Current Trends in Drug Discovery, Development and Delivery (CTD4-2022)

Current Trends in Drug Discovery, Development and Delivery (CTD4-2022)

Edited by

Manikanta Murahari
KL University, College of Pharmacy, India
Email: manikantam@kluniversity.in

Buchi N. Nalluri
KL University, College of Pharmacy, India
Email: buchinalluri@kluniversity.in

and

G. Chakravarthi
KL University, College of Pharmacy, India
Email: chakra_varthi123@kluniversity.in

ROYAL SOCIETY OF CHEMISTRY

This publication is based on peer-reviewed manuscripts from the 2022 Conference on Current Trends in Drug Discovery, Development and Delivery (CTD4-2022) held at KL University, India.

Special Publication No. 358

Print ISBN: 978-1-83767-083-3
PDF ISBN: 978-1-83767-109-0

A catalogue record for this book is available from the British Library

The Royal Society of Chemistry is a charity, registered in England and Wales, Number 207890, and a company incorporated in England by Royal Charter (Registered No. RC000524), registered office: Burlington House, Piccadilly, London W1J 0BA, UK, Telephone: +44 (0) 20 7437 8656.

Visit our website at books.rsc.org

Editorial Contributions

Editors

Dr. Buchi N. Nalluri, M.Pharm., PhD

Dr. Manikanta Murahari, M.Pharm., PhD

Dr. G. Chakravarthi, M.Pharm., PhD

Associate Editors

Dr. A. Anka Rao, M.Pharm., PhD

Dr. M. Narender, M.Pharm., PhD

Dr. P. Rajeswari, M.Pharm., PhD

Dr. Sivadasu Praveen, M.Pharm., PhD

Dr. Naresh Dumala, M.Pharm., PhD

Dr. J. Risy Namratha, M.Pharm., PhD

Dr. M. Kasi Vishwanath, M.Pharm., PhD

Assistant Editors

Mrs. N. Srilakshmi, M.Pharm., (PhD)

Mr. A. Narayana Rao, M.Pharm., (PhD)

Ms. Md. Jahasultana, M.Pharm., (PhD)

Mrs. N. Venkata Naga Jyothi, M.Pharm., (PhD)

Mr. A.V. Surendra, M.Pharm., (PhD)

Mr. K. Ramakrishna, M.Pharm., (PhD)

Mr. T. Daveedu, M.Pharm., (PhD)

Ms. Vineela, M.Pharm., (PhD)

Mr. B. Karan Kumar, M.Pharm., (PhD)

Dr. R. Supriya, PharmD

Dr. P. Lokeswari, PharmD

Dr. K. Sukeerthi, PharmD, (PhD)

Mr. G. Narayana Murthy, M.Pharm

Mrs. T. Prasanna Kumari, M.Pharm

Ms. L Haritha, M.Pharm

Ms. G. Haritha, M.Pharm

Contents

Part 1 Computational Design of Small Molecules for Potential Targets

PART 1

COMPUTATIONAL DESIGN OF SMALL MOLECULES FOR POTENTIAL TARGETS

HIT IDENTIFICATION FROM *GANODERMA LUCIDUM* TO TREAT POLYCYSTIC OVARIAN SYNDROME: AN *IN-SILICO* APPROACH

Manisha Devi[a*], Anbu Jayaraman[a], Parasuraman Pavadai[b], Selvaraj K[c], Damodar Nayak A[a]
[a]Department of Pharmacology, Faculty of Pharmacy, M S Ramaiah University of Applied Sciences, Bengaluru, Karnataka -560054. [b]Department of Pharmaceutical Chemistry, Faculty of Pharmacy, M S Ramaiah University of Applied Sciences, Bengaluru, Karnataka -560054. [c]Department of Department of Biotechnology, Kalasalingam Academy of Research and Education, Krishnankoil, Tamil Nadu -626126.
Corresponding Author: manisha1996.mc1@gmail.com

1 INTRODUCTION

Stein-Leventhal syndrome, often known as Polycystic ovarian syndrome, is a group of gynaecological, endocrine, and metabolic problems that affect women who have reached reproductive age. Due to an abnormally high number of immature eggs growing in the ovarian periphery, polycystic ovaries have a "string of pearls"-like structure. The clinical manifestations are hypergonadotropism, amenorrhea, oligomenorrhea, numerous ovarian cysts, hirsutism, obesity, and they are frequently linked to infertility. Diabetes and cardiovascular disease risk are increased by a number of metabolic abnormalities, including insulin resistance, hyperinsulinemia, poor glucose tolerance, endothelial dysfunction, hypertension, and dyslipidemia.[1]

According to the World Health Organization, 3.4% of women worldwide experienced PCOS in 2012. PCOS is extremely prevalent worldwide, with prevalence rates ranging from 2.2% to 26%. The frequency of PCOS varies from area to region in India. According to a study done by Nidhi et al., girls between the ages of 15 to 18 had a PCOS prevalence of 9.13%.[2] According to Bharthiet al., the prevalence of PCOS in India was 6%, with urban populations having a greater frequency than rural ones.[3]

The main risk factor for PCOS is an excess of androgen. Generally speaking, increasing total free testosterone causes a rise in LH levels. High amounts of androgen also prevent ovarian follicle development and maturation. DHEA levels are also elevated in 50–75% of PCOS-afflicted women, which is a significant contributor to elevated androgen levels. Excess androgen is another factor in the rise in body weight. Moreover, elevated androgen in PCOS women is the primary cause of hirsutism. 50–70% of PCOS-afflicted women have been found to have insulin resistance, while 30–40% of obese PCOS-afflicted women have been shown to have diabetes and impaired glucose tolerance. Insulin resistance and PCOS are two endocrine disorders that are connected. The most prevalent causes of insulin resistance include obesity, a poor diet, and stress. By increasing the production of ovarian testosterone, lowering serum SHBG, and signalling the impending ovulation, insulin resistance causes hyperinsulinemia, which is a significant pathophysiology of PCOS. Women with PCOS and a family history of diabetes have reduced ß - cell function. A significant part of the pathophysiology of the pituitary-ovarian axis is played by IR and hyperinsulinemia.[4] Obesity raises IR, which helps to boost testosterone, decrease the production of gonadotropins, and is also linked to anovulation. Insulin increases free serum testosterone via interacting with LH and lowering SHBG. Hyperinsulinemia amplifies the upregulation of LH receptors brought on by FSH in granulosa cells, which inhibits cell division and follicle development. Due to hyperandrogenism, insulin resistance, and altered intraovarian paracrine signalling, follicular development is disturbed in women with PCOS. Women with PCOS have follicular arrest due to an increase in aromatase expression in granulose cells seen in the antral follicles. High LH promotes theca cell hyperandrogenism, whereas decreasing FSH prevents the maturation and growth of follicles. Follicular arrest in PCOS is associated with peripheral buildup of antral follicles in the ovaries, which results in a polycystic morphology, oligomenorrhea, amenorrhea, anovulation, and subfertility. In addition to being connected to numerous metabolic issues like dyslipidemia,[5] T2DM, and cardiovascular risks, obesity affects 80% of PCOS patients. In PCOS women with insulin resistance, increased caloric intake reduces the ability of adipose tissue stores to grow, which may partially indicate defective adipogenesis. Estrogen aberration, gonadotropin abnormality, hypothalamic-pituitary-adrenal axis abnormality, genetic, and environmental variables are additional risk factors for PCOS.

An eastern fungus native to Europe and parts of China called Ganoderma lucidum (G. lucidum) has a variety of medical uses. It has a huge, black, glossy, and woody outer mushroom. It has been used for more than 2000 years in traditional Chinese medicine and is known as a "herb of spiritual potency". Polysaccharides, triterpenes, amino acids, sterols, proteins, enzymes, vitamins, and peptides are all abundant in them. In-depth pharmacological characteristics of G. lucidum's polysaccharides and triterpenoids are the focus of current research.[6] Throughout the past few decades, more than 380 triterpeniods have been found in the various Ganoderma species. The majority of triterpenoids are a subclass of terpenoids that can be thought of as polymerizing from six isoprene units and have a basic mother nucleus of 30 carbon atoms.[7] Tetracyclic and pentacyclic triterpenoids make up the majority of the triterpenoids that have been found so far, while chain, monocyclic, bicyclic, and tricyclic triterpenoids are also present. It is further subdivided into ganoderic acid, ganodermic acid, ganoderal, lucidenic acid, lucidone, and ganolactone depending on the structure and functional groups. The broad spectrum of pharmacological actions of ganoderic acids include immunomodulation, anticancer, antidiabetic, antioxidant, antiatherosclerotic, antifibrotic, antitumor, analgesic, anti-inflammatory, antinociceptive, antimicrobial, hypolipidemic, hepatoprotective, antiandrogenic, antiproliferative, and estrogenic activity.[8]

A novel technique called computer-assisted drug development (CADD) combines simulation and mathematical modelling to hasten the creation of new drugs. With the specification of future trial experimental designs and the assessment of potential pharmaceutical safety concerns, this method offers a knowledge-based decision-making tool for alternative development methods with projected power and success possibilities.[9] Drug research and discovery are time- and money-consuming processes. On average, it takes 10-15 years and \$500–\$800 million to bring a medicine to market. Homology modelling, molecular docking, virtual high-throughput screening, quantitative structure-activity relationships, and 3D pharmacophore mapping in general are the main CADD components. There are three types of docking techniques: rigid ligand and rigid receptor docking, flexible ligand and flexible receptor docking. This study uses molecular docking, ADMET characteristics, and molecular dynamics to discover the lead moiety for treating polycystic ovarian syndrome using Ganoderma lucidum triterpenoids.

2 MATERIALS AND METHODS

2.1 Protein preparation

The X-ray crystallographic structure of androgen receptor (PDB ID: 1E3G), AMP activated kinase (PDB ID: 2UV7), estrogen receptor α (PBD ID: 4PPS), insulin receptor (PDB ID: 5E1S), mitogen activated protein kinase (PDB ID: 6FXV), insulin like growth factor (PDB ID: 3KR3), estrogen receptor β (PDB ID: 3OLS), and follicle stimulating hormone receptor (PDB ID: 4AY9) was provided by the RCSB Protein Data Bank (http://www.rcsb.org/pdb).[10] Prior to analysis, the protein was cleaned and the missing residues were inserted using Swiss- PDB Viewer v4.1.0. The structure of the protein and the locations of the amino acids from the active areas were further established using the BIOVIA discovery studio version 4.0 software; they were then used for docking investigation.

2.2 Retrieval and preparation of active compounds

About 250 triterpenoids make up the ganoderma lucidum, which are divided into substances like ganoderic acids, ganoderiol, ganoderone, ganolactone, and ganoderal. For this study, clomiphene citrate and metformin, common medications used to treat PCOS, and a collection of 22 ganoderic acids (GAs), were chosen. The data repository (https://cb.imsc.res.in/imppat/home),[11] previously published literature, and PubChem (https://pubchem.ncbi.nil.nih.gov/) were used to gather ganoderic acids and standard drugs. The chosen standard compounds and chosen ganoderic acids were downloaded in SDF format for additional research.

2.3 Molecular docking

This approach is a pivotal part of structural biology research, and it is one of the most often used methodologies in the drug design process. Molecular docking study was conducted using PyRx 0.8 tool[12] and AutoDock Vina programme.[13] Androgen receptor (PDB ID: 1E3G), AMP activated kinase (PDB ID: 2UV7), estrogen receptor α (PBD ID: 4PPS), insulin receptor (PDB ID: 5E1S), mitogen activated protein kinase (PDB ID: 6FXV), insulin like growth factor (PDB ID: 3KR3), estrogen receptor β (PDB ID: 3OLS), and follicle stimulating hormone receptor (PDB ID: 4AY9) were used as the proteins and the selected bioactive components were used as ligands. The Kollam partial charges and Polar hydrogen atoms were also added using PyRx software to the 3D structures. Moreover, "pdbqt" format was used to store the receptor and ligand data in order to determine the docking energy affinities. AutoDock Vina calculated each ligand's energy affinity values using up to 10 different docking sites. The complex's 2D and 3D structures, as well as the amount of hydrogen bonds and non-covalent interactions for each complex, were developed by Discovery Studio Visualizer.[14]

2.4 Molecular dynamics simulation

Molecular dynamic simulation was used to investigate the binding stability, conformation, and interaction processes between the discovered bioactive compounds and receptors.[15] Using the DESMOND v3.6 programme, a molecular dynamic simulation of the ligand-receptor complex was performed. Also, a water model was created at a distance of 10A units from the orthorhombic period boundary using the pre-established SCP water model. In addition, the requisite quantity of counter ions was added to balance the electric charges, and before the molecular dynamic simulation process started, the system's energies were reduced by heating and equilibrium processes. The Nose-Hoover method and the NPT (isothermal-isobaric ensemble) ensemble were utilised in the system production step, which lasted 100ns with time steps of 0.001ps. 300K temperature and 1 atmosphere of pressure were also used. Based on the interactions and dynamical features of the complexes, the best confirmations were chosen.

3 RESULTS AND DISCUSSION

This study used pharmacoinformatics to examine the role of Ganoderma lucidum active ingredients in the prevention and treatment of PCOS. A group of in silico molecular modelling techniques known as pharmacoinformatics are used to test bioactive compounds for binding affinities, pharmacokinetics, and pharmacodynamic characteristics. By enabling researchers to focus the effects of biological and synthetic studies, pharmacoinformatics has enhanced the discovery of bioactive substances. The positive effects of some medications were foreseen through pharmacoinformatics research and then validated in vitro and in vivo. If researchers can understand how substances bind, interact, and inhibit/excite a protein, they may be able to find therapeutic treatments for certain disease states. The purpose of this research work was to inhibit various receptors involved in the pathophysiology of PCOS by using bioactive components found in *Ganoderma lucidum* for the treatment of PCOS.

3.1 Retrieval of bioactive compounds and preparation

In order to find the triterpenoids that are found in Ganoderma lucidum, a literature research was done. Out of more than 250 triterpenoids, the ganoderic acids (GAs) group was chosen as the most easily available bioactive component. For this study, a list of 22 bioactive ingredients and two common medications currently used to treat PCOS were chosen. The active ingredients in Ganoderma lucidum that function as inhibitors of HIV protease and tyrosinase underwent a comparable computational research.

3.2 Molecular docking studies

Molecular docking analysis was carried out to investigate the intermolecular interaction between the bioactive components and target proteins. Twenty two bioactive components and the standard drugs were docked against androgen receptor (1E3G), AMP activated kinase (PDB ID: 2UV7), estrogen receptor α (PBD ID: 4PPS), insulin receptor (PDB ID: 5E1S), mitogen activated protein kinase (PDB ID: 6FXV), insulin like growth factor (PDB ID: 3KR3), estrogen receptor β (PDB ID: 3OLS), and follicle stimulating hormone receptor (PDB ID: 4AY9) employing one target protein at a time using AutoDock Vina to analyse their binding capacity. Following molecular docking, the binding energies of bioactive components were found to be ranging from -5to -8.9 kcal mol^{-1} for androgen receptor. Similarly the binding energies of bioactive components was found to be ranging from -4.7 to -8.1 kcal mol^{-1} for AMP activated kinase, from -4.8 to -8.4 kcal mol^{-1} for estrogen receptor α, from -4.9 to -8.6 kcal mol^{-1} for insulin receptor, from -4.6 to -10.2 kcal mol^{-1} for mitogen activated protein kinase, from -4.3 to -7.3 kcal mol^{-1} for insulin like growth factor, from -5 to -7.4 kcal mol^{-1} for estrogen receptor β, and from -4 to -8.1 kcal mol^{-1} for follicle stimulating hormone receptor respectively illustrated in Table 1.1.

Table 1.1 *Ganoderic acids and standards with their binding affinities against selected target receptors.*

Compound Name	Docking Scores(kcal mol^{-1})							
	1E3G	2UV7	3KR3	3OLS	4AY9	4PPS	5E1S	6FXV
GAA	-8.9	-6.6	-6.3	-6.7	-7.3	-7.9	-7.4	-8.4
GAB	-8.3	-7.6	-6.6	-6.7	-7.1	-7.7	-7.9	-9.4
GAC1	-7.6	-7	-7.3	-6.7	-7.9	-8.1	-7.6	-8.3
GAC2	-8.3	-7.6	-6.5	-6.6	-6.6	-7.9	-8.2	-8.7
GAC6	-7.4	-6.8	-6.7	-6.8	-7.4	-7.4	-8.6	-8.4
GAD	-8.1	-8	-6.5	-7.4	-7.4	-7.8	-7.6	-8.3
GAE	-7.7	-8.1	-6.8	-6.9	-7.6	-8.2	-8.3	-9
GAF	-7.5	-7.1	-6.6	-6.4	-8.1	-7.3	-7.8	-8

GAG	-7.4	-7.8	-7.2	-7.1	-6.9	-7.8	-8	-10.2
GAH	-7.7	-6.7	-6.4	-6.9	-7.4	-7.1	-7.4	-8.7
GAI	-8.5	-7	-6.9	-6.6	-7.4	-7.2	-8.3	-8.2
GAJ	-7.7	-7.1	-7.2	-6.8	-7	-7.4	-8.5	-8.6
GAK	-7.4	-6.9	-6.7	-7.4	-7.5	-7.8	-7.7	-8.4
GAM	-8	-7.3	-6.5	-7.3	-8.1	-8.1	-8.2	-8
GAN	-8.5	-7.6	-6.5	-7	-7.5	-7.6	-8.2	-8.5
GAL	-8.4	-7.2	-6.5	-7	-7.4	-7.2	-8.3	-9.3
GAS	-8.1	-6.7	-6.7	-6.6	-7.5	-7.7	-8.2	-7.9
GAT	-7.3	-8.1	-6.3	-6.6	-6.4	-6.7	-7.5	-8.6
GAAM1	-7.2	-6.8	-6.6	-7	-7.1	-8.1	-8.3	-9.2
GALm2	-8.6	-8.1	-7.2	-7.4	-7.3	-8.4	-8.1	-9.4
GADM	-7.9	-7.5	-7.6	-6.6	-8	-7.3	-7.9	-8.3
Gaα	-7.4	-7	-6.4	-6.5	-6.6	-7.3	-6.7	-8.3
STD1	-5.8	-6.6	-5.2	-6	-5.6	-6	-7.6	-7.3
STD2	-5	-4.7	-4.3	-5	-4	-4.8	-4.9	-4.6

3.2.1 Identification of lead targets. Molecular docking results and the binding energies of selected ganoderic acids along with the standards revealed five targets with better binding affinities which include androgen receptor, AMP activated kinase, estrogen receptor α, insulin receptor and MAPK1. Ganoderic acids exhibited highest binding energies with MAPK1. See Table 1.2.

3.2.2 Interpretation of protein ligand interaction. Ganoderic acid A(GAA) (CID-471002) was selected for the further studies based on easy availability and pharmacological properties. BIOVIA discovery studio visualizer tool was used to anticipate the interactions between Ganoderic acid A and the five selected target receptors. The interactions of GAA with the targets are listed in Table 1.3 and the 3D and 2D interactions with amino acid residues are depicted in Figure 1.1.

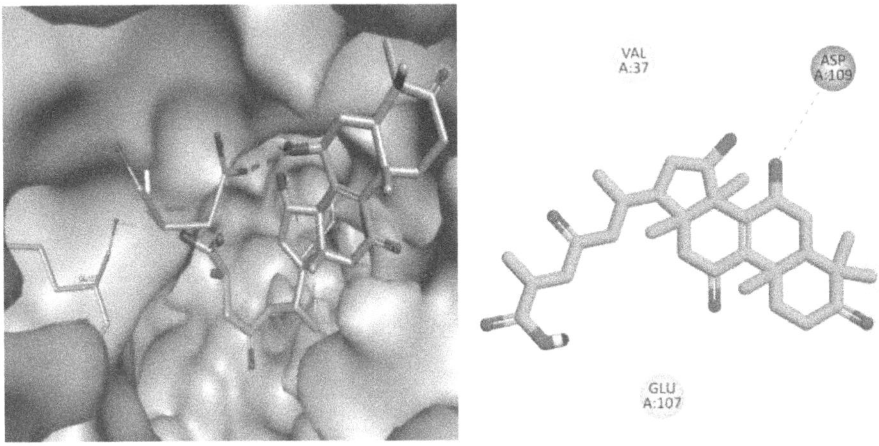

Figure 1.1 *Interaction between GAA and MAPK 1. Left figure represents 3D and right figure represents 2D complex protein-ligand interaction.*

Table 1.2 *Molecular docking targets with highest energies.*

Sl. No.	Molecular docking targets	Ligand Name	Highest binding energies
1	Androgen receptor	GAA	-8.9 kcal mol^{-1}
2	AMP activated kinase	GAE	-8.1 kcal mol^{-1}
3	Estrogen receptor α	GALm2	-8.4 kcal mol^{-1}
4	Insulin receptor	GAC6	-8.6 kcal mol^{-1}
5	MAPK 1	GAG	-10.2 kcal mol^{-1}

Table 1.3 *List of bonding interactions of GAA with target receptors.*

Compound Name	Residues	Amino Acids	Bond Category
Ganoderic acid A with androgen receptor	783A	GLN	Hydrogen bond
	782A	SER	Hydrogen bond
	778A	SER	Hydrogen bond
	779A	ARG	Carbon hydrogen bond
	697A	PHE	Alkyl
	880A	LEU	Alkyl
	883A	LYS	Pi-alkyl
Ganoderic acid A with AMP activated kinase	236A	ARG	Hydrogen bond
	266A	LEU	Alkyl
	273A	PHE	Pi-alkyl

	272A	TYR	Pi-alkyl
Ganoderic acid A with estrogen receptor α	340A	ALA	Hydrogen bond
	348A	ASN	Hydrogen bond
	537A	SER	Hydrogen bond
	535A	PRO	Hydrogen bond
	536A	LEU	Carbon hydrogen bond
Ganoderic acid A with insulin receptor	1097A	ASN	Hydrogen bond
	1080A	ALA	Hydrogen bond
	1000A	ARG	Hydrogen bond
	1002A	LEU	Alkyl
Ganoderic acid A with MAPK 1	107A	GLU	Hydrogen bond
	37A	VAL	Alkyl

3.2 Molecular dynamics simulation

Although being a popular and effective application, protein-ligand docking only provides a static picture of the ligand's binding pose, with the receptor's active site looking just like a photograph. With the inclusion of Newton's conditions of movements, Molecular Dynamics (MD) should be used to simulate the atoms of the particles in the framework as a component of time. A molecular dynamic simulation for the GAA-1E3G complex was performed in a previous work, and simulations for the GAA-2UV7, GAA-4PPS, GAA-5E1S, and GAA-6FXV complexes were performed in the current investigation. Each of the four GAA complexes was subjected to a 100 ns molecular dynamic simulation, with the target proteins' unbound state used to interpret the data. To understand the fluctuations and stability of the protein-ligand complex, root mean square deviation (RMSD) and root mean square fluctuation (RMSF) of receptor atoms were used to analyse the molecular dynamic trajectories.

3.2.1 Root mean square deviation (RMSD. In order to explore the equilibration of molecular dynamic trajectories and assess the stability of the protein-ligand complex system during the simulation process, RMSD is an important parameter. According to the RMSD of GAA-2UV7 complex, variations in protein backbone were observed initially from 1.8 to 5.2nm. At the time intervals of 10ns and 80ns, a stable conformation was established with no significant deviations. However the protein backbone showed fluctuations from 2.4 to 3.6nm. The GAA-4PPS complex RMSD analysis indicated no significant changes in the protein backbone throughout the simulation. The fluctuations were observed only between 1 to 2.7nm. RMSD analysis of GAA-5E1S indicated no significant changes in the protein backbone from 0 to 64ns ranging from 1 to 3nm. However the protein backbone was observed to be unstable after 65ns till the end of the simulation. The GAA-6FXV complex showed variations ranging from 1.2 to 2.8nm in the protein backbone throughout the simulation period as depicted in Figure 2.

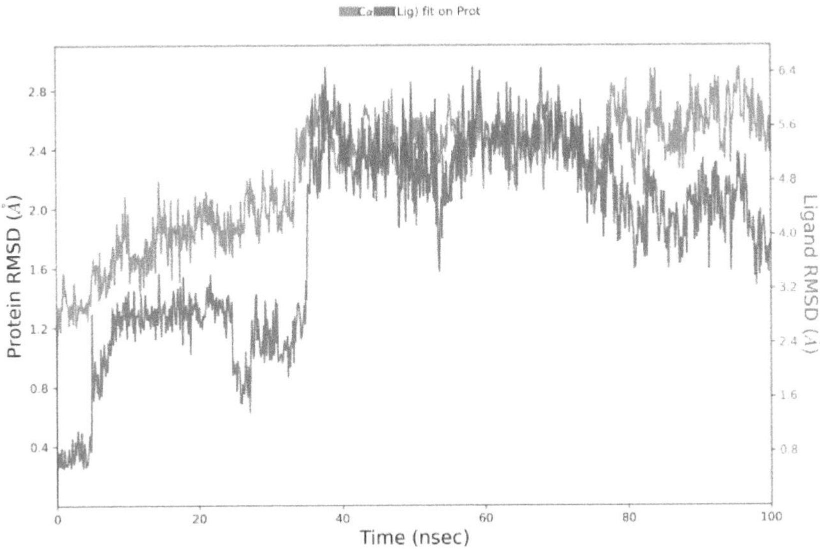

Figure 1.2 *RMSD graph of GAA-6FXV complex.*

3.2.2 Root mean square fluctuation (RMSF). Another important parameter to consider when examining the stability and adaptability of protein-ligand complex systems in simulation is the RMSF. RMSF was used to analyse how the target protein's amino acid residues changed after interacting with a ligand. The protein's Cα atoms' RMSF values were calculated and shown in relation to the residues. The amino acid residues in the under investigation GAA-1E3G complex exhibited no variations during the simulation, and the resulting graph was noted to be smooth. The amino acids of AMPK (2UV7) which interacted with GAA showed fewer fluctuation values during MD simulation, amino acids of estrogen receptor (4PPS) which interacted with GAA showed a smooth graph during MD simulation, amino acids of insulin receptor (5E1S) which interacted with GAA showed minimal fluctuations values during MD simulation and amino acids of MAPK (6FXV) showed no fluctuations and resulted in a smooth graph as depicted in Figure 3.

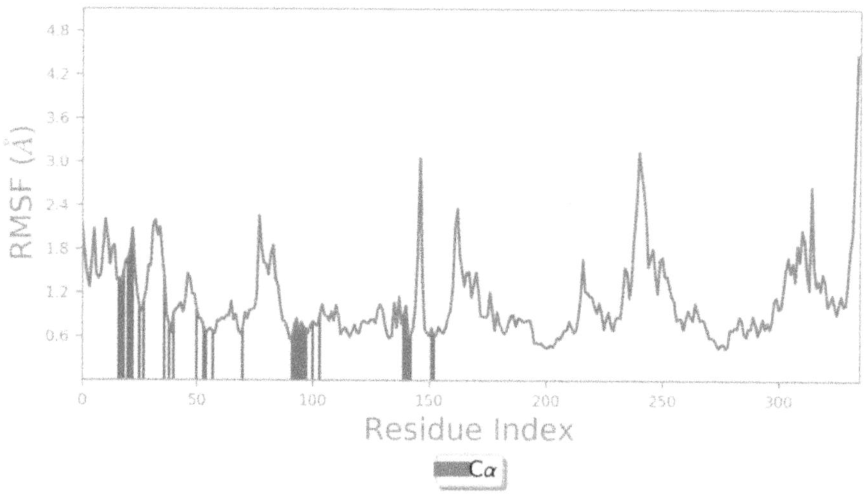

Figure 1.3 *RMSF plot of 6FXV during 100ns of MD simulation.*

3.2.3 Protein ligand interactions. During the MD simulation, protein interactions with the ligand were observed. By hydrogen bonding with the ligand and a water molecule, the ligand was able to bind with the 2UV7 amino acid residue ASP232 in a fraction of 0.39, indicating that 39% of the simulation time was spent maintaining the association. Similar results were obtained with LEU266, ARG236 and VAL238 where the ligand displayed interaction fractions of 0.32, 0.25 and 0.20, respectively. These results suggest that through there is an interaction between GAA and 2UV7 the interactions are not strong enough and sustained for a lesser duration of time. GAA showed 0.7 interaction fraction with 4PPS amino acid residue THR347, 0.68 with ASN3448, 0.66 with PRO535, and 0.44 with GLY344 through hydrogen bonds and water bridges respectively. 38% of the simulation time in the chosen trajectory was spent in ligand interactions with the 4PPS amino acid residue GLY344, and 30% of the simulation time was spent in ligand interactions with PRO535. The results suggested a good quality interaction of GAA along with 4PPS and lasted for 70% of the time. GAA showed 0.35 interaction fraction with 5E1S amino acid residue VAL1010, 0.35 with GLP1096, 0.35 with MEF1079, 0.36 with THR1025, and 0.33 with ALA1080 through hydrogen bonds and water bridges respectively. These results suggest that through there is an interaction between GAA and 2UV7 the interactions are not strong enough and sustained for a lesser duration of time. GAA showed 0.88 interaction fractions with 6FXV amino acid residue ILE29, more than 1 with ASP109, and 0.6 with LYS112 through hydrogen bonding mediated with water molecules respectively as depicted in Figure 4. Figure 5 shows that the ligand interactions with the 6FXV amino acid residue ILE29 took place for 59% of the simulation time in the chosen trajectory, those with ASP165 took place for 36% of the simulation time in the chosen trajectory, and those with GLU69 took place for 33% of the simulation time. The results of both molecular docking and MD simulation suggested a maximum interaction of GAA along with 6FXV which lasted throughout the simulation period.

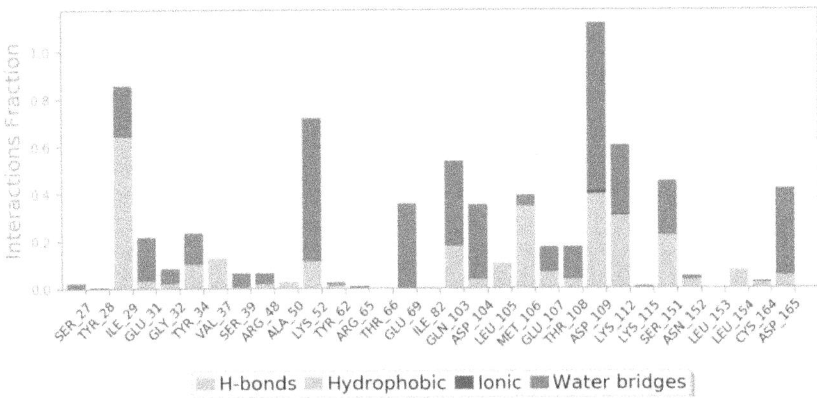

Figure 1.4 *Interacting amino acid residues of 6FXV with GA.*

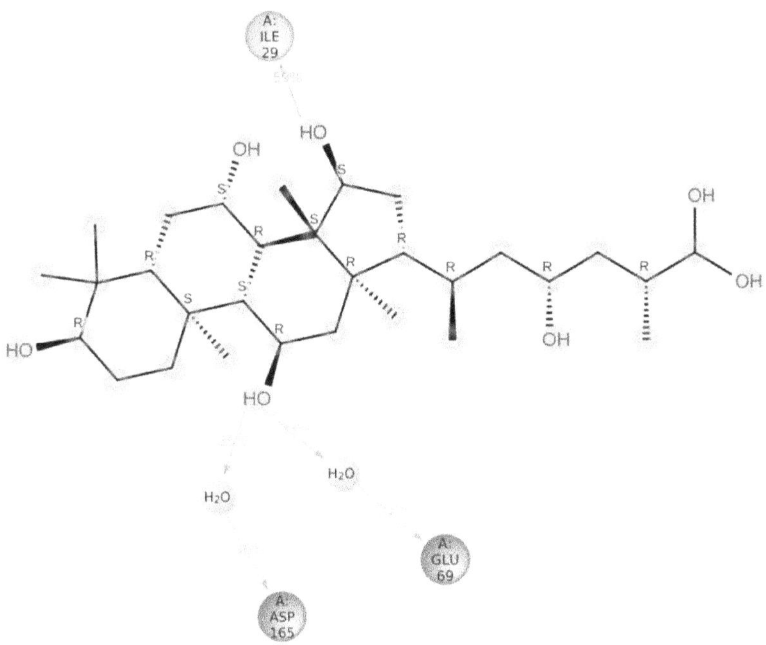

Figure 1.5 *Protein ligand (GAA-6FXV) contacts.*

4 CONCLUSION

PCOS requires a comprehensive pharmaceutical approach for therapy because it is characterised by endocrinological, gynaecological, and metabolic problems. The androgen receptor, AMP activated kinase, oestrogen receptors A and B, follicle stimulating hormone, insuln receptor, and MAPK are a few of the significant receptors involved in the pathophysiology of PCOS. In this study, we investigated how the target receptors of ganoderic acids were inhibited to reverse the pathophysiology of PCOS. This study's thorough pharmacoinformatics-based molecular docking investigation of GAs against eight distinct receptors demonstrated the potential of ganoderic acids for the treatment of PCOS. *In-silico* molecular docking investigations revealed that the selected 22 ganoderic acids showed best binding affinities with MAPK, AMP activated kinase, androgen receptor, estrogen receptor, and insulin receptor. The binding affinities of all the ganoderic acids were better compared to the selected standard drugs used in the treatment of PCOS. Molecular dynamic stimulation experiments revealed the safety profile of Ganoderic Acid A as well as the protein–ligand complex's stability. Molecular simulation was carried out for four ligand receptor complexes including GAA-2UV7, GAA-4PPS, GAA-5E1S, and GAA-6FXV for analysing the parameters like RMSD, RMSF and protein ligand contacts. Molecular dynamic simulation results revealed good interactions of protein and ligand, best one among them being MAPK receptor. MAPK may be the possible pathway for the effective treatment of PCOS using ganoderic acids. To determine the effectiveness and explore the pharmacological activities of GAA in the treatment of PCOS, in vivo and in vitro research is necessary.

5 ACKNOWLEDGMENT

The authors thank the Department of Pharmaceutical Chemistry, Faculty of Pharmacy, M. S. Ramaiah University of Applied Sciences, Bengaluru for providing computational facility.

6 REFERENCES

1. S. Mousa, Brady, S. Mousa, and Mousa. *Drug, Healthcare and Patient Safety*, 2009, 9. DOI:10.2147/dhps.s4388.
2. RV. Bharathi, S. Swetha, J. Neerajaa, JV. Madhavica, DM. Janani, and SN. Rekha. *Middle East Fertility Society Journal*, 2017, **22**(4), 313–316. DOI:10.1016/j.mefs.2017.05.007.
3. M. Gupta, D. Singh, M. Toppo, A. Priya, S. Sethia, and P. Gupta. *International Journal Of Community Medicine And Public Health*, 2017, **5**(1), 95. DOI:10.18203/2394-6040.ijcmph20175603
4. CJ. Glueck, and N. Goldenberg. *Metabolism*, 2019, **92**, 108–120. DOI:10.1016/j.metabol.2018.11.002.
5. P. Jin, and Y. Xie. Gynecological endocrinology, 2018, **34**(4), 272–277. DOI:10.1080/09513590.2017.1395841.
6. MF. Ahmad.*Biomedicine& Pharmacotherapy*, 2018, **107**, 507–519. DOI:10.1016/j.biopha.2018.08.036.
7. MF. Ahmad.*Journal of ethnopharmacology*, 2020, **260**, 113047.

8. DH. Ryu, JY. Cho, NB. Sadiq, JC. Kim, B. Lee, M. Hamayun, TS. Lee, HS. Kim, SH. Park, CW. Nho, and HY. Kim. *Food Chemistry*, 2021, **335**, 127645. DOI:10.1016/j.foodchem.2020.127645.
9. R. Gomeni, M. Bani, C. D'Angeli, M. Corsi, and A. Bye. *European Journal of Pharmaceutical Sciences*, 2001, **13**(3), 261–270. DOI:10.1016/s0928-0987(01)00111-7.
10. HM. Berman. *The Protein Data Bank. Nucleic Acids Research*, 2000, **28**(1), 235–242. DOI:10.1093/nar/28.1.235.
11. K. Mohanraj, BS. Karthikeyan, RPV. Ananth, RPB. Chand, SR. Aparna, P.Mangalapandi, and A. Samal. *Scientific Reports*, 2018, **8**(1), 4329. DOI:10.1038/s41598-018-22631-z.
12. X. Fradera, and K. Babaoglu. *Current Protocols in Chemical Biology*, 2017, **9**(3), 196–212. DOI:10.1002/cpch.27.
13. T. Gaillard. *Journal of Chemical Information and Modeling*, 2013, **58**(8), 1697–1706. DOI:10.1021/acs.jcim.8b00312.
14. UP. Mohan, S. Kunjiappan, PBT. Pichiah, and S. Arunachalam. *3 Biotech*, 2021,**11**(1). DOI:10.1007/s13205-020-02530-9.
15. AK. Kalimuthu, T. Panneerselvam, P.Pavadai, SRK. Pandian, K. Sundar, S. Murugesan, DN. Ammunje, S. Kumar, S. Arunachalam, and S. Kunjiappan. *Scientific Reports*, 2021,**11**(1). DOI:10.1038/s41598-021-01008-9.

FRAGMENT BASED DRUG DESIGN (FBDD) OF AN EXEMPLAR INHIBITOR AGAINST SPIKE GLYCOPROTEIN OF SARS-COV-2

Manjiri S. Shinde[a], Bandoo C. Chatale[a*]

[a]MET Institute of Pharmacy, MET Complex, Bandra (West), Mumbai, India-400050.
* Corresponding Author E-Mail ID: banduchatale@gmail.com

1 INTRODUCTION

1.1 Virus

SARS-CoV-2 has several developing variations, each of which poses a serious threat, making treatment equally difficult. Fragment-Based Drug Design (FBDD) holds a sequential strategy to begin with a pathway of developing a chemical against the virus (FBDD: A classic branch of Computer Aided Drug Discovery).

One needs to understand the virus in order to create an inhibitor. Virus's fundamental composition and replication process play a key role in the construction of fragments. By this approach, it is possible to identify the protein that the virus uses to replicate and cause symptoms to appear in the body.

Potential targets to consider in order to develop inhibitors against the virus's proteins are

1) Spike glycoprotein
2) Envelope protein
3) RNA genome
4) Proteins responsible for exocytosis from host's cell.

1.1.1 The rationale behind selecting spike glycoprotein of the SARS-CoV-2. Despite the fact that numerous possible targets were also identified, including the RNA genome, the spike-ACE-2 interface, and the envelope protein. Spike protein was selected a target for creating fragments in this study since it is in charge of the very first stage of viral replication in the host.[1]

1.1.2 Replication of SARS-CoV-2.

1. Entry of virus from the respiratory pathway
2. Interaction of virus's spike glycoprotein with human-ACE2 protein
3. Endocytosis: Entry into the human cell.
4. Attaches to the host ribosome and starts synthesizing their proteins
5. Translation: Synthesis of virus's proteins
6. Formation of nucleocapsid
7. Exocytosis: Exiting the replication forms of the virus

1.2 Fragment-Based Drug Design

It has always been crucial to pinpoint the specific seed or lead to the receptor. FBDD makes the process simpler as a result. Similar to the pieces of a jigsaw puzzle, fragments must be put together to form a whole. They are the components that make the final product, hence.[2]

The research approach of FBDD:

1) Fragment screening – Screening of virtual library, Identification of fragment scaffolds, multidisciplinary analysis of target protein, establishing guidelines for lead identification.

2) Fragment growth and screening – Virtual molecule generation from FBDD hit fragment growth. The step includes high precision docking studies.

3) Lead Identification – Structural modification of the lead according to medical chemistry point of view.

4) ADME/ Tox prediction: ADMET properties or also known as pharmacokinetic properties for fragments and inhibitor were predetermined using Swiss ADME for drug likeness.

1.2.1 The rationale behind selecting the FBDD approach. FBDD has an edge in arriving at an unambiguous hit molecule compared to the conventional method. Fragments were thus selected on basis of a literature survey of already FDA-approved drugs focusing on antivirals class as the research was established for inhibitory activity against spike glycoprotein. Approved drugs were selected for ease of availability of API and related molecules.[2]

2 OBJECTIVE OF THE STUDY

Screening and filtering top-scoring fragment hit against the spike protein with docking score < -7.0. Developing an exemplar inhibitor against SARS-CoV-2 spike glycoprotein.

3 SOFTWARES/ DATABASES APPLIED

The software with its corresponding purpose has been mentioned in Table 2.1 which was used in study.

Table 2.1 *Software or websites used for study.*

Sr. No.	Software/ website used	Purpose
1.	RSCB Protein Data Bank.	For retrieving the desired protein.[3]
2.	Marvin Sketch.	Drawing structures, calculating molecular properties of a compound.[4]
3.	Zinc Pharmer.	Providing fragments on requesting the query for given compounds.[5]
4.	Chimera.	Visualization software, provides Ramachandran plot for knowing the quality of protein.[6]
5.	AutoDock Vina.	For performing docking studies.[7]
6.	SmiLib.	Combinatorial library for gathering features of all fragments and providing combined molecules with possible chemistry rules valid combinations.[8]
7.	Swiss ADME.	Studying physicochemical properties, pharmacokinetic properties of the molecules/fragments asked.[9]

4 RESULTS AND DISCUSSION

4.1 Protein selection

SARS-CoV-2 spike glycoprotein was isolated from RCSB Protein Data Bank. PDB ID: 6VXX. 6VXX was selected because it followed the following criteria:

1) Having resolution <3.0Å
2) Having no mutations

3) Consisting of no missing parts, which were yet to be discovered. The Ramachandran plot, in which the blue portion is the permissible regions that are present in the green plot if the parameter does not fulfil it illustrates that the whole structure of a protein is still to be determined, was used to evaluate the quality and validity of protein 6VXX. In the following Figure 2.1 it was confirmed that the protein is valid to study on.

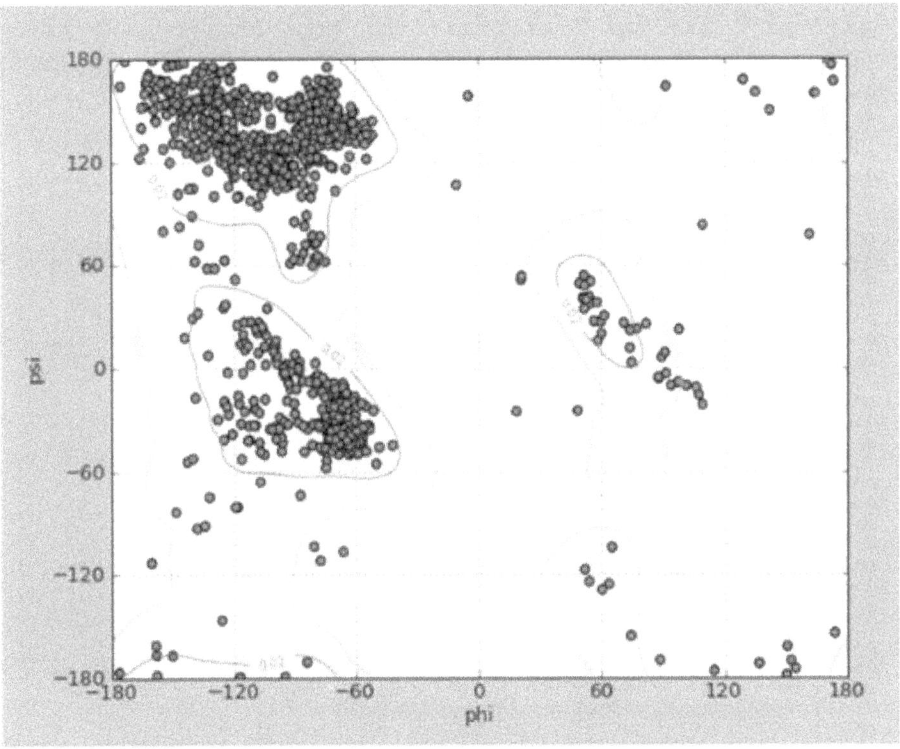

Figure 2.1 *Ramachandran plot for spike protein: 6VXX.*

A to R were the chains assigned to the 6VXX protein. Whereby chains A, B, and C contained ligand-binding sites and were large molecules. NAG was the ligand that was already attached to the protein (2-acetamido-2-deoxy-beta-D-glucopyranose). Considering the macromolecules the fragments and the inhibitor was bound to the pocket between the three macromolecules (between chains A, B, C).

4.2 Fragment selection

Fragments were retrieved from already repurposed drugs from a literature survey, in which the antivirals were docked against spike glycoprotein of the virus with a promising docking score. Same repurposed antivirals were taken into consideration for further fragments development. Each drug in the antiviral section was subjected as a query to retrieve fragments on Zinc Pharmer. Zinc Pharmer online software displayed several fragments for one drug molecule. Filtration of those fragments was further carried out by considering Lipinski's Rule of 5.

Antivirals preferred according to the physicochemical properties were – Saquinavir, Ombitasvir, Asunaprevir, Daclatasvir, Telaprevir, Tipranavir. Several fragments were selected as such and were docked against the spike glycoprotein[10]. The top 10 fragments with good affinities were filtered according to the docking score. Table 2 thus discusses the docking score of those fragments and availability of the same in the market.[10]

Table 2.2 *Retrieved fragment's IUPAC, ZINC ID, Dock score, Fragment source, Availability of fragment in the market if purchased from a licensed vendor(from Zinc Pharmer).*[5,6,7]

Sr no	Fragments	Dock score	Fragment source
1.	N-[2-(4-methylpiperazin-1-yl)phenyl]quinoline-2-carboxamide 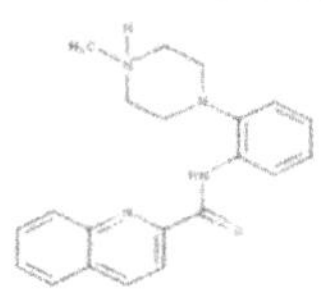	-8.0	Saquinavir
2.	6-[(1R)-1-[[(1R)-2-methyl-1-phenyl-propyl]amino]ethyl]-N4-phenyl-1,3,5-triazine-2,4-diamine 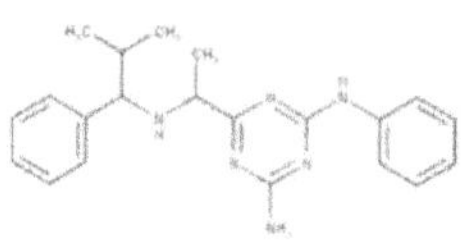	-8.4	Tipranavir
3.	2-[[3-(5-methyl-4H-1,2,4-triazol-3-yl)phenyl]methylamino]-3H-quinazolin-4-one 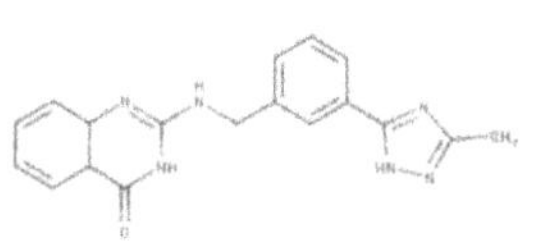	-8.1	Ombitasvir

4.	1H-benzoimidazol-2-ylamino-(1H-benzoimidazol-2-ylimino)methanethiol 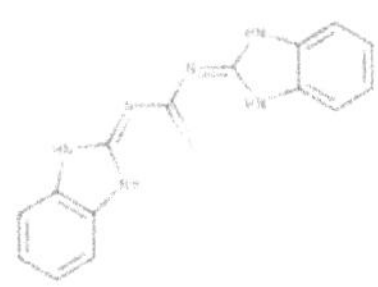	-7.0	Asunaprevir
5.	1-(2-Chlorophenyl)-3-(3-hydroxynaphthalene-2-carbonyl)-urea 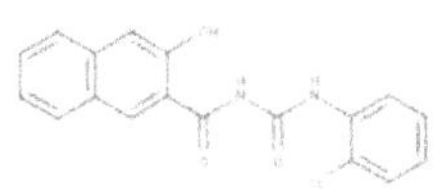	-7.8	Asunaprevir
6.	1-[(1S,2R)-2-(hydroxymethyl)cyclopentyl]-3-(6-methyl-8-quinolyl)urea 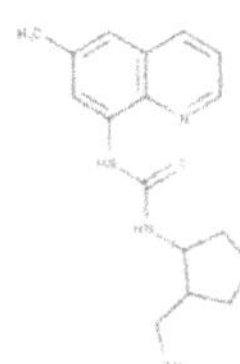	-7.4	Telaprevir
7.	N-[6-(5-acetylamino-3H-benzoimidazol-2-yl)benzothiazol-2-yl]acetamide 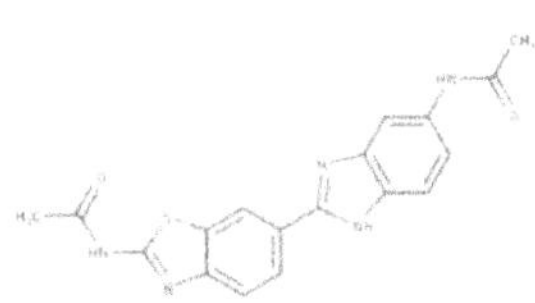	-7.8	Daclatasvir
8.	2-(pyrimidin-2-ylamino)-N-(5-quinolyl)thiazole-4-carboxamide 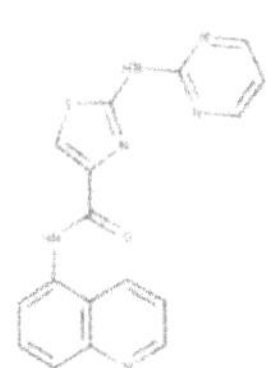	-8.1	Tipranavir

9.	N-(4-ethyl-3-methyl-benzothiazol-2-ylidene)-2-oxo-chromene-3-carboxamide 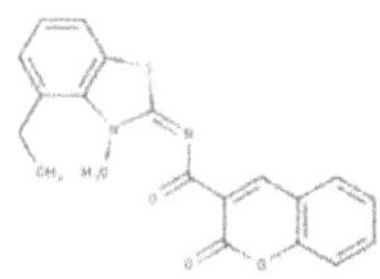	-7.8	Saquinavir
10	N-dibenzofuran-3-yl-N'-(3-methyl-1H-pyrazol-4-yl)oxamide 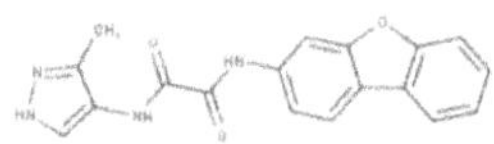	-7.2	Ombitasvir
11	3-(6-ethyl-5-methyl-2,3,3a,6,7,7a-hexahydro-1,3-benzothiazol-2-yl)-1-(4-oxo-1,2,3,4-tetrahydroquinazolin-2-yl)urea 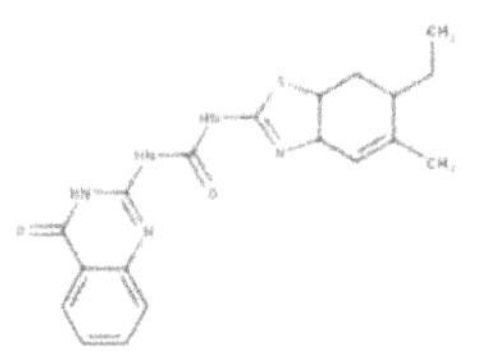	-7.9	New design

4.3 Design of an exemplar inhibitor

Once all 10 fragments had been located and had received favourable docking scores, they were loaded as SMILES file format into the combinatorial library programme SmiLib. A Java-based programme called SmiLib integrates data uploaded in scaffolds, linkers, and building blocks to produce a chemical with the proper chemistry. SmiLib's loaded fragments produced a number of other chemicals. These substances were once more docked with the spike protein. The effective one was then chosen as an exemplar inhibitor.

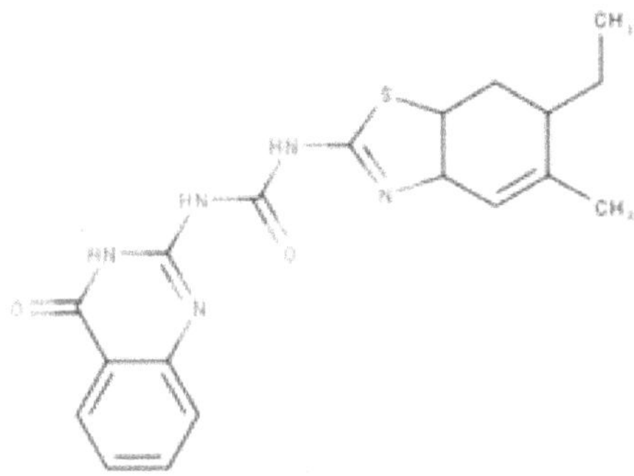

Figure 2.2 *Designed inhibitor; IUPAC: 3-(6-ethyl-5-methyl-2,3,3a,6,7,7a-hexahydro-1,3-benzothiazol-2-yl)-1-(4-oxo-1,2,3,4-tetrahydroquinazolin-2-yl)urea.*

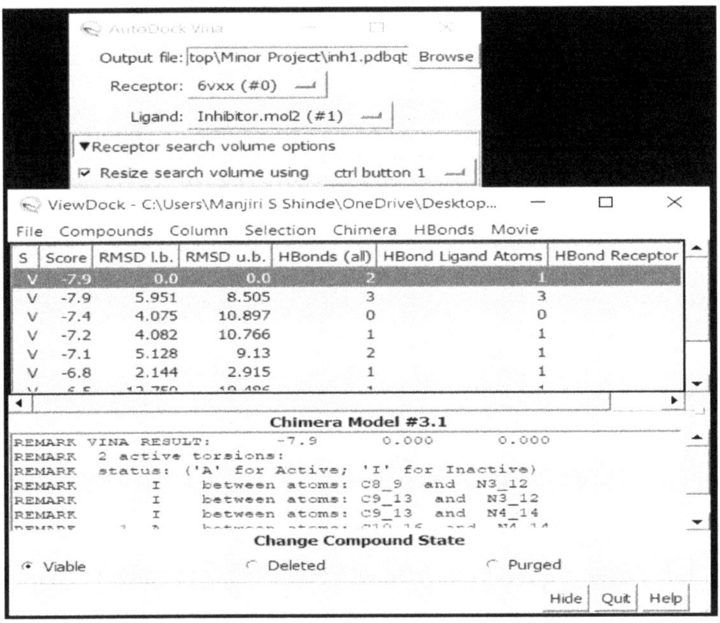

Figure 2.3 *Docking score of designed inhibitor.*

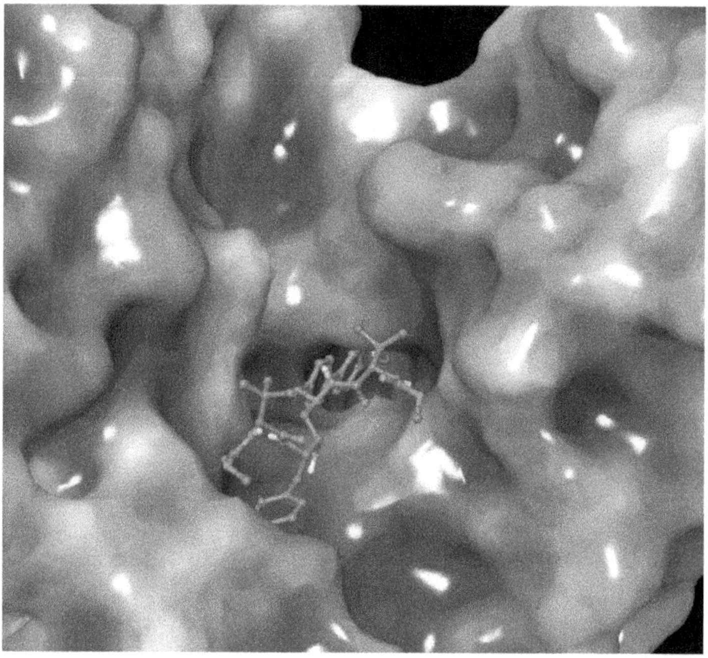

Figure 2.4 *Docked new inhibitor against the receptor.*

Table 2.3 *Pharmacokinetic properties of retrieved fragments whose GI absorption is high and with 0 violations from Lipinski's Rule of 5.*

Fragment number from table 4.2.1	ZINC ID	MW	XLOGP	BBB permeation
1.	06822583	347.43	3.29	Yes
2.	28790361	362.47	4.17	No
3.	89642199	332.36	2.14	No
4.	05071637	308.36	5.76	No
5.	72266059	340.76	2.6	No
6.	92145236	299.37	2.29	Yes
7.	05444072	365.41	2.88	No
8.	56924453	348.38	3.83	No
9.	07385706	364.42	3.83	No
10.	92653488	334.33	2.8	No

4.4 Outcomes of the research

1) Potential fragments against SARS-CoV-2 were developed. These fragments could be undergo further lead optimization as the fragments developed have gone through several filtration steps, thus discarding other hits from the list.
2) Designed an exemplar inhibitor that showed good affinity towards the receptor with a docking score of -7.9 kcal/mol.
3) The Swiss ADME online database was used to assess the fragments' chemical, physical, drug-like qualities, and pharmacokinetic parameters, which further adds value to the design of an exemplary inhibitor.

5 CONCLUSION

FBDD has become an alternative approach to HTS for the generation of lead compounds for drug targets. This approach was applied to establish a new design computationally against SARS-CoV-2. In the proposed data, fragments with a good affinity towards the spike glycoproteins were developed. Thus, further investigation with fragments to develop a different inhibitor can be carried out thus saving time of working/ screening/ filtering with all sets of fragments even in the antivirals section. ADME properties were also determined to know if the proposed molecule can be orally available or not.

6 ACKNOWLEDGMENT

The authors are thankful to Mumbai Educational Trust Institute of Pharmacy (Degree) for providing the facilities to conduct the study.

7 REFERENCES

1. Guo, Y.-R., et al., *The origin, transmission and clinical therapies on coronavirus disease 2019 (COVID-19) outbreak – an update on the status.* Military Medical Research, 2020. **7**(1): p. 11.
2. Bian, Y. and X.-Q. Xie, *Computational Fragment-Based Drug Design: Current Trends, Strategies, and Applications.* The AAPS Journal, 2018. **20**(3): p. 59.
3. H.M. Berman, J.W., Z. Feng, G. Gilliland, T.N. Bhat, H. Weissig, I.N. Shindyalov, P.E. Bourne, *The Protein Data Bank.* Nucleic Acids Research, 2000. **28**: p. 235-242.
4. {{Cherinka}, B.a.A., Brett H. and {S{\'a}nchez-Gallego}, Jos{\'e} and {Brownstein}, Joel and {Argudo-Fern{\'a}ndez}, Mar{\'\i}a and {Blanton}, Michael and {Bundy}, Kevin and {Jones}, Amy and {Masters}, Karen and {Law}, David R. and {Rowlands}, Kate and {Weijmans}, Anne-Marie and {Westfall}, Kyle and {Yan}, Renbin},

Marvin: A Tool Kit for Streamlined Access and Visualization of the SDSS-IV MaNGA Data Set. 2019. **158**.
5. Koes, D.R., Camacho, C.J., *ZINCPharmer: pharmacophore search of the ZINC database.* Nucleic Acids Research 40, 2012.
6. Pettersen EF, G.T., Huang CC, Couch GS, Greenblatt DM, Meng EC, Ferrin TE., *UCSF Chimera--a visualization system for exploratory research and analysis.* J Comput Chem., 2002. **25(13)**.
7. Forli, S., et al., *Computational protein–ligand docking and virtual drug screening with the AutoDock suite.* Nature Protocols, 2016. **11**(5): p. 905-919.
8. A. Schüller, V.H., G. Schneider, *SmiLib v2.0: A Java-Based Tool for Rapid Combinatorial Library Enumeration.* QSAR & Combinatorial Science, 2007.
9. Daina, A., O. Michielin, and V. Zoete, *SwissADME: a free web tool to evaluate pharmacokinetics, drug-likeness and medicinal chemistry friendliness of small molecules.* Sci Rep, 2017. **7**: p. 42717.
10. Patil S, H.J., Ballester PJ, Fattakhova E, DiFlumeri J, Campbell A, et al., *Drug Repurposing for Covid-19: Discovery of Potential Small-Molecule Inhibitors of Spike Protein-ACE2 Receptor Interaction Through Virtual Screening and Consensus Scoring.* ChemRxiv. Cambridge: Cambridge Open Engage, 2020.

VIRTUAL SCREENING OF MILK PROTEINS OF RISPERIDONE FOR TASTE IMPROVEMENT; EXPERIMENTAL VALIDATION OF SODIUM CASEINATE BINDING BY FLUOROMETRY AND TASTE ASSESSMENT BY PALATABILITY

Bandoo C. Chatale[a,c], Laxmi Ananthanarayan[b] and Mariam S. Degani[a*]

[a]*Department of Pharmaceutical Sciences &Technology Institute of Chemical Technology, Mumbai 400019, India,* [b]*Food Engineering and Technology Department, Technology Institute of Chemical Technology, Mumbai 400019, India,* [c]*MET Institute of Pharmacy, MET Complex, Bandra Reclamation, Bandra (West), Mumbai, India- 400050.*
*Corresponding Author E-Mail ID: ms.degani@ictmumbai.edu.in

1 INTRODUCTION

Taste is a key facet in the improvement of oral food products, and oral formulations. Chemical compounds react with taste receptors on the tongue in the oral cavity to produce the experience of taste. The five basic tastes are umami, sourness, sweetness, bitterness, and saltiness;although bitterness is the most challenging taste for oral food products since it is intuitively associated with negative health effects.[1] By using a variety of techniques, including coating the drug particle with sugar, polymers, lipids, gelatin, collagen, protein, ion exchange resins, surfactants, inclusion complexation with β-cyclodextrin, 2-hydroxyl-propyl-β-cyclodextrin, and pro-drugs synthesis, thus taste masking is accomplished physically and chemically.[2]

A biopharmaceutics categorization (BCS) class II medication with high potency, risperidone (RIS), is frequently prescribed to treat schizophrenia. The main drawback of RIS is that it has a bitter taste.[3] Despite the fact that RIS is an old medication, more palatable formulations are clearly needed. The following are a few research to increase palatability: Orodispersible tablets of RIS-cation exchange resin complexes were employed to mask the bitter taste of RIS in ratios of 1:1 for RIS-Kyron T-104, 1:2 for RIS-Indion 204 complex, 1:4 for RIS-Amberlite IRP 64 resin, and 2:1 for RIS-cetyl alcohol complexes.[4]

Milk proteins are generally recognized as safe (GRAS) material, that is why milk-proteins are promising for taste improvement. Previously calcium caseinate together with lecithin was employed for masking the bitter taste of acetaminophen.[5] Milk proteins are branched into two classes mainly, caseins and whey proteins (β-lactoglobulin (β-LG), α-lactalbumin, and lactoferrin).

Caseins are a family of phosphoprotein, principally four primary varieties including αS1, αS2, β and κ-casein.[5] Caseins have an arbitrary and open structure as they contain a lot of proline and few cysteine residues. Owing to this open random coil structure along with high hydrophobic regions, caseins have strong propensity to associate with themselves, other proteins and several other ligands. NaCAS (sodium salt of casein) is produced from acid casein.[5]

The native form of β-LG with 162 amino acid residues, made up of eight antiparallel β-strands called calyx or β-barrel is a conical structure with α-helix placed on the exterior surface of the β-barrel. The native form of α-lactalbumin is made up of 123 amino acids with 4 disulfide bridges, consists of two domains that are coupled by a calcium-binding loop: a large α-helical domain and a small β-sheet domain.[6]

Lactoferrin is an 80-kDa iron-binding glycoprotein containing 700 amino acids and comprises of a single-chain polypeptide with two gobular lobes. [6]

Research work has been reported to understand the interaction between NaCAS and polysaccharides such as NaCAS/gum arabic complexes formed via electrostatic interaction. Chitosan and NaCAS have formed nanoparticles in aqueous solution. Casein was able to form complexes with polyphenols, such as epigallocatechin gallate, gallic acid.[7]

The ability of the protein to bind ligands has been proven by diverse methods that consist of electron spin resonance spectroscopy, equilibrium dialysis, affinity chromatography, and NMR spectroscopy. One of the simplest procedures was perturbation of the intrinsic tryptophanyl (Trp) fluorescence. This effect was particularly noticeable in the cases in which the Trp fluorescence decreases on ligand addition.[8]

The primary criterion for evaluating oral formulations is taste. Various in-vitro and in-vivo methods, including animal studies, electronic tongues, and human taste panels, are used to evaluate how well compositions hide flavours. The renowned human panel for taste assessment approach has drawbacks, including being costly and tedious, vulnerable to bias and human inaccuracy, raising ethical questions about human taste panels, and making it challenging to match adult taste perception to paediatric dosage formulas.[9] The aim of the present research was identifying milk proteins for improvement of taste for RIS which was elucidated through molecular modelling, mainly the molecular docking used for screening milk proteins and casein was found the effective mode of binding with RIS. Subsequently, interaction studied of RIS with casein were performed using induced fit docking (IFD), MMGBSA, and fluorometry. Casein protein structure not elucidated by X-ray crystallography or NMR spectroscopy so homology model building was performed.[10]

Animals have also been used in vivo for taste improvement formulation evaluation as a potential substitute to human taste panels and electronic tongues.[9] The approaches that have been previously reported utilising rats use a laser-based instrument gadget to detect rat licks, but they are expensive.[10] Rats in particular are very helpful for studies on taste perception as they exhibit many of the same physiological and behavioural traits as people. For instance, same taste sensitivity.[10] According to reports, the single bottle test using rat model is an effective method for evaluating the taste of taste-masked solutions.[11]

Taste improvement potential assessment was performed by previously reported rat's palatability study using single bottle test of the trained model of male rats (Albino Wistar strain).[11]

2 METHODS AND MATERIALS

2.1 Materials

RIS was taken gift sample from Cipla Ltd. Mumbai, India. NaCAS was purchased from Alfa aesar (A Johnson Matthey Company).

2.2 Methods

2.2.1 Molecular modelling. The Schrodinger Suite was used for all molecular modeling tasks.

2.2 Structure building

The structure of RIS was drawn utilizing a 2D sketcher of Schrödinger maestro 2017-1. RIS was developed for docking using LigPrep modules where using OPLS2005, the structure was transformed to 3D and its energy was optimized. For RIS at pH 6.8, the least energetic ring conformation was produced in LigPrep using the Epik module and all other default settings.

2.2.2 Protein preparation. 3D structure of α-lactalbumin, β-Lactoglobulin (β-LG), and lactoferrin were recovered from protein data bank (PDB ID 1A4V, 3BLG, 1JW1) and casein homology model were clarified using Protein preparation wizard module[c] wherein bond order was established when missing residues, side chains from amino acids, and hydrogen atoms were added. These proteins subjected to pre-processing, and optimized to create H-bond network, remove waters with less than 3 Å distance, and using OPLS2015 force field minimized to remove any steric clashes.

2.2.3 Homology model building of casein. Sequence Alignment: The protein sequence of casein protein was retrieved from Uniprot Knowledgebase database using accession no. P02666. BLASTp search within the Protein Data Bank (PDB) was used to search the template for homology modelling with best sequence homology with casein. PDB ID 1K0M template was found identical with casein of alignment for homology. 1K0M was the crystal structure of a soluble monomeric form of intracellular chloride ion channel CLIC1 with resolution 1.4 angstroms and which was downloaded from the protein data bank.[12]

Model Building: The homology model of human casein was made with PRIME module[b] using the knowledge-based (faster) composite/chimera modelling protocol (Figure 1). The homology model was subjected to various refinement protocols such as optimization of side-chains and refinement of loops. The entire structure was then energy-minimised using a reduced-Newton energy minimization protocol. The OPLS-2005 all-atom force field was used and solvation energies were treated via the Surface Generalized Born (SGB) continuum solvation model.[13] The constructed casein structure was examined to ensure it was acceptable using the phi/psi distributions of the Ramachandran plot that were acquired using the PROCHECK analysis.[14]

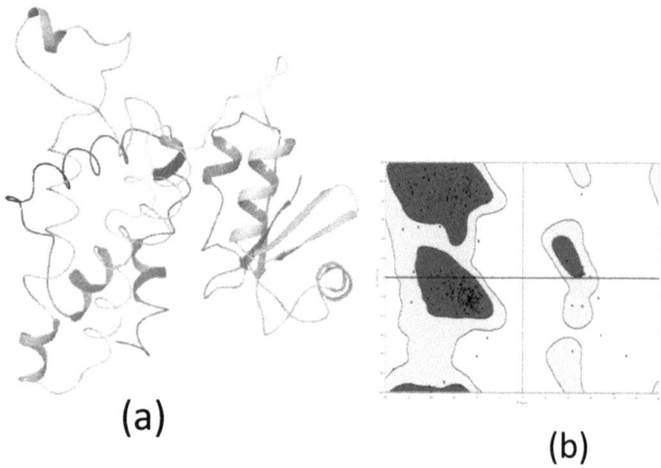

Figure 1 *(a) Homology model casein (b) Ramachandran plot of casein homology model.*

2.2.4 Active site prediction and binding mode prediction using molecular docking of RIS with milk proteins. The fully optimized proteins structures subjected for active site prediction using the SiteMap module was used to characterize the milk proteins components such as the binding pockets and functional residues, SiteMap uses grid points to identify areas of the protein surface that might be appropriate for interacting with ligands. Fully optimized proteins structures of β-LG, lactalbumin, lactoalbumin, lactoferrin, homology model of casein were used for the receptor grid generation using centroid of selected residues which were predicted utilizing SiteMap module[d] having sitescore 1.228, 0.560, 1.194, 1.105 and 1.027 respectively. The optimized RIS was docked to proteins grid using glide module[e] the automated docking program implemented in the Schrodinger package.[15]

2.2.5 Induced Fit Docking (IFD) of RIS with casein. Completely optimized protein, casein structure was utilized for generating the protein grid utilizing centroid of selected residues which were predicted utilizing Site-Map module.[d] The optimized RIS was induced fit docked to protein grid using induced fit docking program[f] implemented in the Schrodinger package. The graphical user interface included within maestro 2017-1 was used to run the IFD protocol. Extended sampling protocol of IFD was utilized for docking the RIS using the Induced fit docking program.

2.2.6 The Molecular Mechanics-Generalized Born Surface Area (MMGBSA) study. The complex of the RIS with casein was subjected to MM-GBSA method using PRIME module[b].

2.2.7 RIS with NaCAS interaction assessment using fluorometry. Potassium chloride, Sodium chloride salts along with disodium hydrogen phosphate, and potassium dihydrogen phosphate were dissolved in double-distilled water to prepare a phosphate buffer with a pH of 6.8, and diluted HCl was used to modify the pH.[15]

A small amount of ethanol (10% v/v) was added to the nearly insoluble RIS to make it soluble for the production of 1 mM. To keep the same components in both solutions and prevent fluorescence interferences, a RIS 0.5 mM solution of NaCAS was made in phosphate buffer 6.8 pH and 10% ethanol was added to this. Using quartz 96 well plates, fluorescence spectrum scanning was done to examine how RIS quenches the intrinsic fluorescence of NaCAS (sodium salt of casein).

Fluorescence quenching studies were conducted using a well plate with a maximum volume capacity of 300 µl per well and an experimental volume maintained at 200 µl per well to maintain a constant path-length and volume. Fluorescence is affected by pH fluctuations, hence phosphate buffer pH 6.8 was employed [39]. For the complexation tests, molar ratios of NaCAS to RIS were taken at 1: 2, 1:4, 1:6, 1:8, 1:12, 1:14, and 1:18. RIS and NaCAS solutions were used as controls. NaCAS's intrinsic fluorescence was quenched by adding RIS (a quencher) in descending concentrations of 100, 200, 300, 400, 600, 700, 800, and 900 M while maintaining a constant concentration of the protein at 50 M. At an excitation wavelength of 295 nm, the fluorescence from each quartz well plate was individually measured over the emission range of 350 to 550 nm.

2.2.8 In-vivo taste assessment and statistical analysis: ANOVA. Screening of taste mask solutions of NaCAS: RIS in the ratios of 1:1, 2:1, and 3:1 were performed using rats of Wistar stain, female sex, in three groups with average weights of 255.7 g, 231.5 g, and 208.8 g respectively. Rats were accommodated in cages includes a continuous 12-hour light/dark cycle, bedding made of corncobs, and a 22 °C ambient temperature. With the prior consent of the Institutional Animal Ethics Committee (IAEC) at Institute of Chemical Technology (ICT), Mumbai, India (protocol number: IAEC/ICT/2017/044), a slightly modified version of the reported approach for rats palatability for taste assessment was used.[11]

Oral lethal dose (LD50) of rat is 56.6 mg/kg as per material data safety sheet of Cayman Chemicals. RIS doses for rats were calculated as per average body weights of seven rats in groups and RIS dose was found 3.2 mg/0.3 kg per body weights. RIS has practically insoluble in water therefore small quantity of ethanol (0.5 % v/v) used for solubilisation of RIS and it was added in 50 mL purified water to form bitter drug, RIS solution. Then 1:1, 2:1, 3:1 molar ratios of NaCAS to RIS were added 50 mL purified water to form NaCAS taste mask RIS solution.

2.2.9 Training paradigm. Seven rats were employed for study. On day 1, rats were given access to water for 1.5 hr. The rats were deprived of water for 22 hrs, to determine response to thirst; in this study they were given access to food. After a period of water deprivation, for 30 mins the rats were supplied with 50ml of drinking water and the amount ingested was recorded. Four days of training were accomplished prior to starting with the research.

Experiment: 24hrs were required for each experiment. In the initial part of the experiment, rats were provided with water for 1.5hr. According to schedule, for 30 min rats were given 50 mL of test solution in drinking bottle itself, followed by 22 hrs water deprivation. Later, for 30mins rats were given free access to 50 mL of test solution in graduated drinking bottles in accordance to the schedule. The bottles were then removed to record the average consumed volume.

RIS-NaCAS solution in 1:1, 1:2, 1:3 molar ratios were analyzed utilizing one way ANOVA (GraphPad Prism 6). Volume of consumed water (control) was compared with test solutions such as RIS solution, sucrose solution and RIS-NaCAS taste masked solution (n=7). The variation in the volume consumed by rats was analyzed statistically by a one-way analysis method ANOVA is used before the Tukey multiple comparison test, in a 1:1 or 1:2 solution for RIS-NaCAS. A difference of $P > 0.001$ was considered statistically non-significant with respect to days defined in the figure legends. For the RIS-NaCAS solution in 1:3, a difference of $P < 0.001$ was regarded to be statistically significant in relation to the days specified in the figure legends (Figure 7, 8, 9).

3 RESULTS AND DISCUSSION

3.1 Molecular docking of RIS with milk proteins

The screening of milk protein for taste masking of bitter drug RIS was performed using molecular modelling technique such as molecular docking. Interaction studies of RIS with sodium caseinate (NaCAS) was studied using molecular modelling techniques including homology modelling, induced fit docking, molecular dynamic (MD) simulation and the Molecular Mechanics-Generalized Born Surface Area (MMGBSA) were used to study interaction of the RIS with NaCAS.

3.2 Ligand binding site prediction

The probable binding sites of the protein, β-LG, lactalbumin, bovin serum albumin and homology model of casein were mapped by Site-Map of module.[d] The probable binding sites of the protein, β-LG, lactalbumin, bovin serum albumin and homology model of casein were mapped by SiteMap Figure 3) and scored in accordance with various criteria such as cavity size, solvent exposure, protein enclosure, the balance between hydrophilicity and hydrophobicity and the degree to which a ligand might donate or accept hydrogen bonds. For β-LG Two sites, site 1 & site 2 (Figure 2). Mapping of entire β-LG revealed that yellow color hydrophobic, blue color H-bond donor region were found on site 1 and site 2 showed the blue color H-bond donor and red color H-bond acceptor regions were found on site 2. Blind docking was used with ligands with known interactions to validate the ligand binding sites on β-LG.

For lactalbumin, mapping of entire lactalbumin revealed that only site 1 was found which were include larger portion of blue colour H-bond donor, medium portion of red color H-bond acceptor and small portion of yellow color hydrophobic region.

Homology model of casein mapping of entire proteins revealed that five sites, site 1, 2, 3, 4 and 5 were found and on which yellow color hydrophobic, blue colour H-bond donor region were found.

3.3 The screening of milk protein for taste masking of RIS using molecular docking

Fully optimized proteins structures of β-LG, lactalbumin, lactoalbumin, homology model of casein were used for the receptor grid generation using centroid of selected residues which were predicted utilizing Site-Map module[d] having sitescore 1.228, 0.560, 1.194, and 1.027 respectively.

Table 1 *RIS-Milk protein interaction analysis.*

RIS dock on milk protein	Docking score	RIS-Milk protein interaction analysis
RIS dock on β-LG	-3.457	Hydrophobic with VAL41, LEU58, ALA86, VAL92, PHE105
RIS dock on lactalbumin	-2.817	polar with GLN54, SER56, THR33, HIS32; Electrostatic with GLU49, ASP102, LYS58, LYS99; hydrophobic with VAL42, ALA106, LEU105, TRP104, TYR103; Pi cation with TYR103
RIS dock on lactoferrin	-0.399	H-bond TYR524; polar with ASN642, ASN638, THR636, ASN551, SER557,SER556, THR527, SER519; Electrostatic with GLU354, GLU635, GLU521, ARG463, LYS520, LYS522, LYS637,ARG531; hydrophobic with PHE641, LEU640,LEU639,TRP560,ALA558, TYR523, TYR524,TYR526; Pi cation with ARG531

RIS dock on casein	-4.454	H-bond with GLN161; polar with GLN161, GLN-164, HIS65, GLN69, GLN161, THR169; Electrostatic with GLH20, GLH19, GLH17, GLU-52; hydrophobic with LEU18, PRO165, ILE-64, LEU18, PHE67, ALA68, PRO86, LEU85, LEU92

RIS docked on β-LG, lactalbumin, lactoferrin, and homology model of casein and docking score were -3.457, -2.817, -0.399, -1.193, -4.454 respectively. The molecular docking was effectively used for screening milk proteins for RIS and casein was found effective mode of binding with RIS.

3.4 The binding mode prediction using IFD of RIS with casein

RIS was induced fit docked on the Site-Map specified binding cavity of Site Map 1 with amino acid residues of casein and it was shown the best overall docking score for the RIS and was finalized as the most probable active site for this protein. The scoring from IFD calculation was the computed GlideScore reflecting the interaction of RIS with protein, casein. The IFD study was used to elucidate how RIS bind to the casein.

The IFD results provided appropriate details regarding the RIS-casein interactions, binding affinity, binding energy, and orientation. The knowledge of the binding location of the RIS on casein was the prerequisite for understanding the molecular level interaction (Figure 2). It exhibits better interactions results into strong binding. It was recognized that protein flexibility incorporated in docking studies to more realistically understanding of casein–RIS binding. Induced fit docking was helpful for recognize the internal cavity of protein as the probable binding site for the drug RIS.

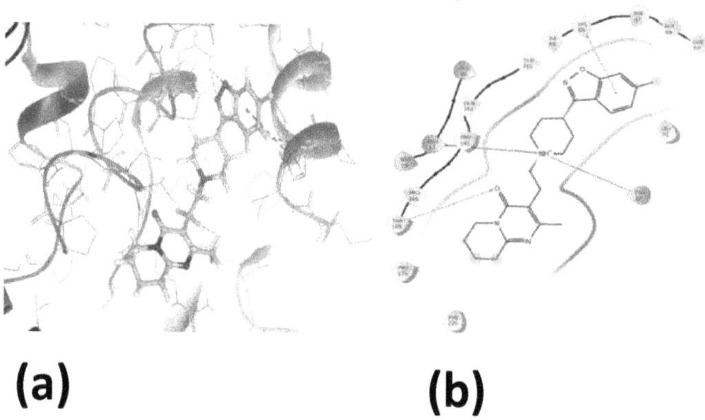

Figure 2 *(a) RIS IFD pose with casein (b) interaction of RIS IFD pose with casein.*

Table 2 *Binding affinity analysis of RIS with casein.*

Ligand	Protein	MMGBSA dGBind (kcal/mol)	MMGBSA dG Bind contribution (kcal/mol)			
			Coulomb	Vdw	Lipo	Solv GB
RIS	Casein	-99.13	-71.24	-30.36	-1153	70.28

3.5 Binding affinity analysis

The MM-GBSA (molecular mechanics-generalized born surface area) is used to estimate relative binding affinity for IFD complex of the RIS with NaCAS. The binding strength of RIS with casein determined by the binding free energy, ΔG_{bind} and its shows binding Affinity (dG binding energy) constant with casein (Table 2).

3.6 Quenching of intrinsic fluorescence of NaCAS protein by RIS

For confirm stability of RIS-casein complex predicted by molecular modelling fluorescence quenching study performed. NaCAS contain tryptophan amino acid which having an excitation at 295 nm and an emission range of longer wavelength of 350 to 550 nm. Intrinsic fluorescence of tryptophanyl amino acid of NaCAS undergoes progressive quenching with incremental addition of the drug, RIS. The phenomenon of fluorescence was described by Stern Volmer equation as:

$$F_0/F = 1 + Kq\,\tau(Q) = 1 + Ksv\,(Q) \quad (1)$$

Here, Intensity of fluorescence without a quencher is F_0. Intensity of fluorescence with a quencher, (here RIS) is F. Stern volmer quenching constant is Ksv. Quenching rate constant is Kq. Lifetime of the emissive excited state of the protein, NaCAS is τ. Molar concentration quencher (here RIS) is Q.

3.7 Fluorometric behaviour of RIS-NaCAS complex

The interactions of RIS with NaCAS were studied by fluorometry. The maximum excitation of NaCAS was located at 295 nm and fluorescence emission was acquired in the range of 350 to 550 nm. The fluorescence emission progressively decreased with incremental addition of RIS, indicating the quenching of fluorescence intensity of β-LG upon binding to RIS. This was attributed to the formation of RIS-NaCAS complex (Figure 3).

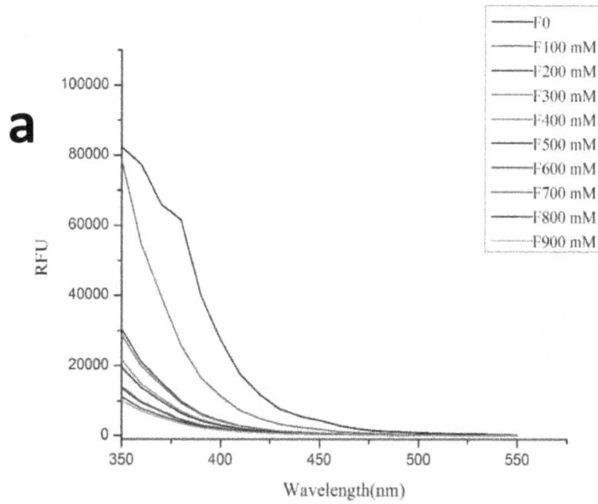

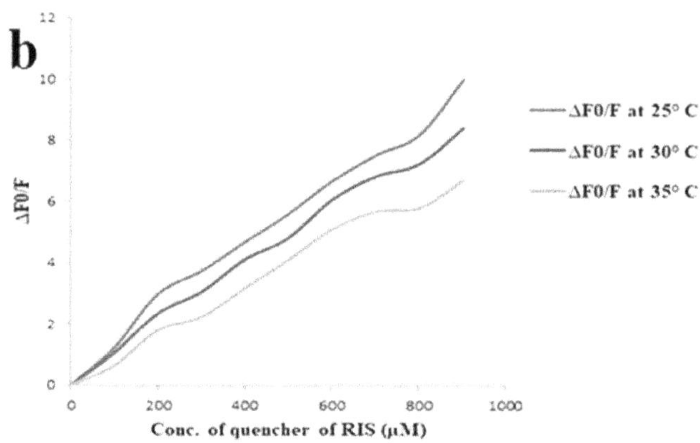

Figure 3 *(a) Fluorometric behaviour of RIS-NaCAS complex (b) Change in fluorescence of protein, sodium caseinate versus concentration of quencher, risperidone at temp 25, 30 and 35 ^{0}C.*

Experiments were performed for quenching of intrinsic fluorescence of NaCAS, by RIS at 25, 30, and 35 °C temperature. In order to determine the mechanism of quenching by RIS on NaCAS, data on fluorescence quenching were evaluated using the Stern-Volmer equation.

As the temperature rises, the slope of the curve declines, obviously confirming static quenching (Figure 3b).

3.8 In-vivo taste assessment

All these three groups of the Wistar rats were provided training as per authorised protocol before performing experiments. After 30 minutes of test solution access, the total amount of test solutions eaten was noted. For experiments, at the end of the 30-minute experiment on day 1, the total amount of water consumed was about 16.7-17.2 mL in three groups, and neutral mouth movements were noticed, thereby suggesting that the rats taste perception was comparatively neutral for water. On the 2nd day of experiment, sucrose-water solution (1 % w/v) was used as a sweet taste solution, at end of 30 min experiments the total average volumes ingested were approximately 14.2-17.7 mL. The volume of the sucrose-water solution consumed indicates that sucrose was palatable to rats. On 3rd day, a bitter drug, RIS solution, was provided, at the end of 30 min experiments the total average volumes consumed were approximately 9.5-10.5 mL, a decrease in consumption of volume of RIS solution observed. Rats displayed alterations in their jaw, smacking, oral grooming, and paw shaking activities, showing that they preferred the taste of sucrose over bitter medicines. The rats were given water again on day 4 to avoid any false-positive or false-negative observations in the succeeding trials because rats have taste perception recall. The total average quantities consumed at the conclusion of the 30-minute experiments were roughly 13.7-17.2 mL. On the 5th, each rat in groups were given sucrose-water solution (1 % w/v), at the end of 30 min experiments total average volumes consumed were approximately 15.1-17.8 mL. On the 6th day, taste mask solutions of NaCAS: RIS 1:1, 2:1, 3:1 ratio were given to each groups. Taste masking potential of sodium caseinate with risperidone in ratios of 1:1, 2:1, 3:1 were accessed, average volumes of total consumption at end of 30 min experiments were approximately 12.5 mL, 13 mL, 17.8 mL and it was found that 3:1 ratio, notable increase in the solution consumption improved as compared with 1:1, 2:1 ratio. After 30 min of access to water, sucrose solution, and taste mask solution of NaCAS: RIS 1:1, 2:1, 3:1 ratios, total volume consumed by rats recorded. The measurement of total average volumes of the test solution consumed, this developed rats model has psychophysical taste perception characteristics (facial/body expression) similar to those observed in children.

3.9.1 Statistical analysis: ANOVA. Significant variances between the test solutions and the control were found using one way ANOVA (Figure 6). When RIS was compared to water, there was a substantial variance in volume consumption (p 0.0001). Rats consumed RIS and RIS-NaCAS in a ratio of 1:3 orally, and this revealed a substantial variance (p 0.001), at the same time volume consumption with RIS-NaCAS of ratios of 1:1 and 1:2 were showed the non-significant variance ($p > 0.001$) compared with RIS. Furthermore RIS-NaCAS solution and sucrose solution demonstrated that the bitterness of RIS was greatly reduced while showing no significant change from control (Figure 4, 5, 6).

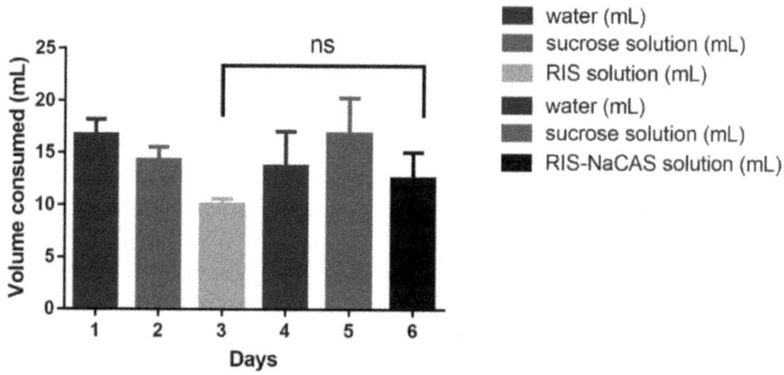

Figure 4 *Test sample volume intake: *(p <0.0001), substantial variance in volume of the RIS solution (day 3) consumed as compared to water (day 1, 4); @(p < 0.0001), substantial variance in volume of the RIS solution (day 3) consumed to sucrose solution (day 2, 5); and # (p <0.001), substantial variance in volume of the RIS solution (day 3) consumed as compared to RIS-NaCAS solution in 1: 1 ratio (day 6) [11] (In vivo taste assessment).*

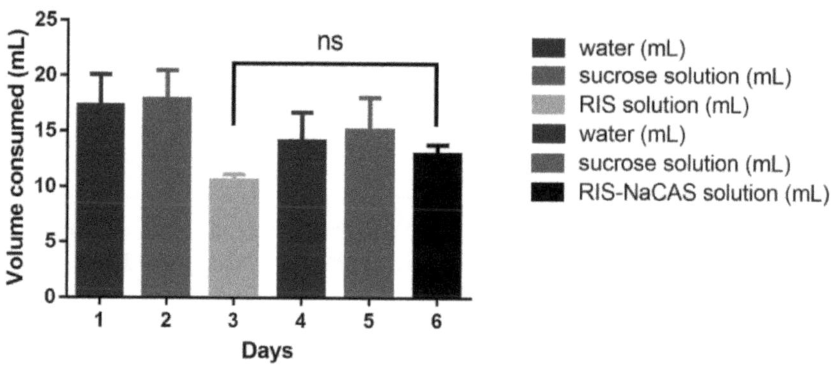

Figure 5 *Test sample volume intake: *(p <0.0001), substantial variance in volume of the RIS solution (day 3) consumed as compared to water (day 1, 4); @(p < 0.0001), substantial variance in volume of the RIS solution (day 3) consumed to sucrose solution (day 2, 5); and # (p <0.001), substantial variance in volume of the RIS solution (day 3) consumed as compared to RIS-NaCAS solution in 1: 2 ratio (day 6) [11] (In vivo taste assessment).*

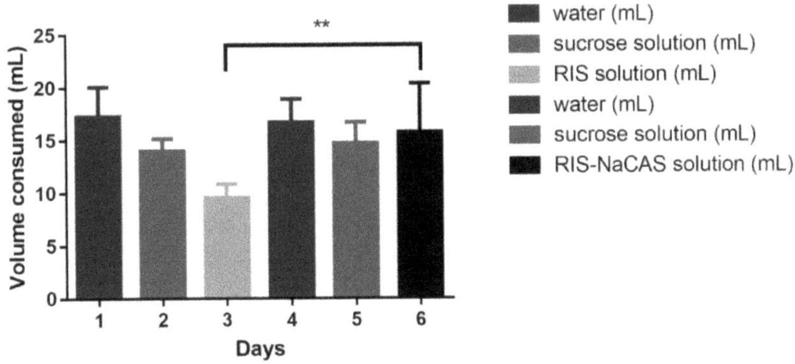

Figure 6 *Test sample volume intake: *(p <0.0001), substantial variance in volume of the RIS solution (day 3) consumed as compared to water (day 1, 4); @(p < 0.0001), substantial variance in volume of the RIS solution (day 3) consumed to sucrose solution (day 2, 5); and # (p <0.001), substantial variance in volume of the RIS solution (day 3) consumed as compared to RIS-NaCAS solution in 1: 3 ratio (day 6)[11] (In vivo taste assessment).*

4 CONCLUSIONS

The fluorescence studies as well as molecular modelling indicate good binding of RIS with NaCAS. The static quenching indicated that a stable complex of RIS with NaCAS is obtained. The previously described single bottle-test rat model was used to measure and compare the volume ingested of the RIS solution and the RIS-NaCAS solutions for the palatability investigation for taste improvement. in molar ratios of 1:1, 2:1, 3:1 and it was noted that 3:1 ratio solution palatability improved in comparison with 1:1, 2:1 ratios thus it was found that complex formation in between NaCAS with RIS in 3:1 ratio was effective for taste improvement of RIS and thus that milk protein sodium caseinate was shown to greatly minimize RIS's bitterness. Therefore, these research could assist to lessen the bitterness of RIS.

5 ACKNOWLEDGEMENTS

Bandoo Chatale thanks to UGC-BSR, India, for the Research Fellowship and Cipla Ltd Mumbai, India for donation of RIS sample.

6 REFERENCES

Notes

[a]LigPrep, Schrödinger, LLC, New York, NY, 2017.
[b]PRIME, Schrödinger, LLC, New York, NY, 2017.
[c]Protein preparation wizard, Schrödinger, LLC, New York, NY, 2017.
[d]Site Map, Schrödinger, LLC, New York, NY, 2017.

[e]Glide, Schrödinger, LLC, New York, NY, 2017.
[f]Induced fit docking, Schrödinger, LLC, New York, NY, 2017.

1. D. V. Smith and R. F. Margolskee, *Scientific American*, 2001, **284**, 32-39.
2. V. D. Wagh and S. V. Ghadlinge, *Journal of Pharmacy Research Vol*, 2009, **2**.
3. M. A. Tawakkul, R. B. Shah, A. Zidan, V. A. Sayeed and M. A. Khan, *Pharmaceutical development and technology*, 2009, **14**, 409-421.
4. D. Shukla, S. Chakraborty, S. Singh and B. Mishra, *Chemical and Pharmaceutical Bulletin*, 2009, **57**, 337-345.
5. C. De Kruif and C. Holt, in *Advanced Dairy Chemistry—1 Proteins*, Springer, 2003, pp. 233-276.
6. C. Holt and L. Sawyer, *Journal of the Chemical Society, Faraday Transactions*, 1993, **89**, 2683-2692.
7. I. Hasni, P. Bourassa, S. Hamdani, G. Samson, R. Carpentier and H.-A. Tajmir-Riahi, *Food Chemistry*, 2011, **126**, 630-639.
8. L. Sawyer, S. Brownlow, I. Polikarpov and S.-Y. Wu, *International Dairy Journal*, 1998, **8**, 65-72.
9. S. Muresan, A. van der Bent and F. A. de Wolf, *Journal of agricultural and food chemistry*, 2001, **49**, 2609-2618.
10. R. K. Palmer, D. Long, F. Brennan, T. Buber, R. Bryant and F. R. Salemme, *PloS one*, 2013, **8**, e72391.
11. B. C. Chatale, and M.S. Degani, *Drug Development and Industrial Pharmacy*, Volume 45, 2019 - Issue 10.
12. J. Sherman, K. W. Rusiniak and J. Garcia, in *Recent developments in alcoholism*, Springer, 1984, pp. 59-79.
13. S.J. Harrop, DeMaere, M.Z., Fairlie, W.D., Reztsova, T., Valenzuela, S.M., Mazzanti, M., Tonini, R., Qiu, M.R., Jankova, L., Warton, K., Bauskin, A.R., Wu, W.M., Pankhurst, S., Campbell, T.J., Breit, S.N., P.M. Curmi, *J Biol Chem* 276: 44993-45000.
14. R. A. Laskowski, M. W. MacArthur, D. S. Moss, J. M. Thornton (1993).. *J. App. Cryst.,* 26, 283-291.
15. B. C. Chatale, L. Ananthanarayan, and M.S. Degani *In Conference on Drug Design and Discovery Technologies, Royal Society of Chemistry* (Vol. 355, p. 319).

IN-SILICO EVALUATION OF PHYTOCHEMICALS AS POTENTIAL ANTIEPILEPTIC DRUGS

Harshini Anand[a*], B. Sushmitha[a], Ritrick Dey[a], M. Akhilan[a], S Shiva[a] and AR. Mahesh[a*]

[a]Department of Pharmaceutical Chemistry, Faculty of Pharmacy, Ramaiah University of Applied Sciences, Bengaluru

Corresponding Author E-Mail ID: mahesh.py.ph@msruas.ac.in, reachharshini.anand@gmail.com

1 INTRODUCTION

Epilepsy is a chronic neurological disorder characterized by recurrent episodes of seizures. The general pathophysiology behind a seizure is linked with excessive and abnormal electrical activity in parts of the brain. [1] The established causes of epilepsy vary with the population. Childhood onset of epilepsy could be due to genetic issues, while epilepsy in adults could be due to history of stroke, traumatic head injury, infection in the brain and brain tumor. [2,3]

Anti-Seizure Drug Therapy (ASD) is the primary method of treatment of epilepsy. Non-pharmacological therapy is rarely used. The goal of ASD is to relieve the symptoms of seizures, with minimal side effects. The drug therapy for epilepsy depends on the type of seizures experienced. [4,5,6,7] The target receptors/ pathways for ASD and classification of antiepileptic drugs, based on their mechanism of action is shown in Table 1.

Table 1.1 *Antiepileptic Drugs and their Mechanisms of Action.* [3]

Decreasing Neuronal Excitability	Reported Examples of Drugs
Blockade of Voltage Gated Sodium Channel	Phenytoin, Carbamazepine, Lamotrigine, Sodium Valproate
Inhibition of Low-Threshold T-Type Calcium Channels	Ethosuximide, Zonisamide, Lamotrigine, Sodium Valproate

Enhancement of the Action of GABA	Reported Examples of Drugs
Increasing GABA Synthesis	Gabapentin, Pregabalin
Blockade of GABA Transaminase	Vigabatrin, Sodium Valproate
Inhibition of GABA Reuptake	Tiagabine
Allosteric Modulation of GABA Receptors	Benzodiazepines, Barbiturates
Blockade of NMDA and/or AMPA Receptors	Reported Examples of Drugs
Blockade of NMDA Receptors	Felbamate
Blockade of AMPA Receptors	Topiramate, Lamotrigine, Phenobarbital

The most used antiepileptic drugs today are Lamotrigine, Gabapentin, Levetiracetam, Phenytoin, Zonisamide and Carbamazepine. [3,4] Anti-Seizure Drug Therapy has made a great progression over the years and is effective in managing symptoms in about 70% of the cases. However, there are various limitations to be overcome. Common side effects linked with most of the antiepileptic drugs include sedation, dizziness, blurred vision, stomach upset and ataxia. [10,11]

A shift to the usage of herbal medicines can be considered to overcome some of the existing limitations. Herbal medicines have been used since times immemorial in countries and continents like India, China, and Africa. They are more economical and have better accessibility. [12,13,14,15]

To optimize their antiepileptic activity, it is necessary to perform *in-vitro, in-vivo* and *in-silico* studies. [16,17,18] This study aims to predict the mechanism of action, pharmacokinetic and toxicological properties of phytochemicals isolated from plants, who's *in-vivo* / *in-vitro* studies have already proven to show anti-epileptic activity. [19]

The current study aims to identify potential receptors for these phytoconstituents' binding and forecast their pharmacokinetic, toxicological, and action mechanisms. To ascertain their probable binding mechanisms, molecular interactions were examined using *in-silico* molecular docking research and validated docking methodologies. To ascertain the pharmacokinetic and toxicological characteristics of phytochemicals, those with favorable docking scores were sent for *in-silico* ADME/T (Absorption Distribution Metabolism Excretion Toxicity) research. Molecular Dynamics simulations were used to validate the study's hit molecule.

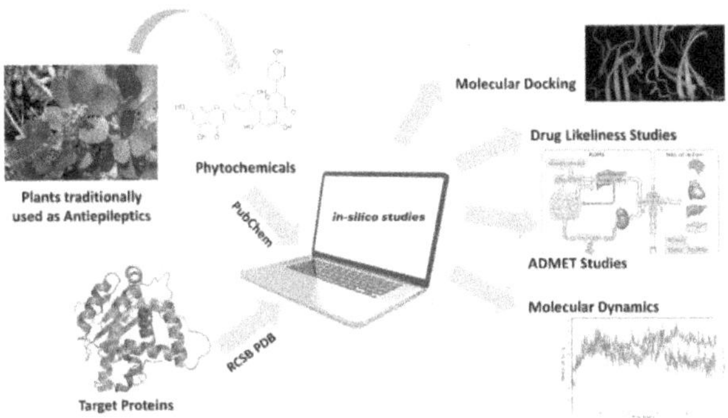

Figure 1 *Graphical Abstract.*

2 METHODOLOGY

Docking studies were carried out to analyse the different types of biomolecular interactions and ligand receptor binding affinities. The docking study was performed on nine proteins namely crystal structure of AMPA Receptor, Carbonic Anhydrase 2, Connexins, GABA A Receptor, GABA B Receptor, HCN1, NMDA Receptor, Synaptic Vesicle Protein and Voltage Gated Sodium Channel proteins with PDB id's: 3R7X, 3F8E, 6L3V, 6D6T, 4MS3, 5U6P, 5EWJ, 4V11, 6AGF respectively. The computational work was performed on Intel(R) Core (TM) i5-6300U CPU at 2.40GHz 2.50 GHz, Installed RAM 8.00 GB (7.35 GB usable).

2.1 Protein preparation

The Crystal structure of AMPA Receptor, Carbonic Anhydrase 2, Connexins, GABA A Receptor, GABA B Receptor, HCN1, NMDA Receptor, Synaptic Vesicle Protein and Voltage Gated Sodium Channel proteins with PDB id's: 3R7X, 3F8E, 6L3V, 6D6T, 4MS3, 5U6P, 5EWJ, 4V11, 6AGF respectively were retrieved from the RCSB Protein Data Bank, All the proteins were prepared by removing the other ligands using Swiss PDB viewer, the prepared proteins were saved in PDB format.

2.2 Identification of Active Pockets on Protein

The web portal "PrankWeb" was used to identify active pockets on the protein. Active Pockets are ranked according to their binding score. The pocket[s] with the highest ranking, namely pocket 1 was identified for use as the binding site for docking. The amino acids constituting the active pocket 1 were noted down.

2.3 Ligand preparation

The 3D structures of the ligands were drawn using chemsketch and uploaded in BIOVIA Discovery Studio Visualizer-2020. Ligand minimization was done and using small molecule wizard in 'SMALL MOLECULE' wizard in BIOVIA Discovery Studio Visualizer-2020 and was saved as a cluster SDF file.

2.4 Docking studies

To reduce false positives and to identify the perfect orientation of ligand within the active site of protein, docking study acquires its importance.

Docking was done using PyRx-Virtual Screening Tool. Converted all ligands to PDBQT in PyRx-Virtual Screening Tool and selected those as ligands in Vina wizard. Loaded the prepared proteins into the PyRx-Virtual Screening Tool and selected it as macro molecule. Calculation of interaction energy [interaction between ligand and receptor], with amino acids involved in binding was carried out.

2.5 Visualization of 3D and 2D Interactions between Ligand and Protein

2.5.1 Visualisation of 3D interactions between Ligand and Protein. Visualization of 3D interactions was done by using PyMOL. PDBQT files of both the ligand and protein were inserted into the PyMOL software and saved as Multifile in PDB format.

2.5.2 Visualization of 2D Interactions between Ligand and Protein. The saved 3D complexes of each ligand with the protein were opened in BIOVIA Discovery Studio to visualize the 2D interactions and identify the amino acids in the protein responsible for binding with the ligand.

2.6 Drug Likeliness Studies

The selected phytochemicals were loaded into DruLiTO in SDF format and the test for drug likeliness was carried out.

2.7 ADME/T Studies

The SMILES of the selected phytochemicals were loaded into Swiss ADME/T [http://www.swissadme.ch/] and the ADME/T properties of the same were recorded.

2.8 Molecular Dynamics

Simulation of Molecular Dynamics [MD] helps visualize the action of Protein-Ligand complexes [PLC] at the target's binding site region under physiological conditions. MD was performed using the Desmond module of Schrödinger developed by D.E Shaw research group [Academic license, Version 2020-1] 1. Through the system's builder panel; the orthorhombic simulation box was prepared with the Simple Point-Charge [SPC] explicit water model in such a way that the minimum distance between the protein surface and the solvent surface is 10 Å. Complexes docked with receptors were solvated with the orthorhombic TIP3P water model 2. Neutralization of the solvated system was accomplished by counter ions and limiting the salt concentration in the physiological system to 0.15 M. The receptor-ligand complex system was designated with the OPLS AA force field 3.

Two seconds of relaxation time was used for the Reversible Reference System Propagator Algorithms [RESPA] integrator 4, Nose-Hoover chain thermostat 5 and Martyna-Tobias-Klein barostat. The final production of MD simulations was performed using the equilibrated system. This MD simulation was set to run for 100 ns with 310 K temperatures at 1.0 bar pressure with NPT [Isothermal-Isobaric ensemble, constant temperature, constant pressure, constant number of particles] ensemble at default settings 6 for relaxation before simulation. The MD simulation was performed with the MD simulation tool, with the simulation time set to 100 ns. Furthermore, the .out file was used to view the trajectories and create a movie. The out. cms file was imported and the movie was exported at a higher resolution [1280x1024] with better quality. The trajectory was written with 1000 frames during the complete MD simulation. The protein backbone frames were aligned to the backbone of the initial frame to better understand the complex's stability during MD simulation. Finally, after loading the .out file and selected the Root Mean Square Deviation [RMSD] and Root Mean Square Fluctuation [RMSF] in the analysis type to oblique, the simulation interaction diagram and the results was analysed.

3 RESULTS AND DISCUSSION

3.1 Molecular Docking Studies

3.1.1 Docking of phytochemicals with 3R7X - AMPA Receptor. Among all the phytochemicals docked with AMPA Receptor, the phytochemicals in Table 3.1 have the best binding affinities. Palasitrin has the highest binding affinity of -7.4, while Perampanel, the standard, has a binding affinity of -5.2.

Table 3.1 *Docking Results of Phytochemicals with AMPA Receptor.*

Ligands	Binding Affinity	Hydrogen Bonds Interactions
Standard – Perampanel	-5.2	B-ARG101,B-ASP231
Palasitrin	-7.4	A-ASP101,A-PHE102,A-ARG231,A-ASN232,
Isocoreopsin	-7	A-GLU97,A-PHE102,A-ARG231,A-ASN232,B-PHE102,B-LYS226
Coreopsin	-6.7	A–GLU97,A–ILE100,A–LYS104,A–LYS226,A–ASN232,A–ASN235,B–GLU97,B-ASP101,B–PHE102,B–LYS104,B–ASN232,B–LEU236

Isobutrin	-6.6	A-GLU97,A-ASP101,A-PHE102,A-LYS104,A-ARG231,B-ASP101,B-LYS104,B-ASN232
Gallic Acid	-6.3	A-PHE102

3.1.2 Docking of phytochemicals with 3F8E - Carbonic Anhydrase 2. Among all the phytochemicals docked with Carbonic Anhydrase 2, the phytochemicals in Table 3.2 have the best binding affinities. Withanolide has the highest binding affinity of -8.5, while Sodium Valproate, the standard, has a binding affinity of -4.2.

3.1.3 Docking of phytochemicals with 6D6T – GABA-A Receptor. Among all the phytochemicals docked with GABA A Receptor, the phytochemicals in Table 3.4 have the best binding affinities. Glycyrrhizin has the highest binding affinity of -11.7, while Clonazepam, the standard, has a binding affinity of -10.7 kcal/mol.

Table 3.2 *Docking Results of Phytochemicals with Carbonic Anhydrase 2.*

Ligands	Binding Affinity	Hydrogen Bonds Interactions
Standard - Sodium Valproate	-4.2	-
Withanolide	-8.5	A-ASN62,A-HIS64,A-ASN67,A-GLN-92
Silydianin	-8	A-HIS64,A-ILE91, A-GLN92
Amentoflavone	-7.9	A-ARG58,A-ASN62,A-HIS64,A-ARG67,A-THR200
Isocoreopsin	-7.5	A-GLN92,A-LYS170,A-GLY171,A-THR200
Agathisflavone	-7.4	A-ASN62,A-HIS64,A-ASN67,A-PHE131

Table 3.4 *Docking Results of Phytochemicals with GABA A Receptor.*

Ligands	Binding Affinity	Hydrogen Bonds Interactions
Standard–Clonazepam	-10.7	D-HIS102,D-SER206
Glycyrrhizin	-11.7	D-HIS102,D-SER159,D-SER205,D-SER206D-THR207,E-ASN60,E-PHE77
Amentoflavone	-10.3	D-SER159,A-GLU189,E-ASN60
Butrin	-10.2	D-HIS102,D-SER205D-SER206,D-207,D-TYR210,E-ASP56,E-PHE77,E-THR142,E-VAL190
Palasitrin	-10.2	D-GLU138,E-ASP56,E-TYR58,E-ASN60
Withaferine A	-10.1	D-LYS156,D-THR207

Table 3.5 *Docking Results of Phytochemicals with GABA B Receptor.*

Ligands	Binding Affinity	Hydrogen Bonds Interactions
Standard - Ethosuximide	-5.1	B-SER128,B-GLU421
Amentoflavone	-11.2	A-CYS103,A-GLU138,A-GLU144,A-THR-203,A-GLU253
Agathisflavone	-10.9	A-SER131,A-GLN197,A-THR198,A-GLU200,
Somniferine	-10.5	A-CYS103,A-PRO105,A-THR199
Urs-12-en-27a, 30 di-oic acid 3-0-a-L-rhamnopyranosyl [1,2] a-L-arabinopyranoside	-10.3	A-GLN197,B-GLU144,B-LYS167,B-GLN206
Glycyrrhizin	-9.9	A-GLY64,A-THR198,A-THR199,A-GLU253,B-SER140,B-GLU144,B-GLN147,B-LYS168

3.1.4 Docking of phytochemicals with 4MS3 - GABA B Receptor. Among all the phytochemicals docked with GABA B Receptor, the phytochemicals in Table 3.5 have the best binding affinities. Amentoflavone has the highest binding affinity of -11.2, while Ethosuximide, the standard, has a binding affinity of -5.1.

3.1.5 Docking of phytochemicals with 5U6P - HCN1. Among all the phytochemicals docked with HCN1, the phytochemicals in Table 3.6 have the best binding affinities. Glycyrrhizin has the highest binding affinity of -10.2, while Phenytoin, the standard, has a binding affinity of -5.1.

Table 3.6 *Docking Results of Phytochemicals with HCN1.*

Ligands	Binding Affinity	Hydrogen Bonds Interactions
Standard - Phenytoin	-5.1	C-TYR501,C-ARG504
Glycyrrhizin	-10.2	B-HIS421,B-LYS422,B-ALA425,B-ASP426,C-ASN455,C-GLN516
Amentoflavone	-9.8	B-LYA422,C-GLN440,C-ARG467,D-ARG126,D-SER204
Silibinin	-9.6	C-LYS531
Withaferine A	-9.6	C-ARG467,C-LYS468,C-GLN516,D-ARG126,D-GLU205
Withanolide	-9.6	C-GLN440,C-ARG560

Table 3.7 *Docking Results of Phytochemicals with NMDA Receptor.*

Ligands	Binding Affinity	Hydrogen Bonds Interactions
Standard - Remacimide	-7.9	A-SER108,A-TYR109,A-ILE133
Agathisflavone	-11.5	A-ILE133,A-THR333,A-ASN334,B-MET207,B-THR233
Amentoflavone	-10.3	A-ILE133,A-THR333,A-ASN334,B-THR233
Isobutrin	-10.1	A-ARG115,A-SER132,A-THR333,B-ASP104,B-GLN110
Isocoreopsin	-9.8	A-ARG115,A-THR333,B-GLN110
Withanolide	-9.3	A-ARG115

3.1.6 Docking of phytochemicals with 5EWJ - NMDA Receptor. Among all the phytochemicals docked with NMDA Receptor, the phytochemicals in Table 3.7 have the best binding affinities. Agathisflavone has the highest binding affinity of -11.5, while Remacimide, the standard, has a binding affinity of -7.9.

3.1.7 Docking of phytochemicals with 4V11 - Synaptic Vesicle Protein 2A. Among all the phytochemicals docked with Synaptic Vesicle Protein 2A, the phytochemicals in Table 3.8 have the best binding affinities. Agathisflavone has the highest binding affinity of -7.2, while Levetiracetam, the standard, has a binding affinity of -4.3.

3.1.8 Docking of phytochemicals with 6AGF - Voltage Gated Sodium Channel. Among all the phytochemicals docked with Voltage Gated Sodium Channel, the phytochemicals in Table 3.9 have the best binding affinities. Urs-12-en-27a, 30 di-oic acid 3-0-a-L- rhamnopyranosyl [1,2] a-L- arabinopyranoside has the highest binding affinity of -12.2, while Levetiracetam, the standard, has a binding affinity of -5.3.

Table 3.8 *Docking Results of Phytochemicals with Synaptic Vesicle Protein 2A.*

Ligands	Binding Affinity	Hydrogen Bonds Interactions
Standard – Levetiracetam	-4.3	A-ARG323,A-ASP415
Agathisflavone	-7.2	A-TYR312,A-ARG323,A-LYS326,A-THR361,A-LEU363,A-GLU411
Palasitrin	-7.1	A-ARG323,A-ASP415,B-ASP82
Amentoflavone	-7	A-TYR312,A-ARG323,A-ASP415,B-ASP82
Silychristine	-6.9	A-TYR312,A-GLY321,A-LYS376,A-ASP415
Butrin	-6.8	A-LYS314,A-HIS316,A-THR361,A-LYS370,A-ASN371,A-ASP415

Table 3.92 *Docking Results of Phytochemicals with Voltage Gated Sodium Channel.*

Ligands	Binding Affinity	Hydrogen Bonds Interactions
Standard-Levetiracetam	-5.3	A-GLU161,A-ARG222
Urs-12-en-27a, 30 di-oic acid 3-0-a-L-rhamnopyranosyl [1,2] a-L-arabinopyranoside	-12.2	A-CYS759,A-LYS1244,A-THR1534
Amentoflavone	-11.2	A-CYS1167,A-SER1283
Glycyrrhizin	-11.1	A-GLN405,A-THR1533,A-SER1535
Somniferine	-11.1	A-SER1535,A-ILE1582
Withanolide	-10.9	A-SER1283

3.2 Drug Likeliness Studies

Table 3.10 summarizes the physicochemical properties of the selected phytochemicals that were found to possess good binding affinities post docking studies. Drug Likeliness or Drugability is often evaluated through Lipinski's Rule of 5. It is observed that most phytochemicals violate only up to three Lipinski's Rules. Violations are most commonly seen in Molecular Weight and H-Bond Donors.

Table 3.10 *Drug Likeliness Studies.*

Ligands	Molecular Weight [Daltons]	Log P	Rotatable Bonds	H - Bond Acceptors	H - Bond Donors	Lipinski's Rule Violations
Agathisflavone	538.464	5.134	3	10	6	3
Amentoflavone	538.464	5.134	3	10	6	3
Betulinic Acid	456.711	7.0895	2	2	2	1
Butrin	596.538	-2.5439	7	15	9	3
Coreopsin	434.397	-0.1218	6	10	7	1

Gallic Acid	170.12	0.5016	1	4	4	1
Glycyrrhizin	822.942	2.2456	7	13	8	3
Isobutrin	596.538	-2.6487	9	15	5	2
Isocoreopsin	434.397	-0.017	4	10	6	1
Palasitrin	594.522	-2.6342	7	10	9	2
Silibinin	482.441	2.3627	4	10	5	0
Silychristine	482.441	2.4045	4	10	6	1
Silydianin	482.441	0.9905	3	10	5	0
Urs-12-en-27α, 30 di-oic acid 3-0-α-L-rhamnopyranosyl [1→2] α-L-arabinopyranoside	764.95	3.6134	6	11	5	2
Withaferine A	470.606	3.3529	3	6	2	0
Withanolide	470.606	3.4954	2	6	2	0

3.3 Prediction of Pharmacokinetic properties

3.3.1 Absorption. Molecules with water solubility values above -4 are said to have considerable water solubility. All the phytochemicals except Withaferine A and Withanolide exhibit reasonable water solubility. Caco2 is an immortal cell line of epithelial colorectal adenocarcinoma cells. It acts as an in-vitro model of intestinal mucosa. A molecule has high $CaCO_2$ permeability if the value is above 0.9. from the table below, Betulinic Acid and Withanolide have high $CaCO_2$ permeability. Agathisflavone, Amentoflavone, Betulinic Acid, Silibinin, Silychristine, Silydianin, Withaferine A and Withanolide largely undergo absorption through the small intestine and through the P-Glycoprotein transporter. However, Agathisflavone, Amentoflavone, Silibinin, Silychristine, Withaferine A and Withanolide act as P-Glycoprotein inhibitors, which may have pharmacokinetic implications for other P-Glycoprotein substrates.

3.3.2 Distribution. Molecules with log VDss values above -0.15 are considered to have good distribution. Coreopsin, Isocoreopsin, Palasitrin, Silychristine, Silydianin, Withaferine A and Withanolide are distributed well in the bloodstream while the other phytochemicals have poor distribution. Fraction unbound reflects the property of the molecules to bind to plasma proteins. All the phytochemicals have considerably low fraction unbound values, indicating that they are highly bound to plasma proteins. BBB Permeability reflects the potential of a molecule to cross the blood-brain barrier. This property is important for drugs intending to act on the CNS. Molecules with logBB above -1 have considerable BBB permeability. From the table below, Amentoflavone,

Betulinic Acid, Silydianin, Withaferine A and Withanolide can permeate the BBB to a considerable extent.

3.3.3 Metabolism. CYP3A4 is a liver enzyme responsible for the metabolism of many drugs. Agathisflavone, Betulinic Acid, Urs-12-en-27α, 30 di-oic acid 3-0-α-L- rhamnopyranosyl [1→2] α-L- arabinopyranoside, Withaferine A and Withanolide act as CYP3A4 substrates and thus get metabolized by the same. The rest of the phytochemicals that are not metabolized, get excreted unchanged. CYP inhibitors can interfere with the drug metabolism and cause drug-drug interactions. From the table below, Silibinin acts as a CYP2C9 and CYP3A4 inhibitor.

3.3.4 Excretion. Total clearance refers to the volume of plasma that is cleared of the drug, per unit time. Agathisflavone, Amentoflavone, Gallic Acid, Isobutrin and Withaferine A undergo considerable clearance, while the other phytochemicals have moderate total clearance values. Organic Cation Transporter-2 [OCT-2] is a renal uptake transporter that allows for renal clearance and disposition of drugs. OCT2 substrates have the potential of interacting with OCT2 inhibitors, which could lead to adverse drug reactions. From the table below, Glycyrrhizin and Withaferine A act as Renal OCT2 substrates.

3.3.5 Toxicity. The toxicity parameters for the selected phytochemicals were evaluated and summarized in Table 3.15. The AMES test is used to assess the mutagenic potential of drug molecules. From the table below, Butrin, Isocoreopsin and Palasitrin were found to possess AMES toxicity. Maximum tolerated dose provides an estimate of the toxic dose and therapeutic index of drugs. A value of less than 0.4 is considered low. From the table below, Butrin, Isobutrin, Palasitrin and 3-0-α-L- rhamnopyranosyl [1→2] α-L-arabinopyranoside have low values of maximum tolerated dose. Betulinic Acid was found to possess hepatotoxicity. All compounds possess a low potential to cause Minnow toxicity.

Table 3.15 *Toxicity Parameters of the phytochemicals.*

Ligands	AMES toxicity	Max. tolerated dose [human] [log [mg/kg/day]]	hERG I inhibitor	Oral Rat Acute Toxicity [LD50] [mol/kg]	Oral Rat Chronic Toxicity [LOAEL]	Hepatotoxicity	Minnow toxicity [log Mm]
Agathisflavone	No	0.438	No	2.507	3.353	No	1.716

Amentoflavone	No	0.425	No	2.489	3.219	No	1.017
Betulinic Acid	No	0.403	No	2.869	2.36	Yes	-1.761
Butrin	Yes	0.236	No	2.512	5.448	No	4.594
Coreopsin	No	0.564	No	2.628	4.035	No	0.757
Gallic Acid	No	0.7	No	2.218	3.06	No	3.188
Glycyrrhizin	No	0.431	No	2.427	4.907	No	4.364
Isobutrin	No	0.28	No	2.516	6.499	No	2.601
Isocoreopsin	Yes	0.474	No	2.543	3.566	No	1.791
Palasitrin	Yes	0.204	No	2.546	5.688	No	4.395
Silibinin	No	0.814	No	2.795	3.505	No	2.441
Silychristine	No	0.612	No	2.614	3.541	No	3.66
Silydianin	No	0.483	No	2.981	3.912	No	2.028
Urs-12-en-27α, 30 di-oic acid 3-0-α-L-rhamnopyranosyl [1→2] α-L- arabinopyranoside	No	-0.002	No	2.622	3.679	No	2.529
Withaferine A	No	-0.695	No	2.779	0.918	No	0.738
Withanolide	No	-0.893	No	2.819	1.653	No	0.604

3.4 Molecular Dynamics

Molecular Dynamics simulations were performed between Amentoflavone and GABA-B Receptor.

3.4.1 Root Mean Square Deviation [RMSD]. Despite protein-ligand docking being extensively employed and proven to be successful, it only presents a static image of the ligand's binding posture in the receptor's active site. Molecular dynamics must be utilized to simulate the atomic dynamics in the system over time after integrating Newton's equations of motion. The top-scoring test complex for binding, amentoflavone, was subjected to an MD simulation for 100 ns in order to better understand the unbound with GABA B receptor (ligand-receptor complex). Using the RMSD, RMSF, RGY, and SASA of the receptor atoms, MD trajectories analysis was utilised to assess the stability and volatility of these complexes. In order to analyse the MD trajectory equilibration and verify the stability of complex systems during the simulation process, RMSD is a crucial measurement. The RMSD of the protein backbone atoms was plotted against time to assess the variations in structural confirmation. There were no variations throughout the first 60 ns of the experiment, but for the remaining 100 ns, oscillations were noted (Figure 2).

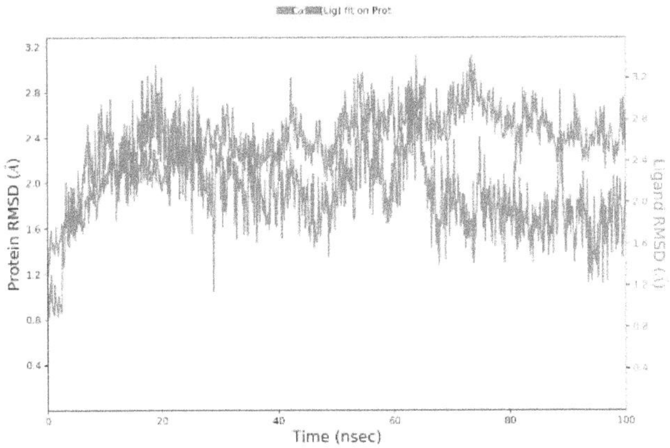

Figure 2 *RMSD of GABA-B Receptor with Amentoflavone.*

3.4.2 Root Mean Square Fluctuation [RMSF]. The Root Mean Square Fluctuation [RMSF] is useful for characterizing local changes along the protein chain. Peaks in the Figure 3 indicate areas of the protein that fluctuate the most during the simulation. GABA-B Receptor residues that interact with Amentoflavone are marked with green-colored vertical bars. RMSF also show fluctuations.

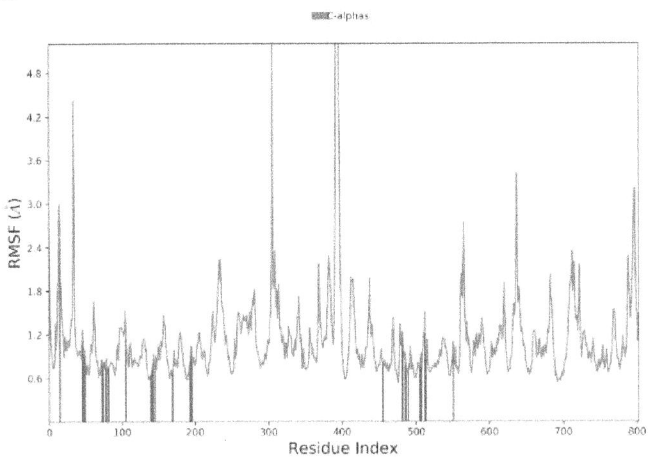

Figure 3 *RMSF of GABA-B Receptor.*

3.4.3 Ligand-Protein 2D Contacts. Figure 4 is a schematic of detailed interactions of the atoms of Amentoflavone with GABA-B Receptor residues. Interactions that occur more than 30.0% of the simulation time in the selected trajectory [0.00 through 100.00 nsec], are shown.

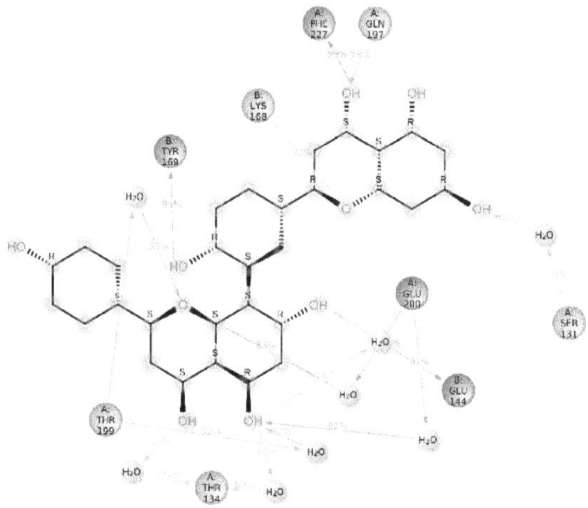

Figure 4 *Amentoflavone GABA-B Receptor 2D Contacts.*

3.4.4 Protein-Ligand Contacts Timeline. Figure 5 represents the residues of GABA-B Receptor which interact with Amentoflavone in each trajectory frame. Some residues make more than one specific contact with the ligand, which is represented by a darker shade of orange, according to the scale to the right of the plot.

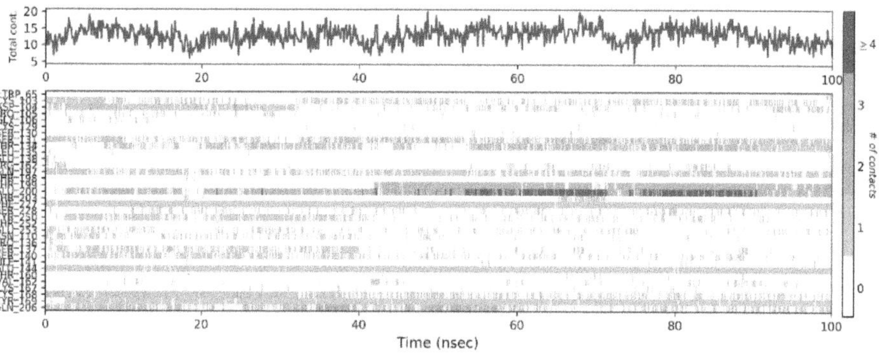

Figure 5 *GABA-B Receptor Amentoflavone Contacts Timeline*

4 CONCLUSION

Among the isolated phytochemical compounds which were reported to be having antiepileptic properties, around 70 molecules were studied for their binding affinities with various proteins responsible in treating epilepsy which includes AMPA Receptor [3R7X], Carbonic Anhydrase 2 [3F8E], Connexins [6L3V], GABA-A Receptor [6D6T], GABA-B Receptor [4MS3], HCN1 [5U6P], NMDA Receptor [5EWJ], SV2A [4V11] and Voltage-Gated Sodium Channel [6AGF]. The compounds with best docking scores were compared to the respective standards and reported.

Among all the phytochemicals screened, Amentoflavone was found to have very good binding affinity with all the above proteins except AMPA Receptor [3R7X], which indicates that Amentoflavone might act through multiple targeted mechanisms.

The compounds with best docking scores were evaluated for drug likeliness studies in which few of them showed slight deviations from Lipinski's Rule. These violations can be considered as prediction and activities should be tested *in-vitro/ in-vivo*, since some of these compounds are reported as active antiepileptic agents.

In-silico pharmacokinetic studies showed that most of the compounds estimated with good binding energies are slightly water soluble, while compounds like Withaferine A and Withanolide are poorly water-soluble. Most phytochemicals show good intestinal absorption. They possess good excretion rates and have limited potential to cause toxicity. The validation of the docking studies of Amentoflavone with GABA-B receptor was performed using molecular dynamics studies. RMSD and RMSF data shows that the complex even though seems to be stable initially, displays some fluctuations after 60 ns which shows that the complex is unstable. There is a scope of testing other complexes for its stability using MD simulations as an extended study.

Since *in-silico* evaluation serves as a preliminary study, further *in-vitro* and *in-vivo* studies are to be done on the above reported phytochemicals to ascertain the efficiency of these molecules as potential antiepileptic drugs.

5 ACKNOWLEDGEMENT

The authors are thankful to Department of Pharmaceutical Chemistry, Faculty of Pharmacy, MS Ramaiah University of Applied Sciences, Bengaluru for providing the facilities to carry the work.

6 REFERENCES

1. Who.int. 2022. Epilepsy. [online] Available at: <https://www.who.int/news-room/fact-sheets/detail/epilepsy> [Accessed 30 May 2022].
2. Nice.org.uk. 2022. Epilepsies: diagnosis and management | Guidance | NICE. [online] Available at: <https://www.nice.org.uk/guidance/cg137> [Accessed 30 May 2022].
3. Sharma, H. and Sharma, K., 2017. Sharma & Sharma's principles of pharmacology. 3rd ed. Hyderabad: Paras Medical Publisher, pp.523-529.
4. J. DiPiro, G. Yee, L. Posey, S. Haines, T. Nolin and V. Ellingrod, Pharmacotherapy. New York [i pozostałe]: McGraw Hill. 2020.
5. NS. Santhosh, S Sinha, and P. Satishchandra, Epilepsy: Indian perspective. Annals of Indian Academy of Neurology, 17(Suppl 1), p.S3. 2014.

6. M. Tsvere, MK. Chiweshe, and N. Mutanana, General side effects and challenges associated with anti-epilepsy medication: A review of related literature. African Journal of Primary Health Care and Family Medicine, 2020, 12(1), 1-5.
7. A. Wahab, Difficulties in treatment and management of epilepsy and challenges in new drug development. Pharmaceuticals, 2010, 3(7), 2090-110.
8. A. Gaitatzis, and JW. Sander, The long-term safety of antiepileptic drugs. CNS drugs, 2013, 27(6), 435-455.
9. M. Hassan Baig, K. Ahmad, S. Roy, J. Mohammad Ashraf, M. Adil, M. Haris Siddiqui, S. Khan, M. Amjad Kamal, I. Provazník, and I. Choi, Computer aided drug design: success and limitations. Current pharmaceutical design, 2016, 22(5), 572-81.
10. TT. Talele, SA. Khedkar, and AC. Rigby, Successful applications of computer aided drug discovery: moving drugs from concept to the clinic. Current topics in medicinal chemistry, 2010, 10(1), 127-41.
11. DE Pires, TL. Blundell and DB. Ascher, pkCSM: predicting small-molecule pharmacokinetic and toxicity properties using graph-based signatures. Journal of medicinal chemistry, 2015, 58(9), 4066-72.
12. SA. Hollingsworth, and RO. Dror, Molecular dynamics simulation for all. Neuron, 2018, 99(6), 1129-1143.
13. JD. Durrant, and JA. McCammon, Molecular dynamics simulations and drug discovery. BMC biology, 2011, 9(1), 1-9.
14. MA. Rogawski, AMPA receptors as a molecular target in epilepsy therapy. Acta Neurologica Scandinavica, 2013, 127, 9-18.
15. L. Ciccone, C. Cerri, S. Nencetti, and E. Orlandini, Carbonic anhydrase inhibitors and epilepsy: State of the art and future perspectives. Molecules, 2021, 26(21), 6380.
16. Y. Noam, C. Bernard, and TZ. Baram, Towards an integrated view of HCN channel role in epilepsy. Current opinion in neurobiology, 2011, 21(6), 873-79.
17. S. Chen, D. Xu, L. Fan, Z. Fang, X. Wang, and M. Li, Roles of N-methyl-D-aspartate receptors (NMDARs) in epilepsy. Frontiers in Molecular Neuroscience, 2021, 333.
18. Y. Ohno, and K. Tokudome. Therapeutic Role of Synaptic Vesicle Glycoprotein 2A (SV2A) in Modulating Epileptogenesis. CNS & neurological disorders drug targets, 2017, 16(4), 463–71
19. DI. Kaplan, LL. Isom, and S. Petrou, Role of sodium channels in epilepsy. Cold Spring Harbor perspectives in medicine, 2016, 6(6), p.a022814.

SCREENING OF PLACENTAL PROTEINS AGAINST PROTEASES FOR THEIR POTENTIAL USE IN OSTEOARTHRITIS: A COMPUTATIONAL APPROACH

Jithu Jerin James[a*], Parasuraman Pavadai[b] and Sandhya K V[a]
[a]Department of Pharmaceutics, Faculty of Pharmacy, MS Ramaiah University of Applied Sciences, Bengaluru, India, [b]Department of Pharmaceutical Chemistry, Faculty of Pharmacy, MS Ramaiah University of Applied Sciences, Bengaluru, India.
Corresponding Author E-Mail ID: jithujerinjms25@gmail.com

1 INTRODUCTION

Osteoarthritis (OA) is a chronic joint condition owing to increasing articular cartilage weakening and deterioration. It is categorized as mild and severe OA based on breakdown of articular cartilage, narrowing of joint spaces and growth of osteophytes. OA occurs in adults (40-60 years old) and causes movement disorders after 65 years of age. Symptomatic osteoarthritis affects an estimated 240 million persons globally. According to National health portal, it is the most frequent joint disease with a prevalence of 22-39% in India.

The mode of treatment includes non-pharmacological (exercise, weight loss, physiotherapy), pharmacological (to reduce pain and inflammation), hyaluronic acid injections, supplements (vitamin D, chondroitin sulphate, glucosamine) and surgery.[1-3]

Risk factors such as age, previous joint injury, stress, obesity or genetics stimulate macrophages within synovium to release proinflammatory cytokines such as tumour necrosis factor α (TNF-α), *Interleukin 1* beta (IL-1β) and Interleukin 6 (IL-6), which then activates synoviocytes to release metalloproteases which results in articular cartilage deterioration.[1,2,4]

One of the common pathways leading to OA includes stress-induced premature senescence (SIPS) induced by reactive oxygen species (ROS) via enhancing endoplasmic reticulum (ER) stress. SIPS is characterised by a cessation in cellular replication, an increase in the production of inflammatory agents and metalloproteinases, and a reduction in the responsiveness of chondrocytes to anabolic growth factors.[5,6]

Human placental tissue repairs cartilage either by regenerating chondrocyte or by altering intracellular signalling pathways that senescence targets.[7] Human placental tissue is rich with biofactors such as proanabolic - basic fibroblast growth factor (bFGF), anti-catabolic – tissue inhibitor of metalloproteases (TIMP) and anti-inflammatory biofactors – *Interleukin 1* receptor antagonist (IL-1Ra).[8,9]

The aim of this study is to screen the placental protein against matrix metalloproteinase -13 (MMP-13) for the treatment of osteoarthritis.

2 METHODS AND RESULTS

2.1 Source and retrieval methods

For the docking study, placental proteins with possible anti-arthritic effects were taken into consideration. The placental tissue contains over forty proteins that are related with cartilage tissue repair.[9] For the study, selected twenty-six proteins and were obtained from the protein data bank (PDB) (http://www.rcsb.org).

2.2 Preparation of target and ligand

Three-dimensional (3D) X-ray crystallographic structure of MMP-13 receptor of less than 2.5 Å resolution was downloaded from the protein data bank (PDB) (http://www.rcsb.org). The PDB ID was 7JU8. For the study, the 3D X-ray crystallographic structure of ligand proteins with a resolution of less than 2.5 were taken from the PDB database. The proteins were cleaned and subjected to minimization using CHARMM-GUI server (http://charmm-gui.org).The files were named protein ID.pdb and saved for further analysis.[10,11]

2.3 Protein-Protein docking

Protein-protein docking was performed by using ZDOCK server (https://zdock.umassmed.edu/) and was evaluated for their binding affinities.[11] The receptor was MMP-13 (PDB ID: 7JU8), and the ligand protein was one of the chosen placental proteins. For each ligand protein, ZDOCK calculated the docking score. After docking, the complexes were loaded into Biovia Discovery Studio 2021 Client.lnk. The Visualizer generated the complex's 3D structures and displayed the number of hydrogen bonds, non-covalent interactions, bond distance and interacting residues for each complex.[12-14]

2.4. Molecular dynamics simulations of docked complexes

Molecular dynamics (MD) simulations were used to assess the conformational stability of the interactions between selected ligand proteins and receptors.[11] The top two docked protein–protein complex (saved as PDB file) was subjected to molecular dynamic simulation using DESMOND v3.6 package.[15] The lead complexes were prepared using the OPLS force field. The complexes were initially immersed in an orthorhombic box at the distances of 10Å units containing SPC water, pH neutralized by adding adequate amount of Na+/Cl− ions. For molecular dynamic runs, the ensemble class was set to NPT (300K and 1 atmospheric pressure) and 100ns at 0.001ps time steps.

3 RESULTS AND DISCUSSION

3.1 Retrieval of ligand proteins and preparation

Twenty-six placental proteins from the PDB database with 3D X-ray crystallographic structures less than 2.5 Å were taken into consideration for docking. The proteins were cleaned and subjected to minimization using CHARMM-GUI server (http://charmm-gui.org).The files were named protein ID.pdb and saved for docking studies.[11]

3.2 Molecular docking studies

Twenty-six ligand proteins were docked against MMP-13 using ZDOCK server to analyze their binding capacity. Following protein-protein docking, the binding energies of ligand proteins was found to be ranging from 1046.272 to 2229.332. The detailed ZDOCK binding score in descending order, were shown in (Table 3.1).

Of the 26 ligand proteins, five proteins were in line with the excellent docking score. The top five complexes were selected for further research, which include metalloproteinase inhibitor-3, plasminogen activator inhibitor 1, vitronectin, fibrillin-1 and Keratin, type II cytoskeletal 1complex with MMP-13. These proteins can inhibit MMPs, have anti-inflammatory effect, anti-apoptosis effect, extracellular matrix – cell interactions and macrophage polarization effect.[9] The 3D structures of top five complexes were shown in (Figure 3.1).

Table 3.1 *ZDOCK score results for ligand proteins.*

Sl. No	Protein name	Protein ID	Docking score
1	Metalloproteinase inhibitor 3	3CKI	2229.332
2	Plasminogen activator inhibitor 1	7AQF	2131.215
3	Vitronectin	6O5E	2126.44
4	Fibrillin-1	1UZJ	2083.607
5	Keratin, type II cytoskeletal 1	6E2J	2015.896
6	Laminin subunit alpha 5	5XAU	2009.504
7	Vimentin	1GK7	1942.767
8	Glia-derived nexin	4DYO	1929.703
9	Collagen alpha 1(II) chain	6HG7	1915.297
10	Lactotransferrin (fragment)	2BJJ	1903.067

11	Transforming growth factor-beta–induced protein ig-h3	7ASG	1889.915
12	Collagen alpha 3(VI) chain	1KTH	1870.947
13	Collagen alpha 1(I) chain	5k31	1814.312
14	Keratin, type I cytoskeletal 10	6E2J	1804.145
15	Alpha-1-antitrypsin	3NE4	1785.138
16	Plectin	1MB8	1774.785
17	Keratin, type I cytoskeletal 14	6JFV	1765.148
18	Laminin subunit beta 1	5XAU	1761.196
19	Keratin, type II cytoskeletal 5	6JFV	1740.909
20	Laminin subunit gamma 1	5XAU	1734.832
21	Actin, cytoplasmic 2	6WK2	1722.963
22	Collagen alpha 2(I) chain	5CVA	1720.229
23	Collagen alpha 1(III) chain	4AE2	1717.219
24	Alpha-2-macroglobulin	2P9R	1707.327
25	Fibronectin	2CG7	1495.959
26	Dickkopf-related protein 1	3S0Q	1046.272

3.3 Interpretation of protein-protein interactions

BIOVIA discovery studio visualizer tool was used to anticipate the interactions between the five selected ligand proteins and MMP-13.

3.3.1 Binding mode of Metalloproteinase inhibitor 3. Metalloproteinase inhibitor-3 (3CKI) showed maximum interaction with MMP-13 with binding score of 2229.332 as compared to other ligand proteins. Metalloproteinase inhibitor-3 formed six electrostatic interactions with LYS234, ASP174, ASP179, ASP231, TYR195 and HSD226, three conventional hydrogen bonds with TYR195, ALA188 and PHE175, two carbon hydrogen bonds with THR110 and TYR176, one Pi-Donor hydrogen bond with PHE189, and six hydrophobic interactions with PRO108, ARG109, LEU111, TYR195, LEU111 and PRO190 active site amino acid residues of MMP-13 respectively showed in (Figure 3.1) and (Table 3.2).

3.3.2 Binding mode of Plasminogen activator inhibitor 1. Plasminogen activator inhibitor 1 (7AQF) showed binding score of 2131.215, and interactions such as one electrostatic with LYS234, three conventional hydrogen bonds with TYR176, TYR195 and TYR176, four carbon hydrogen bonds with PHE175, ASN194, ALA188 and ASP231, and twelve hydrophobic interactions with PRO190, PRO242, LEU111, LEU184, PHE107, two

PHE175, TYR176, TYR195, HSD232 and PRO242 active site amino acid residues of MMP-13 respectively as shown in (Figure 3.1) and (Table 3.2).

3.3.3 Binding mode of Vitronectin. Vitronectin (6O5E) showed three electrostatic interactions with ASP174, ASP198 and PHE175, four conventional hydrogen bond interactions with ASN105, TYR195, PRO193 and GLY173, three carbon hydrogen bond interactions with two ARG109 and ASP174, one Pi-Donor Hydrogen Bond, and fourteen hydrophobic interactions with two PHE107, PHE175, VAL106, ARG109, three PRO190, PRO193, TYR104, PHE189, two TYR195 and PRO177 active site amino acid residues of MMP-13 respectively as depicted in (Figure 3.1) and (Table 3.2).

3.3.4 Binding mode of Fibrillin-1. Fibrillin-1(1UZJ) showed six electrostatic interactions with GLU223, ASP179, three ASP231 and PHE107, formed four conventional hydrogen bonds with LEU111, ALA188 and two ALA186, three carbon hydrogen bonds with HSD187, two PRO190, and nine hydrophobic interactions with VAL106, ARG109, LEU184, three PRO242, PHE107, HSD232 and PRO193 active site amino acid residues of MMP-13 respectively as depicted in (Figure 3.1) and (Table 3.2).

3.3.5 Binding mode of Keratin, type II cytoskeletal 1. Keratin, type II cytoskeletal 1 (6E2J) showed one conventional hydrogen bond with TYR176, one electrostatic interaction with GLU171 and seven hydrophobic interactions with PRO177, two LEU184, TYR104, PHE107, PHE175, and PRO177 active site amino acid residues of MMP-13 respectively as depicted in (Figure 3.1) and (Table 3.2).

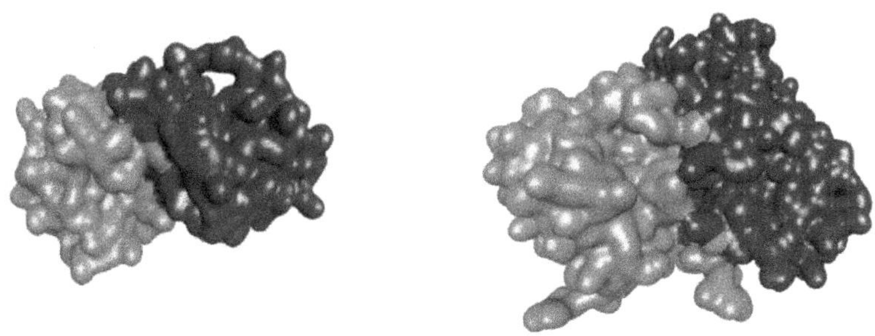

(a) **(b)**

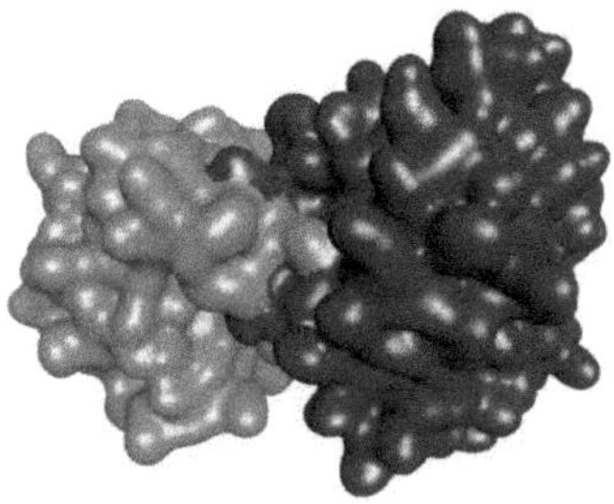

(c)

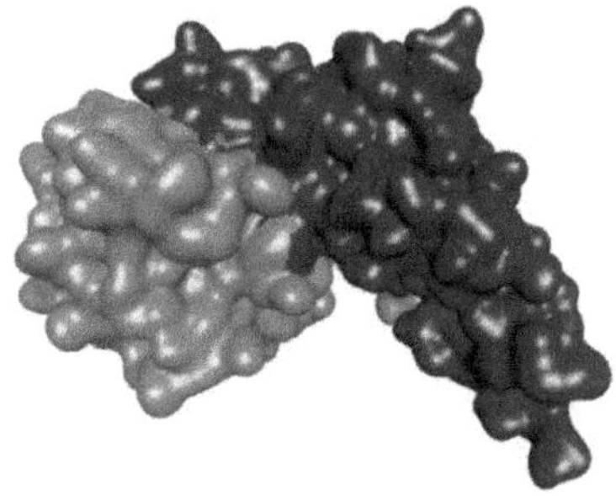

(d)

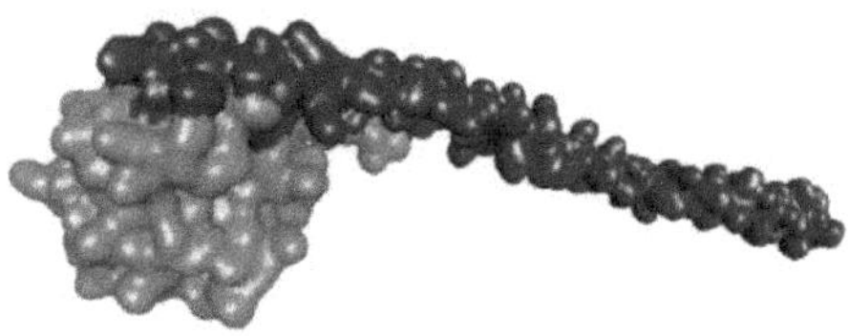

(e)

Figure 3.1 *3D representation of MMP-13 complex with (a) Metalloproteinase inhibitor 3 (b) Plasminogen activator inhibitor 1 (c) Vitronectin (d) Fibrillin-1 and (e) Keratin, type II cytoskeletal 1.*

Table 3.2 *List of bonding interactions between selected 5 ligand proteins with MMP-13.*

Protein name & ID	Interacting residues From	To	Category	Bond distance (A^O)
Metalloproteinase inhibitor 3 (3CKI)	LYS234	ASN121	Electrostatic	4.63543
	ARG20	ASP174	Electrostatic	5.07138
	LYS110	ASP179	Electrostatic	4.30339
	ARG115	ASP231	Electrostatic	3.8693
	LYS45	TYR195	Hydrogen Bond	2.41966
	LYS45	ALA188	Hydrogen Bond	1.68092
	ARG109	PHE175	Hydrogen Bond	1.90045
	THR110	PHE50	Hydrogen Bond	1.48808
	ARG20	TYR176	Hydrogen Bond	3.66592
	ARG20	TYR195	Electrostatic	4.69705
	LYS45	PHE189	Hydrogen Bond	3.49048

	ARG115	HSD226	Electrostatic	4.01533
	MET44	TYR195	Other	4.32022
	PRO108	PRO54	Hydrophobic	4.41556
	ARG109	PRO54	Hydrophobic	4.9471
	LEU111	MET46	Hydrophobic	3.19269
	TYR195	LYS45	Hydrophobic	5.23673
	PHE50	LEU111	Hydrophobic	4.95519
	PHE50	PRO190	Hydrophobic	4.99374
Plasminogen activator inhibitor 1 (7AQF)	LYS234	ASP89	Electrostatic	4.63624
	TYR176	ALA74	Hydrogen Bond	1.90747
	TYR195	GLY70	Hydrogen Bond	3.37391
	LEU78	TYR176	Hydrogen Bond	2.74086
	SER6	PHE175	Hydrogen Bond	2.67966
	LYS69	ASN194	Hydrogen Bond	3.67476
	HSD77	ALA188	Hydrogen Bond	3.45396
	GLY84	ASP231	Hydrogen Bond	2.97603
	ARG115	PHE107	Hydrophobic	3.14198
	PRO190	LYS80	Hydrophobic	4.87948
	PRO242	PRO85	Hydrophobic	3.88765
	ARG118	LEU111	Hydrophobic	3.99988
	PRO366	LEU184	Hydrophobic	5.37588
	PHE107	PRO111	Hydrophobic	5.14724
	PHE175	VAL8	Hydrophobic	4.8214
	PHE175	ALA74	Hydrophobic	4.6375
	TYR176	ALA74	Hydrophobic	5.2244
	TYR195	PRO73	Hydrophobic	4.64294
	HSD232	PRO85	Hydrophobic	3.97393
	TRP86	PRO242	Hydrophobic	4.52441
	ARG450	ASP174	Electrostatic	5.19597
	ARG450	ASP198	Electrostatic	5.25719
	ASN105	GLY471	Hydrogen Bond	2.86312
	TYR195	ILE354	Hydrogen Bond	1.95791
	ASN448	PRO193	Hydrogen Bond	2.73539
	ARG450	GLY173	Hydrogen Bond	2.15972
	ARG109	GLY170	Hydrogen Bond	3.33088
	ASP174	ARG450	Hydrogen Bond	2.59962
	GLY170	ARG109	Hydrogen Bond	1.48805

Vitronectin (6O5E)	ARG452	PHE175	Electrostatic	3.84197
	ARG450	TYR195	Hydrogen Bond	3.07417
	PHE107	TRP469	Hydrophobic	4.86077
	TRP469	PHE107	Hydrophobic	4.48706
	ARG450	PHE175	Hydrophobic	5.10598
	VAL106	PRO190	Hydrophobic	5.33824
	ARG109	LYS186	Hydrophobic	4.46908
	PRO190	ILE354	Hydrophobic	4.91994
	PRO193	ILE354	Hydrophobic	5.31964
	ARG353	PRO190	Hydrophobic	4.44133
	TYR104	PRO473	Hydrophobic	5.21917
	PHE189	ARG450	Hydrophobic	4.40452
	TYR195	ILE354	Hydrophobic	5.17744
	TYR195	ARG450	Hydrophobic	3.96328
	HSD263	PRO190	Hydrophobic	5.13938
	TYR265	PRO177	Hydrophobic	3.54358
Fibrillin-1 (1UZJ)	ARG1632	GLU223	Electrostatic	3.76246
	LYS1621	ASP179	Electrostatic	5.31923
	ARG1644	ASP231	Electrostatic	4.03123
	ARG1644	ASP231	Electrostatic	5.42934
	LEU111	GLU1593	Hydrogen Bond	1.70343
	ALA188	GLN1630	Hydrogen Bond	3.02601
	GLN1630	ALA186	Hydrogen Bond	2.93596
	ARG1632	ALA186	Hydrogen Bond	2.45038
	HSD187	GLN1630	Hydrogen Bond	3.09553
	PRO190	THR1625	Hydrogen Bond	2.70032
	PRO190	PHE1626	Hydrogen Bond	3.25129
	LYS1586	PHE107	Electrostatic	4.74291
	ASP231	PHE1629	Electrostatic	3.26593
	CYS1631	HSD232	Other	5.80321
	VAL106	LYS1586	Hydrophobic	4.72425
	ARG109	VAL1580	Hydrophobic	3.72357
	LEU184	ILE1623	Hydrophobic	5.0994
	PRO242	LEU1639	Hydrophobic	5.44132
	ARG1632	PRO242	Hydrophobic	4.68025
	CYS1633	PRO242	Hydrophobic	4.36023
	PHE107	PRO1590	Hydrophobic	5.48279

	HSD232	ARG1632	Hydrophobic	5.02265
	PHE1595	PRO193	Hydrophobic	5.03869
Keratin, type II cytoskeletal 1 (6E2J)	ILE235	TYR176	Hydrogen Bond	2.8222
	MET225	LEU184	Other	3.25798
	GLU171	TYR230	Electrostatic	3.26221
	PRO177	LEU227	Hydrophobic	4.37853
	LEU184	MET225	Hydrophobic	5.10533
	TYR104	LEU256	Hydrophobic	4.70234
	PHE107	LEU245	Hydrophobic	5.05471
	PHE175	LEU233	Hydrophobic	4.55191
	TYR230	PRO177	Hydrophobic	3.7698
	PHE231	LEU184	Hydrophobic	5.2531

3.4. Molecular dynamic simulation study

The results of a protein-protein docking study showed that, of the 26 ligand proteins that were screened, metalloproteinase inhibitor-3 and plasminogen activator inhibitor 1 displayed the best docking energy for MMP-13, with binding scores of 2229.332 and 2131.215, respectively. The results of a protein-protein docking study showed that, of the 26 ligand proteins that were screened, metalloproteinase inhibitor-3 and plasminogen activator inhibitor 1 displayed the best docking energy for MMP-13, with binding scores of 2229.332 and 2131.215, respectively. These two complexes were subjected for molecular dynamic studies. A 100ns dynamic simulation was carried out for metalloproteinase inhibitor 3-MMP-13 complex, plasminogen activator inhibitor 1-MMP-13complex, and the unbound form of the MMP-13 target protein and the respective results were interpreted. To evaluate the fluctuations, intermolecular interactions, structural changes, and stability of protein-protein complexes, the trajectories of each complex were subjected to parameters like root mean square deviation (RMSD) and root mean square fluctuation (RMSF) of receptor atoms.[15]

Using the RMSD value from the dynamic simulation, the backbone deviation of proteins and the stability of protein-protein complexes were estimated. For this, protein RMSD (Å) is plotted against time in ns. The metalloproteinase inhibitor 3-MMP-13 complex showed high deviation (1.6 to 5Å) up to equilibration (22ns). The system stabilizes upon equilibration, and this trajectory persisted for another 100ns. It suggests that throughout the MD simulation, the complex was quite stable. The RMSD of the metalloproteinase inhibitor 3-MMP-13 complex is depicted in (Figure 3.2).

Initially, the plasminogen activator inhibitor 1-MMP-13 complex showed rise in RMSD till 12 ns ranging from 2 to 4.7 Å. From 13ns to 35ns the system was stable. From 36ns to 45ns complex showed variations in backbone RMSD up to 7.2 Å. The stable conformation was attained from 46ns to 100ns. The RMSD of the metalloproteinase inhibitor 3-MMP-13 complex is depicted in (Figure 3.3).

The RMSD of plasminogen activator inhibitor 1-MMP-13 complex showed more deviations compared to metalloproteinase inhibitor 3-MMP-13 complex. The RMSD analyses confirm that the metalloproteinase inhibitor 3-MMP-13 complex showed more structural stability than plasminogen activator inhibitor 1-MMP-13 complex during the dynamic simulation.

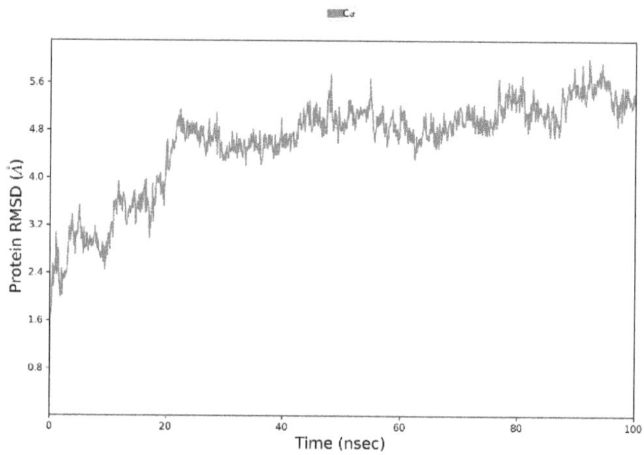

Figure 3.2 *The MD simulation's RMSD plot for metalloproteinase inhibitor 3-MMP-13 complex.*

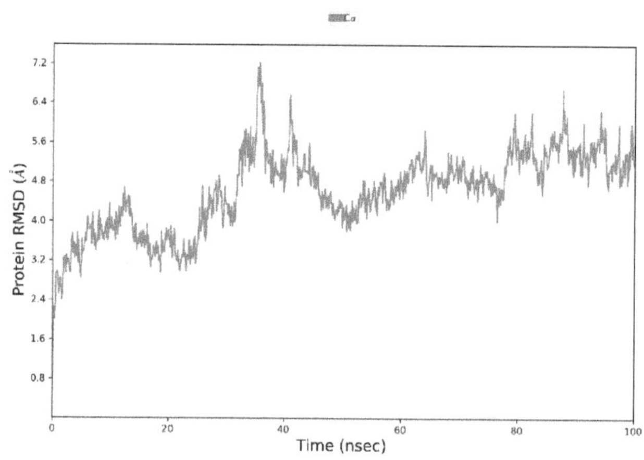

Figure 3.3 *The MD simulation's RMSD plot for plasminogen activator inhibitor 1-MMP-13 complex.*

The flexibility of the complex's during simulation was examined using the RMSF. The proteins Cα atoms RMSF values were computed and shown in relation to the residues. In the examined metalloproteinase inhibitor 3-MMP-13 complex, the amino acid residues showed no fluctuations for the entire simulation and the resulted graph was observed to be smooth. The obtained findings showed, as shown in (Figure 3.4), that binding of the ligand protein did not significantly alter the receptor protein's flexibility. The plasminogen activator inhibitor 1-MMP-13 complex exhibited high fluctuation in the backbone residue positions ranging from 430 to 460 by crossing 8.5Å as depicted in (Figure 3.5). The results suggest that the RMSF of metalloproteinase inhibitor 3-MMP-13 complex showed no fluctuations of the receptor compared to RMSF of plasminogen activator inhibitor 1-MMP-13 complex.

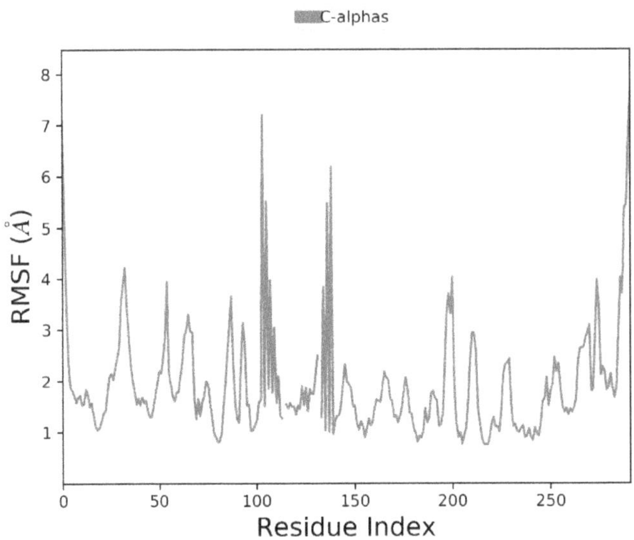

Figure 3.4 *The MD simulation's RMSF plot for metalloproteinase inhibitor 3-MMP-13 complex.*

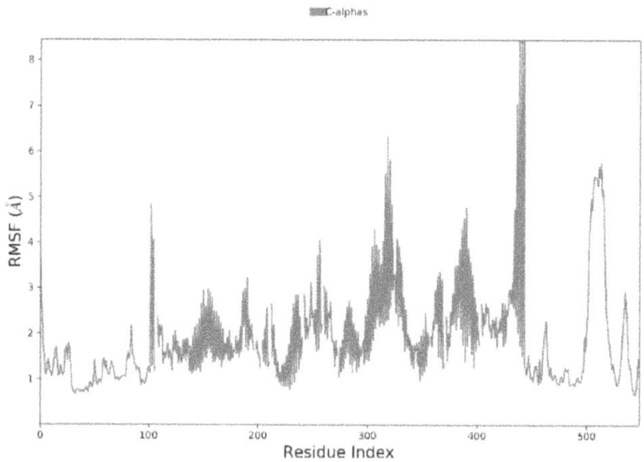

Figure 3.5 *The MD simulation's RMSF plot for plasminogen activator inhibitor 1-MMP-13 complex.*

4 CONCLUSION

The incidence of OA and the related burden are rapidly rising over the world, posing a serious public health problem in the future decades. The present pharmacological therapies for OA are only symptomatic, and their long-term effectiveness and safety are unknown. Emerging therapeutics target matrix-degrading proteases and senescent chondrocytes while also promoting cartilage regeneration, limiting bone remodelling, and reducing local inflammation. Human placental proteins help to prevent cartilage deterioration by inhibiting MMP expression. An insight of receptor inhibition is provided by the present in silico investigation. Here, we have shown how a systematic virtual screening, a protein-protein docking approach, and molecular dynamic studies all contributed to reaching a substantial result.

MMP-13 receptor and twenty-six placental proteins were taken into consideration for in-silico investigations to have a comprehensive approach. By using protein-protein docking, placental proteins that interact with MMP-13's binding pocket have been identified. Protein-protein docking findings based on scoring functions anticipated the strong binders. By molecular dynamic study, identified top two complexes that are stable in computed environment. The *in-silico* experiments lead us to the conclusion that the placental proteins plasminogen activator inhibitor 1 and metalloproteinase inhibitor 3 are appropriate for inhibiting MMP-13 receptor or altering the intracellular signalling pathways that senescence targets in OA.

5 ACKNOWLEDGMENT

The authors are highly grateful to Department of Pharmaceutical Chemistry, Faculty of Pharmacy, MS Ramaiah University of Applied Sciences, Bengaluru, and the management for the constant encouragement and support.

6 REFERENCES

1. A. Latourte, M. Kloppenburg and P. Richette, *Nat. Rev. Rheumatol.,* 2020, 16(12), 673-688.
2. X. Cai, S. Yuan, Y. Zeng, C. Wang, N. Yu and C. Ding, *Front. Pharmacol.,* 2021, 12, 645842.
3. IA. Marino-Martinez, AG. Martinez-Castro, VM. Pena-Martinez, CA. Acosta-Olivo, F. Vilchez-Cavazos, A. Guzman-Lopez, EP. Rodriguez, VJ. Romero-Diaz, JA. Ortega-Blanco, and J Lara-Arias, *Exp. Ther. Med.*, 2019, 17, 11–16.
4. KA. Kimmerling, AH. Gomoll, J. Farr and KC. Mowry, *J. Orthop. Res., 2020,* 38(5), 1141-1149.
5. C. Gwam, C. Ohanele, J. Hamby, N. Chughtai, Z. Mufti, and X. Ma, *Ann. Transl. Med.*, 2019.
6. AK. Emara, H. Anis, NS. Piuzzi, *Ann. Transl. Med.,* 2020, 8(5), 156.
7. NS. Mohamed, WA. Wilkie, EA. Remily, RE. Delanois RE, *Ann. Transl. Med.,* 2020, 8(4), 62.
8. CR. Flannery, SA. Seaman, KE. Buddin, MA. Nasert, EJ. Semler, KL. Kelley, M. Long, J. Favret, A. Pavesio, and RF. Loeser, *Osteoarthritis. Cartilage.*, 2021, 29, 1203–1212.
9. M. Bhattacharjee, JL. Escobar Ivirico, HM Kan, S. Shah, T. Otsuka, R. Bordett, M. Barajaa, N. Nagiah, R. Pandey, LS. Nair and CT. Laurencin, *Proc. Natl. Acad. Sci. U.S.A.*, 2022, 119(4), e2120968119.
10. VN. Saiprabha, D. Giles and UR. Sharanabasava, *Int. J. Pharm. Pharm. Res.*, 2019, 17 (1), 13-24.
11. GSP. Matada, PS. Dhiwar, N. Abbas, E. Singh, A. Ghara, A. Das and SV. Bhargava, *J.Biomol. Struct. Dyn.*, 2022, 40(13), 6183-6192
12. C. Teoh, T. Heidelberg, M. Rizman-Idid, *DARU. J. Pharm. Sci.,* 2011, 19(6), 469-475.
13. AK. Kalimuthu, T. Panneerselvam, P. Pavadai, SRK. Pandian, K. Sundar, S. Murugesan, DN. Ammunje, S. Kumar, S. Arunachalam, S. Kunjiappan, *Sci. Rep.*, 2021, 11(1), 21488.
14. WL. Chong, K. Chupradit, SP. Chin, MM. Khoo, SM. Khor, C. Tayapiwatana, P. Nimmanpipug, W. Thongkum and VS. Lee, *Molecules.,* 2021, 26, 5696.
15. T. Mohankumar, V. Chandramohan, HS. Lalithamba, RL. Jayaraj, P. Kumaradhas, M.Sivanandam, G.Hunday, R. Vijayakumar, R. Balakrishnan, D.Manimaran and N.Elangovan, *Sci. Rep.,* 2020, 10, 599.

IN SILICO ADMET PREDICTION AND MOLECULAR DOCKING STUDIES IN THE DESIGN OF NEW THERAPEUTIC AGENTS TARGETING PIM 1 KINASE

S. Muni Sireesha[a], A. Sejal Mudiraj[a], A. Varmini[a], A. Pallavi [a], Afirin Farahana[a] and T. Saritha Jyostna[a*]

[a]Osmania University, Sarojini Naidu Vanita Pharmacy Maha Vidyalaya, Department of Pharmaceutical Chemistry, 500017, Hyderabad, India.

Corresponding Author E-Mail ID: sirisnvpmv@gmail.com

1 INTRODUCTION

Janus kinase (JAK) / signal transducers and activators of transcription (STAT) pathway controls PIM 1 and interacts with numerous proteins,[1] regulates multiple signaling pathway such as cell growth, cell survival, cell cycle and drug resistance.[2] The PIM1 gene was initially recognized as proviral integration site for Moloney murine leukemia virus. It is a proto-oncogene that translates a serine/threonine kinase which has a dynamic character in oncogenesis.[3] Transcription of pim1 can be triggered by the several interleukines, cytokinines. The inhibitors of pim1 kinase includes flavonoids, cinnamic acid, quinoline 3,4 diones, imidazole (1,2 – b) pyridazines most of which are in clinical trials.[4] Even though these agents have revealed good effectiveness and negligeable toxicities invitro cell lines. Thus pim1 is a promising antitumor agent. Pim 1 is involved in numerous human cancers together with prostate cancer, acute myeloid leukemia and other hemapoietic malignances. Pim 1 kinases are exclusive and different from other kinases.[5]

IN-VITRO cell lines and their strong toxicity investigation is still being accepted by researchers worldwide and there is an extensive possibility to progress novel leads further side quinolone has seemed as an appreciated scaffold in medicinal chemistry maintaining various variety of pharmacological actions.[6] In view of literature we designed a 2-quinoline with substitution of dithiocarbamate at 3rd position. Molecular docking help to project a novel effective inhibitor by mechanism of drug receptor interaction.

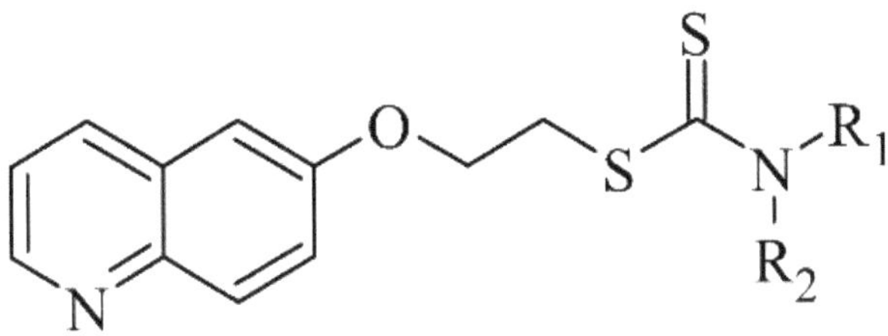

Figure 1 *General Structure of 2-quinoline with substitution of dithio carbamate.*

Table 1 *Structures and IUPAC names of the designed quinoline derivatives.*

S. No.	CHEMICAL STRUCTURE	CHEMICAL NAME
1	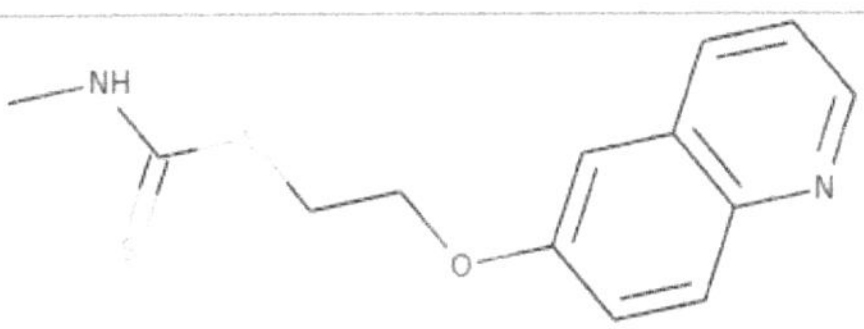	2-[(quinolin-6-yl)oxy]ethyl methylcarbamodithioate
2	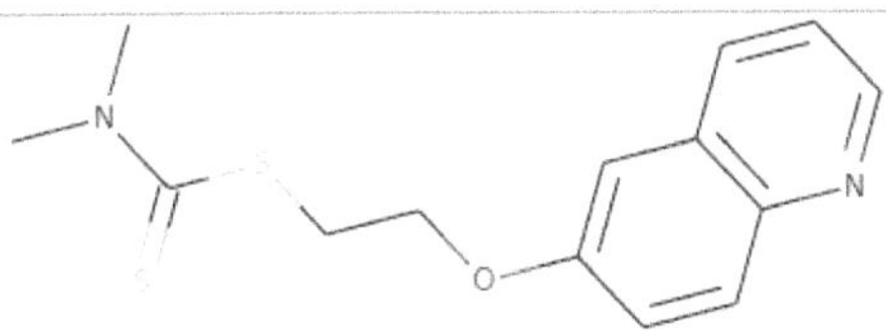	2-[(quinolin-6-yl)oxy]ethyl dimethylcarbamodithioate
3	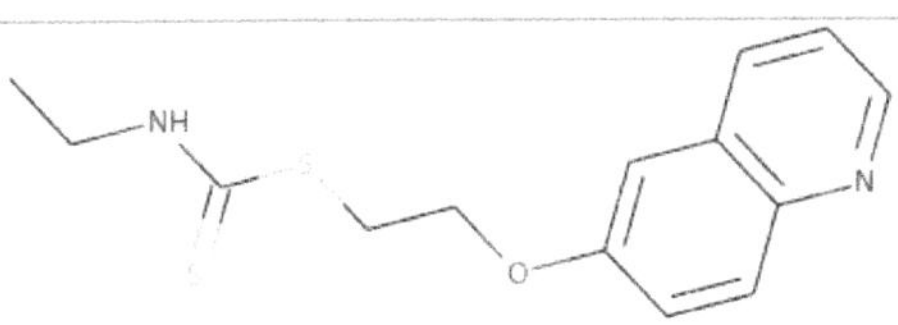	2-[(quinolin-6-yl)oxy]ethyl ethylcarbamodithioate

4	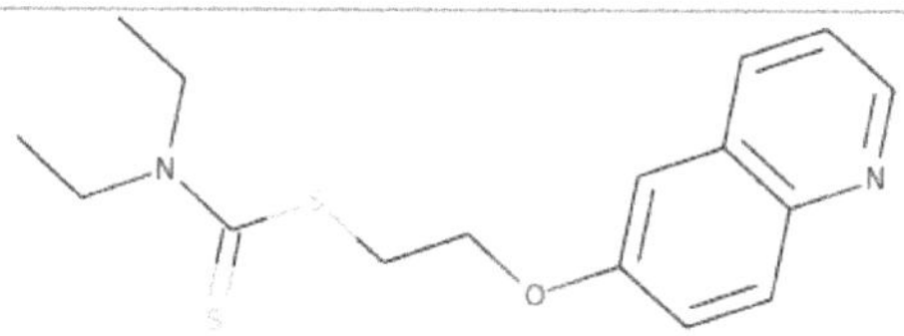	2-[(quinolin-6-yl)oxy]ethyl diethylcarbamodithioate
5	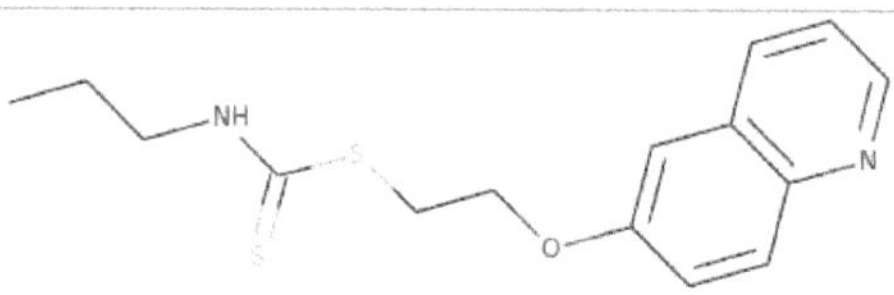	2-[(quinolin-6-yl)oxy]ethyl propylcarbamodithioate
6	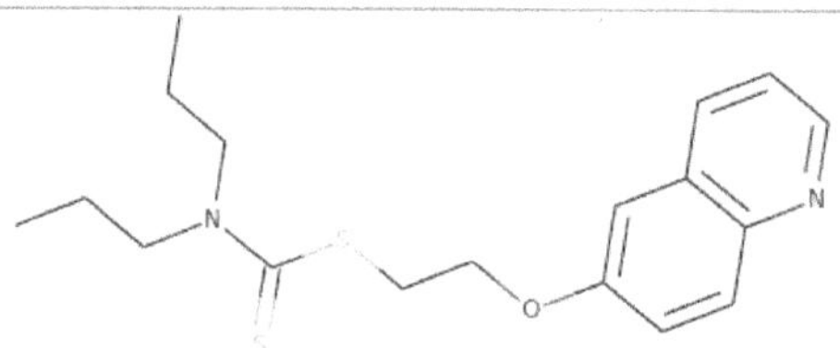	2-[(quinolin-6-yl)oxy]ethyl dipropylcarbamodithioate
7	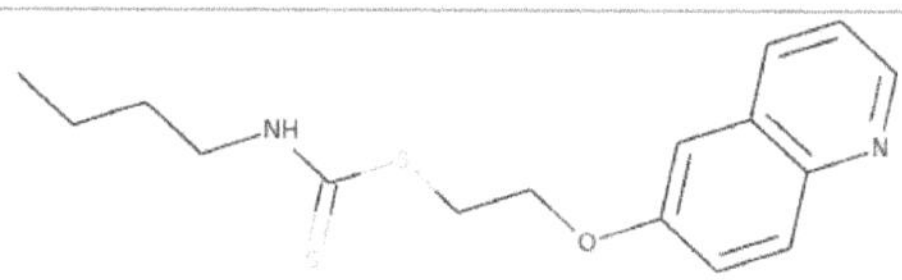	2-[(quinolin-6-yl)oxy]ethyl butylcarbamodithioate
8	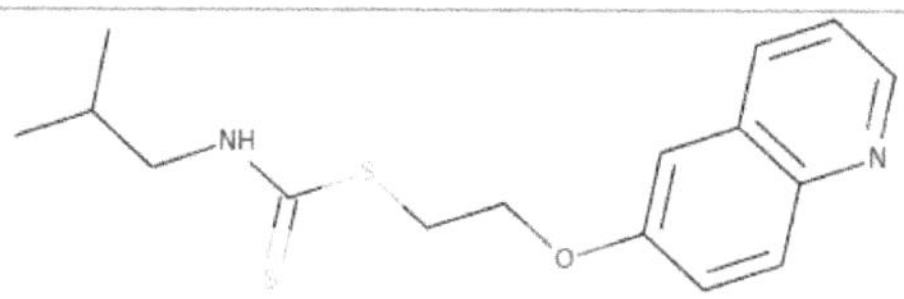	2-[(quinolin-6-yl)oxy]ethyl(2-methylpropyl)carbamodithioate
9	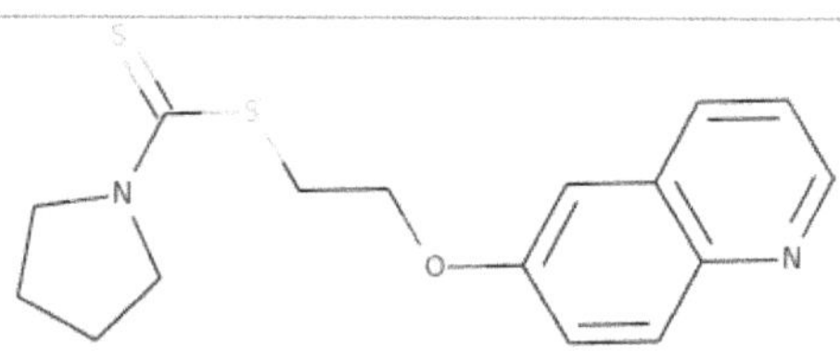	2-[(quinolin-6-yl)oxy]ethyl pyrrolidine-1-carbodithioate

10	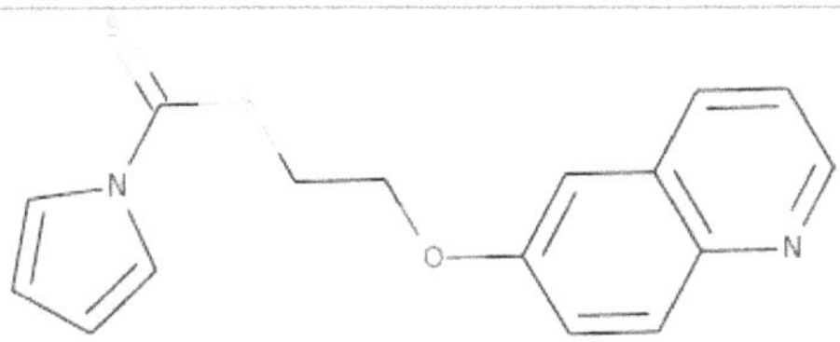	2-[(quinolin-6-yl)oxy]ethyl 1H-pyrrole-1-carbodithioate
11	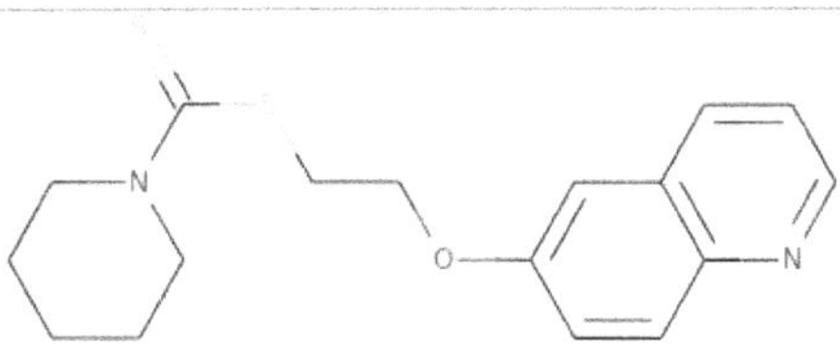	2-[(quinolin-6-yl)oxy]ethyl piperidine-1-carbodithioate
12	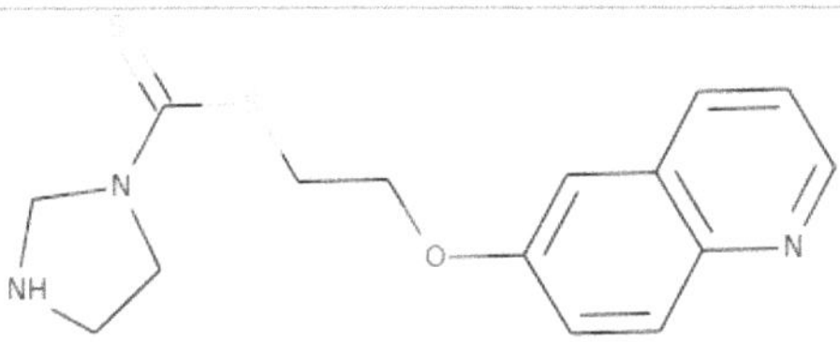	2-[(quinolin-6-yl)oxy]ethyl imidazolidine-1-carbodithioate
13	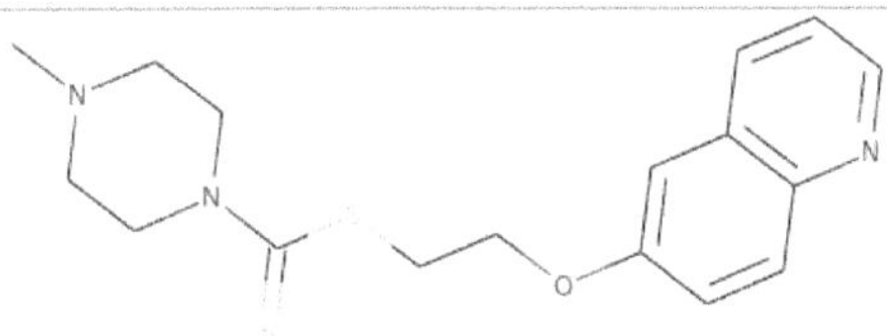	2-[(quinolin-6-yl)oxy]ethyl4-methylpiperazine-1 carbodithioate
14	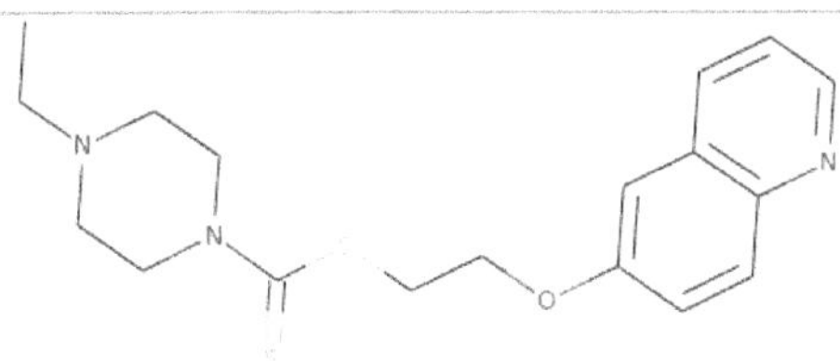	2-[(quinolin-6-yl)oxy]ethyl4-ethylpiperazine-1-carbodithioate
15	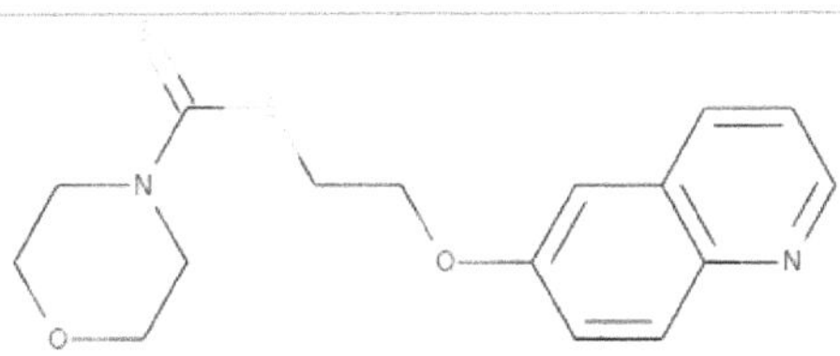	2-[(quinolin-6-yl)oxy]ethylmorpholine-4-carbodithioate

16	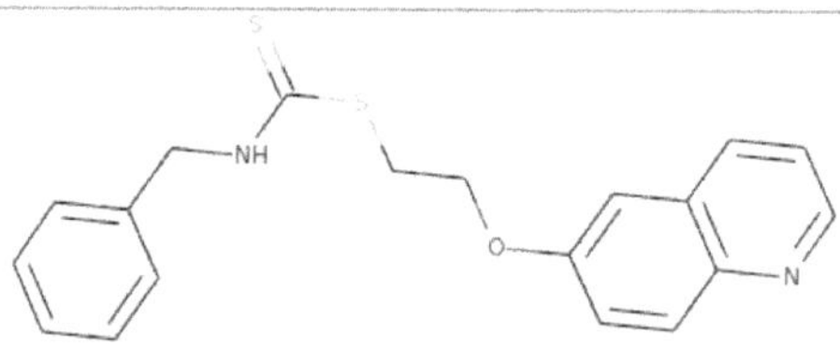	2-[(quinolin-6-yl)oxy]ethyl benzylcarbamodithioate
17	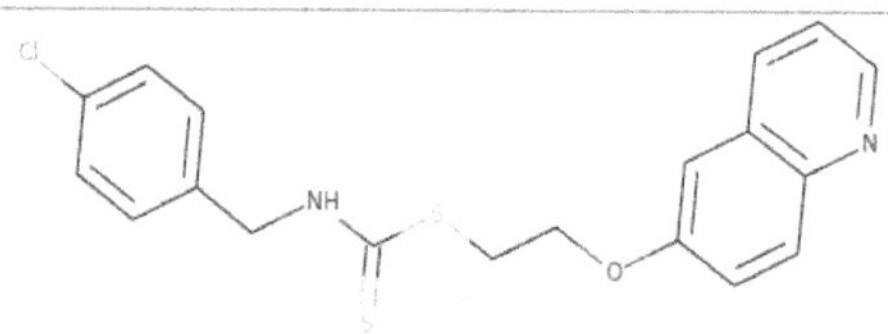	2-[(quinolin-6-yl)oxy]ethyl[(4-chlorophenyl)methyl]carbamodithioate
18	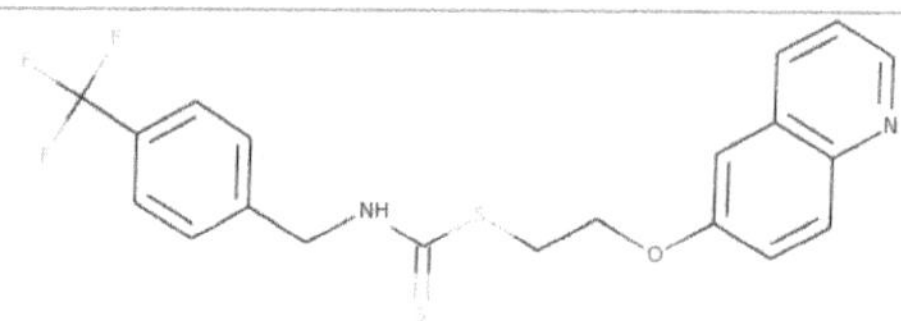	2-[(quinolin-6-yl)oxy]ethyl {[4-(trifluoromethyl)phenyl]methyl}carbamodithioate
19	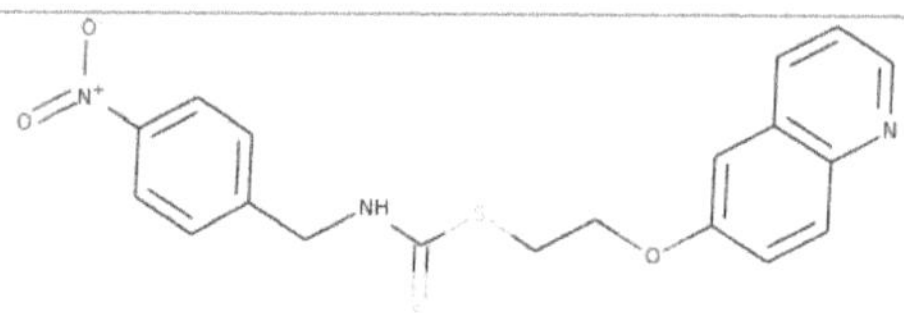	2-[(quinolin-6-yl)oxy]ethyl[(4-nitrophenyl)methyl]carbamodithioate
20	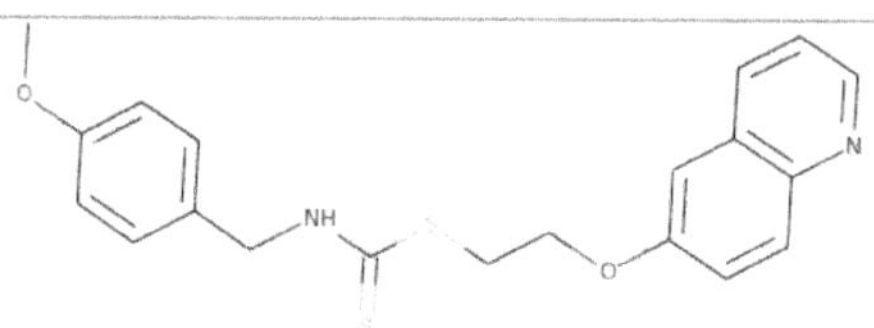	2-[(quinolin-6-yl)oxy]ethyl[(4-methoxyphenyl)methyl]carbamodithioate

2 MATERIALS AND METHODS

In the process of discovery, computational (in silico) facilitates optimization of lead molecules by correlating pharmacokinetic properties and screening of molecules. In this study data sets of 20 molecules have been selected. The software choosed are SWISS ADME (free, cc-by 40 creative commons 4.0 international liscence , swiss institute of bioinformatics, Switzerland), Molsoft (free, Molsoft LLC, US), Molinspiration, PKCSM. All the designed molecules were exposed to Swiss ADME, to estimate pharmacokinetic studies and drug-likeness. This tool is beneficial in calculation of properties and Lipinski violation. Molsoft, is another leading tool in the range of structure prediction, structural proteomics, bioinformatics, cheminformatics, molecular awareness and simulation. log P (O/W) values are projected with this tool. Molinspiration endeavor extended variety of cheminformatics software tools encouraging molecule handling and processing, SMILES and SD files modification, molecular standardization, origination of tautomers, molecule fragmentation, calculation of numerous molecular properties desired in QSAR and molecular modeling, high quality molecule portrayal. This helps to envision GPCR ligand, ion channel modulator, protease inhibitor, kinase inhibition, Nuclear receptor ligand, enzyme inhibitor. PKCSM is an innovative technique for envisaging and optimizing small molecules. It predicts pharmacokinetic properties, human intestinal absorption (HIA %), IN-VITRO skin permeability, IN-VITRO VDss (Human), IN-VITRO Blood brain barrier permeability and toxicity.

2.1 Molecular docking studies

Docking studies was performed to recognize the PIM1 protein structure by means of binding ability of several ligand molecules gone through for this study. PIM1 kinase 3D structure was (PDB ID: 3BGQ) downloaded from Protein Data Bank and saved as a Brookhaven pdb file format. ACD/Chemsketch helps to create the 2D structures of ligands. All the created ligands were cleaned and 3D optimization was done and saved in MDL Molfile format. Later by using Open Babel all the ligands were converted to PDB file format. Docking studies was performed by AutoDock Vina. Around the co-crystallized ligand a grid was established. The co-ordinates (x = -8.298, y = 40.748, z = 0.228) were produced with the help of Pharmit. Prepared pdbqt files for both target and ligands. docking accomplished in the absence of water molecules by adding charges for all 22 molecules (20+2 standard molecules). After docking all the molecules were examined and envisioned in the discovery studio for active site amino acids interactions.[13]

Table 2 *Predicted molecular properties of the designed quinoline derivatives.*

S. No.	Human Intestinal Absorption	Caco-2 cell permeability	Skin permeability (log Kp)	VDss (human) (log L/KG)	Blood-brain barrier permeability (log BB)	Toxicity
1	94.124	1.658	-2.693	-0.021	0.501	NO
2	93.914	1.675	-2.675	0.062	0.248	NO
3	93.705	1.654	-2.715	0.075	0.485	NO
4	93.076	1.666	-2.727	0.256	0.338	NO

5	93.292	1.659	-2.719	0.168	0.454	NO
6	92.741	1.291	-2.836	0.47	0.417	NO
7	92.666	1.659	-2.71	0.269	0.44	NO
8	93.263	1.675	-2.691	0.215	0.451	NO
9	93.925	1.624	-2.801	0.757	0.579	NO
10	96.999	1.775	-2.484	-0.041	0.136	NO
11	93.536	1.627	-2.797	0.806	0.563	NO
12	92.406	1.6	-2.939	1.317	0.546	NO
13	94.219	1.654	-3.085	0.807	0.25	NO
14	93.8	1.27	-3.087	0.898	0.326	NO
15	95.324	1.747	-2.896	0.511	0.454	NO
16	92.531	1.113	-2.747	0.02	0.589	NO
17	89.806	0.749	-2.735	0.011	0.233	NO
18	89.649	1.104	-2.675	0.27	0.542	NO
19	91.414	0.59	-2.704	0.115	-0.857	NO
20	92.643	1.112	-2.737	0.181	0.452	NO
SGI 1776	96.80	24.30	-2.735	0.662	1.18	Mutagenicity
AZI 1208	96.02	3.11	-2.699	1.168	0.82	Mutagenicity

Table 3 *In silico predicted bioactivity properties of the designed quinoline derivatives.*

S. No.	GPCR Ligand	Ion channel modulator	Kinase inhibitor	Nuclear receptor Ligand	Protease inhibitor	Enzyme inhibitor
1	-0.39	-0.73	-0.63	-0.77	-0.68	-0.18
2	-0.28	-0.78	-0.64	-0.74	-0.73	-0.26
3	-0.31	-0.74	-0.58	-0.67	-0.62	-0.17

4	-0.25	-0.82	-0.63	-0.67	-0.67	-0.20
5	-0.21	-0.61	-0.54	-0.58	-0.49	-0.13
6	-0.13	-0.71	-0.54	-0.52	-0.49	-0.16
7	-0.15	-0.61	-0.48	-0.49	-0.40	-0.09
8	-0.21	-0.61	-0.53	-0.53	-0.36	-0.15
9	-0.17	-0.72	-0.56	-0.60	-0.53	-0.25
10	-0.08	-0.28	-0.06	-0.07	-0.30	0.04
11	-0.15	-0.72	-0.54	-0.57	-0.52	-0.14
12	0.04	-0.59	-0.53	-0.95	-0.43	-0.22
13	-0.13	-0.67	-0.43	-0.59	-0.51	-0.15
14	-0.13	-0.69	-0.48	-0.55	-0.51	-0.18
15	-0.24	-0.81	-0.51	-0.62	-0.56	-0.19
16	-0.14	-0.55	-0.35	-0.40	-0.30	-0.11
17	-0.14	-0.54	-0.36	-0.40	-0.33	-0.14
18	-0.07	-0.40	-0.25	-0.21	-0.23	-0.09
19	-0.28	-0.55	-0.45	-0.46	-0.41	-0.21
20	-0.16	-0.55	-0.34	-0.40	-0.31	-0.13
SGI 1776	-0.2	-0.27	-0.13	-0.2	-0.19	-0.21
AZI 1208	-1.26	-1.22	-0.75	-1.01	-1.33	-0.46

Table 4 *Insilico predicted lipinski's properties of the designed quinoline.*

S. No.	M.W. g/mol	H.B. Don.	H.B. Acc.	LogP (o/w)	M.R. cm3/mol	TPSA Å 2
1	278.39	1	2	2.83	80.84	91.54
2	292.42	0	2	3.03	85.74	82.75

3	292.42	1	2	2.95	85.65	91.54
4	320.47	0	2	3.65	95.35	82.75
5	306.45	1	2	3.31	90.45	91.54
6	348.53	0	2	3.96	104.97	82.75
7	320.47	1	2	3.28	95.26	91.54
8	320.47	1	2	3.5	95.26	91.54
9	318.46	0	2	3.27	97.16	82.75
10	314.43	0	2	3.35	92	84.44
11	332.48	0	2	3.63	101.96	82.75
12	319.44	1	3	2.99	99.07	94.78
13	347.5	0	3	3.68	108.77	85.99
14	361.52	0	3	3.84	113.58	85.99
15	334.46	0	3	3.34	98.24	91.98
16	354.49	1	2	3.43	105.33	91.54
17	388.93	1	2	3.73	110.34	91.54
18	422.49	1	5	3.62	110.33	91.54
19	399.49	1	4	3.18	114.15	137.36
20	384.52	1	3	3.62	111.82	100.77
SGI 1776	405.42	1	7	4.27	107.95	54.68
AZ 1208	221.28	1	2	2.42	65.26	71.47

Table 5 *Bond Interactions Between Ligand & Target Protein (3BGQ).*

S. No.	Binding EN(Kcal/mol)	H-bond interactions	Interacting amino acids
1	-6.8	LYS:67	VAL52, ILE104, LEU120, ILE185,
2	-7	LYS:67	LEU44, PHE49, VAL52, ILE104, LEU120, ILE185

3	-7.1	LYS:67; ASP:128	PHE49, VAL52, ILE104, LEU120, PHE130, ILE185
4	-7		PHE49, VAL52, ILYS67, ILE104, LEU120, ILE185
5	-7.2	LYS:67, GLU:171	PHE49, VAL52, ILE104, LEU120, PHE130, ILE185
6	-7	LYS:67	PHE49, VAL52, ILE104, LEU120, ILE185
7	-6.9	-	LEU44, PHE49, VAL52, ALA65, LYS67, ILE104, LEU120, LEU174, ILE185
8	-7	-	PHE49, VAL52, LYS67, ILE104, LEU120, ILE185
9	-7.6	-	LEU44, PHE49, VAL52, ALA65, LYS67, ILE104, LEU120, LEU174, ILE185
10	-7.5	-	LEU44, PHE49, VAL52, ALA65, LYS67, ILE104, LEU120, LEU174, ILE185
11	-8.1	-	LEU44, VAL52, ALA65, LYS67, ILE104, LEU120, ILE185, ASP186
12	-7.2	LYS67	VAL52, LEU120, ASN172, ILE185
13	-8	LYS67	VAL52, LEU120, GLU121, ILE185
14	-7.6	LYS67	LEU44, VAL52, LEU120, ARG122, VAL126, ILE185
15	-7.5	LYS67	LEU44, VAL52, LEU120, ILE185
16	-8.9	LYS67	LEU44, PHE49, VAL52, LYS67, ILE104, LEU120, ASP128, LEU174, ILE185
17	-8.2	-	LEU44, PHE49, VAL52, ALA65, LYS67, GLU171, LEU174, ILE185
18	-8.6	ASP128	LEU44, PHE49, VAL52, ALA65, LYS67, ILE104, LEU120, VAL126, ASN172, LEU174, ILE185, ASP186,
19	-8.2	ASP128	LEU44, PHE49, VAL52, LYS67, ILE104, LEU120, PRO123, VAL126, ASP128, LEU174, ILE185
20	-8.1	-	LEU44, PHE49, VAL52, ALA65, GLU89, ILE104, LEU120, ASN172, ILE185, ASP186
AZD 1208	-9.7	LEU44	VAL52, GLN127, ASP128, LEU174, ILE185

SGI 1776	-10.1	LYS67	LEU44, PHE49, VAL52, ALA65, ILE104, LEU120, GLU121, ARG122, ASP128, LEU174, ILE185, ASP186

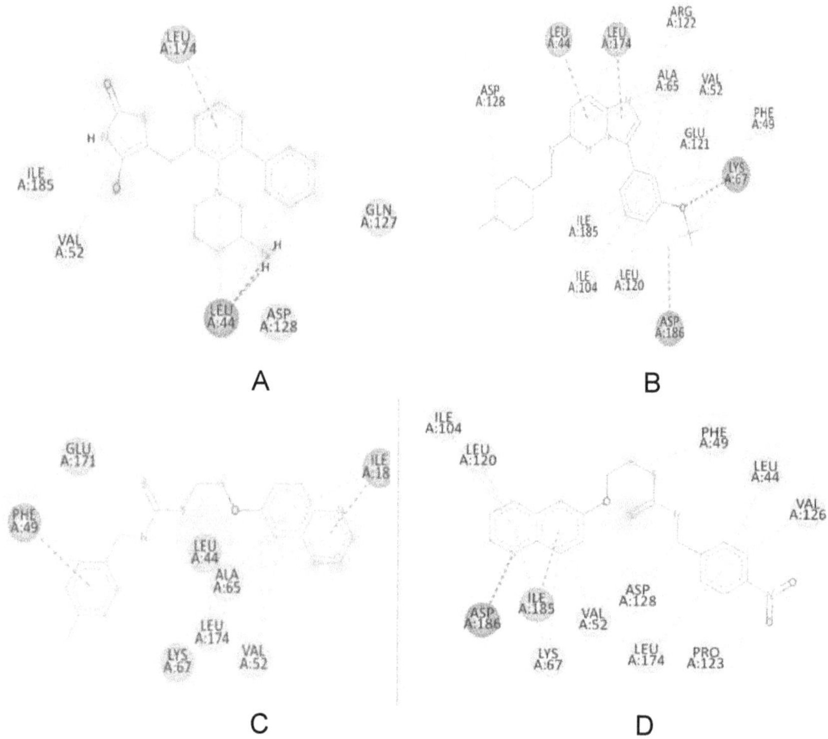

Figure 2 *The docking mode of (A) AZD1208 (B) SGI1776 (C) Compound 16 (D) Compound 18, into PIM1 Kinase (pdb code: 3BGQ) shows hydrogen bonds, hydrophobic interactions pi-sigma, pi-pi staked, pi-alkyl, and alkyl type interactions.*

3 RESULTS AND DISCUSSION

3.1 Molecular properties

The molecular properties of designed compounds (1-20) were predicted and shown in Table 2. All the designed compounds have considerable HIA capacity from 78.59 to 93.80. and are fewer accessible to Caco-2 cells between 0.69-1.54 cm/sec. The invivo blood brain barrier diffusion values are from -0.773 to 0.386 evidencing them as CNS inactive compounds. All the compounds show non-toxic nature. Furthermore, all the properties of designed compounds are compared with the SGI 1776 and AZ 1208 (PIM 1 Kinase inhibitors). The overall estimation has shown that the compounds are protected to the normal cells and suitable anti-tumor agents. The bioactivity parameters of compounds 1-

20 (Table 3) were approximated for all the compounds and related with standards of SGI 1776 and AZ 1208. The outcome shows them as protected compounds with upright cell growth inhibitors and binding capacities. The projected lipinski properties of 1-20 were shown in (Table 4). An orally accessible drug must follow lipinski rule with molecular weight of fewer than 500 daltons, hydrogen bond donors is less than 5, hydrogen bond acceptors is lower than 10, the log P value is lower than 5 and molecular refractivity is within the dimension. The outcomes show that all the molecules follow to Lipinski rule, illustrating good oral absorption. The TPSA was determined and detected that most of the compound's value is less than 140. This ability with respect to the lipinski rule confirm as harmless administrable drugs and authorizes their biological activity.

3.2 Docking

The ligands with binding energies and various bond interactions of all 20 compounds are shown in table 5. The binding energies are in the range of -7.20 to -9.70 KJ/mol for Pim-1 kinase protein. They formed 1–3 hydrogen bonds with LEU 44, LYS 67, GLU 121 and ASP 186 amino acids in the active site. Various types of hydrophobic interactions were also detected together with pi-sigma, pi-pi staked, pi-alkyl, and alkyl types are shown in Figure 6.2. Among the series compound 16 and compound 18 exhibited equivalent dockscore values (-9.7 k.cal/mol) when compared to standard drugs AZD1208 and SGI1776.

4 CONCLUSION

The investigational approaches for the calculation of molecular toxicity are tedious and intense tasks. Due to that, the computational methods are used to progress different approaches for molecular properties prediction. Exploration for the progress of novel PIM1 inhibitors, identified 2-oxo indole as a lead molecule. In the current study, we designed a series of Dithiocarbamate substituted at 3rd position at 2-oxoindole ring. All the designed derivatives were estimated for molecular properties by various softwares. From the insilico studies, all the compounds possess drug-like properties and serve as capable lead molecules for inhibition of PIM1 target. All 20 compounds are docked into active site of PIM1 kinase and compared with standard drugs SGI 1776 and AZD 1208. The docking results specified the compound 16 and 18 perfect fitting into active site of PIM1.The overall outcomes suggests that the effective use of these computational physio chemical properties together with docking interactions may form an integrated background that can be used in lead optimization to develop effective PIM1 inhibitors in future.

5 ACKNOWLEDGMENT

The authors are obliged to Management of Sarojini Naidu Vanita Pharmacy Maha Vidyalaya for providing the services to carry the work.

6 REFERENCES

1. A.S.Green, Catherine Lacombe, Patrick Mayeux, Didier Bouscary and Jerome Tamburini, *Blood,* 2012,**119**,1791-1792.

2. H. Arrouchi, W. Lakhlili, A. Ibrahimi, *Bioinformation*, 2019, **15(1),** 40– 45.

3. L. Broccatelli, E. Salphati, J. Plise, A. Cheong, M.L. Gobbi, I. Lee, Aliagas, D.Brown. *Drug Discoverv Today,* 2003, **8**,1067-1077.

5. L.S Chen, S. Redkar, P.Taverna, J.E.Cortes and V.Gandhi, *Blood,* 2011, **118(3),** 693-702.

6. M. Congreve, C.W.Murray, T.L.Blundell, *Drug Discov.Today*, 2005, **10**, 895-907.

7.J.A.DiMasi, R.W.Hansen, H.G.grabowski, *J. Health Economics*, 2003, **22**,15l-185.

8. J. Drews, *Drug Discov. Today*, 2003, **8**, 411-420.

9. M.M.J. Haddach, M.K. Schwaebe, F.Pierre, S.E. O'Brien, C.T.J. Borsan, N.Raffaele, *ACS Med ChemLett,* 2012, **3**, 135–9.

10. S. Holder, M. Zemskova, C. Zhang, M. Tabrizizad, R. Bremer, J.W.Neidigh, M.B. Lilly, *Mol Cancer Ther*, 2007, **6**, 163–72.

11. W.L. Jorgensen, *Science*, 2004, **303**, 1813-181.

12. E.K. Keeton, K. McEachern, K.S.Dillman, S. Palakurthi, Y. Cao, M.R. Grondine, S. Kaur, S. Wang ,Y. Chen, A. Wu, M. Shen, F.D. Gibbons, M.L. Lamb, X. Zheng, R.M. Stone, D.J. Deangelo, L.C. Platanias, L.A. Dakin, H. Chen, P.D. Lyne, D. Huszar. *In 53rd ASH Annual Meeting*, 2011.

13. O. Kim, T. Jiang, Y. Xie, Z. Guo, H. Chen and Y. Qiu, *Oncogene* 2004, **23**, 1838–1844

14. G. Klebc, *Drug Discovery today*, 2006, **11**, 580-594

15. M. Lee, K.H. Lee, A. Min, J. Kim, S. Kim, H. Jang, J.M. Lim, S.H. Kim, D.H. Ha, W.J. Jeong, K.J. Suh, Y.W.Yang, T.Y. Kim, D.Y.Oh, Y.J. Bang, S.A. Im, *Cancer Res Treat*, 2019, **51(2),** 451-463.

DRUG REPURPOSING FOR PRIMARY LATERAL SCLEROSIS: A MOLECULAR DYNAMIC STUDY

Shannon D Almeida[a], Akash Kumar Pandey[a], Mohamed Mutahir[a], Prakash Pai[a], SM Prakruthi[a], Shivaraj Gouravakkanvara[a], Rishita Gosh[a], Damodar Nayak A[a], Parasuraman Pavadai[b*]
[a]Department of Pharmacology, Faculty of Pharmacy, M S Ramaiah University of Applied Sciences, Bengaluru-560054, Karnataka, India, [b]Department of Pharmaceutical Chemistry, Faculty of Pharmacy,M S Ramaiah University of Applied Sciences, Bengaluru-560054, Karnataka, India.
Corresponding Author E-Mail ID: pvpram@gmail.com

1 INTRODUCTION

Primary Lateral Sclerosis (PLS) is a slow and selective neurological disorder that affects the central nervous system of adults. It leads to continuous muscle stiffness and subtle loss of mobility through the development of corticobulbar dysfunction.[1] The main differences from amyotrophic lateral sclerosis (ALS) are that PLS has no effect on lower motor neurons, no muscle atrophy or spinal motor neuron degeneration.[2] PLS usually appears spontaneously after the age of 50 and advances slowly over years, if not decades. Discomfort typically starts in the legs, but it can also start in the tongue or hands. Clumsiness, weight loss, weakness, and leg stiffness are all possible symptoms. Other common symptoms include spasticity (spontaneous muscle loss caused by stretching that depends on the velocity of stretching), dragging of the foot, and speaking and swallowing difficulties due to facial muscle involvement. Respiration may also be a risk factor in later stages of the disease, requiring atypical respiratory support in individuals who develop respiratory failure. As seen in patients with Babinski syndrome, hyperreflexia is a significant component of PLS. Some person's exhibit emotional instability in one area and urgency in another, and patients with PLS occasionally have moderate cognitive alterations seen in neuropsychology examinations, particularly in managerial function assessments.[3]

There is no single test that can confirm a diagnosis of PLS. In fact, because this ailment can resemble the symptoms and signs of other neurological disorders including multiple sclerosis and ALS, it requires a battery of testing to rule out other infections. Blood tests, Electromyogram, MRI, Nerve conduction study, Spinal piercing are few of the tests performed for the diagnosis of PLS. Doctors may delay a diagnosis for 3 to 4 years since early ALS may resemble PLS until further symptoms develop a few years later.[4]

There are no medications available to prevent, treat, or reverse primary lateral sclerosis. Treatment is centred on symptom alleviation and maintenance.[5] Hence Drug repositioning/Drug repurposing can be used as an alternative methodology for drug discovery, and this 'poly-pharmacology' can be a therapeutic requirement for complex diseases. This study focuses on repurposing of FDA approved drugs against Transactive response DNA binding protein of 43 kDa (TDP43) protein of PLS disease by molecular docking and dynamic studies.

2 METHODS AND RESULTS

2.1 Preparation of protein

For the X-ray crystallographic structure of TAR DNA Binding protein [TDP43] (PDB-7Q8V), the RCSB Protein Data Bank (PDB: http://www.rcsb.org/pdb)[6] was utilised. Before beginning the analysis, the protein structure was cleaned and missing residues were added using the Swiss-PDB Viewer v4.1.0 (https://spdbv.unil.ch/). The file was stored as 7Q8V.pdb and was used for further investigation. We used BIOVIA Discovery Studio Visualizer version 4.0 software to determine protein structure and amino acid locations from active areas for use in molecular docking studies.

2.2 Active compounds retrieval and preparation

We discovered that around 137 FDA-approved compounds were effective against the TDP43 protein (PDB -7Q8V), but only the top ten were used for studies due to their high affinity towards the target protein and low free energy, and the process was carried out using (https://sibiolead.com/) and other plant constitutional drugs were discovered from Indian Medicinal Plants, Phytochemistry, and Therapeutics (IMPPAT) (https://cb.imsc.res.in/imppat/home),[7] previously published studies and public database PubChem (https://pubchem.ncbi.nlm.nih.gov/).

2.3 Identification of the binding location

A binding site is a portion of a macromolecule or a section of a macromolecule that permits the target molecule to directly participate in forming a specific combination with another chemical molecule to advance a chemical process. The binding of the ligand to the particular position of the chemical molecules plays a critical function in illness treatment. By providing adequate and favourable catalytic regions, the compounds are able to create enough contact sites to initiate fast interactions with target enzymes. All probable active binding sites of targeted drugs were found for further study using the Prank Web (https://prankweb.cz/) website. Once the active site of the protein was determined, a receptor grid was generated using the PyRx tool.

2.4 Molecular docking

Molecular docking has evolved into an essential method in drug discovery. The purpose of molecular docking is to use computing methods to anticipate the ligand-receptor complex structure. Docking is done in two steps: first, ligand conformations in the active site of the protein are sampled, and then these conformations are ranked using a scoring function. The molecular docking analysis was carried out using the PyRx 0.8 tool[8] and the AutoDockVina program.[9] The ligand was chosen from the FDA-approved chemical list (https://sibiolead.com/), and the target receptor was chosen as TDP43 (PDB-7Q8V).PyRx software was used to add the polar hydrogen and kollman partial charges to the 3D structure. The receptor and ligand data were saved in ".pdbqt" format to calculate out the molecular docking energy affinities (kcal mol1). AutoDockVina estimated the energy affinity values of each ligand for up to ten possible docking locations. The impacts of energies were determined using ligand conformation at the active binding site and the RMSD between the original and subsequent structures. The amount of hydrogen bonds and non-covalent interactions in each complex were computed using Discovery Studio Visualizer, which produced visualisations of features, compounds, and interactions (2D and 3D).[10]

2.5 Molecular dynamics simulation

The molecular dynamic studies are a type of computer stimulation that allows us to analyse the physical motions of atoms and chemical compounds in relation to their ligand interactions and conformation modes with receptors, allowing us to assess the binding stability.[11] DESMOND v3.6 package[12] was used to perform molecular dynamics experiments on the chosen ligand TDP43 complex files. The chosen ligands topology was obtained from the PRODRG server.All of the complexes have their systems prepared. To begin with molecular dynamic stimulations, the vacuum was reduced for 5000 steps using the steepest descent technique. A simple point charge (SPC) water model was used to solve the complicated structure in a cubic periodic box of 0.5 nm. The complex system was then maintained at an adequate salt concentration of 0.15 M by injecting a sufficient amount of Na+ and Cl counterions. For the last run, each complex was given a simulation time of 100 ns from the NPT (Isothermal-Isobaric, constant number of particles, pressure, and temperature) equilibration.[13, 14] The Desmond simulation software was used to do the trajectory analysis of root mean square deviation (RMSD) and root mean square fluctuation (RMSF).

Table 1 *Top ten compounds and their binding affinity against TDP43.*

SL NO.	Molecule name	Binding score (kcal mol−1)
1	Indinavir	-10
2	Adapalene	-9.5
3	Alclometasone	-9.3
4	Amsacrine	-9.3
5	Diflorasone	-9.3
6	Medrysone	-9.1
7	Doxycycline	-9
8	Flunisolide	-9
9	Lorazepam	-9
10	Nifedipine	-9

Ten active compounds were found to have lower binding energies (≤-9 kcal mol^{-1}) with the target protein and the values are displayed in Table 1. These 10 active compounds were submitted to further testing.

3 DISCUSSIONS

3.1 Protein–ligand interactions are interpretation

The BIOVIA Discovery studio visualizer tool was used to visualise the interactions between the 10 ligands selected and TDP43. It was discovered that compound Indinavir interacted better with TDP43, having a binding energy of -10 kcal mol^{-1}.

Table 2 *Chemical interactions of top ten compounds with TDP43 protein.*

Compounds	Residues, Amino acid	Distance (Å)	Bond category
Indinavir	45A, PHE	3.55	Hydrophobic interaction
	45A, PHE	3.34	Hydrophobic interaction
	48A, ILE	3.88	Hydrophobic interaction
	61A, ALA	3.71	Hydrophobic interaction
	73A, VAL	3.11	Hydrophobic interaction
	109A, LEU	3.39	Hydrophobic interaction
	110A, GLN	3.62	Hydrophobic interaction
	179A, LEU	3.73	Hydrophobic interaction
	63A, LYS	3.49	Hydrogen bond
	77A, GLU	2.85	Hydrogen bond
	176A, ASP	3	Hydrogen bond
	156A, LYS	4.42	Pi cation interaction
Adapalene	40A, ILE	3.38	Hydrophobic interaction
	48A, ILE	3.45	Hydrophobic interaction
	61A, ALA	3.33	Hydrophobic interaction
	109A, LEU	3.41	Hydrophobic interaction
	110A, GLN	3.86	Hydrophobic interaction
	175A, LEU	3.9	Hydrophobic interaction
	175A, LEU	3.05	Hydrophobic interaction

	175A, LEU	3.41	Hydrophobic interaction
	154A, ASD	2.43	Hydrogen bond
	154A, ASD	2.48	Hydrogen bond
	156A, LYS	4.86	Salt bridge
Aclometasone	40A, ILE	3.69	Hydrophobic interaction
	40A, ILE	3.86	Hydrophobic interaction
	113A, ASN	3.76	Hydrophobic interaction
	175A, LEU	3.91	Hydrophobic interaction
	175A, LEU	3.89	Hydrophobic interaction
	63A, LYS	2.53	Hydrogen bond
	113A, ASN	2.46	Hydrogen bond
	116A, ASP	1.9	Hydrogen bond
	176A, ASP	2.77	Hydrogen bond
Amsacrine	40A, ILE	3.56	Hydrophobic interaction
	48A, ILE	3.7	Hydrophobic interaction
	61A, ALA	3.96	Hydrophobic interaction
	109A, LEU	3.74	Hydrophobic interaction
	110A, GLN	3.99	Hydrophobic interaction
	175A, LEU	3.93	Hydrophobic interaction
	110A, GLN	2.4	Hydrogen bond
	156A, LYS	2.49	Hydrogen bond
	158A, SER	2.43	Hydrogen bond
	158A, SER	2.54	Hydrogen bond
	159A, ASN	3.2	Hydrogen bond
Diflorasone	40A, ILE	3.7	Hydrophobic interaction
	40A, ILE	3.81	Hydrophobic interaction
	40A, ILE	3.86	Hydrophobic interaction
	113A, ASN	3.79	Hydrophobic interaction
	175A, LEU	3.91	Hydrophobic interaction
	63A, LYS	2.58	Hydrogen bond
	111A, GLY	1.94	Hydrogen bond
	113A, ASN	2.4	Hydrogen bond
	116A, ASP	1.94	Hydrogen bond
	176A, ASP	2.54	Hydrogen bond
Medrysone	40A, ILE	3.69	Hydrophobic interaction
	40A, ILE	3.62	Hydrophobic interaction
	40A, ILE	3.66	Hydrophobic interaction
	61A, ALA	3.61	Hydrophobic interaction
	175A, LEU	3.6	Hydrophobic interaction
	175A, LEU	3.85	Hydrophobic interaction
	63A, LYS	2.67	Hydrogen bond
	176A, ASP	2.86	Hydrogen bond

Doxycycline	48A, ILE	3.71	Hydrophobic interaction
	175A, LEU	3.46	Hydrophobic interaction
	175A, LEU	3.8	Hydrophobic interaction
	110A, GLN	2.22	Hydrogen bond
	111A, GLY	2.06	Hydrogen bond
	113A, ASN	3.35	Hydrogen bond
	113A, ASP	3.57	Hydrogen bond
	116A, ASP	1.81	Hydrogen bond
	116A, ASP	4.46	Salt bridge
Flunisolide	40A, ILE	3.93	Hydrophobic interaction
	40A, ILE	3.66	Hydrophobic interaction
	40A, ILE	3.86	Hydrophobic interaction
	61A, ALA	3.98	Hydrophobic interaction
	109A, LEU	3.9	Hydrophobic interaction
	30A, LYS	2.56	Hydrogen bond
	50A, GLU	2.48	Hydrogen bond
	63A, LYS	3.3	Hydrogen bond
Lorazepam	40A, ILE	3.81	Hydrophobic interaction
	40A, ILE	3.71	Hydrophobic interaction
	48A, ILE	3.6	Hydrophobic interaction
	175A, LEU	3.31	Hydrophobic interaction
	175A, LEU	3.62	Hydrophobic interaction
	176A, ASP	3.68	Hydrophobic interaction
	63A, LYS	2.92	Hydrogen bond
	108A, GLN	2.68	Hydrogen bond
	110A, GLN	2.17	Hydrogen bond
	175A, LEU	2.37	Hydrogen bond
Nifedipine	45A, PHE	3.67	Hydrophobic interaction
	47A, GLU	3.66	Hydrogen bond
	63A, LYS	3	Hydrogen bond
	64A, VAL	3.12	Hydrogen bond
	110A, GLN	2.39	Hydrogen bond
	113A, ASN	2.39	Hydrogen bond
	164A, ARG	3.47	Hydrogen bond
	175A, LEU	2.71	Hydrogen bond

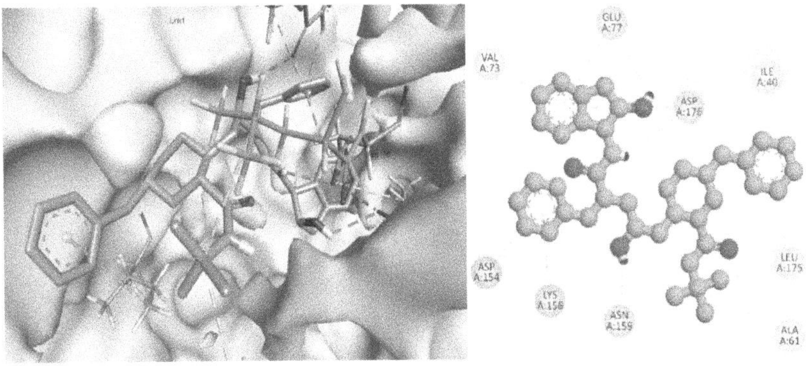

Figure 1 *The interaction of INDINAVIR and TDP43 was shown. The left side represents complex protein–ligand interaction in 3D, whereas the right side represents complex protein–ligand interaction in 2D.*

3.3 Molecular dynamics simulation

MD simulations were used to examine the stability, measure intermolecular interactions, and validate the interaction between the 10 FDA-approved medicines and the TDP43 protein receptor. A model simulating identical physiological circumstances was created. For a time span of 100 ns, various thermodynamical conditions such as temperature, pressure, density, and volume were applied. Complex adjustments to the wielding DESMOND PACKAGE were estimated under the applied conditions. Using the ensembles, the entire system was annealed and equilibrated. The last production stage also looked at structural modifications to drug receptor complexes. Furthermore, MD trajectories analysis was performed using unique parameters such as root mean square deviation (RMSD), root meansquare fluctuation (RMSF), conformational modification of ligands, and intermolecular interactions of these complexes to analyse the stability and structural changes at each level of these complexes

3.3.1 Root mean square deviation (RMSD). Checking the RMSD values is one of the most critical analyses that should be performed. The Root Mean Square Deviation (RMSD) is used to calculate the average change in displacement of a set of atoms with respect to a reference frame for a given frame. It is computed for each frame of the trajectory. It demonstrates the stability of the complex, as well as the precise instant when the structure of the complex has altered, providing a rapid suggestion regarding the time period that has to be focused on. The RMSD of the protein backbone atoms was plotted against time to determine structural conformation alterations. Initially, the backbone RMSD of the 7q8v-Indinavir and 7q8v-Amsacrine complexes varied from 0.6 to 2.4 Å until 45 ns. The stable conformation was achieved at a time range of 50 to 100 ns, with slight variances in the results.

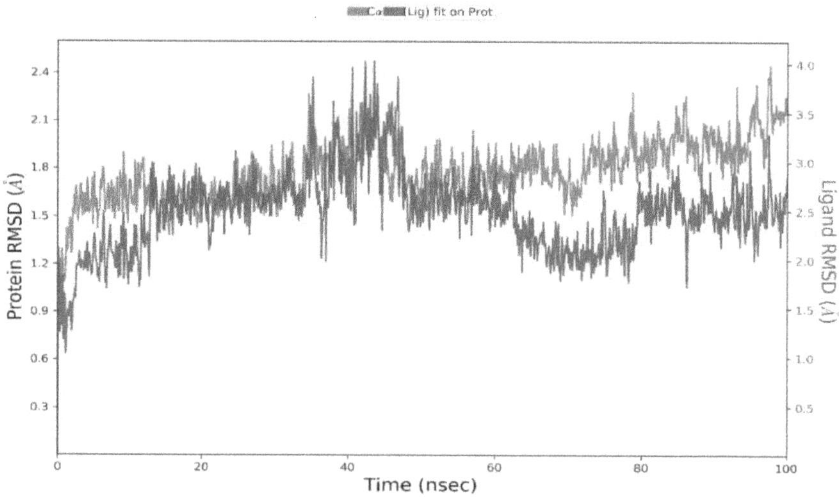

Figure 2 RMSD study plot of Indinavir-TDP43 complex.

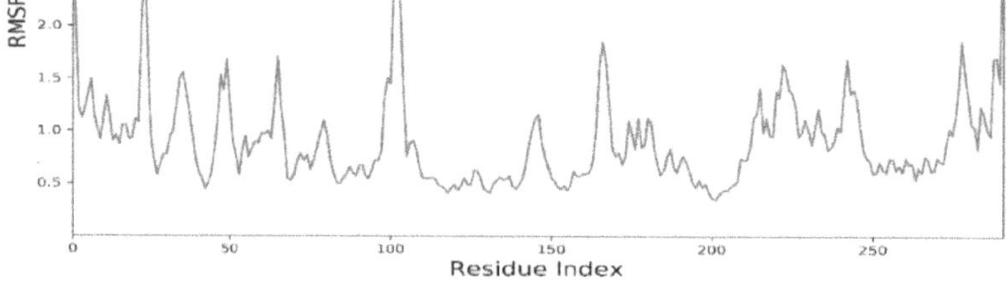

Figure 3 *RMSF study plot of Indinavir-TDP43 complex.*

3.3.2 Root mean square fluctuation (RMSF). A very helpful analysis, that helps you understand what each residue is doing, is the root mean square fluctuation

(RMSF) value. This value is the average distance of the movement of the α-carbon of each residue during the simulation time. In a protein or an enzyme, some residues play key roles in the inhibition or the activation mechanism of that protein. Therefore, it is necessary to check their properties in the simulation, and by analysing their RMSF values, it can be found whether they have been affected by the ligand or not. On this plot, peaks indicate areas of the protein that fluctuate the most during the simulation. Typically, you will observe that the tails (N- and C-terminal) fluctuate more than any other part of the protein. Secondary structure elements like alpha helices and beta strands are usually more rigid than the unstructured part of the protein, and thus fluctuate less than the loop regions.

3.3.3 Protein ligand contacts/ interactions. Inspecting all formed interactions between a ligand and the residues inside a binding site can help judge whether a ligand is potential or not. There are numerous direct and indirect ways to analyse these interactions. The 2D interaction diagram is an example of direct ways, and it is a tremendous tool to comprehensively study these interactions. The first step to analyse these interactions is to know their strength and their importance. For example, hydrogen bonds are one of the strongest non-covalent interactions, and monitoring their number and their positions is very helpful.

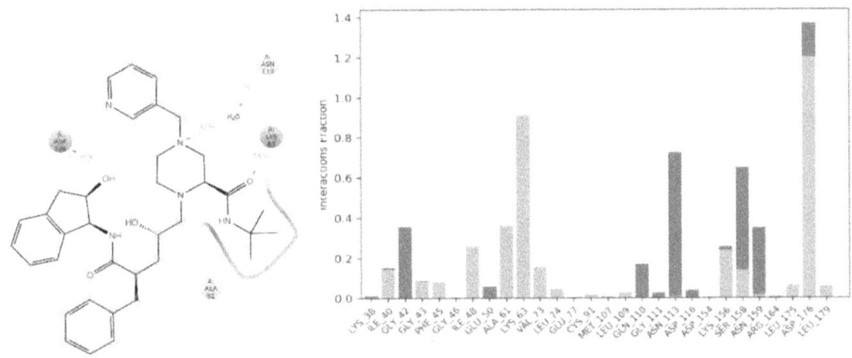

Figure 4 *Interaction of ligand with protein. The left side represents INDINAVIR-TDP43 contacts, whereas the right side represents Interacting amino acid residues of INDINAVIR with TDP43.*

The major interaction fractions of the ligand Indinavir with the amino acid residues were H-bond interaction with an interaction fraction of 1.4 with ASP_176 and 0.9 with LYS_63, Water bridges with an interaction fraction of 0.7 with ASN_113 and Hydrophobic interaction with an interaction fraction of 0.4 with ALA_61. There was no such significant ionic interaction seen between the indinavir and amino acid residues of TDP-43.

4 CONCLUSION

The project includes data related to Primary Lateral Sclerosis and its possible treatment through in silico molecular docking. This includes repurposing of FDA approved drugs and evaluation of its potency towards the disease target, TDP43 protein Thus, molecular docking has been carried out using SIBio Lead software to observe the interaction of the protein with ligand and measure various interactions between them such as hydrogen bonding, hydrophilic interactions etc which gives the information regarding the strength and affinity of the bonds in the complexes. The affinity of all drugs towards the protein was observed by docking, among which Indinavir (FDA approved drug) was found to be drug with highest binding affinity towards the target receptor. Molecular Dynamic (MD) studies of individual drugs were studied. MD studies revealed the protein-ligand stability. Hence these drugs can be evaluated further in the animal model to establish a drug for treating PLS.

5 ACKNOWLEDGMENT

The authors thank the Department of Pharmaceutical Chemistry, Faculty of Pharmacy, M. S. Ramaiah University of Applied Sciences, Bengaluru for providing computational facility.

6 REFERENCES

1. MK. Floeter, and R. Mill. *Amyotrophic lateral sclerosis*, 2009, **10**(5-6), 339-346.
2. RJ. Caselli, BE. Smith, and D. Osborne. *Neurology*, 1995, **45**(11), 2005-2009.
3. S. Chan, P. Kaufmann, DC. Shungu, and H. Mitsumoto. *Neuroimaging Clinics*, 2003, **13**(2), 307-326.
4. MR. Turner, and K. Talbot. *Practical Neurology*, 2020, **20**(4), 262-269.
5. MK. Floeter, D. Warden, D. Lange, J. Wymer, S. Paganoni, and H. Mitsumoto. *Amyotrophic Lateral Sclerosis and Frontotemporal Degeneration*, 2020, **21**(1), 67-73.
6. HM. Berman. *The Protein Data Bank. Nucleic Acids Research*, 2000, **28**(1), 235–242. DOI: 10.1093/nar/28.1.235.
7. K. Mohanraj, BS. Karthikeyan, RPV. Ananth, RPB. Chand, SR. Aparna, P. Mangalapandi, and A. Samal. *Scientific Reports*, 2018, **8**(1), 4329. DOI: 10.1038/s41598-018-22631-z.
8. X. Fradera, and K. Babaoglu. *Current Protocols in Chemical Biology*, 2017, **9**(3), 196–212. DOI: 10.1002/cpch.27.
9. T. Gaillard. *Journal of Chemical Information and Modeling*, 2013, **58**(8), 1697–1706. DOI: 10.1021/acs.jcim.8b00312.
10. UP. Mohan, S. Kunjiappan, PBT. Pichiah, and S. Arunachalam. *3 Biotech*, 2021, **11**(1). DOI: 10.1007/s13205-020-02530-9.

11. AK. Kalimuthu, T. Panneerselvam, P. Pavadai, SRK. Pandian, K. Sundar, S. Murugesan, DN. Ammunje, S. Kumar, S. Arunachalam, and S. Kunjiappan. *Scientific Reports*, 2021, **11**(1). DOI: 10.1038/s41598-021-01008-9.
12. S. Kunjiappan, M. Sankaranarayanan, BK. Kumar, P. Pavadai, E. Babkiewicz, P. Maszczyk, EG. Mrowka, S. Arunachalam, SRK. Pandian, V. Ravishankar, S. Baskararaj, S. Vellaichamy, L. Arulmani, and, T. Panneerselvam. *Nanotechnology*, 2020, **32**(9), 095101. DOI: 10.1088/1361-6528/abc57e.
13. WL. Jorgensen, J. Chandrasekhar, JD. Madura, RW. Impey, and ML. Klein. *The Journal of Chemical Physics*, 1983, **79**(2), 926–935. DOI: 10.1063/1.445869.
14. Y. Hu, and SB. Sinnott. *Journal of Computational Physics*, 2004, **200**(1), 251–266. DOI: 10.1016/j.jcp.2004.03.019.

ARTIFICIAL NEURAL NETWORK PROCESS USED IN GLOBAL DRUG DISCOVERY AND DEVELOPMENT

Grandhi Sai Meena*, Koushik Yetukuri

Department of Regulatory Affairs, Chalapathi institute of pharmaceutical sciences, lam, Guntur, Andhra Pradesh, 522034

*Corresponding Author E-Mail ID: gsaimeena8589@gmail.com

1 INTRODUCTION

In the 1940s, McCulloch and Pitts made an effort to develop a mathematical model of the human brain. After the perceptron underwent significant development, Rosenblatt published the first technique for pattern recognition using a two-layer ANN.Early in the 1970s, ANNs were first used to discover new drugs.Due to Rumelhart and McClelland's widespread adoption of the parallel distributed processing notion, neural networks gained enormous popularity in the middle of the 1980s[1]. One of the branches of Artificial Intelligence (AI) that is most frequently employed in the advancement of medicine is artificial neural networks (ANN). An alternative name for neural networks is ANN. These are the computing systems inclined by the biological nervous system. It is built on a network of interconnected components known as artificial neurons. Similar to a biological neuron, the ANN is also having the interconnection nodes. The ANN is a simple mathematical function. This model has three sets of rules. Those are multiplication, summation and activation. In order to predict the properties of chemical compounds for drug discovery, ANNs are used. It has been demonstrated mathematically that an ANN may be used to roughly predict the link between any chemical property and its structure. Drug research and design are routinely done using artificial neural networks (ANNs). In terms of both structure and capacity, ANN is similar to the human brain. The neurons that make up the brain are able to store, retrieve, and link bits of information (see Figure 1).

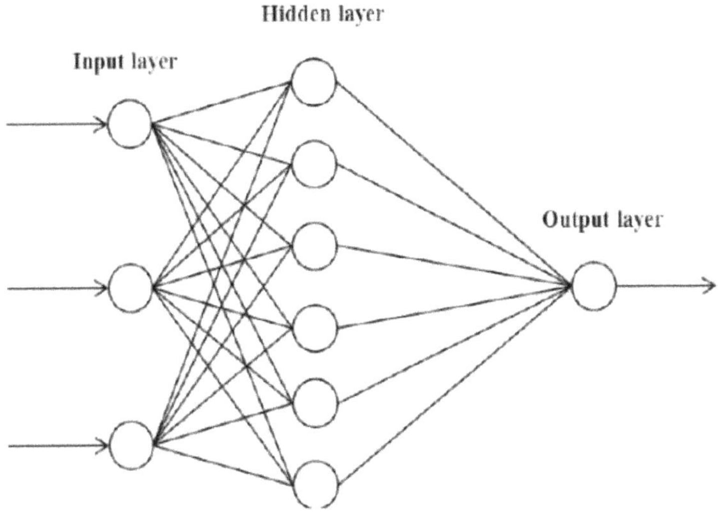

Figure 1 *Structure of ANN.* [2]

Standard data (proper information) is supplied to the input layer during learning (as a mathematical value), which produces output value. A comparison is made between the output value and the necessary (proper, standard) output value, which is also included in the standard data set. The ANN modifies its internal logical rules and factors based on the difference between these two values to produce the desired output value. Since drug discovery is a Natural Science, it can be challenging to understand, but once understood, it might be simple to put into practice[3]. Because ANN is a result of formal science, it is simple to understand yet could be challenging to use for drug discovery. If used properly, ANN can greatly simplify a variety of human chores and issues. However, "How, the ANNs have been used to the drug discovery problem" is crucial to producing successful results. ANNs can be used to solve any sophisticated drug development challenges when the human brain is no longer able to be helpful[3].

2 METHODS & RESULTS

ANNs are a virtual approach for predicting the effects of a medicine on the body, with the ultimate goal of enhancing patient care by expediting pharmaceutical research. Pharmacologists and computational biologists are involved in the interdisciplinary and difficult field of the development of ANNs for drug disposal. The advantage of ANNs is that they can scale back on the requirement for standard pharmacology, including animal testing, which saves time and money on drug development and discovery research. Software can be used to design novel pharmaceuticals as well as enhance existing ones. However, more research and methods are developing, and ANNs' accuracy in drug disposition is still being developed[4]. It is envisaged that high throughput technology that can provide & "omics" the data would help this field of study expand. The use of biological knowledge acquired in this postgenomic era has created a brand-new discipline called "systems pharmacology". New forms of neural networks are developed every week. Based on the network's architecture and connectivity pattern, ANNs are separated into feedforward and recurrent networks. The five primary types of neural networks are as follows:

- Multilayered perceptron/backpropagation networks
- Kohenen neural networks
- Counter propagation networks
- Bayesian neural networks
- Recurrent neural network (RNN)
-

2.1 Multilayered perceptron/ backpropagation networks

The multilayer perceptron is the most used neural network model. These networks use the backpropagation algorithm and are supervised. The error collected after every step is backpropagated throughout the network, significantly reducing the error.

2.2. Kohenen neural networks

Due to the nature of unsupervised networks, the property under investigation is not used during training. The points from a multidimensional space are projected into a 2D plane via the Kohenen network[5].

2.3 Counter propagation networks

Counter propagation (CPG) networks are supervised networks trained using Kohenen's training technique. In CPG networks, only the input layer is used when calculating distance, whereas during the adaptation phases, both the input and output layers' weights are changed.

2.4 Bayesian neural networks

Statistical pattern identification and data analysis using neural networks that is consistent with the Bayesian perspective.

2.5 Recurrent neural network (RNN)

The connections between the Processing Errors (PEs) in an RNN create a directed cycle. This controls the program's dynamic temporal behaviour. RNNs have the ability to process inputs using internal memory. Recurrent neural networks with Elman neural networks process information in a conventional feedforward manner prior to applying the learning algorithm. The context units constantly keep a copy of the prior values that were present in the hidden units during the learning process.

Any type of neural network can be defined using the transfer functions of their

- Neurons
- The connection formula
- The Learning formula

2.6 Neurons

The building block of the ANN created to mimic the function of the biological neuron is the artificial neuron. As create the output for that neuron, the incoming signals, referred to as inputs, are first merged and then multiplied by the connection adjustment[6]. The sigmoid function is the most widely utilized transfer function, while the activation function is the weighed sum of the neuron's inputs (see Figure 2).

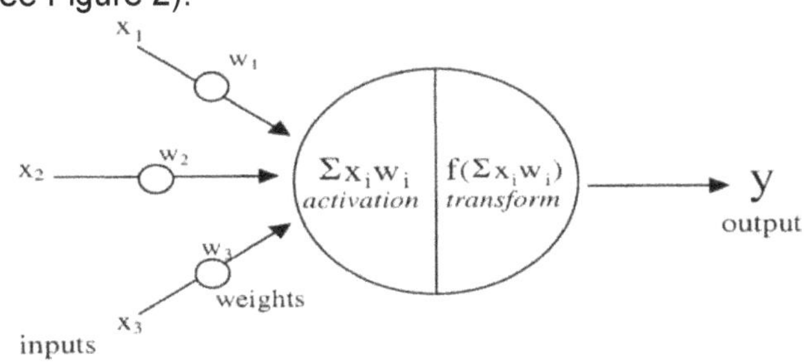

Figure 2 *Model of Sigmoid Function in ArtificialNeuron[6].*

2.7 The connection formula

The functioning of the artificial neural network is significantly influenced by how the neurons are connected to one another. Artificial neurons can accept inputs that are either excitatory or inhibitory, just like "actual" neurons can. The following neuron's summing mechanism subtracts in response to inhibitory inputs while adding in response to excitatory inputs. The term for this is lateral inhibition. Another form of connection is feedback, in which the output of one-layer travels back to the input of a succeeding layer or to the same layer. Depending on whether a network has feedback connections (see Figure 4) or not, there are two different styles of design that can be distinguished. The previous output values of feedforward (see Figure 3) architecture are not remembered since there is no connection from the output to the input neurons. Neurons in the feedback architecture are connected from the output to the input. When attempting to reduce training error, each neuron has an additional input weight that gives it one more degree of freedom[6]. Such a network maintains a memory of the prior state such that the subsequent state is dependent not only on the input signals but also on the networks past states.

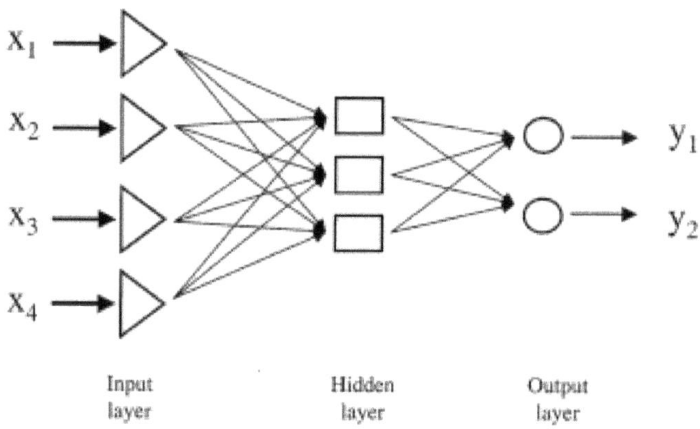

Figure 3 *Feedforward network*[6].

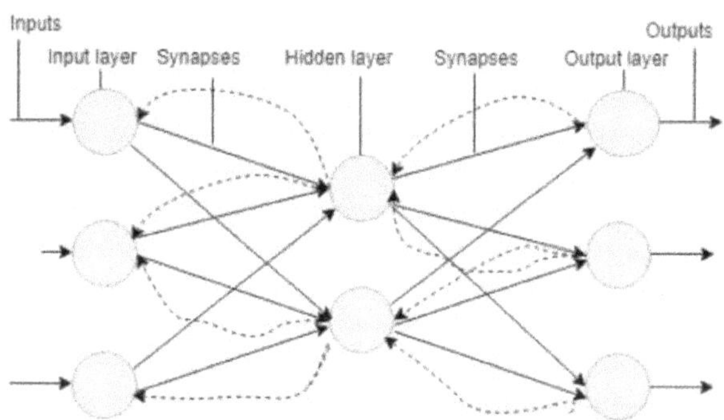

Figure 4 *Feedback network*[7].

2.8 The learning rule

In general, learning includes acting upon and adapting to change as and when the environment does. An artificial neural network (ANN) is a complex system, or more precisely, we can say that it is a complex adaptive system, which can modify its internal structure in response to the information that passes through it[8]. Learning in ANN implies that a processing unit is capable of modifying its input/output behaviour in response to the change in environment because it is a sophisticated adaptive system. When a specific network is built, the fixed activation function and input/output vector improve the significance of learning in ANN. We must now modify the weights in order to alter the behaviour of the input and output. It might be described as the process of discovering similar characteristics among samples belonging to the same class in order to understand how to categorize data from samples into different groups. For example, we have some training samples with special features for ANN training and other testing samples with additional special features for ANN testing.

Although there are many alternative learning rules (see Figure 5), the Delta rule or Back-propagation rule is the one that is most frequently applied. By repeatedly adjusting the weights, a neural network is learned to map a collection of input data[6]. The ANN's recognition skills depend on the use of weighted linkages. To maximize the weights across neurons, input data is sent forward through the network. By backward propagating the error during the training or learning phase, the weights are optimized. To reduce the deviation between the predicted and target values, the ANN reads the input and output values from the training data set and modifies the value of the weighted links. Throughout numerous training cycles, the prediction error is reduced until the network reaches the desired level of accuracy.

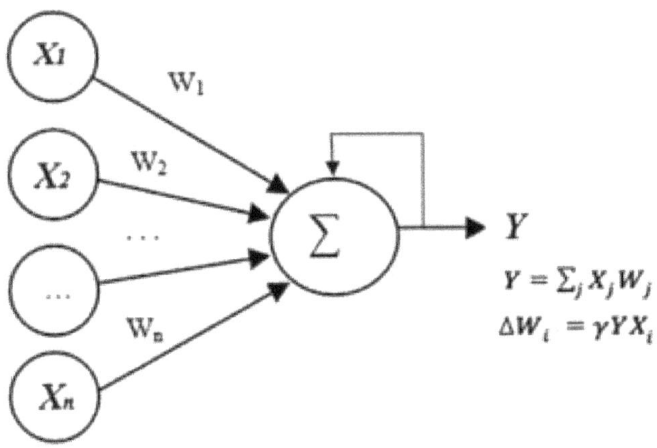

Figure 5 *Learning rule in ANN*[9].

Applications of ANN in Drug Discovery and Development (see Figure 6)

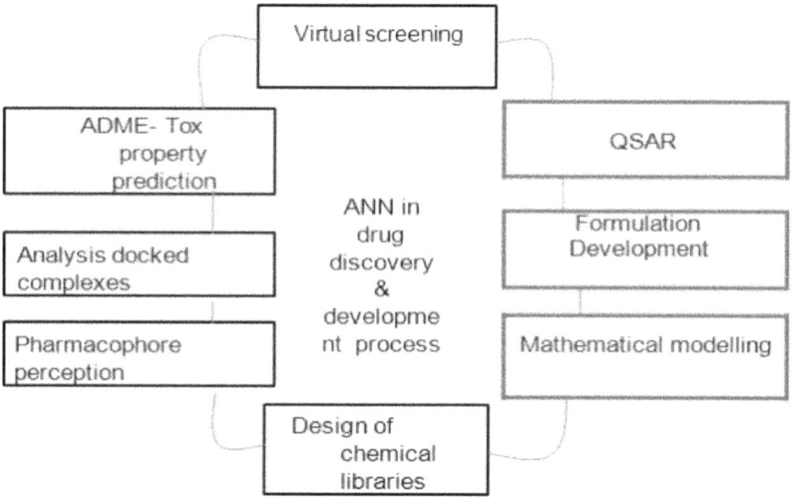

Figure 6 *Applications of ANN in drug discovery and development.*

ANNs are well-known for being a potent tool for simulating diverse nonlinear interactions and have been used to solve many issues of great complexity[10]. Generally, classification or pattern recognition, prediction, and modelling can be used to group together a variety of ANN applications.

ANNs are used for modelling drugs, designing dosages, predicting protein structure and function, modelling pharmacokinetics and Pharmacodynamics, interpreting analytical data, and interpreting in vitro/in vivo correlations[5]. ANNs are highly helpful in data analysis of pharmacological research because of their capacity to identify patterns from complex sets of analytical data, including non-linear correlations. For instance, using the entire spectrum during the identification process, ANNs can be used to analyze the spectral data of multi component samples like mixes to quantify the concentrations in the mixture.

Furthermore to being used more frequently for docking, QSAR studies, and the prediction of ADME-Tox properties, ANNs are also increasingly used for virtual screening (VS), also known as high-throughput screening (HTS), and the evaluation of the properties of the ligand in terms of their pharmacophoric features. In biological systems, nonlinearity is common. Particularly helpful for data mining are ANNs[5]. To uncover patterns in the data, data mining technologies including decision trees, genetic algorithms, regression lines, and clustering algorithms are utilized. The ANN's capacity to handle nonlinear interactions gives it an edge over other statistical methods. A system also includes a number of independent and dependent variables. Even if some essential variables are absent or there is insufficient information available for those variables, ANNs can still provide output. ANNs, which are based on sophisticated algorithms, can handle numerous variables simultaneously and dynamically evolve to deal with more challenging challenges.

3 DISCUSSION

3.1 Limitations of ANN

Variability of the training dataset: The training dataset for ANNs must be sufficiently diverse. The most frequent issue with ANNs is the lack of diversity in the sample dataset. The test dataset must be correctly predicted by the ANN, which requires the appropriate training dataset.

Issues in the standard dataset: The standard dataset should be error-free because errors there could affect the test format's results.

Overfitting: The ANN's size and the amount of PEs involved are also considered. There shouldn't be any under training if ANN is the right size.

Overtraining: The ANN should not be over trained since otherwise it will try to become accustomed to the output.

The researcher must be well aware of the issue that ANN is intended to address.

Due to their capacity to understand complicated trends and patterns in datasets and models, ANNs may be especially helpful in the drug discovery process.

Table 1 *Differentiating the biological neuron and artificial neuron.*

Feature	Biological Neuron	Artificial Neuron
Structure	Not completely understood	Very easy to learn, and it automatically generates while learning
Speed	Less; in milliseconds	Very fast; in MHz
Vision	Superior; 1 Brain > 1000 super computers	Depend on learning parameters & data[1]; can be superior to brain.
Intelligence	Depend on individual exposure	Depend on learning
Arithmetic calculation ability	Comparatively less	Very high and customizable
Knowledge	Gain by learning, education	Gain by learning from input data set
Information processing	By neurons, a structural and functional unit; unidentified process	by the structural and operational component known as a "Processing Element"; strange mechanism
Working	processing in parallel; capacity to handle multiple	processing in parallel; handling multiple tasks
	tasks concurrently; Perceptual	concurrently; Perceptual

4 CONCLUSION

In this article we examined recent advances in the use of neural networks for drug discovery, including virtual screening and the development of QSAR models to predict activity profiles and drug-target interactions, binding constants with respect to various targets, drug selectivity, inhibition constants for various enzymes, toxicity profiles, ADMET and physicochemical properties, and toxicity profiles. Clearly, Ann models are incredibly useful instruments for the process of developing the most snipping medication delivery systems. It's important to take into account the growing interest in neural networks used in pharmacological analysis. Better equipment is being used in analytical laboratories, including instruments that can digitally record thermograms, voltammograms, chromatograms, and spectrograms.

5 ACKNOWLEDGEMENT

The author is thankful to Chalapathi institute of pharmaceutical sciences, for providing the facilities to carry out the research work.

6 REFERENCES

1. Patel, J. Science of the Science, *Curr. Drug Discov. Technol.* **10**, 2–7 (2013).
2. Jin, C., Jang, S., Sun, X., Li, J. & Christenson, R. *J. Civ. Struct. Heal. Monit.* **6**, 545–560 (2016).
3. Baskin, I. I., Winkler, D. & Tetko, I. V. *Expert Opin. Drug Discov.* **11**, 785–795 (2016).
4. Puri, M. *et al. Artificial Neural Network for Drug Design, Delivery and Disposition* (Elsevier Inc., 2016). doi:10.1016/B978-0-12-801559-9.00001-6.
5. Mandlik, V., Bejugam, P. R. & Singh, S. *Artificial Neural Network for Drug Design, Delivery and Disposition* (Elsevier Inc., 2016). doi:10.1016/B978-0-12-801559-9.00006-5.
6. Agatonovic-Kustrin, S. & Beresford, R. *J. Pharm. Biomed. Anal.* **22**, 717–727 (2000).
7. Abiodun, O. I. *et al. Heliyon* **4**, e00938 (2018).
8. Kanda, T. & Ishiguro, H. *Human-Robot Interaction in Social Robotics* 297–343 at https://doi.org/10.1201/b13004-7 (2012).
9. Asadi, R., Kareem, S. A., Asadi, S. & Asadi, M. *Artif. Intell. Eng. Des. Anal. Manuf. AIEDAM* **31**, 30–54 (2017).
10. Wesolowski, M. & Suchacz, B. *J. AOAC Int.* **95**, 652–668 (2012).

SCREENING OF SOME NOVEL ISOXAZOLES AGAINST *C. ALBICANS* FOR THEIR POTENTIAL USE AS ANTIFUNGALS: A COMPUTATIONAL APPROACH

Judy Jays[a*], P Parasuraman[a], J Saravanan[b]

[a]Department of Pharmaceutical Chemistry, Faculty of pharmacy, M.S.Ramaiah University of Applied Sciences, Bengaluru, India, [b]Department of Pharmaceutical Chemistry, PES college of pharmacy, Bengaluru, India.

Corresponding Author E-Mail ID: judyjays.py.ph@msruas.ac,in

1 INTRODUCTION

Candida albicans, an adjustable fungal pathogen is the most prevelant Candida species responsible for causing candidiasis in humans. Candidiasis comprises a horde of fungal infections, including invasive fungal infections (Candidemia), where most patients are immunocompromised.[1,2] Hence, the success of the treatment depends more on the effectiveness of the antifungal agent. Antifungal resistance represents a major challenge for treating invasive fungal infections due to the limited arsenal of systemic antifungal drugs.[3] One of the biggest challenges to successful clinical treatment is azole resistance among Candida and Aspergillus species, followed by multi drug resistance among Candida species.[4] Many synthetic isoxazole (See Figure 1) derivatives displayed a wide spectrum of pharmacological activities including antibacterial and antifungal activities.[5,6] With this background, novel isoxazoles were designed with the objective of obtaining more potent antifungals.

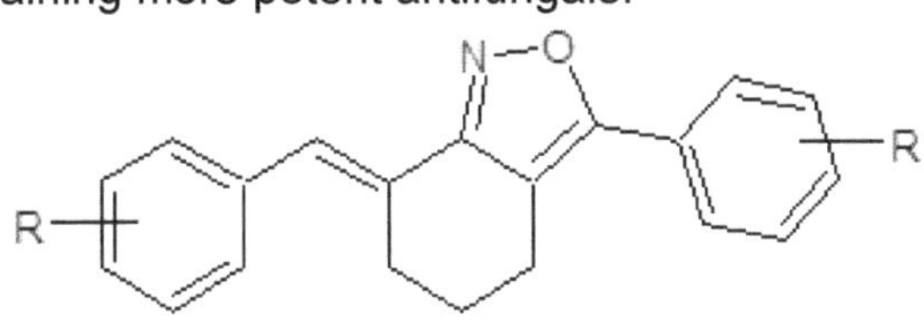

Figure 1 *General Structure of Isoxazole.*

Molecular docking studies are used to explain the binding of ligands with the target proteins. This helps in optimization of the ligand structure. All the designed isoxazoles were docked onto the active site of the crystal structure of selected antifungal enzymes. The mechanism of interaction of the novel isoxazoles were studied on two target proteins-N-myristoyl transferase of C. albicans (PDB ID: 1IYK) and dihydrofolate reductase of C. albicans (PDB ID: 4HOE).[7]

N-myristoyl transferase (NMT) catalyzes the transfer of the rare fatty acid myristate from myristoyl-CoA to the N-terminal glycine of substrate proteins, and is found only in eukaryotic cells. NMT activity is essential for vegetative growth. Recent N-myristoyltransferase inhibitors show *in vivo* antifungal activity and are promising selective fungal N-myristoyltransferase inhibitors.[8,9]

Dihydrofolate reductase catalyzes the reduction of dihydrofolate into tetrahydrofolate, and is critical for the biosynthesis of purines, thymidylate and some amino acids. These are important for cell proliferation and cell growth. As the DHFR enzyme is found in both humans and microorganisms, the ideal antifungal agent should selectively inhibit the fungal enzyme.[10] Fungal DHFR is different from the human enzyme.[11] Hence, inhibitors of fungal DHFR may be useful as antifungal agents.[12]

2 METHODS AND RESULTS

2.1 Molecular Docking Studies

Molecular modelling was carried out on PC operating with 64-bit workstation, 3.70 GHz Intel® Core i3- 6100 processor with 4GB RAM using CentOS Linux operating system. Simulation studies were carried out using Ligprep, protein preparation Wizard and GLIDE (Grid –based Ligand Docking with energetics) modules implemented in Maestro 11 (Schrodinger LLC).

All the proposed derivatives were docked onto the target proteins identified from the literature. Molecular docking process was carried out as per standard protocol of Glide manual.[13]

2.1.1 Preparation of ligands [14]. The ligands were drawn in 2D and converted to 3D. Molecules were optimized and minimized with LigPrep module as per standard protocol. The implementation was carried out with the graphical user interface (GUI) of Maestro software package, using the OPLS_2005 forcefield. Protonation states at biologically relevant pH were correctly assigned. The optimized molecules were taken up for the molecular docking studies.

2.1.2. Preparation of protein. X-ray crystallographic structures of target proteins were downloaded from the Protein Data Bank (PDB extracted from the Brookhaven Protein Database http://www.rcsb.org/pdb) and used for docking studies. "Protein Preparation Wizard" tool of Maestro was used to fix common inaccuracies like protonation or missing side chains and loops. Unwanted water molecules were removed, H-atoms were added, disulphide bonds were created and side chains were fixed. Energy refinement and energy minimization was carried out.

2.1.3. Receptor grid generation of target protein. The optimized protein with co-crystallized ligand was taken to generate a 3D (20 x 20 x 20 A°) grid at the active site of the target protein as per the standard protocol of glide manual. The co-crystallized ligand molecule is removed and the prepared ligand is docked in its place. Receptor grid generation allows defining the position and size of active site for ligand docking

2.1.4. Molecular docking [15]. GLIDE module of Maestro was used to study the docking of the derivatives on the target proteins. Glide was run on flexible docking mode where the protein is rigid and ligand is flexible. GLIDE docking and scoring methods were used to identify the binding pattern and interactions of title compounds at the active site of target proteins. The pharmaceutically relevant protein-ligand complexes with known binding affinities are described and the correct ligand-protein docked pose is viewed. GLIDE uses a series of hierarchical filters to search for possible locations of the ligand in the binding-site of a receptor. The binding of the derivatives was estimated using a variety of scoring functions that have been compiled into the single score known as GScore. An empirical scoring function, it is comprised of terms that account for the physics of the binding process encompassing lipophilic term, hydrogen bond terms, coulomb-vdW energies, etc. It also includes terms to account for hydrophobic enclosure.

GScore = 0.065*vdW+0.130*Coul +Lipo + Hbond + Metal + buryP + rotB +site

(VdW=Van der Waals energy, Coul= Coulonb energy; Lipo=lipophilic term; Hbond=hydrogen bonding term; Metal=metal binding term; buryP=penalty for buried polar groups; rotB=penalty for freezing rotatable bonds; site=polar interactions in the active site).

2.2 Molecular Dynamics

The binding stability, conformation, and interaction mechanisms between the protein and ligand were determined using molecular dynamic simulation [16]. The ligand-receptor complex was subjected to molecular dynamic simulation using DESMOND v3.6 package.[16] Furthermore, using the preset SCP water model, a water model was developed at a distance of 10A units of orthorhombic period boundary.[17] Further, the electric charges were neutralized by adding the necessary number of counter ions, and before the molecular dynamic simulation process began, the system decreased their energies through heating and equilibrium process. The system production step lasted 100ns, with time steps of 0.001ps; 300K temperature and 1 atmospheric pressure were used with Nose-Hoover method and the NPT (isothermal-isobaric) ensemble.[16,17] The best confirmations were chosen based on the interactions and dynamical features of the complex.

3 DISCUSSIONS

3.1 Molecular docking results

Docking of title compounds with target proteins was carried out in both Standard Precision (SP) and Extra Precision (XP) modes of Glide. The XP results are discussed as it provides a better correlation between good poses and good scores. It includes additional terms over SP scoring function and is a more complete treatment of some of the SP terms like scoring of H-bonds, detection of buried polar groups, etc.

Glide uses the Emodel scoring function to select between protein-ligand complexes of a given ligand and the Gscore function to rank-order compounds to separate compounds that bind strongly (actives) from those that don't (inactives). The Emodel scoring function is primarily defined by the protein-ligand coulomb-vdW energy. GScore is an empirical scoring function designed to maximize separation of compounds with strong binding affinity from those with little to no binding ability. Docked poses were ranked according to their docking scores. Usually, good ligand affinity for the receptor may be expected for low GScore.

3.1.1 Molecular docking with N-myrystyl transferase of C. albicans. Results for the docking of the title compounds with NMT are given in Table 1. XP mode has given significant results with good Glide scores and Emodel. The docking score for the ligands varied from -4.459 to -7.857, while the GScore for docking of the standards miconazole, ketoconazole, and clotrimazole with N-myristoyltransferase was -7.065, -6.357 & -3.889 respectively. Compound 2d (2-hydroxy derivative) with GScore -7.857 and Emodel -53.534 was ranked the highest among the derivatives. Two hydroxyl groups of the compound have formed H-bonding interaction with HIS 227 and PHE 414, while the phenyl ring has formed pi-pi stacking interaction with PHE 240 and HIS 227 at the active site of the protein (Fig. 2). Most of the compounds have shown similar binding interactions with PHE 117 at the active site indicating the significant role played by the amino acid in binding with the receptor.

Table 1 *Molecular docking results of isoxazole derivatives with N-myristyl transferase (PDB ID: 1IYK).*

Comp Code	R	SP mode		XP mode		
		GScore	Emodel	GScore	Emodel	Interactions at active site
2a	H	-6.779	-49.437	-5.557	-46.885	Phe 117, Phe 339
2b	Furfuryl	-5.73	-41.201	-5.377	-41.143	Nil

2c	4-hydroxy	-6.502	-54.586	-6.432	-49.178	Phe 117, Try 225, Thr 211
2d	2-hydroxy	-7.071	-52.787	**-7.857**	**-53.534**	Phe 240, His 227, Phe 414
2e	4-methyl	-6.027	-47.689	-5.75	-45.79	Phe 117, His 227, Phe 339
2f	4-ethyl	-6.584	-51.909	-6.194	-52.108	Phe 117
2g	4-nitro	-6.441	-58.929	-4.828	-60.379	Phe 117, Phe 240, Thr 211, Asp 412
2h	3-nitro	-6.486	-61.743	-5.695	-63.115	Phe 240, Asp 412
2i	4-dimethyl amino	-5.844	-53.817	-5.025	-53.345	Phe 117, Phe 339
2j	4-chloro	-6.439	-50.007	-4.837	-52.994	Phe 117
2k	3-chloro	-6.424	-54.749	-4.459	-52.047	Phe 117, His 227, Phe 339
2l	2,4-dichloro	-5.746	-52.831	-4.7	-56.428	Phe 117, Phe 240
2m	2,6-dichloro	-5.495	-44.89	-5.382	-44.232	Phe 240
2n	3-methoxy	-6.418	-55.189	-5.678	-50	Phe 117, His 227, Asp 412
2o	4-methoxy	-5.761	-49.606	-5.581	-46.162	Phe 117
2p	3,4-dimethoxy	-6.461	-57.78	-5.647	-52.574	Phe 117, His 227, Phe 339
2q	3,4,5-trimethoxy	-7.117	-66.932	-6.997	-59.011	Phe 117, Try 225
2r	2-nitro	-6.205	-59.123	-4.865	-56.511	Phe 240, Asp 412
2s	3-methoxy-4-hydroxy	-5.684	-58.713	-4.659	-52.456	Phe 117, His 227
2t	3-ethoxy-4-hydroxy	-5.399	-56.432	-4.875	-53.379	Phe 117
Miconazole		-5.736	-68.807	-7.065	-66.578	Leu 451, Tyr 354
Ketoconazole		-7.503	-88.869	-6.357	-79.190	Phe 240, Asp 412
Clotrimazole		-5.372	-30.007	-3.889	-66.518	Try 225

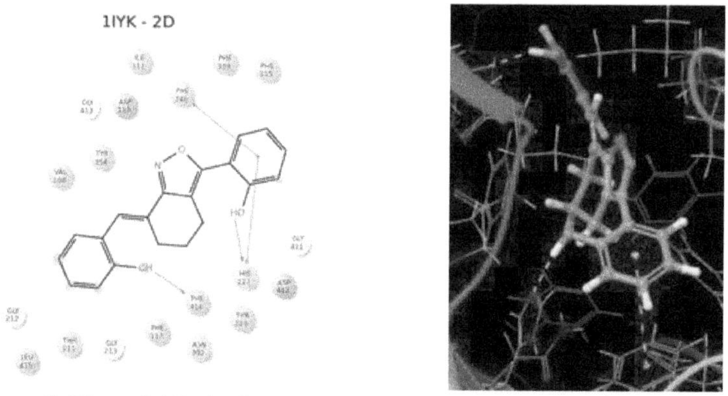

Figure 2 2D and 3D docked pose of Compound 2d with NMT of C. albicans.

3.1.2 Molecular docking with DHFR of C. albicans. Docking results of the ligands with DHFR are given in Table 2. Gscore for the derivatives varied from -4.125. to -6.779 against the enzyme. While the GScore for docking of the standards Miconazole, Ketoconazole, and clotrimazole was -7.334, -6.020 & -4.154 respectively. The results, given in Table 2 rated compound 2d (2-hydroxy derivative) having GScore -6.779, Emodel -61.309 with highest score. The ligand displayed H-bonds with ILE112 at the active site (Fig.3). Oddly, the compounds having similar scaffold have exhibited interactions with different amino acids at the active site, while some compounds have not formed any interactions.

Table 2 *Molecular docking results of isoxazole derivatives with Dihydrofolate reductase (PDB ID: 4HOE).*

Comp Code	R	SP mode		XP mode		
		GScore	Emodel	GScore	Emodel	Interactions at active site
2a	H	-5.961	-47.856	-5.737	-45.153	Tyr 118, Ala 11
2b	Furfuryl	-6.108	-44.075	-5.507	-37.101	Phe 36
2c	4-hydroxy	-7.199	-58.21	-6.572	-52.346	Glh 32
2d	2-hydroxy	-7.311	-62.796	**-6.779**	**-61.309**	Ile 112
2e	4-methyl	-6.722	-46.396	-6.605	-54.876	Phe 36
2f	4-ethyl	-5.995	-42.746	-6.658	-55.227	Nil
2g	4-nitro	-6.344	-56.715	-5.487	-53.572	Arg 72, Ala 115
2h	3-nitro	-6.51	-55.52	-5.33	-63.715	Nil
2i	4-dimethyl amino	-5.451	-45.99	-5.189	-54.172	Nil
2j	4-chloro	-6.244	-46.132	-6.657	-53.643	Nil
2k	3-chloro	-6.08	-48.305	-5.702	-56.119	Nil

2l	2,4-dichloro	-6.058	-48.828	-6.462	-63.894	Nil
2m	2,6-dichloro	-5.633	-44.899	-6.266	-41.351	Nil
2n	3-methoxy	-5.976	-45.38	-6.536	-44.538	Tyr 118,Glh 32, Ala 11
2o	4-methoxy	-5.732	-44.734	-6.219	-37.267	Phe 36, Ala 11
2p	3,4-dimethoxy	-5.779	-53.193	-6.191	-52.976	Nil
2q	3,4,5-trimethoxy	-6.934	-64.359	-6.19	-60.504	Nil
2r	2-nitro	-6.22	-53.29	-5.45	-56.159	Nil
2s	3-methoxy-4-hydroxy	-4.51	-55.904	-4.274	-49.778	Nil
2t	3-ethoxy-4-hydroxy	-5.87	-57.205	-5.012	-53.183	Nil
Miconazole		-6.680	-57.104	-7.334	-63.555	Phe 36
Ketoconazole		-7.582	-67.082	-6.020	-64.858	Ala 115, Glh 32
Clotrimazole		-4.718	-28.043	-4.145	-34.119	Phe 36, Ile 112

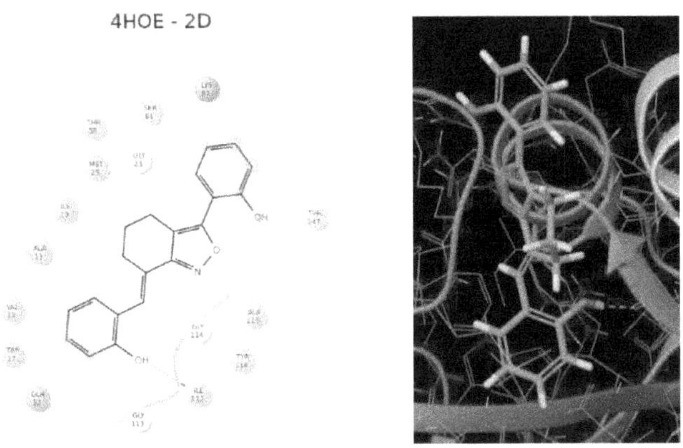

Figure 3 *2D and 3D docked pose of Compound 2d with DHFR of C. albicans.*

3.2 Results of Molecular dynamics

The molecular dynamic simulation was carried out for compound 2d with N-myristyl transferase and Dihydrofolate reductase proteins.

The MD trajectory events of 2d-N-myrystyl transferase complex showed that protein RMSD fluctuated initially up to 5.4 Å and retain its stability thereafter till the end with 5.4 Å. Ligand RMSD was stable, and fluctuation were found to be at 4 Å till 35 ns and retained its stability with 1.8 Å till the simulation period (Fig. 4a). RMSF analysis of all complex shows no major fluctuation due the binding of amino acids with the key functional groups of ligands (Fig. 4b). The protein-ligand contacts of 2d-N-myrystyl transferase revealed that amino acid residue GLU109(40%), PHE240 (43%), HIS227(42%) and ASP412(54%) contributes maximum interaction with 2d compound for its stability and activity (Fig. 4c).

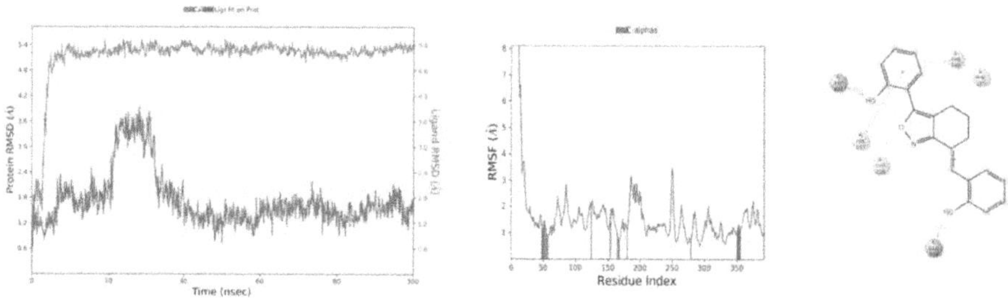

4a 4b 4c

Figure 4 *a) RMSD graph of 2d-N-myrystyl transferase complex; b) RMSF plot of 2d-N-myrystyl transferase complex;* ***c)*** *Protein-Ligand contact of 2d-N-myrystyl transferase complex.*

The MD trajectory events of 2d-Dihydrofolate reductase complex showed that protein RMSD fluctuated stabilized up from 6 Å to 7 Å throughout the simulation period. Ligand RMSD was stable, and fluctuation were in between 2.2 to 4.1 Å (Fig. 5a). RMSF analysis of all complex shows no major fluctuation due the binding of amino acids with the key functional groups of ligands (Fig. 5b). The protein-ligand contacts of 2d- Dihydrofolate reductase revealed that amino acid residue ILE112 (42%), ALA115(45%), TYR118 (79%) and ASP146(63%) contributes maximum interaction with 2d compound for its stability and activity (Fig. 5c).

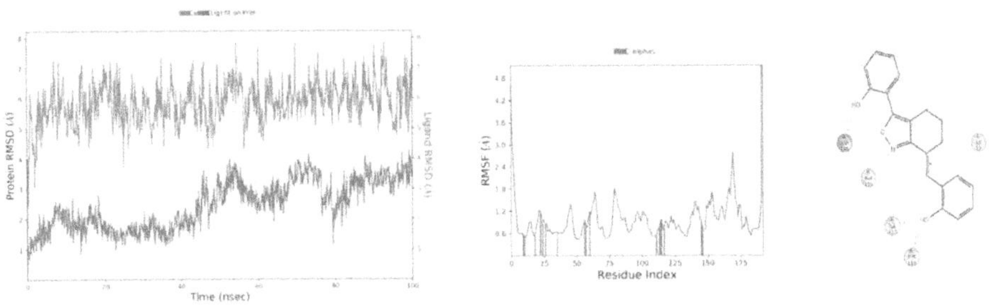

5a 5b 5c

Figure 5 *a) RMSD graph of 2d-Dihydrofolate reductase complex; b) RMSF plot of 2d-Dihydrofolate reductase complex; c) Protein-Ligand contact of 2d-Dihydrofolate reductase complex.*

Molecular dynamic study reveals compound 2d was found to be better in binding and stable with Dihydrofolate reductase protein.

4 CONCLUSION

Molecular docking studies were performed for twenty isoxazole derivatives against two antifungal target proteins of *C. albicans*- N-myristiyl transferase and Dihydrofolate reductase.

The results obtained from molecular docking of title compounds with N-myristyl transferase of *C.albicans* are quite promising. Based on the docking studies, it can be concluded that compound 2d (2-hydroxy derivative) is predicted to have significant binding affinity for N- myristyl transferase. The study suggests that the compounds are specific in binding at the active site of the essential enzyme of *C. albicans*, and they may be used as leads for further structural modifications to arrive at potential anti-fungal agents.

5 ACKNOWLEDGMENT

The authors are thankful to Pharmacological Modelling and Simulation Center, M S Ramaiah University of Applied Sciences for providing the facilities.

6 REFERENCES

1. People at Risk for Invasive Candidiasis. cdc. Gov. February 20, 2022. http://www.cdc.gov/fungal/diseases/candidiasis/invasive/risk-revention.html
2. B.J. Kullberg, M.C. Aerndrup. *N Engl J Med*, 2015, **373**, 1445.
3. NP. Wiederhold, *Infect Drug Resist.* 2017,**10**, 249.
4. DS. Perlin, R. Rautemaa-Richardson, A. Alastruey-Izquierdo, *Lancet Infect Dis.* 2017,**17**, 383.
5. J. Zhu, J. Mo, HZ Lin, Y Chen, HP Sun, Bioorg *Med Chem* 2018, **26**, 3065.

6. KV. Chikkula, S. Raja. *Int J Pharm Pharm Sci*, 2017, **9**, 13.
7. J. Judy, S. Mohan and J. Saravanan, *chemm.*, 2019, **3**, 442.
8. KK. Prasad, M.P. Toraskar, V.J. Kadam *Mini Rev. Med. Chem.*, 2008, **8**:142
9. C. Zhao, S. Ma *ChemMedChem.*, 2014, **9**, 2425.
10. J.H. Chan, J.S. Hong, L.F. Kuyper, D.P. Baccanari, S.S. Joyner, R.L.Tansik, C.M. Boytos, S.K. Rudolph *J. Med. Chem.*, 1995, **38**, 3608.
11. M. Whitlow, A.J. Howard, D. Stewart, K.D. Hardman *J. Biolog. Chem.*, 1997, **272**, 30289.
12. J.H. Chan, D.P. Baccanari, R.L. Tansik, C.M. Boytos, K.R. Sharon, A.D. Brown *et. al. J. Het. Chem.*, 1997, **34**,145.
13. Beginning with Maestro. Schrödinger software release. LLC, New York, NY, 2015.
14. Schrödinger: LigPrep software, version 2.5. New York, NY: LLC; 2011.
15. Glide, version 6.7, User Manual. Schrödinger, LLC, New York, NY, 2015
16. AK. Kalimuthu, T. Panneerselvam, P. Pavadai, SRK. Pandian, K. Sundar, S. Murugesan, *et.al. Scientific Reports*, 2021, **11**, 21488.
17. S. Kunjiappan, M. Sankaranarayanan, B.K. Kumar, P. Pavadai, E. Babkiewicz, P. Maszczyk, *et al. Nanotechnology*, 2020, **32**, 095101.

EVALUATION OF *IN-SILICO* AND *IN-VITRO* ANTHELMINTIC ACTIVITY OF *HEYDOTIS ASPERA* (HEYNE EX ROTH) USING *PHERETIMA POSTHUMA*

D S N B K Prasanth [a*], Risy Namratha Jamullamudi [b], Suneetha Achanti [c]
[a] Department of Pharmacognosy, KVSR Siddhartha College of Pharmaceutical Sciences, Vijayawada – 520010, Andhra Pradesh, India, [b] Pharmaceutical Chemistry Research Division, K L College of Pharmacy, Koneru Lakshmaiah Education Foundation, Vaddeswaram, India, [c] Department of Pharmaceutical Analysis, KVSR Siddhartha College of Pharmaceutical Sciences, Vijayawada – 520010, Andhra Pradesh, India.
Corresponding author email id: dsnbkprasanth@gmail.com

1 INTRODUCTION

Helminth-related diseases are often chronic in nature. Humans with helminthiasis are infested by pinworms, roundworms, and tapeworms. Parasite-borne diseases cause morbidity, resulting in conditions such as onchocerciasis and schistosomiasis. Developing countries have reported more worm infections owing to inadequate hygienic conditions. Herbal medicines are preferred because they are affordable and have fewer side effects than synthetic compounds. To test the anthelmintic properties of Pheretima posthuma, an adult Indian earthworm with similar anatomy and physiology to roundworm parasites in human intestines was chosen for testing [1].

There has been growing interest in natural drugs as potential therapeutic agents. Clinically, drugs are natural products or pharmacophores of natural substances. This shows the importance of drugs derived from natural sources in drug discovery [2]. Plants have been used as medicines for as long as mankind exists. Traditionally, Indian medicine has relied on plants [3].

Heydotis aspera (Heyne ex Roth) syn. Oldenlandia aspera (*Heyne ex Roth.*) (Rubiaceae) is a slender erect herb with trachomatous branches. The leaves are whorled, sessile, linear-lanceolate, and have an acute apex. The flowers were purple with branched racemes. The capsules were loculicidal. It is commonly known as the sanna parapatam. It is extensively used in the Indian traditional system of medicine as an Anticancer, Anti rheumatism, febrifuge, anthelmintic, anti-tuberculosis, expectorant, and antiasthma [4].

In this study, we assessed the antihelminthic potency of the methanolic extract of H. aspera and identified molecular interactions between different phytoconstituents and beta-tubulin.

2 MATERIALS AND METHODS

2.1 Collection, identification, and extraction of plant material

Whole H. aspera plants were collected in 2020 from the Seshachalam Forest, Tirumala, Andhra Pradesh, and authenticated via a botanical taxonomist by Dr. K. Madhava Chetty, Sri Venkateswara University, Tirupati, Andhra Pradesh, India. In brief, the plant was dried and pulverized to a coarse powder, followed by maceration with methanol for three days. The dark greenish powder extract was collected in airtight glass bottles and refrigerated before use for analysis and other assays.

2.2 Preliminary phytochemical screening

In accordance with standard procedures, methanol extracts of H. aspera (MEHA) were screened for preliminary phytochemical identification of major chemical constituents. [5-7]

2.3 GC-MS Analysis

GC-MS analysis of MEHA was performed using an Agilent 5977 MSD gas chromatograph, following the procedure of Mater H Mahnashi et al. [8-10]

2.4 Anthelmintic activity

2.4.1 Collection of Earthworms. Indian earthworms were collected from muddy soil in Pamaru Village, Krishna District, Andhra Pradesh, India. Using normal saline, fecal matter was removed from the worms.

2.4.2 Invitro anthelminthic activity. A total amount of 30 earthworms had collected and were divided up into five groups, each made up of 6 worms. Various concentrations (5, 10, 15, and 20 mg/ml) of the methanol extract and the standard (mebendazole) were prepared at 20 mg/ml in 10 ml. We washed the earthworms in regular water before they were placed in a 10 ml Petri dish. Earthworm motility was observed shortly after release into Petri dishes as soon as the launch time was stated. When earthworms showed no activity except for trembling, the time of paralysis was observed, and the time of death was observed when earthworms did not relocate excessively after shivering or when showered with hot water (40–50 °C). [11, 12]. All results are expressed as the mean ± SEM of six animals in each group.

2.5 Molecular Docking

Docking studies of GC-MS-eluted phytocompounds were carried out using AutoDock Vina and Discovery studio Biovia 2020 software to determine the interaction between the ligand and target protein. The crystal structure of β-tubulin (1SA0) [13] was obtained from the Protein Data Bank. The physicochemical properties of the ligands were evaluated using Lipinski's Rule of Five. [14-16].

2.6. ADMET Analysis

The pharmacokinetic properties of ADMET ligands must be examined to establish their functions in the body. AdmetSAR was used to study the ADMET ligand inheritance.

3 RESULTS AND DISCUSSION

3.1 Preliminary Phytochemical Analysis

Preliminary phytochemical screening of the methanol extract and results are presented in Table 1.

Table 1 *Preliminary phytochemical analysis of methanol extract of H. aspera.*

Phytoconstituents	Method	MEHA
Flavonoids	Shinoda Test	+
	Zn. Hydrochloride test	+
	Lead acetate Test	+
Volatile oil	Stain test	+
Alkaloids	Wagner Test	+
	Hager's Test	+
Tannins & Phenols	$FeCl_3$ Test	+
	Potassium dichromate test	+
Saponins	Foaming Test	+
Steroids	Salkowski test	+
Carbohydrates	Molish test	+
Acid compounds	Litmus test	-
Glycoside	Keller-Killani Test	+
Amino acids	Ninhydrin test	+
Proteins	Biuret test	+

3.2 GC-MS Analysis and Drug Likeliness

Eight compounds were identified by analyzing the GC-MS data of H. diffusa (Figure 1 and Table 1). Lipinski's rule was applied to all eight compounds. In addition to Berberine, Androsta-1,4-diene-3,17-dione, and Glafenine, the rest of the compounds violate Lipinski's rule in terms of logp value, and based on SwissADME results, they all have satisfactory "Abbott Bioavailability Score (> 0.1)" (Table 2). See also Table 3 for the drug-likeness properties of phytochemicals of H. diffusa based on Lipinski's rule.

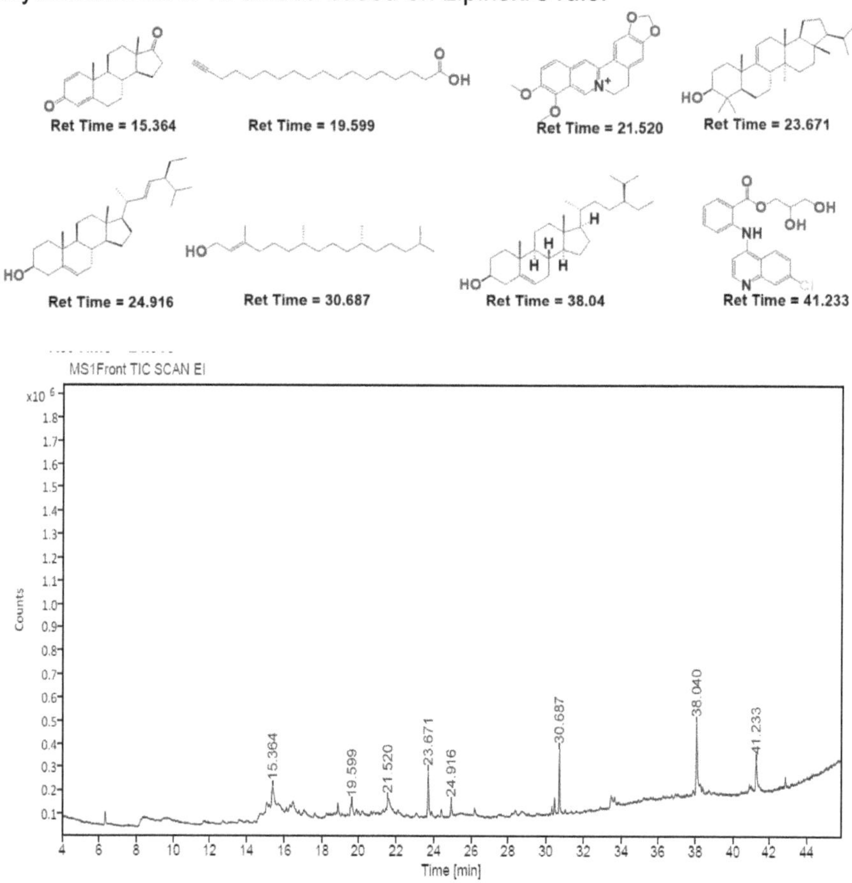

Figure 1 *Representative image of the most abundant compounds identified in MEHA using GC-MS analysis.*

Table 2 *GC-MS profiles of the compounds identified from methanolic extract of H. diffusa.*

Retention Time	**Name of the compounds**	**Base (m/z)**	**Area**	**% Area**	**Mol. Weight (g/mol)**
15.364	Androsta-1,4-diene-3,17-dione	211.1	539885.001	3.89	284.4
19.599	17-Octadecynoic acid	81	730024.449	5.26	280.4
21.52	Berberine	336.2	37785698.95	26.92	336.4
23.671	Isoarborinol	130	1693212.601	12.24	426.7
24.916	Stigmasterol	55	875751.763	6.31	412.7
30.687	Phytol	71.1	2012424.813	14.5	296.5
38.04	Gamma-sitosterol	55.1	3620976.785	26.09	432.7
41.233	Glafenine	225	600951.685	4.33	372.8

Table 3 *Drug-likeness properties of phytochemicals of H. diffusa based on Lipinski's rule.*

Compounds	**Lipinski Rule**				**Lipinski Violations**	**Bioavailability Score**
	MW	**Mlogp**	**HBA**	**HBD**		
	<500	**≤4.15**	**<10**	**<5**	**≤1**	**>0.1**
Berberine	336.1	2.19	4	0	0	0.55
Androsta-1,4-diene-3,17-dione	260	3.41	2	0	0	0.55
Glafenine	355	1.98	6	0	0	0.55
Phytol	256	5.25	1	0	1	0.55
17-Octadecynoic acid	248	4.57	2	0	1	0.85
Stigmasterol	371.1	6.62	1	0	1	0.55
Isoarborinol	426.4	6.92	1	1	1	0.55
Gamma-sitosterol	414.7	6.73	1	1	1	0.55

3.3 Anthelmintic activity

In contrast to mebendazole, the extract displayed a significant anthelmintic activity. As the concentrations increased, P. posthuma began to lose motility and became fatal. The methanol extract in concentrations 5, 10, 15, and 20 mg/ml showed paralysis of worms within 10.16±0.22, 7.34±1.22, 5.66±0.55 and 3.51±0.23 min and death with 34.62±2.51, 28.17±3.51, 18.22± 3.21 and 9.15±1.22 min depending on the concentration. The standard drug, Mebendazole, showed paralysis at 1.85±0.31 min and death after 2.78±0.11 min at 20 mg/ml concentration (Figure 2 and 3).

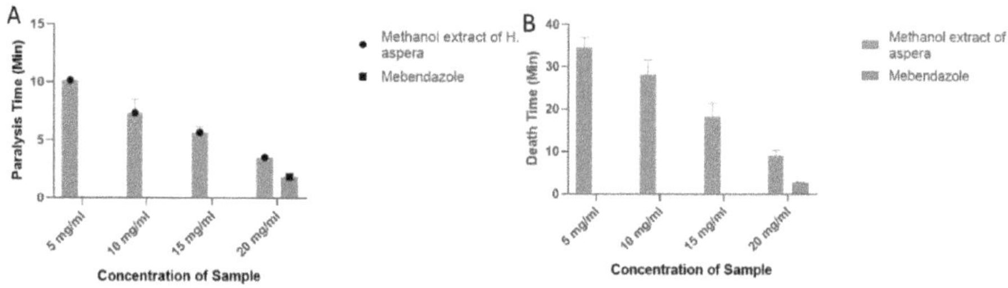

Figure 2 *Paralysis time and death time of standard mebendazole and various concentrations of MEHA.*

3.4 Molecular Docking

Docking simulations of eight identified compounds using GC-MS for β-tubulin revealed promising binding affinities, and the order of binding energy was Androsta-1,4-diene-3,17-dione (HA1) > glafenine (HA2) > stigmasterol (HA3) > berberine (HA4) > γ-sitosterol (HA5) > phytol (HA6) >17-Octadecynoic acid (HA7) > isoarborinol (HA8). The binding energies of HA1, HA2, HA3, HA4, HA5, HA6, HA7, and HA8 are exhibited -7.4, -6.6, -6.3, -5.7, -5.4, -4.6, -4.2, and -3.4 kcal/mol, respectively. Androsta-1,4-diene-3,17-dione had the best docking rating of -7.4, which showed hydrophobic interactions of ALA A:12 with the β-tubulin enzyme. Interestingly, none of the phytocompounds exhibited any electrostatic interactions. Surprisingly, with the exception of berberine and glafenine, the remaining compounds interacted with ALA A (12 amino acids) either by hydrogen bonds or by hydrophobic interactions. This clearly indicates that ALA A:12 amino acids play a crucial role in the inhibition of β-tubulin. The standard mebendazole exerts a binding energy of -6.7 kcal/mol and interacts with amino acids in β-tubulin via hydrogen, electrostatic, and hydrophobic interactions. The complex protein–ligand interactions are shown in Table 4 and Figure 3.

Table 4 *Binding energies and interactions of potential A compounds with β-tubulin (PDB ID:1SA0).*

Ligand	**Binding energy (Kcal/mol)**	**Hydrogen bond Interactions**	**Hydrophobic bond Interactions**
Berberine	-5.7	THR A:145 (3.70)	-
Androsta-1,4-diene-3,17-dione	-7.4	-	ALA A:12 (6.66)
Glafenine	-6.6	ALA A:12 (3.70), GLU A:183 (5.18)	ALA A:99 (5.08), ALA A:180 (6.72)
Phytol	-4.6	ALA A:100 (4.27)	ALA A:12 (4.85)
17-Octadecynoic acid	-4.2	ASP A:98 (4.64)	ALA A:12 (5.82), ALA A:180 (6.95)
Stigmasterol	-6.3	-	ALA A:12 (4.47), ALA A:99 (6.50)
Isoarborinol	-3.4	-	ALA A:12 (5.17)
Gamma-sitosterol	-5.4	-	ALA A:12 (6.89), ALA A:180 (6.63)
Mebendazole	-6.7	ASN A:101 (3.90), GLU A:183 (5.69)	SER A:178 (5.72), VAL A:177 (6.03), GLU A:183 (6.40)

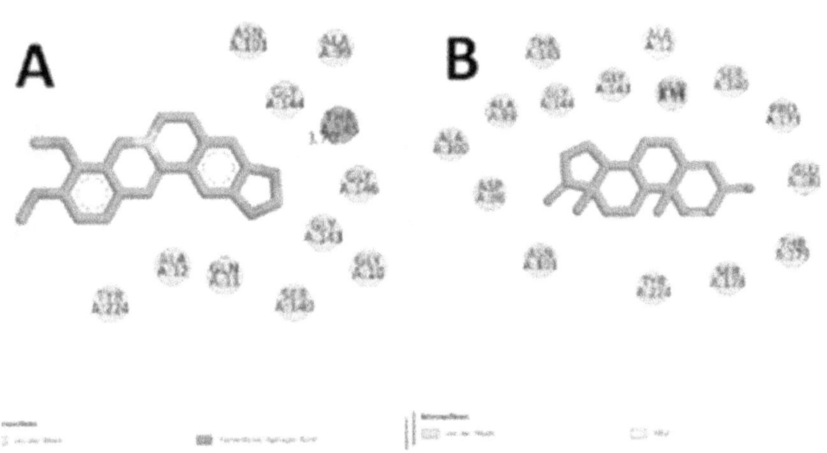

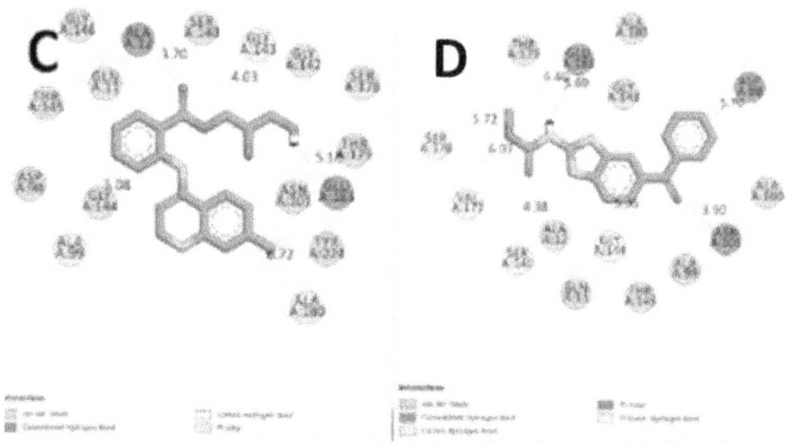

Figure 3 *2D interactions of GC-MS-eluted phytoconstituents of MEHA with β-tubulin (PDB:1SA0). A) Berberine, B) Androsta-1,4-diene-3,17-dione, C) Glafenine, D) Mebendazole.*

3.5 ADMET Analysis

The ADMET attributes of the ligands were studied using the Protox-ii server (https://tox-new). charite. de/protox_II/). Table 5 shows the predicted ADMET properties of the selected phytocompounds.

Table 5 *ADME/T analysis of phytoconstituents eluted from MEHA. Active indicates A; Inactive indicates IA.*

Phytocompound	Protox-II						
	LD50 (mg/kg)	**Predicted Toxicity Class**	**Hepatotoxicity**	**Carcinogenicity**	**Immunotoxicity**	**Mutagenicity**	**Cytotoxicity**
Berberine	200	3	IA	A	A	A	A
Androsta-1,4-diene-3,17-dione	2920	5	IA	A	A	IA	IA
Glafenine	1486	4	A	IA	A	IA	IA
Phytol	5000	5	IA	IA	IA	IA	IA

17-Octadecynoic acid	1600	3	IA	IA	IA	IA	IA
Stigmasterol	890	4	IA	IA	A	IA	IA
Isoarborinol	2000	4	IA	IA	A	IA	IA
Gamma-sitosterol	890	4	IA	IA	A	IA	IA

According to the in-silico toxicity prediction through Protox-II, compounds berberine and 17-Octadecynoic acid are toxic, and the rest of the compounds comes under the category of harmful. None of the compounds were non-toxic.

4 CONCLUSION

In the present study, the potent anthelmintic effect of MEHA was observed against P. posthuma. This suggests that H. aspera is a viable alternative drug source for controlling worm infections. Molecular docking experiments suggest that Androsta-1,4-diene-3,17-dione, glafenine, and stigmasterol are potential ligands for anthelmintic activity and interact with β-tubulin. In addition, in vivo experiments must be performed to confirm the results obtained in vitro.

5 REFERENCES

1. M. B. Manke, S. C. Dhawale and P. G. Jamkhande, *Asian Pacific Journal of Tropical Disease*, 2015, **5**, 175-180.
2. I. O. Ademola and J. N. Eloff, *Afr J Tradit Complement Altern Med*, 2011, **8**, 134-139.
3. K. Sriti, A. Sahoo and C. Chakraborti, *Int Res J Pharm*, 2011, **2**, 133-136.
4. K. Madhava Chetty, K. Sivaji and K. Tulasi Rao, *Andhra Pradesh, India*, 2008, **169**, 201.
5. M. S. Gokhale and C. Kokate, *Practical pharmacognosy*, Editora Record, 2008.
6. K. R. Khandelwal, *Practical pharmacognosy*, Pragati Books Pvt. Ltd., 2008.
7. T. E. Wallis, *Practical pharmacognosy*, J. & A. Churchill, 1925.
8. H. Oberacher, G. Whitley and B. Berger, *Journal of mass spectrometry : JMS*, 2013, **48**, 487-496.
9. M. H. Mahnashi, Y. S. Alqahtani, B. A. Alyami, A. O. Alqarni, M. Ayaz, M. Ghufran, F. Ullah, A. Sadiq, I. Ullah, I. U. Haq, M. Khalid and H. C. A. Murthy, *Evidence-based complementary and alternative medicine : eCAM*, 2022, **2022**, 7924171.
10. O. D. Sparkman, *Journal of the American Society for Mass Spectrometry*, 2005, **16**, 1902.

11. P. A. Chander, H. Y. Sri, N. B. Sravanthi and U. V. Susmitha, *Asian Pacific Journal of Tropical Disease*, 2014, **4**, S233-S235.
12. K. Mahato, B. B. Kakoti, S. Borah and M. Kumar, *Asian Pacific Journal of Tropical Disease*, 2014, **4**, S259-S262.
13. D. S. N. B. K. Prasanth, S. P. Panda, A. L. Rao, N. Teja, V. B. N. Vani, T. Sandhya and P. B. B. Rao, *Indian Journal of Pharmaceutical Education and Research*, 2020, **54**, S251-S260.
14. O. Trott and A. J. Olson, *J Comput Chem*, 2010, **31**, 455-461.
15. G. M. Morris, R. Huey, W. Lindstrom, M. F. Sanner, R. K. Belew, D. S. Goodsell and A. J. Olson, *J Comput Chem*, 2009, **30**, 2785-2791.
16. M. M. Jaghoori, B. Bleijlevens and S. D. Olabarriaga, *Journal of computer-aided molecular design*, 2016, **30**, 237-249.

IDENTIFICATION, MODELLING AND DOCKING STUDIES ON SERT LIKE MYCOBACTERIAL PROTEIN AS A POSSIBLE ANTI TB DRUG TARGET

Manal NK Hamem, PV Sangeeta Guruvelli, Sony Priya Kurati, Murali Krishna Kumar Muthyala*

Pharmaceutical Chemistry Research Lab, AU College of Pharmaceutical Sciences, Andhra University, Visakhapatnam, Andhra Pradesh, 530003, India.

Corresponding Author E-Mail ID: profmmkau@gmail.com

1 INTRODUCTION

Tuberculosis (TB), one of the oldest and most lethal human illnesses, is caused by the bacteria *Myobacterium tuberculosis* (Mtb). Mtb is a significant infectious killer, killing over 1.2 million people each year.[1] It is a re-emerging pathogen, due to the development of multiple-drug resistant (MDR) and extensively drug resistant (XDR) strains.[2] Mtb shows unique abilities to protect itself from toxic environments and host immunity by using its lipid-rich cell wall and other chemical sensing mechanisms.[3] It is also able to survive inside the highly acidic, enzyme rich environment of a macrophage. Here, Mtb is found to produce chemical messengers, including cAMP to hack into the host protection mechanisms.[4]In view of its ability to hibernate inside host tissue, it can also be hypothesized that Mtb continuously feeds on host biochemistry, like any other pathogenic bacteria and also monitors via sensing host chemical signalling molecules.[5]

Recently, several research groups, in their pursuit to repurpose approved drugs for anti Tb therapy, found potential activity in non-antibiotics including antipsychotics, antidepressants and antihistaminic drugs.[6] Thioridazine has already been put to use successfully to treat XDR TB on terminally ill patients.[7] For these drugs, synergism is observed with existing classes of anti TB drugs but the mechanism of action is not clearly established. For thioridazine, it is suggested that the drug alters biochemistry of Mtb by altering intracellular pH, oxidation/reduction potential and also nutrient exchange mechanisms. Recently, we reported potent inhibitory activity of sertraline (antidepressant drug) against *M. tuberculosis* H37Rv (MIC 1.6 μg/ml).[8]

This hints to the assumption that, as a part of adaptation to host environment, the Mtb is presumed to be using host neurochemicals, including dopamine, GABA, serotonine and histamine for various purposes.[9] This is possible only if Mtb shall carry proteins (Cytosolic enzymes, transmembrane proteins, ion channels etc) to use specific neurochemical, for tuning its biochemistry qualitatively and quantitatively. It is reported that, Mtb can undergo vertical evolution via gene fusion events to produce chimeric genes to perform specific biological functions, including secreting specific chemicals or synthesize proteins.[10]

Serotonin (5HT) is an endogenous autacoid that can modulate activity in visceral sensory nerves. Circulating 5-hydroxytryptamine (5-HT) is produced mainly in the enterochromaffin cells of the intestine. 5-HT is also, however, locally released from pulmonary neuroendocrine cells and neuroepithelial bodies distributed throughout the airways. Further, several studies proved that pulmonary endothelial cells take majority of circulating supply of 5HT outside CNS.[11] Secretion of large amounts of 5-HT from these cells occurs in response to airway hypoxia, stress, allergy and infection.[12]

Inspired by the fact that Mtb preferentially attacks pulmonary tissue and potent anti TB activity of sertraline (selective SERT inhibitor), we pursued the problem of finding possible presence of sodium dependant serotonin reuptake protein in mycobacterial proteome using bioinformatics tools and use them for virtual screening to identify hits from Drug Bank database.

2 METHODS AND RESULTS

2.1 Query selection

Serotonin Reuptake Transporter SERT mediates rapid clearance of synaptic and circulating 5-hydroxytryptamine (serotonin) (5-HT), thereby playing a critical role in the spatiotemporal fine-tuning of 5-HT signalling.[13] The serotonin transporter (SERT) terminates serotonergic signalling through the sodium and chloride-dependent reuptake of neurotransmitter into presynaptic neurons. SERT is a target for antidepressant and psychostimulant drugs, which block reuptake and prolong neurotransmitter signalling. Sertraline acts as Selective Serotonin Reuptake Inhibitor (SSRIs) due to that we selected sodium dependent Serotonin Reuptake receptor (SERT) as query for performing the first step of our approach for alignment with Mtb. Proteins (see Figure 2) using BLAST.[14]

2.2 Softwares, databases and servers used

1. Computer requirements: Laptop computer, desktop computer or computer workstations with Windows operating systems.
2. Databases: Drug bank, PubChem, Gene bank, protein data bank and UniProt .
3. Bioinformatics tools: BLAST and Clustal Omega
4. Homology Model: Swiss model and I-Tasser servers.
5. Validation tools: PROCHECK, Verify_3D, ProSA
6. Docking softwares: MGLTools 1.5.6 and VINA, AUDocker.
7. Visualisation tool: PyMol

All the experimental data of Serotonin Reuptake Transporter was collected from Uniprot database (Entry Id P31645). The amino acids in fasta format is used as a query in protein blast. For the BLAST (blastp) alignment, *Mycobacterium tuberculosis* was chosen as microorganism; Non-redundant protein sequence, Uniprot KB/Swissprot, Protein Data Bank proteins as standard database. For scoring BLOSUM62 matrix is used. In this study, the mycobacterial proteins, SGC57945.1 (identity-24.8%; similarity-45%), CNI52794.1 (identity-26.9%; similarity-43%) CNJ20565.1 (identity-26.9%;similarity-43%) and MBR7538981 (identity-33.7%; similarity-55%) were obtained as hits. See Figure 1.

Select for downloading or viewing reports	Description	Scientific Name	Query Cover	E value	Per. Ident	Acc. Len
Select seq emb|SGC57945.1|	Na+-dependent transporters of the SNF family [Mycobacterium tuberculosis]	Mycobacterium tuberculosis	65%	5e-29	24.88%	438
Select seq emb|CNI52794.1|	Na+-dependent transporters of the SNF family [Mycobacterium tuberculosis]	Mycobacterium tuberculosis	35%	1e-16	26.99%	300
Select seq emb|CNJ20565.1|	Na+-dependent transporters of the SNF family [Mycobacterium tuberculosis]	Mycobacterium tuberculosis	35%	2e-16	26.99%	300
Select seq gb|MBR7538981.1|	sodium-dependent transporter [Mycobacterium tuberculosis]	Mycobacterium tuberculosis	12%	2e-09	33.75%	102

Figure 1 *We represented the alignment in new presentation way based on the coverage area of the hits on the query by using different colour for each protein.*

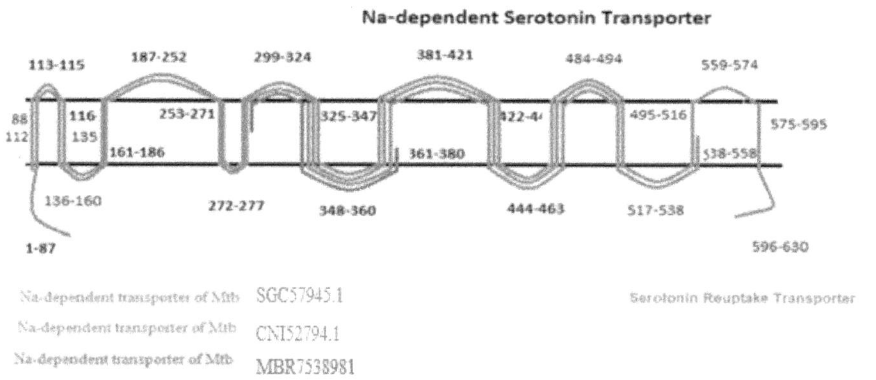

Figure 2 *Showing the alignment of the query in blue with hits in red, yellow and green.*

2.3 Selection of Mtb protein for homology modelling

Mtb is reported to possess essential transmembrane proteins that mediate nutrient uptake and metabolite efflux.[15] The transporters, required for the uptake of essential nutrients, enable mycobacteria to persist in the harsh intracellular host environments by scavenging nutrients. They thus represent potential new drug targets. Characterizing unique mycobacterial transporters provides another route to discover novel therapeutics to treat TB.[16] Despite considerable progress in the development of genetic methods for mycobacteria, the function of many transporters is still unknown. It has long been known that Na-symports are the principle and sometimes the only form of secondary transporters in alkalophilic and thermophilic bacteria.[17]The presence of primary Na-pumps in many pathogens indicates that they could use sodium motive force for solute uptake. Indeed, most of them encode Na-dependent symporters for alanine, proline, and several other amino acids.[18] The SGC57945.1 protein obtained in the alignment study is a sodium dependant transporter having maximum coverage area of 65% with SERT and less E-value (5e-29) which made it best selection for further work of modelling, docking and virtual screening.

2.4 Homology modelling of 3D structure protein

The FASTA sequence of the protein (Id: SGC57945.1) was retrieved from the Genebank. The sequence was then screened for similarity search against all the proteins deposited in Protein Data Bank using template and similarity search of I-Tasser server tool and Swiss model server.The 3D model of the protein was generated by using the default protocol available in the respective web servers.

2.5 Model validation

The models obtained from both servers is found same with RMSD less than 2A°. The obtained 3D model of the protein was evaluated using PROCHECK and Verify_3D available from the Structural Analysis and Verification Server (SAVES) and ProSA server. The PROCHECK server analyses the stereochemical quality of the protein and gives an assessment of overall quality of the structure compared to well refined structures of similar resolution, which is given by the Ramachandran (RC) plot. The accuracy of a protein 3D structure can be determined by Verify_3D, which tests the quality of protein structures by comparing them with the 3D profiles of correct protein structures having high scores. ProSA is a widely used tool to check potential errors in 3D models of proteins which are obtained either from experimental elucidation, from protein engineering or theoretical models. The program gives a z-score plot and energy plot for the protein. The energy of the structure is evaluated using a distance based and solvent exposure potential of protein residues. The z-score is indicative of overall model quality of the protein 3D structure. The ProSA energy plot determines the local quality of the 3D model by plotting energy as a function of amino acid sequence position. Positive values correspond to erroneous parts, whereas negative values indicate a stable and good model.

2.6 Active site prediction

The binding of the ligand in the active site of the protein is brought about by the micro environment, which is created by folds of the 3D structure of protein into cavities or pockets. It is necessary to identify the active site of the protein where the ligands can bind. In the present study, active site of the protein was identified using CASTp (Computed Atlas of Surface Topography of proteins) server and P2Rank server a new open-source software package for ligand binding site prediction from protein structure.

2.7 Molecular Docking Study

Molecular docking studies are used to understand the interaction between a ligand and a target protein at the atomic level. Before the start of the docking process, we subjected the protein and ligand for preparation using MGLTools. After defining active pocket region using GRID parameters, VINA was implemented with the help of AUDocker GUI [19] to perform virtual screening Drug Bank database. The top 10 hits were presented in the Table 7.

3 DISCUSSIONS

The FASTA sequence of selected protein (Accession no. SGC57945.1) was obtained from Gene bank database. The conserved domains of the protein predicted using BLAST server are shown in Figure 3.

The template search result from BLAST predicted the Apo-form of PDB code: 4US3_Ato have the highest identity of 40% and E value of 8e-63, the JPred3 server predicted an E value of 1e-89 for the protein 4US3_A.

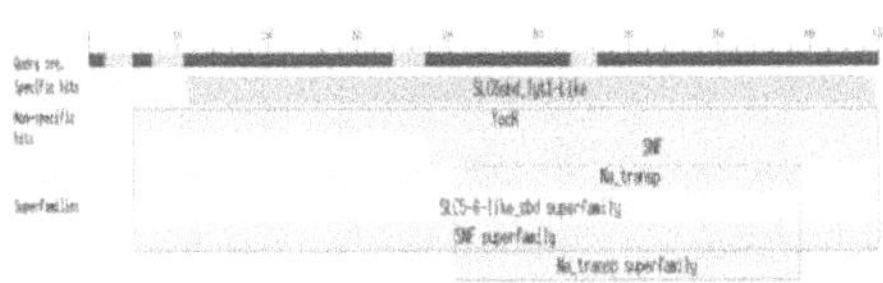

Figure 3 *Conserved regions (blue) on the query sequence and coverage of selected hits.*

Table 1 *Showing servers used to identify templets.*

S. no.	Name of the server	Parameters considered for template selection	E value	PDB code of template protein
1	BLAST	Sequence position specificity	7e-64	4US3_A
2	JPred3	Secondary structure prediction, solvent accessibility and coiled-coil region prediction	1e−89	4US3_A

The E values with the corresponding templates are given in Table 1. The protein sequence alignment with the template protein 4US3_A shows a 62% sequence similarity by using Clustal Omega shown in Figure 4. Therefore, the Crystal Structure of the bacterial NSS member MhsT in an Occluded Inward-Facing State was taken as the template for model generation based on the sequence alignment and predicted E values.

Figure 4 *Alignment of 4US3 protein and SGC57945.1 Mtb protein by Clustal Omega. The stars and dots represent identity and similarity respectively.*

3.1 Secondary structure prediction

The secondary structure of Mtb Na-dependent protein (see Figure 5) which consists of twenty four α-helices and nineteen β-turn (Tables 2 and 3). The residues corresponding to the secondary structure which are predicted using ProFunc (Prediction of protein function from 3D structure) of PDB sum web server. The α-helices in the secondary structure of Mtb Na-dependent transporter protein form an embellishment pattern which is characteristic of Na-dependent transporter (see Figure 5).

Figure 5 *3D structure model of Mtb Na-dependent transporter protein.*

Table 2 *Showing nineteen beta turns of the secondary structure.*

No.	**Beta turn**	**Sequence**
1	Arg80-Ala83	RHWA
2	His81-Trp84	HWAW
3	Gly117-Asn120	GELN
4	Leu119-Thr122	LNIT
5	Asp123-Gln126	DSSQ
6	Asn137-Trp140	NPWW
7	Lys206-Gly209	KLTG
8	Ile225-Gly228	IGLG
9	Ser238-Lys241	SDNK
10	Glu275-Ala278	EANA
11	Gly279-Leu282	GPGL
12	Leu294-Gly297	LPGG
13	Gly358-Ser361	GPLS
14	Leu360-Val363	LSDV
15	Thr364-Gly367	TLGG
16	Leu365-Lys368	LGGK
17	Gly395-Arg398	GWSR
18	Ile408-Leu411	ITGL
19	Leu411-Gln414	LKRQ

Table 3 *Twenty-four α-helices represent the secondary structure.*

No	**Start**	**End**	**No. of Residues**	**Sequence**
1	Ile12	Ala19	8	ILAAAGSA
2	Ile22	Asn34	13	IGAIWKFSYVAAN
3	Gly38	Val50	13	GAFLVVFLLFSFV
4	Gly52	Thr66	15	GLAVLLAENILGSST
5	Ala70	Met77	8	AVNAFKKM
6	Trp84	Ala115	32	WIGVIGVFSSWCIYSFYSVVGGWTI GYTVMAA
7	Gln126	Ser136	11	QLTGIFTRFIS
8	Trp140	Ala158	19	WPVIAHLVFAGMTCFVVLA
9	Arg162	Ser188	27	RGLEKSVKVMMPLLFIIMIVLIIVGIS
10	Ser192	Leu200	9	SSAGLKLFL
11	Phe204	Lys206	3	FSK
12	Gly211	Leu223	13	GVLDALGLAFFSL
13	Ile229	Ala235	7	IHITYSA
14	Ile243	Ala272	30	IANSSMWVVILSCMVCVLAGLMIFP ALAAA
15	Leu282	His293	12	LTFMTMPVYFAH
16	Gly297	Phe331	35	GSVLAVTFFVLLLMAALTSAISLLEH IVAYVQMRF
17	Arg335	Ser354	20	RRRAGLVVTASIMLMGIPVS
18	Leu355	Phe357	3	LSF
19	Ile370	Ser379	10	IFDILDYLTS
20	Leu382	Gly395	14	LMPLFGIAVCLIFG
21	Ala400	Glu406	7	ANALIPE
22	Gln414	Leu417	4	QSLL
23	Leu418	Arg421	4	LVWR
24	Val423	Val433	11	VGPICIGVILV

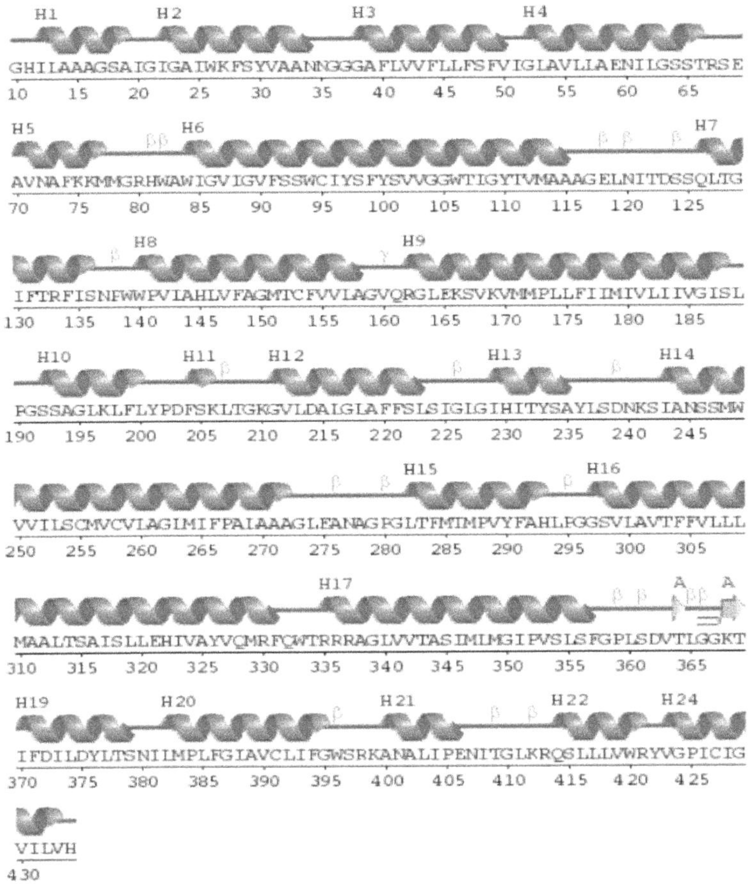

Figure 6 *The secondary structure of Mtb Na-dependent transporter protein shown in helices and sheets predicted by PDBsum. The red line represents β-hairpin motif.*

The protein core consists of one antiparallel β-strand (see Figure 6). There is a β-hairpin motif between the two residues in β strand 1 (The364-Leu365) to the strand 2 residues (Lys368–THR 369).

3.2 Structural evaluation

3.2.1 Model validation. The Ramachandran plot of Mtb Na-dependent transporter protein is shown in Figure 7 obtained from SAVES server. The plot statistics are given in Table 4. The plot statistics show that 93.2% of amino acid residues are in most favoured region, 6.4% in allowed region and none of the residues in the generously allowed with0.3% in disallowed regions. Therefore, the generated 3D model is a good-quality model in terms of stereochemical quality.

Table 4. *Ramachandran plot statistics of Mtb Na-dependent transporter protein.*

Description	**No. of amino acids**	**Percentage**
Residues in most favored red color regions	341	93.2%
Residues in additional allowed yellow color regions	24	6.4%
Residues in generously allowed regions beige	0	0.0%
Residues in disallowed white color regions	1	0.3%
non-glycine and non-proline residues	366	100.0%
Number of end-residues (excl. Gly and Pro)	1	
Number of glycine residues (shown as triangles)	44	
Number of proline residues	14	
Total number of residues	425	

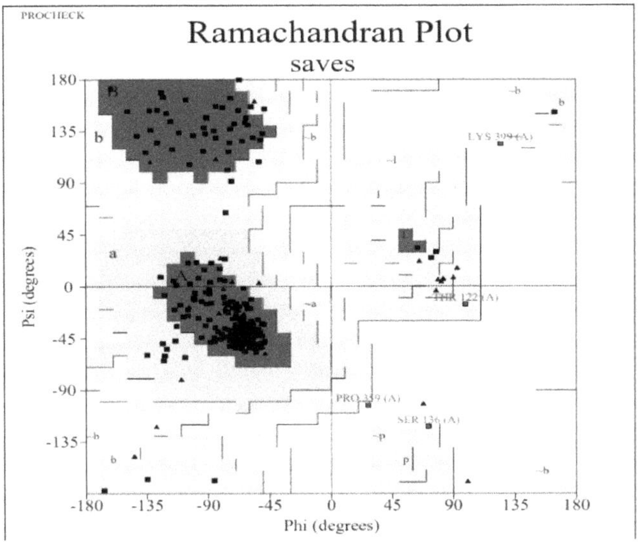

Figure 7 *Ramachandran Plot by SAVES server.*

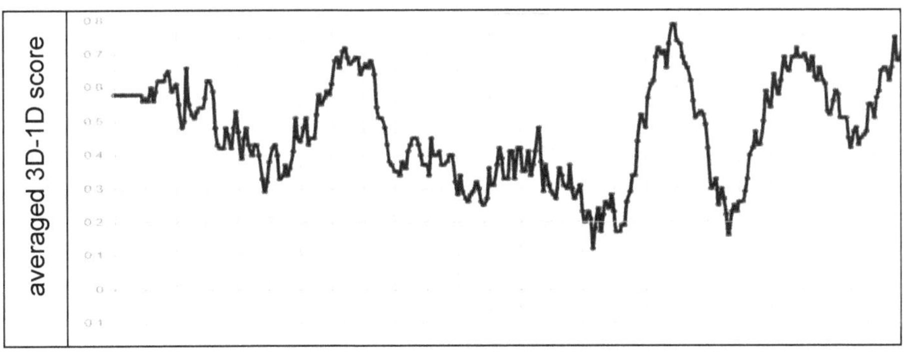

Figure 8 *Verify-3D Plot of Mtb Na-dependent transporter.*

The Verify_3D plot of Na-dependent transporter protein (see Figure 8) shows that 90.59% of the residues have an averaged 3D-1D score≥0.2. A good-quality structure requires at least 80% of the amino acids to have a score≥0.2 in the 3D/1D profile. The Mtb protein was submitted to ProSA webserver, and a z-score of − 5.32 was obtained. The plot (see Figure 9) shows z-scores of all protein chains in PDB determined by X-ray crystallography (light blue colour region) and by NMR spectroscopy (dark blue region).

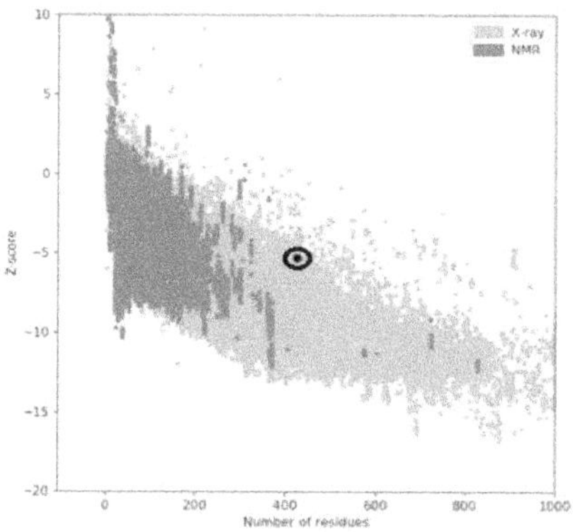

Figure 9 *Z-plot of Mtb Na-dependent transporter protein obtained ProSA webserver. the black spot encircled corresponds to Mtb protein.*

ProSA energy plot for the model shows maximum residues within negative region of energy, which indicate that the built model for Mtb Na-dependent transporter protein is reasonably stable.

3.2.2 Active site identification. The binding site predictions for Mtb Na-dependent transporter protein from CASTp and P2RANK servers which perform two pockets are given in table 5. The active site of Mtb protein generated using CASTP (see Fig 10). Both these servers predicted an active site that generated using CASTp. Both these servers predicted an active site that consists of His11, Ala14, Gly17-Ser18, Glu69, Gln161-Arg162, Leu164-Glu165, Val168, Ile229, Thr232-Tyr233, Tyr236, Thr314, Ile317-Ser318, Glu321-His322 Ala325, A335. The sequence alignment (Table 6) clearly showed that the conserved amino acids for recognition of sodium, 5HT and sertraline present in human SERT protein also are present in the predicted active pocket of Mtb protein. This further proved our hypothesis about presence of serotonin binding region on mycobacterial proteins. The active site of Mtb protein was generated and considered for active site prediction and virtual screening.

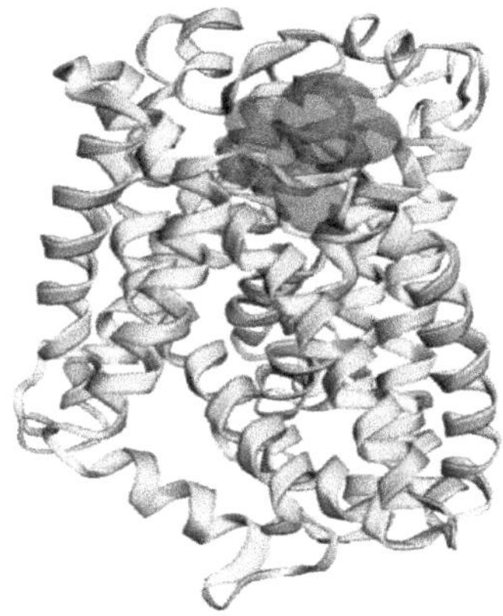

Figure 10 *Active site on by CASTP server.*

Table 5 *Servers used in binding site prediction and binding sites residues.*

S. No	Server	Binding site residues
1.	CASTP	His11, Ala14, Gly17-Ser18, Glu69, Gln161-Arg162, Leu164-Glu165, Val168, Ile229, Thr232-Tyr233, Tyr236, Thr314, Ile317-Ser318, Glu321-His322 Ala325, A335
2.	P2RANK	
	Pocket 1	His11, Ala14, Gly17-Ser18, Glu69, Gln161, Leu164-Glu165, Ile229, Thr232-Tyr233, Tyr236, Thr314, Ile317-Ser318, Glu321-His322 , Arg335
	Pocket 2	Trp26-Lys27, Tyr30-Val31, Asn34-Asn35, Leu213-Asp214, Leu218, Gly279, Pro280, Lys368, Asp372, Asp375-Tyr376, Asn380

Table 6 *Comparison of SERT and Mtb Na-dependent transporter protein binding site amino acids, Red colour amino acids represent matching amino acids.*

	Serotonin binding site	Sertraline binding site	Na-binding site
Serotonin Reuptake Receptor Na-dependent transporter	Y95,A96,D98,L99,G100,N101,W103,R104, Y175,Y176,I179,W182,F335,S336,F341,V343,P403,F407,S438,A441, G442	Leu99, G100, W103, R104, Y176, I179, F335	G94, A96, V97, N101, S336, N368, L434, D437, S438
Mtb Na-dependent transporter	S18, A19, G21, I22, G23, A24, W26, K27, F221, S222, L227, I229, P280, F284, S315, S318, L319	L13,A14, G17, I20, W26, K27, Y30,V31,V168, M172,L218,P280,F284,D375	G17,A19, I20,A24, S222, A311, T314, S315

3.3 Molecular docking analysis

Structure-based virtual screening was carried out for Mtb Na-dependent transporter protein using DrugBank database. The DrugBank database (version 5.1.9, released on 2022-01-03) contains 14,758 drug entries including 2,723 approved small molecule drugs; 1,542 approved biologics (proteins, peptides, vaccines, and allergenics), 132 nutraceuticals and over 6,696 experimental (discovery-phase) drugs. The experimental and FDA approved drugs were subjected to ligand preparation. The stereoisomers with specific chirality were retained to generate five conformers per ligand. The generated conformers were subjected to molecular docking. Results of the docking study of the active drugs into the active site of protein revealed different molecular interactions exhibiting hydrogen bonding, π interactions and hydrophobic interactions between the drugs ligands and Mtb Na-dependent transporter protein.

Results of docking simulations showed 665 drugs having several structurally diverse ligands bound with the receptor pocket with scores ranging from −10.2 to −5.2 kcal/mol. The resulting docked complexes were visually inspected, analysed and prioritised based on score. A sample of top 10 drugs are given in Table 6 prioritised based on docking score. The protein ligand interactions were visualised using PyMol.

3.3.1 Results of docking simulation. (See Table 7). In the docking simulations resulted in identification of mosapride, a 5HT4 agonist, as the top hit with docking score of -10.2 kcal/mol. The top 100 hits belong mainly to antifungals, antidiabetic, cardiotonic, antipsychotics,antidepressant, antiinflammatory and antihyperlipidemic drugs. Among these, azole antifungals are already proven to possess clinically useful anti TB activity.[20]Antipsychotics like pimozide (-9.3kcal/mol) has shown potent anti TB activity.[21] Efforts are on foot to obtain remaining drugs and screen them for anti TB activity.

Table 7. *Results of the top 10 drugs obtained in the docking study.*

S.No	Drug	Score (Kcal/mol)
1	Mosapride	-10.2
2	Itraconazole	-9.7
3	Carpipramine	-9.4
4	Sertindole	-9.4
5	Fluspirilene	-9.3
6	Metildigoxin	-9.3
7	Penfluridol	-9.3
8	Pimozide	-9.3
9	Deslanosede	-9.2
10	Ketoconazole	-9.0

4 CONCLUSION

During our research on Mtb adaption environments in human hosts, we found out a new drug repurposing strategy. By aligning Serotonin Reuptake

Receptor amino acids with Mtb proteins, a novel Mtb target belonging to the Na-dependent transporter family was found after hit selection. Further modelled and virtually screened with drug data bank. The results showed 665 drugs scores ranging from −10.2 to −5.2 kcal/mol. The top 100 drug exhibit diversity in pharmacological activity, including cardiotonic, antipsychotics, antidepressant, anti-inflammatory, antifungals, antidiabetic and antihyperlipidemic drugs. Some drugs already proved having anti-mycobacterium activity like antidepressant and antifungal drugs. We working to find out the mycobacterium activity of the other drugs by invitro screening.

5 REFERENCES

1.World Health Organization. 2020. Global tuberculosis report 2019.World Health Organization, Geneva, Switzerland.

2.Seung KJ, Keshavjee S and Rich ML. *Cold Spring Harb Perspect Med*, 2015, **5,** 1

3.Jarlier V and Nikaido H. FEMS *Microbiol Lett*. 1994, **123**, 11.

4.Rahman SA, Singh Y, Kohli S, Ahmad J, Ehtesham NZ, Tyagi AK and Hasnain SE. *mBio* 2014, **5,** 02020.

5.Meena LS, Rajni. *FEBS J.* 2010, **277**, 2416.

6.Diane Ordway, Miguel Viveiros, Clara Leandro, Rosario Bettencourt, Josefina Almeida, Marta Martins, Jette E. Kristiansen, Joseph Molnar, and Leonard Amaral. *Antimicrob Agents Chemother*, 2003, **47**, 917.

7. Amaral L, Viveiros M. *Antibiotics*, 2017, **14**, 3.

8.Sangeeta GPV, Risy Namratha J and Muthyala Murali Krishna Kumar. *IJC-B*, 2018, **57**, 1295.

9.Cambier CJ, Falkow S and Ramakrishnan L. *Cell*, 2014, **159**, 1497.

10. Gallant J, Mouton J, Ummels R, Ten Hagen-Jongman C, Kriel N, Pain A, Warren RM, Bitter W, Heunis T and Sampson SL. *NAR Genom Bioinform*, 2020, **18**, 1.

11. MacLean MR, Herve P, Eddahibi S and Adnot S. *Br J Pharmacol*, 2000, **131**, 161.

12. Nau F Jr, Miller J, Saravia J, Ahlert T, Yu B, Happel KI, Cormier SA and Nichols CD. *J Physiol Lung Cell Mol Physiol*, 2015, **308**,

191.

13.Glenda Mac Queen, Leslie Born and Meir Steiner. *CNS Drug Reviews*, 2001, **7,** 1.

14. Altschul SF, Madden TL, Schaffer AA, Zhang J, Zhang Z, Miller W and Lipman DJ. *Nucleic Acids Res*, 1997, **25**, 3389.

15. Lodish H, Berk A, Zipursky SL, Matsudaira P, Baltimore D and Darnell J. *In Molecular cell biology*, WH Freeman, New York, NY, 4th ed, 2000.

16. Cook GM, Berney M, Gebhard S, Heinemann M, Cox RA, Danilchanka O and Niederweis M. *Adv MicrobPhysiol*, 2009, **55**, 81.

17. Heyne RI, De Vrij W, Crielaard W and Konings WN. *J. Bacteriol*, 1991, **173**, 791.

18. Comparisons Claudia C. Hase, Natalie D. Fedorova, Michael Y. Galperin, and Pavel A. Dibrov. *Microbiology and Molecular Biology reviews*, 2001, **65**, 353.

19. Sandeep G, Nagasree KP, Hanisha M, Kumar MM. BMC *Res Notes*, 2011, **25**, 445.

20. Imperiale BR, Cataldi ÁA and Morcillo NS. *Rev Argent Microbiol*, 2017, **49**, 332.

21. Heemskerk MT, Korbee CJ, Esselink JJ *et al.*,. *Sci Rep*, 2021, **11**, 19634.

IN SILICO DESIGN OF NATURAL COMPOUND-DERIVED NOVEL INHIBITORS AGAINST RdRP OF SARS-CoV-2

Sanjay Madala[a], S S V Kiran K[a], Burra V L S Prasad[a*]
[a]Center for Advanced Research and Innovation in Structural Biology of Diseases, K L E F University, Vaddeswaram, Andhra Pradesh 522 502, India
*corresponding author email: dr.prasad.bvls@gmail.com

1 INTRODUCTION

The month of November 2019 marked the emergence of a new strain of coronavirus, SARS-CoV-2, which caused havoc and panic around the world. This resulted in the announcement of Coronavirus disease 2019 (COVID-19) as a pandemic by the World Health Organization (WHO) from the beginning of 2020. [1,2] COVID-19 had infected millions of people and caused about 6,500,000 deaths as of August 2022 (https://covid19.who.int/). SARS-CoV-2 belongs to the family of coronaviridae and shares significant similarity with its family members, SARS-CoV and MERS-CoV, both of which are highly pathogenic. But unlike these two, SARS-CoV-2 is highly transmissible among humans with a low death rate of ~3.5%. [3,4]

The knowledge about the structure and organization of SARS-CoV-2 genome paves a way to develop drugs that target the virus and combat against it. The 29.8kb ss-RNA contains 14 open reading frames (ORFs) out of which the ORF-1a and ORF-1ab, present towards the 5' end of the RNA, encodes information to synthesize 15 of 27 non-structural proteins (nsps). [5] These nsps include a chymotrypsin-like domain, PL-pro domain, an RNA-dependent RNA polymerase domain (RdRP), a helicase domain. [6] The RdRP, also known as Nsp12, is a 932 amino acid protein enzyme necessary for SARS-CoV-2 genome duplication and transcription. Nsp7 and 8 together with Nsp12 are needed for the replication and transcription. This core protein also includes two zinc ions which provide structural stability. [7] An analogy to understand the structure of RdRP is to consider it as a right-handed cup, with palm, finger and thumb domains acting as major sites, besides having NiRAN domain, interface domain and a β-hairpin like domain (see Table 1). [8] There are seven binding site motifs (A-G) located in the palm and finger domains. [9] These domains and the motifs (see Figure 1) are shown to play an important role in the RNA polymerase activity. [10]

Currently, there are 11 WHO approved vaccines available as a means of protection against SARS-CoV-2. However, there is always a need to develop drugs against COVID-19 disease for those who already are infected and need immediate cure. Remdesivir is one such antiviral and nucleotide analogue drug that was approved by FDA for emergency use to treat COVID-19 infected patients. [11] This was originally a therapeutic used against Ebola virus. This drug was found effective by targeting the RNA synthesis mechanism in the RdRP complex. It intervenes in the replication process by stalling the addition of new nucleotides, creating a barrier to disrupt the RNA translocation. [12]

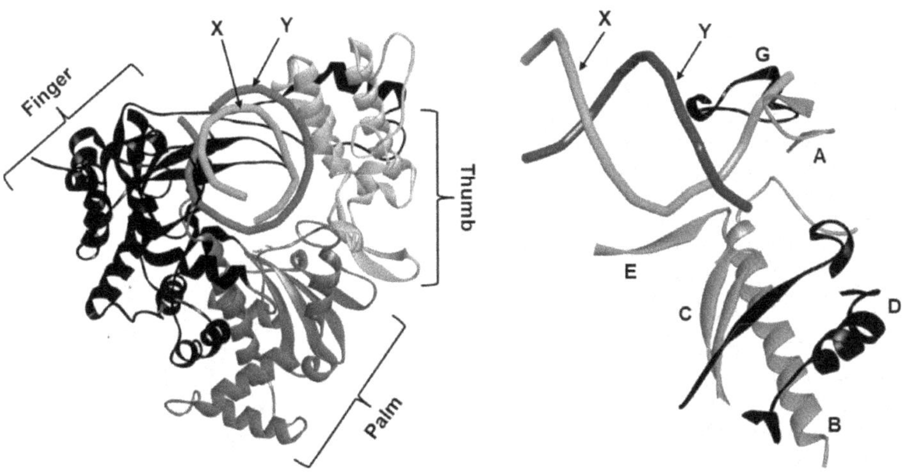

Figure 1 *3D representation of domains (left figure) and motifs (right figure).*

Table 1 *Domain and Motifs in RdRP core structure.*

Type		**Residue Number**
Domain	**Finger**	366-581 & 621-679
	Palm	582-620 & 680-815
	Thumb	816-932
Motif	**A**	611-626
	B	678-710
	C	753-767
	D	771-796
	E	810-820
	F	544-557
	G	499-514

The National Cancer Institute (NCI) natural product repository is the world's unparalleled database maintained by the US, containing approximately 230,000 terrestrial, marine and microbes extracts around the world. Its diverse range of products aids in discovering and designing drugs against the emerging diseases. The advancements in the field of computational drug discovery and designing minimized the problems of performing multiple assays and compatibility issues on the available compounds that are in small quantities (https://dtp.cancer.gov/organization/npb/introduction.htm). With these improvements the NCI introduced the NCI Program for Natural Product Discovery (NPNPD) to create libraries of the available natural products based on their crystal structures. These libraries or databases are free to access around the world aiding in the immediate use for High Throughput Virtual Screening (HTVS) and screening of novel compounds. [13] About 250,000 natural compounds are currently enlisted in the database. The present work is an attempt to screen and obtain novel inhibitory natural compounds that target the RdRP of SARS-CoV-2 efficiently.

2 MATERIALS AND METHODS

2.1 Protein Preparation

7BV2, the PDBID, containing the 3D structure of SARS-CoV-2 RdRP bound with remdesivir monophosphate was obtained from RCSB PDB (www.rcsb.org). This structure was further processed using the protein preparation wizard of the Schrödinger suite (version 2020-2) to ensure structural correctness and high confidence for later steps of drug designing.

2.2 Ligand Preparation

The National Cancer Institute (NCI) natural compound database was downloaded from the internet that contained about 250 thousand molecules in SDF file format (https://cactus.nci.nih.gov/download/nci/index.html). The Ligprep module from the Schrödinger suite was used to clean the structures which removed unwanted molecules, optimized geometries and neutralized charged groups, besides various ionization states and tautomer were generated.

2.3 Docking

The docking protocol was performed using the Ligand Docking Module of Schrödinger Glide v8.7. Initially the potential ligand binding sites were predicted using Schrodinger's SiteMap module. Individual grids were generated for the top three outputs of the SiteMap module using the Receptor Grid Generation module of the Schrödinger suite. High-Throughput Virtual Screening (HTVS) was used for primary screening of ligands with the predicted RdRP binding site.

Further the obtained top contenders were processed by redocking again with RdRP using the standard precision and extra precision docking (SP and XP docking). The ligands with favorable energies were considered for the later steps.

2.4 ADMET Predictions

Absorption, Distribution, Metabolism, Excretion and Toxicity (ADMET) are the important pharmacokinetic properties to consider for a drug to be declared as safe. ADMET includes parameters like molecular weight (MW), various types of solvent accessible surface area (SASA, FISA, FOSA, and PISA), volume, octanol/water partition coefficient prediction (log Po/w), prediction of solubility in aqueous phase (log S), IC50 prediction for HERG channel blockage (log HERG), brain-blood barrier partition value prediction (log BB), skin permeability prediction (log Kp) and human serum albumin binding prediction (log Khsa). These were predicted using the Qikprop module of Schrödinger suite. Toxicity prediction was performed using the Toxicity (extensible) module of Discovery Studio. Ames Mutagenicity, NTP Rodent Carcinogenicity, FDA Rodent Carcinogenicity, Weight of Evidence Carcinogenicity, Skin Irritancy and Skin Sensitization models were used for the toxicity prediction. [14,15]

2.5 Molecular Dynamics

The Desmond molecular dynamics package available in Schrödinger suite was used to perform simulation runs for the top leads. The docked protein-ligand complex was placed in the center of an orthorhombic box of size 10*10*10 Å. The box was filled up with water as the solvent with the TIP4P model. Further the system was neutralized with Na+ and Cl- ions maintaining a concentration of 0.15M. The whole system was applied with the OPLS force field. A short minimization process was done which was followed by a 100 picoseconds (ps) heating step with the temperature set at 100 K. The final production run was carried out for a time period of 40 nanoseconds (ns) and a snapshot of the trajectory was taken at every 10 ps interval which resulted in an output of 4000 frames. Nose-Hoover thermostat was used to maintain the temperature at 300 K and Martyna-Tobias-Klein barostat was used to control the system pressure at 1.01 bars. Post analysis of the simulations were performed to examine the RMSD and RMSF values of the protein and the protein-ligand complex.

2.6 Binding Free Energy (ΔG) Calculation

The binding free energy for shortlisted protein-ligand complexes was calculated using MM/PBSA protocol in Discovery studio visualizer 2021 with the applied CHARMm force field. About 500 frames were generated with a simulation run of 1000 ps and a snapshot at every 2 ps.

3 RESULTS AND DISCUSSION

3.1 Virtual Screening and Docking

The protein preparation wizard of Schrödinger suite generated a clean protein structure by removing all unnecessary atoms and water molecules, besides, adding required amino acid residues that were missing in the raw protein structure file. Initial screening of the NCI database with the use of modified Lipinski rule of five (with MW from 500 to 600 Da) produced 39,944 leads. Subsequently, 1081 compounds were obtained with the use of HTVS docking protocol. SP and XP docking protocol was used to further screen the obtained 1081 leads yielding the top seven lead compounds, with IDs NCDID_135_136, NCDID_135_142, NCDID_135_144, NCDID_135_149, NCDID_135_153, NCDID_135_170, and NCDID_142. The glide score of these seven novel leads ranged between -11.39 to -8.49 kcal/mol which were significantly higher than remdesivir -7.59 kcal/mol. [16] It is noteworthy to specify that arginine 249 (Arg249) and proline 677 (Pro677) residues of RdRP formed h-bond with all of the seven lead molecules making them highly conserved. Besides h-bonds, most of the lead ligands also formed π-stacking interactions between the pyrrole ring of ligand and Arg349 of RdRP (see Table 2). NCDID_135_144 had the highest number of h-bond interactions along with its high XP gscore of -11.39 kcal/mol.

Table 2: *Glide score, hydrogen bond and Pi stacking bond number of screened lead compounds.*

Ligand	XP gscore (kcal/mol)	Number of h-bonds	Number of Pi stackings
NCDID_135_144	-11.39	10	0
NCDID_135_142	-10.80	8	1
NCDID_142	-9.75	8	1
NCDID_135_136	-9.70	7	1
NCDID_135_153	-9.26	6	0
NCDID_135_170	-8.79	8	2
NCDID_135_149	-8.49	9	1

3.2 ADMET Predictions

The obtained seven initial leads were evaluated for their safety by ADMET predictions. The Molecular weight of the leads were between 510 - 530 Da. All the lead molecules were observed to be within compatible range for all the ADME properties, except for those that describe the blood-brain barrier. In this case, all the lead compounds were found to exhibit the same characteristic of not being able to cross the blood-brain, which can be signified from the resultant values of log BB. This character is advantageous as targeting RdRP does not require the leads to move to the brain. Slight deviation was observed in the values of FISA with all the leads except NCDID_142 being higher than the optimum range (see Table 3).

The TOPKAT predictions (see Table 4) done using Discovery Studio proved that all the leads were non-mutagenic and non-carcinogenic and hence can be considered safe for in-vivo analysis. NCDID_142 was observed to have mild properties of skin irritancy which was the only deviation among all the predicted properties.

3.3 Molecular Dynamic Studies and Binding Free Energy (ΔG) Calculations

The structural conformation of the protein throughout the simulation can be interpreted by the protein RMSD plot. The RMSD value for RdRP was maintained between 1.5 to 3.2 Å in the four protein-ligand complex simulations. The C and N terminals of a protein will have the highest degree of freedom due to the high flexibility. This was the reason for high RMSF values (see Figure 3) for terminal residues in RdRP. Fluctuations in RMSF of the protein from residue 360 to 450 was observed in all simulations. Interestingly, these residues belong to the finger domain which were predicted to be flexible to all the sliding of RdRP over the template RNA, hence explaining the RMSF fluctuations. The RMSD plots (see Figure 2) describe the stability of ligands with respect to protein. NCDID_135_142 and NCDID_135_153 were observed to be more stable, unlike NCDID_135_136 and NCDID_135_144 which had significantly high RMSD value compared to protein RMSD signifying that they were moving away from the binding cavity of RdRP. At least one intramolecular h-bond was present (see Figure 5) resulting in the stability of the structure of ligands. This explains the steady RMSD of the ligands (see Figure 4) with reference to its conformation.

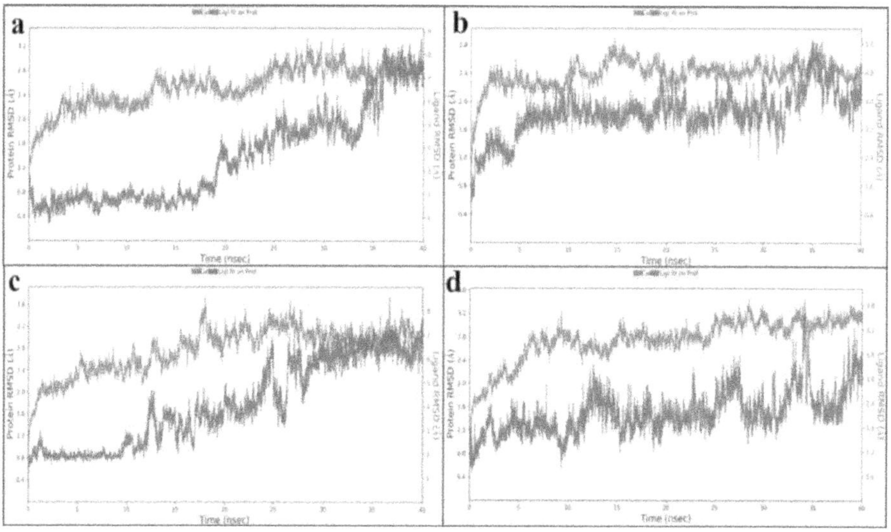

Figure 2 *RMSD Plots for Cα and Ligand fit to Cα in a) NCDID_135_136; b) NCDID_135_142; c) NCDID_135_144; d) NCDID_135_153.*

Desmond interprets four types of molecular interactions which include h-bond, hydrophobic, ionic and water bridges from the simulation. All protein-ligand simulations (see Figure 6), except NCDID_135_144, had these four types of interactions in their simulation. The four lead molecules were found to have conserved interactions with Arg349, Glu350 and Pro677. These ligands formed a conserved hydrogen bond with Pro677 which was located just before the start of motif B. This interaction was assumed to hinder the flexibility of the motif B and in turn the finger domain making it harder for the RdRP to slide along the template RNA and synthesize new RNA strand. Besides, the hydrophobicity of the protein surface demonstrates that the hydrophilic nature of remdesivir makes it hard to enter its deep-seated binding pocket compared to the novel ligands which were hydrophobic and have easily accessible binding cleft (see Figure 7). The binding free energy analysis was utilized to further optimize them. Based on the obtained energies, which were in range of -16.31 to -38.29 kcal/mol, they were ranked as NCDID_135_144 > NCDID_135_153 > NCDID_135_142 > NCDID_135_136.

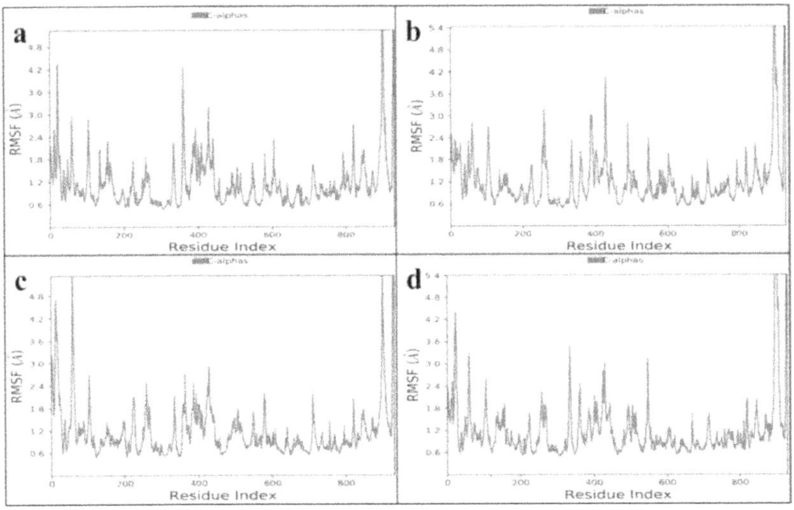

F

Figure 3 *RMSF Plots of protein backbone in a) NCDID_135_136; b) NCDID_135_142; c) NCDID_135_144; d) NCDID_135_153.*

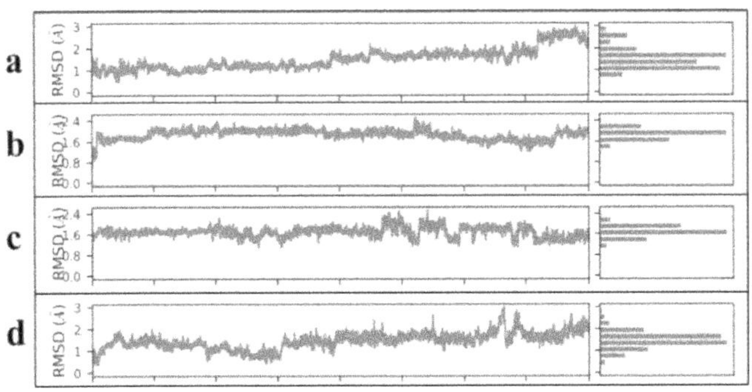

Figure 4 *RMSD Plots for ligand in a) NCDID_135_136; b) NCDID_135_142; c) NCDID_135_144; d) NCDID_135_153.*

Table 3: *ADME predictions of the novel compounds using Qikprop.*

Title	MW	SASA	FOSA	FISA	PISA	volume	QPlog					
							Po/w	S	BB	HERG	Kp	Khsa
NCDID_135_144	528.56	905	376.02	375.23	153.76	1627.61	-0.64	-2.62	-5.04	-3.73	-6.26	-1.2
NCDID_135_142	528.56	914.87	398.19	358.27	158.41	1628.57	-0.53	-3.18	-4.92	-3.88	-5.93	-1.22
NCDID_135_136	528.56	910.43	381.77	369.88	158.78	1636.36	-0.55	-3.01	-4.99	-3.79	-6.15	-1.19
NCDID_135_153	528.56	913.87	433.1	332.63	148.14	1618.21	0.77	-4.16	-4.57	-3.44	-5.59	-1.02
NCDID_135_170	527.58	920.08	386.98	370.37	162.72	1647.2	-1.5	-1.39	-4.72	-4.6	-8.11	-1.4
NCDID_135_149	528.56	882.67	371.1	350.83	160.74	1604.15	-0.57	-2.41	-4.61	-3.58	-5.79	-1.23
NCDID_142	512.56	893.54	414	279.84	199.7	1608.55	0.41	-3.16	-3.78	-3.89	-4.44	-1.02

Table 4: *Toxicity prediction of novel compounds using Discovery Studio Suite.*

Title	Female Mouse NTP	Male Mouse NTP	Female Mouse FDA	WOE Prediction	Ames Prediction	DTP Prediction	Skin Irritancy	Skin Sensitization
NCDID_135_136	Non-Carcinogen	Non-Carcinogen	Non-Carcinogen	Non-Carcinogen	Non-Mutagen	Non-Toxic	None	None
NCDID_135_142	Non-Carcinogen	Non-Carcinogen	Non-Carcinogen	Non-Carcinogen	Non-Mutagen	Non-Toxic	None	None
NCDID_135_144	Non-Carcinogen	Non-Carcinogen	Non-Carcinogen	Non-Carcinogen	Non-Mutagen	Non-Toxic	None	None
NCDID_135_149	Non-Carcinogen	Non-Carcinogen	Non-Carcinogen	Non-Carcinogen	Non-Mutagen	Non-Toxic	None	None
NCDID_135_153	Non-Carcinogen	Non-Carcinogen	Non-Carcinogen	Non-Carcinogen	Non-Mutagen	Non-Toxic	None	None
NCDID_135_170	Non-Carcinogen	Non-Carcinogen	Non-Carcinogen	Non-Carcinogen	Non-Mutagen	Non-Toxic	None	None
NCDID_142	Non-Carcinogen	Non-Carcinogen	Non-Carcinogen	Non-Carcinogen	Non-Mutagen	Non-Toxic	Mild	None

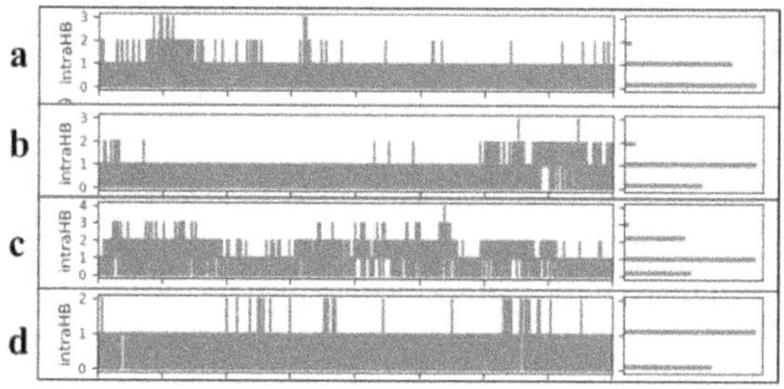

Figure 5 *Intramolecular hydrogen bond plots of ligand in a) NCDID_135_136; b) NCDID_135_142; c) NCDID_135_144; d) NCDID_135_153.*

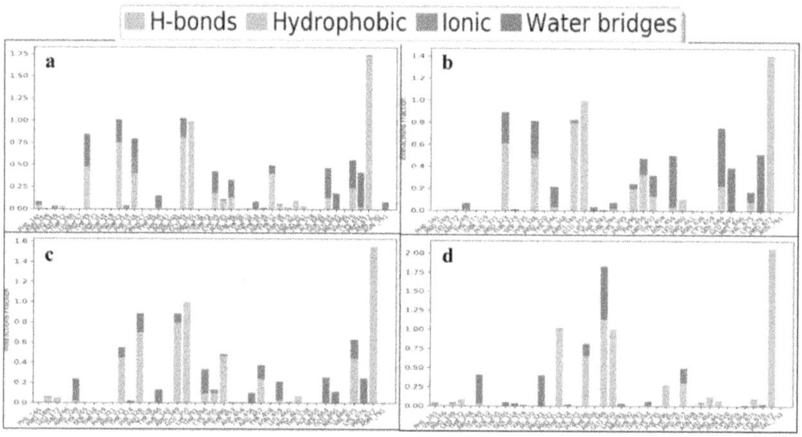

Figure 6 *Protein-Ligand interaction plots in a) NCDID_135_136; b) NCDID_135_142; c) NCDID_135_144; d) NCDID_135_153.*

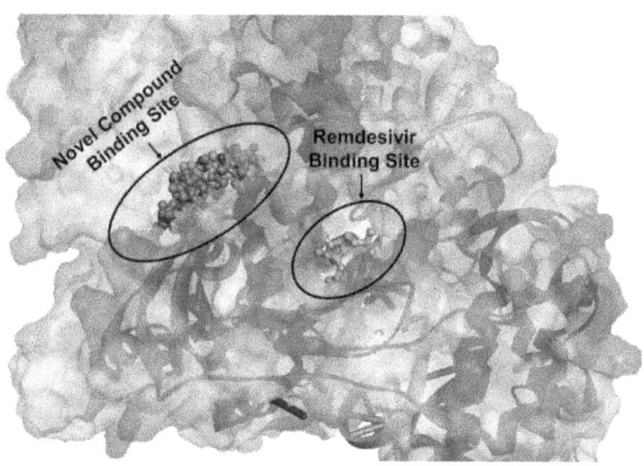

Figure 7 *Hydrophobic surface representation of RdRP (Top view) bound with remdesivir and novel leads.*

4 CONCLUSIONS

This work resulted in two novel lead molecules, from a library of 250 thousand natural compounds, that can serve as potential drugs against SARS-CoV-2 RdRP. The primary screening and processing of the entire database revealed seven leads with binding affinities of -11.39, -10.8, -9.75, -9.7, -9.26, -8.79 and -8.49 kcal/mol that were considerably higher than the binding affinity of remdesivir (-7.59 kcal/mol) with RdRP. These novel compounds are binding to the binding pocket that is better accessible compared to remdesivir which is binding a cavity that is deep inside. The novel accessible binding pocket is not only unique but also provides a better mechanism and opportunity to inhibit RdRP. These leads are binding between the Finger and Palm domains hindering the further sliding of the RdRP over the template RNA, thereby stalling the viral replication and hence further SARS-CoV-2 infection. All the four leads are observed to have a highly conserved interaction with Pro677, which is an important residue contributing to the formation and function of motif B of RdRP which resides at the junction of base of finger domain and palm domain. NCDID_135_153 and NCDID_135_142 are considered as potential leads based on the careful interpretation of the results obtained from docking, ADMET and simulation studies. These advantages of the identified novel compounds demand for immediate in vitro studies.

5 ACKNOWLEDGMENT

BVLS gratefully acknowledges the Indian Government funding agencies, DST-SERB-CRG, ICMR-ISRM and DST-NSM for their resource support through the project grants. BVLS acknowledges PARAM Yukti Facility under the National Supercomputing Mission (NSM), Government of India at the Jawaharlal Nehru Centre for Advanced Scientific Research (JNCASR), Bangalore.

6 CONFLICT OF INTEREST

The authors confirm that this article content has no conflicts of interest.

7 REFERENCES

1 Z. Wang, B. Yang, Q. Li, L. Wen and R. Zhang, *Clin Infect Dis*, 2020, **71**, 769–777.

2 D. Wu, T. Wu, Q. Liu and Z. Yang, *Int J Infect Dis*, 2020, **94**, 44–48.

3 M. Bchetnia, C. Girard, C. Duchaine and C. Laprise, *J Infect Public Health*, 2020, **13**, 1601–1610.

4 M. Fonzo, C. Bertoncello and V. Baldo, *Pathog Glob Health*, 2020, **114**, 347–348.

5 A. Wu, Y. Peng, B. Huang, X. Ding, X. Wang, P. Niu, J. Meng, Z. Zhu, Z. Zhang, J. Wang, J. Sheng, L. Quan, Z. Xia, W. Tan, G. Cheng and T. Jiang, *Cell Host Microbe*, 2020, **27**, 325–328.

6 J. Ahmad, S. Ikram, F. Ahmad, I. U. Rehman and M. Mushtaq, *Heliyon*, , DOI:10.1016/J.HELIYON.2020.E04502.

7 R. N. Kirchdoerfer and A. B. Ward, *Nat Commun*, , DOI:10.1038/S41467-019-10280-3.

8 S. Venkataraman, B. V. L. S. Prasad and R. Selvarajan, *Viruses*, , DOI:10.3390/V10020076.

9 S. G. Itoh, S. Tanimoto and H. Okumura, *Chem Phys Lett*, , DOI:10.1016/J.CPLETT.2021.138819.

10 Y. Gao, L. Yan, Y. Huang, F. Liu, Y. Zhao, L. Cao, T. Wang, Q. Sun, Z. Ming, L. Zhang, J. Ge, L. Zheng, Y. Zhang, H. Wang, Y. Zhu, C. Zhu, T. Hu, T. Hua, B. Zhang, X. Yang, J. Li, H. Yang, Z. Liu, W. Xu, L. W. Guddat, Q. Wang, Z. Lou and Z. Rao, *Science*, 2020, **368**, 779–782.

11 D. Focosi, F. Maggi, S. McConnell and A. Casadevall, *Antiviral Res*, , DOI:10.1016/J.ANTIVIRAL.2022.105247.

12 G. Kokic, H. S. Hillen, D. Tegunov, C. Dienemann, F. Seitz, J. Schmitzova, L. Farnung, A. Siewert, C. Höbartner and P. Cramer, *Nat Commun*, , DOI:10.1038/S41467-020-20542-0.

13 C. C. Thornburg, J. R. Britt, J. R. Evans, R. K. Akee, J. A. Whitt, S. K. Trinh, M. J. Harris, J. R. Thompson, T. L. Ewing, S. M. Shipley, P. G. Grothaus, D. J. Newman, J. P. Schneider, T. Grkovic and B. R. O'Keefe, *ACS Chem Biol*, 2018, **13**, 2484–2497.

14 J. C. Dearden, *J Comput Aided Mol Des*, 2003, **17**, 119–127.

15 E. Aiassa, J. P. T. Higgins, G. K. Frampton, M. Greiner, A. Afonso, B. Amzal, J. Deeks, J. L. Dorne, J. Glanville, G. L. Lövei, K. Nienstedt, A. M. O'connor, A. S. Pullin, A. Rajić and D. Verloo, *Crit Rev Food Sci Nutr*, 2015, **55**, 1026–1034.

16 M. Hosseini, W. Chen, D. Xiao and C. Wang, *Precis Clin Med*, 2021, **4**, 1–16.

IN SILICO SCREENING, ANALYSIS AND DESIGN OF POTENTIAL INHIBITORS AGAINST RSMD FOR TREATING TUBERCULOSIS

Mangal S Kadam[a], Burra V L S Prasad[a*]
[a] Centre for Advanced Research and Innovation in Structural Biology of Diseases, K L E F University, Vaddeswaram, Andhra Pradesh, 522502, India
[*]Corresponding Author E-Mail ID: dr.prasad.bvls@gmail.com

1 INTRODUCTION

Mycobacterium tuberculosis (M.tb) causes TB and has ever since remained as a major global health burden because of increased rate of antibiotic resistance, an alert raised by WHO. The target, RsmD-like RNA methyltransferase from Mycobacterium tuberculosis (M.tb) (PDB ID:6AIE) was solved at 1.71Å (current best resolution) from our lab. [1] N2 G966 16S rRNA methyltransferase (RsmD) belongs to the classic Rossman fold according to SCOP classification which uses S-Adenosine-L-Methionine (SAM a.k.a AdoMet) as a methyl donor hence belongs to SAM-dependent methyltransferases (Type-I methyltransferases) family. Structurally, RsmD has 3 helices (labeled H1-3, H4-6) on either side of a β-sheet like a sandwich, containing seven strands with the 7th strand antiparallel to the other strands, typically represented as 3214576 where the numerical values stand for the labeled strand number in a sequence starting from N-terminal. SAM-dependent methyltransferases have a well conserved catalytic domain. [2] However, the mode of binding of cofactor and substrate in the active site and the mechanism of methyl transfer is still elusive.

The in silico drug discovery and design are indispensable protocols of the drug development life cycle which has proven its importance by demonstrating substantial reduction in the overall time, resources and money for a novel drug to reach the market. [3] Computational drug discovery process helps to generate potential lead compounds by using various computational tools that are developed incorporating the respective algorithms and relevant biological and chemical knowledge about the molecules and their interactions and pathways. [4] In silico drug discovery and design methods also provide the opportunity to modify, repurpose and improvise the existing drugs to achieve better efficacy. The computational tools help in discarding the unwanted, irrelevant, toxic, and unsafe compounds. [5] The lead compounds that are thus computationally discovered and validated are proceeded for in vitro, in vivo and phase trials eventually to launch the molecules as drugs against specific diseases in the market.

The classic strategy of designing and or discovering a small molecule that competitively binds (with higher relative binding affinity) to the catalytic pocket is implemented. The small molecules are designed to prevent SAM (cofactor) and or substrate binding to RsmD thereby rendering RsmD dysfunctional resulting in the continuance of drug sensitivity and failure of protein translation in the Mycobacterium which results in the death and hence a potential cure to TB through RsmD as the new target. In the current study, the attempts and the steps followed to discover novel lead compounds from the ASINEX nucleoside mimetic database (https://www.asinex.com/nucleoside-mimetics), are described. Three novel lead compounds are successfully discovered as potential RsmD inhibitors, paving the way for novel modes of curing TB.

2 MATERIALS AND METHODS

To arrive at well characterized and in silico validated set of lead compounds, the broad computational drug discovery and designing protocol involves protein and ligand preparation, binding site prediction, molecular docking that includes high throughput virtual screening (HTVS), binding free energy calculations, absorption, distribution, metabolism and excretion (ADME) compliance, toxicity and safety studies and molecular dynamics simulation (MDS) studies. [6,7] Each step in the process is provided as an independent software module having adjustable parameters that help optimize the output, as close as possible to the in vivo system, from the respective module. The parameters available were customized carefully to obtain optimal output suitable for the succeeding steps of the drug discovery protocol. The licensed version of Discovery Studio (DS) 2020, an integrative structural biology and bioinformatics analytic platform was used throughout the study (Biovia Discovery Studio 2021 Accelrys, Inc., San Diego, CA, USA). [8]

2.1 Protein Preparation

For the target (RsmD) preparation, the high resolution RsmD crystal structure was retrieved from RCSB Protein Data Bank (PDB ID: 6AIE). One of the most important steps is checking the structural quality and integrity of the target protein and ligand. To address the same, DS provides Protein Report of Structure protocol under the sub-module of Protein Report & Analysis which is under the Macromolecules module. The protein report, when generated, contains the information associated with missing residues, missing atoms, bond distance, angle and torsional angle inconsistencies and relevant links to the modules and tasks which fix the inconsistencies, if any. Hence, the protein report was generated and checked for any 3D structural inconsistencies. Two loops (from THR21 to THR25), (from PHE185 to VAL 188), residues ARG22, PRO23, THR21, THR24, LEU184, PHE185, ALA186, ASN187, VAL188 and side chain atoms of TYR122 were found missing. At first, water molecules were deleted from the structure as there was no role of water in the catalysis. This was followed by the Prepare Protein sub-module with tasks such as Clean Protein, Insert Loop and others. On running these tasks on the target protein, the missing residues and atoms were fixed and added. Loop modeling and side chain optimization were also performed. Using the Automatic Preparation sub-module, the hydrogens were added, terminal atoms were renamed to prepare the target protein compliant and ready for further processing by other protocols. CHARMm-based energy minimizer was used to perform the initial energy minimization and structural optimization of the protein. Once the protein preparation was done the next step of ligand selection and preparation was performed as explained below.

2.2 Ligand Preparation

It is important to prepare the ligands correctly while studying the receptor-ligand interactions. The ASINEX database has a library of nucleoside mimetic compounds which were downloaded in SDF (Structure Data File) format for further processing. The compounds were screened through Lipinski Rule of Five and Veber's rule with an additional filter that restricts the range of the molecular weight of the compounds to 350 Da to 450 Da. To check the quality and integrity of the compounds, DS provides the *Prepare Ligand* submodule under the *Small Molecules* module. The ligand preparation includes processes such as changing ionization, generation of tautomer and isomers, fixing bad valences and generating 2D or 3D coordinate parameters using the protocols such as *Change Ionization, Generate Tautomer, Generate Isomers, Fix Bad Valencies, Generate Coordinates* and others. The filtered ligands were subjected to the ligand preparation that now included isomers and tautomer. The successfully prepared 3D ligand structures were minimized using the *Minimization* protocol under the *Small Molecule* module of DS. CHARMm-based energy minimizer with 2000 steps of Smart Minimizer was used for ligand minimization, which contains a combination of Steepest descent and Conjugate Gradient algorithms. The prepared target and the ligands were considered for the next step of binding site prediction and molecular docking studies as described below.

2.3 Defining Binding Site

One of the most important tasks of Structure based Drug Designing (SBDD) or Computer Aided Drug Designing (CADD) is identifying or predicting the ligand binding site. [9] Binding site is a part of the 3D structure of protein that allows the binding of a small molecule called ligand through non-covalent interactions. Hence recognizing the binding site is of great importance and an essential step in drug discovery and design. This is achieved by using the *Define and Edit Binding Site* module of DS. It provides three options to define the binding sites: a) the receptor cavity based, b) from PDB SITE record based, c) from current/custom selection. In the current study, the binding site was predicted using the receptor cavity option. It predicted 4 binding sites out of which the first binding site, matching the catalytic active site of RsmD and other Rossman-

Fold methyltransferases described in relevant literature, was selected. [10] Coincidentally, when compared to other three predicted binding sites, the first binding site had the volume to accommodate molecules of similar MW of SAM (S-adenosyl L methionine) (around 399 Da), the universal natural cofactor of methyltransferases. Even after many CDOCKER based docking trials using the same set of ligands and receptor, the results could never be reproduced. It was realized after critical observations that the centroid of the binding site sphere was changing resulting in the inconsistent docked poses after every run. On fixing the centroid coordinates (X: 32.941843, Y: 9.792891, Z: 21.325793) and the radius of the sphere (10.5 Å), the results could be reproduced.

2.4 Molecular Docking

Molecular docking is a computational method that predicts affinity, orientation, and interaction of a ligand in the binding site of a receptor. The molecular docking was performed using CDOCKER of DS. In this study, to dock ligands into a defined binding site the CDOCKER (CHARMm-based molecular dynamics) sub-module, which is part of the Discovery Studio, was used. [11] CDOCKER uses simulated annealing with 2000 heating steps and 5000 cooling steps for generating ligand conformational variants. Further, using Molecular Dynamics (MD) based energy minimization best conformations and poses are selected and displayed by CDOCKER. The forcefield which was used for atom typing was CHARMm which is used for general purpose all-atoms, proteins, nucleic acids and organic molecules. In addition, CDOCKER provides many options to modify various parameters such as number of hits per ligand, generation of random conformations, refinement of ligand orientations and others, to obtain optimized CDOCKER energy. The resulting CDOCKER energy is the first indicator used to judge and rank ligand receptor complexes. In our study, for getting the best receptor-ligand interactions we set the value of Random Conformations to 10, Orientation to Refine to 10 leaving the rest of parameters to respective default values. After docking, CDOCKER gives an integer value that indicates the energy of the ligand receptor binding. The convention used in CDOCKER is that the higher positive values of CDOCKER energy and CDOCKER interaction energies indicate better, stable, and favorable ligand receptor complexes. On docking the ligands, generated from the previous ligand preparation step, a smaller subset of ligands docked successfully. These ligand poses were ranked based on the CDOCKER energy in descending order that gave the best lead compounds on the top from which, the top favorable compounds were selected for the ADME and Toxicity prediction module.

2.5 ADMET Predictions

As a part of the drug discovery and development process the study of pharmacokinetic parameters viz., absorption, distribution, metabolism, excretion, and toxicity (ADMET) properties related to human physio-anatomical compatibility has become mandatory which help us to screen the drug candidate based on their human safety, physio-anatomical relevance and efficacy which are necessary for regulatory approval. To avoid the problem in the development process it is essential to check and study ADMET properties that allow us to remove compounds with unfavorable ADMET properties. Essentially aqueous solubility in water at 25°c, ability to penetrate blood-brain barrier, inhibition of the ubiquitous enzyme, cytochrome P-450 2D6 (CYP2D6) enzyme and ability in binding to carrier proteins in the blood are predicted. The study also provides information about the ease of absorption of the compounds into the human intestine and the toxicity of the compounds to liver (hepatocytes). The ADMET values for each ligand were predicted using the ADMET submodule. The top ligands from the CDOCKER step were passed for ADMET compliance. These resulting molecules were sorted in the descending order as per the binding energies. This sorted list was subjected for the safety tests using the TOPKAT module which is described below.

2.6 Toxicity Prediction (TOPKAT)

The main feature of drug analysis is safety. Computational toxicity prediction plays a major role in decreasing time to market, judge human health risk, lower the animal experiments, and plan design of downstream chemical development process. TOPKAT uses the QSTR (quantitative structural toxicology relationship) model for evaluating different standards of toxicity, to predict carcinogenicity, mutagenicity, lethal dose, skin irritancy and sensitization and environmental toxicity. [12] To determine the safety of compounds, the USFDA (Food and Drug Administration) and NTP (National Toxicology Program) standard toxicity risk properties of compounds were predicted by applying the TOPKAT module of Discovery Studio. The compounds were predicted and assessed for their carcinogenicity and mutagenicity as compared to the USFDA rodent carcinogenicity and Ames mutagenicity.

The compounds are passed through Weight of Evidence (WOE) predictions. WOE evaluates all the considered ligands with the relevant evidence for the risk, skin irritancy and skin sensitization. [13] All compounds are listed with their Mouse Female NTP model, Rat Female NTP model and Mouse Male FDA model. The sorted compounds with good ADMET properties falling within standard acceptable ranges were chosen for toxicity prediction using the *TOPKAT* Prediction submodule. The final compounds with best binding energies, non-carcinogenic and safe were taken for MDS studies as described below.

2.7 Molecular Dynamics Simulations (MDS)

In computational studies, the MDS plays a very important role where the ligand and receptor interactions are simulated, mimicking the *in vivo* atomic environment of a molecular experiment as closely as possible. It is through the results of MDS, we interpret the simulated data and extend our understanding about the natural stability and affinity of the ligand receptor complexes. [14] The best ligand-RsmD complexes selected from the previous step were subjected to MDS. GROMACS v.2021 was used for the simulation studies. [15] MDS provides the flexibility and many options to simulate the protein-ligand interactions in different conditions such as various temperatures, pressures and time that mimic the real-life molecular system. MDS gives the advantage of doing an experiment *in silico* which is otherwise hard and very costly to perform in real-life experiments.

MDS run was performed for the native and docked complexes. The pre-processing of protein and ligand were done to generate topology, solvation, charge correction and ligand modification (if any). After successful pre-processing, energy minimization was carried out to reduce aberrations in the structural parameters of the system, making it stereo-chemically stable. This step was followed by heating the system that makes atomic vibrations according to the simulation temperatures. The whole system was optimized for pressure and volume to maintain the stability. GROMACS v.2021 (*GROningen MAchine for Chemical Simulation*) was used to perform simulation for the top potential lead compounds. The topology of ligand-bound structures was generated with the amber99sb force field and solvated in a dodecahedron periodic boundary box of 10.0x10.0x10.0 nm^3. This box was then solvated with approximately 5485 TIP3P water models. To neutralize the system having a net charge, counter ions (Na^+ and Cl^-) were added by replacing the water molecules. A minimization step was done by using the Steepest Descent method to optimize the bond angles, bond lengths, and torsion angles. To study electrostatic interactions Particle-Mesh Ewald (PME) method was used and Van der Waals interactions were truncated with a default cut-off of 0.9 nm. NPT (Number of particles, Pressure and Temperature) and NVT (Number of particles, Volume and Temperature) canonical ensemble calculations were performed at 50000 steps each. The system was heated to 300 K using a Berendsen thermostat, the pressure was maintained of 1 bar with coupling time 0.1 ps. The final production run was carried out for 100 ns. The embedded packages of GROMACS were used to generate the plots for Root Mean Square Deviation (RMSD), Root Mean Square Fluctuations (RMSF), Hydrogen bond, Radius of gyration (Rg) and Interaction energy for the complexes for interpretation and insights.

3 RESULTS

Virtual screening, docking, human physio-anatomical compatibility studies, ligand interaction analysis and molecular dynamics simulations were performed to screen, evaluate, assess their potential to be the lead compounds for further invitro and other tests, eventually to treat the disease. The binding affinities, molecular conformations, and the stability of the selected compounds were investigated.

High throughput virtual screening of the 4507 ASINEX nucleoside mimetic compounds resulted in 2840 molecules. On ligand preparation that includes tautomer and isomer generation resulted in a total of 3480 nucleoside mimetics. These were docked in the selected binding site using CDOCKER that resulted in only 1019 poses. The top 10% (100 compounds) based on best CDOCKER energy were selected from these poses having favorable binding energy that ranged from 20 kcal/mol to 34 kcal/mol. The selected 100 compounds were subjected to ADMET compliance which resulted in 34 compounds. The compounds passed through ADMET were subjected to TOPKAT and top 3 compounds were finalized for further MDS studies considering their safety level, favorable binding energies and number of conserved hydrogen bonds. The flowchart is shown in Figure 1.

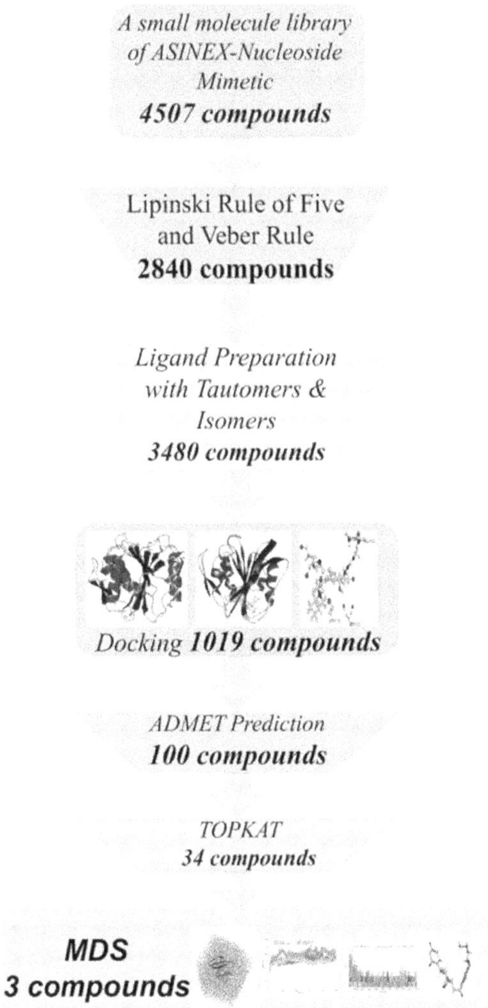

Figure 1 *Flowchart to find potential lead compounds.*

3.1 ADMET Prediction

The 34 compounds passed through ADMET were sorted based on their binding energies and the top 5 were picked for further studies (see Table 1). The 34 compounds were found with a good absorption level for the human intestine and blood-brain-barrier (BBB) penetration level. The 34 compounds were also found to be non-inhibitors of cytochrome P450 2D6 (CYP2D6) which is involved in drug metabolism. All the selected compounds were found to be good with absorption, BBB penetration and solubility as shown in Table 2.

Table 2 gives the ADME compliant values for the top 5 compounds. The compounds passed through ADME descriptors were subjected to the TOPKAT prediction module for safety studies.

Table 1 *CDOCKER scores and binding energies of ASXNA compounds.*

Compound Name	-CDOCKER (kcal/mol)	-CDOCKER Interaction (kcal/mol)	Binding Energy of Complex (kcal/mol)
ASXNAID_1651	30.2187	53.3098	-151.98
ASXNAID_2037	29.9142	58.1144	-144.4
ASXNAID_3215	33.8561	56.2497	-128.42
ASXNAID_3187	32.8009	44.9954	-111.52
ASXNAID_3218	27.624	52.2905	-100.11

3.2 Toxicity Prediction

Among the top 5 compounds, ASXNAID_2037 showed Carcinogen alert for Rat- female -NTP Prediction and multi- carcinogen alert for Mouse male FDA animal model. Similarly, ASXNAID_1651 showed multi-carcinogenic alert for Mouse Male FDA model as shown in Table 3.

However, all compounds were found safe when followed WOE (Weight of Evidence) predictions (see Table 3). WOE evaluates all the considered relevant evidence for a risk assessment. [13] Though ASXNAID_1651, ASXNAID_2037 show highest binding energies, -151.98 and -144.40 Kcal/mol respectively, the top 2 molecules were discarded from further studies. The top 3 compounds, ASXNAID_3215, ASXNAID_3187 and ASXNAID_3218 were considered to run MDS.

3.3 Molecular Docking and Ligand Interaction Studies

All five compounds (Table 1) had good binding energies and were found to bind in the conserved SAM binding pocket. The top 2 mentioned above have been discarded since they were predicted to be carcinogenic and mutagenic. The remaining 3 compounds, ASXNAID_3215, ASXNAID_3187 and ASXNAID_3218 had relatively good binding energies; -128, -111 and -100

Kcal/mol, respectively. The number of hydrogen bonds formed between the receptor and ligand were analyzed. Hydrogen bonds play a very important role in the stability, mechanics, and dynamics of a protein. Hydrogen bonds were found to be highest for ASXNAID_3218 with a total of 11, followed by ASXNAID_3215 with 7 hydrogen bonds and ASXNAID_3187 had only 5 hydrogen bonds. The detailed individual atomic interactions are shown in Figure 2 a-c and Tables 4 a-c. As can be seen from the tables, all the hydrogen bonds can be categorized as strong hydrogen bonds.

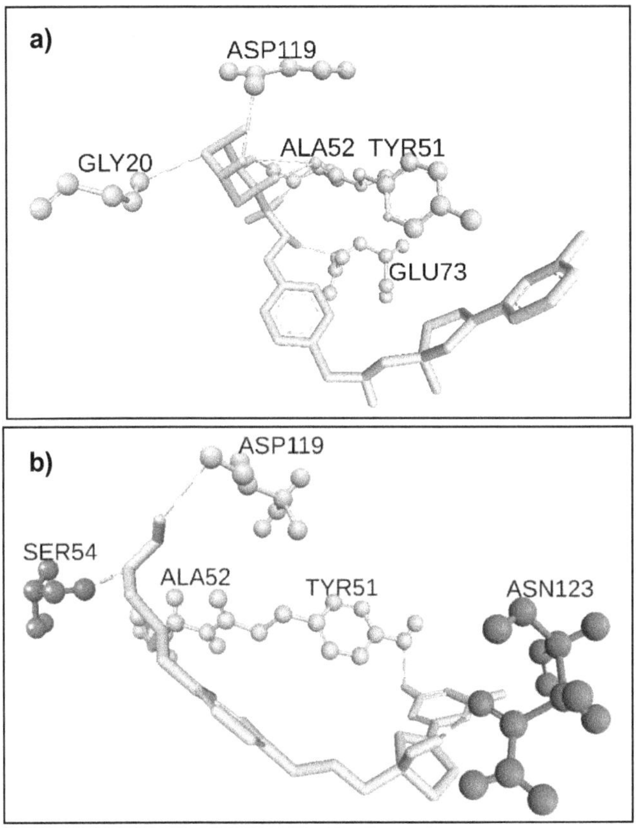

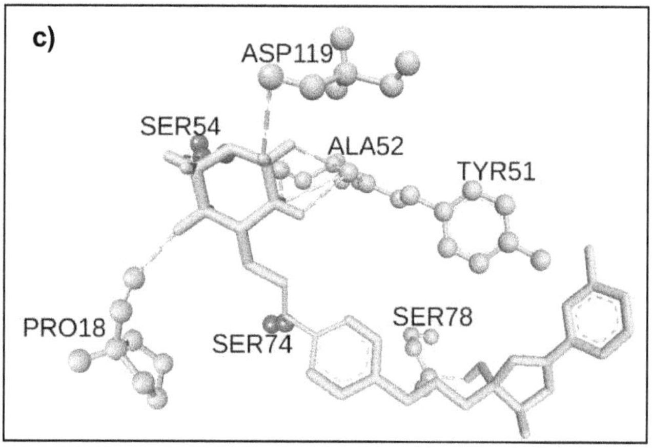

Figure 2 *a) Interaction diagram of ASXNAID_3215; b) Interaction diagram of ASXNAID_3187; c) Interaction diagram of ASXNAID_3218. Ligand is represented as stick model; SER54 and SER74 interactions were observed in SAM and its analogs; ASN123 new residue interaction was not listed in the 25 amino acids; other interacting amino acids are unique to the current ligands.*

Table 2 *Solubility level 0-4, where 0 (extremely low), 1 (very low), 2 (low), 3 (good), 4 (optimal); Blood-brain-barrier 0-4 (Very high penetrant), 1 (High), 2 (Medium), 3 (Low), 4 (Very Low); CYP2D6 prediction (0- Non-Inhibitor, 1-Inhibitor); Absorption Level in human intestine (0-2), where 0 is good absorption 1 (moderate) 2 (low) 3 (very low); No. of rotatable bond, No. of Acceptor, No. of Donor. AlogP - logarithm of the partition coefficient between n-octanol and water (Must be less than five).*

Compound Name	**Molecular Weight**	**Solubility Level**	**BBB Level**	**CYP 2D6 prediction**	**Absorption Level**	**Rotatable Bonds**	**Acceptors**	**Donors**	**AlogP**
ASXNAID_1651	432.55	3	3	0	0	8	9	5	-1.468
ASXNAID_2037	410.52	3	3	0	0	6	4	3	0.548
ASXNAID_3215	428.54	3	3	0	0	9	6	3	-0.303
ASXNAID_3187	386.48	3	3	0	0	10	6	2	1.649
ASXNAID_3218	427.53	3	3	0	0	9	7	2	1.258

Table 3: *Toxicity properties of leads predicted using Discovery Studio Suite. (Note: C – Carcinogen; SC – Single-Carcinogen; MC – Multi-Carcinogen; NC – Non-Carcinogen; NM – Non-Mutagen)*

Compound Name	Mouse female NTP Prediction	Rat female NTP Prediction	Mouse Male FDA	WOE Prediction	Ames Prediction	Skin Irritancy	Skin Sensitization
ASXNAID_1651	NC	NC	MC	NC	NM	None	None
ASXNAID_2037	NC	C	MC	NC	NM	None	None
ASXNAID_3215	NC	NC	NC	NC	NM	None	None
ASXNAID_3187	NC	NC	NC	NC	NM	None	None
ASXNAID_3218	NC	NC	NC	NC	NM	None	None

Table 4 a) *List of Hydrogen bonds between RsmD-ASXNAID_3187.*

S. No	Residue No.	Residue name	Atom	Distance (Å)	Atom	Residue Name
1	54	SER	HG	2.22	O	ASXNAID_3187
2	123	ASN	OD1	2.28	H	ASXNAID_3187
3	52	ALA	O	3.01	H	ASXNAID_3187
4	119	ASP	OD2	2.87	H	ASXNAID_3187
5	51	TYR	O	2.82	H	ASXNAID_3187

Table 4 b) *List of Hydrogen bonds between RsmD-ASXNAID_3215.*

S. No	Residue No.	Residue name	Atom	Distance (Å)	Atom	Residue Name
1	20	GLY	HN	2.81	O	ASXNAID_3215
2	51	TYR	O	2.6	H	ASXNAID_3215
3	51	TYR	O	2.51	H	ASXNAID_3215
4	73	GLU	OE1	2.51	H	ASXNAID_3215
5	51	TYR	O	2.99	H	ASXNAID_3215
6	52	ALA	O	2.48	H	ASXNAID_3215
7	119	ASP	OD1	2.95	H	ASXNAID_3215

Table 4: c) *List of Hydrogen bonds between RsmD-ASXNAID_3218.*

S. No	Residue No.	Residue name	Atom	Distance (Å)	Atom	Residue Name
1	54	SER	O	2.54	HG	ASXNAID_3218
2	74	SER	N	2.12	HG	ASXNAID_3218
3	78	SER	O	2.27	HG	ASXNAID_3218
4	74	SER	OG	1.82	H	ASXNAID_3218
5	18	PRO	O	2.34	H	ASXNAID_3218
6	51	TYR	O	2.83	H	ASXNAID_3218
7	52	ALA	O	2.33	H	ASXNAID_3218
8	51	TYR	O	3.04	H	ASXNAID_3218
9	54	SER	OG	2.76	H	ASXNAID_3218
10	119	ASP	OD2	2.89	H	ASXNAID_3218
11	51	TYR	O	2.78	H	ASXNAID_3218

Table 4: d) *List of Hydrogen bonds between RsmD-SAM.*

S. No	Residue No	Residue Name	Atom	Distance (Å)	Atom	Residue Name
1	29	ARG	HH12	2.14	N2	SAM
2	54	SER	H	2.31	O2	SAM
3	78	SER	HG	1.88	O5	SAM
4	26	ASP	OD1	2.34	H21	SAM
5	19	ARG	O	2.15	H23	SAM

3.4 Molecular Dynamics Simulation

The MD simulation is studied for investigating conformational stability, conformational changes, internal movements and receptor-ligand interactions between the receptor and ligand complexes. MD simulation was performed for RsmD-ligand complexes over 100 ns under physiological environment to record conformational changes, its stability, and interactions of ligands with amino acid residues over time. MD trajectories were used to interpret and draw the standard RMSD, RMSF, Rg, interaction energy and hydrogen bond plots.

3.4.1 RMSD analysis. RMSD plot provides the information about the structural changes happening at the atomic resolution over atomic time scales i.e., picoseconds (ps) and nanoseconds(ns) and hence correlates to the stability of the system. To understand the stability of receptor and or receptor-ligand complexes in a systematic time frame, RMSD is the key factor. There are no significant changes in the structural conformation of native protein throughout simulations. An average RMSD value of native protein was 0.21 nm. An average RMSD value of 0.20 nm was observed for the RSMD-ASXNAID_3215 complex, 0.19 nm for RSMD-ASXNAID_3187 and 0.25 nm for RSMD-ASXNAID_3218 complex. Overall, the ligands did not influence the structural conformation and stability of the protein when compared with the native protein (see Figure 3 a).

3.4.2 RMSF analysis. To identify the flexible and rigid regions of the protein structure RMSF is an essential parameter. It is a standard measure related to the deviation of the particle from its original position. RMSF can also be used to understand the flexibility of the backbone of the protein structures and ligands. To understand the protein structure and ligand complex showing each atom's mean displacement, RMSF was analyzed. In the case of RsmD, the highest fluctuation between 20 to 30 residues was observed. For the RSMD-ASXNAID_3215 complex, residue fluctuation was observed between 20 to 30 residues, 52 to 56, 164 to 176, whereas for RSMD-ASXNAID_3187 complex 16 to 28, from 120-136 residues and for the RSMD-ASXNAID_3218 complex 18 to 28, from 116 to 136 residues showed fluctuations. The average RMSF value for the native protein was 0.24 nm. The average RSMD-ASXNAID_3215 complex showed 0.050 nm, RSMD-ASXNAID_3187 showed 0.069 nm and RSMD-ASXNAID_3218 showed 0.035 nm. It is interesting to note that the RSMD-ASXNAID_3218 showed the least fluctuation (see Figure 3 b).

3.4.3 Radius of gyration (Rg). To understand the compactness and folding properties of the protein and protein-ligand complexes Rg is a useful parameter. It is also useful to demonstrate the influence of a ligand in a protein-ligand complex applying changes in it. Higher R_g value specifies protein molecules with loose packing, whereas lower R_g value specifies tight packing of the protein structure. In case of ligands bound to RsmD protein, an average R_g value for RSMD-ASXNAID_3215 complex showed 0.031 nm, RSMD-ASXNAID_3187 showed 0.046 nm and RSMD-ASXNAID_3218 showed 0.079 nm. The radius of gyration for native protein was 0.016 nm was observed, which could be attributed to conformational shifts resulting in changes in the secondary structure of the protein during the MD simulations (see Figure 3 c).

3.4.4 Intermolecular H-bonding. To understand the binding affinity of ligands with the protein molecule intermolecular H-bonding is a crucial parameter.

Formation or distortion of hydrogen bonds is a principal component during the MD simulation. Large number of H-bonds present between the protein and ligand signifies a stronger binding affinity. In the case of the RSMD-ASXNAID_3215 complex showed 5, RSMD-ASXNAID_3187 showed 7 and RSMD-ASXNAID_3218 showed 11 hydrogen bonds were formed. Among top 3 lead compounds RSMD-ASXNAID_3218 shows strong interactions in the binding site of RsmD with a significant number of hydrogen bonds (see Figure 3 d).

3.4.5 Interaction Energy. To understand the interaction between receptor and ligand interaction energy parameters plays a crucial role. RSMD-ASXNAID_3187 and RSMD-ASXNAID_3215 showed favorable interactions, however RSMD-ASXNAID_3218 maintained the highest favorable interactions throughout the 100 ns simulation time (see Figure 3 e).

a)

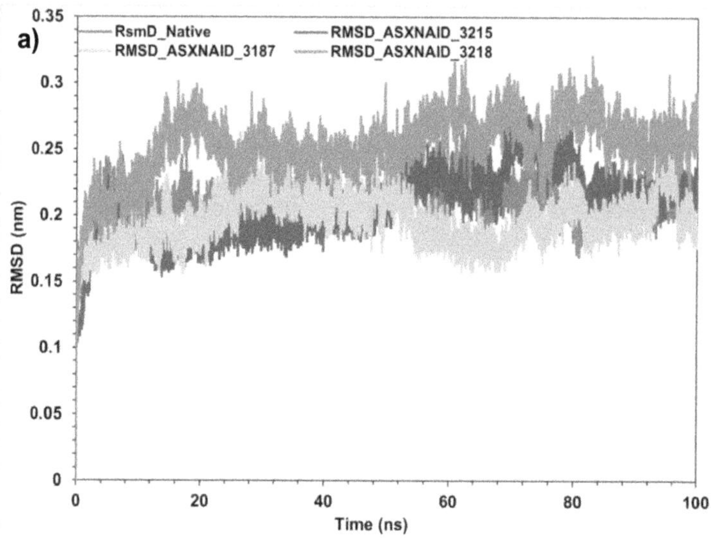

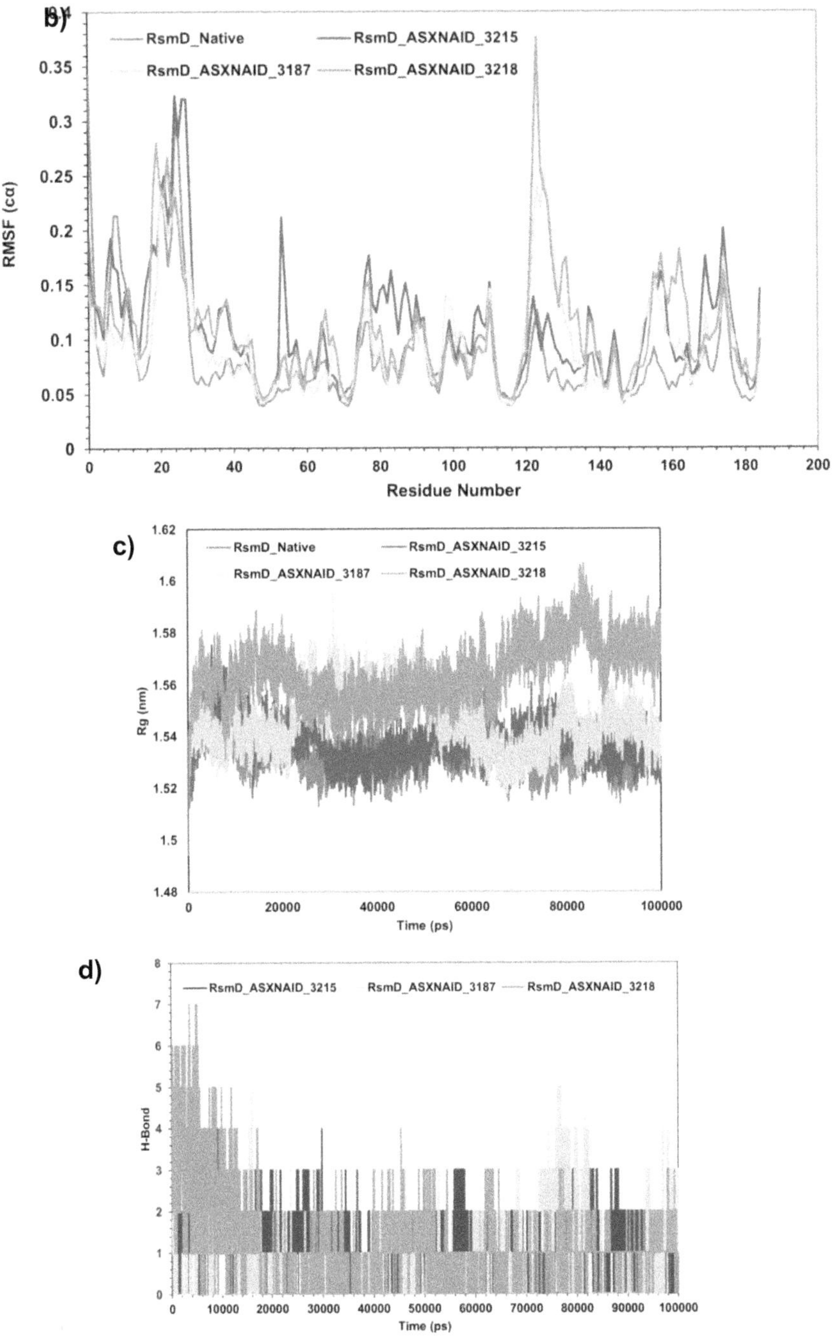
b)
RsmD_Native
RsmD_ASXNAID_3215
RsmD_ASXNAID_3187
RsmD_ASXNAID_3218
RMSF (cα)
Residue Number
c)
RsmD_Native
RsmD_ASXNAID_3215
RsmD_ASXNAID_3187
RsmD_ASXNAID_3218
Rg (nm)
Time (ps)
d)
RsmD_ASXNAID_3215
RsmD_ASXNAID_3187
RsmD_ASXNAID_3218
H-Bond
Time (ps)

e)

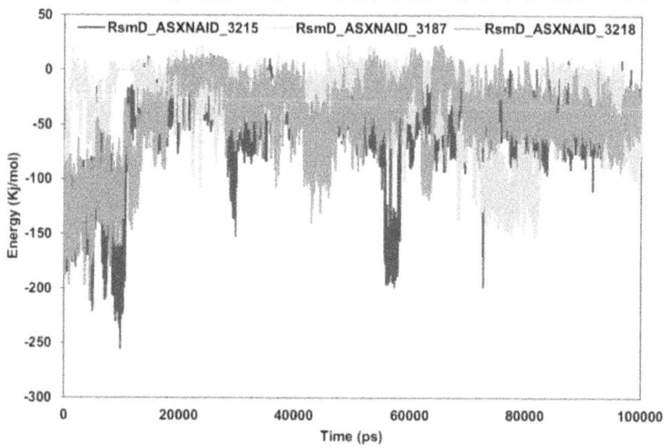

Figure 3 *Molecular dynamics of RsmD bound to ASXNAID a) Root mean square deviation (RMSD); b) Root mean square fluctuations (RMSF); c) Radius of gyration (R_g); d) number of hydrogen bonds formed; e) Interaction Energy. (Apo enzyme (blue), RSMD-ASXNAID_3215 (red), RSMD-ASXNAID_3187 (yellow), RSMD-ASXNAID_3218 (green)).*

4 DISCUSSIONS

RsmD is one of the thirteen known 16S rRNA methyltransferases which methylates N2 position of Guanine 966 nucleotide of the 16S rRNA of the 30S subunit. [16] It has been recently shown that proper protein synthesis and antibiotic resistance is primarily because of rRNA methylation. [17] RsmD is one of these rare examples, when a housekeeping methyltransferase provides antibiotic resistance. It has been revealed in both antibiotic producing strains and pathogenic bacteria causing diverse infections. [17] RsmD modifies m2 G966 nucleotide residue present in tetracycline binding site. [18] The deletion of RsmD in E. coli increases susceptibility to tetracycline. [19]

In our recent review article, we have analyzed more than three hundred methyltransferase-SAM structures and have shown that RsmD active site is surrounded by ~25 amino acids within a spherical radius of 4Å. Further studies of the properties of the 25 amino acids revealed that twelve amino acids belong to non-polar category (Val16, Pro17, Pro18, Pro23, Leu57, Ala56, Ala52, Tyr51, Pro121, Gly55, Gly20, Gly53) indicating hydrophobic disposition that is suitable for SAM binding, six to polar (Thr2, Thr21, Thr24, Thr25, Ser 78, Ser54), three to positively charged (Arg19, Arg22, Arg29) and four to negatively charged (Asp26, Asp75, Glu73, Asp119) residues.

There are no crystal structures of RsmD, SAM and or substrate bound complexes till now. All the crystal structures of methyltransferase-SAM complexes were analyzed to obtain the consensus and conserved interactions. This knowledge gained about the SAM - methyltransferase interactions were used to deduce the RsmD-SAM binding sites and their potential interactions. With SAM as a natural binding cofactor of RsmD for methyl transfer and the interaction studies, four novel SAM analogs with improved binding free energies were designed as potential leads against RsmD [20]. The docking and simulation studies of SAM-RsmD complex revealed that SAM is interacting with five amino acids (SER54, ARG29, SER78, ASP26, ARG19) of the 25 amino acids with one strong hydrogen bond each as shown in the table 4 d. [20] As can be seen from tables 4 a-c, the three nucleoside mimetic compounds are having 5, 7 and 11 hydrogen bonds with the receptor indicating these compounds have much better binding energies, more specificity and new interaction partners than the SAM and its analogs discussed elsewhere [20].

On comparing the interactions of the three ASXNA ligand complexes and SAM analog complexes, it was observed that there are interactions which are a) seen in ASXNA ligands as well as in SAM analogs (see Figure 2), b) seen only in ASXNA ligands and not seen in SAM analogs and c) new interactions not listed in the 25 amino acids (see ta 3.1). SER54 and SER78 were interacting with the ASXNA ligands similar to SAM analogs indicating their possible importance in stabilizing the binding. The green-colored residues represent those amino acid interactions which are unique to the ASNX ligands, however, they are one among the 25 amino acids already identified in the active site region of methyltransferases. The residues which are unique are TYR51, ALA52, ASP119, GLY20, PRO18 and GLU73. ASN123 is the new interacting residue which is not among the 25 amino acids. ASN123 is a new residue interaction because of the tilting of the ligand towards S4 C-terminal end instead toward H4 and S1.

The current goal of discovering novel molecules against RsmD resulted in three highly potent small molecules which are predicted to be safe with acceptable ADME properties showing good docking scores with higher binding energies. The three lead compounds are also learnt to be easy to synthesize. The ASXNAID_3215, ASXNAID_3187 and ASXNAID_3218 are computationally verified and now need to be tested and evaluated for the in vitro and in vivo studies. However, among these three the ASXNAID_3218 is considered more potent than the other two and will be the priority for the in vitro and in vivo studies and testing.

5 CONCLUSIONS

As 16S rRNA methyltransferases are considered as a novel target against tuberculosis to overcome the problem of drug resistance caused by methylation. Our target is one of them that is RsmD which transfers the methyl group to N2 G966. The strategy to inhibit the function of RsmD is explored, resulting in three potent lead compounds which need to be evaluated and tested for their inhibition activity. The chemical environment of the binding pocket of RsmD consists largely of hydrophobic amino acids. The 4507 compounds from ASINEX nucleoside mimetics library three compounds, ASXNAID_3215, ASXNAID_3187, ASXNAID_3218 showed promising binding energies, docking scores and stable and stronger interactions with RsmD. The three leads are learnt to be easy to synthesize. The top 3 potent lead molecules i.e., ASXNAID_3215, ASXNAID_3187, ASXNAID_3218, are computationally verified and now need in vitro and in vivo evaluation and testing.

6 ACKNOWLEDGMENT

BVLS acknowledges ICMR (ISRM/12(7)/2019) for the financial support.

7 CONFLICTS OF INTEREST

The authors confirm that this article content has no conflicts of interest.

8 FUNDING

BVLS gratefully acknowledges ICMR (Indian Council of Medical Research) for providing the financial support (Grant No. ICMR: ISRM 12(07)/2019).

9 REFERENCES

1 S. Venkataraman, A. Dhankar, K. M. Sinha, P. Manivasakan, N. Iqbal, T. P. Singh and B. V. L. S. Prasad, 2018.

2 G. Maravić, J. M. Bujnicki, M. Feder, S. Pongor and M. Flögel, *Nucleic Acids Res.*, 2003, **31**, 4941–4949.

3 B. Shaker, S. Ahmad, J. Lee, C. Jung and D. Na, *Comput. Biol. Med.*, 2021, **137**, 104851.

4 J. L. Dahlin and M. A. Walters, *Future Med. Chem.*, 2014, **6**, 1265–1290.

5 S. Ekins, *Computational Toxicology: Risk Assessment for Pharmaceutical and Environmental Chemicals*, John Wiley & Sons, 2007.

6 S. K. Saxena, *High-Throughput Screening for Drug Discovery*, BoD – Books on Demand, 2022.

7 S. Mohane Coumar, *Molecular Docking for Computer-Aided Drug Design: Fundamentals, Techniques, Resources and Applications*, Academic Press, 2021.

8 D. S. Biovia, *San Diego: Dassault Systèmes.*

9 S. J. Campbell, N. D. Gold, R. M. Jackson and D. R. Westhead, *Curr. Opin. Struct. Biol.*, 2003, **13**, 389–395.

10 P. V. Sergiev, A. A. Bogdanov and O. A. Dontsova, *Nucleic Acids Res.*, 2007, **35**, 2295–2301.

11 G. Wu, D. H. Robertson, C. L. Brooks 3rd and M. Vieth, *J. Comput. Chem.*, 2003, **24**, 1549–1562.

12 E. Gonzalez, S. Jain, P. Shah, N. Torimoto-Katori, A. Zakharov, Đ.-T. Nguyễn, S. Sakamuru, R. Huang, M. Xia, R. S. Obach, C. E. C. A. Hop, A. Simeonov and X. Xu, *Drug Metab. Dispos.*, 2021, **49**, 822–832.

13 E. Aiassa, J. P. T. Higgins, G. K. Frampton, M. Greiner, A. Afonso, B. Amzal, J. Deeks, J.-L. Dorne, J. Glanville, G. L. Lövei, K. Nienstedt, A. M. O'connor, A. S. Pullin, A. Rajić and D. Verloo, *Crit. Rev. Food Sci. Nutr.*, 2015, **55**, 1026–1034.

14 M. A. F. Nasution, E. P. Toepak, A. H. Alkaff and U. S. F. Tambunan, *BMC Bioinformatics*, 2018, **19**, 419.

15 J. Lemkul, *LiveCoMS*, , DOI:10.33011/livecoms.1.1.5068.

16 M. R. Salaikumaran, V. P. Badiger and V. L. S. P. Burra, *Protein J.*, 2022, **41**, 97–130.

17 I. A. Osterman, O. A. Dontsova and P. V. Sergiev, *Biochemistry* , 2020, **85**, 1335–1349.

18 L. Jenner, A. L. Starosta, D. S. Terry, A. Mikolajka, L. Filonava, M. Yusupov, S. C. Blanchard, D. N. Wilson and G. Yusupova, *Proc. Natl. Acad. Sci. U. S. A.*, 2013, **110**, 3812–3816.

19 R. J. Nichols, S. Sen, Y. J. Choo, P. Beltrao, M. Zietek, R. Chaba, S. Lee, K. M. Kazmierczak, K. J. Lee, A. Wong, M. Shales, S. Lovett, M. E. Winkler, N. J. Krogan, A. Typas and C. A. Gross, *Cell*, 2011, **144**, 143–156.

20 M. R. Salaikumaran and B. V. L. S. Prasad, *Lett. Drug Des. Discov.*, , DOI:10.2174/1570180819666220616105517.

TOXICITY AND BREAST CANCER RISK PREDICTION STUDIES FOR CARBAMATES: BIOINFORMATICS APPROACH

Daveedu Thathapudi[a], Naresh Dumala[b*], Raja Jayarao Yendluri[c], Manohar Vadela[a], Nagaraju Bandaru[b]

[a]Department of Biotechnology, Vikrama Simhapuri University, Nellore-524320, Andhrapradesh, India; [b]Department of Pharmacy, Koneru Lakshmaiah Education Foundation, Vaddeswaram, Guntur-522302, Andhra Pradesh, India;[c]Dr. Samuel George Institute of Pharmaceutical Sciences, Markapur-523316, Andhrapradesh, India.

*Corresponding author: nareshpharma2020@kluniversity.in

1 INTRODUCTION

Pesticides are the chemical substances that possess rodenticide, insecticide, fungicide, and nematicide activities. Due to these properties, they are commonly used for controlling the pests in various fields. Depending on the chemical nature, these pesticides are grouped as carbamates, pyrethroids, organophosphorus agents, neonicotinoids, and triazines. As they are being used excessively in the agricultural, domestic, and industrial applications and also because of their high persistence these pesticides are considered as potential environmental pollutants.[1] Humans, animals and non-target species are getting exposed to these pesticides preferably by the oral route through intake of food or water that is contaminated by those harmful chemicals.[2-6] Exposure to the pesticides is major health concern due to their well-established toxic risks. These unintentional or occupational exposures perhaps the reason for triggering neurodegenerative, metabolic, developmental, reproductive, and cancerous like conditions. Most of the toxic effects caused by the pesticides were linked with their ability to initiate either agonistic or antagonistic responses towards neurological endocrine receptors, blocking ion channels, inhibiting enzymes like acetylcholine esterase otherwise disturbing the redox balance or inducing oxidative stress.[7-9]

In silico analysis is one of the simplest approaches that is comprehensively used in studying drug efficacy during the drug developmental process.[10] Molecular Docking is the technological software that gives information about the ligand and receptor molecular interactions.[11] In this study, selected carbamate pesticides were chosen to adjudicate their ability to interact with the human breast cancer gene [BRCA1]. The BRCA1 gene is located in the 17th chromosome long arm at 17q21 and this BRCA1 possesses 24 exons that range about 80 kb.[12] Initially, BRCA1 was mapped in 1990 and cloned in the year of 1994.[13] BRCA1 makes interaction with enormous proteins that participates in cellular events like cell cycle progression, cellular pathways, gene regulation, ubiquitination, and also in the condition of DNA damage in response.[14] BRCA1 is a unique DNA Damage Response [DDR] protein that involves in DNA activation and also as a check point for repair.[15] Further BRCA1 also binds to the p53 protein and RAD51 complex leading to the formation of both homologous recombination and Non-Homologous end joining that plays a significant role in the DNA damage repair process.[16] Any deficiency in the BRCA1 gene may lead to abnormalities in the duplication of the centrosome, G2/M checkpoint, S-phase checkpoint and spindle checkpoint.[17] The most important feature of the BRCA1 gene-associated breast cancer is that it directly comprises the complex of aneuploidy than malignant tumours without having any mutations in the Breast cancer susceptibility gene1.[18]

Considering the above key points, the present study was intended to address the questions: a) on the general toxicity and carcinogenic profile of 72 carbamates, and b) find the number of carbamates interacting with BRCA1 gene. It was followed by their protein targets identification and providing preliminary information on cancer concerns of the carbamate exposure.

2 METHODOLOGY

2.1 In silico prediction of the toxicity profile of carbamates

In the current study, 72 carbamates were chosen, and they were subjected to analyze their possible carcinogenic effect. The general toxicity risks associated with pesticides/ carbamates were predicted using the pkCSM platform, a computational simulation model. Toxicological assessment was performed for mutagenicity, Aquatic toxicity [minnow toxicity], acute & oral toxicity, hepatotoxicity, and skin surface toxicity. Based on the preliminary investigation results obtained from the PKcsm tool out of the 72 carbamates selected 10 compounds found to be carcinogenic hence those 10 carbamates were carefully chosen for further study. They were Allethrin [Pubchem ID: 11442], Butacarb [Pubchem ID: 17564], Carbaryl [Pubchem ID: 6129], Camphechlor [Pubchem ID: 58167], Diallate [Pubchem ID: 5284376], Fenoxycarb [Pubchem ID: 51605], Mobam [Pubchem ID: 14119], Propoxur [Pubchem ID: 4944], Triallate [Pubchem ID: 5543], Toxaphene [Pubchem ID: 5284469]. Above compounds, 3D structures were retrieved from the PubChem in their SDF format and converted into the PDB, using the software Open Babel version 2.3.2[19].

2.1.1 Protein sequence retrieval. The sequence of the BRCA1 gene [human] was retrieved from the Uniprot ID: P38398.[20] The retrieved sequence was used for further investigation by subjecting the sequence to similarity search protein with 100% similarity [4Y2G] was selected as a template for building the homology modelling. For molecular docking studies, a reliable protein structure is required, hence the modelled structure of the BRCA1 receptor was analyzed in terms of its reliability. The stability of the protein was analyzed using the Ramachandran plot statistics and there were [94.2%] of residues are present in the allowed region. The ProSA analysis indicated that the BRCA1 receptor built from the UniProt sequence is similar to the PY2G protein and it was modelled using the software Modeller 9V8.

2.1.2 Protein structure Quality evaluation. Quality evaluation was done to the modelled protein using platforms like PROCHEK, Protein structure analysis [ProSA], Protein quality predictor [ProQ], discrete optimized protein energy [DOPE] score, and finally superimposition of target and template. Further, modelled BRCA1 protein was given to the galaxy refine to verify the torsional angle deviation and to measure the angle bonds and their length, mismatches in the sequence, missing atoms, level of precision in fold prediction, water molecule arrangements in their right position and if any extra atoms and errors in their chirality.

2.1.3 Active site prediction. Region of active sites in the BRCA1 gene was identified using the Plip web server[21] which shows the residues involved in the active site of the protein molecule and also predicts the volume and area of the active site. Further, residues in the active site are mapped to the structure to evaluate where the residues are located.

2.2 Molecular docking studies

The molecular docking technique is widely used to study the interaction between the macromolecules [Proteins] and small molecules [Ligands] molecular docking was done using Auto Dock Vina, before running the docking process, the format of macromolecule was converted from PDB to PDBQT and ligands format also turned from SDF to PDBQT, in Auto Dock Vina gasteiger partial charges and hydrogen atoms were added to the macromolecule. Docking of the carbamates [ligands] to the macromolecules were focused on the specified sites. The interaction energy/binding energy between ligand and receptor are measured in terms of Kcal/mol and those energy levels were shown in table 1.

2.2.1 Visualization of Binding interactions. Interaction between receptor and ligand in the form of Hydrogen bonding and Van Der Waals forces were seen using the Visualizer software, Accelrys Discovery Studio,[22] and figures were shown below.

2.2.2. Residues participating in active sites. Active sites involved in the BRCA1 macromolecule were determined using the PLip web server, those residues are listed as follows serine (1655), glycine (1656), leucine (1701), and lysine (1702).

2.2.3. Protein target identification and annotation. Initially, selected carbamates interacting protein targets were found individually using the database of STITCH, where in humans [Homo sapiens] related proteins were selected with a high score of confidence, excluded the data mining of the above 10 interactions. The targeted proteins interacting with carbamates were derived from the database of STITCH and were carefully reconnoitered for the Gene Ontology [GO] term for identifying the protein biological functions, further, those affected pathways concerning the cancer were also identified using the PANTHER database 11.1.[23]

3 RESULTS AND DISCUSSION

pkCSM studies of toxicological assessment for the 72 carbamates indicated that out of 72 at least 10 compounds from this category of carbamates could induce different mutagenicity. Based on the *in silico* toxicological studies performed on these 10 carbamates such as butacarb, camphechlor, diallate, triallate could induce the skin sensitization, on other hand other carbamates like allethrin , fenoxycarb, mobam, propoxur, carbaryl and cloethocarb does not showed any difficulties related to skin sensitization. Range of pesticide toxicity is largely depends on the toxicodynamic properties of those pesticides, hence the examination for those ten carbamates were selected for the further studies. Commonly, human intestinal absorption and their ability to cross the blood-brain barrier, distribution, metabolism and excretion properties provides valuable information in predicting the pesticidal toxicity. All the ten carbamates were well absorbed through the intestine. Similarly, most of the carbamates from the selected ten have passed through blood brain barrier except one carbamate i,e camphechlor (shown in Figure 1).

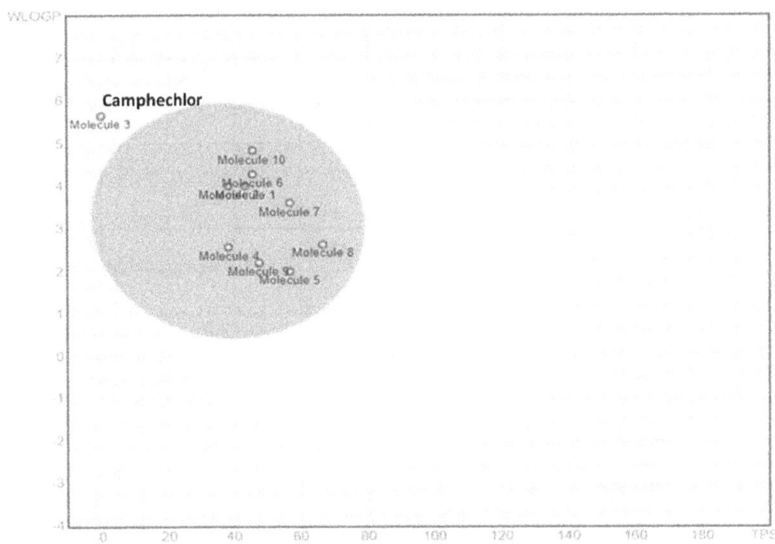

Figure 1 *Potentiality of Carbamates intestinal absorption and penetration into the brain. (Molecule 3, Camphechlor), white egg represents intestinal absorption while yellow circle denotes brain penetration.*

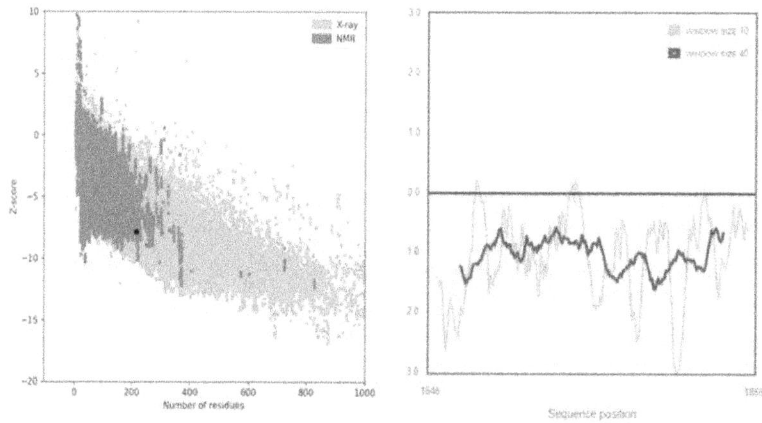

Figure 2 *Protein structural analysis (PROSA) plot of BRCA1 protein.*

Based on this ADME study we can conclude that except camphechlor all the carbmates could potentially enters into the brain and might result in the neurological complications. Homology modeling of BRCA1 protein and its quality analysis was analyzed using ProSA Z- plot and ProSA e- plot and they showed the good quality. Further, residues in the protein and their allotment in the favoured and unfavourable region were analyzed and their images were shown in Figure 2.

Molecular docking was performed for those 10 compounds which are likely cancer-causing carbamates, various binding energies for those molecules with the BRCA1 gene were obtained, and those affinities were measured in terms of Kcal/mol. These binding affinities are shown in Table 3.1.

Table 3.1 *Binding Energy values of carbamates with BRCA1.*

Sr. No.	Name of the Ligand	Binding Energy (kcal/mol)	Binding residues of BRCA1
1.	Propoxur	-4.6	Arg1699,Leu1701
2.	Triallate	-4.4	Gln1857
3.	Carbaryl	-5.7	Leu1701,Asn1774
4.	Alletrin	-5.8	Arg1695
5.	Mobam	-4.9	Leu1701
6.	Butacarb	-5.9	Arg1699,Leu1701
7.	Diallate	-5.8	Gln1857
8.	Fenoxycarb	-4.0	Gln1798

The lowest binding energy indicates the stronger binding affinity between the receptor (BRCA1) and ligands (carbamates) (See Figure 3). To screen their ability to cause breast cancer, Auto dock vina platform 4.0 was chosen. Carbamates were docked with the human BRCA1 gene using Autodock vina4.0. Docked results were visualized in Chimera[24] and Discovery studios. In the present study, out of the screened 10 carbamates for the molecular docking, except 2 carbamates such as Cloethocarb and camphechlor, remaining all 8 compounds showed binding interactions with the BRCA1, and those interactions were shown in the Table 3.1. Where the carbamate, propoxure made the hydrogen bond binding interactions using with leucine -1701, triallate made three hydrogen bonds with two amino acids such as leucine-1854 and with glycine-1825, carbaryl with leucine-1701, allethrin with the amino acids of both leucine-1701 and glycine-1656. Mobam made the single hydrogen bond interactions with leucine-1701, butacarb made two hydrogen bonds with glycine-1656 and leucine-1701, fenoxycarb two hydrogen bonds with leucine-1705 and lysine-1711 and finally diallate two hydrogen bond interactions with glycine-1825 and leucine-1854.

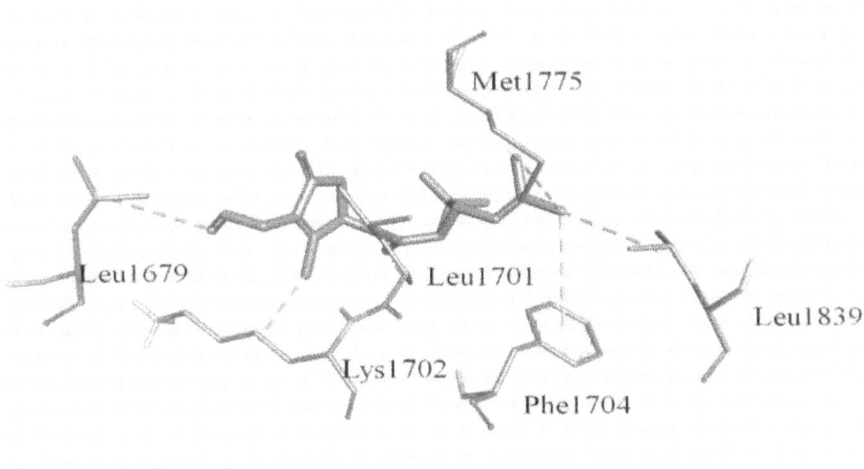

Figure 3 *Interaction of BRCA1 gene with carbamate (Allethrin) interactions obtained using Discovery Studio (Green amino acids are in Van Der Waal forces; Violet amino acids are in π-π stacking).*

The carbamates interacted with protein targets and their biological process implications were illustrated in the Figure 4.

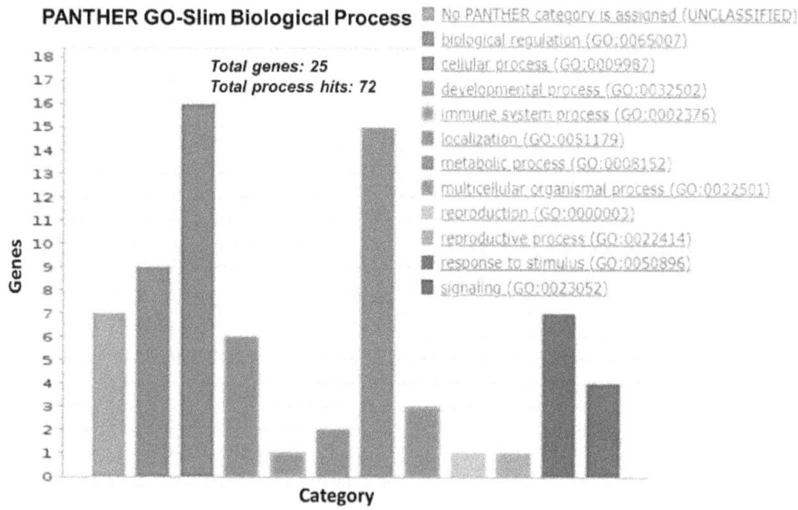

Figure 4 *Targeted proteins involved in the biological process.*

Most importantly from Kyoto Encyclopaedia of Genes and Genomes (KEGG) pathway of panther database shown us different pathways that were affected by the carbamates and it has been illustrated in the Figure 5.

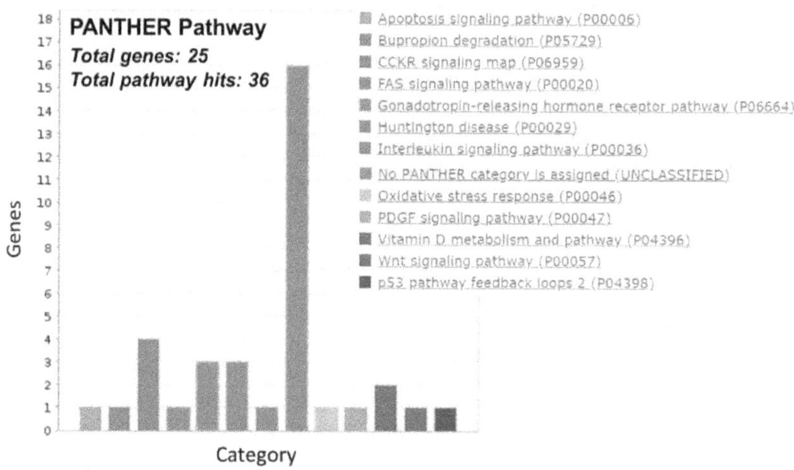

Figure 5 *Targeted signaling pathways effected by the carbamates.*

Interestingly, KEGG pathway indicated the some of the pathways which were linked with the cancer incidence were chiefly effected. Those cancer connected pathways were the WNT signaling pathway and P53 signaling pathway. It is well known that p53 is one of the most recognized proteins that participate in cancer incidence, its role in suppressing the tumour is most likely compromised by disturbing functions of genes or mutating environmental pollutants or by radiation. Some of the carbamates in this study were involved in targeting the p53 chemicals, which indicates, their role in inducing breast cancer.[25, 26]

4 CONCLUSION

The results of molecular docking analysis demonstrated that carbamates such as allethrin, butacarb, carbaryl, diallate, fenoxychlor, mobam, propoxur and triallate compounds made the molecular interactions with BRCA1 gene, which provides a clue for their ability to cause breast cancer. From the present molecular docking studies, we hypothesize that the selected chemicals can harm the women's health by targeting the BRCA1 gene and might induce breast cancer, at least in part targeting the p53 and WNT signaling pathway. Furthermore, the present study acts as a platform to understand the plausible mechanisms underlying the toxic effects of selected chemicals on breast cancer incidence. However, we cannot conclude based on mere in silico studies hence to support this notion, in vivo studies involving molecular mechanisms should be conducted. Nevertheless, this study provides invaluable information on preliminary understanding about the toxic effect of carbamates.

5 ACKNOWLEDGEMENTS

The authors are thankful to management, Koneru Lakshmaiah Education Foundation for their encouragement and support for conducting this study.

6 CONFLICT OF INTEREST

All the authors declare that they have no conflicts of interest.

7 REFERENCES

1. C. A. Damalas and I. G. Eleftherohorinos, *International journal of environmental research and public health*, 2011, **8**, 1402-1419.
2. D. Hamilton, Á. Ambrus, R. Dieterle, A. Felsot, C. Harris, B. Petersen, K. Racke, S. S. Wong, R. Gonzalez and K. Tanaka, *Pest Management Science: formerly Pesticide Science*, 2004, **60**, 311-339.
3. S. Mostafalou and M. Abdollahi, *Archives of toxicology*, 2017, **91**, 549-599.
4. K. S. Schafer and S. E. Kegley, *Journal of Epidemiology & Community Health*, 2002, **56**, 813-817.
5. C. K. Winter, *Reviews of environmental contamination and toxicology*, 1992, 23-67.
6. K.-H. Kim, E. Kabir and S. A. Jahan, *Science of the total environment*, 2017, **575**, 525-535.

7. S. Mostafalou and M. Abdollahi, *Toxicology and applied pharmacology*, 2013, **268**, 157-177.
8. C. N. Pope, *Journal of Toxicology and Environmental Health Part B: Critical Reviews*, 1999, **2**, 161-181.
9. D. M. Soderlund, *Archives of toxicology*, 2012, **86**, 165-181.
10. S. Zhang, *Drug Design and Discovery*, 2011, 23-38.
11. X.-Y. Meng, H.-X. Zhang, M. Mezei and M. Cui, *Current computer-aided drug design*, 2011, **7**, 146-157.
12. Y. Miki, J. Swensen, D. Shattuck-Eidens, P. A. Futreal, K. Harshman, S. Tavtigian, Q. Liu, C. Cochran, L. M. Bennett and W. Ding, *Science (New York, N.Y.)*, 1994, **266**, 66-71.
13. J. M. Hall, M. K. Lee, B. Newman, J. E. Morrow, L. A. Anderson, B. Huey and M.-C. King, *Science (New York, N.Y.)*, 1990, **250**, 1684-1689.
14. E. M. Rosen, S. Fan and Y. Ma, *Cancer letters*, 2006, **236**, 175-185.
15. R. Roy, J. Chun and S. N. Powell, *Nature Reviews Cancer*, 2012, **12**, 68-78.
16. L.-s. Teng, Y. Zheng and H.-h. Wang, *Journal of Zhejiang University SCIENCE B*, 2008, **9**, 85-89.
17. C.-X. Deng, *Nucleic acids research*, 2006, **34**, 1416-1426.
18. M. Tirkkonen, O. Johannsson, B. A. Agnarsson, H. Olsson, S. Ingvarsson, R. Karhu, M. Tanner, J. Isola, R. B. Barkardottir and Å. Borg, *Cancer research*, 1997, **57**, 1222-1227.
19. S. Salentin, S. Schreiber, V. J. Haupt, M. F. Adasme and M. Schroeder, *Nucleic acids research*, 2015, **43**, W443-W447.
20. U. Consortium, *Nucleic acids research*, 2015, **43**, D204-D212.
21. D. Studio, *Accelrys [2.1]*, 2008.
22. A. Daina and V. Zoete, *ChemMedChem*, 2016, **11**, 1117-1121.
23. P. D. Thomas, M. J. Campbell, A. Kejariwal, H. Mi, B. Karlak, R. Daverman, K. Diemer, A. Muruganujan and A. Narechania, *Genome research*, 2003, **13**, 2129-2141.
24. E. F. Pettersen, T. D. Goddard, C. C. Huang, G. S. Couch, D. M. Greenblatt, E. C. Meng and T. E. Ferrin, *Journal of computational chemistry*, 2004, **25**, 1605-1612.
25. J. X. Chen, Y. Zheng, M. West and M.-s. Tang, *Cancer research*, 1998, **58**, 2070-2075.
26. G. P. Pfeifer, M. F. Denissenko, M. Olivier, N. Tretyakova, S. S. Hecht and P. Hainaut, *Oncogene*, 2002, **21**, 7435-7451.

IN SILICO PHARMACOKINETIC AND PHARMACODYNAMIC SCREENING METHODS OF PHYTOL; AN ANTITUBERCULAR LEAD DISCOVERY STUDY

Fathima Asra[a, b], Daveedu Thathapudi[a], Fathima Ajra[b], Sukeerthi Katari[a], Naresh Dumala[a] *

[a]Department of Pharmacy, Koneru Lakshmaiah Education Foundation, Vaddeswaram, Guntur-522302, Andhra Pradesh, India, [b]Nimra Institute of Medical Sciences, Nimra Nagar, Ibrahimpatnam, Vijaywada-521456, Andhra Pradesh, India.

*Corresponding email address: nareshpharma2020@gmail.com

1 INTRODUCTION

Tuberculosis (TB) is among the serious infectious diseases of global concern for which the *Mycobacterium tuberculosis* (MT) is the causative agent that affects primarily the lungs of an infected person. Even after announcing this TB as a remains as a health emergency with global impact by the World Health Organization (WHO), the MT remained as deadly pathogen over decades of time.[1] In the 2020 year, it was estimated like 89 to 109 million people fell sick because of MT infection worldwide of which there were 11 lakh children. Overall 8 countries which include Bangladesh, China, India, Indonesia, Nigeria, Pakistan, South Africa and the Philippines accounted for two-thirds of all TB cases worldwide. Around 15 lakh people were died because of TB in the year 2020. Globally, the rate of new cases of tuberculosis decreased by 11% from 2015 to 2020 (for one lakh people from 142 in 2015 to 127 in 2020), which include a decrease of 1.9% from 2018 to 2019.[2] New medication combinations have recently demonstrated strong potential to dramatically reduce the length of time needed to cure TB.[3] However, relatively few novel chemical compounds are being created to combat this worldwide health concern. Hence, it is a necessity for novel chemical components which may be developed as possible TB-fighting compounds.

Phytol (3,7,11,15-Tetramethyl-2-hexadecen-1-OL), is an acyclic diterpene alcohol that is found in chlorophyll (See Figure 1). Phytol is a typical precursor in the production of synthetic variants of vitamin E and vitamin K1. Phytol is the most common acyclic diterpene found in the environment since it is a component of chlorophyll, and it may be generated by virtually all photosynthetic organisms such as algae, cyanobacteria, and plants.[4] In recent years, studies have been conducted on Phytol to study their beneficial effects and reports are available for this compound that are showing anticancer, anti-inflammatory, antimicrobial, antidiuretic, antidiabetic, antioxidant, and anticholinesterase abilities.[5, 6] However, very limited studies pertaining to phytol and its beneficial role in the TB are available. Knowing the phytol has excellent antibacterial and antioxidant properties, this study was intended to explore the its possible role in controlling the TB using *in silico* approaches.

The key objectives of this study are, 1) to explore the toxicokinetic properties of phytol and 2) to identify the binding ability of phytol with the key targets using molecular docking.

2 METHODS

2.1 In silico toxicokinetics screening

Absorption related parameters such as percentage of intestinal absorption, permeation through the skin, distributing patterns of the phytol through CNS, BBB (Blood brain barrier) and steady state volume of distributions were estimated.

The metabolism site of the phytol in the body was predicted by a web-service (Site of metabolism prediction (SOMP) for *in silico* metabolism site prediction.[7] For estimation of sites for metabolism, 5 important human cytochrome P450s (CYP3A4, CYP2C9, CYP1A2, CYP2D6 and CYP2C19) were examined for drug-like compounds. UGT-catalyzed glucuronidation sites were also included in the training set. The forecast is as per the PASS (Prediction of Activity Spectra for Substances) technology[8], and LMNA descriptors. Finally, the excretion of the phytol molecule was anticipated by pkCSM software[9].

2.1.1 Toxic and adverse effects prediction. The possible toxicity risk of phytol was predicted for mutagenicity, tumorigenecity, irritation and gonadal toxicity using DataWarrior[10], Cardiotoxicity and Skin sensitization were estimated by pkCSM and Endocrine toxicity was also predicted using Endocrine disruptome software.[11]

2.1.2 In silico acute rat toxicity of Phytol. LD50 values for rats with various administration routes (oral, intraperitoneal (IP), intravenous (IV), and subcutaneous (SC)) were predicted by GUSAR software (Institute of Biomedical Chemistry of the Russian Academy of Medical Sciences).[12]

2.1.3. Molinspiration bioactivity score. The bioactivity properties were calculated using Molinspiration tools, Molinspiration Cheminformatics, Slovensky Grob, Slovak Republic.[13]

2.2 Molecular docking study

Molecular docking investigations were conducted by Molegro Virtual Docker (MVD, Molegro ApS, Aarhus C, Denmark).

2.2.1 Preparation of target proteins. The proteins that were targeted for docking studies present study are isocitrate lyase enzyme (PDB ID: 1F8I), Mycobacterial DHFrase (PDB ID: 2CIG), and Mycobacterial DNA Gyrase (PDB ID: 4BAE). The 3D structures of DHFrase, DNA Gryase, Isocitrate lyase enzyme were extracted from the protein data database in .pdb file format. The target for docking study was selected as A-Chain of DHFrase, B-Chain of DNA Gryase and A-Chain of Isocitrate lyase enzyme. PDB files often lacks denotative hydrogen assignments, and the PDB file format cannot provide the bond order information. Hence, the MVD software was used to assign appropriate bonds, bond ordering, hybridization, and charges. MVD's cavity detection algorithm was used to calculate the potential binding sites of targets.

2.2.2 Preparation of ligand. ChemDraw Ultra 8.0 software (Cambridge Soft Corporation, Cambridge) was used to create the 3D structures, which were then stored in. mol format following energy reduction. In the workspace, the ligands are imported and prepared them.

2.2.3 Molegro Virtual Docker's docking search algorithms and scoring functions. The docking studies were conducted by MVD (See Figure 2). In a quick and flexible manner, docking with MVD provides the most likely configuration of ligand binding to a macromolecule. MolDock is a molecular docking algorithm, based on a novel heuristic search technique that fuses cavity prediction with differential evolution. MolDock's docking scoring function is an extension of the piecewise linear potential (PLP) that incorporates additional hydrogen bonding and electrostatic components. To boost docking accuracy even more, a re-ranking scoring function is implemented, which determines the most promising docking solution from the docking algorithm's solutions.[14] The parameters that were set for the molecular docking was, radius of 25 Angstroms was defined for the binding site on the protein that extended in the X, Y, and Z directions around the selected cavity. The docking scores of the Phytol were compared against the standard drugs, Succinic acid - a selective Isocitrate lyase enzyme inhibitor, INH- a DHFrase enzyme inhibitor, and Ciprofloxacin - a DNA Gryase enzyme inhibitor.

3 RESULTS AND DISCUSSION

3.1 In silico toxicokinetic screening

3.1.1 Absorption. Intestinal absorption is essential to elicit the biological activity of any orally administered drugs. It is predicted by the physicochemical properties responsible for the simple diffusion of a drug. Observation of *in silico* prediction reveals that Phytol absorbs orally.

Table 1 *Various parameters related to the permeability of Phytol. Permissible ranges: LogKp in negative: impermeable to skin; LogKp in positive: permeable to skin.*

Intestinal absorption (%)	P-gp substrate	P-gp I Inhibitor	P-gp II Inhibitor	Skin permeation (LogKp)
90.71	No	No	Yes	-2.576

3.1.2. P-glycoprotein substrate and Skin permeation. By eliminating medications absorbed in the intestines back into the gut lumen, P-glycoprotein protects the organism against hazardous xenobiotics. Drug removal from the kidneys and liver into urine and bile, respectively. Maintaining the blood-brain barrier's integrity. P-glycoprotein-1, also known as multidrug resistance protein 1, is a cell membrane protein that pumps several foreign chemicals out of cells. Phytol is not a substrate of P-glycoprotein hence the bioavailability and duration of action will not be affected by P-glycoprotein.

Skin permeability coefficient (Kp) property of Phytol was predicted by pkCSM software by multiple linear regression model to predict the possibility of various routes of administration and to design suitable dosage forms. It was found that Kp linearly correlated with molecular size and lipophilicity (R2 = 0.67). The molecules with more negative log Kp (with Kp in cm/s) values are less skin permeant. Phytol gained logKp in negative (Table 1), hence it may be impermeable to skin. Transdermal route of administration is selected based on the ability of Skin permeation of a drug. These results indicting that the Phytol may not be suitable for Transdermal patches. However, complimenting with *in vivo* experimental evidences may help in drawing firm conclusions.

3.1.3. Distribution. Using the pkCSM prediction method, we have calculated Blood–brain barrier permeation which measures a percentage of drugs that can permeate the brain. The results obtained for phytol is shown in Table 2.

Table 2 *Various parameters related to the distribution of Phytol. Permissible ranges: BBB permiation: (<0.1= low, 0.1-2 = medium, >2.0 = high absorption to CNS), VDss: Steady state volume of distribution.*

BBB Permeability (log BB)	CNS Permeability (log PS)	VDss in human (log L/kg)
0.806	-1.563	0.468

The percentage of BBB permeation was found to be no permeation to Blood–brain barrier. Steady state volume of distribution is one of the important pharmacokinetic parameter to predict the pattern of distribution and duration of drug action. CNS Permeability and Steady state volume of distribution of Phytol was found to be poor.

3.1.4. Metabolism. Biotransformation is one of the enzymatic processes of drug disposition which may leads to termination of therapeutic and toxic effects. There are five isoforms of cytochrome enzymes mainly involved in the biotransformation reaction. The interactions of Phytol with these enzymes were ranked based on their Delta P values and same are presented in the Tables 3 and 4.

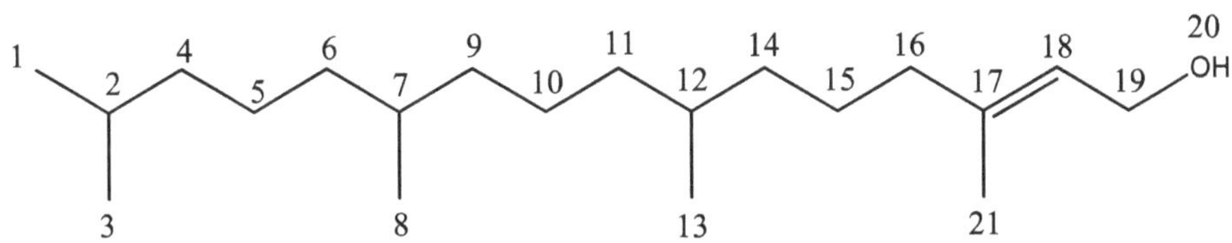

Compound CID: 5366244, MF: $C_{20}H_{40}O$ MW: 296.5g/mol,
IUPAC Name: (E)-3,7,11,15-tetramethylhexadec-2-en-1-ol,
Isomeric SMILES: CC(C)CCCC(C)CCCC(C)CCC/C(=C/CO)/C.

Figure 1 *Structural formula of Phytol with numbering of atoms.*

Metabolism impacts the drug efficiency and toxicity and is classified into two phases: Phase I (oxidation, hydrolysis, and reduction) and Phase II (conjugation). The Cytochromes P450 (CYP) family is the primary enzymes of Phase I that metabolize the majority of medicines. The major reaction of Phase II is glucuronidation, which is performed by UDP-glucuronosyltransferase (UGT) and acts as a clearance mechanism for medications from a variety of therapeutic classes. Prediction of sites of metabolism (SOM) is critical throughout the drug development process to predict potential 'metabolic hot spots' in lead compounds that may be changed to produce more metabolically stable molecules. The SOM prediction may be used to estimate a compound's activity and toxicity, including potential metabolites.

Table 3 *Effects of UGT, CYP3A4 and CYP2D6 enzymes on Phytol during biotransformation process.*

enzyme ***UGT***			enzyme ***CYP3A4***			enzyme ***CYP2D6***		
Atom No	Rank	DeltaP	Atom No	Rank	DeltaP	Atom No	Rank	DeltaP
20	1	0,861	19	1	0,859	18	1	0,582
18	2	-0,896	2	2	0,603	19	2	0,394
17	3	-0,882	7	3	0,424	5	3	0,282
16	4	-0,821	12	3	0,424	10	3	0,282
14	5	-0,777	10	4	0,159	16	4	0,167

Table 4 *Effects of CYP2C9, CYP2C19 and CYP1A2 enzymes on Phytol during biotransformation process.*

enzyme **CYP2C9**			enzyme **CYP2C19**			enzyme **CYP1A2**		
Atom Number	Rank	DeltaP	Atom Number	Rank	DeltaP	Atom Number	Rank	DeltaP
5	1	0,613	2	1	0,678	19	1	0,904
10	1	0,613	19	2	0,598	2	2	0,564
19	2	0,580	18	3	0,580	7	3	0,140
2	3	0,553	4	4	0,550	12	3	0,140
7	3	0,553	17	5	0,444	20	4	-0,986
12	3	0,553	7	6	0,376	21	5	-0,681
4	4	0,506	12	6	0,376	18	6	-0,585
15	5	0,475	5	7	0,241	14	7	-0,523

Site of metabolism predictor is a freely available web server that predicts the sites of metabolism (SOM) for five major P450 and UGT isoforms. It was created using a Bayesian technique using tagged multilevel neighborhoods of atoms descriptors to describe the structures of over 1000 metabolized xenobiotics based on chemical structural formulas and assessments of'structure-SOM' connections. The server predicts SOMs catalyzed by cytochrome P450 isoforms 1A2, 2C9, 2C19, 2D6 and 3A4 as well as enzymes from the UDP-glucuronosyl transferase family. The average invariant prediction accuracy computed for the training and evaluation sets (using leave-one-out cross-validation) is 0.9 and 0.95, respectively. Delta P was determined for each non-hydrogen atom in the chemical structure that was supplied to the prediction, and the atoms are listed in descending order of Delta P.[15] According to the top-ranking atom estimation, 15 SOM were predicted. The possible enzymes may be Cytochromes P450 3A4, 1A2 and 2D6 for phase 1 reaction and UDP-glucuronosyl transferase (UGT) for UGT glucuronidation in phase 2 was ranked as number one with high delta value.

3.1.5. Excretion. Clearance is defined as the rate at which a drug is eliminated from plasma (mg/min) divided by the drug's concentration in plasma (mg/mL). The body's overall capacity to remove a drug from plasma is the sum of renal clearance, and all other tissues involved in clearence. Organic cation transporters translocate endogenous and exogenous substances of cationic nature and, therefore, play an important role in the detoxification of exogenous compounds. *In silico* prediction reveals that phytol would be eliminated from the body by various excretory organs like kidney, liver. Total clearance for Phytol was determined as 1.686 log mL/min/kg. There was no renal organic cation transporter -2 substrate was involved in the clearance of Phytol.

3.2 *In silico* toxicity risk assessment

A toxic risk alert indicates that the drawn structure may be harmful according to the specified risk category. The major toxicity risk like, irritating effects, reproductive effects, mutagenicity and tumorigenicity were assessed by employing the DataWarrior v04.04.04. The toxicity potential was graded as high risk, medium risk and low risk.[10] Phytol is free from major toxicity risks mentioned above. Furthermore, Phytol is free from cardiotoxicity.

Skin sensitization is a complicated immunological reaction that has a significant influence on one's quality of life and capacity to function. Computational QSAR models offer a quick screening method and important information for assessing chemical toxicity. Pred-Skin web app version 1.0 is a free to use tool for predicting the skin sensitization abilities for chemical compounds. The outputs of this online software when analyzing a drug are, (i) binary predictions of human and murine skin sensitization potential; (ii) multiclass prediction of murine skin sensitization; and (iii) probability maps demonstrating the projected contribution of chemical fragments. Phytol is found to be non-sensitizer.

3.3 Acute Rodent Toxicity

Any chemicals that have not previously been present in the human diet to any great extent, the corresponding chemicals are considered to be novel with respect to human consumption. Therefore, the principle focus of the acute rodent toxicity study of chemical is to assess the safety of chemicals to human consumption as drug. LD50 values for rats with different routes of administration like oral, IV, IP and SC were predicted by General unrestricted structure-activity relationships (GUSAR) online software. Based on the appropriate training sets represented as MDL SDfiles with chemical structure data and quantitative endpoint information, GUSAR software was developed to create QSAR models. Phytol is nontoxic to oral and IP routes, but in case of IV and SC route, fall under the class of 4 and 5, respectively (Table 5). LD50 cut-off range for Class 4 drugs and Class 5 drugs are 300-2000 and 2000-5000 mg/kg body weight, respectively.

Table 5 *In silico prediction of LD50 values and acute rodent toxicity classification of Phytol by OECD project.*

Oral rat toxicity		**Acute Rodent Toxicity Classification**		
Acute toxicity (LD_{50}) mol/kg	Chronic toxicity (LOAEL) log mg/kg bw/day	Oral & IP	IV	SC
1.607	1.043	Non Toxic	Class 4	Class 5

LD_{50} cut-off range (mg/kg body weight): Class 1 drugs: 0-5, Class 2 drugs: 5-50, Class 3 drugs: 50-300, Class 4 drugs: 300-2000 and Class 5 drugs: 2000-5000.

3.3.1 Endocrine toxicity. Predicting a compound's endocrine disruption potential is challenging, but important in toxicity prediction studies. There is a simple, and new tool called Endocrine Disruptome that runs on an open source platform called Docking interface for Target Systems (DoTS) for this purpose. Molecular docking is conducted using AutoDock Vina and 18 nuclear receptor crystal structures are docked to 18 compounds: The predicted values are expressed in three different colours. They are green which shows that the predicted molecule has low probability of binding, red colour are with high probability of binding and yellow colour with intermediate probability of binding. Phytol secured low probability of binding (Table 6), hence it may be free from adverse events mediated by endocrine disturbances.

Table 6 P*rediction of endocrine receptors mediated adverse effects of Phytol.*

AR	G: -6.9	**LXR β**	G: -7.6
AR an	Y: -6.8	**MR**	Y:-7.0
ER α	G: -7.2	**PPAR α**	G: -6.3
ER α an	G: -6.6	**PPAR β**	G: -7.4
ER β	G: -7.2	**PPAR γ**	G: -1.7
ER β an	G: -7.1	**PR**	G: -7.6
GR	G: -6.4	**RXR α**	G: -7.6
GR an	G: -5.5	**TR α**	Y: -8.1
LXR α	G: -7.4	**TR β**	Y: -8.4

AR, Androgen receptor; ER α, Estrogen receptor alpha; ER β, Estrogen receptor beta; GR, Glucocorticoid receptor; LXR α, Liver X receptor alpha; LXR β, Liver X receptor beta; MR, Mineralocorticoid receptor; PPAR α, Peroxisome proliferator-activated receptor alpha; PR, Progesterone receptor; RXR α, Retinoid X receptor alpha; TR α, Thyroid hormone receptor alpha; TR β, Thyroid hormone receptor beta. G, Green colour: The predicted molecule has low probability of binding; Y, Yellow colour: intermediate probability of binding.

3.4 *In silico* quantitative prediction of bioactivity score by Molinspiration tool

Molinspiration assists by providing free on-line cheminformatics services for calculating crucial molecular characteristics and predicting bioactivity score for GPCR ligands, ion channel modulators, kinase inhibitors, nuclear receptor ligands, protease inhibitors, and other enzyme targets.

Table 7 *Molinspiration bioactivity Score of Phytol.*

Target proteins	Molinspiration bioactivity Score of Phytol
GPCR ligand	0.11
Ion channel modulator	0.16
Kinase inhibitor	-0.32
Nuclear receptor ligand	0.35
Protease inhibitor	0.00
Enzyme inhibitor	0.31

The larger the value of the score is, the higher is also probability that the particular molecule will be active. On the basis of the data presented in Table 7, Phytol has the highest molinspiration bioactivity score of 0.35 for nuclear receptor and 0.31 for enzyme. From these studies, it was evident that Phytol may act as enzyme inhibitor as well as nuclear receptor inhibitor.

3.5 Molecular docking study

In the present *in silico* analysis to evaluate the inhibitory effect of Phytol on isocitrate lyase enzyme, DHFRase, and DNA Gyrase, molecular docking study using molegro virtual docker was performed. Phytol found to have a good docking score of -119.779 (isocitrate lyase enzyme), -162.24 (DHFRase) and -126.82 (DNA Gyrase), respectively.

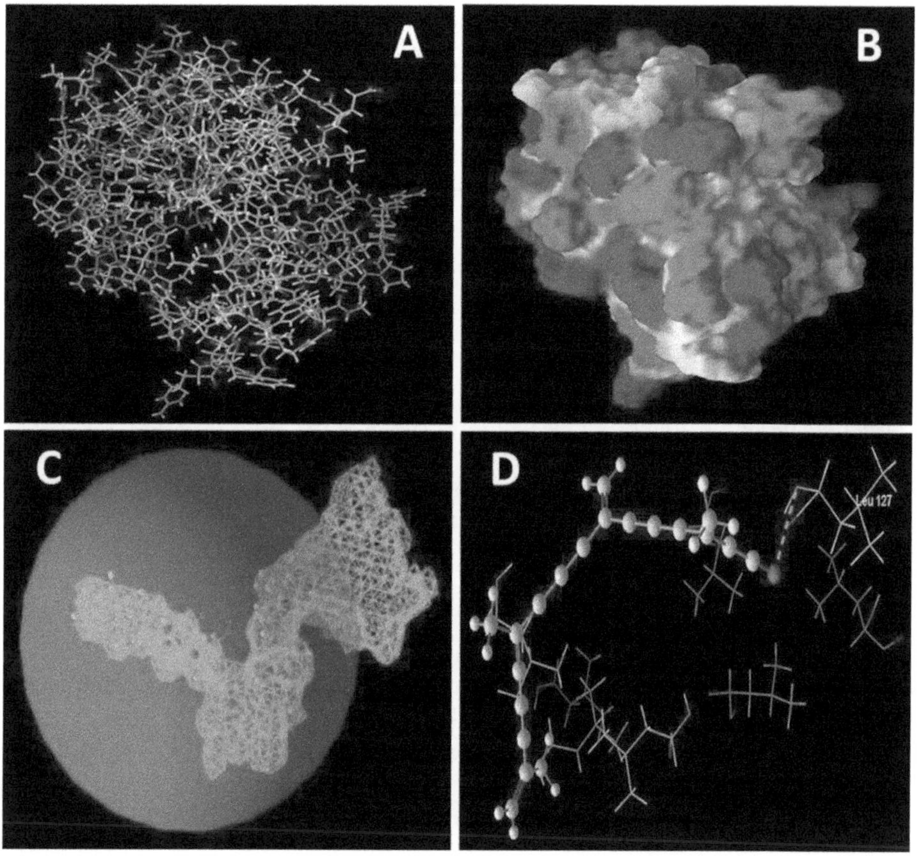

Figure 2 A. *Visualization of Mycobacterial DHFRase (2CIG) enzyme by MVD. 2B. Visualization of Mycobacterial DHFRase (2CIG) enzyme with surface by MVD. 2C. Visualization of Grid generated around the binding pocket in the Mycobacterial DHFRase (2CIG) enzyme by MVD. 2D. 2D plot of ligand protein (mycobacterial DHFRase (2CIG) Enzyme) interaction: Hydrogen bond interaction of Phytol with Leu-127 by MVD.*

Isocitrate lyase enzyme involved in the biosynthetic pathway of ATP in the Mycobacterium tuberculosis during dormant condition. Inhibition of Isocitrate lyase may lead to unavailability of ATP causes death of *Mycobacterium tuberculosis*. DHFRase, and DNA Gyrase enzyme inhibition interfere with nuclear division and bacterial replication. High docking score reveals that Phytol may avidly binds with these enzymes and produces antitubercular activity (Table 8).

Table 8 *Ligand – Target interaction parameters by MVD.*

Ligand	Target Protein (PDB ID) Moldock Score		
	Isocitrate lyase (1F8I)	*DNA Gyrase* (4BAE)	*DHFRase* (2CIG)
3,7,11,15-tetramethylhexadec-2-en-1-ol	-119.779	-126.82	-162.24
Succinic Acid (Co-crystalized ligand)	-76.5715	-	-
Ciprofloxacin (Co-crystalized ligand)		-124.357	
(4R)-Isonicotinic-Acetyl-Nicotinamide-Adenine Dinucleotide (Co-crystalized ligand)	-		-204.656

4 CONCLUSION

In the current study, the phytol was adjudicated for its possible anti tubercular activity, using *in silico* methods, and the obtained results were satisfactory in terms of its toxicokinetics profile and the binding interactions with different targets of *Mycobacterium tuberculosis* bacteria were promising. However, the detailed investigation should be conducted to test this hypothesis *in vivo* and *in-vitro* studies.

5 ACKNOWLEDGEMENTS

The authors are thankful to management, Koneru Lakshmaiah Education Foundation for their encouragement and support for conducting this study.

6 CONFLICT OF INTEREST

All the authors declare that they have no conflicts of interest.

7 REFERENCES

1. K. Patra, S. Batabyal, K. Mandal, D. Ghose and J. Sarkar, *Clinical Epidemiology and Global Health*, 2022, 101031.
2. W. H. Organization, 2021.
3. D. S. Reddy, A. Sinha, A. Kumar and V. K. Saini, *Archiv der Pharmazie*, 2022, **355**, 2200214.
4. N. Seddek, M. Fawzy, W. El-Said and M. Ahmed, *Global NEST Journal*, 2019, **21**, 329-337.
5. M. Das and M. Himaja, *Int J Pharm Pharm Sci*, 2014, **6**, 592-594.

6. G. A. Ismail, S. F. Gheda, A. M. Abo-shady and O. H. Abdel-karim, *Food Science and Technology*, 2019, **40**, 681-691.
7. A. Rudik, A. Dmitriev, A. Lagunin, D. Filimonov and V. Poroikov, *Bioinformatics*, 2015, **31**, 2046-2048.
8. S. Parasuraman, *Journal of pharmacology & pharmacotherapeutics*, 2011, **2**, 52.
9. D. E. Pires, T. L. Blundell and D. B. Ascher, *Journal of medicinal chemistry*, 2015, **58**, 4066-4072.
10. T. Sander, J. Freyss, M. von Korff and C. Rufener, *Journal of chemical information and modeling*, 2015, **55**, 460-473.
11. K. Kolšek, J. Mavri, M. Sollner Dolenc, S. Gobec and S. Turk, ACS Publications, 2014.
12. A. Lagunin, A. Zakharov, D. Filimonov and V. Poroikov, *Molecular informatics*, 2011, **30**, 241-250.
13. E. P, *Molinspiration Cheminformatics*, 2014.
14. R. Thomsen and M. H. Christensen, *Journal of medicinal chemistry*, 2006, **49**, 3315-3321.
15. J. Kirchheiner and J. Brockmöller, *Clinical pharmacology and therapeutics*, 2005, **77**, 1-16.

PART 2

DEVELOPMENT AND VALIDATION OF DRUGS AND DOSAGE FORMS BY ANALYTICAL AND BIOANALYTICAL TECHNIQUES

DEVELOPMENT AND VALIDATION OF HPLC METHOD FOR QUANTIFICATION OF PARECOXIB IN PARECOXIB INJECTION

Haranatha Babu Balanagu[1], Tatavarti SVS Gopala Krishna[2], Ramesh Cheruvu[2], Buchi N. Nalluri[1]*,

[1]Department of Pharmacy, Koneru Lakshmaiah Education Foundation, Vaddeswaram, AP, pin 522302, India. [2]Alphamed Formulations Private Limited, Hyderabad, Telangana, 500101

*Corresponding Author E-Mail ID: buchinalluri@kluniversity.in

1. INTRODUCTION

Parecoxib sodium is an amide prodrug of Valdecoxib, a nonsteroidal anti-inflammatory drug (NSAID) used for the management of post operative pain and also used in the treatment of acute renal colic. It is marketed as Powder for Solution for Injection prepared by Lyophilization process. Parecoxib sodium is an amide prodrug of valdecoxib. Chemical name of parecoxib is N-[[4-(5-methyl-3-phenyl-4 isoxazolyl)phenyl] sulfonyl] propanamide, sodium salt having molecular formula of $C_{19}H_{17}N_2O_4SNa$ with a molecular weight of 392.40 g/mol. No chiral centers are present in Parecoxib Sodium drug substance. Hence, Parecoxib sodium drug substance does not exhibit potential isomerism. Parecoxib sodium is a freely water soluble, slightly in tetrahydrofuran and completely insoluble in methylene chloride.

Literature search indicated that, compendial monograph are not available for Parecoxib sodium drug substance and Parecoxib sodium Injection in the IP, USP and BP. A very few research articles were available on UPLC and HPLC methods and bio-analytical methods for the estimation of Parecoxib. Most of the analytical literature is about bio-analytical methods for measuring Parecoxib in plasma samples and pharmacodynamic studies of Parecoxib Injection. Shaikh et al. made and tested a new RP-HPLC method for measuring

Parecoxib in human plasma and drug formulations. The method used an internal standard (Ibuprofen) to measure how much Parecoxib was in the sample[1]. Saccomanni et. al developed a LC-MS method for the simultaneous quantification of the parecoxib and valdecoxib in canine plasma[2]. Chen et al. developed UPLC-MS/MS method for the determination of parecoxib and valdecoxib in rat plasma samples[3]. Jain evaluated application of intravenous parecoxib on relief of acute post-surgical pain[4]. A very few references are available for the HPLC based quantification of Parecoxib. Crane et al developed HPLC method for the stability evaluation of reconstituted Parecoxib for Injection with commonly used diluents[5]. Dhananjay et al. developed a method and validated stability indicating UPLC-MS/MS method for the estimation of Related Compounds of Parecoxib Sodium in Parecoxib Sodium Injection[6]. Hence, in the present research, it was proposed to develop and validate a new RP-HPLC method for the routine analysis of Parecoxib in Quality Control laboratories.

2. MATERIALS

Parecoxib as a gift sample was received from M/s MSN Organics Pvt. Ltd., India. Disodium hydrogen phosphate anhydrous was a gift sample from Alphamed Formulations Pvt. Ltd., Hyderabad, India. Acetonitrile and Methanol were HPLC grade purchased from Ranchem and all the other chemicals used for this study were of analytical grade.

2.1 Methods

2.1.1 Instrumentation. HPLC system with PDA detector (Make & Model: Waters e 2695 with PDA detector; Software: Empower) was used in the quantification of Parecoxib in Parecoxib Injection.

2.1.2 Chromatographic conditions. The column was a reversed phase Symmetry Fluoro-phenyl column(150mm×4.6mm;3.5μ; Waters Corporation). The gradient elution mode with a Mobile Phase-A (pH 3.0 buffer: Methanol::30:70) and Mobile Phase-B (Water: Acetonitrile::10:90) at a flow rate 1.2 mL/min were used. Parecoxib was detected at 215 nm[5] with a retention time of 18 min. See Table 1.

Table 1: *Gradient Program of Mobile phase used in Assay analysis and Retention times of API and impurities.*

Gradient Program of Mobile phase		
Time (minutes)	**% Mobile Phase-A**	**% Mobile Phase-B**
0.01	98.0	2.0
20.0	80.0	20.0
32.0	50.0	50.0
35.0	50.0	50.0
35.1	98.0	2.0
58.0	98.0	2.0

2.1.3 Preparation of solutions:

Diluent: Purified water

0.9% Sodium chloride Solution. About 0.9 g of sodium chloride was accurately weighed and put into a 100 mL volumetric flask. 30 mL of water for injection was added, and the mixture was stirred well to dissolve the salt. Added enough water to make the right volume for injection and mixed well.

Preparation of Standard Solution. About 42 mg of Parecoxib sodium standard was accurately weighed and put into a 100 mL volumetric flask. Then, 30 mL of diluent was added, and the flask was shaken with a sonicator to dissolve the standard. Diluted to the right volume with diluent and mixed well.

Preparation of Test Solution. Dissolved 2.0 mL of 0.9% sodium chloride solution in a sample vial using ultrasonication. The solution from five separate vials was combined. The solution was transferred into a 250 mL volumetric flask, diluted to volume with diluent, and thoroughly mixed.

Preparation of Placebo Solution. 5 mL of placebo solution was placed into a 250 mL volumetric flask, the volume was filled with diluent, and the solution was mixed well.

Preparation of pH 3.0 Buffer. 1.15 g of Ammonium dihydrogen phosphate and 3.0 g of 1-Octane sulfonic acid sodium salt were weighed and put into a beaker with 1000 mL of water, where they dissolved. We used diluted phosphoric acid to get the pH to 3.0±0.05, and then filter the solution through a 0.45 membrane filter.

Preparation of Mobile Phase-A. A mixture of pH 3.0 buffer and methanol that was 70:30 by v/v was made.

Preparation of Mobile Phase-B. Water and acetonitrile were mixed together in a 10:90 v/v ratio.

Blank. Diluent (i.e., purified water) was used as blank.

Analytical Procedure. Separately, 10 µl of the blank solution, the placebo solution, the standard solution, and the test solution were each injected into the chromatographic system and left to run for a total of 58 minutes with the mobile phase gradient program. During the procedure sample temperature of 25°C was maintained and a column temperature of 50°C was maintained.

2.2 Method Validation:

As per ICH requirements, an HPLC technique for the measurement of Parecoxib was designed and validated for important validation parameters.

2.2.1 Specificity and System suitability. Specificity of an analytical method is its capacity to distinguish between the analyte and other components of a complex combination. Separate injections of 10 µl solutions of blank, standard, sample, and placebo were used to evaluate the method's specificity.

Following system suitability criteria as considered during validation of HPLC method for quantification of Parecoxib Injection.

1. The column efficiency as calculated for the Parecoxib sodium peak in the standard solution must not be less than 2000 theoretical plates.
2. The tailing factor should not exceed 2.0 for the same.
3. The percent relative standard deviation (%RSD) of the area response of the Parecoxib sodium peaks from five replicate injections of standard solution should not exceed 2.0.

2.2.2 Forced Degradation. Forced degradation purposefully degrades the sample to test the HPLC method's capacity to quantify an active ingredient and its breakdown products without interference. Drug substance and drug product were forced to undergo acid, base, oxidative, thermal, hydrolytic, photolytic, and metal degradation during method validation. This study identifies drug substance and drug product degradation routes, develops and validates stability-indicating technique, and distinguishes degradation products in formulations related to drug substances from those related to non-drug substances (e.g., excipients).

2.2.3 Linearity. Parecoxib working standard was used to prepare solutions from 80% to 120% of target concentration to test the method's linearity. Injected three injections from each concentration into the chromatographic system under identical conditions. Triplicated all solution peak area responses were recorded.

2.2.4 Precision. Repeated sample solution injections determined technique precision. For procedure precision, injected six replicates of sample solutions into the chromatographic system (repeatability). Repeatability results were derived using %RSD.

2.2.5 Accuracy. Recovery studies at three concentration levels (80%, 100%, and 120%) were used to determine the assay method's precision. Prepared solutions in triplicate by adding placebo to Parecoxib at 80% and 120% of the intended concentration. To achieve a level of 100 percent, six placebo-drug mixtures were spiked with Parecoxib pharmacological material. For each replicate sample, the percentage recovery and percent RSD were computed.

2.2.6 Range. Using linearity, precision, and accuracy data, analytical method range can be determined.

2.2.7 Solution stability. By testing standard and sample solutions, the Solution Stability was assessed. Preparation and injection of standard and sample solutions into the chromatographic system. Standard and sample solutions were stored in the refrigerator and at room temperature, then evaluated after 24 and 48 hours relative to the fresh standard, respectively.

2.2.8 Robustness. The robustness of an analytical method denotes its dependability throughout routine application, i.e., its capacity to remain unaffected by minor, but deliberate alterations in method parameters, such as flow rate, pH of the mobile phase, and column temperature.

2.2.9 Mobile phase stability. It was evaluated by using initial day mobile phase on different days for the analysis of freshly prepared standard solution.

3. RESULTS AND DISCUSSION

3.1 Method Development and Optimization

Physical and chemical parameters of Parecoxib sodium were extracted from the scientific literature. The HPLC technique was created to determine reversed phase HPLC-PDA chromatographic conditions, such as sample preparation, detection wavelength, and mobile phase. In order to accomplish this, a series of tests were conducted by adjusting the ratio of Mobile Phase-A to Mobile Phase-B and optimizing the chromatographic conditions on the Water's Xselect CSH Fluoro-phenyl (150x4.6mm)3.5µm column. The mobile phases Mobile Phase-A (pH 3.0 buffer: Methanol:30:70) and Mobile Phase-B (Water: Acetonitrile:10:90) in gradient elution mode with a flow rate of 1.2 mL/min, injection volume 10 µl, run time 58 min, and column temperature 50°C at wavelength (λ) 215 were chosen as the optimal conditions for the entire study, where Parecoxib was eluted with symmetrical peak

3.2 Method Validation

3.2.1 Specificity and System suitability. Specificity was assessed by comparing the chromatograms of mobile phase blank, placebo solution, standard solution, and sample solution, each injected into the HPLC system at 10 µl. The results show no interference from coeluting peaks at Parecoxib sodium's retention time. The pure analyte peak proved the method's specificity. To ensure system suitability for analysis, parameters were monitored. Hence, standard preparations and % RSD were used to determine system accuracy. USP tailing (0.92), plate count (30534), and % RSD (0.50%) were likewise good. See Figure 1 and Table 2.

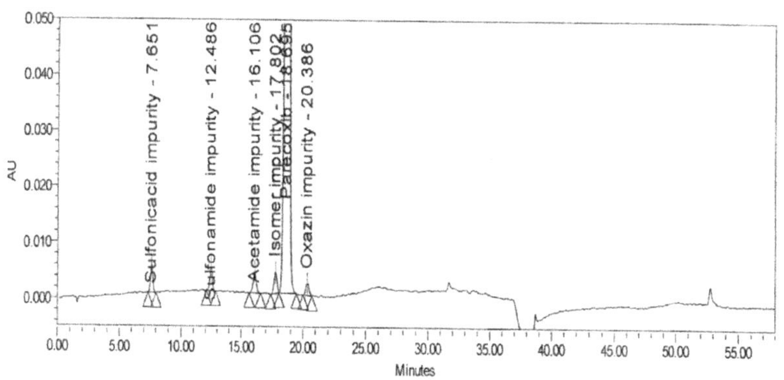

Figure 1: Chromatogram of spiked sample solution

Table 2: *Results of Precision and System Suitability*

Parameter	Acceptance Criteria	Results#
Theoretical Plates	NLT 2000	30534
Tailing Factor	NMT 2	0.92
% RSD	NMT 2%	0.50
#: *Data from five replicate injections*		

3.2.2 Forced Degradation. Forced degradation investigations on parecoxib API, drug product, and placebo samples demonstrated the method's specificity and stability-indicating power. Hydrogen peroxide, acid hydrolysis, base hydrolysis, heat, and photolytic stress were used to establish test solution specificity. The stressed samples were then diluted and chromatographed. Parecoxib peak purity is below the purity standard. See Tables 3–4. The peak purity data of Parecoxib peak for every degradation sample demonstrates that the peak was homogenous and had no co-eluting peaks, showing stability-indicating and specificity.

Table 3: *Forced degradation study results of Parecoxib API.*

Degradation Mechanism	Degradation Condition	% Assay	% Total Impurities	Purity angle	Purity threshold	Mass Balance
Undegraded sample	---	100.0	0.05	0.048	0.295	--
Acidic degradation	Heat for 2 hrs on water bath maintained at 90°C in 0.1 N HCl Solution	84.4	10.6	0.045	0.287	95.0
Base degradation	Heat for 2 hrs on water bath maintained at 90°C in 0.1 N NaOH Solution	97.3	1.7	0.080	0.335	99.0
	Heat for 2 hrs on water bath maintained at 90°C in 1.0 N NaOH Solution	70.9	23.3	0.038	0.263	94.2

Peroxide degradation	Kept on benchtop for 2 hrs in 3 % H_2O_2 Solution	88.0	5.1	0.101	0.270	93.1
	Kept on benchtop for 5 hrs in 3 % H_2O_2 Solution	75.2	14.8	0.045	0.266	90.0
Thermal degradation	Kept in an oven at 105°C for 48 hrs	98.8	0.05	0.060	0.289	98.9
	Kept in an oven at 105°C for 144hrs	98.8	0.04	0.040	0.328	98.8
Photolytic degradation	Exposed to 1.2 million lux hrs / 200 watts/m^2	100.1	0.09	0.063	0.295	100.2
	Exposed to 2.4 million lux hrs / 400 watts/m^2	96.1	0.04	0.113	0.308	96.1
Humidity degradation	Exposed to 90 % RH for 168 hrs	97.4	0.08	0.049	0.281	97.5
	Exposed to 90 % RH for 336 hrs	96.2	0.04	0.146	0.302	96.2
Hydrolytic degradation	Heat on water bath for 2 hrs at 90°C	97.9	0.10	0.070	0.285	98.0
	Heat on water bath for 6 hrs at 90°C	97.9	0.23	0.093	0.345	98.1
	Heat on water bath for 24 hrs at 90°C	97.5	0.70	0.067	0.331	98.2
Metal degradation	Kept on benchtop for 72 Hrs with $FeCl_3$	97.7	0.08	0.052	0.331	97.8

Table 4: *Forced degradation study results of Parecoxib drug product.*

Degradation Mechanism	**Degradation Condition**	**% Assay**	**% Total Impurities**	**Purity angle**	**Purity threshold**	**Mass Balance**
Undegraded sample	---	102.6	0.04	0.049	0.304	--
Acidic degradation	Heat for 2 hrs on water bath maintained at 90°C in 0.1 N HCl Solution	85.6	12.9	0.04	0.288	96.0

Base degradation	Heat for 2 hrs on water bath maintained at 90°C in 0.1 N NaOH Solution	99.7	1.3	0.049	0.342	98.4
	Heat for 2 hrs on water bath maintained at 90°C in 1.0 N NaOH Solution	77.9	19.1	0.040	0.274	94.5
Peroxide degradation	Kept on benchtop for 2 hrs in 3 % H_2O_2 Solution	94.3	6.1	0.040	0.284	97.9
	Kept on benchtop for 5 hrs in 3 % H_2O_2 Solution	78.3	14.5	0.032	0.273	90.5
Thermal degradation	Kept in an oven at 105°C for 48 hrs	98.7	1.8	0.052	0.294	98.0
	Kept in an oven at 105°C for 144hrs	96.3	2.1	0.045	0.324	95.9
Photolytic degradation	Exposed to 1.2 million lux hrs / 200 watts/m^2	99.9	0.06	0.064	0.296	97.4
	Exposed to 2.4 million lux hrs / 400 watts/m^2	99.6	0.50	0.039	0.335	97.6
Humidity degradation	Exposed to 90 % RH for 168 hrs	100.5	0.06	0.064	0.295	98.0
	Exposed to 90 % RH for 336 hrs	101.1	0.04	0.042	0.329	98.6
Hydrolytic degradation	Heat on water bath for 2 hrs at 90°C	100.2	0.11	0.063	0.296	97.8
	Heat on water bath for 6 hrs at 90°C	100.0	0.21	0.048	0.343	97.7
	Heat on water bath for 24 hrs at 90°C	101.6	0.70	0.058	0.361	99.7
Metal degradation	Kept on benchtop for 72 Hrs with $FeCl_3$	98.1	0.05	0.048	0.328	95.7

3.2.3 Linearity. The analytical method's linearity was tested to ensure that test findings are directly proportional to analyte concentration in samples within a specific range. Hence, different concentrations of Parecoxib standard working solution were produced and injected into the chromatographic apparatus to record chromatograms. Plotting peak regions versus concentrations determined detector response linearity. The correlation coefficients for parecoxib and placebo matrix are >0.999 and 0.999, respectively, indicating a linear relationship between concentration and peak response.

3.2.4 Precision. Six sample solutions were injected into the chromatograph to test HPLC technique precision. Six sample preparations had RSD values below 2%. The method's low RSD indicates precision.

3.2.5 Accuracy. The devised HPLC method for Assay test parameter accuracy reflects closeness of agreement between the expected value and the obtained value. Calculating analyte recovery % yielded it. Recovery experiments at three concentration levels by spiking the Parecoxib drug material to placebo at 80%, 100%, and 120% of target concentration assessed the HPLC method for assay accuracy. The quantities recovered and % RSD were obtained from the analyte peaks [Table 5]. Analyte recovery is 98.9-101.8 with an RSD of 0.06–1.32%. It's accurate and precise.

Table 5: *Results of Accuracy.*

Sample Concentration	**% Recovery**	**Mean % Recovery**	**SD**	**%RSD**
80%	99.5	99.9	0.40	0.40
	99.9			
	100.3			
90%	101.8	100.3	1.33	1.32
	99.5			
	99.5			
100%	99.4	99.6	0.24	0.25
	99.4			
	100.0			
	99.7			
	99.4			
	99.7			
110%	99.4	99.2	0.26	0.27
	99.3			
	98.9			
120%	100.2	100.1	0.06	0.06
	100.1			
	100.1			
Overall Mean % Recovery	99.8	---	---	---
Overall SD	0.62	---	---	---
Overall % RSD	0.63	---	---	---

3.2.6 Range. Linearity, precision, and accuracy determined analytical method range. The devised HPLC method quantifies Parecoxib in samples from 80 to 120%, with good precision, accuracy, and linearity.

3.2.7 Solution stability. Assay results were 100.9% to 102.1%, similarity factor for standard was 0.98–1.02, and % assay difference was less than 2.0%, demonstrating satisfactory stability of the standard and sample solutions for 48 hr at room temperature and refrigerated storage. See Table 6.

Table 6: *Difference in % Assay during solution stability.*

Time in Hours / Storage Condition	% Assay	Difference
Initial	101.5	--
About 24 hrs at Room Temperature Condition	100.9	0.6
About 24 hrs at Refrigerated Condition (2-8°C)	102.1	0.6
About 48 hrs at Room Temperature Condition	101.8	0.3
About 48 hrs at Refrigerated Condition (2-8°C)	101.6	0.1

3.2.8 Robustness. The developed HPLC method for assay test parameter's robustness was tested by examining how minor HPLC circumstances affected system suitability parameters. Robustness testing indicated that moderate changes to method conditions such mobile phase pH, column, and flow rate are acceptable [Table 7]. In all variations, Parecoxib peak area RSD from five replicate injections of the standard solution was less than 2.0%. The tailing factors and theoretical plates were likewise acceptable.

Table 7: *Results of Robustness.*

<table>
<tr><th colspan="6">System Suitability for Robustness</th></tr>
<tr><th colspan="2">Parameter</th><th colspan="2">Theoretical plates</th><th>Tailing Factor</th><th>% RSD of peak area response</th></tr>
<tr><td colspan="2">System suitability as per test procedure</td><td colspan="2">30413</td><td>1.06</td><td>0.11</td></tr>
<tr><td rowspan="2">Flow rate (± 10 %)</td><td>From 1.2 to 1.32 mL/min</td><td colspan="2">29578</td><td>1.18</td><td>0.16</td></tr>
<tr><td>From 1.2 to 1.08 mL/min</td><td colspan="2">37139</td><td>1.23</td><td>0.28</td></tr>
<tr><td rowspan="2">Column temperature (± 5°C)</td><td>From 50°C to 55°C</td><td colspan="2">26412</td><td>1.08</td><td>0.34</td></tr>
<tr><td>From 50°C to 45°C</td><td colspan="2">28056</td><td>1.26</td><td>0.20</td></tr>
<tr><td rowspan="2">Mobile phase pH(± 0.2)</td><td>From pH 3.0 to 3.2</td><td colspan="2">34213</td><td>1.15</td><td>0.45</td></tr>
<tr><td>From pH 3.0 to 2.8</td><td colspan="2">32039</td><td>1.20</td><td>0.24</td></tr>
<tr><th colspan="6">Robustness</th></tr>
<tr><th colspan="2">Parameter</th><th colspan="3">% Assay</th><th>Overall % RSD</th></tr>
<tr><td colspan="2" rowspan="2">Precision as per test procedure</td><td>100.7</td><td>101.3</td><td>101.0</td><td rowspan="2">--</td></tr>
<tr><td>101.5</td><td>101.2</td><td>100.8</td></tr>
<tr><td></td><td>From 1.2 to 1.32 ml/min</td><td>100.6</td><td>100.4</td><td>100.6</td><td>0.37</td></tr>
</table>

<table>
<tr><td>Flow rate (± 10 %)</td><td>From 1.2 to 1.08 ml/min</td><td>99.9</td><td>100.6</td><td>100.5</td><td>0.48</td></tr>
<tr><td rowspan="2">Column temperature (± 5°C)</td><td>From 50°C to 55°C</td><td>100.3</td><td>100.9</td><td>100.8</td><td>0.36</td></tr>
<tr><td>From 50°C to 45°C</td><td>101.1</td><td>100.4</td><td>100.4</td><td>0.39</td></tr>
<tr><td rowspan="2">Mobile phase pH(± 0.2)</td><td>From pH 3.0 to 3.2</td><td>100.9</td><td>100.7</td><td>100.5</td><td>0.32</td></tr>
<tr><td>From pH 3.0 to 2.8</td><td>100.2</td><td>100.6</td><td>100.4</td><td>0.43</td></tr>
</table>

3.2.9 Mobile phase stability. The peak regions of Parecoxib from five replicate injections of the reference solution were within 2.0% RSD. Tailing factors and theoretical plates were likewise within acceptable limits, showing two-day mobile phase stability.

4. CONCLUSION

On the basis of the obtained results, the developed and validated HPLC method for the Assay test parameter of Parecoxib sodium is straightforward, accurate, precise, and indicative of linear stability, and may therefore be used for routine quality control analysis. The parameters of the analytical technique and the solvents of the mobile phase produced excellent resolution for Parecoxib sodium. The approach was validated according to ICH standards. The stability-indicating HPLC technique devised suggests that Parecoxib degrades under acidic and basic environments. Moreover, it undergoes oxidative and thermal deterioration. The approach is robust enough to reproduce precise and accurate results under a variety of chromatographic conditions. The validation data satisfies the acceptance criteria for the validation parameters in accordance with the current ICH recommendations.

5. ACKNOWLEDGMENT

The authors are thankful to Alphamed Formulations Private Limited, Hyderabad for providing the facilities to carry the work.

6. REFERENCES

1. Shaikh SM, Manjunatha DH, Seetharamappa J, Kandagal PB. *Analytical letters*. 2007;**40**(15):2925.
2. Saccomanni G, Giorgi M, Del Carlo S, Manera C, Saba A, Macchia M.. *Analytical and bioanalytical chemistry*. 2011;**401**:1677.
3. Chen M, Sun W, Wang Z, Huang C, Hu G, Chen Y, Wang L. *BMC Pharmacology and Toxicology*. 2020;**21**(1):1.
4. Jain KK. *Expert Opinion on Investigational Drugs*. 2000;**9**(11):2717.

5. Crane IM, Mulhern MG, Nema S. *Journal of clinical pharmacy and therapeutics*. 2003;**28**(5):363.
6. Dhananjay, P. D., Ravi Kumar, B., Sridhar, Y., Prof. Sriramulu, *IOSR J Pharm. Bio. Sci.*, 2017, **12**(4), 58.
7. ICH Guideline, "*Validation of analytical procedures: text and methodology*," in Proceedings of International Conference on Harmonization, Topic Q2 (R1), Geneva, Switzerland, November 2005.
8. FDA, CDER, Beers, and Donald, *Analytical Procedures and Methods Validation for Drugs and Biologics Guidance for Industry*, FDA, Silver Spring, MD, USA, 2015.

QUANTIFICATION OF DARIDOREXANT IN HUMAN PLASMA USING RP-HPLC: A BIOANALYTICAL METHOD DEVELOPMENT AND VALIDATION

Bonagiri Pavani[a], Narender Malothu[*b], D S N B K Prasanth[c], Chakravarthi Guntupalli[b]

[a]Research Scholar, Department of Pharmacy, Koneru Lakshmaiah Education Foundation, Vaddeswaram, AP, India; [b]Department of Pharmacy, Koneru Lakshmaiah Education Foundation, Vaddeswaram, AP, India; [c]Department of Pharmacognosy, KVSR Siddhartha College of Pharmaceutical Sciences, Vijayawada, AP, India-520010.

***Corresponding author:**narendermalothu@gmail.com

1. INTRODUCTION

A medical condition called insomnia involves having issues falling and staying asleep.[1] There are plenty of opportunities to sleep, but the quality or length of the sleep is subpar. According to estimates, over one-third of adult Americans experience insomnia.[2] Benzodiazepines, nonbenzodiazepines, GABA-A modulators, sedative antidepressants (like trazodone and amitriptyline), and over-the-counter sleeping pills, which often contain first-generation sedative antihistamines and are sometimes combined with an analgesic, are the most common medications used to treat insomnia.[3-5] These treatments are thought to affect neurotransmitters involved in promoting sleep.[5] The antagonism of the orexin system has recently been discovered as a potential novel mechanism for modulating sleep. Neurons in the lateral part of the hypothalamus in the brain make orexin A and orexin B, which are neuropeptides.[6,7] It is believed that these neuropeptides cause wakefulness because they interact to orexin receptors 1 and 2. Several experimental results support the idea of the function of the orexin system.[8] Because of the interest in the orexin system, orexin receptor antagonists have been synthesised and tested as possible therapies for insomnia. Many of these drugs have already gone through clinical trials.[9] Daridorexant (DXT, Figure 1a), an orexin receptor 1 and 2 antagonist, was recently approved by the FDA for the treatment of insomnia.[10, 11]

Despite the prevalence of sedative-hypnotics in forensic toxicology investigations, there have been few studies documenting the DXT examination of biological fluids. Following a 50 mg oral dosage, the highest concentration is often greater than 950 ng/mL. A very sensitive test is necessary as forensic toxicology laboratories frequently have to detect a drug several hours after a single administration. Thus, a valid test for DXT assessment in human plasma was required to assist the therapeutic medicine's commercialization and impending clinical studies of the drug, enabling the characterization of the compound's clinical pharmacokinetics. The present study describes a human plasma DXT test that begins with liquid-liquid extraction and ends with RP-HPLC with PDA detection. This is, to the best of our knowledge, the first such technique to emerge in the literature. Suvorexant was used as an internal standard (IS), as it possesses structural similarities to the analyte (Figure 1 b).

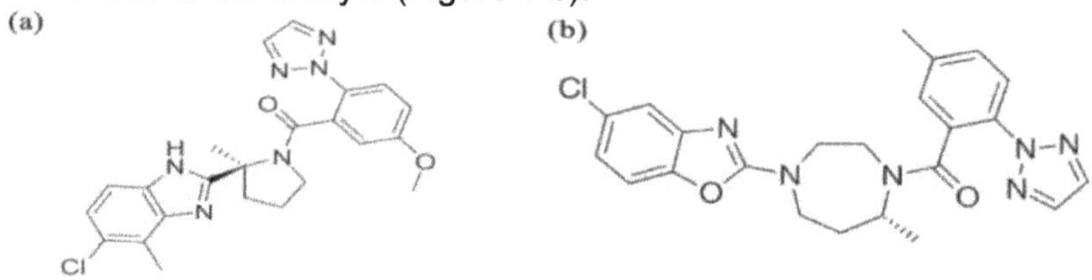

Figure 1. Chemical structures of (a) DXT; (b) Suvorexant.

2. MATERIAL AND METHODS

2.1 Materials

DXT and Suvorexant were collected from Maruthi Drugs Pvt. Ltd., Hyderabad, India as gifted samples. Methanol, acetonitrile (HPLC grade), n-butyl chloride, sodium acetate (99.9%) and formic acid have been bought from Sigma-Aldrich (Milwaukee, WI, USA). Dipotassium EDTA human plasma seemed to be purchased from Bruhada Blood Bank, Kadapa, India. A Millipore Milli-Q plus system (Bedford, MA, USA) has been used to start generating deionized water.

2.2 Instrumentation

A HPLC was provided by Waters LC (Milford, Massachusetts, USA). A quaternary gradient system (600 controller), as for degasser (waters, model AF), PDA detector (waters, model 2998), as well as an auto sampler, have all been included in the structure (waters, framework 717 plus). The data must have been handled by employing inspire pro (Waters, Milford, MA, USA).

2.3 Optimized Chromatographic conditions

The Inertsil ODS column (250 x 4.6 mm, 5μ) has been used for the chromatographic separation and has been preserved at 25°C. An isocratic mobile phase of 0.5 mm sodium acetate buffer pH 3.5: methanol (78: 22 v/v) has been moved through to the article at an air flow of 0.5 mL/min. The isosbestic wavelength of recognition of such compounds has been 225 nm.

2.4 Preparation of standards

A DXT stock solution of 100 µg/mL was created and serially diluted to yield standard working solutions of 0.5, 1, 2, 4, 7, 10, 17, and 20 µg/mL. A stock solution of 25 µg/mL internal standard was produced and diluted to a working solution of 8 µg/mL. During the mobile phase, all solutions were prepared. The solutions were kept in ordinary amber glass bottles at 4 °C.

About 50 µL of working standard stock and 50 µL of working IS solution in 100 µL of human Dipotassium EDTA plasma were combined to make plasma standards. These standards were used to measure clinical plasma samples with concentrations ranging from 0.25 to 10 µg/mL. Clinical samples that had a concentration that was too high for the calibration range were diluted with plasma and tested again.

2.5 Preparation of Quality control (QC) samples

A separate DXT stock solution was produced in the mobile phase at a concentration of 50 µg/mL. This solution was then diluted to produce solutions containing 0.25, 1.5, 10, and 17 µg/mL. These solutions were diluted further in QC samples with concentrations of 0.13 µg/mL (LLOQ), 0.75 µg/mL (low), 5.00 µg/mL (mid), and 8.50 µg/mL (high). 0.5 mL of each solution was transferred to a 50 mL volumetric flask, which was then completely filled with human plasma. 300 µL aliquots of plasma QC solutions were put into 3.6 mL polypropylene cryotubes. The tubes were then sealed and stored at -20 ° C.

2.6 Extraction procedure

Calibration and QC samples were collected and seemed to be fully ready by fortifying 0.5 mL of plasma with the appropriate volume of DXT working standard inside a 10 mL small glass round bottom glass centrifuge tube. It was indeed controlled to increase to a final concentration of 8 µg/mL. Following fortification of the plasma, 1 mL of formic acid and 2.5 mL of n-butyl chloride has been added, respectively. The samples were put on a rotating mixer for 5 min to achieve homogeneity. Furthermore, the samples were centrifuged at 3000 rpm for 5 min. An organic layer was removed and placed in a single conical glass container, which was then faded to dryness with nitrogen (50°C). The extracts were reconstructed with 30 µL of mobile phase, of which 20 µL had been implanted in and out of RP-HPLC.

2.7 Bioanalytical method validation

A method validation protocol has been prepared based on the guidelines of the USFDA.[12,13] Linearity, carryover, recovery, the matrix effect, precision, accuracy, and stability all seemed to confirm the exact relationship. Calibration curves were attained by spiking the blank matrix with a known concentration of DXT to provide concentrations of 0.25, 0.5, 1, 2.25, 3.63, 5, 8.5 and 10 µg/mL. The impact of carryover has even been evaluated by analyzing drug-free plasma (n =6) on the basis of such a blood survey spike as well as the 0.25 and 10.00 µg/mL.A lack of positive response or a response less than LLOQ in the smoke free fusion following its spiked samples at the highest concentration was assessed as the concentration not directly affected by the carryover effect. Its calibration plot of absorption spectra versus concentration (µg/mL) of DXT had been plotted. For such a calibration plot, the regression variable must have been calculated, and the concentrations of added variety must have been interlaced first from regression parameters. Sample concentration was calculated through a simple regression, just using the template Y = mx + b, under which y = peak area, x = concentration of standard in µg/mL, m = the slope of both the curve and b=the intercept with the y-axis. Recovery (%) must have been evaluated[14].

Within-run and between-run accuracy and precision seem to have been assessed through production QC samples but rather determined through it copy analysis using four determinations of different initial concentrations: LLOQ (0.13 µg/mL), low QC (0.75 µg/mL), medium QC (5.00 µg/mL), and high QC (8.50 µg/mL).

The stock solution of DXT (100 µg/mL) and thus IS (50 µg/mL), as well as high and low QC tests, were performed at room temperature conditions for an extended period of time - still having situations (-20°C), but rather free-thaw stability. Short-term stability determinations were obtained without thawing QC specimens but rather by keeping them at room temperature for 24 h, whereas long-term stability was evaluated by storing the entire tests for 30 days at -20 °C. Freeze-thaw stability has been reviewed after each cycle for something like a maximum of three cycles.

3. RESULTS AND DISCUSSION

The criteria that would optimize the applicability of future HPLC experiments for metabolite identification were determined by analyzing the makeup of the mobile phase. Formic acid and orthophosphoric acid were evaluated at various concentrations (0.1%, 0.5%, and 1%). Two different acids were used to change the pH of the mobile phase. At concentrations between 0.1 and 0.5%, the retention times of DXT and IS were achieved at 2.56 and 6.55 min, respectively. However, a greater concentration of acid-induced impurities in the matrix masked the analytes. In terms of chromatographic separation and analyte detection using the PDA detector, orthophosphoric acid has been determined as the best candidate, with a mobile phase flow rate of 0.5 mL/min offering excellent peak clarity and separation. These settings were developed for the Inertsil ODS column (250 x 4.6 mm, 5μ) with a mobile phase of 0.5 mM sodium acetate buffer (pH 3.5): methanol (78:22 v/v). The analytes were detected at 225 nm, which is the isosbestic point.Suvorexant was primarily evaluated as an IS for the following three reasons: (1) it has a comparable molecular structure to DXT; (2) it has similar physicochemical attributes to DXT; and (3) it has great recovery. Furthermore, the DXT has been shown to be stable at -20 °C for at least 6 months, even after three days of room temperature freeze-thaw cycles. The DXT and Suvorexant peaks, fortunately, were well resolved and did not overlap. DXT and IS had retention durations of 2.56 and 6.55 min, respectively. Figure 2a-f depicts representative blank, IS, HQC, MQC, LQC, and LLOQ chromatograms.

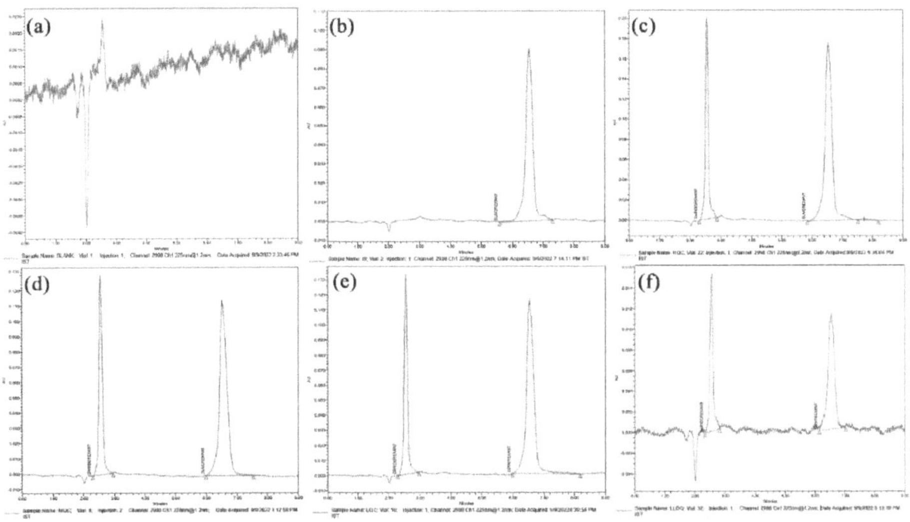

Figure 2. Representative chromatograms of (a) Blank; (b) IS (8 μg/mL); (c) HQC (8.50 μg/mL); (d) MQC (5.00 μg/mL); (e) LQC (0.75 μg/mL); (f) LLOQ (0.13 μg/mL).

3.1 Optimization of extraction method

Recovery has been judged to use ether/toluene (1:1), *n*-butyl chloride or MTBE. In a 10 mL round underside glass centrifuge tube, 0.5 mL of plasma had been fortified with this was to achieve a final concentration of 8 μg/mL. In terms of independently associated, serum was indeed fortified, as was the appropriate volume of DXT required to work conventionally prior to liquid-liquid extraction. In all tubes, 1 mL of formic acid and 2.5 mg/μL of appropriate solvent seemed to be incorporated. Whole samples were kept on a rotary mixer for a few minutes before being centrifuged at 3000 rpm for 5 min. Its organic surface seemed to be eliminated but rather located in a cylindrical glass tube, and non-extracted samples were collected from DXT to achieve the desired concentration of 5 μg/mL. Extracts have been vaporised to dryness under nitrogen (50°C), reconstituted as both 30 µL of organic phase and evaluated using RP-HPLC. Samples were collected and obtained using MBTH were just not inoculated as a result of their own cloudy appearance. Recoveries seemed to be derived through site evaluation of a corresponding peak region (drug/IS) of harvested but rather non-extracted extracts such as quadruplicate. Following collection of *n*-butyl chloride as that of the crude extract, this same infusion capacity was optimized (20 µL). Peak shape, abundance, and signal-to-noise (S/N) ratios have all been investigated. Table 1 provides the outcomes of extraction recovery.

Table 1. % recovery of various extraction solvents.

Extraction solvent	% Recovery	
	DXT	IS
ether/toluene (1:1)	87.37 ± 4.93	88.49 ± 2.94
n-butyl chloride	98.53 ± 3.95	97.65 ± 4.83
MTBE	90.58 ± 4.93	91.74 ± 3.59

3.2 Method validation

Calibration curves were constructed by graphing the ratio of peak areas to concentrations within the working range (0.25-10 g/mL). In the tested ranges of DXT (n = 3) and IS (n = 3), significant linearity (r2 = 0.9991) was attained, according to what is expected in biological fluids. It was found that the DXT (n = 3) and IS (n = 3) tested ranges exhibited acceptable linearity (r2 = 0.9991), which was expected for biological fluids. The signal-to-noise ratio was used to calculate the Limit of Detection (LOD) and Limit of Quantitation (LOQ). The predicted LOD was then the concentration determined for three times this noise based on the peak height slope. The peak-to-peak noise across the retention time and duration of analytes (DXT) for a blank sample was assessed. This estimated LOD is then verified by injecting this diluted solution with samples to confirm a peak height that is three times the noise of the baseline. For LOQ and LOD, the usual S/N was 10:1 and 3:1, respectively (Table 2). Precision and accuracy were within the proposed criterion (RSD 15%).

Chromatograms of different batches of blank matrices (plasma from four distinct batches) were compared to those from spiked plasma solutions and test samples to ensure that the other co-eluting components were not causing misleading or interfering findings. It was discovered that peaks produced by the matrix did not interfere with DXT and IS under optimal chromatographic conditions. DXT and IS (both n = 10) showed typical retention times of 2.56 and 6.55 min, respectively. Despite extraction, plasma samples may still contain large amounts of residual material, such as oils, lipids, and transition metals, which might deposit on the chromatography column. The retention times for the column are subsequently shortened by these materials, which are also very difficult to remove from the column. Regardless, no discernible change in retention time or peak area, as well as a loss in column capacity with time, has been found. By using PDA analysis, the methods selectivity was also validated; no substance that co-eluted with the target analytes was found. These circumstances are crucial for analyzing in vivo models where the co-elution of metabolites may obscure the DXT and/or IS peaks.

The recovery rates for six replicates of each quality control sample were calculated in order to assess the accuracy of the procedure. Recovery rates for DXT ranged from 92.66 to 115.38%. The range of the CV values (%) was 2.18 to 15.32%. The approach's reproducibility was verified using stability testing. The analytes were stable in human plasma (CV% 5%) after 30 days of long-term storage, 24 h of short-term storage, and three cycles of freezing and thawing (Table 3). These results suggested that analyte storage in plasma samples at -20 °C was appropriate and that frequent pharmacokinetic analyses would not encounter any stability problems. These results demonstrate that the method can precisely quantify DXT.

Table 2. Summary of validation data for DXT.

Parameter	DXT
Linearity range (μg/mL)	0.25-10
Calibration equation	y = 0.9973x - 0.0247
Correlation coefficient (r^2)	0.9991
LLOQ (μg/mL)	0.13
LOD (μg/mL)	0.045
ULOQ (μg/mL)	10.00
Accuracy (Mean %)	100.09
Precision (Mean %CV)	7.54
Selectivity	Selective

Table 3. Summary of stability studies for DXT.

Stability	QC Level	Mean Measured concentrations (μg/mL); (n=6)		% Change	%CV	% Mean stability*
		Comparison sample	Stability sample			
Benchtop	HQC	8.53	8.85	3.67	1.19	103.67
	LQC	0.73	0.76	4.09	4.35	104.09
Freeze-Thaw	HQC	8.51	8.55	0.45	0.64	100.45
	LQC	0.74	0.78	5.84	3.04	105.84
Auto sampler	HQC	8.53	8.30	2.66	0.75	97.34
	LQC	0.74	0.73	0.34	1.57	99.66
Short term	HQC	8.39	8.38	-0.11	0.14	99.89
	LQC	0.73	0.78	6.79	4.09	106.79
Long term	HQC	8.23	8.17	-0.75	1.21	99.25
	LQC	0.70	0.74	5.04	11.8	105.04

4. CONCLUSION

Sedative-hypnotic medications are prominent in clinical and forensic toxicology studies, but no study reporting the analysis of DXT in human plasma using RP-HPLC-PDA has been published to date. The RP-HPLC-PDA analytical approach presented in this work provides selective, sensitive, and accurate DXT analysis without the requirement for costly cleaning procedures, solvent-intensive extractions, or expensive instruments. This is the first time HPLC-PDA has been used to detect DXT in biological matrices. The lower limit of quantification of (0.13 µg/mL) suggests that the approach might be beneficial for drug assessment even at lower dosages. Furthermore, just 0.5 mL of plasma is needed, which is ideal for DXT pharmacokinetic studies in smaller animals (eg, cats). With this method, a range of DXT concentrations in plasma was found that fit with amounts that have been suggested as therapeutic for both humans and animals.

5. ACKNOWLEDGEMENTS

We gratefully thank the Bruhada blood bank, Kadapa, India for the support of drug-free human plasma and Management of Vasavi institute of Pharmaceutical Sciences, Kadapa for providing instrumentation and infrastructure facility to carry out this work. Authors are also thankful for the management of KL University for their support.

6. REFERENCES

1. C.M. Morin, C.L. Drake, A.G. Harvey, A.D. Krystal, R. Manbe, D. Riemann, and K. Spiegelhalder, *Nat Rev Dis Primers*, 2015, DOI: 10.1038/nrdp.2015.26
2. E.S. Ford, T.J. Cunningham, W.H. Giles, and J.B. Croft, *Sleep Med*, 2015, **16**, 372. DOI: 10.5665/sleep.4684
3. O. Bruni, M. Angriman, M.G. Melegari, and R. Ferri, *Expert Opin Pharmacother*, 2019, **20**, 2257, DOI: 10.1016/j.sleep.2021.02.003
4. B.X. Lou, and M. Oks, *Clin Geriatr Med*, 2021, **37**, 401,DOI: 10.1016/j.cger.2021.04.003
5. D.N. Neubauer, J. Minkel, and A.D. *Krystal, Psychiatry*, 2015, 2193.

A. Flores, R. Saravia, R. Maldonado, and F. Berrendero, *Trends Neurosci*, 2015, **38**, 550.

6. Y. Han, K. Yuan, Y. Zheng, and L. Lu, *Neurosci Bull*, 2020, **36**, 432.

7. A.J. Roecker, C.D. Cox, and P.J. Coleman, *J Med Chem*, 2016, **59**, 504.DOI: 10.1021/acs.jmedchem.5b00832
8. C. Muehlan, C. Vaillant, I. Zenklusen, S. Kraehenbuehl, and J. Dingemanse, *Expert Opin Drug Metab Toxicol*, 2020, **16**, 1063.
A. Markham, Drugs, 2022, 82, 601.
9. J.B.R. Dos Santos, M.R.R. Da Silva, *Eur J Clin Pharmacol*, 2022, **78**, 1749.
10. B.M. Ishaq, L.S. Reddy, G.M. Basha, M. Sreenivasulu, C.M. Chetty, and H.A. Ahad, *J Chem Soc Pak,* 2020, **42**, 171.
11. Guidance for Industry, Bioanalytical Method Validation, US Department of Health and Human Services, Food and Drug Administration Centre for Drug Evaluation and Research (CDER), Centre for Veterinary Medicine (CVM), (2001) BP. http://www.fda.gov/cder/guidance/index.htm. (Accessed on 5th May 2022).
12. Z. Han, X. Liu, Y. Ren, L. Luan, and Y. Wu, *J Sep Sci*, 2010, **33**, 1923.

A UV-SPECTROPHOTOMETRIC METHOD FOR SIMULTANEOUS ESTIMATION OF ABACAVIR AND LAMIVUDINE IN THEIR PHARMACEUTICAL DOSAGE FORMS

Jahasultana Mohammed, Narender Malothu*, Narayana Murthy Ganta, Nestory Mbilinyi, Uma Maheswari J, Monzir Idris

Department of Pharmacy, Koneru Lakshmaiah Education Foundation, Vaddeswaram, AP, India.

**Corresponding Author:*

Dr M. Narender, narendermalothu@gmail.com

1. INTRODUCTION

A strong nucleoside analogue reverse transcriptase inhibitor (NRTI) called Abacavir (ABA) is a familiar drug used in the treatment of HIV and AIDS. It is a structured synthetic carbocyclic nucleoside. This has an absolute configuration of 1S:4R on the cyclopentene ring. In the body, ABA sulphate breaks down to ABA, its free base. ABA has activity against Type - 1 Human Immunodeficiency Virus (HIV-1)[1]. The active metabolites produced by phosphorylating ABA compete with one another to be incorporated into viral DNA. They function as a DNA synthesis chain terminator and competitively inhibits HIV reverse transcriptase enzyme. Cellular enzymes transform ABA intracellularly into carbovir triphosphate, an analogue of deoxyguanosine-5'-triphosphate (dGTP), the active metabolite[2]. By combating with the natural substrate dGTP and incorporating itself into viral DNA, carbovir triphosphate inhibits the action of HIV-1 reverse transcriptase (RT). The lack of a 3'-OH Group in the incorporated nucleotide, which is required to generate the 5′ to 3′ phosphodiester linkage essential for DNA Chain elongation, leads viral DNA growth to terminate.

A zalcitabine analogue, lamividine (LAM) is a reverse transcriptase inhibitor. It is employed to treat hepatitis B and HIV-1 (HBV). The nucleoside reverse transcriptase inhibitor lamivudine is incorporated into the viral DNA.

LAM prevents the production of viral DNA and is effective against HIV-1 and hepatitis B (HBV). It can produce active metabolites when it is phosphorylated, and these metabolites combat for viral DNA incorporation. LAM metabolites competently limit the HIV reverse transcriptase enzyme's function by DNA incorporation, serving as a chain terminator for DNA synthesis. 5' to 3'-phosphodiester linkage is responsible for DNA chain elongation, but is not formed by inserted nucleoside analogues because they lack a 3'-OH group. The active 5'-triphosphate metabolite of LAM, a synthetic nucleoside analogue, is phosphorylated inside of cells (L-TP). DNA chain termination results the incorporation of this nucleoside analogue into viral DNA by HIV reverse transcriptase and HBV polymerase[3-4]. See Figure 1.

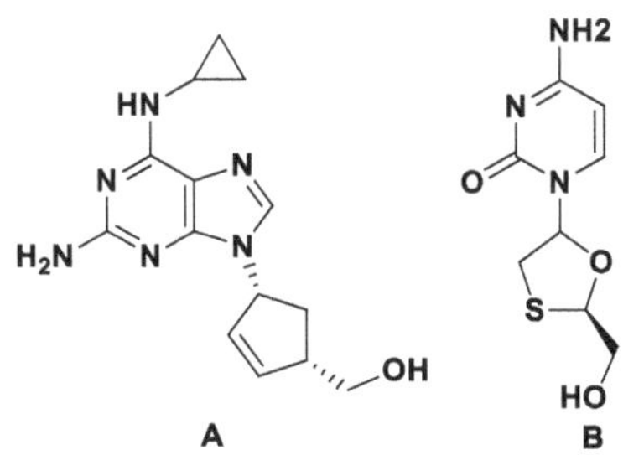

Figure 1: Chemical structures of ABA (A) and LAM (B).

According to a review of the relevant literature, there are a few analytical techniques that have been published for quantifying ABA and LAM both on their own and in combination with other medications. These methods make use of UV-Visible spectroscopy[5-8], HPLC[9-12], and HPTLC[13-16]. UV-spectrophotometry was used to determination of ABA and LAM drugs in combined dosage forms. Although only a few techniques have been reported so far. Because of its widespread availability, versatility, and user-friendliness, UV-spectrophotometric analysis is often used for the quality control of medicines in conventional laboratories. This is mostly owing to the fact that it is an effective method. As a result, an effort has been made to estimate both medications in a combined dose form using a straightforward quantitative approach known as the Simultaneous Equation Method. In this research, the techniques displayed the simplicity and sensitivity towards determination of pharmaceuticals in formulations, having been refined by relevant measures and verified in accordance with ICH standards. In addition, the methods were shown to be effective.

2. METHODS AND MATERIALS

2.1 Reagents and apparatus

We acquired the API ABA as free samples from Taj Pharmaceuticals in Hyderabad. The API LAM was got as free samples from Mylan Pharmaceuticals in Hyderabad. All the other chemicals used in this study were Analytical Grade

(AR) and bought from Merck India Limited, Mumbai and SD Fine Chemicals, Mumbai. All absorbance measurements were made with a UV-Visible spectrophotometer (Eppendorf Biospectrometer Basic) with a fixed 2 nm spectral bandwidth and 10 mm quartz cells.

2.2 Determination of absorption maxima

Based upon solubility and compatibility in absorption for both drugs in the UV region, a mixture of acetonitrile and water in *50:50 %v/v* was selected as solvent diluent. Standard solutions containing 10µg/mL of each ABA and LAM was prepared in the diluent (Acetonitrile and water), approximately 3.0mL of solution in 10mm cuvette scanned under UV region 400- 200nm using UV-Visible spectrophotometer (LAB India) with fixed slit width of 2.0nm.

2.3 Preparation of standard solution (100 µg/ mL)

These was prepared by accurately weighing 10 mg of ABA and LAM each into two separate 100 mL volumetric flask containing 70 mL of solvent diluent and sonicated for 20 min, and then the make up the volume to the mark with the same solvent.

2.4 Method development

2.4.1 Preparation of sample solution. Weighed accurately about 20 tablets of Abamune-L and the tablet powder equivalent to 10 mg of each ABA and LAM into 100 mL was transferred into standard volumetric flask containing 70 mL of solvent diluent. The solution was sonicated for 10 minutes, and the volume was made up to the mark to get 100µg/mL concentration of each analyte. The resulted solution was filtered using 0.45µ Millipore nylon membrane and scanned under UV- spectrophotometer.

2.4.2 Method: Simultaneous Equation Method. From the two absorbing species in the sample (X = ABA and Y = LAM) each of which absorbs at the maximum wavelength (λ_{max}) of different from the other, henceforth the amount or concentration of both drugs was determined by the SEM technique (Vierodt's method). The following data is required for the estimation,

- aX_1, aY_1 are the absorptivity of ABA and LAM at 218 nm.
- aX_2, aY_2 are the absorptivity of ABA and LAM at 274 nm.
- A_1, A_2 are the absorbances of both samples at 218 and 274 nm.

As the sample contains ABA and LAM and their concentration would be taken as C_{AMA} and C_{LAM}, respectively. Following equations were considered based upon the data obtained that at 218 and 274nm, the absorbance is the sum of the individual ABA and LAM absorbances.

$$C_{ABA} = \frac{A_2 aY_1 - A_1 aY_2}{a\,X_2 a\,Y_1 - a\,X_1 a\,Y_2} \quad (1)$$

$$C_{LAM} = \frac{A_1 aX_2 - A_2 aX_1}{a X_2 a Y_1 - a X_1 a Y_2} \quad (2)$$

$$\% \text{ Amount ABA } (C_X) \text{present (assay)} = \frac{C_X}{\text{amount taken}} \times 100 \quad (3)$$

$$\% \text{ Amount ABA } (C_Y) \text{present (assay)} = \frac{C_Y}{\text{amount taken}} \times 100 \quad (4)$$

2.5 Method validation:

After achieving the method development, the validation studies were conducted for parameters such as linearity, precision, accuracy, robustness, ruggedness, LOD and LOQ as per ICH Q_2 (R_1) guidelines[17-18].

2.5.1 Specificity. **Preparation of standard solution of LAM (1000 µg/ mL):**

The standard solution was prepared by placing 10 mg of LAM in a 10 mL volumetric flask with 3 mL of solvent diluent and sonicating it for about 20 minutes. The solvent was then used to get the volume up to the mark. From the aforementioned, 1 mL was pipetted into a 10 mL volumetric flask and diluted to 100 g/mL concentration. Pipetting 1 mL of the aforesaid 100 g/mL concentration solution into a 10 mL volumetric flask filled with solvent diluent to the mark yielded a 10 g/mL concentration.

Preparation of standard solution of ABA (1000 µg/mL):

The 10 mg of ABA that was precisely weighed and placed in a 10 mL volumetric flask with 3 mL of solvent diluent and sonicated for around 20 min to make the standard solution (1000 µg/mL). Further, the similar procedure given under LAM was employed to achieve the standard solution.

Preparation of working standard solutions of ABA and LAM:

The final concentration of 10 µg/mL was achieved by pipetting off individual portions of 1 mL from the 100 µg/mL solution of ABA and LAM into two separate 10 mL volumetric flasks and brought up the volume with diluent.

Preparation of sample solution:

Add a few mL of diluent, dissolve, and then add the remaining diluent to make the final volume (62.5 g/mL of ABA and 25 µg/mL of LAM). This mixture was designated as sample stock solution after being ultrasonically agitated for 10 min. To obtain the final concentration of 31.25 µg/mL of ABA and 12.5 µg/mL of LAM, pipette out 1 ml from the sample stock into 10 mL volumetric flask and filter through a 0.45 membrane filter. The volume is then brought up to 10 mL with diluent.

3 *Linearity.* Freshly prepared solutions of six concentrations of ABA and LAM were assessed for linearity by measuring their absorbances at 218 and 274 nm. After recording the absorbances, calibration curves against

concentration were plotted. The standard solution was generated by precisely weighing 10mg of both ABA and LAM in a 10 mL volumetric flask with 5mL of methanol, sonicating the mixture for about 30 min, and then adjusting the volume to concentration 1000 µg/mL with diluent. From this 100 µg/mL concentration solutions were prepared further.

2.5.3 Accuracy.

Accuracy 50%:

2.5 mL of each ABA and LAM standard solutions were pipette out separately into a clean standard 20 mL volumetric flask. To each flask 5 mL of sample was added and the flask was diluted with solvent and filtered through 0.45 µ filter.

Accuracy 100%:

5 mL of each ABA and LAM standard solutions were pipette out separately into a clean standard 20 mL volumetric flask. To each flask 5 mL of sample was added and the flask was diluted with solvent and filtered through 0.45 µ filter.

Accuracy 150%:

7.5 mL of each ABA and LAM standard solutions were pipette out separately into a clean standard 20 mL volumetric flask. To each flask 5 mL of sample was added and the flask was diluted with solvent and filtered through 0.45 µ filter.

2.5.4 Precision:

It is the degree of agreement among individual test results. The repeatability of precision was established in compliance with ICH recommendations. Six replicate injections of a standard solution into the chromatographic system were used to establish the system precision despite maintaining the optimal Spectroscopic conditions. A total of six concentrations of the drug product were produced and injected into the spectroscopic apparatus to acquire method precision.

2.5.5 Robustness:

Method capacity to remain unaffected by small parameter variation like wavelength, temperature, pH etc and it is determined during development of the analytical procedure using spectroscopic system

2.5.6 Ruggedness

The degree of test findings reproducibility acquired by analysing the same samples under various test conditions like different analysts, instruments, lots of chemicals, etc., was the measure of an analytical method's robustness. A sample solution containing 10 g/mL was generated in triplicate, and under various environmental conditions, their relative absorbances were evaluated. The % RSD of the outcomes to the absorbances was reported.

2.5.7 Detection Limit (DL) and Quantitation Limit (QL). The ratio Signal to noise (S/N) is commonly related to this. Analysts generally employ S/N ratio of 2:1 or 3:1 for DL, whereas the QL is frequently regarded to require a S/N of 10:1. The peak-to peak noise, which is determined from the injection of blank plasma, is calculated by dividing the signal, which is typically measured from the base line to the peak apex.

$$DL = \frac{3.3\,\sigma}{S} \quad (5)$$

$$QL = \frac{10\,\sigma}{S} \quad (6)$$

Where, σ - Standard deviation from response S - Slope from calibration curves.

4 RESULTS AND DISCUSSION

This work deals with the establishment of the UV Spectrophotometric method for the simultaneous estimation of ABA and LAM in the tablet dosage forms.

3.1 Selection of detection wavelength

The absorption maximum for ABA was found at 218nm whereas 274nm for LAM. The isoabsorptive point was observed at 230nm (Fig 2). At these wavelengths, it gave the maximum compatibility to the proposed methods. The blank spectrum indicated that the solvent system is not interfering the absorbance of analyte. Further method development protocols for the assay of ABA and LAM (Abamune-L tablets) were initiated based on method development guidelines and literature.

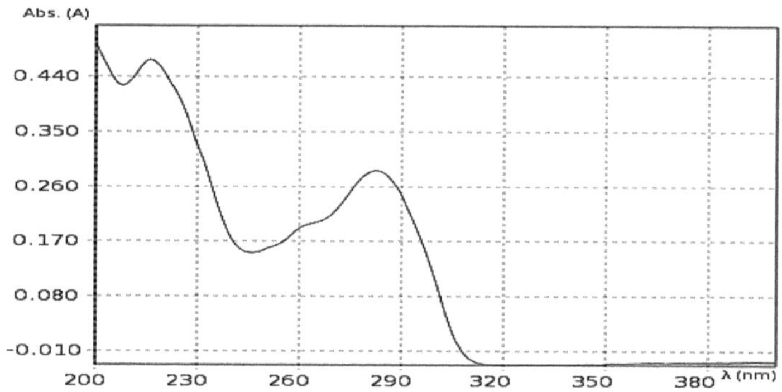

Fig 2: Absorption spectrum of ABA and LAM.

3.2 Method: Simultaneous Equation Method (SEM)

The standard solutions of drugs were prepared separately in appropriate concentrations. The absorptivities values of both drugs aX_1, aX_2 (ABA) and aY_1, aY_2 (LAM) were calculated using standard solution at the concentration of 0.001g /100 mL. The sample solution equivalent to standard amount (as per Label claim) was prepared and scanned to measure absorbances A_1 and A_2 at $\boldsymbol{\lambda_{max}}$ of ABA ($\boldsymbol{\lambda_2}$ = 218 nm) and LAM ($\boldsymbol{\lambda_2}$ = 274 nm), respectively (Table 3.1). The sample solution absorbance was found to be A_1 = 1.257 (at $\boldsymbol{\lambda_1}$ = 218 nm) and A_2 = 0.524 (at $\boldsymbol{\lambda_2}$ = 274 nm). The amount of drug/ analyte present was found to be 61.25 µg/mL (98.12 %) of ABA and 25.2 µg/mL (100.8 %) of LAM, which was found as per label claim amount.

Table 3.1 *The absorbance values in Simultaneous equation method.*

S. No.	Analyte	λ_{max} of ABA 218 nm	λ_{max} of LAM 274 nm
1	ABA	1.257	0.524
2	LAM	0.342	0.276
3	Sample	0.466	0.245

3.3 Method validation:

3.3.1 Specificity. As test procedure both **s**tandard and sample solutions were prepared and analyzed Spectrophotometrically and the absorption spectrum Figure 3 that showed no interferences of excipients.

Blank determination: Acetonitrile: water (1:1) was scanned in UV region under optimized conditions. Blank spectrum showed no interference of solvent in absorbance.

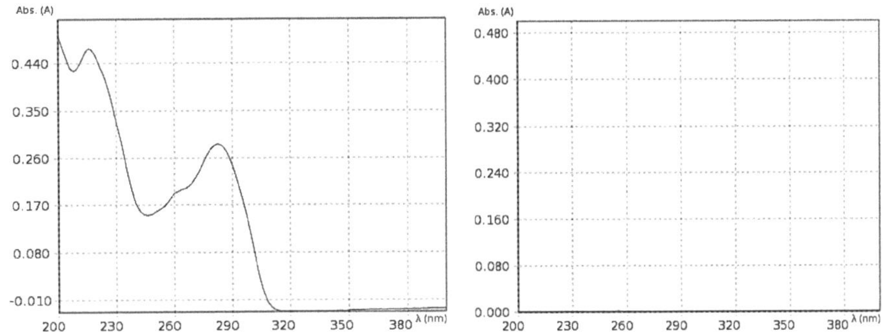

A B

Figure 3: Absorption spectrum of formulation (A), the absorption spectrum of Blank (B).

3.3.2 Linearity. Six concentrations of ABA and LAM were freshly prepared and measured the absorbance at 218 nm (λ_1) and 274 nm (λ_2). The overlain spectrum was drawn for six concentrations individually. The calibration curve was plotted against concentration v/s absorbances (Figure 4). Each solution was scanned three times and linearity was evaluated by linear-regression analysis. Both the drugs showed good linear relationship calibration curve plot with regression equation of y = 0.0795x for ABA and y = 0.0518x for LAM (Table 3.2). The Correlation coefficient (r^2: ABA = 0.999 and LAM = 0.992) found was within prescribed range. Hence the method was linear in the concentration ranging from 5-30 µg/mL.

Table 3.2 *Linearity data of ABA and LAM (n=3)*

Analyte	Concentration (µg/mL)	Absorbance	Linear regression Equation
ABA	2.5	0.183	y = 0.0795x r^2 = 0.9986
	5	0.384	
	7.5	0.571	
	10	0.762	
	12.5	0.983	
	15	1.244	
LAM	2.5	0.195	y=0.0518x r^2 = 0.9924
	5	0.315	
	7.5	0.428	
	10	0.534	
	12.5	0.632	
	15	0.730	

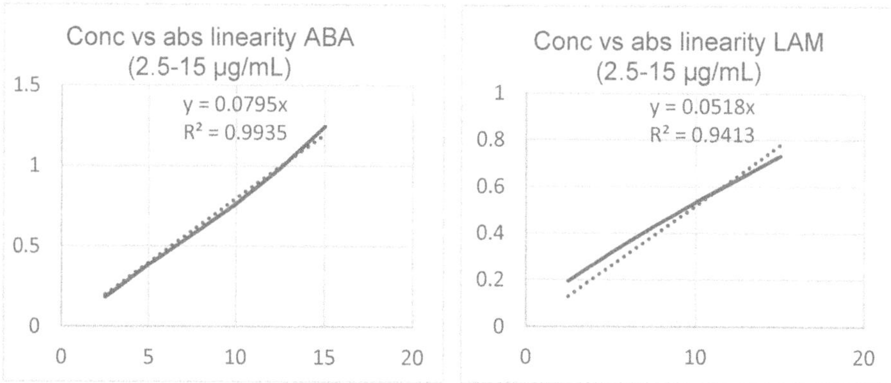

A **B**

Figure 4 Calibration curve for ABA (2.5-15 µg/mL) (A), Calibration curve for LAM (2.5-15 µg/mL) (B).

3.3.3 Accuracy. The accuracy was tested by three concentration levels: 50, 100, and 150% of the labelled claim (Figure 5). Three determinations at 218 nm (ʎ1) and 274 nm (ʎ2) were undertaken for each level, and the results were recorded. Percent analyte recovered by the suggested procedure was used to measure accuracy. The method showed mean recovery of 98.7-99.7 % for ABA and for LAM (Table 3.3). The % RSD was found to be less than 2.0 %. By these results showed the proposed method was accurate.

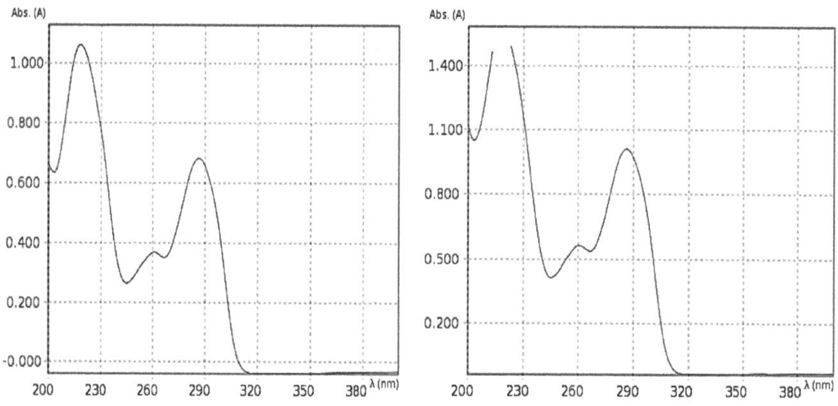

A **B**

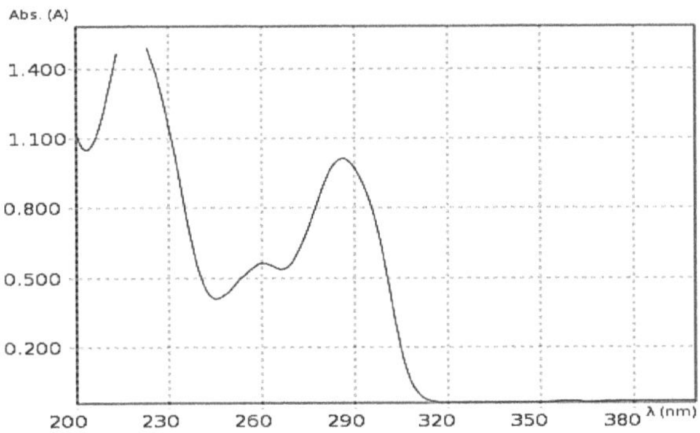

C

Figure 5 Accuracy level 100% (A), Accuracy level 150% (B), Accuracy level 50% (C).

Table 3.3 *Accuracy data of ABA and LAM (standard addition method) (n=3).*

	Spiked level	**Absorbance**	**% Recovery**	**SD**	**% RSD**	**Mean Recovery**
ABA	50 % 50 % 50 %	0.339 0.404 0.402	101.0 100.0 100.0	0.001	0.324	99.7 %
	100 % 100 % 100 %	0.826 0.812 0.806	100.3 98.5 100.8	0.002	0.324	99.7 %
	150 % 150 % 150 %	1.064 1.064 1.077	99.1 99.1 99	0.0026	0.21	98.7 %
LAM	50 % 50 % 50 %	0.270 0.269 0.279	102.0 101.6 102.0	0.21	0.73	101.8 %
	100 % 100 % 100 %	0.515 0.514 0.517	99.0 99.7 100.8	0.032	0.81	99.8 %
	150 % 150 % 150 %	0.685 0.681 0.741	99.9 99.6 99.5	0.014	0.35	99.6 %

3.3.4 Precision. The tablets' analysis yielded low% RSD readings, which demonstrated the method's high degree of precision. Table 3.4 shows the absorbances at 218 nm (λ_1) and 274 nm (λ_2) for six replicates of standard solutions of ABA and LAM under spectroscopic conditions. The method precision predicted by assaying the ABA and LAM as per the protocols shown under system precision (Table 3.5).

The % assays and % RSD of assay were calculated with proposed method which showed the results were within limit, hence the method called to be passes repeatability.

Table 3.4 *System Precision data (n = 3).*

S. No.	Concentration (µg/mL)	Absorbance	
		λ1	λ_2
1	10	1.267	0.773
2	10	1.262	0.776
3	10	1.279	0.781
4	10	1.275	0.784
5	10	1.278	0.784
6	10	1.279	0.785
Mean	-	1.273	0.781
S. D	-	0.007	0.005
%RSD	-	0.563	0.632

Table 3.5 *Method Precision data (n = 3).*

S. No.	Concentration (µg/mL)	Absorbance		Assay (%)	
		λ1	λ_2	ABA	LAM
1	10	0.446	0.251	99.8	99.9
2	10	0.436	0.254	99.1	100.1
3	10	0.440	0.257	99.5	101.6
4	10	0.445	0.248	99.6	99.5
5	10	0.441	0.247	99.5	99.4
6	10	0.445	0.254	99.6	100.1
Mean	-	0.442	0.252	99.5	100.1
S.D	-	0.004	0.004	-	-
% RSD	-	0.88	1.54	-	-

3.3.5 Robustness. The following changes were made in the spectroscopic system to determine the effect of deliberate variations in the optimized spectroscopic conditions like wavelength and temperature etc. When the detection wavelength was varied the % RSD found was 1.106 for ABA and 1.106 for LAM (Table 3.6). Variation in temperature for 25°C ± 5°C not showed any major changes in absorbance's. The % RSD was found to be 1.102 (Table 3.7), which shows the method was robust by minimal variation in wavelength (λ) and temperature (°C).

Table 3.6 *Robustness data change in wavelength of ABA and LAM (n = 3).*

S. No	Change in Wavelength	Absorbance of ABA at 218nm	% RSD	Absorbance of LAM at 274 nm	% RSD
1.	Low (216 nm)	0.801	1.106	0.578	1.106
2.	Original (218 nm)	0.814		0.579	
3.	High (220 nm)	0.807		0.570	

Table 3.7 *Robustness data change in diluent ratio of ABA and LAM (n = 3).*

S. No	Change in diluent ratio	Absorbance of ABA at 218 nm	% RSD	Absorbance of LAM at 274 nm	% RSD
1.	Low (45:55)	0.715	1.102	0.737	1.102
2.	Original (50:50)	0.814		0.579	
3.	High (55:45)	0.918		0.592	

Note: *Diluent: Acetonitrile: water (50:50 %v/v).*

3.3.6 Ruggedness. At these variable conditions (test analysts, laboratory conditions and instrument) the % RSD of the results analogous to the absorbance's were within acceptance criteria i.e., 1.102 and 1.106 for both ABA and for LAM.

3.3.7 DL and QL. According to experimental methods, S/N ratios of 2:1 or 3:1 for DL and 10:1 for QL are typically used. In order to produce the standard solutions, the concentration was reduced along the linearity range until S/N ratios of 3:1 for QL and 10:1 for DL could be reached. For each concentration, the absorbance was measured three times. Table 3.8 shows the determined DL for ABA and LAM.

Table 3.8 *DL and QL data of ABA and LAM.*

Parameters	ABA (μg/mL)	LAM (μg/mL)
DL	0.047	0.025
QL	0.142	0.076

The method gives good resolution between both the drug analytes (ABA and LAM) with a short analysis time. The method was validated for system suitable parameters according to ICH guidelines were found to be within limits were postulated in the following Table 3.9.

Table 3.9 *Results in method validation.*

Parameters	Results		Acceptance limits
	ABA	LAM	
Specificity	-	-	No interference
Linearity	r^2 = 0.9923	r^2 = 0.9825	r^2 = 0.998
Accuracy	99.7 %	100.3 %	98-102%
Precision	0.82	1.58	RSD < 2%
Robustness	< 1.5%	< 1.5%	RSD < 2%
LOD (µg/mL)	0.047	0.025	S/N should be more than 3:1
LOQ (µg/mL)	0.142	0.076	S/N should be more than 10:1
Assay	99.3%	99.8%	98-102%

4 CONCLUSION

Using spectrophotometric techniques, a straightforward simultaneous equation method that is accurate, sensitive, and precise has been developed for the purpose of estimating ABA and LAM in the pharmaceutical dosage form. At a temperature of 37°C ± 5°C with a relative humidity of 51%, the acetonitrile:water (50:50%)v/v solvent system was utilized. The wavelengths at which the absorbance was found to be at its highest for ABA and LAM were 274 nm and 218 nm, respectively. The validation criteria for all of the parameters that were evaluated have been met by the proposed method, which has also been found to be reliable, straightforward, and speedy. According to the guidelines provided by ICH Q2(R1), the system suitability parameters, including precision, linearity, accuracy, robustness, and stability, were evaluated and found to be satisfactory.

5 ACKNOWLEDGMENTS

We thank the Principal, KL College of Pharmacy, KL Deemed to be University, Vaddeswaram for their co-operation and also providing facilities to conduct the experiments.

6 CONFLICT OF INTEREST

Authors declare no conflict of interest.

7 REFERENCES

1. Nagulwar VP, Bhusari KP. Int J Pharm Sci Res 2012;**3**(7):2159.

2. Anantha Kumar D, Naveen Babu MN, Seshagiri Rao JVLN, Jayathirth Rao V. Rasayan J Chem 2010;3(1):94.

3. Devmurari VP, Patel RB. Int J Pharm Sci Res 2010;1(7):82.

4. Sharma P, Kosanam S, Rao SS. J Drug Deliv Ther 2021;11(2-S):14.

5. Sriveni T, Naveen V, Rupa VS, Renuka A, Porika S, Akiful Haque M, et al. Int J Pharm Qual Assur 2021;12(1):57.

6. Patil BR, Bhusnure OG, Paul BN, Ghodke AY, Mulaje SS. Int J Drug Regul Aff 2018;2(2):78.

7. Behera S. J Anal Bioanal Tech 2012;03(06).

8. Grewal IK, Singh S, Arora S, Sharma N. Plant Arch 2020;20:3265.

9. Arun R, Anton Smith A. Int J Res Pharm Sci 2010;1(3):321.

10. Kumar PR, Prakash PB, Krishna MM, Yadav MS, Deepthi CA. E-Journal Chem 2006;3(3):142.

11. Calcium KA. Eurasian J Anal Chem 2010;5(1):89.

12. V. Juyal , M. Chaudhary , P. Kumar G. Gnanarajan PKY. Met. J Pharm Res 2008;1(2):182–7.

13. Joshi RS, Pawar NS, Sawant RL. Lat Am J Pharm 2010;29(7):1226.

14. Srilatha D, Nasare M, Nagasandhya B, Prasad V, Diwan P. ISRN Spectrosc 2013;2013:1.

15. Noorbasha, K., & Nurbhasha, S. Future Journal of Pharmaceutical Sciences, 2020, **6**(1), 1.

16. Godela, R. Future Journal of Pharmaceutical Sciences, 2020, **6**(1), 1.

PHARMACOKINETIC PROFILE OF NOVEL CURCUMIN-METFORMIN CONJUGATE IN RATS BY LC-MS/MS

Vijaybhanu Perumalsamy[a*], Harish Kumar D R[a] and Sarasija Suresh[b]

[a]Department of Pharmaceutical Chemistry, Faculty of Pharmacy, M.S. Ramaiah University of Applied Sciences, Bengaluru, Karnataka, India.

[b]Project Director, Institute for Drug Delivery and Biomedical Research, Bengaluru, Karnataka, India.

Corresponding Author E-Mail ID: vijaybhanu.p@gmail.com

1 INTRODUCTION

Despite progress in cancer treatment, the disease remains one of the greatest threats to human health. It is the second leading global cause of death after cardiovascular diseases. There are limitations associated with conventional chemotherapeutic agents, including their random distribution in the body and poor selectivity. There is an increasing need for better therapeutic approach in the treatment of cancer with drugs of higher efficacy and less drawbacks. Conjugation of two or more anticancer drugs through a covalent bond is an effective approach amid different drug design strategies for the development of new drugs in cancer therapy.[1] The hybrid conjugates synthesized by this approach may result in synergistic effect with enhanced anticancer efficacy, selectivity and minimal side effects.[2]

Curcumin, a well-known natural compound obtained from turmeric, is known to possess diverse biological activities including anticancer.[3, 4] Based on animal and human studies, it has been reported that curcumin is devoid of significant toxicity.[5, 6] The poor oral bioavailability of curcumin which is due to its low aqueous solubility and chemical stability, limits its clinical applicability despite its outstanding safety profile.[7, 8] Several strategies have been explored to overcome this limitation including physical and chemical modification of curcumin.

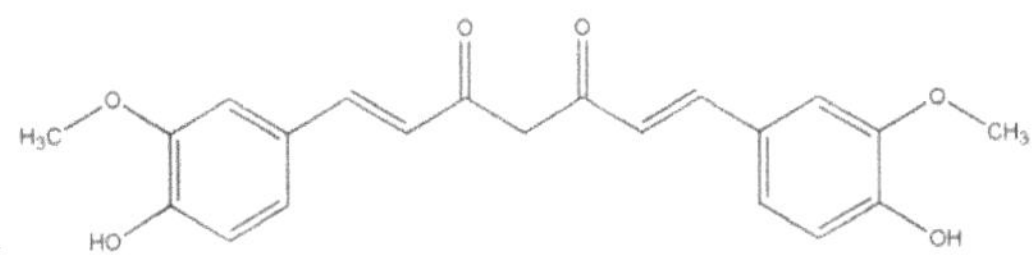

Figure 1a. Structure of Curcumin.

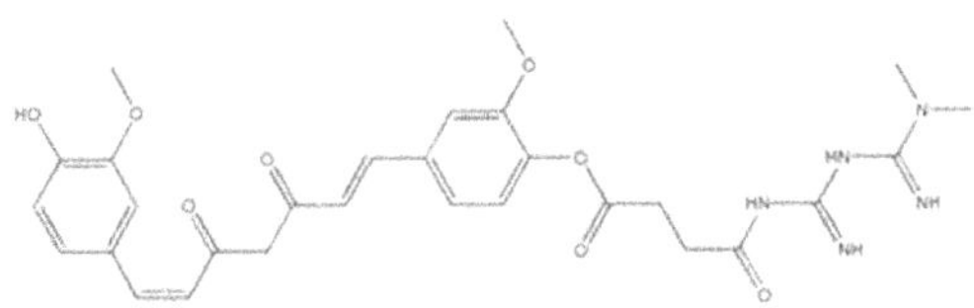

Figure 1b. Structure of Curcumin-Metformin conjugate.

Metformin has been reported to exhibit anticancer potential and was found to act by activation of AMPK (AMP activated protein kinase) pathway.[9] Curcumin and metformin in combination has been found to decrease cell proliferation and tumour development *in vitro* and *in vivo.*[10] Conjugation of curcumin and metformin as a single moiety increases the steric hindrance around curcumin which may stabilize it against chemical and enzymatic degradation thereby resulting in its enhanced bioavailability.[11] Curcumin was conjugated with metformin through amide linkage using succinic anhydride as linking agent and dicyclohexyl carbodiimide as coupling agent.[12] The synthesized conjugate may enhance the bioavailability of curcumin.

A highly sensitive analytical method for quantifying small amount of plasma after oral administration is in high demand. Despite the fact that many methods have been developed, their applicability in *in vivo* pharmacokinetic studies is limited. The present study was designed to develop LC-MS/MS method for quantification of curcumin in plasma and examine the pharmacokinetic profile of curcumin-metformin conjugate in Sprague-Dawley (SD) rats by applying the validated method for quantitative analysis of curcumin in rat plasma.

2 METHODS AND RESULTS

2.1 Materials

Curcumin with purity of 97% was purchased from Biogenuix Med systems Pvt Ltd (New Delhi, India). Curcumin D6 of purity 99.77%, used as internal standard in LC-MS/MS, was procured from Clearsynth (Mumbai, India). Acetonitrile, formic acid, methanol, diethyl ether, dichloromethane, sodium acetate and sodium tetra borate were procured from Merck Limited (Mumbai, India).

β-glucuronidase was purchased from Sigma Aldrich (USA).

2.2 Animals

Pharmacokinetic studies were carried out in male SD rats weighing between 180-200 g. The animals were checked for their body weight and the skin surface was superficially examined for abnormalities before randomization into groups. The animals were housed at a temperature of 25°C (±2°C) with relative humidity of 50 - 60% under the conditions of 12 hr light and dark sequence. The study protocols were in accordance with the guidelines after approval from institutional animal ethics committee (IAEC).

2.3 Animal experiments

SD rats were orally administered with curcumin and conjugate suspensions in 0.5% carboxy methyl cellulose (CMC) at doses of 130 mg/rat and 40 mg/rat respectively. Initially free curcumin (130 mg) and curcumin-metformin conjugate (40 mg) suspensions were prepared in 0.5% CMC separately. The dose was administered at 10 mL/kg body weight in 0.5% CMC suspension, orally (200 g of rat was administered with 2.0 mL of drug suspension).

Two groups of animals with eight animals each were used for the study. Sparse sampling approach was followed so that not more than three samples were collected from each animal. Approximately 400-500 μL of blood was drawn from the retro-orbital plexus and placed into micro centrifuge tubes pretreated with heparin, at the intervals of 0.5, 1, 2, 4, 6, 8, 12, 24 hr. Blank plasma was collected from control animal dosed with vehicle. The blood samples were then centrifuged at 4000 rpm for 10 mins to separate the plasma which was kept at–20°C.

2.4 LC-MS/MS analysis

2.4.1 Calibration curve standards and quality control samples preparation. The calibration curve (CC) standards, quality control (QC) samples of curcumin and internal standard were prepared from the diluted stock solutions. Curcumin was dissolved in methanol and 2 mg/mL standard stock solution was prepared. The stock solution of curcumin was appropriately diluted with methanol to prepare working solutions. Eight levels of calibration curve standards of curcumin in the range 1002.362 ng/mL to 1.010 ng/mL were prepared by serial dilution of curcumin in blank plasma. QC samples at four concentration levels (lowest-LLOQ, low-LQC, middle-MQC and high-HQC) were prepared in blank plasma using methanol as diluent in the range of 753.989 to 1.004 ng/mL.

2.4.2 Sample preparation. Curcumin was isolated from plasma using liquid-liquid extraction method. The frozen plasma samples were allowed to thaw in a water bath at room temperature and vortexed to evenly mix the contents. 50 µL of plasma and 10 µL of internal standard were added. Then 15 µL of beta glucuronidase and 100 µL of 0.1 M solution of sodium acetate were added and subsequently incubated for 1 hr at 60°C. 100 µL of 0.1 M borax and 1 mL of diethyl ether: dichloro methane (80:20) were added and shaken for 15 mins. 1 mL of supernatant organic layer was transferred to tubes and dried in vacuum concentrator followed by reconstitution in 100 µL mobile phase. 20 µL of processed sample was used for LC-MS/MS analysis.

2.4.3 Chromatographic conditions. Agilent 1200 Series equipped with a quaternary pump was used for analysis. The separation was achieved using Agilent HPH C18, 4.6*100mm, 5µm column. Mobile phase employed for analysis was the mixture of formic acid (0.2%) and acetonitrile (40:60% v/v). Flow rate was maintained at 0.7 mL/min. The temperature of auto sampler and column oven were maintained at 10ºC and 50°C respectively. The volume of sample injected was 20 µL with the total chromatographic run time of 4 mins. Agilent 6470 triple quadrupole mass spectrometer equipped with electrospray positive ionization was used for mass spectroscopic detection. Quantitation was performed in multiple reaction monitoring (MRM) mode. Optimised mass spectrometric conditions are shown in Table 1.

Table 1. *Mass spectrometric parameters for curcumin and internal standard.*

Parameters	Curcumin	Curcumin D6
MRM transition		
Parent ion (m/z)	369.1	375.1
Daughter ion (m/z)	176.9	179.9
Dwell time (m sec)	200	200
Declustering potential (V)	70	70
Entrance potential (V)	10	10
Collision energy (V)	20	20
Collision cell exit potential (V)	5	5

2.4.4 Data analysis. A graph was plotted with concentration of curcumin in plasma and corresponding time in hr on y and x axis respectively. Pharmacokinetic parameters of curcumin in rat plasma including maximum concentration -Cmax, time taken to reach maximum concentration -Tmax, area under the curve from time zero to observed time-$AUC_{0\text{-}t}$ or infinity $AUC_{0\text{-}\infty}$ and elimination half-life-$t_{1/2}$ were determined using WinNonlin software.

3 DISCUSSIONS

3.1 Method Development

A sensitive and selective analytical method for the quantitative analysis of curcumin in rat plasma was developed and validated with curcumin D6 as internal standard. The chromatographic conditions like column, mobile phase composition, flow rate, pH, temperature, injection volume and splitting of sample into ion source were optimised to achieve the desired selectivity, sensitivity and symmetric peak shapes of analyte and internal standard. Feasibility of various short length C18 columns were tested for optimizing the chromatographic conditions with various composition of mobile phases along with different flow rates. Also suitability of organic solvents and buffers with variable pH range was ensured. As the mobile phase, a mixture of 0.2% formic acid: acetonitrile (40:60) was found to be effective.

The correlation coefficient for curcumin was determined to be 0. 9959 for the concentration range of 1.0 ng/mL-1002.36 ng/mL. Regression equations' average slope and intercept were 0.00376 and 0.002312, respectively. The calibration curve for curcumin is shown in Figure 2.

Quadrupole full scans were performed in positive ion detection mode to optimise the ESI conditions for curcumin and curcumin D6. Detection of curcumin was performed using MRM mode to monitor precursor to product ion transitions of *m/z* 369.1 → 177 for curcumin and *m/z* 369.1 → 145 for IS. Figure 3 shows mass spectra of curcumin with precursor and product ion.

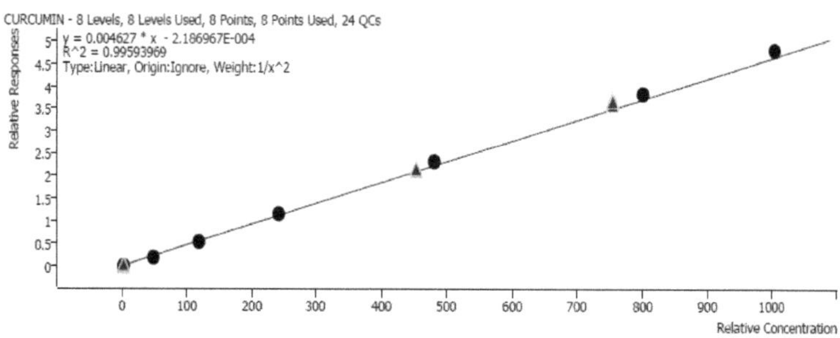

Figure 2. Calibration curve for curcumin.

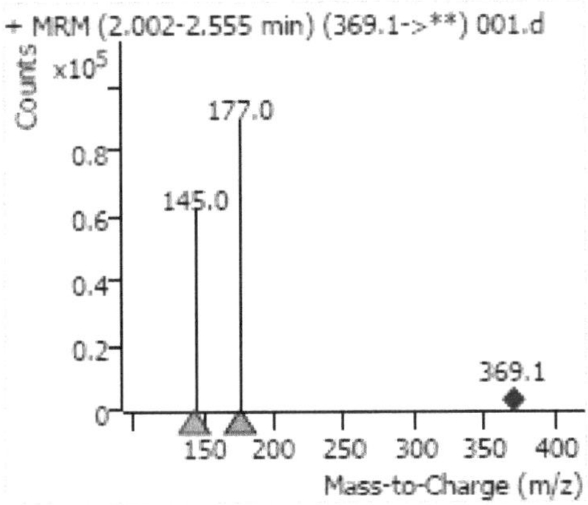

Figure 3. Mass spectra of curcumin with parent and product ion.

3.2 Method Validation

The method was validated in accordance with FDA guidelines for linearity, precision, accuracy, specificity, selectivity and recovery.[13]

3.2.1 Selectivity. It was evaluated by comparing the response of extracted blank plasma with that of extracted LLOQ samples processed with internal standard. The percentage interference was well within the limits for both analyte and internal standard and percentage CV was 9.33. Direct interference from endogenous compounds was not observed in blank matrix samples at the retention time of curcumin. The retention time obtained for curcumin and IS was 2.231. The representative chromatograms of blank plasma and plasma sample with IS and LLOQ of curcumin are shown in Figure 4, 5 and 6.

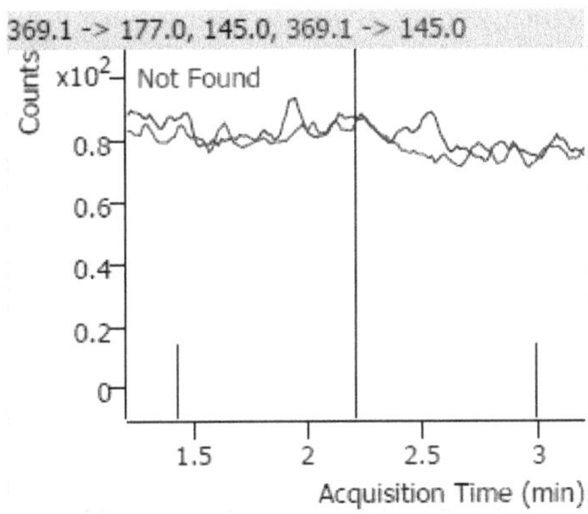

Figure 4. Chromatogram of blank plasma sample.

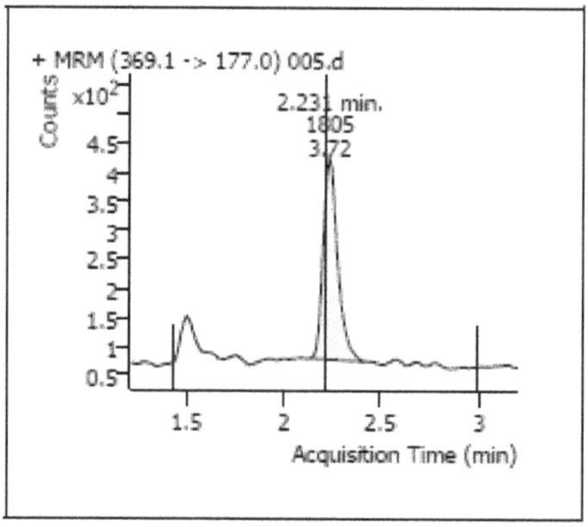

Figure 5. Representative chromatogram of LLOQ of curcumin.

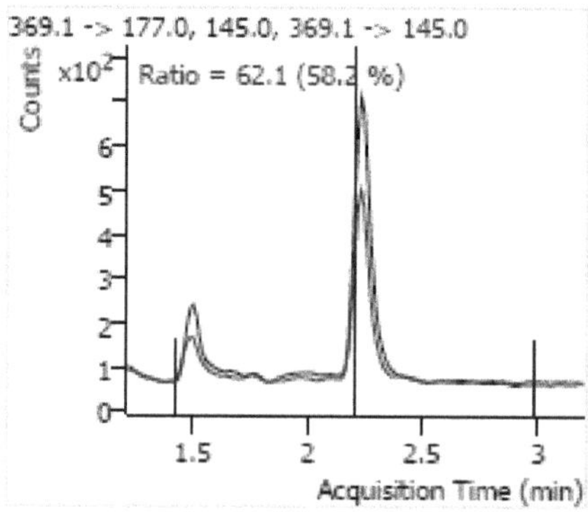

Figure 6. Representative chromatogram of curcumin (black) and IS (blue).

3.2.2 Linearity. Calibration curves for curcumin were constructed over the concentration range of 1.002 ng/mL – 1002.362 ng/mL, with eight calibration standards. It was found to be linear with best fit with correlation coefficient of 0.9959. The reported range of precision for calibration curve standards was 0.29% to 1.94%, while their accuracy ranged between 90.04% - 97% during the course of validation (Table 2).

Table 2. *Concentration of curcumin in calibration standards.*

Calibration standards ID	Concentration in ng/ml	Mean± SD*	CV%
A	1.000	1.049±0.009	0.84
B	2.002	1.812±0.035	1.94
C	48.113	43.320±0.416	0.96
D	120.283	119.916±1.281	1.07

E	240.567	252.520±0.730	0.29
F	481.134	493.082±4.933	1.00
G	801.889	831.340±3.682	0.44
H	1002.36	1039.738±11.109	1.07

*Represented as mean±standard deviation of three observations (n=3)

3.2.3 Precision and Accuracy. It was estimated by analysing QC samples within each run and in different runs. It was studied at four QC concentrations of curcumin as depicted in Table 3. The intraday precision and accuracy of the method at four QC concentration levels (1.004 ng/mL, 2.714 ng/mL, 452.393 ng/mL and 753.989 ng/mL, n=6) were found to be 3.83, 2.78, 0.33.0.99% and 96.00, 86.23, 102.70, 105.13% respectively. The interday precision and accuracy of the method at QC (n=6) were found to be 5.16, 2.79, 0.55, 1.01% and 95.32, 86.96, 102.08, 105.54% respectively. The results were found to be within the acceptable limits.

Table 3. *Intra and inter batch precision and accuracy.*

Concentration of QC sample (ng/mL)	Intra-day		Inter-day	
	Precision (% CV)	Accuracy (%)	Precision (% CV)	Accuracy (%)
LLOQ QC (1.004)	3.83	96.00	5.16	95.32
LQC (2.714)	2.78	86.23	2.79	86.96
MQC (452.393)	0.33	102.70	0.55	102.08
HQC (753.989)	0.99	105.13	1.01	105.54

3.2.4 Recovery. The mean, standard deviation (SD) and coefficient of variation (% CV) of peak area for extracted and un-extracted sample at different QC levels were calculated. The % recovery of extracted quality control samples was assessed by comparing their mean peak area with that of un-extracted samples. Also, the overall mean recovery, SD, and % CV were calculated. The percent mean recovery at HQC, MQC, LQC levels were 52.50%, 60.05%, and 65.62% respectively and the overall recovery was 59.39% (within the acceptable limits of ≤ 110.0 %) as tabulated in Table 4.

Table 4. *Recovery of curcumin from rat plasma.*

Concentration of QC sample (ng/ml)	% Mean recovery	Overall % CV	Overall % recovery
LQC (2.714)	65.62		
MQC (452.393)	60.05	11.09	59.39
HQC (753.989)	52.50		

3.3 Pharmacokinetic Studies

The plasma concentration of curcumin over time following single oral administration of curcumin-metformin conjugate (200 mg/kg) and curcumin (650 mg/kg) in rats is depicted in Figure 7. Table 5 summarizes the pharmacokinetic parameters estimated following the administration of free curcumin and conjugate orally to rats.

Two groups of rats were used for the study. One group of animals were given orally with curcumin suspension in 0.5% CMC at a dose of 650 mg/kg. The other group of animals were administered orally with conjugate suspension in 0.5% CMC at a dose of 200 mg/kg. Curcumin dose used was based on previous studies.[14] The conjugate dose was determined on the basis of equivalent dose of metformin at 100 mg/kg.[15]

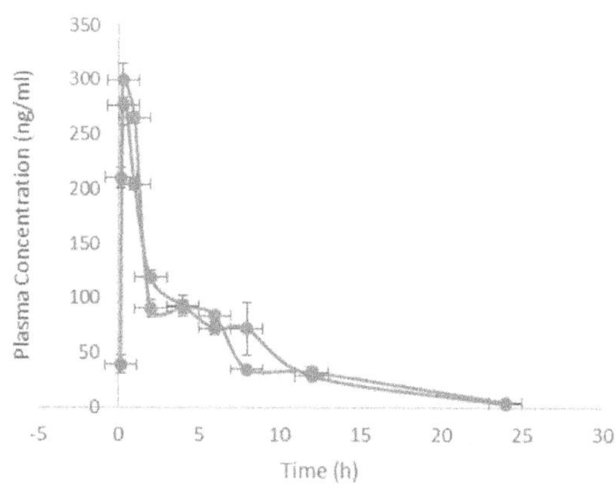

Figure 7. *Concentration versus time profile of curcumin in rat plasma, All points are expressed as mean ± SD (n=3)*

Table 5. *Pharmacokinetic parameters of curcumin and conjugate.*

Pharmacokinetic parameters	Curcumin	Curcumin conjugate
Cmax (ng/ml)	276.32±18.39	299.38±15.85
T max (hr)	0.30	0.30
$AUC_{0\text{-}t}$ (ng ml^{-1} hr^{-1})	1278.54±61.04	1268.44±80.72
$AUC_{0\text{-}\infty}$ (ng ml^{-1} hr^{-1})	1299.56±59.45	1305.61±81.14
$t_{1/2}$ (hr)	3.88±0.48	4.92±0.28
Ke (hr^{-1})	0.18±0.02	0.14±0.01

In case of rats administered with 130 mg of curcumin orally, the peak plasma concentration was found to be 276.32 ±18.39 ng/mL in 30 mins and then decreased to about 3.72 ng/mL in 24 hr of dosing. For the conjugate given orally at a dose of 40 mg, the peak plasma concentration was about 299.38 ±15.85 ng/mL in 30 mins and then decreased to about 5.21 ng/mL in 24 hr. Tmax was found to be 30 mins in case of both curcumin and the conjugate. The half-life of curcumin was found to be 3.88 ±0.48 hr whereas for that of conjugate it was observed to be 4.92 ±0.28 hr. The conjugate exerted slightly longer half-life than curcumin. The mean values obtained for $AUC_{0\text{-}\infty}$ were found to be 1299.56 ± 59.45 ng/mL/hr and 1305.61 ±81.14 ng/mL/hr for curcumin (650 mg/kg) and conjugate (200 mg/kg) respectively. The amount of curcumin present in curcumin-metformin conjugate (100 mg/kg or 20 mg/rat) is significantly less than that of pure curcumin (650 mg/kg or 130 mg/rat). The mean value for $AUC_{0\text{-}\infty}$ for curcumin (20 mg/rat) present in curcumin-metformin conjugate was found to be 1305.61 ±81.14 ng/mL/hr. The mean value calculated for $AUC_{0\text{-}\infty}$ for pure curcumin (130 mg/rat) was found to be 8486.47 ng/mL/hr but the value obtained was 1299.56 ± 59.45 ng/mL/hr. So upon comparing the mean values of $AUC_{0\text{-}\infty}$, it was observed that curcumin-metformin conjugate increased the bioavailability of curcumin by 6.5 times of free curcumin.

4. CONCLUSION

A sensitive and selective LC-MS/MS method was developed and validated to quantify curcumin from rat plasma using labelled IS curcumin D6. The chromatographic conditions like column, mobile phase composition, flow rate, pH, temperature, injection volume and splitting of sample into ion source were optimised through different trials to achieve the desired selectivity, sensitivity and symmetric peak shapes of analyte and internal standard. The developed bio analytical method was validated in accordance with FDA guidelines for linearity, precision, accuracy, specificity, selectivity and recovery. The method validation parameters were within the acceptance criteria. This method was successfully applied for pharmacokinetic studies of curcumin-metformin conjugate in SD rats.

The pharmacokinetic profile of the synthesized conjugate was studied in SD rats and compared with that of free curcumin. Free curcumin and curcumin-metformin conjugate suspensions were administered by oral route to two groups of animals at a dose of 130 mg/rat and 40 mg/rat respectively. Retro-orbital sampling was done at regular time intervals and the plasma was analyzed for curcumin by liquid chromatography with tandem mass spectrometry. The pharmacokinetic parameters like Cmax, Tmax, AUC and $t_{1/2}$ were determined. The results showed that conjugation of curcumin with metformin had enhanced the bioavailability of curcumin, which can be related to the increased stability of conjugated curcumin.

5. ACKNOWLEDGEMENT

The authors are thankful to Faculty of Pharmacy, M. S. Ramaiah University of Applied Sciences and Wipro Life Sciences Lab, Bengaluru for providing the facilities to carry out the research work.

6. ETHICAL APPROVAL

Animal procedures were carried out in accordance with the guidelines from institutional animal ethics committee (IAEC), approval number-BITS-Hyd/IAEC/2021/04.

7. FUNDING

The authors have not received any finding for this study.

8. REFERENCES

1. S. S. Panda, Q. L. Tran, P. Rajpurohit, G. G. Pillai, S. J. Thomas, A. E. Bridges, J. E. Capito, M. Thangaraju and B. L. Lokeshwar, *Pharmaceuticals (Basel, Switzerland)*, 2022, ***15*(4)**, 451.
2. M. Hasan, R. K. Leak, R. E. Stratford, D. P. Zlotos, and P. A. Witt-Enderby, *Pharmacology Research & Perspectives*, 2018, **6(4)**, e00417.
3. B. C. Jordan, B. Kumar, R. Thilagavathi, A. Yadhav, P. Kumar and C. Selvam, *Chemical Biology & Drug Design,* 2018, **91(1)**, 332.
4. C. D. Mock, B. C. Jordan and C. Selvam, *RSC advances*, 2015, ***5*(92)**, 75575.
5. R. A. Sharma, C. R. Ireson, R. D. Verchoyle, *Clinical Cancer Research,* 2001, **7**, 1452.
6. A. L. Cheng, C. H. Hsu, J. K. Lin, M. M. Hsu and C. Y. Hsieh, *Anticancer Research,* 2001, ***21***, 2895.
7. P. Anand, A. B. Kunnumakkara, R. A. Newman and B. B. Aggarwal, *Molecular Pharmaceutics.*, 2007, **4**, 807.

8. W. Liu, Y. Zhai, X. Heng, F. Y. Che, W. Chen, D. Sun and G. Zhai, *Journal of Drug Targeting*, 2016, **24(8)**, 694.
9. M. Zakikhani, R. Dowling and I. G. Fantus, *Cancer Research*, 2006, **66(21)**, 10269.
10. R. R. Falah, W. H. Talib and S. J. Shbailat, *Therapeutic Advances in Medical Oncology*, 2017, **9(4)**, 235.
11. M. H. Teiten, M. Dicato and M. Diederich, *Molecules*, 2014, **19(12)**, 20839.
12. P. Vijaybhanu, D. R. Harish Kumar and S. Suresh, *Journal of Young Pharmacists,* 2022, **14(1)**, 73.
13. R. J. Meesters and S. Voswinkel, *Journal of Applied Bioanalysis*, 2018, **4(3)**, 67.
14. S. K. Jain, M. S. Gill, H. S. Pawar, and S. Suresh, *Indian Journal of Pharmaceutical Sciences*, 2015, **76(5)**, 415.
15. J. T. Cheng, C. C. Huang, I. M. Liu, T. F. Tzeng, and C. J. Chang, *Diabetes*, 2006, **55(3)**, 819.

DEVELOPMENT AND VALIDATION OF A RP-HPLC METHOD FOR THE SIMULTANEOUS DETERMINATION OF TWENTY RELATED SUBSTANCES OF SULFAMETHOXAZOLE AND TRIMETHOPRIM IN INJECTION DOSAGE FORM

K. Rajasekhar[a], Ch. Gangu Naidu[a,b*], Ch. Naga Sesha Sai Pavan Kumar[a*], K. Varaprasada Rao[c], Y. Srinivasa Rao[c]

[a]Division of Chemistry, Vignan's Foundation for Science, Technology & Research, Vadlamudi, [b]Division of Chemistry, Vignan's Institute of Information Technology, Visakhapatnam, [c]Department of Pharmaceutical analysis, Vignan Institute of Pharmaceutical Technology, Visakhapatnam
*Corresponding Authors E-Mail: naiduiict@gmail.com, pavaniict@gmail.com

1. INTRODUCTION

The component mixture of five parts of the antibiotic sulfonamide sulfamethoxazole and one part of the antibacterial medication trimethoprim is listed on the product's label as a typical treatment for a variety of bacterial infections.[1] Together, trimethoprim (TM) and sulfamethoxazole (SM) produce nucleic acids in two distinct steps. Trimethoprim inhibits dihydrofolate reductase, the enzyme that changes dihydrofolic acid into its tetrahydro form. This metabolic process is necessary for the production of the folate coenzymes, which are crucial for the biosynthesis of thymine, purine, serine, and methionine.

Para-aminobenzoic acid and sulfamethoxazole compete to exert their antibacterial effects. Teva-Trimel, an antibacterial medication with a broad spectrum of activity, contains trimethoprim and sulfamethoxazole. Sulfamethoxazole and trimethoprim, two of its two main antibacterial ingredients, work together to inhibit many different bacterial species. Given how frequently this product is used in medicine, methods for figuring out what's in pharmaceutical dosage forms are a regular analytical need in quality control of the pharmaceutical sector. Numerous methods have been suggested for the identification of trimethoprim and sulfamethoxazole in the mixture, including UV spectroscopy with multivariate calibration. [2,3] FT-IR spectrometry [4], amperometry [5], voltammetry [6], HPTLC [7,8], or HPLC. [9,10]

The United States Pharmacopoeia uses reverse phase HPLC and related substance by TLC, in accordance with the British Pharmacopoeia's official pharmacopoeial procedures for sulfamethoxazole and trimethoprim detection in pharmaceuticals a combination of UV spectrophotometry and diazotization titration. [11] The relevant ingredient is nevertheless quantified using the TLC technique. There is no method described in the literature for simultaneously quantifying SM and TM-related substances. Measurement of related chemicals is not especially efficient using the TLC method. A novel related substance process for SM and TM injection using efficient chromatographic technology has to be created and verified in order to replace the current TLC approach.

2. MATERIALS AND METHODS

2.1 Availability of data and materials.

Sulfamethoxazole and Trimethoprim Injection (Figure 1), USP is sold over the counter and is made by Teva Pharmaceuticals in India. Sulfamethoxazole and trimethoprim are each contained in a 5 mL dose of 400 mg each. Related contaminants are offered in the EP and USP catalogues. HPLC-grade acetonitrile, methanol, and orthophosphate from Merck (Darmstadt, Germany). Prior to usage, water was cleaned using a Millipore Milli-Q water-purification system (Bedford, MA, USA) and a Durapore 0.45 m membrane filter (Millipore, Dublin, Ireland).

2.2 Instrumentation.

A Waters (Milford, MA, USA) Alliance HPLC system with a photodiode array detector was used for the HPLC analysis. Software called Waters Empower was in charge of this system. Employed for chromatographic separation is Kromasil 100-5-C18, 5.0 m, (4.0 x 250) nm, manufactured by Nouryon, column part number M05CLB25.

2.3 Chromatographic Conditions.

Using a straightforward gradient run programme, the separation of the analytes was accomplished on an Alliance HPLC system Kromasil 100-5-C18 column kept at a temperature of 30°C. The mobile phase's flow rate was set to 0.8 mL/min, and its composition was 0.08%. As the aqueous component in mobile phase A, ortho-phosphoric acid buffer was used. As the organic component in mobile phase B, acetonitrile: methanol (80:20 v/v) was utilised. Acetonitrile and water (5:95 v/v) was used as the diluent for API Samples and impurities.

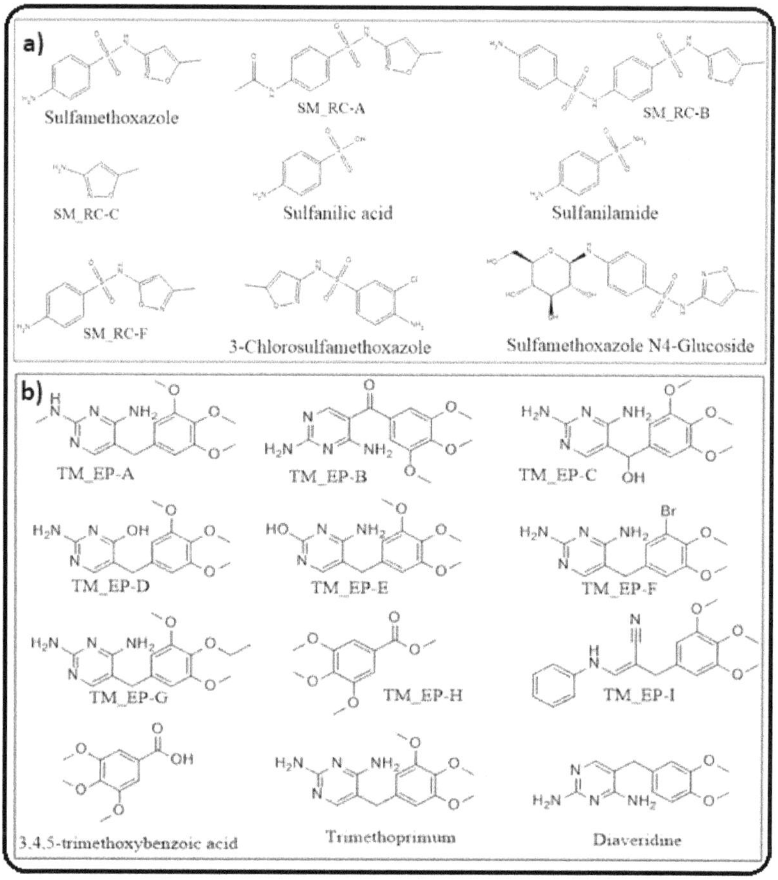

Figure 1 Chemical structures of (a) SM and related impurities (b) TM and related impurities.

The HPLC was configured with the following gradient programme: (T/% B): 0.1/5, 2/5, 10/10, 30/15, 70/60, and 75/60. To keep the sample stable, a sampler cooler set at 5 ^{0}C was employed. Before usage, a 0.45 µm Millipore filter was used to filter the mobile phase. A PDA detector tuned to 210 nm wavelength was used to achieve the detection. 10 µL of the sample injection volume were used.

2.4 Standard stock, sample and spike sample preparation.

For SM and TM-related impurities, a different stock solution of impurities was prepared. After being thoroughly mixed and diluted to volume with diluent, an accurately weighed quantity of each individual contaminant was dissolved in diluent to create the stock solutions. This produced a concentration of each impurity of 0.1 mg/mL for the two stocks, respectively.

By weighing and transferring around 2.2 mg of Sulfamethoxazole RC-F into a 100 mL volumetric flask, diluting the volume with diluent, the System Suitability Solution (0.16 mg/mL of Sulfamethoxazole, 0.032 mg/mL of Trimethoprim, and 0.00032 mg/mL of Sulfamethoxazole RC-F) Then, 0.16 mL of this solution was placed into a 10 mL volumetric flask, and 1 mL of System suitability stock solution was added and diluted with diluent to the desired volume.

In order to create the sample solution, which contained 1.6 mg/mL of SM and 0.32 mg/mL of TM, 1 mL of the injection test sample was transferred into a 50 mL volumetric flask, and the remaining volume was filled with diluent.

Sample solution for spiking robustness (1.6 mg/mL of sulfamethoxazole and 0.32 mg/mL of trimethoprim, all impurities at 1% to test concentrations): Transferred 4 mL of each Sulfamethoxazole Impurity mix Stock solution and Trimethoprim Impurity mix Stock solution, each diluted to volume with diluent, into a 50 mL volumetric flask together with 1 mL of Sample solution.

3. RESULTS AND DISCUSSION

3.1 Method development and optimization.

The chromophore functional groups in SM, TM, and all related impurities are all active for UV absorption, hence an HPLC linked to a UV/PDA detector was chosen as the analytical approach for quantifying SM and TM-related impurities. HPLC is a reliable equipment, and the cost of analysis would be lower and more practical in quality control labs. The main excipient in STI is benzyl alcohol, which is also a UV-active substance. Based on their spectral data, the isosbestic point for all of the contaminants was chosen to be 210 nm.

3.1.1 Related Substances. The technique must be able to identify and quantify both known and unidentified SM and TM impurities in pharmaceutical products. Ideal range for it to fall inside is between reporting threshold and 1.5 times specified limit. The characteristics of the method, such as specificity, precision, linearity, and accuracy, should be clearly defined in Tables 1 to 4 in order for measurements to be within 20% of the true value for impurity levels below 0.15%, with a probability of 80%, and within 15% of the true value for impurity levels above 0.15% , with a probability of 90%.

3.1.2 Selection of Method Conditions. The TLC test for organic impurities is described in the current USP monograph for Sulfamethoxazole and Trimethoprim Injection. Additionally, there is no method to measure relevant contaminants in the drug items Sulfamethoxazole and Trimethoprim Oral solution and Sulfamethoxazole and Trimethoprim tablet. According to a literature search, this medication product does not have any processers that can measure all the contaminants listed in the EP and USP catalogues. So, new method development for STI-related substance operations began in earnest.

There are a total of 22 compounds (see Fig. 1) in the Sulfamethoxazole and Trimethoprim Injection that were chosen for this investigation, of which Sulfamethoxazole and associated impurities make up 9 and Trimethoprim and related impurities make up 12 together with one excipient. Few chemicals are polar, few are mid-polar, and few are non-polar, making this approach crucial.

As per physicochemical properties, the pKa values for SM and TM were 1.6 and 5.7, respectively. Based on this, think about utilizing three different buffers for the initial approach development: ortho-phosphoric acid, potassium dihydrogen phosphate, and dipotassium hydrogen phosphate. Solvents used as acetonitrile:methanol %v/v (50:50). The stationary phases C18, C8, and phenyl, which range from non-polar to polar, were selected.

For up to 70 minutes, a straightforward linear gradient from 5 to 70% was used. The column temperature was set to be at room temperature. Based on the aforementioned permutations and combinations, all the experiments were run, and the ortho-phosphoric acid as mobile phase A and a mixture of acetonitrile and methanol as mobile phase B as the best experiment were chosen.

By altering technique parameters such column temperature and gradients, it was possible to further tune settings in order to obtain the necessary system suitable characteristics, such as resolution, runtime, and capacity factor.

Critical factors including buffer concentration, flow rate, greater organic ration of gradient programme, and column temperature are used to test the durability of the finalizing the process.

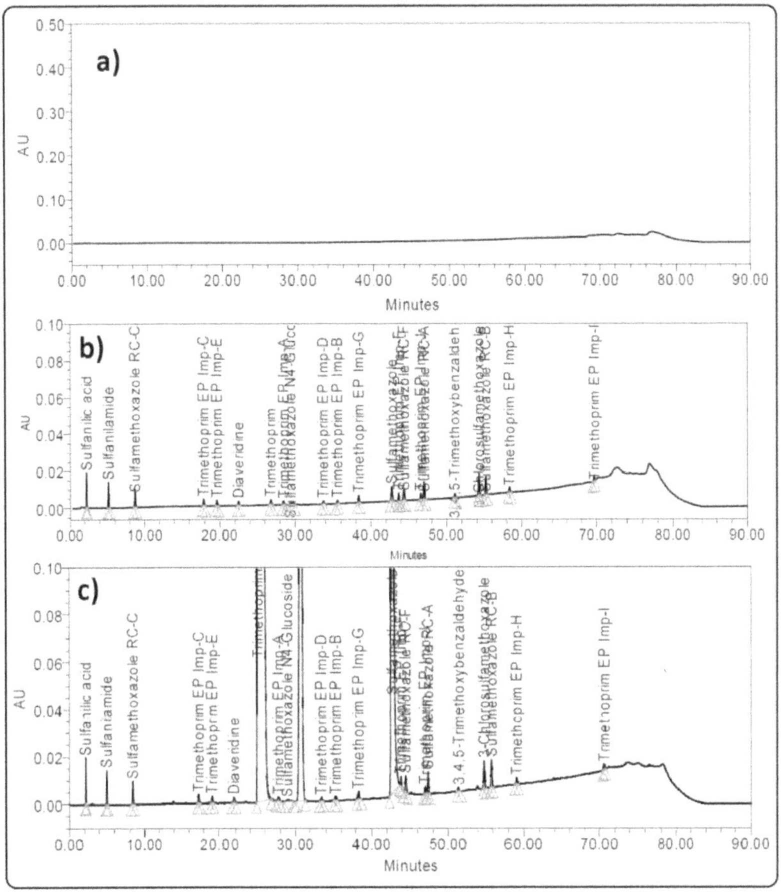

Figure 2 Chromatogram of (a) blank (b) LOQ solution (c) LOQ spiked sample solution.

3.2 Method Validation

According to "ICH Q2 (R1)" criteria, the improved method underwent technical validation.[12,13] The system's appropriateness criteria complied with the standards for acceptance. The resolution between each individual peak was higher than 2.0, TM EP-A & SM N4 gluconate, TM EP-J & SM RC-A is NLT 1.5, and the relative retention duration of the impurity peak was comparable. The relative standard deviation for LOQ solution of replicate injections was less than 5.0 percent for all the peaks. Figure 2 displays the LOQ solution, LOQ spiked sample solution, and the Blank.

3.2.1 System suitability test. The LOQ standard solution was injected into the chromatograph six times, and the chromatograms were recorded, as part of the system suitability test. The resolution between SM and SM RC-F should be greater than 2, the relative standard should not be greater than 5%, and system suitability solution was injected.

3.2.2 Specificity. All of the contaminants and medication product were infused to assess the method's specificity. Diluents (blank), STI sample, all listed contaminants, excipient, and LOQ spiked sample solutions were injected onto an HPLC in separate solutions. Peak purity were assessed using the Empower software in spiked sample solutions determined with a PDA detector under ideal chromatographic conditions for all the compounds and were found to be better (purity angle should be less than purity threshold value). This showed that no other peaks were co-eluting with the analytes and showed the method's capability to assess the analyte of interest unambiguously in the presence of potential interference.

For every molecule under investigation, baseline resolution was attained. Table 1 shows the retention time, RRTs, resolution and peak purities for each component in the spiked solution.

Table 1 *Selectivity and sensitivity data.*

Compound Name	t_R (min)	RRT	Rs	Purity angle	Purity threshold	s/n
Sulfanilic acid	2.24	0.08	-	1.00	3.32	494
Sulfanilamide	5.15	0.19	23.84	0.87	1.89	352
Sulfamethoxazole RC-C	8.65	0.32	18.72	6.24	12.66	247
Trimethoprim EP Imp-C	17.9	0.67	37.53	10.64	22.03	85
Trimethoprim EP Imp-E	19.64	0.73	5.93	9.37	19.79	71
Diaveridine	22.49	0.84	8.73	11.29	22.72	56
Trimethoprim	26.77	1	12.45	0.13	0.36	63
Trimethoprim EP Imp-A	28.47	1.06	4.46	21.12	46.59	48
Sulfamethoxazole N4-Glucoside	29.38	1.1	2.04	9.48	19.18	18
Trimethoprim EP Imp-D	33.69	1.26	8.73	14.08	30.55	37
Trimethoprim EP Imp-B	35.56	1.33	5	7.29	14.40	39
Trimethoprim EP Imp-G	38.39	1.43	8.79	15.76	27.05	82
Sulfamethoxazole	42.79	1.6	14.53	6.02	7.00	177
Trimethoprim EP Imp-F	43.71	1.63	3.45	6.65	12.68	91
Sulfamethoxazole RC-F	44.39	1.66	2.74	1.03	2.08	157
Trimethoprim EP Imp-J	46.59	1.74	7.6	8.57	17.34	74

Sulfamethoxazole RC-A	47.01	1.76	1.92	1.05	2.25	226
3,4,5-Trimethoxybenzaldehyde	51.08	1.91	15.63	6.64	13.39	39
3-Chlorosulfamethoxazole	54.34	2.03	12.54	1.22	2.58	288
Sulfamethoxazole RC-B	55.27	2.06	3.77	0.86	1.85	240
Trimethoprim EP Imp-H	58.4	2.18	11.96	9.15	18.02	67
Trimethoprim EP Imp-I	69.64	2.6	39.99	2.80	5.76	62

t_R – Retention time; RRT – Relative retention time; RRF - Relative response factor; Rs- Resolution; s/n- signal to noise ratio

3.2.3 Precision. Six STI samples were constituted by spiking impurities to their limit of quantification (LOQ) of evaluated concentration (0.15 g/ml), blending the solution to obtain each impurity's LOQ concentration, and then using a standard solution made at a concentration of 1.6 g/ml for all SM-related impurities and 0.32 g/ml for TM-related impurities. It pumped a test sample. A different lot column, a different instrument manufacturer, and several analysis days were also used to test the intermediate precision. Information from precision studies is presented in Table 2.

Table 2 *Precision data.*

Component Name	Standard solution Precision (% RSD)	Intermediate precision (% RSD)
Sulfanilic acid	1.1	1.7
Sulfanilamide	1	0.8
Sulfamethoxazole RC-C	1.3	0.9
Trimethoprim EP Imp-C	3.7	2.4
Trimethoprim EP Imp-E	2.8	1.9
Diaveridine	0.7	2.2
Trimethoprim	2.6	3.3
Trimethoprim EP Imp-A	1.5	3.1
Sulfamethoxazole N4-Glucoside	1.7	1.3
Trimethoprim EP Imp-D	3	2.4
Trimethoprim EP Imp-B	3.3	3.2
Trimethoprim EP Imp-G	2.9	3.7

Sulfamethoxazole	1.5	0.6
Trimethoprim EP Imp-F	3.4	1.7
Sulfamethoxazole RC-F	1.3	1.8
Trimethoprim EP Imp-J	1.8	1
Sulfamethoxazole RC-A	1.4	1.9
3,4,5-Trimethoxybenzaldehyde	3.2	2.1
3-Chlorosulfamethoxazole	1	0.6
Sulfamethoxazole RC-B	1.2	1
Trimethoprim EP Imp-H	2.6	2
Trimethoprim EP Imp-I	2.2	2

3.2.4 Accuracy. A series of solutions, five levels of recovery experiments were carried out, and the developed method's precision was assessed. Limit of quantification (LOQ) levels of 50, 100, 120, and 150 percent (a hypothetical concentration of LOQ to 1%) contaminants were spiked during test preparation. The remaining level solutions were created and injected thrice, and the LOQ, 100 and 150% were prepared and injected six times. The average results are shown in Table 3.

Table 3 *Accuracy data.*

Component Name	LOQ	50%	100%	120%	150%
Sulfanilic acid	101.33±0.69	100.88±0.90	100.49±0.41	100.22±0.12	100.88±0.34
Sulfanilamide	100.74±0.74	100.38±0.40	99.59±0.77	99.85±0.14	99.52±0.29
Sulfamethoxazole RC-C	100.36±1.00	99.82±0.34	100.38±0.41	99.92±0.56	100.59±0.34
Trimethoprim EP Imp-C	100.24±0.37	100.58±0.86	100.50±0.24	100.20±0.35	100.16±0.18
Trimethoprim EP Imp-E	99.53±0.89	100.33±0.53	99.93±0.82	99.87±0.71	100.26±0.32
Diaveridine	99.43±0.64	100.79±1.12	100.07±0.55	99.57±0.62	100.28±0.44
Trimethoprim EP Imp-A	100.14±0.58	100.38±0.59	100.21±0.21	99.89±0.35	99.70±0.23
Sulfamethoxazole N4-Glucoside	100.39±0.43	100.34±1.42	100.22±0.59	100.50±0.67	100.85±0.15
Trimethoprim EP Imp-D	99.84±1.13	100.25±0.75	100.03±0.29	99.89±0.59	99.93±0.46
Trimethoprim EP Imp-B	100.21±0.69	100.64±0.56	100.75±0.47	100.08±0.39	99.74±0.14
Trimethoprim EP Imp-G	99.92±0.73	99.94±0.85	99.79±0.21	99.88±0.43	100.02±0.46

Trimethoprim EP Imp-F	99.24±1.13	100.86±1.05	100.89±0.49	100.20±0.65	100.31±0.33
Sulfamethoxazole RC-F	100.77±2.92	99.62±1.40	100.10±0.98	99.97±0.37	99.95±0.13
Trimethoprim EP Imp-J	99.11±2.85	100.53±0.71	99.45±0.18	99.86±0.46	100.08±0.51
Sulfamethoxazole RC-A	100.26±0.81	100.70±0.55	100.53±0.30	100.14±0.38	100.29±0.32
3,4,5-Trimethoxybenzaldehyde	99.58±2.34	99.86±0.53	99.43±0.11	99.70±0.24	99.67±0.32
3-Chlorosulfamethoxazole	100.71±0.71	100.28±0.68	100.37±0.25	100.17±0.54	100.19±0.72
Sulfamethoxazole RC-B	99.70±1.92	100.59±0.97	100.39±0.67	100.26±0.89	100.05±0.31
Trimethoprim EP Imp-H	100.88±0.30	100.66±1.32	100.00±0.55	100.24±0.56	100.13±0.13
Trimethoprim EP Imp-I	100.04±0.96	99.99±0.52	100.18±0.70	100.06±0.35	100.48±0.29

3.2.5 Sensitivity. The method's sensitive limit of detection (LOD) for lowest peak was found to be around 0.02% based on the established signal to noise ratio. The limits of quantitation (LOQs) for SM, TM, and their impurities were established by inoculating a variety of diluted solutions with specified amounts at a signal-to-noise ratio of 10:1. For more information, see Table 1.

3.2.6 Linearity. The linearity curves of SM, TM, and associated impurities were assessed after the calibration standards (0.1, 0.25, 0.5, 0.75, and 1% w.r.t. to test concentration) were prepared in triplicate injection. All calibration curves were made by graphing the peak area against the concentration of each analyte, and excellent agreement was found for all analytes. In order to calculate the departure from zero, these curves were used to establish correlation coefficients (r), relative response factors, and % y-intercept.

3.2.7 Solution stability. The goal of the solution stability procedures was to determine how long a sample solution would remain stable in typical and conceivably occurring scenarios during sample collection, handling, and analysis. A freshly manufactured LOQ and LOQ spiked sample solution were injected into the system every 4 hours up to 24 hours to complete this activity. The first experiment was carried out at room temperature (about 22°C), and it was discovered that this temperature had a substantial impact on the quantitative determination (% change from freshly made) of spiking sample solution.

In establishing the sampler cooler temperature to 5°C, the identical experiment was carried out again. Under these circumstances, it was discovered that all the chemicals were stable in both LOQ and LOQ boosted sample solution. At a sample temperature of 5 °C, Trimethoprim EP Impurity-H is stable for up to 12 hours. As a result, the solutions that were created for the validation were light-protected and could be used for up to 12 hours.

3.2.8 Robustness study. The robustness study's objective is to assess the proposed analytical approach and determine whether the outcomes are reliable even under slightly different conditions. According to screening and optimization experiment knowledge, critical method parameters such as column temperature between 27 and 33°C, flow rate between 0.7 and 0.9 mL/minute, and buffer concentration between 0.6 and 1.0% were chosen for this evaluation and helped to pinpoint the analytical method's failure point. Several experiments were conducted, and it was found that every resolution complied with the requirements for system appropriateness.

However, the essential pair of TM EP-A and SM N4 Glucoside negatively impacted buffer concentration, flow rate, and column temperature. On SM N4 Glucoside, Benzyl alcohol had a moderately negative impact on buffer concentration, a positive impact on column temperature, and a negative impact on flow rate. On the other pair, TM EP-J and SM RC-A, the column temperature was negatively impacted, the flow rate was positively impacted, and the buffer concentration was moderately negatively impacted.
Finally, this study demonstrated how a system can continue to work even when little changes are made.

4. CONCLUSION

The quantification of SM, TM, and associated impurities in liquid pharmaceutical dosage form, a straightforward, reliable, and focused RP-HPLC approach was devised. Regarding 22 substances, all potential relevant impurities, API, and excipients were taken into account in accordance with the individual pharmacopoeia monographs for SM and TM and the labelling on drug products. Reverse phase HPLC technology, affordable analytical technique, was used to separate all of these components. To accomplish the specified system appropriateness criteria, the method was created using a systematic methodology. By altering important procedure parameters, The approach has proven to be reliable. The technique was verified in accordance with ICH guidelines and it was established that it may be used to identify and quantify relevant STI substances. Finally, the technique was used to assess the impurity levels in the various sources of STI samples and demonstrated its efficacy. Since it contains SM, TM, and their contaminants, it is appropriate for quality control examination of complicated pharmaceutical preparations.

5. REFERENCES

1. A.Sorayya, P.Gharbani, Iran. *Journal of Chemistry and Chemical Engineering*, 2013, **32**,1.
2. M.E.W. Hassouna, *Analytical Letters,* 1997, **30,** 2341.
3. M.H. Givianrad, M. Saber-Tehrani, P. Aberoomand-Azar, M. Mohagheghian, *Spectrochimica Acta Part A*, 2011, **78**, 1196.

4. K. Kargosha, S.H. Ahmadi, *Analytical Letters*, 1999, **32,**1613.
5. Ch. Gangu Naidu, R. Nageswara Rao, N. Narendra Varma, *Analytical Chemistry Letters,* 2019, **9**, 535.
6. L.S. Andrade, R.C. Rocha-Filho, Q.B. Cass, O. Fatibello-Filho, *Electroanalysis,* 2009, **21**,1475.
7. D.H. Shewiyo, E. Kaale, P.G. Risha, B. Dejaegher, J. Smeyers-Verbeke, Y. Vander Heyden, *Journal of Chromatography A*, 2009, **1216**, 7102.
8. K. Datta, S.K. Das, *Journal of liquid Chromatography*, 1988,**11**, 3079.
9. A.S. Amin, M.F.E. Shahat, R.E. Edeen, M.A. Meshref, *Analytical Letters*, 2008, **41**, 1878.
10. A.U. Kulikov, A.G. Verushkin, L.P. Loginova, *Chromatographia*, 2005, **61**, 455.
11. G. P. Tahan, S. C. Machado, E. C. Malaguti, P. P. Maia, S. Rath, I. Martins, *Eclética Química*, 2018, **40**, 32.
12. The United States Pharmacopoeia, National Formulary, United States Pharmacopoeia Convention, Rockville, 2022, **44**, 39.
13. International Conference on Harmonization, ICH Guidelines, Q2(R1) Validation of analytical procedures, International Conference on Harmonization, Geneva, 1996.

A SENSITIVE AZEOTROPIC SOLVENT SCREENING HEAD SPACE GAS CHROMATOGRAPHIC METHOD FOR TERIFLUNOMIDE DRUG SUBSTANCE

Bhagavan Rajesh Babu Koppisetty[b*], Rajendra Prasad Yejella[a], A. Krishna Manjari Pawar[a], Srinivasa Rao Yarraguntla[b], Vasudha Dadi[b], Challa Gangu Naidu[c].

[a]Department of Pharmaceutical Analysis, College of Pharmaceutical Sciences, Andhra University, Visakhapatnam, [b]Department of Pharmaceutical analysis, Vignan Institute of Pharmaceutical Technology, Visakhapatnam, [c]Division of Chemistry, Vignan's Institute of Information Technology VIIT(A), Visakhapatnam,

*Author for correspondence: Email: koppisettybrbabu@gmail.com

1. INTRODUCTION

Teriflunomide(TFM) is an immunomodulatory agent taken orally It is applied to the treatment of multiple sclerosis relapse forms. It is the leflunomide drug's active metabolite. It is a class-2 compound according to the biopharmaceutical categorization system (BCS). The drug is insoluble in water, essentially soluble in acetone, and just marginally soluble in ethanol. It works by selectively inhibiting the reversible mitochondrial enzyme dehydrooroate dehydrogenase, which is necessary for the production of denovo pyrimidine. Chemically speaking, TFM is also known as (Z)-2-cyano-3-hydroxy-but-2-enoic acid-(4-trifluoromethylphenyl) amide.[1,2,3,4] Leflunomide is produced when 4-(trifluoromethyl)aniline reacts with 5-methyl-1,2-oxazole-4-carbonyl chloride in toluene to create TFM. TFM is produced by the subsequent alkaline hydrolysis of leflunomide in methanol. According to their level of toxicity, solvents were categorised into classes I, II, and III in the ICH guideline for residual solvents. Solvents classified as Class-I should be avoided because they may be human carcinogens. Solvents in the Class-II category should be used sparingly since they may be hazardous in an irreversible way. Solvents in the Class-III category are less hazardous and pose less threat to human health. The toluene and methanol used in synthesis were classified as class-2 solvents by ICH, whose use is limited in pharmaceutical products due to their irreversible toxicity such as teratogenicity or neurotoxicity. The concentration limits of class-II solvents, methanol and toluene in pharmaceutical substance are 3000 ppm and 890 ppm respectively according to ICH.[5]

There were reported works for estimation of solvent residuals in various pharmaceutical drug substances by use of head space gas chromatographic technique.[6,7,8,9] There was no reported work for estimation of solvent residuals in drug TFM. The present study focuses on determination of solvent residuals in the drug product TFM. The novel synthetic scheme of TFM involves the use of toluene and methanol during the stages of condensation and alkaline hydrolysis respectively (Figure 1). The methanol and toluene being the residual solvents in the drug TFM were determined qualitatively and quantitatively by the use of Head Space Gas Chromatographic (HSGC). HSGC is used for determining the concentration limits of residual solvents because of its highly sensitive and very high separation efficiency.[10,11,12,13] The validated method was used for estimation of this azeotropic solvent mixture in TFM drug substance.[14,15] LOQs obtained in this study are satisfactory for the detection of ICH concentration limits

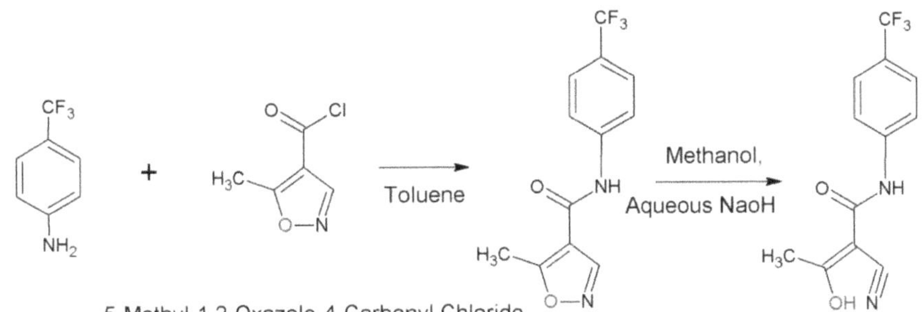

5-Methyl-1,2-Oxazole-4-Carbonyl Chloride

4-(Trifluoromethyl)Aniline Leflunomide Teriflunomide

Figure 1 *A novel synthetic pathway of TFM*

2 MATERIALS AND METHODS

2.1 Chemicals and Reagents

The bulk drug TFM was provided as a gift sample by a biophore india pharmaceuticals pvt. ltd. The HPLC grade solvents dichloromethane, ethanol, methanol, toluene, ethylacetate, N,N-dimethyl formamide and acetic acid were purchased from merck(mumbai, India)

2.2 Experimental

2.2.1 Instrumentation. The analysis was carried out using a flame ionisation detector (FID) and head space sampler (7694E) equipped Agilent 7890A GC system. The data handling system used was empower for data integration. Analytical balance (Sartorius CP224S, Germany). Gas chromatographic column used was DB-624, length of 30m, 0.53mm i.d, 3.0µm film thickness, fused silica capillary column.

2.2.2 Chromatographic Conditions. The HSGC optimized parameters are tabulated in Table 1.

Table 1 *Optimized HSGC* parameters.

Carrier gas	:	Nitrogen
Carrier gas Flow rate	:	2. mL/min
Oven Temp.		
Initial oven temp.	:	40°C
Initial holding time	:	8 min
Temperature rate	:	25°C/min
Final oven temp.	:	240°C
Analysis time	:	10 min
Injector Parameters		
Injector temp.	:	180°C
Split ratio	:	1:7
Data sampling rate	:	5.0
Detector		
Type	:	FID
Detector temp.	:	260°C
Zero air flow	:	300.0 mL/min
Hydrogen flow	:	30.0 mL/min
Makeup flow (N_2)	:	25.0mL/min
Zone Temperatures		
Vial temp.	:	90°C
Loop temp.	:	100°C
Transfer line temp.	:	110°C
Event times:		
GC cycle time	:	26.0 min
Thermostatic time	:	10.0 min
Pressurization time	:	0.2 min
Loop fill time	:	0.2 min
Injection time	:	1.0 min
Loop equilibration time	:	0.1 min
Vial parameters:		
Shake mode	:	Low
Vial pressure	:	14.5 psi
Run time	:	26 min

2.2.3 Standard Preparation. The standard was prepared by weighing and adding 300.0 mg of methanol and 89.0 mg of toluene to a calibrated flask that held 100 mL of the solution and contained 10 mL of N,N-dimethyl formamide (diluent). To create the standard solution, 4.0 mL of this solution were diluted to 100 mL with diluents before performing the serial dilution. A 20 mL head space vial was filled with 2.0 mL of the standard preparation and immediately sealed.

2.2.4 Sample Preparation. The samples were prepared in triplicate by adding 2.0 mL of N,N-dimethyl formamide diluent and 80.0 mg of TFM sample to a 20 mL head space vial before being immediately sealed.

2.2.5 Evaluation of System Suitability. The Diluent N,N-dimethyl formamide was injected as a blank, followed by standard for evaluating the system suitability for the developed GC method. Based on the retention times found in a standard chromatogram, the sample preparations were injected in duplicate to identify each solvent peak in the sample solution.

2.3 Method Validation

2.3.1 Specificity Solutions. For assessing the specificity, separate solutions containing 180.0 µg/mL of ethyl acetate, 180.0 µg/mL of ethanol, 120.0 µg/mL of methanol, 25.0 µg/mL of dichloromethane, 36 µg/mL of toluene, and 40.0 mg/mL of TFM were prepared. In the diluent, a spiked solution was prepared with 40.0 mg/mL TFM, 180.0 µg/mL ethyl acetate, 180.0 µg/mL ethanol, 120.0 µg/mL methanol, 25.0 µg/mL dichloromethane, and 36 µg/mL toluene. The prepared solutions were analyzed to find their resolution and retention times (R_T).

2.3.2 Precision. The precision of system was established by injecting of standard solution for six times (n=6) at specified limit of 100.0% with respect to the working concentration for determining system, method and intermediate precision.

2.3.3 Linearity, Accuracy and Range. For methanol and toluene solution, the linearity of the GCHS technique was demonstrated from the limit of quantification to 150.0% of the specification limit. The recommended correlation coefficient (R) is NLT 0.98. On a TFM sample solution laced with methanol and toluene, the accuracy test was run. The inherent amount of the individual solvent was taken into account. The method accuracy was estimated by using three solutions containing TFM samples spiked with the methanol and toluene at approximately 50.0%(n=3), 100.0%(n=3), and 150.0%(n=3) of the working concentration. The percentage recovery calculated should be in the range of 80.0% to 120.0%. Range of the method was determined from the linearity and accuracy data.

2.3.4 Limit of Detection (LOD) and Limit of Quantification (LOQ). A 100.0 ppm level solution of methanol and toluene in diluent was made and repeatedly injected for six injections to determine the LOD and LOQ. For each listed solvent, the S/N ratio values were computed. The LOD and LOQ solutions were created based on the average S/N ratio values received from the 100.0 ppm level solution.

2.3.5 Robustness. The robustness of the method was determined by analysing a sample of the TFM standard solution with altered column oven temperature and carrier gas flow parameters. The actual level for carrier gas flow was 2.0 mL/min, the low level was 1.8 mL/min, and the high level was 2.2 mL/min. The three levels of column temperature fluctuations were Actual level (40°C), Low level (35°C), and High level (45°C).

2.3.6 Solution stability. Standard solution, sample solution, and spiked sample solution were all used to test the stability of the solution. The solutions were stored at room temperature (22–27°C), and from the beginning to 48h, the stability of the solutions was checked at various intervals.

3. RESULTS AND DISCUSSION

3.1 System suitability data. The standard solution was analysed for system suitability for the validation parameters of specificity, precision at LOQ as per the proposed method. The %RSD for peak area of each specified solvent in system suitability solution were determined for six replicate injections. The %RSD was less than 2, for peak areas of the six replicate injections (n=6) during each validation parameter.

3.2 Specificity. This study has shown that all specified solvent peaks were well resolved from each other and there is no interference in the blank sample at the R_T of specified solvent of methanol and toluene peaks (Figure 2). Therefore the method was selective for the determination of residual solvents by HSGC in TFM drug substance. The results of specificity summarizing retention time, R_T ratio and resolution are tabulated in Table 2.

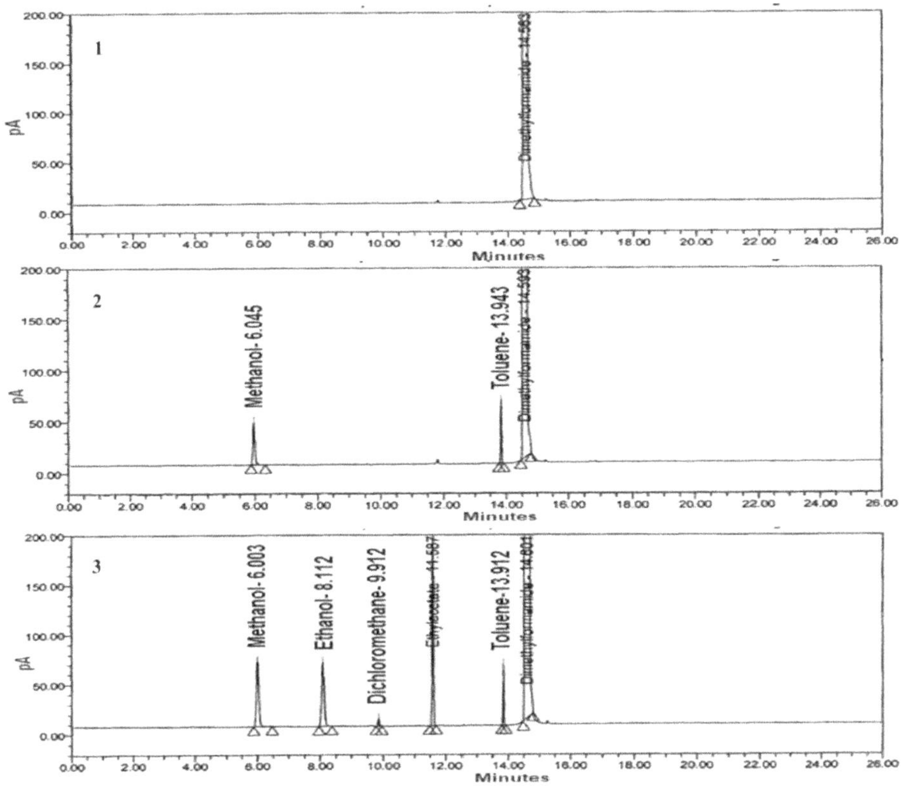

Figure 2 *Chromatograms of 1) Blank 2) Standard solution 3) Spiked solution*

Table 2 *Specificity Study.*

Peak name	R_T[min]	R_T ratio	Resolution
Methanol	6.003	0.41	---
Ethanol	8.112	0.55	15.45
Dichloromethane	9.912	0.68	14.92
Ethyl acetate	11.587	0.79	20.19
Toluene	13.912	0.95	33.70
Dimethyl formamide	14.601	1.00	7.58

3.3 Precision. The system precision showed %RSD for peak area of six replicate injections (n=6) of methanol and toluene in standard solution as1.95% and 1.89% respectively at 100% working concentration.The method precision showed %RSD for recoveries of six replicate injections of each specified solvent of methanol and toluene as 1.71% and 1.95% respectively. The intermediate precision showed %RSD for recoveries for each specified solvent methanol and toluene were listed in Table 3.

Table 3 *Results of intermediate precision for methanol and toluene.*

Sample (n=6)	Day(1) and Column (1)	Day(2) and Column (2)	Day(1) and Column (1)	Day(2) and Column (2)
	Methanol(ppm)		Toluene(ppm)	
1	3089.02	2897.64	953.48	874.12
2	3156.34	2847.84	948.32	902.24
3	3104.87	2947.97	958.56	852.96
4	3180.39	2831.34	936.71	871.23
5	3118.41	2815.13	972.16	880.78
6	3171.28	2850.74	989.24	860.38
Average (ppm)	**3136.71**	**2865.11**	**959.74**	**873.61**
±SD	**37.71**	**49.12**	**18.57**	**17.19**
% RSD	**1.20**	**1.714**	**1.93**	**1.96**

3.4 Linearity. The method linearity for methanol and toluene were 0.65 – 180 µg/mL and 0.24 – 54 µg/mL respectively.At each level, triplicate determinations were performed for getting average peak area. Results obtained are tabulated in **Table-4** and linearity equation for the peak area versus concentration for each specified solvent methanol and toluene was Y= 0.063x +1.163 and Y=0.145x-0.152 respectively with a correlation coefficient (R) 0.99.

Table 4 *Linearity for methanol and toluene.*

Level	Methanol			Toluene		
	Conc. [µg/mL]	**Conc. [ppm]**	**Peak Area (n=3)**	**Conc. [µg/mL]**	**Conc. [ppm]**	**Peak area (n=3)**
LOQ	0.65	16.14	2.4	0.24	6.12	1.16
30 %	36	902.4	56.98	11	264.72	38.36

50%	60	1503.12	98.34	18	446.3	64.47
100%	120	3005.24	192.14	36	898.35	129.66
120%	144	3609.25	232.65	43	1084.38	160.15
150%	180	4510	287.32	54	1336.45	194.12

3.5 Accuracy. The recovery studies were performed at following percentage levels 50, 100 and 150.The % recovery and %RSD for Methanol recoveries were in the range of 91.25 to 96.84 and 1.41 to 2.40 respectively. The % recovery and %RSD for toluene recoveries were in the range of 96.11 to 100.61 and 1.41 to 2.79 respectively. The triplicate(n=3) determinations were performed at each accuracy level for recovery studies of methanol and toluene.

3.6 Range. The range of the method was from LOQ to 150.0% with respect to specification for methanol was 0.65µg/mL– 180µg/mL and for toluene 0.24 µg/mL –53.94µg/mL.

3.7 LOD. Based on the average S/N ratio obtained from 100ppm level solution, the LOD solution was prepared for all specified solvents and injected. The S/N ratio and LOD values for methanol was 5.9:1 and 5.5ppm respectively. The S/N ratio and LOD values for toluene was 4.3:1 and 2.0ppm respectively.

3.8 LOQ. Based on the average S/N ratio obtained from 100ppm level solution, the LOQ solution was prepared for all specified solvents and injected. The S/N ratio and LOQ values for methanol was 12.2:1 and 16.3ppm respectively. The S/N ratio and LOQ values for toluene was 10.9:1 and was 6.0ppm respectively.

3.9 Robustness. The system suitability results obtained when changing the method deliberately was used to prove the robustness of the method for determining the specified solvents. The %RSD was less than 2 for robustness parameter of low column flow rate 1.8 mL/min , high column flow rate 2.2 mL/min, low oven temperature 35°C and high oven temperature 45°C.

3.10 Solution stability. In solution stability study there was no significant variation in the ppm levels of specified solvents upto 48h at room temperature for the solutions of standard, sample and spiked sample. The percentage recoveries for standard solution were observed in the range of 94.82 to 102.12 during the stability period. The variation of the ppm levels of specified solvents in sample solution was below the detection limit. The percentage recoveries for spiked solutions were observed in the range of 95.78 to 101.65 during the stability period. The standard and sample solutions, according to the findings, are stable for 48 hours at room temperature (22–27°C).

4. CONCLUSION

The purpose of determining azeotropic residuals, methanol, and toluene in TFM drug material by HSGC, a designed and verified robust analytical method is helpful. The USP concentration limits in ppm for residual solvents of methanol and toluene were 3000 and 890 respectively. The LOQ values for methanol and toluene were 16.3ppm and 6.0ppm respectively, which are good for the detection of ICH limits. The accuracy is established through the recovery studies at the three levels. The solution stability studies performed showed the stability of the bench work samples for 48 h. The validated method can be used for determination of azeotropic residual solvents methanol and toluene in TFM drug substance.

5. AKNOWLEDGEMENTS

The author K. B. R. Babu is thankful to Vignan Institute of Pharmaceutical Technology, Visakhapatnam for providing laboratory facilities for performing experimental work.

6. REFERENCES

1. L.P. Kotra, J. Park, *Comprehensive Medicinal Chemistry III*, Elsevier, 2017.
2. A. O. Bar, A. Pachner, F. V. Menguy, J. Kaplan, H. Wiendl, *Drugs*, 2014, 74(6) , 659.
3. M. D. Wiese, A. Rowland, T. M. Polasek, M. J. Sorich, C. O'Doherty, *Expert opinion on drug metabolism & toxicology*, 2013, *9*, 1025.
4. J. Oh, & P. W. O'Connor, *Neurology. Clinical practice*, 2013, *3*, 254.
5. ICH, *Impurities : Guideline for Residual solvents,* International Conference on Harmonization, 2016.
6. A. P. Panovska, J. Acevska, G. Stefkov, K. Brezovska, R. Petkovska, A. Dimitrovska, *Journal of Chromatographic Science*, 2016, 54, 103.
7. S. Ramanjaneyulu, K. M. V. Narayana Rao, P. Shyamala, *Journal of Chromatographic Science*, 2022, 60, 117.

8. B. Sivagami., K. S. Ramya, R. Chandrasekar, M. Niranjan Babu, *International Journal of Pharmaceutical Sciences and Nanotechnology,* 2022, 15.
9. C. S. Slater, M. J. Savelski, R. P. Hesketh, E. Frey, In Proceedings of the American Chemical Society 10th Green Chemistry and Engineering Conference, Washington, DC, USA, June, 2006.
10. K. Robards, D. Ryan, *Principles and Practice of Modern Chromatographic Methods,* Academic Press, Cambridge, 2, 2022.
11. N. Kumar, G. G. John, *Journal of Chromatography A*, 1994, 667, 235.
12. H. Hopfer, *Encyclopedia of Dairy Sciences,* Academic Press, Cambridge, 3, 2022.
13. B. V. Ioffe, A. G. Vitenberg, B. V. Ioffe, *Head-space analysis and related methods in gas chromatography,* Wiley, New York, 1984.
14. United States Pharmacopiea, 24th Revision, Asian edition, United States Pharmacopieal convention, Rockville, 2000.
15. ICH, *Validation of Analytical procedures,* International Conference on Harmonization, IFPMA, Geneva, 1994.

A NOVEL FOURIER TRANSFORM INFRARED SPECTROPHOTOMETRIC METHOD FOR QUANTITATIVE ANALYSIS OF IRBESARTAN AND HYDROCHLORTHIAZIDE IN FIXED DOSE FORMULATIONS

N. Raghavendra Babu [a, b], Narender Malothu[b], Y. Padmavathi[a], A. Narmada[a], P. Ravi Kumar[a]

[a]Department of Pharmaceutical Analysis, G. Pulla Reddy Pharmacy, Mehdipatnam, Hyderabad, Telangana-500028, India.

[b]Department of Pharmacy, Koneru Lakshmaiah Educational foundation, Vaddeswaram, Andhra Pradesh-522502, India.

*Corresponding Author E-Mail ID: nayakaraghavendrababu@gmail.com

1 INTRODUCTION

IRB is a non-peptide angiotensin II antagonist with antihypertensive properties. It inhibits angiotensin II ability to bind to the angiotensin I receptor in a selective and competitive manner.[1] The adrenal cortex is stimulated by angiotensin II to produce and secrete aldosterone. It may reduce the excretion of potassium and sodium. The angiotensin II receptor blocker IRB is used either alone or in combination with other medications to treat hypertension and diabetic nephropathy. IRB has been linked to a small number of unusual cases of acute liver injury and has a low rate of transient blood aminotransferase increases. The IUPAC name of IRB is 2-butyl-3-[[4-[2-(1H-tetrazol-5-yl) phenyl] benzyl]- 1,3-diazaspiro [4,4] non-1-en-4-one; Its molecular formula and molecular weight were found to be $C_{25}H_{28}N_6O$ and 428.23 g/mol respectively. The structure of IRB is given in Figure1.

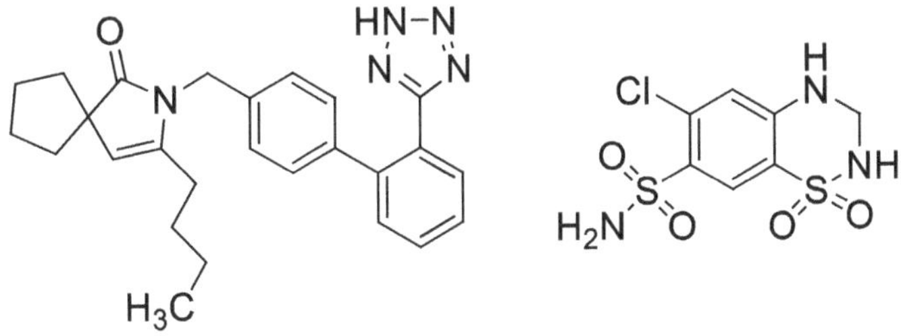

Figure 1. *Structure of IRB and HTZ*

HTZ is a short acting thiazide diuretic and is used in the treatment of hypertension and oedema. The metabolite of this drug appears to bind and accumulate preferentially in red blood cells. The kidneys play a significant role in the excretion of this drug. The physiologic effect of HTZ is by means of Increased Diuresis. The IUPAC name of HTZ is 6-chloro-3,4-dihydro-2H-1,2,4-benzothiadiazine-7-sulfonamide 1,1-dioxide. Its molecular formula and molecular weight are $C_7H_8ClN_3O_4S_2$ and 297.74g/mole respectively.[2] The structure of HTZ is represented in Figure 1.

Separately or in combination, various analytical techniques have been reported in the scientific literature for the estimation of IRB and HTZ.FTIR spectroscopic methods were reported for both the drugs individually.[3-4] Different UV spectrophotometric methods were reported for IRB,[5] whereas a fourth-order derivative UV spectroscopy technique was reported for HTZ. Multicomponent techniques such as simultaneous equation and absorbance ratio methods have been developed for combination of two drug formulations.[6] A generous literature is available on stability-indicating liquid chromatographic methods, HPLC methods for quantification of IRB and HTZ in combined dosage forms.[7-11] The HPTLC method for estimating IRB and HTZ has been published in the literature.[12] A hyphenated technique involving HPLC and tandem mass spectroscopy is also reported for estimation of both the drugs.[13] In light of this, a novel FTIR spectroscopy approach is developed in the current study for the quantification of IRB and HTZ in pharmaceutical dosage forms and pure samples. The FTIR spectroscopy is one of the most significant analytical techniques for scientists. The fundamental advantage of FTIR spectroscopy over other techniques is that all compounds exhibit absorption, allowing for both qualitative and quantitative analysis. Additionally, non-destructive IR spectroscopy allows for in situ and remote measurements of almost any sample, independent of its physical state and without the need for extensive sample preparation. Additionally, without interference from drug excipients, IR-spectroscopic technologies offer a quick, easy, and less time-consuming procedure for assaying the active drugs in their pharmaceutical dosage forms.

Because it doesn't require much solvent, it also reduces the environmental issues associated with industrial chemical waste.[14]

2 MATERIALS AND METHODS

2.1 Chemicals and Reagents Used

IRB and HTZ pure form were obtained from Alembic pharmaceuticals Ltd, Hyderabad and Abbott laboratories pharmaceutical products Ltd, Hyderabad respectively. Potassium bromide of AR grade was obtained from SD fine chemicals limited (SDFCL), Mumbai and Methanol of HPLC grade was obtained from Merck specialities private limited, Mumbai. IROVEL H tablets (Sun Pharmaceuticals Ltd, Hyderabad) were obtained from local market.

2.2 Instrumentation

FTIR instrument- Shimadzu 8400S composed of DLATGS detector, IR solution software and UV-Visible spectrophotometer Shimadzu LC 20AT UV probe 2.43 software was used in the analysis.

2.3 Development of analytical method

Method Development

KBr press were used for sampling solids. FTIR method was developed using fourier transform infrared spectrophotometer.

Selection of measurement mode

Quantitative IR spectra of IRB and HTZ were collected in absorbance mode. Absorbance of peaks at various wave numbers corresponding to various functional groups was determined.

Selection of apodization

The "apodization" function was used to generate the power spectrum by first transforming the interferogram using the Fourier transform. The apodization function influences both the resolution and signal-to-noise ratio of spectra. The noise on the scale becomes more noticeable when the resolution is increased. It was decided to use the apodization function "Happ-Genzel."

Selection of detector

The DLATGS detector of the standard configuration is chosen.

2.3.5 Selection of sampling technique

Geometric mixing was used to prepare pressed pellet concentrations for quantitative estimation. Geometric mixing produced sample pellets. Using FTIR sample cells, sample pellets were generated and analyzed.

The Line plots of IBR and HTZ were plotted by sampling the data obtained from the various FTIR transform infrared spectra by taking wavenumbers on X-axis and absorbance values of respective spectra on Y-axis for the estimation of linear increase in absorbance with respect to wavenumber. IBR Peaks at 3446 cm^{-1}, 2958 cm^{-1}, 1733 cm^{-1}, 1458 cm^{-1}, and 1072 cm^{-1} correspond to N-H stretching, methyl C-H asymmetric stretching, C=O stretching, methylene C-H symmetric bending, and C-N stretching, respectively. IBR quantitative analysis was conducted on the 1733 cm^{-1} group, whose clear, intense peak grew linearly as the concentration increased. Peaks at 3361 cm^{-1}, 1310 cm^{-1}, and 1185 cm^{-1} correspond to N-H stretching, SO_2 symmetric stretching, and SO_2 asymmetric stretching, respectively, in the HTZ IR spectra. For the quantitative study of HTZ, the 3361 cm^{-1} group was chosen because it had a clear, strong peak that grew linearly as the concentration increased.

2.4 Method optimization

The quantization band should be chosen from an area of the spectrum where there is no absorption by other possible sample components. It was decided to use the 400–4000cm^{-1} frequency range. The maximum number of scans chosen was 45 in order to get a higher S/N ratio. To enhance peak separation, a resolution of 8 cm^{-1} was used. Beer-law Lambert's was seen at concentrations of IBR and HTZ between 60 and 300 μg/mg and 5 and 25 μg/mg, respectively.

2.4.1 Optimized analytical method. Standards stocks of IBR and HTZ in KBr: With the help of geometric mixing, 100 mg of dried KBr were combined with the precisely weighed 60 mg of the IBR.This forms the stock of 600μg/mg and for HTZ accurately weighed 5 mg of HTZ 100 mg of dried KBR was mixed to form the stock 50 μg/mg. Mixing should be properly done so that each pellet formed contains uniformly distributed drug.

2.4.2 Preparation of working standard. 5, 10, 15, 20, and 25 mg of IBR were carefully measured from the stock (600 μg/mg of IBR), diluted to 50 mg with dried KBr, and finally concentrated to 60, 120, 180, 240, and 300 μg/mg of IBR, respectively. The drug and dry KBr were appropriately mixed to ensure homogeneous mixing. The final concentration of 5, 10, 15, 20, and 25 μg/mg of HTZ was achieved by properly weighing 5, 10, 15, and 25 mg of HTZ from the stock (50 μg/mg of HTZ) and diluting them to 50 mg with dried KBr. The drug and dry KBr were appropriately mixed to ensure homogeneous mixing.

2.5 Method validation

2.5.1 Linearity and range. Each of the working standard 60, 120, 180, 240, 300 µg/mg of IBR and 5, 10, 15, 20, 25 µg/mg HTZ were prepared and analyzed in FTIR spectroscopy. These standard solutions absorbance of the peaks at 1733 cm^{-1} and 3361 cm^{-1} was noted. Concentration and absorbance were plotted on standard calibration curves. Regression analysis was used to demonstrate linearity, and the regression equation and coefficient of determination are presented.

2.5.2 Limit of Detection and Limit of Quantitation. Limit of detection (LOD) and Limit of quantification (LOQ) were used to determine the developed method's sensitivity for measuring IBR and HTZ. The standard deviation approach was used to calculate the LOD and LOQ. From the calibration curve created for the linearity parameter, standard deviation and slope were obtained.

2.5.3 Sandell's sensitivity. The lowest concentration of IBR (60µg/mg) and HTZ (5µg/mg). Calculate the sandell's sensitivity using the following formula.
Sandell's sensitivity(Л) = Concentration (µg/100mg) × 0.001/absorbance value.

2.5.4 Precision. By reporting repeatability and interday precision, the method's accuracy was determined. An analytical method's precision is typically evaluated using the SD or percentage relative standard deviation (%RSD) of a set of measurements. The precision was determined by using repeatability and interday precision.

2.5.5 Repeatability. By analysing six replicates of the concentrations of 180 µg/mg of IBR and 15 µg/mg of HTZ, respectively, repeatability was determined. The % RSD was calculated.

2.5.6 Interday precision. By analysing three replicates of samples with varied concentrations (120 µg/mg, 180µg/mg, 240 µg/mg for IBR, and 10 µg/mg, 15 µg/mg, 20 µg/mg for HTZ) for two consecutive days, the interday precision of the established method was evaluated. The % RSD was calculated on day 1 and day 2.

2.5.7 Accuracy. Calculating IBR and HTZ's % Recovery using the conventional addition approach served as the basis for the accuracy analysis. A pre-quantified test mixture of IROVEL H tablet extract contain IBR (144,180,216µg/mg) and HTZ (12, 15, 17µg/mg) was mixed with known standards of IBR and HTZ (180 and 15µg/mg). By measuring absorbance and inserting the results into the regression equation of the calibration curve, the percentage of recovery was computed. The % RSD was calculated at each level.[15]

2.5.8 *Assay.* After determining the tablets average weight, twenty IROVEL H tablets, each of which contained 150 mg of IBR and 12.5 mg of HTZ, were crushed and triturated. The tablet powder that was necessary for one tablet was transferred to an Eppendorf tube and dissolved in methanol at this point. After having it vortexed for two minutes, it was then centrifuged at 5000 rpm for 10 min. After that, the supernatant that was generated was collected and allowed to evaporate overnight. It was possible to gather the residue (approximately 172.5 mg when weighed). After that, the entire residue was grinded with 50 mg of KBr to produce a pellet with a concentration of 15 μg/mg, and this pellet was then scanned using the absorbance mode. The following formula (equation 1) can be used to determine the amount of each drug that is contained within the tablet.

$$\text{Assay}=\frac{Concentration\ (\mu g\ per\ mg)\ X\ dilution\ factor\ X\ average\ weight\ of\ the\ tablet}{Weight\ of\ the\ table\ powder\ taken\ (mg)\ X\ Label\ Claim}\ X\ 100 \qquad \ldots\ldots(i)$$

3 RESULTS AND DISCUSSION

3.1. Development and Optimization of the method

By using the KBr pellet method, the FTIR spectra of pure IBR and HTZ are provided in (Figure 2&3.) IBR and HTZ both exhibited strong vibrations at 1733 cm^{-1} and 3361 cm^{-1}, respectively, which are caused by carbonyl (C=O of amide) and amine (N-H) groups. This function is utilised for drug quantification. The Optimized FTIR conditions are given in the (Table 1). Between concentration and absorbance, a linear and proportionate relationship was seen in the 60–300 μg/mg range for IBR and the 5–25 μg/mg range for HTZ and the spectra is shown in (Figure 4&5). This is used to quantify drugs in both pure and formulation form. In the IR spectra of solid pellets of drug combinations, the C=O (carbonyl amide) group peak for IBR and the NH (amine) group peak for HTZ were used to calculate all of the absorbance values.

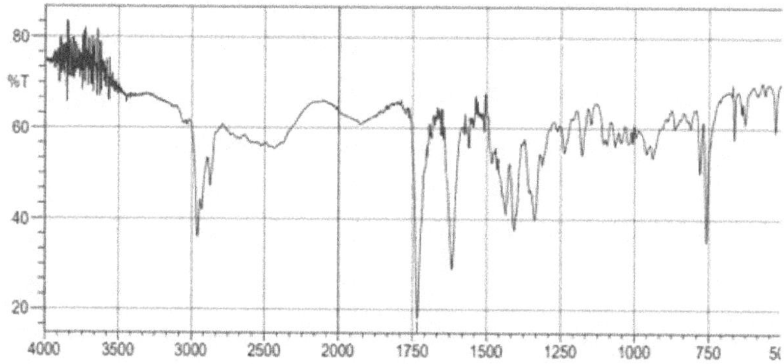

Figure 2. *FTIR spectrum of IBR in transmission mode*

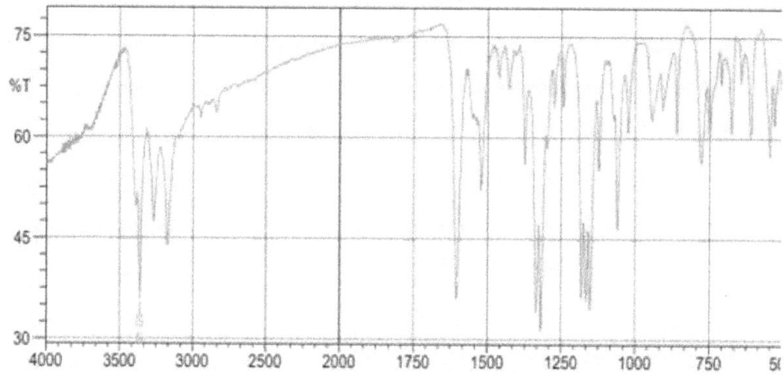

Figure 3. *FTIR spectrum of HTZ in transmission mode*

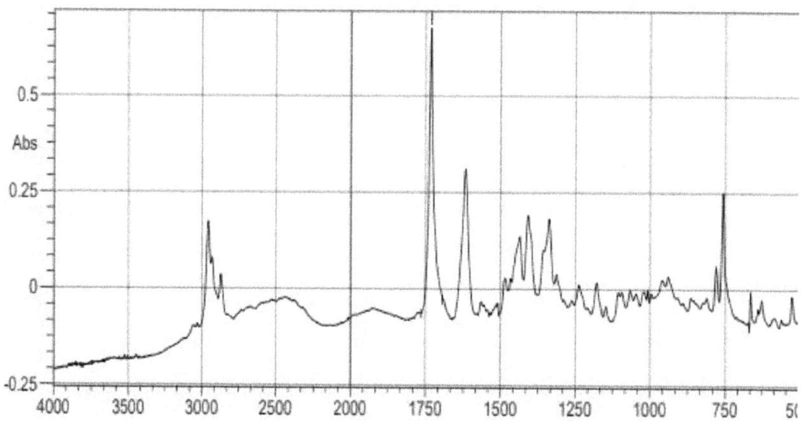

Figure 4. *FTIR spectrum of IBR standard 60μg/mg*

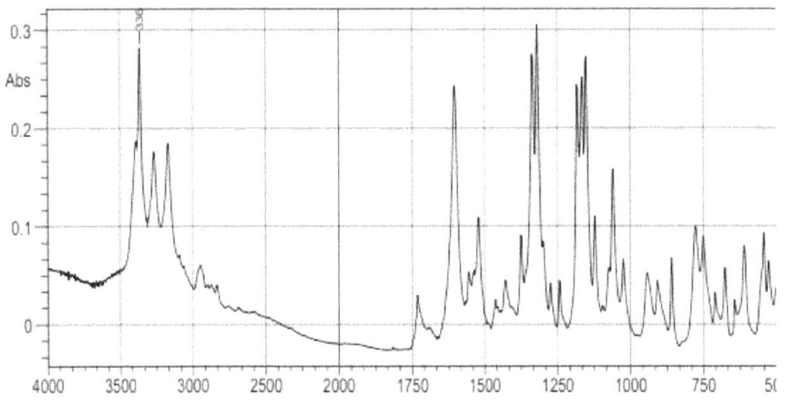

Figure 5. *FTIR spectrum of HTZ standard 5µg/mg*

Table 1. *Optimized FTIR conditions.*

Measurement mode	Absorbance mode
Apodization	Happ-Genzel
Method of making pellets	Direct mixing method
Final weight of pellets	50mg
Frequency range	400-4000cm^{-1}
Peak selection	1733cm^{-1} for IBR 3361cm^{-1} for HTZ
Number of scans	45
Resolution	8.0cm^{-1}

3.2 Validation of the method

Validation was done for the developed FTIR spectroscopy method as per ICH Q2 (R1) guidelines and all the parameters were found to be within acceptable limits.

3.2.1 Linearity

The calibration curve obtained between concentration and absorbance was subjected to linear regression analysis to determine the method's linearity. The method was linear for both the analytes over the range investigated 60-300µg/mg for IBR and 5-25µg/mg for HTZ. The linear regression equation was found to be y = 0.0033x + 0.402 for IRB with R^2 value of 0.9993, while the equation of HTZ was found to be y = 0.0164x + 0.1979 with R^2 value of 0.9996 given in (Table 2) and (Figure 6).

Table 2. *Standard calibration curve data for IBR and HTZ.*

S. No	IBR		HTZ	
	Concentration µg/mL	Absorbance* at 1731cm^{-1}	Concentration µg/mL	Absorbance* at 3361cm^{-1}
1	60	0.5991	5	0.2821
2	120	0.8126	10	0.3580
3	180	0.9940	15	0.4454
4	240	1.233	20	0.5280
5	300	1.4027	25	0.6076

*Average of three determinations

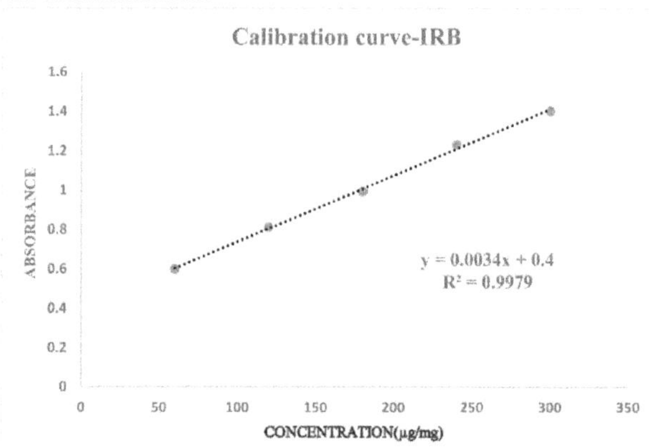

(a)

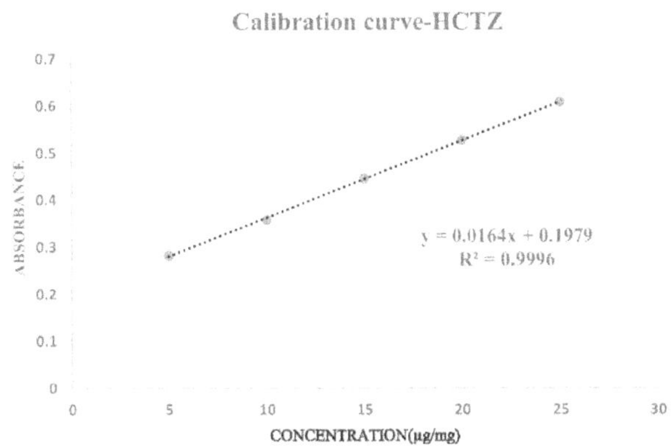

(b)

Figure 6. *(a) Standard calibration curve for IBR and (b) Standard calibration curve for HTZ.*

3.2.2 Sensitivity. The LOD and LOQ values of IBR were calculated out to be 4.526 and 13.715µg/mL, respectively. The LOD and LOQ values of HTZ were calculated out to be 0.278 and 0.844 µg/mL respectively. This indicates the sensitivity of the method. LOD and LOQ are determined based on the standard deviation of the response and the slope.

3.2.3 Sandell's sensitivity. Sandell's sensitivity was calculated for both the drugs at their lowest concentrations. IBR (60µg/mg) shown sandell's sensitivity at 0.1001µg/cm^2 and HTZ (5µg/mg) shown sandell's sensitivity at 0.017 µg/cm^2. These results revealed that the developed method was sensitive.

Table 3. *Sandell's sensitivity data for both IBR and HTZ.*

Name of the drug	Sandell's sensitivity value (µg/cm^2)
IBR	0.1001
HTZ	0.017

3.2.4 Precision. The reproducibility and interday precision of the developed analytical method were reported. The following table 4 displays the repeatability data for six replicates of standard IBR and HTZ pellets. The results revealed that %RSD values were within the limits. Hence the developed method was repeatable.

Table 4. *Repeatability data of IBR and HTZ.*

Concentration (µg/mg)		Absorbance		Mean ± standard deviation (n=6)		% RSD	
IRB	HTZ	IRB	HTZ	IRB	HTZ	IRB	HTZ
180	15	1.0056	0.3720	1.005±0.003063	0.375±0.002836	0.30	0.75
180	15	1.0095	0.3797				
180	15	1.0015	0.3742				
180	15	1.0016	0.3725				
180	15	1.0064	0.3752				
180	15	1.0056	0.3786				

3.2.5 Interday precision. The following table 5 display the interday precision results for three replicates of each IBR concentration (120 µg/mg, 180 µg/mg, and 240 µg/mg) and HTZ concentration (10 µg/mg, 15 µg/mg, and 20 µg/mg). The findings indicate that the% RSD values were within the ranges. As a result, the developed method was accurate. See also Table 6.

Table 5. *Interday precision data of IBR.*

Concentration (µg/mg)	Absorbance*		Mean ± standard deviation (n=3)		%RSD	
	Day 1	Day 2	Day 1	Day 2	Day 1	Day 2
120	0.6752	0.6815	0.674±0.001037	0.683±0.002759	0.15	0.40
	0.6734	0.6863				
	0.6752	0.6815				
180	0.9945	0.9805	0.996±0.001631	0.984±0.007018	0.16	0.40
	0.9968	0.9836				
	0.9978	0.9885				
240	1.0056	1.0016	1.005±0.004018	1.004±0.00256	0.39	0.25
	1.0095	1.0064				
	1.0015	1.0056				

*Average of three determinations

Table 6. *Interday precision data of HTZ.*

Concentration (µg/mg)	Absorbance*		Mean ± standard deviation (n=3)		%RSD	
	Day 1	Day 2	Day 1	Day 2	Day 1	Day 2
10	0.3602	0.3605	0.358±0.004558	0.359±0.000859	1.27	0.23
	0.3624	0.3590				
	0.3536	0.3591				
15	0.4773	0.4721	0.477±0.000994	0.472±0.002737	0.2	0.5
	0.4768	0.4704				
	0.4787	0.4757				
20	0.5801	0.5875	0.585±0.00775	0.582±0.005333	1.3	0.9
	0.5942	0.5820				
	0.5815	0.5768				

*Average of three determinations.

3.2.6 Accuracy. Calculating the percentage recovery of the drugs HTZ and IBR using the standard addition method, respectively, was used for the reliability research. The % of recovery was calculated by measuring absorbance and adding the data into the regression equation of the calibration curve. Table 7 below displays the results for the recovery data for IBR and HTZ (Figure 7). See Table 8 for assay results of marketed tablets.

Table 7. *Recovery data for IBR and HTZ drug product.*

Spike level	Absorbance*		Concentration recovered		% Recovery	
	IRB	HTZ	IRB	HTZ	IRB	HTZ
80%	1.440	0.450	314	15	96%	55%
100%	1.661	0.535	381	20	105%	66%
120%	2.087	0.791	510	36	128%	112%

*Average of three determinations

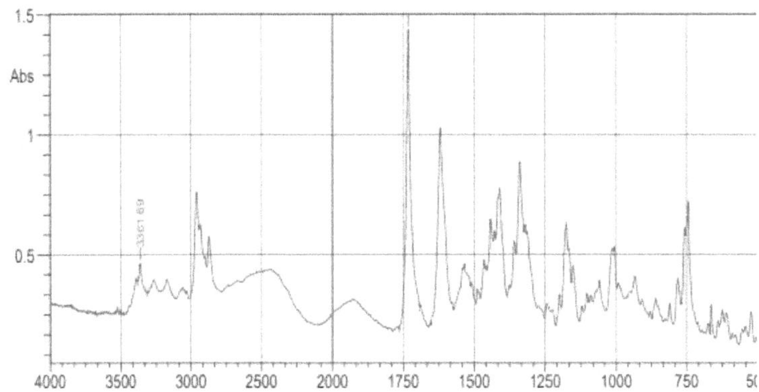

Figure 7. *FTIR spectrum of IBR & HTZ recovery at 80 % level*

3.3 Assay

Table 8. *Assay results of marketed tablets.*

Brand name	Name of the drug	Functional groups	Absorbance*	Label claim mg)		% Purity
				Actual	found	
Irovel H	IBR	1733cm^{-1}	0.623	150	148	98%
	HTZ	3361cm^{-1}	0.586	12.5	12.48	99%

*Average of three determinations.

3.4 Comparison of FTIR and UV methods

Student t- test each the IBR and HTZ linearity findings have been calculated using both of these methods. The outcomes of two different procedures were compared statistically, and the findings revealed a significant difference between the approaches at a significance level (α) of 5%. (tcalculated > tcritical).

Table 9 *Statistical data for t-test of IBR and HTZ linearity.*

Method	Mean of linearity of IBR	Mean of linearity of HTZ	Standard deviation of IBR	Standard deviation of HTZ	Size of sampl es
FTIR	$\bar{x1}$= 1.00434	$\bar{x1}$ = 0.4442	$s_1{}^2$= 0.24920	$s_1{}^2$ = 0.10069	n_1 = 5
UV	$\bar{x2}$ = 0.951	$\bar{x2}$ = 0.3186	$s^2{}_2$= 0.02024	$s^2{}_2$ = 0.0168	n_2 = 5

Hypothesis: The two analytical methods to determine linearity are not significantly different. (ii)

$$H_0 : \mu = \mu_0 \text{------} \quad \text{(ii)}$$

$$Against\ H_1 : \mu \neq \mu_0$$

The t test for difference in means was chosen because the population variances were unknown and the samples were small. The population variances were assumed to be normal, and the test statistic t was calculated using the provided formula:

$$t = \frac{x_1 - x_2}{\sqrt{\frac{(n_1-1)s_1^2+(n_2-1)s_2^2}{n_1=n_2-2} X \sqrt{\frac{1}{n_1}+\frac{1}{n_2}}}} \text{------} \quad \text{(iii)}$$

P value (Probability of rejection) = 0.05 (two – tailed)

$t_{critical}$ (0.05) = 2.3060

Degrees of freedom (df) = n1+ n2−2; (5+5−2) = 8

For HTZ;

$t_{calculated}$ =0.8193

$t_{critical}$ (0.05) =2.3060

Degrees of freedom (df) = n1+ n2−2; (5+5−2) = 8

Our hypothesis was two-sided, so we used a two-tailed test to determine the rejection regions at a 5% level. Using a table of distribution for 8 degrees of freedom, the results were as follows.

R: | t | < 2.306

The observed t value was 0.2297 and 0.8193 (tcalculated < tcritical), which is within the range where our hypothesis is rejected. Therefore, we conclude that the two approaches to determining the linearity of IRB and HTZ are considerably different and reject our hypothesis that the two approaches are not significantly different.

4 CONCLUSION

All compendium pharmacopoeias state that FTIR spectroscopy is frequently used for qualitative analysis and for identifying functional groups in both raw materials and finished products.

For the analysis of IBR and HTZ employing solid pellet technique, the FTIR spectrophotometric method was devised. The findings of a statistical comparison between this and UV showed that the newly discovered approach was significantly different. As a result, its applicability is good. It can be used as an alternative to the official procedures because it met all validation standards in a variety of concentrations. It is appropriate for quality control of both pure and commercial solid dosage forms, and comparable techniques can be developed for other drug categories to estimate their presence in formulations.

5 REFERENCES

1. Pub chem of Hydrochlorthiazide [Online] https://pubchem.ncbi.nlm.nih.gov/compound/irbesartan.
2. Pub chem of Hydro chlorothiazide [Online] https://pubchem.ncbi.nlm.nih.gov/compound/hydrochlorothiazide
3. Khalid AA, AbuSeada HH, Mohammed WN and Adel MA, *Anal. Chem. Int. J,* 2017, **1**, 1-0.
4. Wadher SJ, Kalyankar TM, Puranik MP, and Jayshri S, *International Journal of Medi Pharm Research*, 2016, **2**(01), 32-41.
5. Biyani R and Yadav KS, *Journal of Applied Spectroscopy*, 2019, **86**(5), 934-41.
6. Vetuschi C, Giannandrea A, Carlucci G and Mazzeo P, *Il Farmaco*, 2005, **60**(8), 665-70.
7. Kumar JS and Annapurna MM, *Pharm. Methods*, 2015, **6**, 120-5.
8. Ali TA, Mohamed GG, Aglan AA and Heakal FE, *J. Anal.Chem*, 2016, **44**(1): e1601-8.
9. Pradhan KK, *Int J Pharm Sci Res*, 2016, **7**(11):4531.
10. Albero I, Rodenas V, Garcıa S and Sanchez-Pedreno C. *J Pharm Biomed Anal*, 2002, **20**, 29(1-2), 299-305.

11. Alanazi AM, Abdelhameed AS, Khalil NY, Khan AA and Darwish IA, *Acta Pharm*, 2014, **64**(2), 187-98.
12. Rane VP, Patil KR, Sangshetti JN, Yeole RD and Shinde DB, *J Chromatogr Sci,* 2010, **48**(7), 595-600.
13. Khodke AS, Potale LV, Bothara KG and Damle MC, *Pharmaceutical methods*, 2010, **1**(1):39-43.
14. Qiu X, Wang Z, Wang B, Zhan H, Pan X and Xu RA, J. *Chromatogr. B, 2014*, **957**, 110-5.
15. International Conference on Harmonization of Technical Requirements for Registration of Pharmaceuticals for Human Use. International Conference on Harmonization, Q2 (R1): Validation of Analytical Procedures: Text and Methodology. India. 2005

DEVELOPMENT AND VALIDATION OF RP-HPLC-PDA METHOD FOR THE ESTIMATION OF FLUOXETINE HYDROCHLORIDE IN BULK, MOUTH DISSOLVING FILMS AND IN DISSOLUTION SAMPLES

Jahasultana Mohammed, Buchi N. Nalluri*,

Department of Pharmacy, Koneru Lakshmaiah Education Foundation, Vaddeswaram, AP, India.

*Corresponding Author:

Buchi N. Nalluri, Department of Pharmaceutics, K L College of Pharmacy, KLEF, Vaddeswaram, Guntur, AP, India. buchinalluri@kluniversity.in

1. INTRODUCTION

Fluoxetine hydrochloride (FLX), a first-class agent of antidepressant drug. Chemically FLX is methyl ({3-phenyl-3- [4-(trifluoro methyl) phenoxy] propyl}) amine (Figure 1), a white powder soluble in water. Fluoxetine acts as anti-depressant, by inhibiting the reuptake of serotonin at neuronal membrane. This improves the actions of serotonin on 5HT1A autoreceptors and thus is also known as selective serotonin-reuptake inhibitor (SSRI). Premenstrual tension, bulimia nervosa, panic disorder, anxiety disorders, particularly obsessive-compulsive disorders (OCD), and depression are the main conditions, where FLX is used to treat[1-3].

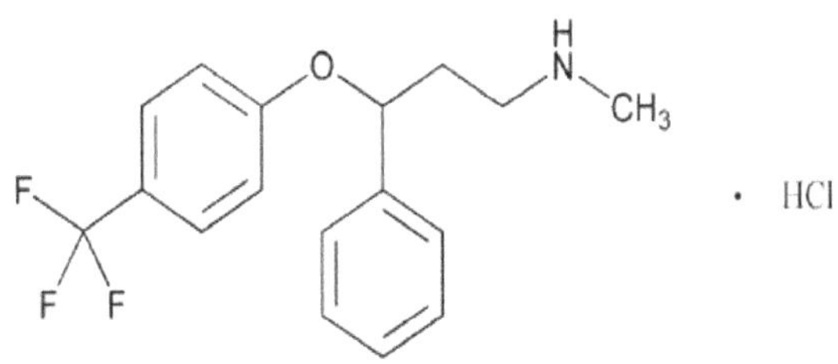

Fig. 1: Structure of Fluoxetine Hydrochloride

The Literature survey states that only few RP-HPLC-PDA methods were reported so far for the estimation of FLX individually[4-5] and in combination with other drugs like Alprazolam, Olanzapine etc., includes HPLC[6-10], Stability indicating RP-HPLC method[11], HPTLC[12-13], UV Spectrophotometry and LC-MS/MS method[14-17] for the FLX determination in bulk, pharmaceutical dosage forms and in plasma samples[18-23]. Despite of this, there are no RP- HPLC-UV/PDA methods were reported for the FLX determination, MDFs in dissolution samples. Here an attempt was made to establish a validated RP-HPLC-PDA method for the FLX determination in bulk, MDFs in dissolution samples as per ICH recommendations[24].

2. MATERIALS AND METHODS

2.1 Instruments

The samples are analysed on a Shimadzu Prominence HPLC system assembled with binary pumps LC-20AD, auto sampler SIL-20ACHT, degasser DGU-20A3, and SPD-M20A photodiode array detector (PDA). After the sample injection, the data was analysed and processed. An Inertsil ODS3 column (150 × 4.6mm, 5μ) was used for during chromatographic analysis

2.2 Chemicals

Fluoxetine as a gift sample was procured from Lifeline pharma, Vijayawada, India. HPLC Grade chemicals like Milli Q water, methanol, and acetonitrile were purchased from Merck in Mumbai, India. Analytical grade Ammonium acetate was procured from Loba Chemi, in Mumbai, India. Fluoxetine mouth dissolving films (MDFs) used for the analysis were in-house prepared containing 20mg of API.

2.2.1. Preparation of mobile phase

Mobile phase was accurately prepared by mixing the composition 10mM Ammonium acetate: acetonitrile in a 58:42v/v ratio and filtered. For about 20 minutes, it was sonicated (degassed). So, prepared mobile phase was placed on HPLC. In isocratic mode, 1.3 mL/min of mobile phase was pumped from the appropriate solvent reservoir. The injection volume of mobile phase was 20μL. Eluents were detected at 226nm using PDA detector.

2.2.2. Preparation of stock and standard solutions

10 mg of FLX were accurately weighed and placed into a clean and dry 10 mL volumetric flask. To that 5ml of methanol was added and sonicated for 2 min. Methanol was used to make the volume up to the required level (1mg/mL). To plot a standard calibration curve, stock solutions were serially diluted at concentrations ranging from 2 to 10μg/mL.

2.3. Preparation of in-house formulation (Mouth dissolving films)

The mouth dissolving films utilized in the research were developed in-house. In a beaker, 5g of FLX was properly weighed and dissolved in a solvent mixture of water and methanol. The other ingredients were added one at a time and blended well. Finally, polymers such as HPMC or Sodium CMC were added and properly mixed. To eliminate entrapped air bubbles, the mixture was sonicated for 5 minutes. The mixture was cast on a glass plate using a wet film applicator (Paul N. Gardner Company Inc. USA) at a thickness of 30mm (750m). The resulting films were dried at a temperature of 45°C in a hot air oven for about 60 minutes. The films dried were then peeled off from the glass plate, trimmed into proper sizes, and kept in a desiccator until they were needed.

2.4 *Method validation*

According to the ICH Q2 (R1) guidelines[24], the developed approach was validated for various method validation parameters.

2.3.1 Specificity. By contrasting sample chromatograms of the FLX standard and MDFs to placebo chromatograms the specificity was found to be satisfactory. Peak purity tests were performed to identify the chromatographic peak from the 3D plots was not due to more than one component.

2.3.2 Linearity. Linearity was determined by preparing different concentrations ranging from 2-10µg/mL (n=3). The resultant data was statistical analysed to characterise slope, intercept, and regression equations.

2.3.3 Precision. The precision was achieved by repeatability. Six replicates of the homogeneous FLX standard sample (6µg/mL) and MDFs samples were injected. Peak area and retention times on chromatograms were statistically analysed to ensure the closeness of the data.

2.3.4 Accuracy. To evaluate this parameter, standard addition and recovery calculations are utilized. Accuracy was determined at three different concentration levels like 80%, 100% and 150% with FLX MDFs samples (6µg/mL) that were spiked with the FLX standard concentration. This samples were injected in triplicate. The %RSD and standard deviation were calculated.

2.3.5 Limit of Detection (LOD) & Limit of Quantification (LOQ). LOD and LOQ for FLX samples was determined by the response of standard deviation (SD) and the calibration plots slope. They were calculated using below formulas:

$$LOD = \frac{3.3\ x\ \sigma}{S} \quad (1)$$

$$LOQ = \frac{10\ x\ \sigma}{S} \quad (2)$$

Where,

σ = Standard deviation of y-intercepts, and
S = mean slope obtained from FLX calibration curve

2.3.6 Robustness. Robustness was evaluated by changing the experimental conditions intentionally. From the above altered conditions, the chromatographic critical parameters (tailing factor, resolution, number of theoretical plates) with respect to change in flow rate ±0.2 mL/min was analysed to know the impact of flow rate on results.

2.3.7 System suitability. This was evaluated by injecting same homogeneous concentration in increased injection volumes of 10-50µL. From the chromatograms the system suitability parameters with respect to change in injection volume was analysed and percentage relative standard deviation was calculated.

2.3.8 Assay of the in-house FLX formulation. Approximately 1cm^2 containing 1.1 mg of FLX mouth dissolving film was taken. This was placed in a 10mL volumetric flask. To that 5mL of methanol (HPLC grade) was added and shaken for a minute. The final volume was made using methanol (0.11mg/mL) and filtered using 0.45µ nylon membrane filter. Further dilution was made using artificial saliva to get 6µg/mL concentration. The preparation of three different individual samples injected onto the column. The amount of FLX present in the MDFs was calculated by comparing the FLX standard peak area with that of formulation sample. The formula below was used to calculate the percentage assay of the samples

$$\% \text{ ASSAY} = \frac{A_T}{A_S} \times \frac{W_S}{D_S} \times \frac{D_T}{W_T} \times \frac{\text{AVG.WT}}{\text{LABEL CLAIM}} \times 100 \tag{3}$$

Where,

W_S = Working standard weight (mg);

W_T = Sample weight.

A_T = Peak area acquired from FLX test preparation.

A_S = peak area achieved from FLX standard preparation.

DS = Standard solution dilution factor.

D_T = Sample dilution factor (mg).

2.3.9 Dissolution Studies. The in-house dissolution method (DISSO 8000, Lab India) described below was followed to evaluate the in-vitro release of prepared FLX MDFs. The dissolution parameters include 500ml artificial saliva dissolution medium with pH 5.7, and USP type II paddle dissolution apparatus. The speed at 75 rpm was maintained. The temperature was kept constant at 37±0.5°C throughout the test. Each FLX 20mg film of approximately 4.6 X 4.6

cm^2 dimensions was transferred to a watch glass and covered with wire mesh made of nylon. After that the film assembled with watch glass was then placed in the dissolution flask. A 5 mL of aliquot samples at regular time intervals were withdrawn, constantly being replaced with fresh 5 mL of dissolution medium. The samples were filter using 0.45µm nylon disk filter and diluted as needed. The dissolution samples were analysed by above mentioned HPLC method. From the chromatographic data, the percentage FLX release were evaluated.

2.3.10 Filter compatibility study. The compatibility of dissolution samples with (0.45µm) nylon filter and (0.45µm) polyvinylidene di-flouride (PVDF) filters was tested. The FLX standard was prepared with dissolution medium and filtered. Chromatograms of FLX standard samples, both in the filtered and unfiltered were injected and chromatographic data was analysed.

3. RESULTS AND DISCUSSIONS

A review of the literature found no published validated HPLC-UV/PDA methods for the measurement of FLX MDFs in dissolution samples. Therefore, the goal of current research work was to establish a simple, cost-effective RP-HPLC-PDA method for the evaluation of FLX in bulk, MDFs and in dissolution samples

3.1 Method development

First trial (Initial) was carried out on an Inertsil ODS3 column (150 x 4.6 mm, 5 µ) with a mobile phase mixture of 10mM Ammonium acetate:acetonitrile in the ratio of 50: 50 v/v. 1mL/min flow rate was maintained and 10mM Ammonium acetate was used as diluent. At a R_T of 3.83 min FLX peak was eluted. In the second trial mobile phase mixture composition was modified to 55:45v/v and all the chromatographic conditions were maintained constant as mentioned above. At a R_T of 4.112 mins FLX peak was eluted but a broad peak was observed. In next trial, the flow rate was altered to 1.2mL/min and all the other LC conditions were maintained constant. At a R_T of 3.521min, but fronting was seen. See Figure 2.

In the subsequent trial, the mobile phase mixture composition of 10mM Ammonium acetate: acetonitrile was changed to the ratio of 60: 40v/v. 1.3mL/min flow rate was maintained. At R_T of 5.516 mins, FLX peak was eluted but there was base line drift and poor peak symmetry Therefore, in the next trial mobile phase mixture composition of 10mM Ammonium acetate:acetonitrile was changed to the ratio of 58: 42 v/v, using artificial saliva as diluent. FLX peak eluted at R_T 4.248 min, in this trial tailing factor was under 2, peak symmetry was satisfied, and fronting was not seen. For quantitative analysis, wavelength was set to 226nm. Fig. 3(A, A1) shows the UV spectra and peak purity curve.

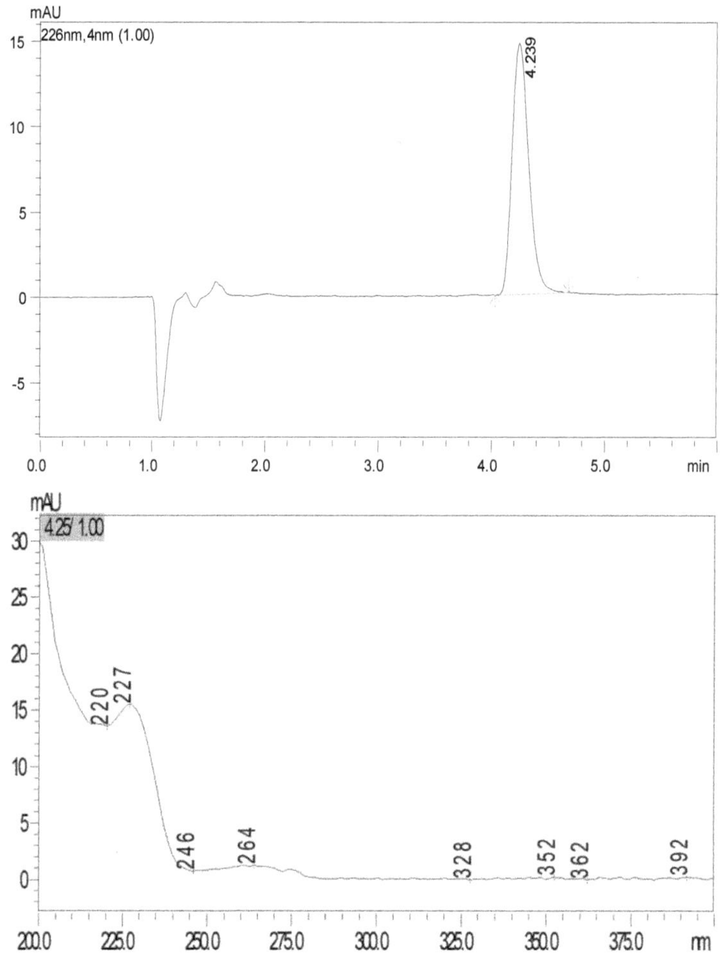

Fig. 2 Chromatogram of FLX

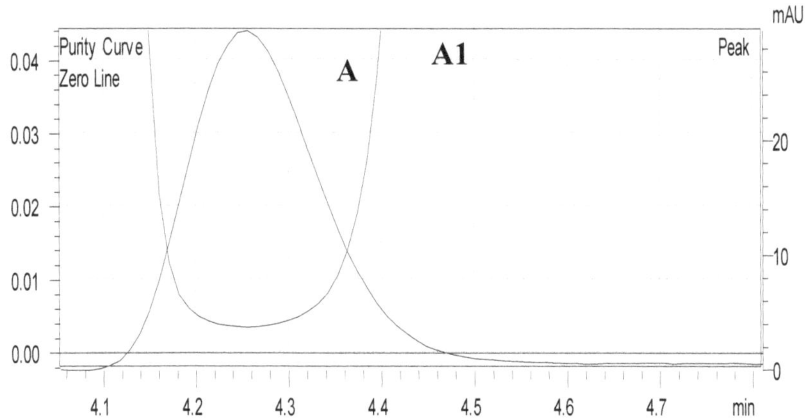

Fig. 3 Absorption Spectrum of FLX (A) and Peak purity curve of FLX (A1).

3.2 Method validation:

The following parameters are determined for validation in accordance with ICH-Guidelines.

3.2.1 Specificity. To examine the interferences, the placebo, diluent sample and standard solutions were individually analysed. Fig.4 showed the base shifted overlay of chromatographic data, with no interfering peaks along with FLX. This demonstrates that the analyte was pure at its peak. The excipients in the formulation of MDFs showed no interfere with the analyte. FLX standard and sample peak purity indices was more than 0.999, supports this parameter.

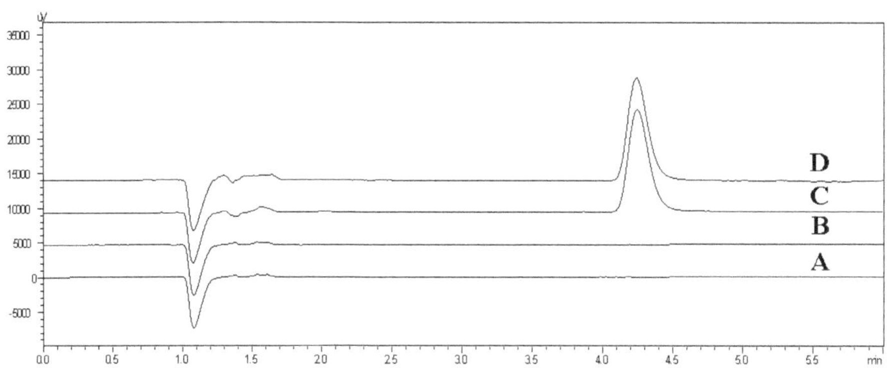

Fig. 4 (I) Typical chromatograms of A) Blank; B) Placebo; C) Standard and D) Sample.

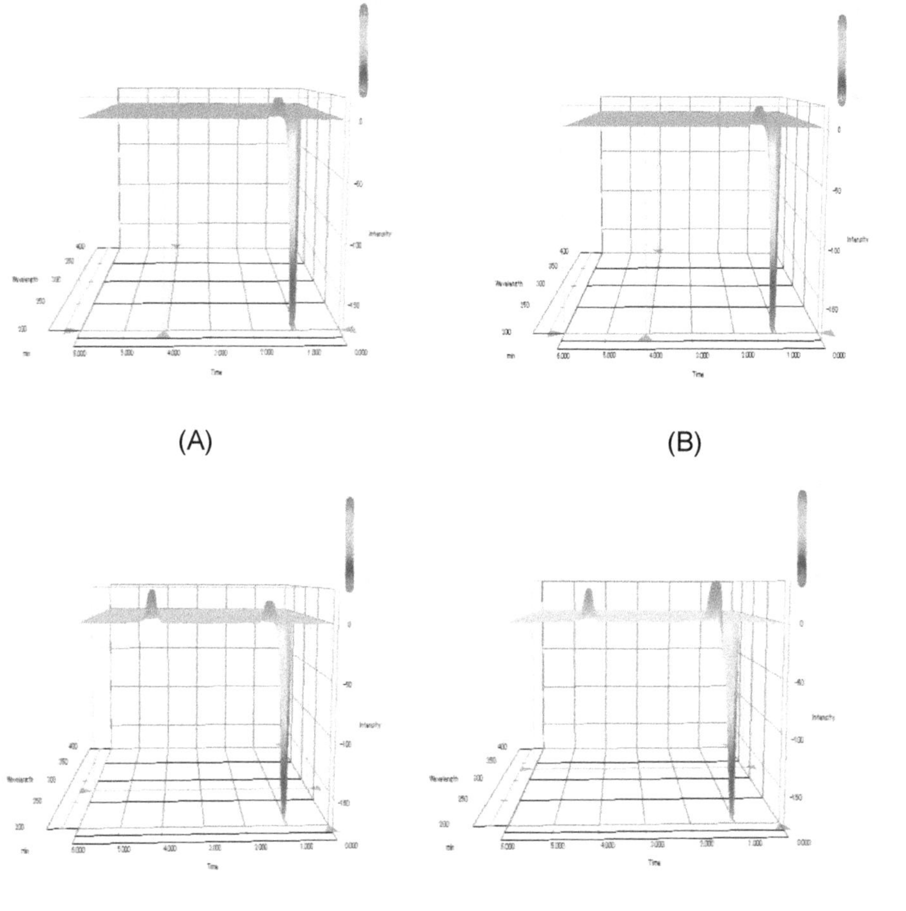

(A) (B)

(C) (D)

Fig. 4(II) 3D plots of Blank (A), Placebo (B), Standard (C), Sample (D).

3.2.2 Linearity. This was determined at 80 to 120% of test concentration. Peak area and concentrations were used to perform the square regression analysis. The correlation coefficient (R^2) was 0.999, shows a linear response curve. From the calibration curve, slope and Y-intercept was calculated and given in Table1.

3.2.3 Precision. Six replicates of the homogeneous FLX standard sample (6µg/mL) and MDFs samples were injected, and the results was tabulated in Table 1. The % RSD value was less than 2, demonstrating the closeness of the values to each other.

3.2.4 Accuracy. The FLX standard to the test sample were analysed and the percent recovery was calculated to be 99.63%. The %RSD was less than 2, indicates this as an accurate method and obtained results were tabulated in Table 1.

3.2.5 Assay. By the proposed technique the FLX assay in MDFs was carried out. The average of three assay samples was about 99.1%. These results indicate that the established RP-HPLC-PDA method can be successfully used for the FLX assay analysis in both bulk and various dosage forms.

3.2.6 Limit of Detection (LOD) and Limit of Quantification (LOQ). The statistical analysis of the calibration curves, was used to calculate LOD and LOQ. The LOD and LOQ was obtained at 0.118µg/mL and 0.558µg/mL concentrations respectively.

Table 1: Linearity, Accuracy, Precision and Assay Results for FLX.

Parameters		FLX
Linearity (n=3)	**Linearity Range**	2-10 µg/mL
	Regression equation	y = 27377x – 8072.1
	Regression coefficient (r^2)	r^2 = 0.999
Accuracy (n=3)	**At % level**	
	80	97.13(0.31)
	100	101.46(0.15)
	120	100.33(1.84)
	% RSD	
Precision (n=6) System Precision	**Peak area**	152322
	(%RSD)	(1.10)
Assay (n=3)	**% Amount found (±SD)**	99.1 (0.78)

3.2.7 Robustness. From the chromatographic data, there was no markable which shows that the established RP-HPLC-PDA method approach was reliable. The robustness parameters were observed and tabulated in Table 2.

Table 2: FLX Robustness data.

Chromatographic parameters	Retention time (min)	Theoretical plates (N)	Theoretical plates
Mobile Phase composition (v/v)			
57:43	5.186	4335.146	1.393
55:45	4.239	3649.974	1.363
52:48	3.547	2849.673	1.335
Flow rate (mL/min)			
0.8	4.958	3503.198	1.388
1.0	4.239	3649.276	1.363
1.2	3.694	3609.276	1.375

3.2.8 System suitability. From the chromatographic data, the retention time (0.19), tailing factor (1.01) and theoretical plate number (0.45) are obtained, indicates the %RSD was less than 2. This indicates the experiment conditions were best suited for the analysis of FLX *invitro* dissolution samples.

3.2.9 Stability of the analytical solution. Stability of samples was obtained by analysing the samples stored under refrigeration (8±1°C) at various time intervals up to 24 hrs. Compared with zero- time interval solution, the % variation in assay results at various time intervals were less than 2, indicates that the solutions were stable for a period of 24 hrs while stored at 8°C.

3.2.10 Filter compatibility study. FLX standard and dissolution samples were filtered and observed for change in assay. From the analytical data, it was considered that nylon filters were more suitable when compared with PVDF filters.

3.2.11 Dissolution analysis of the FLX, MDFs. The dissolution studies of FLX, MDFs was determined, and it was found that 100% of drug released within in 100sec. This shows that this method can be used for the analysis of FLX, MDFs. Figure 5 shows the In-vitro drug release profile of FLX in MDFs formulations.

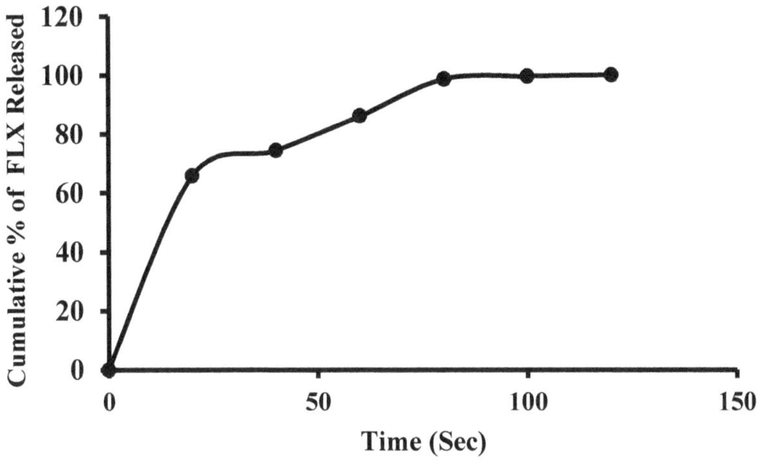

Fig: 5 *In vitro* dissolution profile of FLX from MDFs.

4. CONCLUSION

In this work, a RP-HPLC-PDA method for determining fluoxetine in bulk and its mouth dissolving films was developed and validated. Separation of prepared samples was achieved with a Inertsil ODS3 column (150 × 4.6mm, 5µ) and 58:42v/v (10mM Ammonium acetate: acetonitrile) as mobile phase at flow rate of 1.3 mL/min. The UV wavelength detection was carried out at 226nm. The established method is simple, precise, and rapid. The proposed technique was carried out as per the guidelines of International Conference on Harmonisation (ICH). From the results it was found that this method is applicable for routine analysis of FLX estimation in MDFs and in dissolution samples using both isocratic and binary mode of elution.

Results of method validation parameters like linearity, precision, accuracy, and specificity, showed that they were within acceptable limits. FLX can be determined without use of interference for diluents and placebos. The in-vitro drug release studies of FLX in dissolution medium (artificial saliva) was found to simple and accurate without the need of any other pre-treatment. During in-process QC testing, this method can be used successfully for the determination of *in-vitro* dissolution studies of fluoxetine in mouth dissolving films.

5. ACKNOWLEDGEMENT

The authors are grateful to Lifeline Pharma, Vijayawada, for the Fluoxetine Hydrochloride gift sample, as well as to the college of pharmacy,

Koneru Lakshmaiah Education Foundation Education, Guntur, for providing with the laboratory space needed to conduct the research.

6. REFERENCES

1. Tripathi KD; Essentials of Medical Pharmacology, Jaypee Brother's Medical publisher (P) Ltd; 6th edition; 2008, 446-447.
2. H.P. Rang,MM Dale, JM Ritter, RJ Flower,G Henderson.Rang and Dale's Pharmacology, Elsevier Churchill Livingstone , Elsevier Inc,spain, 7th ed, 2012, 567-568.
3. Indian Pharmacopoeia; Ministry of Health and Family Welfare, Published by Indian Pharmacopoeia Commission; 2010, Vol.II, 1037-1040.
4. Bonde C.G, Saravanam V.S and Agasti L, Ware K, *Asian j of chemistry*; 2008, **20,** 799.
5. Reena paghadal, Mayank Bapna, Jigisha Patel, Hiren Jain, Bhavik Jani. *J of chem and pharm res.*, 2014, **6**, 1166.
6. Shubhanjali Shuklab, Pankaj Kumara , N. S. Hari Narayana Moorthya , Sushant Kumar Srivastava , Piyush Trivedia , Radhey Shyam, Srivastava, Eurasian, *J. Anal. Chem.,* 2010, **5**, 239.
7. Venkateswara Reddy B, Suresh reddy J, Sreeramulu G, Kanumula V, *Chromatographia,* 2007, **66**, 111.
8. Darwish, Ibrahim A; *J of AOAC Int.,* 2005**, 88,** 38-45(8).
9. V. Pranitha, R.Saraswathi, Uma Maheshwar Rao V, Ajitha A. *Int J of Pharm Res & Anal.*,2014, **4**, 281.
10. Shah. R C, Suhagia. N B, Shah. J N, Patel. R D, and Patel. M N, *Indian J Pharm Sci.*, 2008, **70**: 251.
11. Sejal Patel* and N. J. Patel, *Indian J of PharmSci.*,2009, **71**, 477.
12. Mahmoud A. Tantawy, Nagiba Y. Hassan, Nariman A. Elragehy, Mohamed Abdelkawy, *J of Advanced Res.,* 2013, **4**, 173.
13. Duverneuil, Charlottede la Grandmaison, Geoffroy Lorinde Mazancourt, Philippe; Alvarez, Jean-Claude, *Therapeutic Drug Monitoring*: 2003, **25**,565.
14. Sutherlan F.C.W, Badenhorst D, A.D. de Jager, T. Scanes, H.K.L. Hundt, K.J. Swart, A.F. Hundt, *J of Chromatography A*,2001, **914**, 45.
15. Kong M Li, Murray R Thompson, Iain S M.C. Gregor, *J of Chromatography B*,2004, **804**,319.

16. Shanmugam Gopinath, Ramadass Satiesh Kumar, Sekar Alexander, Palanisamy Danabal; *Biomed Chromatography*, 2012, **26**, 1077.

17. Kristoffersen L, Bugge A, Lundanes E, Slordal L, *J of Chromatography B: Biomed Sci and Appl.*,1999,**734**, 229.

18. Manuela T. Maya, Constantino R. Domingos, Teresa M Guerreiro, Jose A. Morais, *J Pharm Biomed Anal.,* 2000,**23**, 989.

19. S.L. Bonde, R.P. Bhadane, Avinash Gaikwad, S.R. Gavali, D.U. Katale, A.S. Narendiran, *J of Pharm& Biomed Anal.,* 2014, **90,** 64.

20. Gennaro.C M, Abrigo C, Angelino S, Albert U, Bogetto F, Maina G, Prolo P and Ravizza L, *J of liquid chromatography & related tech.*, 1997, **20**, 3017

21. Zhongzhou Shen,Sui Wang, Ray Bakhtiar, *Rapid communication in MS*, 2002, **16**,332.

22. Niyazi Yilmaz , Yalcin Ozkan , Sibel A. Ozkan , Inci Biryol & Hassan Y. Aboul-Enein, *J of Liquid Chromatography & Related Tech.,* 2000, **23**, 1699.

23. Sunandana.B, Sushmitha.K, Buchi N. Nalluri; *J of Applied Pharm Sci.,*2013, **3**, 126.

24. International Conference on Harmonization of Technical Requirements for Registration of Pharmaceuticals for Human use. Validation of Analytical procedures Q2 (R1); 2007.

A HIGH-PERFORMANCE LIQUID CHROMATOGRAPHIC METHOD FOR DISSOLUTION TESTING OF LURASIDONE TABLETS

Paras Jain, N V D Prasad Ketha, Buchi N. Nalluri*

Department of Pharmacy, KLEF Koneru Lakshmaiah Education Foundation, Vaddeswaram, AP, pin 522302 India.

*Corresponding Author E-Mail ID; buchinalluri@kluniversity.in

1 INTRODUCTION

Lurasidone Hydrochloride (LH) (See Figure 1) is a central dopamine D_2, and serotonin Type 2 (5HT2A) receptor antagonist used in the treatment of schizophrenia and bipolar depression. The weakly basic drug is benzothiazole derivatives and exhibits strong pH dependent, poor solubility across the physiological pH range, and thus exhibit poor bioavailability. Taking LH with food improves the bioavailability and hence approved product LATUDA® Tablets are to be taken with food. A novel supersaturating amorphous solid dispersion of LH and polymer, having improved solubility was prepared and formulated in a tablet dosage form. A suitable two-stage dissolution methodology was adopted for the characterization of drug release and an HPLC-UV analytical method was developed for quantification of the drug release.

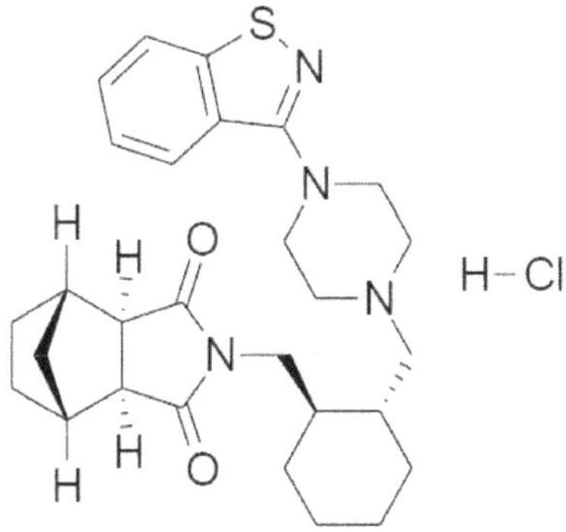

Figure 1 *Chemical Structure of Lurasidone Hydrochloride*

A thorough review of the literature was carried out to enumerate the reported analytical method for LH. A literature survey reveals the determination of Lurasidone hydrochloride in biological fluids and rat plasma, methods such as LC/MS/MS [1-2], UV spectrophotometric[3], and HPLC[4-7]. None of the literature reports a suitable analytical method for quantification of drug release in a two-stage dissolution method to simulate fasted state gastric transit of the lurasidone tablet hence a sensitive HPLC-UV spectroscopic method was developed for the purpose and validated as per ICH guidance.

2 MATERIALS AND METHODS

2.1 Chemicals and Reagents

Acetonitrile (Merck, HPLC Grade), Methanol (Merck, HPLC Grade), Ammonia (Merck, ACS grade), Sodium Hydroxide (Merck, Emplura), Hydrochloric acid (Merch, GR), Orthophosphoric acid (Merck, Emplura), Disodium Hydrogen Phosphate (Merck, Emplura), and Citric Acid are all reagents and solvents of chromatographic grade used for HPLC. Water (HPLC Grade from Milli-Q system) is (Merck, AR). Using LH from MSN Laboratories, the samples of lurasidone hydrochloride tablets were made in the lab. MSN Laboratories provided the working standard of LH.

2.2. Chromatographic conditions

For the creation of a highly sensitive HPLC Method for LH quantification from dissolution samples, the Agilent HPLC system (Agilent Technologies, Germany) was used with a C.D.S software controller and includes a quaternary pump, column heater, sample chiller, and diode array detector. A Waters X-bridge C18 150*4.6 mm, 3.5m was used to produce a symmetrical peak form at a regulated column temperature of 25°C (Part No: 186003034). With a flow rate of 1.5 mL/min, a mobile phase component of A and B in the ratio of 30:70, and a run period of 14 minutes, an isocratic elution was chosen. The components of mobile phase A are a 0.2% ammonia solution in water with a pH of 8.5, and the components of mobile phase B are acetonitrile and water in a 98:2 ratio. 25 µL aliquots of sample solutions were injected, and the results were observed at a 232 nm wavelength.

2.3 Solutions Preparation

2.3.1 Diluents. Three diluents that are needed for testing of both acid and buffer stage dissolution samples were defined; diluent-1 comprises a 50:50 ratio of Acetonitrile and Methanol, and diluent-2 is a triple component mix of 0.05N HCl solution: Acetonitrile: Methanol in the ratios 50:25:25, and diluent-3 is prepared by dissolving 2.8 g of disodium hydrogen phosphate and 2.9 g of citric acid into 1000 mL of water, further adjusted the pH to 3.8 by using orthophosphoric acid or 2N sodium hydroxide solution.

2.3.2 Preparation of standard solutions. A standard stock solution of LH was prepared by dissolving 25mg of standard in 25mL of diluent-1 to obtain a 1mg/ml stock solution. To prepare the standard for Acid Stage dissolution; further, the stock solution is diluted to get a concentration of 12 µg/mL for LH prepared in diluent-3 and for Buffer stage Standard solution, the stock solution is diluted to get a concentration of 12 µg/mL for LH prepared in diluent-2.

2.4. Dissolution Procedure

2.4.1. Acid stage. Dropped the tablet into each of the six dissolution vessels containing a 500mL of 0.1N HCl solution each, and started the dissolution test with USP apparatus-2 using a rotational speed of 75 after ensuring the temperature in each dissolution bowl attends 37°C + 0.5°C. At the specified time, withdraw about 10 mL of the aliquot from each dissolution vessel and replenish the same with fresh media. Filter the sample through a 0.45 µm PVDF (Make: Millipore) filter. Collect the sample after discarding 3mL of the initial filtrate. Further 3mL of the above solution was diluted to 10ml with diluent-3 and mixed well.

2.4.2. Buffer stage. After completion of the acid stage dissolution, added 500 mL of buffer stage media to the acid stage media in each dissolution vessel and resumed the dissolution with the same parameters. At the specified time interval withdraw about 10 mL of the aliquot from each dissolution vessel and replace the same volume with dissolution media. Filter the sample through a 0.45 µm PVDF filter and collected the sample after discarding 3 mL of the initial filtrate. Further 3.0 ml of the above solution was diluted to 10 ml with diluent-2 and mixed well.

3 RESULTS AND DISCUSSIONS

3.1. Chromatographic Method

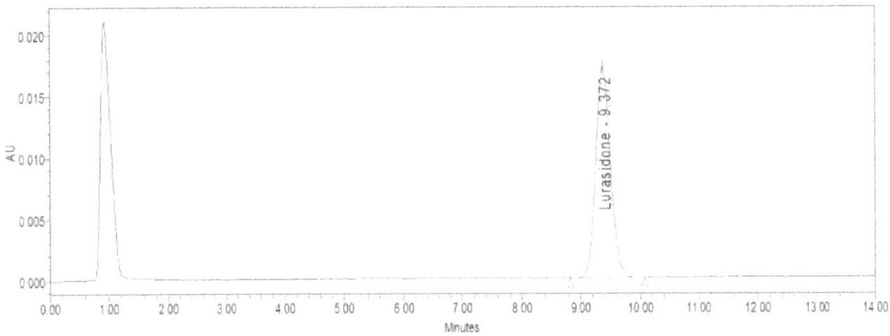

Figure 2 *Typical chromatogram of LH in acid stage dissolution.*

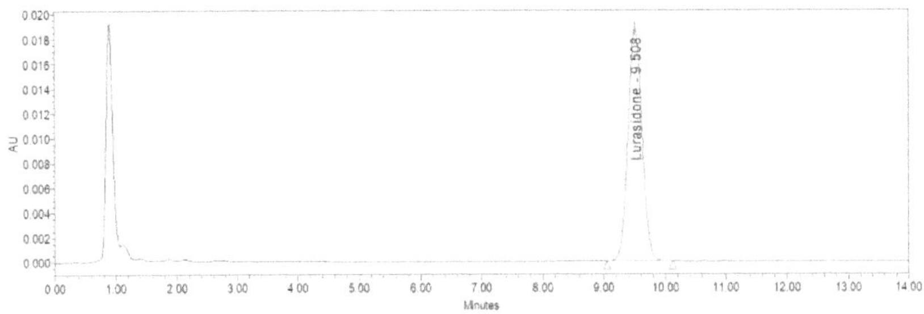

Figure 3 *Typical chromatogram of LH in buffer stage dissolution.*

One of the main objectives of the study is to develop a linear, selective, robust, accurate, and reliable chromatographic method to quantify the LH in its pharmaceutical compounds. A pH-stable resilient water manufacturers' X-Bridge column was chosen for the fundamental separation and quantification of LH due to the difficulties of changing the pH of the solutions for acid and base samples in a single run for comprehensive dissolution profile analysis. Given that LH has a pKa value of 7.6, a mobile phase buffer solution was chosen to adhere to the fundamental ionization principles of avoiding maintaining a mobile phase pH close to the target compound's pKa value, where peak distortion or peak shifting may be seen. The quantification wavelength is selected based on the spectral maxima of LH. To achieve the minimum quantification requirements of the area and peak symmetry, the sample injection volume was chosen to account for as little as 10% of the release profiles of LH from its label claim at the first dissolving point in a time-dependent profile estimation. Figure 2 and Figure 3 despite the chromatograms of LH in acid and Buffer stage dissolutions. A run time of 14 minutes was defined by giving a chance of robustness of the method with preferable variations of 10% changes from each chromatographic parameter.

3.2. Method Validation

The dissolution method has been verified for suitability, specificity, precision, accuracy, linearity, robustness, and solution stability. From the validation data, a set of system appropriateness requirements for the approach have been created.

The minimum level at which the technique was found to be sensitive to LH was 1.2μg/mL. 1.2 μg/mL to 18.0 μg/mL was the range for the method's development, which was based on accuracy and linearity tests. The solutions passed a 24-hour stability test while being stored at room temperature. Overall, based on the results, the strategy has confirmed that the instrument is appropriate for its intended use.

3.2.1 Selectivity. In order to demonstrate that the analyte response is unaffected by potential interferences and endogenous components that coelute with LH in actual samples, the method's selectivity was tested. It was determined by comparing the retention durations of the appropriate standard solutions, placebo, and diluent chromatograms.

3.2.2 Linearity. The linearity for LH was examined using calibration standards ranging from 1.2 to 18.0 μg/mL prepared in the diluent. The standards were injected in triplicate, and the chromatograms were recorded under ideal chromatographic circumstances. The results of the correlation coefficient were around 0.99, and the percentage bias at 100% concentrations was as low as 1.35. Since the data approach is linear. Table 1 refers to the Linearity data and Figure 4 is a demonstrated Graph of linearity data.

Table 1 *Linearity of LH.*

No	The concentration of LH (μg/mL)	Average Peak Area
1	1.2	26497
2	6.0	137854
3	12.0	277074
4	14.4	332704
5	18.0	395021

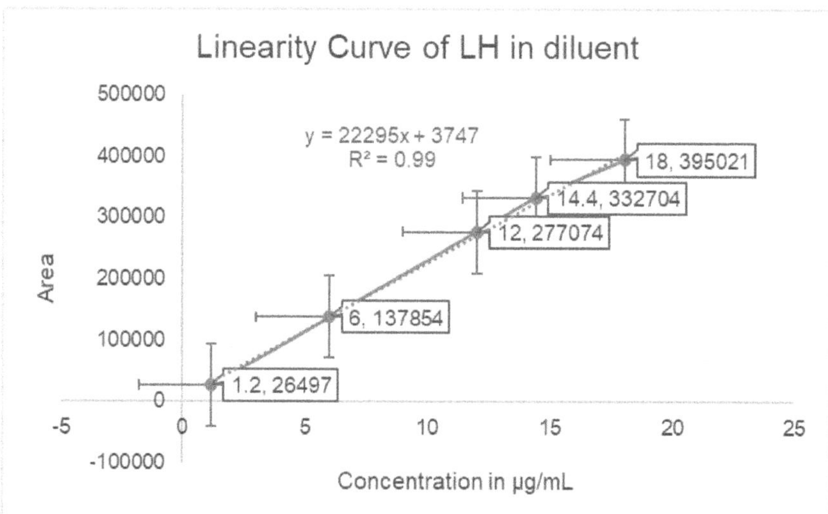

Figure 4 Linearity Graph of *LH in diluent.*

3.2. 3 Precision and accuracy. A Six units dissolution was performed in order to show the precision of the dissolution method and the results of each time point were found with the acceptance criteria of the % RSD for % drug release of six units should not be more than 10.0%. The summary of results was tabulated in Table 2.

The accuracy of the method was expressed in recovery studies. It was performed by the standard addition method, i.e., by adding a known amount of drug to the real sample. The recovery experiments were performed at three different QC levels. The % mean-recovery, SD, and % CV were calculated and reported in Table 3.

Table 2 *Precision Data.*

Time (minutes)	Stage	Unit No % Drug Release						Average	% RSD
		1	2	3	4	5	6		
5	Acid	37	35	37	35	37	35	36	3.04
	Buffer	91	87	88	90	89	84	88	2.55
10	Acid	49	49	49	48	49	47	48	1.77
	Buffer	94	94	95	95	95	92	95	1.34
20	Acid	60	59	60	59	60	57	59	2.04
	Buffer	98	97	98	98	98	96	98	0.98
45	Acid	71	69	71	69	69	68	69	1.83
	Buffer	102	102	103	102	101	99	102	1.47
60	Acid	74	73	74	74	73	70	73	2.2
	Buffer	103	103	105	104	103	101	103	1.41

Table 3 *Accuracy Data.*

No	Actual concentration(μg/mL)	Average recovered concentration (μg/mL)	Percentage recovered (%)
1	1.2	1.225	102.0
2	12.03	11.62	96.6
3	18.04	17.71	98.1

3.2.4 System Suitability. The computation of system suitability parameters is a crucial step in the method development and validation process since it guarantees the system's high efficiency. Chromatographic specifications for six replicate analyses of the LH standard in both acid (A) and buffer stage (B) dissolution conditions, the following variables were tracked: % RSD, USP Plate Count, and USP tailing, data summarised in Table 4

Table 4 *System Suitability Data.*

% RSD		USP Plate Count		USP Tailing of LH	
A	B	A	B	A	B
0.8	1.2	4300	10238	1.4	1.1

3.3.5 Robustness and Ruggedness. The robustness of the developed HPLC method was determined by altering experimental conditions such as flow rate (±0.2 mL/min), pH (±0.2), and column temperature (±5°C) to assess the deviation from the optimized chromatographic conditions.

Table 5 *Robustness Data.*

Condition	% RSD of LH		USP Plate Count of LH		USP Tailing of LH	
	A	B	A	B	A	B
CT - 30°C	1.2	1.3	3702	8792	1.5	1.2
CT - 35°C	0.8	1.2	4300	10238	1.4	1.1
CT - 40°C	1.6	1.2	5672	14982	1.3	1.1
F -1.0mL/Min	1.3	1.4	3298	7895	1.4	1.3
F -1.2mL/Min	0.8	1.2	4300	10238	1.4	1.1
F -1.4mL/Min	0.4	0.9	5234	15432	1.3	1.1
pH -8.0	0.9	1.1	3124	9834	1.5	1.3
pH -8.5	0.8	1.2	4300	10238	1.4	1.1
pH -9.0	1.1	1.1	4980	14356	1.3	1.1

CT – Column Temperature; F -Flow Rate; RSD – Relative Standard Deviation
A -Acid Stage; B-Buffer Stage; USP – United States of Pharmacopeia

4. CONCLUSION

For the intended objective, the devised method for analysing quantitatively dissolution samples of lurasidone hydrochloride has been evaluated. The findings of the statistical analysis demonstrate that the suggested approach has good precision and accuracy. The procedure was fully validated, yielded positive results for each of the parameters assessed for method validation, and was free from interference from any of the additional active components or formulation additions. According to the study's findings, the developed method was found to be rapid, simple, accurate, linear, sensitive, economic, and reproducible. It follows that this technique may be used for the typical product testing analysis of lurasidone HCl in dissolution samples of Lurasidone Hydrochloride Tablets.

5 ACKNOWLEDGMENT

The project and the necessary research facilities were generously provided by Slayback pharma India LLP, Hyderabad, India.

6 REFERENCES

1. Koo, T. S., Kim, S. J., Lee, J., Ha, D. J., Baek, M., & Moon, H., *Biomedical Chromatography*, 2011, **25**(12), 1389.

2. Chae, Yoon-Jee & Koo, Tae-Sung & Lee, *Kyeong-Ryoon.*, *Chromatographia*, 2012, **75** (19-20), 1117

3. *Nikita, M., Jignesh, P., & Mandev, P., International Journal of Research In Pharmacy & Science*, 2012, **2** (2), 44.

4. *Joshi, Nirav and Nehal J. Shah. Pharma Science Monitor*, 2012, **3**(4), 2886.

5. Polawar AR and Damle MC., *International Journal of Research in Pharmacy & Science*, 2014, **4**, 327-332.

6. Chhabda, Pawanjeet., *IJPRD*, 2013, **5**, 103.

7. Balaji M. *IJPRD*, 2013, **5**(09), 103.

8. Validation of Analytical procedure: *Text and Methodology Q2(R1),* International Conference on Harmonisation of Technical Requirements for registration of pharmaceuticals for human use, November 2005.

DEVELOPMENT AND VALIDATION OF STABILITY INDICATING RELATED SUBSTANCES METHOD FOR MONTELUKAST SODIUM AND LEVOCETIRIZINE DIHYDROCHLORIDE COMBINED TABLETS BY RP-HPLC

K Ranjith[a*], D.Rama Sekhara Reddy[b] and TEGK Murthy[c]

[a]Department of Pharmaceutical Chemistry and Analysis, Bapatla College of Pharmacy, Bapatla, Andhra Pradesh, India.

[b]Department of Chemistry, Krishna University, Rudravaram, Machilipatnam, Krishna District, Andhra Pradesh.

[c]Department of Pharmaceutics, Bapatla College of Pharmacy, Bapatla, Andhra Pradesh, India.

Corresponding Author E-Mail ID: ranjith.kapu@gmail.com.

1. INTRODUCTION

Montelukast sodium (MS)[1,2,3] is a leukotriene antagonist which competitively antagonises cysLT1 receptor mediated bronchoconstriction, increased vascular permeability and recruitment of eosinophils. It is indicated for asthma, exercise induced bronchospasm, allergic rhinitis and urticaria.

Levocetirizine dihydrochloride (LD)[4,5,6] is a third generation non sedative antihistamine. It is an active enantiomer of cetirizine. Major effects of levocetirizine are mediated via selective inhibition of H1 receptors. It is indicated for the treatment of seasonal and perennial allergic rhinitis and chronic idiopathic urticaria.

An RP-HPLC method was reported to quantify impurities present in MS active pharmaceutical ingredient (API) [7]. An RP-HPLC method was reported to determine related substances (RS) in LD oral solution[8]. An ultra-high performance liquid chromatographic method was reported to determine RS present in MS and LD combined tablets [9]. Impurities in API mainly depend on raw materials and steps involved in its process of synthesis. Impurities in drug product depend on API, excipients and steps involved in manufacturing. The impurities found in a formulation made by one manufacturer differ from those made by another manufacturer because different formulation manufacturers utilize various APIs, excipients, and manufacturing processes. Therefore, the procedures that have already been disclosed are not quite appropriate for determining RS. This need is the driving force for the development of new methods by analytical research departments for determination of RS in pharmaceutical industry. As most quality control laboratories are equipped with HPLC systems owing to their high performance at affordable cost, Authors were attempted to develop RP-HPLC based gradient method. The present study reports a stability indicating analytical method based on RP-HPLC mode to quantify RS in normal and stressed MS and LD combined tablets.

2. METHOD DEVELOPMENT AND VALIDATION

2.1. Chemicals, solvents and working reference standards

Hydrochloric acid, sodium hydroxide, and hydrogen peroxide-AR grade were purchased from Loba Chemie Pvt.Ltd. Trimethylamine and Methanol –HPLC grade were purchased from E. Merck (India) Ltd. Glacial acetic acid-AR and ACS grade were purchased from Rankem Chemicals. Ammonium acetate-HPLC grade was purchased from Himedia Laboratories. Working reference standards (WRS) of MS (potency: 98.3%w/w), LD (potency: 99.4%w/w), montelukast sulphoxide impurity (potency: 77.0%w/w) and montelukast styrene impurity (potency: 99.3%w/w) were collected as gratuitous samples from Sapco Pharma (P) Ltd, Chennai.

2.2. Sample

MS and LD 10 mg/ 5 mg combined tablets-development stage were provided by Sapco Laboratories (P) Limited, Chennai as gift for this study.

2.3. Preparation of mobile phase

MP-A: 3.85 gm of ammonium acetate was dissolved in 1000 mL of HPLC grade water and adjusted pH to 5.5 with the addition of acetic acid and passed through 0.45 µ millipore nylon filter to remove undissolved particles if any and dissolved gases.
MP-B: Methanol and MP-A (90:10) %v/v were combined.

2.4. Chromatographic conditions

Development of method to accurately estimate RS and subsequently its validation were performed on HPLC make: Prominence with PDA detector SPD-M20A, Prominence HPLC with UV-Visible SPD-20A software: LC Solutions and Waters Alliance with 2695 Separations module and 2996 PDA detector software: Empower-2. All systems were equipped with 100 µL injection loop, auto sampler. Chromatographic conditions (See Table 1.) and gradient programming were set (See Table 2.).

Table 1. *Chromatographic conditions.*

Chromatographic parameters	
Column	Hypersil ODS (150 x 4.6mm), 5µ
Rate of flow	1.0 mL/min
Wavelength for detection	238 nm
Volume of injection	10 µl
Temperature for column	40°C
Runtime	75 mins

Table 2. *Gradient programming.*

Time in mins	%	
	MP-A	MP-B
0	45	55
10	45	55
30	30	70
50	20	80
60	20	80
65	45	55
75	45	55

Note: % - Percentage.

2.5. Preparation of solutions

2.5.1. Sample solvent. Methanol and water were accurately measured and combined (90:10) % v/v.

2.5.2. Diluent. Sample solvent was used as diluent.

2.5.3. MS standard stock solution (26 µg/mL). 26.01 mg of MS WRS was weighed and transferred exactly into a 100 mL volumetric flask, added 70 mL of sample solvent and sonicated for 10 mins. Volume was made up to the mark with sample solvent. 5 mL of the above solution was pipetted into a 50 mL volumetric flask and made up to the volume with sample solvent.

2.5.4. MS standard solution (Reference solution, 2.6 µg/mL). 5 mL of MS standard stock solution was pipetted into a 50 mL volumetric flask and adjusted the volume up to the mark with sample solvent.

2.5.5. LD standard stock solution (12.5 µg/mL). 12.56 mg of LD WRS was weighed and transferred exactly into a 100 mL volumetric flask, added 70 mL of sample solvent and sonicated for 10 mins. Volume was made up to the mark with sample solvent. 5 mL of the above solution was transferred into a 50 mL volumetric flask and made up to the volume with sample solvent.

2.5.6. LD standard solution (1.25 µg/mL). 5 mL of LD standard stock solution was pipetted into a 50 mL volumetric flask and adjusted the volume up to the mark with sample solvent.

2.5.7. Montelukast sulphoxide impurity standard stock solution (974 µg/mL). 19.48 mg of montelukast sulphoxide WRS was accurately weighed and transferred into a 20 mL volumetric flask, added 5 mL of sample solvent and sonicated for 10 mins. Volume was adjusted to 20 mL with sample solvent.

2.5.8. Montelukast sulphoxide impurity standard solution (9.74 µg/mL). 1 mL of montelukast sulphoxide impurity standard stock solution was transferred into a 100 mL volumetric flask and diluted to volume with sample solvent.

2.5.9. Montelukast styrene impurity standard stock solution (276 µg/mL). 5.52 mg of montelukast styrene WRS was accurately weighed and transferred into a 20 mL volumetric flask, added 5 mL of sample solvent and sonicated for 10 mins. Volume was adjusted to 20 mL with sample solvent.

2.5.10. Montelukast styrene impurity standard solution (2.76 µg/mL). 1 mL of montelukast styrene impurity standard stock solution was transferred into a 100 mL volumetric flask and diluted to volume with sample solvent.

2.5.11. Placebo solution. Placebo powder that is equivalent to 50 mg of MS was accurately weighed and transferred into a 100 mL volumetric flask, added 50 mL of sample solvent and sonicated for 5 mins.30 mL of sample solvent was further added and sonicated for 20 mins at a temperature not exceeding 10°C with intermittent shaking. Volume was adjusted to 100 mL with sample solvent and filtered using 0.45µ millipore nylon filter.

2.5.12. Sample solution. Weight of 20 tablets was determined accurately and calculated average weight of the tablet. Tablets were made into a fine powder using mortar and pestle. Powder that is equivalent to 50 mg of MS was accurately weighed and transferred into a 100 mL volumetric flask, added 50 mL of sample solvent and sonicated for 5 mins.30 mL of sample solvent was further added and sonicated for 20 mins at a temperature not exceeding 10°C with intermittent shaking. Volume was adjusted to 100 mL with sample solvent and filtered using 0.45µ millipore nylon filter.

2.5.13. Spiked sample. Tablet powder that is equivalent to 50 mg of MS was accurately weighed and transferred into a 100 mL volumetric flask, added 50 mL of sample solvent and 1 mL of each impurity standard stock solution and sonicated for 5 mins.30 mL of sample solvent was further added and sonicated for 20 mins at a temperature not exceeding 10°C with intermittent shaking. Volume was adjusted to 100 mL with sample solvent and filtered using 0.45µ millipore nylon filter.

2.5.14. System suitability. Reference solution was used to evaluate system suitability of the method. Acceptance criteria for evaluation of system suitability were %RSD of six standard peak areas should be not more than 5.0, theoretical plates for standard peak should be not less than 2000 and tailing factor for standard peak should be not more than 2.0.

%w/w of known Impurity =

$$\frac{At}{As} \times \frac{Ws}{100} \times \frac{5}{50} \times \frac{5}{50} \times \frac{100}{Wt} \times \frac{P}{100} \times \frac{A}{L} \times \frac{586.18}{608.18} \times \frac{1}{RRF} \times 100 \quad \ldots\ldots \quad (1)$$

Where,

At = Area count of known impurity peak in the chromatogram of sample solution
As = Average Area count of standard peak in the chromatograms of reference solution
Ws = Weight of MS WRS taken in mg
Wt = weight of tablet powder taken in mg
P = Purity of MS WRS
L = Label claim of montelukast in mg
A = Average weight of the tablet in mg
RRF = Relative response factor of impurity

586.18 = Montelukast molecular weight

608.18 = MS molecular weight

%w/w of unknown impurity =

$$\frac{At}{As} \times \frac{Ws}{100} \times \frac{5}{50} \times \frac{5}{50} \times \frac{100}{Wt} \times \frac{P}{100} \times \frac{A}{L} \times \frac{586.18}{608.18} \times 100 \quad \ldots\ldots\ldots \quad (2)$$

Where,

At = Area count of unknown impurity peak in the chromatogram of sample solution
As = Average Area count of standard peak in the chromatogram of reference solution
Ws = Weight of MS WRS taken in mg
Wt = Weight of table powder taken in mg
P = Purity of MS WRS
L = Label claim of montelukast in mg
A = Average weight of the tablet in mg
586.18 = Montelukast molecular weight

608.18 = MS molecular weight

% w/w of Maximum Single Unknown Impurity (MSUI) is calculated using highest area unknown impurity peak (i and ii). % w/w of total impurities is the sum of % w/w of total known impurities and % w/w of total unknown impurities.

2.6. Method Validation

Newly developed method to quantify RS was validated subsequently to demonstrate its suitability for intended purpose following ICH guidelines [10,11].

2.6.1. Specificity. Specificity was demonstrated by injecting sample solvent, placebo solution, standard solutions, individual impurity standard solutions, sample solution and spiked sample solution into the chromatographic system. As a part of specificity, sample degradation was initiated by exposure of sample to acidic, basic, oxidative, thermal, light and hydrolytic stress conditions.

2.6.2. Limit of quantitation (LOQ). LOQ values of MS, LD, montelukast sulphoxide and montelukast styrene were estimated by signal to noise ratio method preparing and injecting the solutions at lower concentration into chromatographic system.

-Montelukast and levocetirizine LOQ solution. 3.41 mg of MS WRS and 2.02 mg of LD WRS were accurately weighed and transferred into a 100 mL volumetric flask, added 30 mL of sample solvent and sonicated for 10 mins. Volume was adjusted to 100 mL with sample solvent to prepare LOQ stock solution. 0.5 mL of above LOQ stock solution prepared was transferred into a 50 mL volumetric flask. Volume was adjusted to 50 mL with sample solvent to prepare LOQ solution.

-Montelukast sulphoxide and montelukast styrene LOQ solution. 16.62 mg of montelukast sulphoxide WRS and 10.01 mg of montelukast styrene WRS were accurately weighed and transferred into a 100 mL volumetric flask, added 30 mL of sample solvent and sonicated for 10 mins. Volume was adjusted to 100 mL with sample solvent. 1 mL from the above solution was transferred into a 50 mL volumetric flask and volume was adjusted to 50 mL with sample solvent to prepare LOQ stock solution. 1mL of above LOQ stock solution prepared was transferred into a 10 mL volumetric flask and volume was adjusted to 10 mL with sample solvent to prepare LOQ solution.

2.6.3. Limit of detection (LOD). LOD values of MS, LD, montelukast sulphoxide and montelukast styrene were estimated by signal to noise ratio method preparing and injecting the solutions at lower concentration into chromatographic system.

-Montelukast and levocetirizine LOD solution. 11.01 mg of MS WRS and 6.02 mg of LD WRS were accurately weighed and transferred into a 100 mL volumetric flask, added 30 mL of sample solvent and sonicated for 10 mins. Volume was adjusted to 100 mL with sample solvent. 0.5 mL of the solution prepared as above was transferred into a 50 mL volumetric flask and volume was adjusted to 50 mL to prepare LOD stock solution.1 mL of above LOD stock solution prepared was transferred into a 10 mL volumetric flask and volume was adjusted to 10 mL with sample solvent to prepare LOD solution.

-Montelukast sulphoxide and montelukast styrene LOD solution. 5.55 mg of montelukast sulphoxide WRS and 3.30 mg of montelukast styrene WRS were accurately weighed and transferred into a 100 mL volumetric flask, added 30 mL of sample solvent and sonicated for 10 mins. Volume was adjusted to 100 mL with sample solvent. 1 mL of the solution prepared as above was transferred into a 50 mL volumetric flask and volume was adjusted to 50 mL with sample solvent to prepare LOD stock solution. 1mL of above LOD stock solution prepared was transferred into a 10 mL volumetric flask and volume was adjusted to 10 mL with sample solvent to prepare LOD solution.

2.6.4. Linearity.

MS and LD standard solution. 50%, 75%, 100%, 125% and 150% standard solutions were prepared by transferring 2.5 mL, 3.75 mL, 5.0 mL, 6.25 mL and 7.5 mL each of MS standard stock and LD standard stock solutions accurately into different 50 mL volumetric flasks respectively and volumes were adjusted to 50 mL with sample solvent.

Montelukast sulphoxide and montelukast styrene impurity standard solutions. 50%,75%,100%,125% and 150% were prepared by transferring 0.5 mL, 0.75 mL, 1.0 mL, 1.25 mL and 1.5 mL each of montelukast sulphoxide and montelukast styrene impurity standard stock solutions accurately into different 100 mL volumetric flasks respectively and volumes were adjusted to 100 mL with sample solvent.

A series of standard solutions ranging from LOQ to 150% of specification were injected into chromatographic system. A graph was constructed by taking in µg/mL-concentration on x-axis v/s peak area on y-axis. Slope and intercept were calculated. Correlation between x and y values was indicated by correlation coefficient.

2.6.5. Precision. Reference solution was prepared following test procedure and six repeated injections of predetermined volume from same reference solution vial were introduced into the chromatographic system to evaluate system precision. Mean, standard deviation (SD) and % relative standard deviation (%RSD) for peak areas from six reference chromatograms were calculated. Six different sample solutions were prepared and injected to evaluate method precision. Six different samples of same batch were prepared and analysed separately by another analyst using instrument of different make and same column of different lot on different day to evaluate intermediate precision. Mean, SD, %RSD and 95% confidence interval for % w/w of known impurities, unknown impurities, maximum single unknown impurity and total impurities were calculated.

2.6.6. Accuracy. Known impurities were added to samples at 50%, 100% and 150% of specification limit and injected in triplicate to calculate mean %

recovery. Accuracy for LOQ level was performed by spiking montelukast sulphoxide and montelukast styrene impurities into the placebo. Spiked placebo was injected in triplicate to calculate mean % recovery.

2.6.7. Robustness. Robustness of the developed method was demonstrated by deliberately making changes in chromatographic parameters. System suitability was evaluated under following conditions.

- *Flow rate (+/-10%)*
- *Column oven temperature (+/-5°C)*
- *Organic content (+/-10%)*

2.6.8. Solution stability. Reference and sample solutions prepared as per developed method and maintained at 25°C. Solutions were injected into the chromatographic system at initial, 24 hr and 48 hr to evaluate their stability.

3. RESULTS AND DISCUSSION

Method development [12,13,14] was initiated with mobile phase-A (pH 5.5 ammonium acetate buffer) and mobile phase-B (acetonitrile: mobile phase-A - (90:10) % v/v) using gradient mode but analyte and impurity peaks were eluted closely with less resolution. Acetonitrile in mobile phase-B was replaced by methanol to increase selectivity and different trials were conducted varying gradient programming and temperatures to optimize chromatographic conditions. 40°C temperature (See Table 1) and Gradient programme (See Table 2) were finalized due to good resolution of impurity peaks from each other and other peaks with good symmetry. %RSD of peak areas from six replicate injections 0.2, theoretical plate count more than 7800 and tailing of reference peak below 1.25 indicates meeting of system suitability. Optimised method was subjected to validation to demonstrate its use for intended purpose. Blank, placebo, standard solutions (See Figure 1. and Figure 2), impurity standard solutions (See Figure 3. and Figure 4), sample (See Figure 5) and spiked sample (See Figure 6.) were injected as part of specificity. Standard peaks and impurity peaks in individual solutions, sample and spiked sample were proved to be pure. Degradation of sample in forced manner was performed by stressing samples with different degradation agents (See Figure 7. and Figure 8.) to generate maximum number of impurities which may or may not result in accelerated stability batches of hold time stability. These studies assured applicability of developed method for analysis of stability batches. Purity angle less than purity threshold in Empower 2 software indicates all peaks were pure. Maximum extent of degradation was achieved at acidic, oxidative and photolytic degradation conditions (See Table 3.). Signal to noise ratio method was selected to establish LOQ and LOD. Signal to noise values obtained around 10 for LOQ for MS-0.340 µg/mL, LD-0.200 µg/mL, montelukast sulphoxide-0.332

µg/mL and montelukast styrene-0.199 µg/mL indicate ability of the method to quantify impurities existing in very low quantities. Signal to noise values obtained around 3 for LOD for MS-0.110 µg/mL, LD-0.060 µg/mL, montelukast sulphoxide-0.111 µg/mL and montelukast styrene-0.066µg/mL indicate ability of the method to detect impurities existing in very low quantities. These LOQ and LOD values indicate high sensitivity of the developed method. Standard solutions ranging from LOQ to 150% of specification concentration were prepared and injected to demonstrate linear response of analyte with respect to concentration. Linearity plots were constructed for standards and known impurities. Obtained correlation coefficient (r^2 >0.9900) indicated linear response of the developed procedure. RRFs were established for both known impurities to prevent errors in their quantification. RRFs calculated for montelukast sulphoxide and montelukast styrene were 0.94 and 1.09. % RSD values below 5 for system precision, below 10 for method precision and below 10 for intermediate precision exhibit high precision of the developed method. Accuracy studies were performed in concentration range of LOQ to 150% to ensure recovery of analyte. Mean % recovery values 98.3-101.1 for montelukast sulphoxide impurity and mean %recovery values 96.6-102.5 for montelukast styrene impurity indicates high recovery of analyte from its matrix. Theoretical plates and tailing factor obtained after deliberately varying flow rate, column oven temperature and organic content were within system suitability limits. This indicates robustness of the developed method.

Stability study was performed at room temperature (25°C) to assess stability period of reference and sample solutions. Reference and sample solutions were found to be stable up to 48 hr.

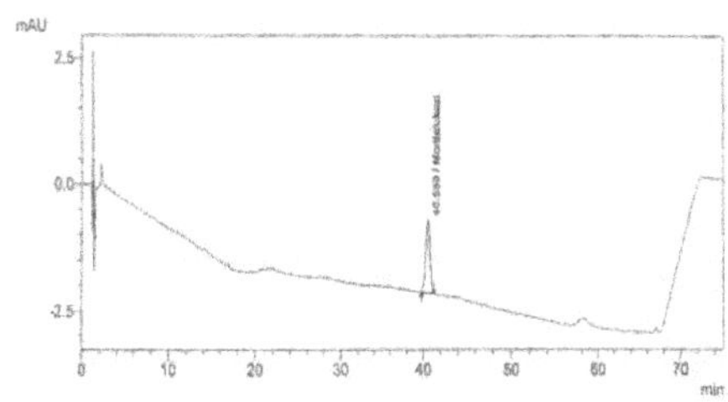

Figure 1. MS standard

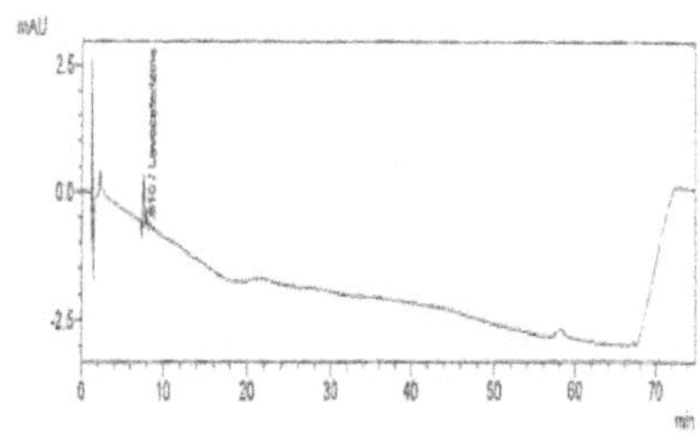

Figure 2. LD standard

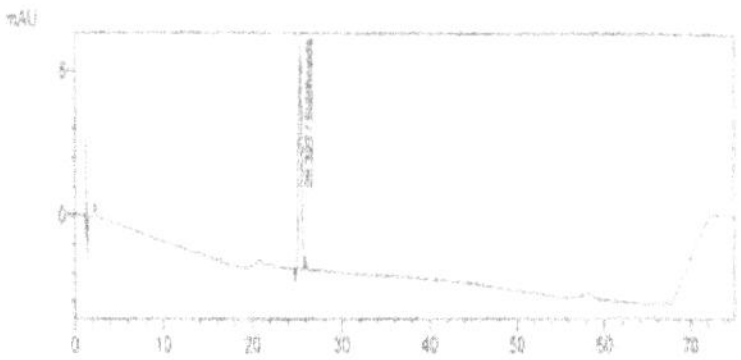

Figure 3. Montelukast sulphoxide impurity standard

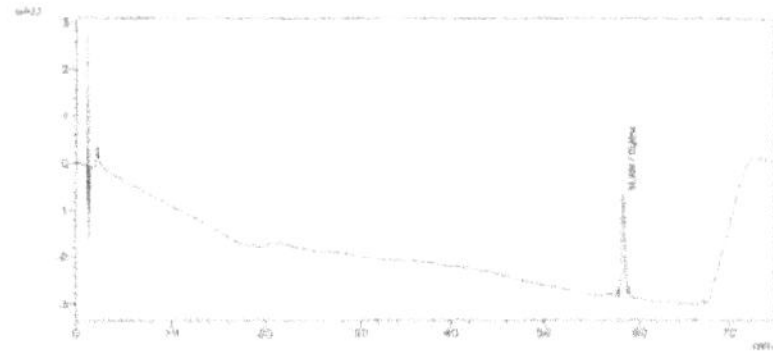

Figure 4. montelukast styrene impurity standard

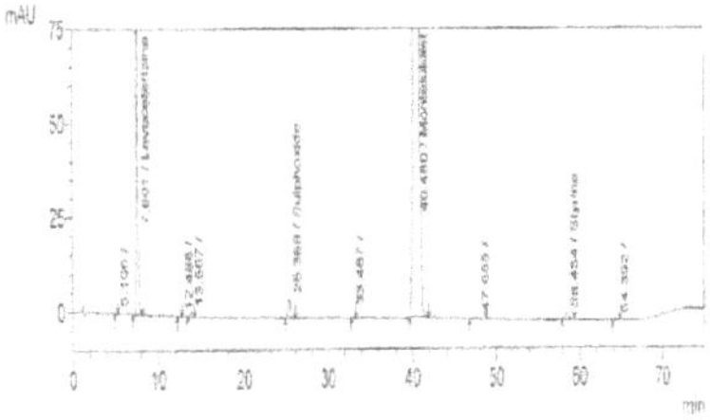

Figure 5. Sample

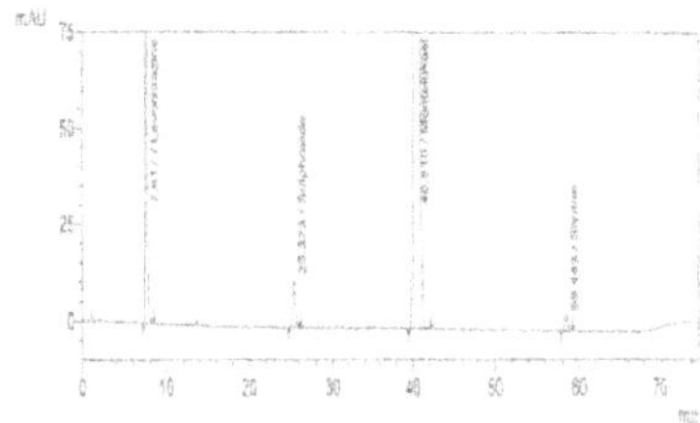

Figure 6. Spiked sample

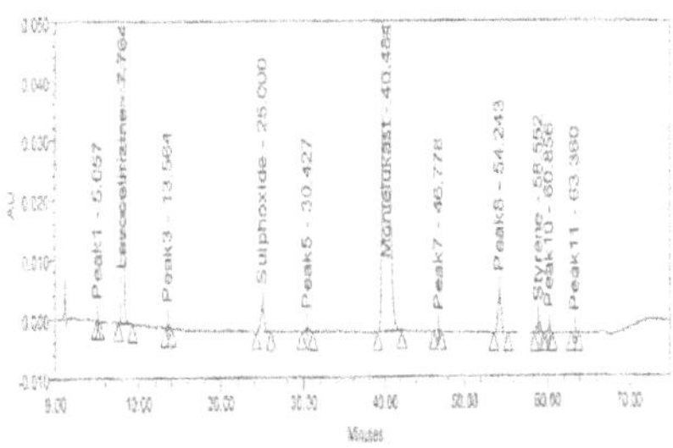

Figure 7. Acid stressed sample

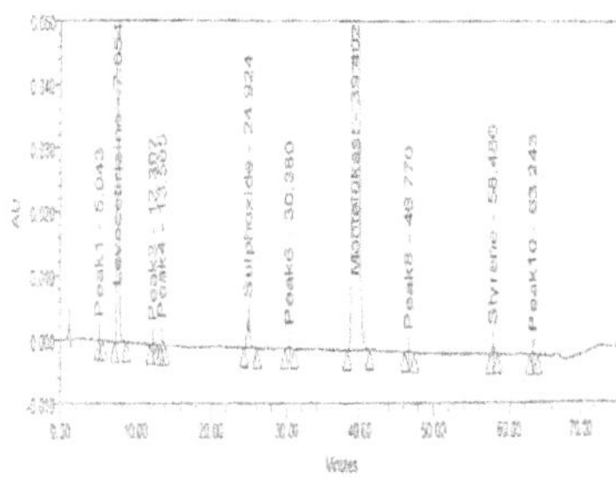

Figure 8. *UV stressed sample*

Table 3. *Forced degradation study data.*

Degradation mechanism	%Weight /weight of				
	Sulphoxide	styrene	total unknown impurities	MSUI	total impurities
Undegraded Sample	1.32	0.25	0.11	0.11	1.68
Acid/5mL of 0.5N HCl stressed for 30 mins at 80°C	1.61	0.44	2.53	2.02	4.58
Base/5mL of 0.05N NaOH	1.61	0.30	0.56	0.18	2.47

stressed for 30 mins at 80°C					
Peroxide/5mL of 1.5%H_2O_2 for 10 mins at 80°C	2.93	0.34	0.79	0.31	4.06
UV/254nm ,168 hr	1.54	2.65	0.26	0.15	4.45
Thermal/105°C for 48 hr	1.46	0.33	0.33	0.13	2.12
Humidity 90%RH,168 hr	1.55	0.33	0.23	0.13	2.11

4. CONCLUSION

A new RP- HPLC method was developed and it was validated following ICH guidelines. As this method was applicable successfully for degraded samples it is claimed to be stability indicating. This procedure can be recommended for quantification of related substances present in newly produced and stability batches of MS and LD combined tablets in quality control laboratory.

5. RECOMMENDATION

As one of the volatile buffers, ammonium acetate buffer was employed as buffer component of movable phase in this approach, this method can be recommended to use for the separation and characterisation of unknown impurities by LC-MS technique.

6. ACKNOWLEDGEMENTS

The author thanks Mr. Brahmaji Valiveti, Chief Strategic Officer of Ingenus Pharmaceuticals, LLC, New Jersey, USA, and Mr. G. Sai Suneel Chand, Director of Sapco Laboratories (P) Limited, Chennai, India, for their support and encouragement.

7. REFERENCES

1. https://go.drugbank.com/unearth/q?query=montelucast&button=&searcher=drugs.

2. https://www.accessdata.fda.gov/scripts/cder/ob/results_product.cfm?Appl_Type=N&Appl_No=020829 #21388.

3. https://www.accessdata.fda.gov/scripts/cder/ob/results_product.cfm?Appl_Type=N&Appl_No=021409 #21857.

4. https://go.drugbank.com/drugs/DB06282.

5. https://www.accessdata.fda.gov/scripts/cder/ob/results_product.cfm?Appl_Type=N&Appl_No=209089 #41953.

6. https://www.accessdata.fda.gov/scripts/cder/ob/results_product.cfm?Appl_Type=N&Appl_No=209090#41954.

7. A. Rashmitha N, B. Raj T, C. Srinivas CH, D. Srinivas N, E. Ray UK, F. Sharma HK and G. Mukkanti K. *E-Journal of chemistry*, 2010, **7(2)**, 555.

8. A. Akula V, B. Sinha BN and C. Seok HJ. *Indian J Pharm Educ Res*, 2018, **52(4)**, 684.

9. A. Kumar SB, B. Jyothi G and C. Rao SB. A. *Indo American Journal of Pharmaceutical Sciences* ,2018, **5(3)**, 1368.

10. International Conference on Harmonisation of Technical Requirements for Registration of Pharmaceuticals for Human, Validation of Analytical Procedures: Text and Methodology Q2(R1), ICH Harmonised Tripartite Guideline, 4th Version, 2013.

11. Validation of Analytical Procedures: Text and Methodology Q3(BR2), Step 4, ICH Harmonised Tripartite Guideline, 2013.

12. A. LR Snyder, B. JJ Kirkland and C. JL Glach. *Practical HPLC method development*, John Wiley & Sons, Inc., New York, 2nd Edition, 1997.

13. A. A Satinder and B. MA Karen. *Handbook of isolation and characterization of impurities in pharmaceuticals*, Academic Press, California, 1st Edition, 2003.

14. A. Y Kazakevich and B.R Lobrutto. *HPLC for Pharmaceutical Scientists*, John Wiley & Sons, Inc.,New Jersey, 2007.

DEVELOPMENT OF RP-HPLC-PDA METHOD VALIDATION FOR THE SIMULTANEOUS ESTIMATION OF ZIDOVUDINE, LAMIVUDINE, AND NEVIRAPINE IN BULK AND DOSAGE FORMS AND IN DISSOLUTION SAMPLES

Potu. Kalyani[a*], Buchi N. Nalluri[b], M. Chinna Eswaraiah[a], Tata Prasanna Kumari[b]

[a]Anurag Pharmacy College, Ananthagiri, Telangana- 508206, India.

[b]Department of Pharmacy, Koneru Lakshmaiah Education Foundation, Vaddeswaram, AP, India.

***Corresponding Author**: kalyanipotu8@gmail.com.

1 INTRODUCTION

The fixed-dose combinations of nucleoside reverse transcriptase inhibitors (NRTIs), such as LMV and ZDV, protease inhibitors, and non-nucleoside reverse transcriptase inhibitors (NNRTIs), NVP are effective in the first-line treatment of HIV infection by acting at various stages of viral replication, and all three drugs (Figure. 1) are approved official by I.P.[1-4]

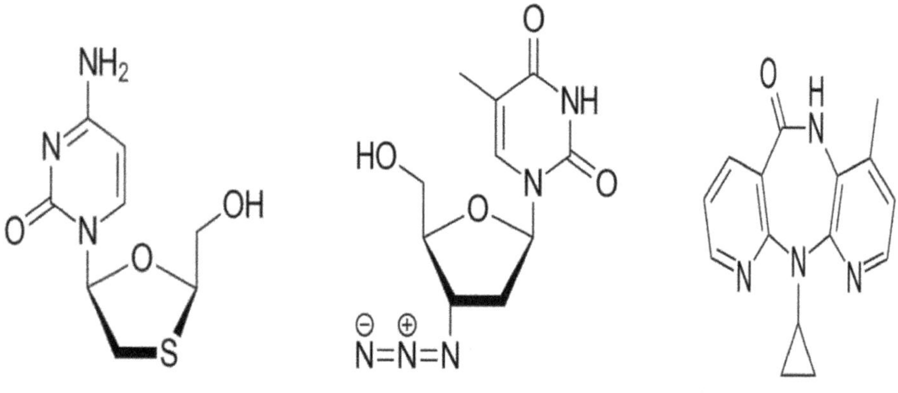

(A) (B) (C)

Figure 1 Structures of LMV (A); ZDV (B); NVP (C).

A review of the scientific literature reveals that only a handful of RP-HPLC techniques for the quantitative estimation of LMV, ZDV, and NVP individually,[5-7] and in combination were established for the analysis.[8-14] On the other hand, no validated HPLC-UV/PDA methods published for the simultaneous determination of LMV, ZDV, and NVP in samples of dissolution. As a result of this, the purpose of this investigation was to establish a validated RP-HPLC-PDA technique for the LMV, ZDV, and NVP determination in bulk, dosage forms, and in dissolution samples by ICH guidelines.[15]

2 MATERIALS AND METHODS

2.1 Chemicals

LMV, ZDV, and NVP were gift samples from Hetero drugs, Hyderabad, India. The HPLC-grade methanol and water were acquired from Merck in Mumbai, India. From Loba Chemi in Mumbai, India, ammonium acetate of analytical grade was acquired. The local pharmacy sold film-coated Duovir-N tablets (Cipla, India) containing a combination of NVP (200mg), ZDV (300mg), and LMV (150mg).

2.2 Analytical method development and optimization

2.2.1 Instrumentation Chromatographic conditions. The data was processed using LC solutions on a Shimadzu Prominence HPLC system equipped with a DGU-20A3 degasser, LC-20AD binary pumps, SIL-20 autosampler, and SPD-M20A PDA detector. The chromatographic analysis was conducted on an Inertsil ODS3 column (150×4.6mm,5µ) for separation. In isocratic, the mobile phase was sonicated and pumped at 1 mL/min flow rate. 10 µL injection volume was maintained at room temperature throughout the analysis. Eluents were analyzed with a PDA detector at 270nm.

2.2.2 Preparation of sample and standard stock solutions. Separate stock of LMV, ZDV, and NVP were prepared by properly weighing 10 mg of each compound and transferring it to a 10 mL volumetric flask, containing 5 mL methanol and sonicated for one minute to solubilize. The volume was topped off with methanol to achieve 1 mg/mL concentration. Using appropriate dilution of stock solution, calibration standards in concentrations 5 to 25µg/mL were prepared.

2.3 Method Validation

The below approaches were validated under ICH Q2 (R1) requirements for analytical method validation.

2.3.1 Specificity. In this investigation, the specificity was determined by contrasting the drug substance, and product chromatograms with 3D plots of the blank and placebo samples. The method's specificity is determined by the lack of peaks other than of excipients or impurities.

2.3.2 Linearity. At a minimum of five concentrations, the linear relationship was assessed. A series of standard dilutions of LMV, ZDV, and NVP were produced and injected in triplicate over a concentration 5-25μg/mL. Using a plot of peak regions, linearity is tested, with proper statistical methods, and slope, intercept, regression (R^2) were calculated.

2.3.3 Precision. Precision, express the closeness of data between a set of measurements obtained by injecting identical samples under the specified conditions. Repeatability, intermediate precision, and reproducibility are three levels of precision that can be measured.

2.3.4 Accuracy. To evaluate this parameter, standard addition and recovery calculations are utilized. Accuracy was determined at three different concentrations i.e., 80%, 100%, and 120% that were spiked with the standard concentration. These samples were injected in triplicate and calculated for %RSD, standard deviation.

2.3.5 Limit of Detection & Limit of Quantification. Calibration curves were used to calculate LOD and LOQ. They were calculated using below formulas:

$$LOD = \frac{3.3\ x\ \sigma}{S} \quad (1)$$

$$LOQ = \frac{10\ x\ \sigma}{S} \quad (2)$$

Where,

σ = Standard deviation of y-intercepts, and

S = mean slope obtained from FLX calibration curve

2.3.6 Robustness. It was evaluated by changing the experimental conditions intentionally. From the altered conditions, the chromatographic interpretation parameters such as tailing factor, resolution, number of theoretical plates for change in flow rate ±0.2 mL/min were analysed to know the impact of flow rate on results.

2.3.7 System suitability. This was evaluated by injecting the same homogeneous concentration in increased injection volumes of 10-50μL. From the chromatograms, the system suitability parameters for change in injection volume were analysed and the percentage relative standard deviation was calculated.

2.3.8 Assay. The average weight of twenty film-coated Duovir-N tablets (Cipla, India) containing a combination of NVP (200mg), ZDV (300mg), and LMV (150mg) was calculated. Using a mortar and pestle, the tablets were finely powdered. A 35.1mg powder mixture containing 6.6mg of NVP, 9.9mg of ZDV, and 5mg of LMV was weighed, transferred to a 10mL volumetric flask, containing 5mL methanol and sonicated for one minute. The contents were brought up to the concentration of 3.51mg/mL, and serial dilutions were performed to get the concentration of 5-25µg/mL. The solution prepared above was filtered and diluted with 15mM ammonium acetate to achieve a concentration 15µg/mL. Three distinct samples were prepared and injected individually. The drug present in formulation was determined by comparing the sample area to that of the standard. By using the above formula the assay percentage was calculated:

$$\%\ \mathbf{ASSAY} = \frac{A_T}{A_S} \times \frac{W_S}{D_S} \times \frac{D_T}{W_T} \times \frac{\text{AVG.WT}}{\text{LABEL CLAIM}} \times \mathbf{100} \qquad (3)$$

Where,

W_S = Working standard weight (mg).

W_T = Sample weight.

A_T = Peak area acquired from LMV, ZDV, and NVP test preparation.

A_S = peak area achieved from LMV, ZDV, and NVP standard preparation.

DS = Standard solution dilution factor.

D_T = Sample dilution factor (mg).

2.4. Dissolution Studies

By using 900mL of 0.1N Hydrochloric acid (pH1.2) as the dissolution medium and a USP type-II (paddle) apparatus, *in-vitro* dissolution studies were conducted (DISSO 8000, Lab India). For the test, a speed of 50 rpm, and a temperature of 37±0.5°C were maintained. The tablets were placed directly in the dissolution vessels. At predetermined intervals, 5mL samples were withdrawn and replaced with 5mL of fresh dissolution medium to maintain sink conditions. The filtered (0.45m nylon disk filter) samples were diluted and analyzed using the HPLC method described previously. The percentage of LMV, ZDV, and NVP releases was determined by comparing the peak area of the test samples to that of the standard.

2.5 Filter compatibility study

The disso samples with both the nylon and PVDF filters (0.45m) was evaluated. The LMV, ZDV, and NVP standards in the dissolution medium were prepared and filtered. Injecting filtered and unfiltered standard samples and observing the resulting chromatograms.

3 RESULTS AND DISCUSSIONS

The work aims to develop a validated method with RP-HPLC-PDA for the simultaneous determination of LMV, ZDV, and NVP in bulk dosage forms and dissolution samples by ICH guidelines.

3.1 Method development

For testing the optimal elution of LMV, ZDV, and NVP, various columns, mobile phases, and diluents were utilized. Initial tests were conducted using an Inertsil ODS3 column (150×4.6mm,5µ) with mobile phase (15mM ammonium acetate and methanol (50:50v/v)) at a flow-rate 1mL/min. In this experiment, the analyte (LMV) peaked at the same time as the solvent. In the subsequent test, the mobile phase composition was altered (52:48v/v) while all other conditions remained the same as in the previous trial. There is no change in the elution time of LMV under these LC conditions. In the subsequent experiment, which was conducted under the same LC conditions but with a 55:45v/v mobile phase composition, the LMV peak was separated from the solvent peak. Optimal chromatographic separation was achieved by using an Inertsil ODS3 (150×4.6mm,5µ) column and a mobile phase (55:45v/v) at a flow rate of 1.0mL/min. Under these conditions, sharp peaks of LMV, ZDV, and NVP were eluted at 2.240, 3.887, and 8.153 min, respectively, and the tailing factor, peak symmetry, and resolution were all within the acceptable range. The wavelength was set to 270 nm, which allowed for improved reproducibility without interference. See Figure 2.

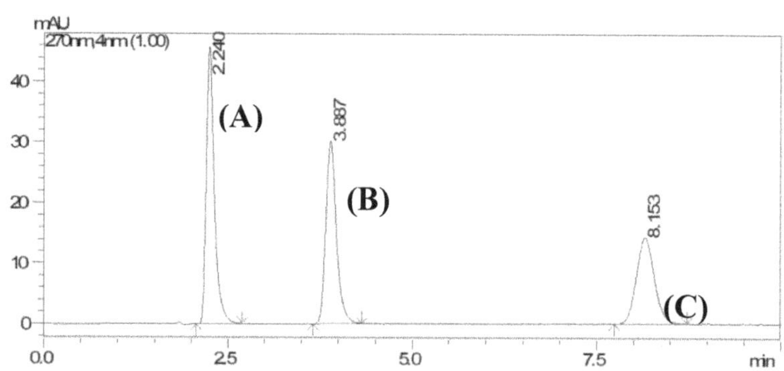

Figure 2 Chromatogram of LMV (A), ZDV (B), and NVP (C).

3.2 Analytical Method Validation

3.3.1 Specificity. By contrasting the analytical data of the drug substance and formulations, the specificity was determined. Following this, placebo, diluent, sample and standard solutions were analyzed separately to examine interferences. Based on the chromatograms base-shifted overlay in Figure 3, concludes that there were no co-eluting or interfering peaks with LVM, ZVD, and NVP. It demonstrates that the analyte's peak is pure and the formulation excipients did not interfere with the analyte. The standard and sample peak purity indices were greater than 0.999, proves the specificity.

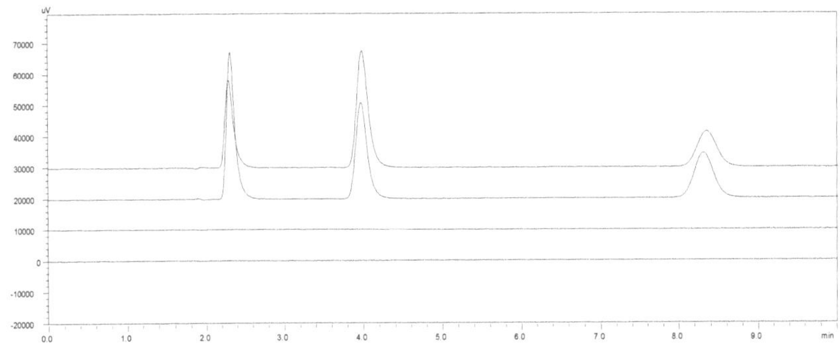

Figure 3(I) Overlay of chromatograms of A) Blank; B) Placebo; C) Standard and D) Sample.

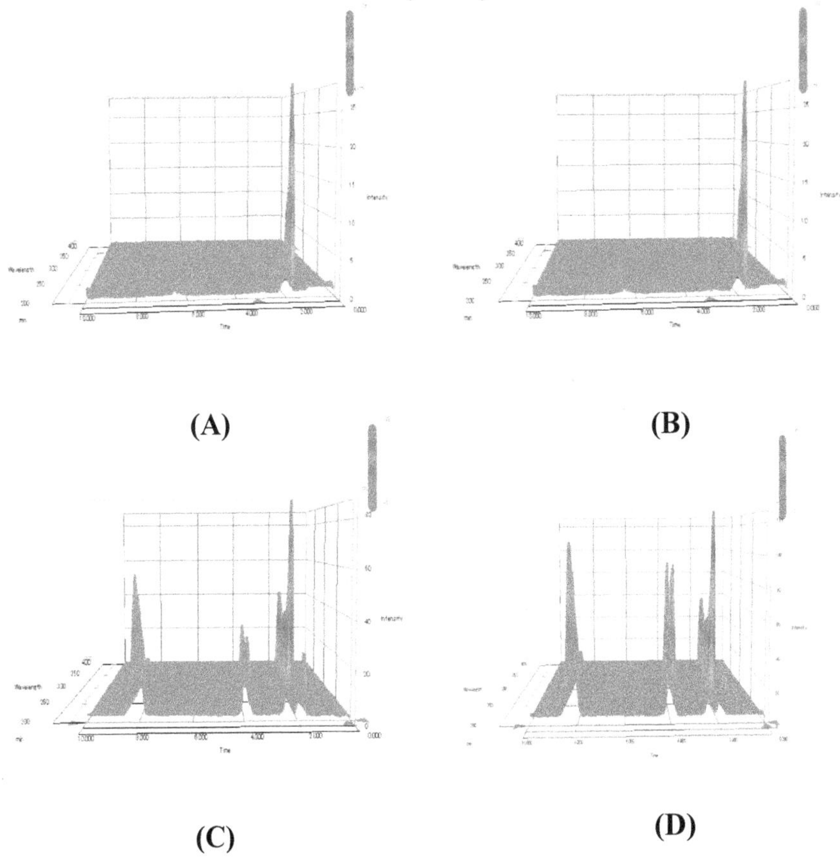

Figure 3(II) 3D plot of Blank (A); Placebo (B); Standard (C); Sample (D).

3.3.2 Linearity. Peak area and concentrations were analyzed using least squares regression. It was determined that the correlation coefficient (R) was 0.999, indicating a linear response over the examined range. Table 1 lists the slope and Y-intercept.

3.3.3 Precision. In terms of repeatability, precision studies were conducted. The repeatability was evaluated using six concentration replicates at the 15μg/mL level, and the results are presented in Table 1. The % RSD was within limits, indicating the proximity of the data values and the accuracy of the method.

3.3.4 Accuracy. It was determined that the percent recovery of the standard added to the sample was 98.10 for LMV, 102.4 for ZDV, and 98.30 for NVP, and as the % RSD was within limits the method was considered as accurate. Table 1 displays the obtained recovery outcomes.

3.3.5 Limit of Detection and Limit of Quantification. The LOD and LOQ was found to be 0.296μg/mL, 0.899μg/mL for LMV and 0.105μg/mL, 0.320μg/mL for ZDV and 0.113μg/mL, 0.343μg/mL for NVP respectively.

3.3.6 Assay. The assay of LMV, ZDV, and NVP in tablets were calculated in three determinations, which were 99.9%, 99.2%, and 100.43%. These results indicate that the current RP-HPLC-PDA method is suitable for the determination of LMV, ZDV, and NVP.

Table 1 Validation data for LMV, ZDV & NVP.

Parameters		**LMV**	**ZDV**	**NVP**
Linearity (n=3)	**Range** **Regression equation** **Regression coefficient(R^2)**	5-25μg/mL y = 118215x - 111216 R^2 = 0.9993	5-25μg/mL y = 106574x - 104510 R^2 = 0.997	5-25μg/ml y = 16550x - 12174 R^2 = 0.9968
Accuracy (n=3)	80 100 120	97.91 (0.15) 99.92 (1.60) 96.66 (1.20)	110.6 (1.57) 99.38 (0.33) 97.5 (1.33)	97.48 (0.98) 99.39 (0.69) 98.11 (1.64)
Precision (n=6) System Precision	**Std Peak area (%RSD)**	355468.8 (0.41)	305875.8 (0.170)	240816.5 (0.243)
Percent Assay (n=3)	**Mean (± SD)**	99.9 (0.1)	99.21 (0.15)	100.43 (0.05)

Table 2 Robustness of LMV, ZDV & NVP.

Parameters	**LMV**			**ZDV**			**LMV**		
Mobile Phase (v/v)	**RT* (min)**	**TP$ (N)**	**TF#**	**RT* (min)**	**TP$ (N)**	**TF#**	**RT* (min)**	**TP$ (N)**	**TF#**
57:43	2.29	1900.22	1.71	4.21	3454.09	1.32	9.57	5463.78	3.16
55:45	2.24	1767.93	1.68	3.88	3218.45	1.34	8.15	5136.09	1.10
52:48	2.19	1683.63	1.67	3.57	2892.07	1.38	6.74	4522.92	2.07
Flowrate (mL/min)									
0.8	2.80	1655.01	1.57	4.89	3229.18	1.29	10.19	5543.33	1.09
1.0	2.24	1767.93	1.68	3.88	3218.45	1.34	8.15	5136.09	1.10
1.2	1.88	1917.62	1.66	3.29	3122.81	1.31	6.94	4702.17	1.09

***Retention time (min); $Theoretical plates (N); #Tailing factor (#).**

3.3.7 Robustness. There were no significant modifications in the chromatograms, indicating that the developed method is reliable. Observations were made of parameters including resolution, capacity factor, and selectivity. The values are presented in Table 2.

3.3.8 System suitability. The system suitability RSD values for retention time (2.240, 3.388, and 8.153), tailing factor (1.5, 1.2, and 0.9), and theoretical plate number (1.4, 1.3, and 1.7) of LMV, ZDV, and NVP are less than 2, indicating that the present LC conditions were suitable for the analysis three combination dissolution samples (LMV, ZDV, and NVP).

3.3.9 Stability of the solution. The stability of samples was analysed at various time-intervals up to 24 hours under refrigeration (8±1°C). By comparing to the zero-time interval, the % variation in assay values at all time intervals was less than 2, indicating that the solutions were stable for 24 hours at 8°C.

3.4 Filter compatibility study

Standard solution and dissolution samples were filtered, analyzed, and the assay variation was computed. From the analytical data, it was considered that nylon filters were more suitable when compared with PVDF filters.

3.5 Dissolution analysis

For the *in-vitro* dissolution, previously validated method was utilized. The percentage of drug release was more than 75% of the labeled amount of LMV, ZDV & NVP dissolved in 45min, indicating that the current method can be applied for the routine analysis.

4 CONCLUSION

The simultaneous determination of LMV, ZDV, and NVP in bulk and tablet dosage forms in dissolution samples required the development of a new RP-HPLC-PDA technique. This method was followed under the guidelines established by the International Council for Harmonisation (ICH) Q2B, and it was observed to be suitable for use in routine analysis for the estimation of LMV, ZDV, and NVP in tablet and bulk dosage. The findings of linearity, precision, accuracy, and specificity were all shown to fall within a predetermined range. This method allows for the selective quantification of LMV, ZDV, and NVP without introducing any interference from diluent or placebo. It was possible to make an accurate estimation of the amount of LVM, ZDV, and NVP in dissolution samples by employing this method, which utilized compendial dissolution conditions. Furthermore, this method can be utilized in the determination of LMV, ZDV, and NVP quantity in bulk and dosage forms, as well as in the analysis of dissolution samples.

5 ACKNOWLEDGMENT

The authors are thankful to Hetero drugs, Hyderabad, for the gift samples of LMV, ZDV, and NVP and thanks to Anurag pharmacy college, Ananthagiri for providing the laboratory facilities to carry out the research work.

Conflict of interest: Nil.

Funding:

1)Potu Kalyani Assistant professor, Department of Pharmaceutical Analysis, Anurag pharmacy college, E-mail ID: kalyanipotu8@gmail.com.

2) Anurag pharmacy college, Ananthagiri, Telangana- 508206, India.

Author contributions:

Conception and design, acquisition of data analysis, and interpretation of data drafting or revising the article.

6 REFERENCES

1. Tripathi KD; Essentials of Medical Pharmacology, Jaypee Brother's Medical publisher (P) Ltd; 6th edition; 2008.

2. H.P. Rang, MM Dale, JM Ritter, RJ Flower,G Henderson, Rang and Dale's Pharmacology, Elsevier Churchill Livingstone , Elsevier Inc,spain, 7th ed, 2012.

3. Indian Pharmacopoeia; Ministry of Health and Family Welfare, Published by Indian Pharmacopoeia Commission; 2010, Vol.II.

4. USP 27, The United State Pharmacopeial Convention, Asian 27thedition, 2005, 1064,1961.

5. Chidambaram Saravanan, Manni Venkatachari Kumudhavalli, Ramalingam Kumar, Vijaya Kumar Latha, Balasundaram Jayakar; Int. res. j. of pharm, 2010, **1**(1) 314.

6. Anantha Kumar. D, Naveen Babu M.V, Seshagiri Rao J.V.L.N. and Jayathirtha Rao.A, Rasayan J Chem., 2010, 94.

7. Sandhya. K, Nagi Reddy. N, Raja Manohar, Rajashekar. P, Indo American J of Pharm. Res.,2007;**2**(1):39.

8. Predrag Djurdjevic, Aleksandra Laban, Slavko Markovic & Milena Jelikic-Stankov, J of Pharm Tec & Drug Res., 2007;2649.

9. Dharmaraj.S. S. Santhosam, Adiyaman.E, and Senthilkumar.M; Int. J of pharm and chem sci., 2013, **2**(4), 1883.

10. Hari Krishnan. N, Gunasekaran.V, Roosewelt.C, Vijaya Kumar Meher, Kalaivani.K and Ravichandiran.V, Asian J of Chem., 2008, **4** (20), 2551.

11. Sock lingam Anbazhagam, Narayanreddy indumathy, Pitchaimuthu shanmugapandiyam, Seshaiah, Krishnan Sridhar, J of Pharm and Biomedical Analysis,2005, **39**(3), 801.

12. Navaneethan.G, Karunaharan.K, and Elango. K.P, J of food and drug analysis, 2012, **20**(2), 446.

13. Rajani Kumar.V, Phani Bhushana Reddy. B, Ravi Kumar.B, K. Sreekanth, Kilaru Naveen Babu, J of Chromatography 2013; 9.

14. Geetha, Ramachandran, Hemanth Kumar. A.K, Kumara swami.V, Soumya Swami Nathan, J of Chromatography,2006, **843**(2), 339–344.

15. International Conference on Harmonization of Technical Requirements for Registration of Pharmaceuticals for Human use. Validation of Analytical procedures Q2 (R1); 2007.

PART 3

NOVEL DRUG DELIVERY SYSTEMS

DESIGN, DEVELOPMENT AND CHARACTERIZATION OF EBERCONAZOLE PRONIOSOMES

Swarupa Arvapalli and Anka Rao Areti

[a]Department of Pharmaceutics, KL College of Pharmacy, Koneru Lakshmaiah Education Foundation, Vaddeswaram, AP, India.

*Corresponding Author E mail ID: rupa03arvapalli@gmail.com

1 INTRODUCTION

Mycosis known as Fungal infections are most common infections, but systemic infection could be lethal to hosts. On an average one billion people suffer from fungal infections across the world. Fungal infections could infect superficially or subcutaneously or systemically in different regions of body [1]. Fungal infections expand through air, or by breathing fungal spores or by direct contact through wounds or cuts on infected skin. People undergoing chemotherapy or suffering through immunodeficiency disorders like AIDS are more susceptible to fungal infections [2]. In recent times, COVID 19 infections led to use of more steroids decreasing patient's immunity and increasing susceptibility to mucromycosis (black fungus)[3]. Treatment regimen for prolonged time period leads to resistance toward antifungal agents [4–6].

Liposomes, transferosomes and ethosomes examples of vesicular systems mainlycontain phospholipids, provide some pros like biodegradable, biocompatible, high entrapment efficiency, provide targeted controlled and prolonged release with enhanced drug delivery across the skin. They contribute few cons due to existence of phospholipids like variation in purity of phospholipids, costly, chemically instable, inclined to oxidative degradation and hydrolysis, increased production price as the lift up of phospholipids needs inert/cryogenic atmosphere. These systems addedly display drawbacks like sedimentation, fusion of vesicles during storage, drug leakage from the vesicular formulations, scale up and sterilization issues, strenuous to obtain products with equal properties, less half-life, acceptable for parenteral administration but cannot combat in residence of bile salts when given via mouth.

Primarily, niosomes provide advantages as same as liposomes. It predominantly comprises of non-ionic surfactants that are cost effective, inert and stable than phospholipids. It won't need any conditions specially during their preparation (low temperature & inert atmosphere), can combat in presence of bile salts in- *vivo*, enhanced half-life, serves as solubilizing agent for less soluble drugs. Surfactant properties modifies the structure of stratum corneum and make them loose and more permeable to drugs. In view of these facts, niosomes are usually chosen over other vesicular systems in drug delivery system. Orderly to meliorate stability of niosomes they are produced as proniosomes.

"Proniosomes are parched, ebullient formulations of surfactant-coated carrier, which can be rehydrated by brief agitation in hot water to form a multi lamellar niosome suspension suitable for administration by oral or other routes". The parched powder form offers flexibility, unit dosing, in which the proniosome powder is provided in capsule as well as compressed into tablets could be beneficial. Proniosomes deprive cons of niosomes namely physical instability such as aggregation, fusion and leaking. Stability of parched proniosomes is thought to be more stable than a pre manufactured niosome formulation.

Eberconazole (EBZ) categorized under broad-spectrum imidazole derivative, topically administered, helps in dermatophytosis, candidiasis, and pityriasis treatment. It belongs to BCS classification class II and, because of its declined solubility, it needs increased dosing, which leads to toxicity. EBZ-loaded niosomes were prepared using Span20 and cholesterol by thin-film hydration technique. The formulated niosomal preparations were characterized for *in-vitro and in-vivo* evaluation methods.

2 METHODS AND MATERIALS

2.1 Reagents and Chemicals Used

Eberconazole is provided by SURA LABS, Dilsukhnagar,Hyderabad as gift sample, sorbitan monooleate (Span 80) is bought from Loba Chemie, Mumbai, chloroform, coconut oil is Purchased from SD Fine Chem Ltd., Mumbai, cholesterol,methanol is Purchased from Merck Limited, Mumbai, dicetyl phosphate (DCP) is Purchased from Himedia, Mumbai.

2.2 Determining absorption maxima

10 µg/ ml solution containing drug was prepared in 5.5 phosphate buffer UV spectrum was taken using Lab India Double beam UV/VIS spectrophotometer (Lab India UV 3000+). The prepared drug mixture is scanned in the region of 200 – 400 nm.

2.3 Standard graph construction

100 mg of Eberconazole was solubilised in 100 mL of pH 5.5 phosphate buffer to make a concentration in 1mg/mL (1000µg/mL) 1 ml was with drawn and diluted to 100 ml with pH 5.5 phosphate buffer to give a concentration of 0.01 mg/ml (10µg/ml). Aliquots of 0.2 ml, 0.4 ml, 0.6 ml, 0.8 ml, 1 ml, were

withdrawn by using pipette in 10 ml volumetric flask and make up the volume to the mark with pH 5.5 phosphate buffer to bring out concentration of 2,4,6,8 and 10 µg/ml respectively. The absorbance of all concentrations was observed at respective (λ_{max}) i.e., 252 nm.

2.4 Formulation of Proniosomes

Multilamellar Proniosome were made by the thin-film hydration technique. Approximately weighed amounts of drug, surfactant (Span), and CHOL were made to dissolve in chloroform in a round-bottom flask . DCP is mixed into each formulation as a negative charge-inducing agent. The chloroform was evaporated at 60°C under reduced pressure using a rotary flash evaporator at various RPM. After chloroform evaporation, the flask was placed under vacuum overnight in a nitrogen atmosphere to erase residual solvent. The thin films are then hydrated with 6 ml of phosphate buffered saline (PBS), pH 5.5 and the formulations were Sonicated 3 times at 50 Hz in a bath- sonicator for 15 min with 5 min interval between successive times. Vesicle suspensions were also Sonicated for 5 min.[7] See Table 1.

Table 1 *Formulation composition of Eberconazole loaded Proniosomes.*

Ingredients (mg)	F1	F2	F3	F4	F5	F6	F7	F8	F9
Eberconazole	10	10	10	10	10	10	10	10	10
Cholesterol (mg)	10	20	30	10	20	30	10	20	30
Span 80(mg)	1.5	2	2.5	3	3.5	4	4.5	5	5.5
Dicetyl phosphate (DCP)	0.5	1	1.5	2	2.5	3	3.5	4	4.5
Coconut oil (mg)	3	3	3	3	3	3	3	3	3
Methanol (mL)	10	10	10	10	10	10	10	10	10
Chloroform(mL)	5	5	5	5	5	5	5	5	5

2.5 Characterization of prepared Proniosome

2.5.1 Particle Sizes, PDI, Zeta Potential. The average particle length and polydispersity index (PDI), dynamic light scattering (Delta Nano C, Beckman counter) is used to that's a degree of the distribution of Proniosome, and Zeta capability becomes anticipated on the premise of electrophoretic mobility under an electric powered field, the use of zeta Sizer Nano ZS (Malvern Instruments, UK). Distilled water is used for diluting samples for measurement of particle size and poly dispersibility Index at an angle of 165^0c. 1 part of sample and 40 parts of distilled water used for measuring zeta potential, PDI, Average particle size. Measurements are done in triplicate[8].

2.5.2 Drug content. UV- Visible spectrophotometer is used to assay the content of Eberconazole in prepared proniosomes. 10 ml methanol is used to dissolve 100 mg proniosome and shaken for 5 mins. 10 ml of methanol is taken to dissolve one ml of above solution. 252 nm wavelength range in using UV-visible spectrophotometer (Lab India 3200). Is set to measure absorbance of aliquots withdrawn[9].

2.5.3 Yield of Proniosome. Collect and weigh completely dried proniosomes and calculate the yield by using formula[10].

$$Percentage\ yield = \frac{Total\ weight\ of\ noisomes}{\text{total weight of drug +weight of added materials}} \times 100 \quad (1)$$

2.5.4 Entrapment Efficiency. Entrapment Efficiency (EE) of the Eberconazole loaded Proniosome changed into determined by measuring the awareness of uninterrupted drug in an aqueous medium by centrifugation method. The Proniosome had been centrifuged during a high-space cooling Centrifuge (C-24.Remi) the usage of nano step centrifuge tubes with ultra-filter out having a relative molecular mass cutoff 100KD (Pall existence sciences-India) at 5000rpm for 15min at 4°c, and therefore separate the supernatant. The quantity of Eberconazole inside the supernatant changed into determining the usage of a UV-Visible spectrophotometer (U-1800, Hitachi) at lambda max 252 nm after suitable dilution[11-12].

The percent entrapment efficiency (%) changed into calculated by means of the usage of the subsequent formula:

$$\% \text{ EE} = \frac{Total\ Drug\ content - Free\ drug}{Total\ drug\ content} \times 100 \quad (2)$$

2.5.6 In- vitro release studies. Method namely Static dissolution is utilized in the *in vitro* release study of different Proniosome formulations of Eberconazole. Set up the apparatus by introducing an open end tube aligned vertically to a beaker containing the 50 ml of PBS pH 5.5 which acts as receptor compartment. To resemble as a barrier inside the body cellophane membrane is tied on the open end tube on one side so that only the tip touches the surface of buffer solution. When the whole study is carried the reaction conditions were standardized. Fill with PBS pH5.5 the receptor compartment and 1 ml vesicular dispersion of Proniosome Eberconazole in the donor compartment. Place the beaker on a magnetic stirrer for a definite study time and at regular intervals with draw samples. The fresh buffer was replaced for each withdrawal into the beaker and collected samples were diluted using pH5.5. Spectrophotometrically Analyze using UV-visible spectrophotometer at 252nm the percentage of drug release in the receptor medium[13].

2.5.7 Scanning electron microscopy. Scan the atomic surface of Eberconazole using The scanning electron microscope (SEM) optimized Proniosome which was view under electron microscope for shape and size. The layered sample surface morphology was examined by using SEM (Hitachi, Japan). Manually dispersed the small amount of powder onto a carbon tab (double adhesive carbon coated tape) adhered to an aluminum stubs. Coat these sample stubs with a thin layer (30Å) of gold by employing POLARON-E 3000 sputter coater. Examine the samples by SEM and photograph under various magnifications with direct data capture of the images onto a computer[14].

2.5.8 Fourier Transform Infrared (FTIR) spectroscopy. FTIR studies are done formulations to find out the possible drug and the excipients interactions when they are prepared. Pure drug and optimized formulation FT IR analysis were carried out using an FT IR spectrophotometer (Bruker FT-IR – GERMANY)[15].

3 RESULTS AND DISCUSSION

3.1 Determination of absorption maxima

The lambda max of Eberconazole was deemed to be 252 nm.

3.2. Calibration curve construction

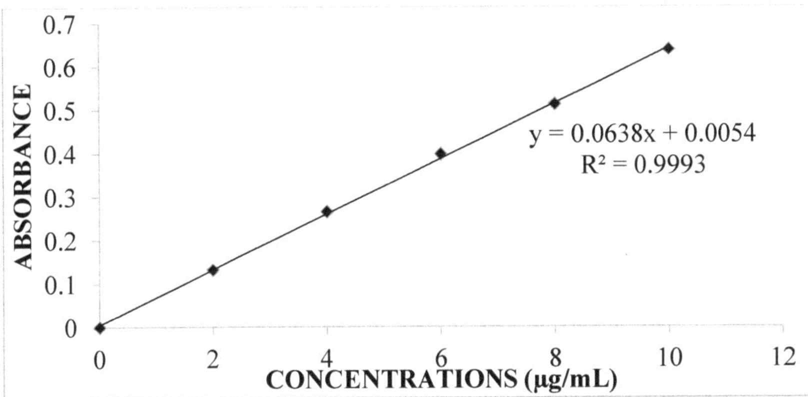

Figure 1: *Standard calibration curve of Eberconazole.*

Eberconazole standard graph was plotted as per experimental process and its linearity is shown in Table 2 and Figure 1. Eberconazole showed good linearity and obeys "Beer- Lamberts" law with R^2 of 0.999 in standard graph.

3.3 Characterization of Proniosome

Table 2 *Percentage yield, Drug Content, Entrapment Efficiency of all Proniosome formulations.*

FORMULATION	Entrapment Efficiency (%)	Drug Content (%)	Percentage yield (%)
F1	93.36	87.69	85.93
F2	89.51	91.47	88.24
F3	86.89	93.73	94.15

F4	90.14	87.26	91.82
F5	94.62	91.18	93.19
F6	97.78	97.09	95.77
F7	83.06	79.61	78.09
F8	91.10	84.78	83.14
F9	93.39	90.42	89.93

Increasing cholesterol concentration increases the entrapment efficiency. Proniosome with lipid composition (Span 80 and cholesterol) in a quantity of 4 mg: 30 mg which gave maximum EE% was chosen for further investigations.

The prepared Proniosome formulations F1, F2, F3, F4, F5, F6, F7, F8 and F9 were evaluated for drug content. As shown in Table 2, the percentage of drug content was deemed to be 87.69%, 91.47%, 93.73%, 87.26%, 91.18%, 97.09%, 79.61%, 84.78% and 90.42% respectively. Among the four formulations, F6 showed maximal drug content as of 97.09% and F6 showed maximal Percentage yield as of 95.77%.

Table 3 *Particle Sizes, PDI, Zeta Potential of all Proniosome formulations.*

FORMULATION	**Particle Size (μm)**	**PDI**	**Zeta Potential (mV)**
F1	7.69	0.85	-23.26
F2	5.63	0.72	-21.49
F3	4.91	0.56	-26.82
F4	6.98	1.36	-13.39
F5	6.85	0.82	-23.14
F6	5.53	0.89	-31.52

F7	8.39	0.36	-23.97
F8	12.29	1.63	-27.15
F9	14.75	1.44	-29.67

The particle size in Proniosome dosage forms other from bioavailability influencing plays an vital role. Obtained vesicle diameter for F6 formulation with diameter of 5.53 μm respectively. (See Table 3).

Sonicated vesicles with low amount of DCP has showed a narrower size distribution and high quantity DCP were smaller. Poly dispersity index value for Span 80 formulation with DCP was 4 for F6 formulation. The enhanced size by high quantity of DCP may be because of separation increased proniosome vesicles bilayers.

Zeta potential results reveal that Span 80 Proniosome possess negative charge at pH 5.5 indicating that a weak electrostatic repulsive force exist in Proniosome bilayer. Also, the inclusion of DCP in Span 80 Proniosome found to have increased the zeta potential. Particles with zeta potential close to zero have been found less phagocytable in comparison with charged particles. The nature and density of charge on the surface of Proniosome influence the extent of biodistribution as well as interaction and uptake of Proniosome by target cells. F6 formulation highest zeta potential and it had good stability.

3.4 *In-vitro* Drug release

Table 4 *In vitro dissolution studies of F1-F9 Proniosome formulations.*

Time (hour)	F1	F2	F3	F4	F5	F6	F7	F8	F9
0	0	0	0	0	0	0	0	0	0
1	23.81	31.16	45.05	28.19	32.24	48.86	10.23	16.14	21.41
2	31.26	38.02	51.61	34.34	36.52	53.51	19.56	24.94	26.75
3	40.34	45.70	55.92	43.96	51.09	66.32	25.49	31.37	34.14
4	45.65	53.94	63.86	55.65	62.90	71.51	33.94	36.19	40.43
6	50.08	65.46	70.73	62.24	73.82	78.06	41.37	45.24	46.18
8	56.11	69.28	77.41	73.19	76.33	83.99	46.59	53.83	55.27
10	63.24	87.15	91.64	82.33	85.61	94.81	52.04	65.14	68.54
12	68.19	91.46	93.23	87.10	93.52	97.35	58.95	68.72	76.78

In vitro studies of selected formulations were carried out in PBS pH 5.5 (See Table 4). It was clearly observed from the data that *in vitro* drug release of Proniosome containing Span 80 of different quantities was sharply increased up to 12 hr. Maximum drug release that is, 97.35%, was reported in case of Proniosome containing 4 mg Span 80. This might be due to Span 80 has monounsaturated alkyl chain. F6 formulation considered as optimized formulation (97.35%).

Optimised formulation F6 was kept for release kinetic studies. The above graph depicts the formulation F6 shows Peppas release kinetics mechanism.

3.5 Scanning Electron Microscopy

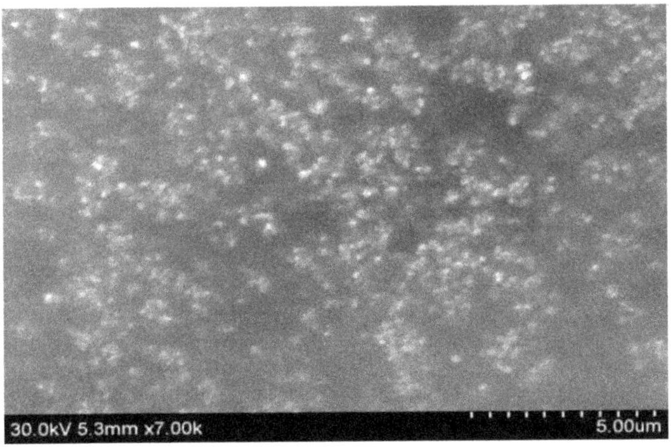

Figure 2 Eberconazole Proniosome optimised formulation (F6).

SEM studies of optimized formulation F6 shown in Figure 2 showed that the Eberconazole - loaded Proniosome with a smooth surfaceand spherical shape.

3.6 FT-IR Studies

FTIR spectra of the drugs and the optimised formulation were recorded. The FTIR spectra of pure Mirtazapine drug, optimised formulation shown in the below figures respectively.

Drugs are also present in the physical mixture, which indicates that there is no interaction between drug and the polymers, which confirms the stability of the drug.

There was no disappearance of any characteristics peak in the FTIR spectrum of drug and the polymers used. This shows that there is no chemical interaction between the drug and the polymers used. The presence of peaks at the expected range confirms that the materials taken for the study are genuine and there were no possible interactions as shown in Figures 3 and 4.

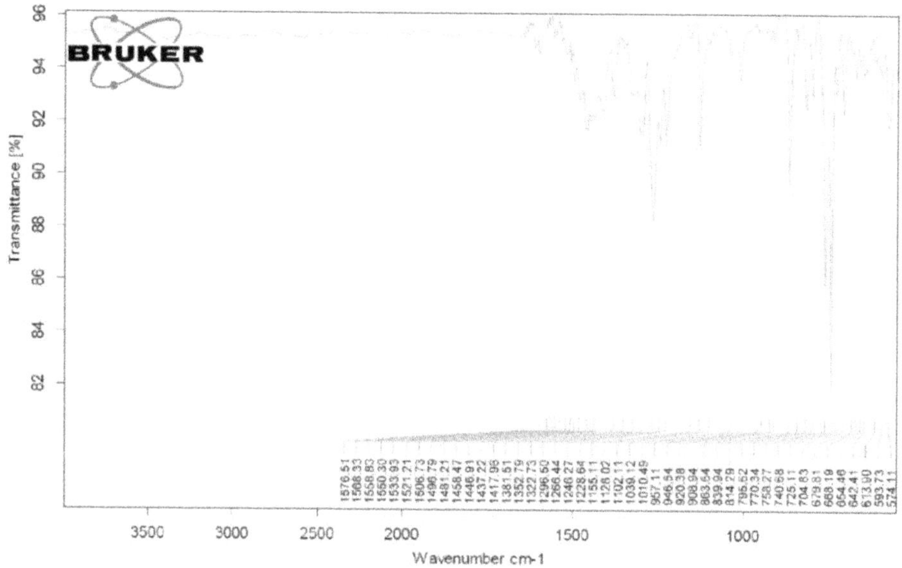

Figure 3 FT-IR graph of Eberconazole.

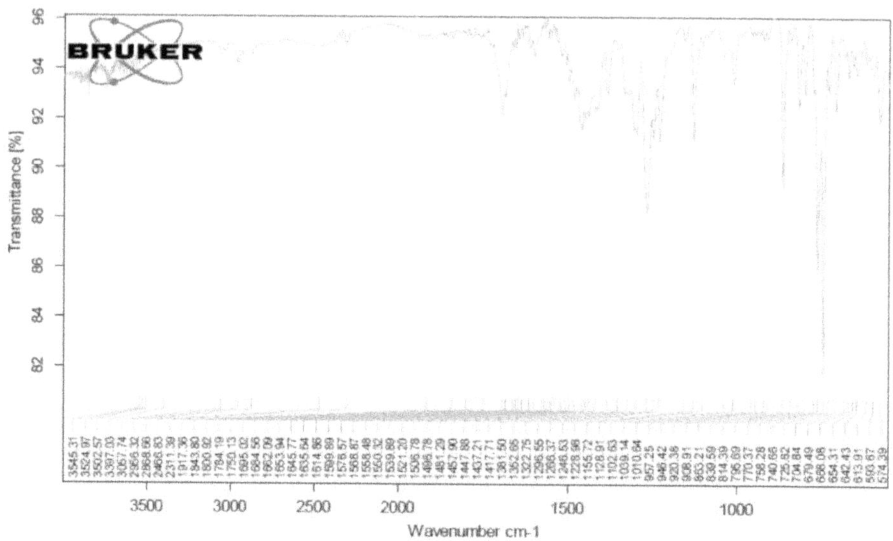

Figure 4 FT-IR graph of optimized formulation.

4 CONCLUSION

The present study demonstrated the successful preparation of Eberconazole Proniosomes by Proniosomes prepared from thin-film hydration method and evaluation. Formulation F6 has proven improved bioavailability, sustained release compared to other formulations. Based on this the Eberconazole Proniosomes containing span 80 and cholesterol considered as optimal formulation.

5 ACKNOWLEDGMENT

The authors are thankful to Joginpally B.R Pharmacy College, Hyderabad for providing facilities in carrying this research.

6 REFERENCES

1. BijuSS,TalegaonkarS,MishraPR,KharRK.Vesicularsystems:an overview.Indian J PharmSci2006;68(2):141-53.
2. Hua S. Lipid–based nano-delivery systems for skin delivery of drugs and bioactives. FrontPharmacol2015;6:1-5.
3. VermaP,PathakK.Therapeuticsandcosmeceuticalpotentialofethosomes:a n overview. JAdv PharmTechnol Res 2010;1(3):274-82.
4. UdupaN.NiosomesasDrugCarriers.In:JainNK,editors.Controlledand Novel Drug Delivery.India: CBS Publishers andDistributors;1997, p. 292-303.
5. J.N.Khandare, G.Madhavi, B.M. Tamhankar, "Niosomes NovelDrug DeliverySystem"TheEasternPharmacist– March,1994; 61.
6. S.P. Vyas and R.K. Knar, "Targetted and controlled Drug Delivery" Ist Reprint 2004; 173, 230, 259-260.
7. N.K.Jain,"ControlledandNovelDrugDelivery"2004;301, 295.
8. VyasandKhar,Targetedandcontrolleddrugdelivery-NovelCarrierSystems, CBSPublishersandDistributors,NewDelhi,Firsteditionreprint2004,249-279.
9. MahaleNB,ThakkarPD,MaliRG,WalunjDR,ChaudhariSR.Niosomes: novelsustainedrelease non-ionic stablevesicularsystems-anoverview,Adv Colloid Interface Sci 2012;183:46-54.

10. Arunothayanun. P., Bernard, M.-S., Craig, D.Q.M., Uchegbu, I.F., Florence, A.T. The effect of Processing Variables on the Physical Characteristics of Non-Ionic Surfactant Vesicles (Niosomes) Formed from a Hexadecyl Diglycerol Ether. *Int. J. Pharm.,* 2000, 201,7–14.
11. Arunothayanun. P., Uchegbu. I. F., Craig, D. Q. M., Turton, J. A.,Florence A.T.*In vitro / in vivo* Characterization of Polyhedral Niosomes. *Int. J. Pharm.,* 1999, 183, 57–61.
12. Uchegbu, I. F., and Florence, A. T. Nonionic Surfactant Vesicles (Niosomes): Physical and Pharmaceutical Chemistry, *Adv Coll Interf Sci., 1995,* 58,1-55.
13. Uchegbu, I. F., and Vyas, S.P. Nonionic Surfactant Based Vesicles (Niosomes) in Drug Delivery, *Int. J. Pharm.* 1998, 172,33–70.
14. Uchegbu, I. F., Double, J. A., Turton, J. A., and Florence, A. T. Distribution, Metabolism and Tumoricidal Activity of Doxorubicin Administered in Sorbitan Monostearate (Span 60) Niosomes in the Mouse. *Pharm. Res., 1995,* 12, 1019-1024.
15. Baillie A. J., Florence A.T., Hume L.R.,. Muirhead, G.T., Rogerson A. The Preparation and Properties of Niosomes Non-Ionic Surfactant Vesicles. *J. Pharm. Pharmacol.*, 1985, 37,863–868.

DESIGN AND EVALUATION OF MUCOADHESIVE IN SITU LIPOSOMAL GEL FOR SUSTAINED OCULAR DELIVERY OF DIFLUPREDNATE USING TWO STEPS FACTORIAL DESIGN

Ajay Kumar BACHUPALLY [a], Anka Rao ARETI[a*]

[a] KL College of Pharmacy, Koneru Lakshmaiah Education Foundation, Vaddeswaram, Guntur, Andhra Pradesh, India.

* ankarao@kluniversity.in

1. INTRODUCTION

Liposomes are membrane-like vesicles containing aqueous compartments. The lipid vesicle's aqueous compartment can hold polar chemicals like inorganic ions and macromolecules, while non-polar medicines can bond to the membrane. By changing lipid content, liposomes can vary in trapping capacity, permeability, and surface charge[1]. These traits give liposomes transportability.

A liposome is a lamellar lipidic structure that encloses an aqueous phase in balance. Lipids in a saline aqueous solution generate it spontaneously[2]. Thus, liposomes are artificial tiny spheres made of the same phospholipids as cell membranes, which may accommodate water-soluble material in the aqueous area and lipophilic stuff in the lipid section[3].

Liposomes are categorized by size, bilayer count, and multi-membrane structure (Figure 1).

I Multilamellar vesicles (MLVs): These are vesicles with five or more concentric lamellae, ranging in size from 0.5 to 20 pm. Oligo-lamellar liposomes or paucilamellar vesicles have a few concentric lamellae[4].

ii. Small unilamellar vesicles (SUVs) are 20-100pm liposomes with a single bilayer.

iii. Large unilamellar vesicles (LUVs) are 100-1000 nm bilayer liposomes.

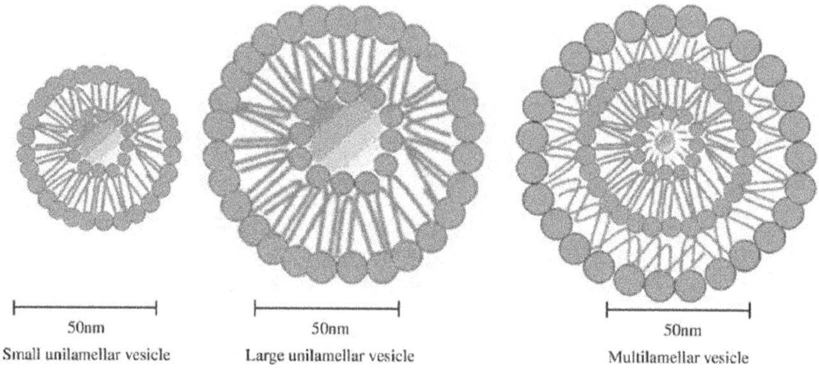

Figure 1 Schematic representation of various types of Liposomes.

2. METHODS AND RESULTS

2.1 Chemicals and Reagents Used

Drug substance Difluprednate was received as a gift sample from Etico Life sciences Pvt Ltd. Cholesterol, Soy phospholipids, Gellan gum, and xanthan gum were received from pharmonix as samples. All other chemicals and excipients used were of analytical and pharmaceutical grade[7].

2.2 Formulation of Difluprednate liposomal in-situ gel formulation

Gellan gum and xanthan gum were used to make topical liposomes. Both gelling agents (1 g) were combined in distilled water with 10 gm propylene glycol for 60 min at 800 rpm (88 g), then neutralized by adding triethanolamine drop-wise (pH 5.5). Using an electrical mixer (25 rpm for 2 min), drug-containing liposomes were blended into 1% (w/v) gelling agents, and the optimum batch's liposomes diffused into the gel, resulting in a 0.125% w/w drug concentration. Experimental Design[9,10,11]

In deionized water, gellan gum powder was dissolved with constant stirring at 500 RPM and 70 ^{0}C (Table 1). The obtained gellan gum solution was cooled to room temperature before being added with constant stirring along with the prescribed amounts of Xanthan Gum and Poloxamer 188, chelating agents, and preservatives. The pH was then adjusted to 4.4 to 5.5 using tri-ethanol amine. The improved liposomal formulation would then be continuously stirred after sterilization by filtering using pyrogen-free cellulose filters with a 0.22 m pore size. The formulations were put into amber-colored glass vials that had been autoclaved for 20 minutes at 121°C and 15 pressure to sterilize them. Before usage, the sterilized liposomal in-situ gel was kept at 4–8 C in a refrigerator [12].

Table 1 *In-situ gel formulation*

	Low	High
Gellan Gum (%w/w)	0.25	0.75
Xanthan Gum (%w/w)	0.25	0.75
Poloxamer 188 (%w/w)	0.5	1
Chelating agent	q.s	q.s
Preservatives	q.s	q.s
Phosphate buffer	q.s	q.s

2.3 Characterization of liposomes

2.3.1 Microstructure and Morphology of lipid nanoparticles. Photon correlation spectrum (PCS) and the Zeta sizer were used to measure the particle size and polydispersity index of drug-loaded liposomes (Nano-ZS, Malvern Instruments, UK). Before doing the analysis, 1 mL of the liposome's solution was diluted with double-distilled water and mixed to avoid multiple scattering. Transmission electron microscopy (TEM) and scanning electron microscopy (SEM) were used to study the shape of the surface (SEM). After the right amount of dilution, a drop of the diluted liposomal suspension was stained with uranyl acetate, put on a copper grid covered with a carbon film, and looked at with a transmission electron microscope (TEM) after drying. For SEM imaging, the lyophilized samples were stuck to a small piece of brass with double-sided tape. Sputtering in an argon environment was used to coat the sample with a gold-palladium alloy. For high-quality SEM pictures, sputtering was used for about two minutes to give the sample a uniform coating[13].

2.3.2 Determination of Entrapment Efficiency and Drug-Loading Capacity. The % EE of drug-loaded liposomes was found by separating the drug that wasn't in a liposome from the drug-loaded liposomes by spinning at 6000 g at 4C for 20 minutes. The above HPLC method was used to measure how much of the unencapsulated DRUG was in the supernatant. The following equations were used to figure out the EE percentage and the real and theoretical drug loading capacities:

$$\%\ \text{Encapsulation efficiency} = \frac{\text{Total drug minus free drug}}{\text{Total drug}}\ x\ 100 \quad (1)$$

2.3.3 *In-vitro Drug Release Study.* Examining in vitro drug release tests of produced liposomes was done using the dialysis bag diffusion method. The drug release assays were carried out using phosphate buffer (pH 7.4). A pre-activated cellulose dialysis bag (Mol. cutoff 12,000; HIMEDIA, Mumbai, India) was used to place and seal the drug-loaded liposomal dispersion, which was equal to 5 mg of medication. The dissolution vessels, which housed the dialysis bag and were stirred at 100 rpm, held 500 mL of dissolving liquid and were maintained at 37 2 °C. The receptor compartment's airtight barrier prevented the dissolving liquid from evaporating. At predetermined intervals of 15 minutes, 30 minutes, 1, 2, 4, 8, 12, 24, and 48 hours, receptor samples were obtained and replaced with the same volume of the fresh dissolving medium. The methanol-dilution samples were filtered through a Millipore membrane (0.22 m), and then the samples were spectrophotometrically analysed.

2.3.4 Stability study. After three months of storage at 20°C, +4°C, and 25°C, the liposomes with the drugs inside were tested to see how stable they were. Depending on how long the samples were stored, the particle size distribution and how well the drugs were wrapped up were measured.

2.3.5 Evaluation of drug Loaded liposomal Gel

2.3.5.1 Clarity and Visual Appearance. The look and clarity of the formulations were evaluated visually using the visual assessment method on white and black backgrounds. Any undesirable particles or turbidity dispersed in the solution were also examined in the formulations.

2.3.5.2 Measurement of pH. The pH of ophthalmic preparations must be taken into account carefully. so as not to irritate the patient's eye. The ophthalmic formulation's pH should range from 5 to 7.4. The pH of the novel formulations was measured using a digital pH meter (Elico Pvt. Ltd. India). An electrode was positioned on top of each formulation and allowed to equilibrate for one minute to ascertain the pH. The pH of the formulations may be altered using 0.1 M NaOH to prevent eye harm.

2.3.5.3 In-vitro Gelation. To find gelling systems that can be employed in situ, the gelling capacity of every formulation that has been generated has been evaluated. A vial containing 2 ml of freshly created pH 7.4 simulated tear fluid was used to measure the gelation time (seconds) and visually analyze the system to determine its ability to gel. The LSNG formulation's viscosity and gelation characteristics were considered. Slow and fades rapidly; No gelation at all. Instantaneous and brief gelation takes place. In an instant, there is gelation, and it lasts for a long time[14].

2.3.5.4 Rheology study. As much as possible, ophthalmic medicines should be used to reduce the "false plasticity" of the tear film before the cornea [27,28].Since the ocular shear rate is quite high, it is best to use viscoelastic fluids that have a high viscosity at low shear rates and a low viscosity at high shear rates. In the best formulation, an increase in shear stress caused an increase in angular velocity, which was shear-thinning (pseudoplastic rheology). After the viscosity measurements were done in solution, 0.5 M NaOH was added to bring the pH up to a healthy level of 7.4. (non-physiological pH). During the measurements, which were done three times with a Brook field viscometer, the spindle speed was changed from 5 to 60 rpm.

3 RESULTS AND DISCUSSION

3.1 Solubility

Solubility investigations were carried out using a cyclone mixer (REMI CM 101, India) in various solvents utilizing the equilibrium solubility method. The table 2 shows the results of solubility tests in various solvents and buffers medium.

Table 2 *Solubility profile of Difluprednate.*

Solvent	Solubility
Water-	0.0097 mg/ml
Chloroform	soluble
Methanol	soluble
Ethanol	soluble
Diethyl ether	Slightly soluble
DMSO	250mg/ml
1.2 pH buffer	insoluble
6.8 pH buffer	Insoluble
7.4pH buffer	Insoluble

3.2 FTIR- Fourier Transform Infrared Spectroscopy

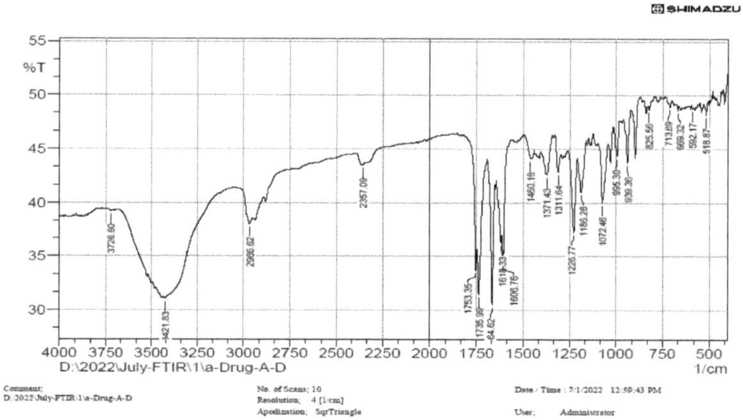

Figure 2 FTIR spectra of pure drug Difluprednate.

Description: The FTIR spectrum of pure difluprednate typically shows several distinct absorption bands in the mid-infrared region (4000-400 cm^-1). The most prominent peaks in the FTIR spectrum of difluprednate are typically observed at around 1700 cm^-1, which corresponds to the stretching vibration of the C=O bond in the ester group (Figure 2). In addition, other peaks may be observed in the FTIR spectrum of difluprednate, including absorption bands at around 1600 cm^-1, which correspond to the stretching vibrations of the aromatic ring, and peaks at around 3000 cm^-1, which correspond to the stretching vibrations of the C-H bonds in the aliphatic chains. Overall, the FTIR spectrum of pure difluprednate provides important information about the molecular structure of this compound and can be used to identify and characterize difluprednate in various pharmaceutical formulations.

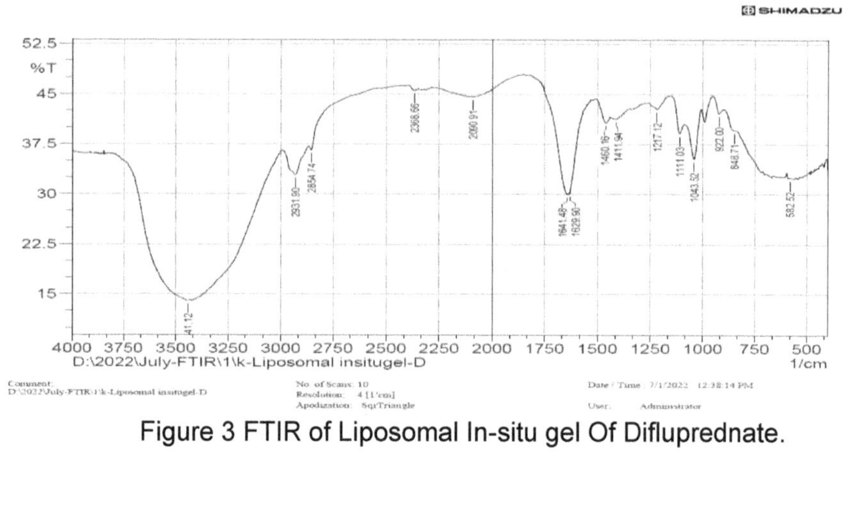

Figure 3 FTIR of Liposomal In-situ gel Of Difluprednate.

Description: The spectra typically show a broad band in the region of 3400 -3200 cm^-1, which corresponds to the O-H stretching vibration of water present in the gel. A sharp peak at around 1620 cm^-1 corresponds to the C=O stretching vibration of difluprednate, indicating the presence of ester functionality (Figure 3).

Another prominent peak is observed in the region of 1460-1470 cm^-1, which corresponds to the bending vibration of the C-H bond in the methylene group present in the molecule. A peak at around 1250 cm^-1 corresponds to the C-O-C stretching vibration of the ether group, which is also present in the difluprednate molecule. Overall, the FTIR spectra of in-situ gel of difluprednate show characteristic peaks corresponding to the various functional groups present in the molecule, which can be used to identify the composition and structure of the gel.

The chosen liposomal formulation, which was the optimum formulation, demonstrated high EE (75.64%), suitable VS and PDI (110.3 nm and 0.285), and ZP (38.6 mv). It also contained 20% penetration enhancer and a 4:1 ratio of lecithin to cholesterol. The improved formulation was used in a design to examine the impact of concentration on the development of in situ mucoadhesive gel using three levels of gellan gum and two levels of Carbopol 934.

4 CONCLUSION

Selected liposomal in situ gel G2 (0.25 percent gellan gum and 1 percent Carbopol 934) had the highest viscosity (3,864 cps) at physiological conditions and the lowest viscosity (156 cps) at normal conditions. It gave off the least amount (15.36%) after 30 minutes and the most (69.82%) after 24 hours. The chosen liposomal in-situ gel will be tested for irritation and pharmacokinetics in animals (G2). When compared to the product on the market, Difluprednate did not cause irritation, had a higher concentration of Difluprednate in aqueous humour, and had effects that lasted longer than the liposomal formulation.

5 ACKNOWLEDGEMENTS

The authors are thankful to KL College of Pharmacy (KL University), Vaddeswaram and Etico Life Sciences Pvt. Ltd., Hyderabad for providing the facilities to carry the work and for timely scientific insights.

6 REFERENCES

1. Nathan Congdon, Benita O'Colmain, Caroline C W Klaver, Causes and prevalence of visual impairment among adults in the United States, Arch Ophthalmol. 2004

2. Jost B Jonas, Rupert R A Bourne, Richard A White, Visual impairment and blindness due to macular diseases globally: a systematic review and meta-analysis, American Journal of Ophthalmology, 2014;158(4):808–15.

3. Frank G Holz, Erich C Strauss, Steffen Schmitz-Valckenberg, Geographic atrophy: clinical features and potential therapeutic approaches, Ophthalmology.2014;121(5):1079–91.

4. Katherine J. Wert, Jonathan H. Lin, Stephen H. Tsang. General pathophysiology in retinal degeneration, Cell-Based Therapy for Retinal Degenerative Disease. 2014;53:33–43.

5. Noriko Yoshida MD, Yasuhiro Ikeda MD, Clinical evidence of sustained chronic infammatory reaction in retinitis pigmentosa. Ophthalmology, 2013;120(1):100–5.

6. Matthew K Zabel, Lian Zhao, Yikui Zhang, Microglial phagocytosis and activation underlying photoreceptor degeneration is regulated by CX3CL1-CX3CR1 signaling in a mouse model of retinitis pigmentosa. Glia.2016;64(9):1479–91.

7. Hussein O Ammar, H A Salama, M Ghorab, Nanoemulsion as a potential ophthalmic delivery system for dorzolamide hydrochloride, AAPS PharmSciTech. 2009;10(3):808–19.

8. Masazumi Yamaguchi, Shin-ichi Yasueda, Akiharu Isowaki, Formulation of an ophthalmic lipid emulsion containing an anti-infammatory steroidal drug, difuprednate. International Journal of Pharmaceutics 2005;301(1–2):121–8.

9. Korenfeld M. Difluprednate: changing the landscape of ocular pharmacology. Expert Review of Ophthalmology, 2008;3(6):619-25.

10. Fujino A, Ohtu S, Shibata K. Studies on the metabolic fate of difluprednate (DFBA), Metabolism in rats and rabbits after subcutaneous administration. Pharmacometrics 1985;29:713-23.

11. Tajika T, Shirasaki Y, Kimura M. Ocular distribution and metabolism after instillation of difluprednate ophthalmic emulsion in rabbits, The 2007 Annual Meeting of the Association for Research in Vision and Ophthalmology (ARVO). Poster B744, Program 2654. Fort Lauderdale, FL, USA, 6-10 May 2007.

12. Senju Pharmaceutical Co.,Ltd.,Osaka, Japan. Phase IIb clinical study of difluprednate 0.05% vs betamethasone 0.1% in the treatment of postoperative inflammation. April 2004.

13. Ohji M, Tano Y, Hida T. Efficacy and safety results of a phase III study of difluprednate ophthalmic emulsion 0.05% in postoperative inflammation. The 2007 Annual Meeting of the Association for Research in Vision and Ophthalmology (ARVO). Poster B807, Program 3903. Fort Lauderdale, FL, USA, 6-10 May 2007.

14. Yamaguchi M, Yasueda S, Isowaki A, Yamamoto M, Kimura M, Inada K, Ohtori A Formulation of an ophthalmic lipid emulsion containing an anti-inflammatory steroidal drug, difluprednate. International journal of pharmaceutics 2005 Sep 14;301(1-2):121-8.

FORMULATION AND EVALUATION OF DESONIDE LOADED NIOSOMES PREPARED BY FILM HYDRATION TECHNIQUE USING BOX-BEHNKEN DESIGN

Megha Shyam MATLAPUDI[a], Anka Rao ARETI[a*]

[a] Department of Pharmaceutics, KL College of Pharmacy, KL University, Vaddeswaram, Guntur, Andhra Pradesh, India.

* ankarao@kluniversity.in

1. INTRODUCTION

Drug delivery systems play a crucial role in delivering the medicament to the target site.[1] However, due to some drawback of the conventional dosage forms, novel drug delivery systems have come into picture. Novel drug delivery systems or NDDS are designed to have specific characteristics to meet the drug delivery requirements. One such NDDS is Niosome (See Figure 1). Niosomes are nanoparticulate vesicles that comprise of lipidic bilayer membrane stabilized by non-ionic surfactant. [2] Lipidic layer usually is Cholesterol or Phosphotidyl choline; and the commonly used non-ionic surfactants may be sorbitan ester/fatty acids, Polysorbates, Poloxamers, Sorbitan esters such as SPAN® 60, SPAN® 20, SPAN 80®, etc. [1,2,3] These nanoparticulate vesicles have better safety tolerance and stability compared to other NDDS such as Liposomes. [1,2] They also tend to deliver the drug to the respective site of action. Due to their enhanced entrapment efficiency, there is a possibility for dose reduction and improved drug tolerance due to minimized side effects. [2]

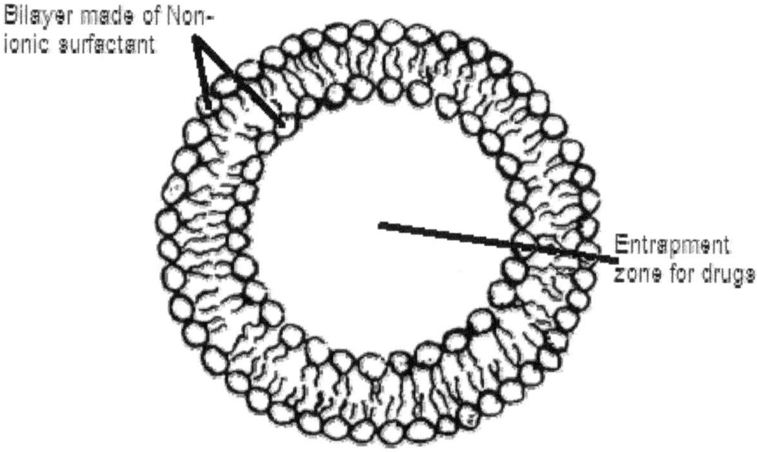

Figure 1 Representation image of Niosome.

In this study, DOE was employed as a systematic methodology for planning the experiments. A Box-Behnken design (a design for response surface methodology) was used to establish the interaction between the factors and their effects on the respective responses. [4] Desonide was selected as the drug of choice as corticosteroids are used for major skin conditions such as Psoriasis, Atopic dermatitis, etc. [5] The corticosteroids when applied to skin may pose adverse effects on long term usage such as thinning of skin, burning, skin oozing, flaking, peeling, to name a few. [6] Hence, Desonide entrapment into Niosomes would provide a stable drug delivery vehicle with reduced side effects, improved efficacy at lower doses, better stability and importantly improved skin permeation.[7]

2. METHODOLOGY

2.1. Chemicals and Reagents Used

Desonide was received from Orbicular Pharmaceutical Technologies as a gift sample. SPAN 80 and Cholesterol and Carbomer 940 were received from Orbicular Pharmaceutical Technologies as excipient gift samples for this study. Other ingredients and reagents were procured.

2.2. Formulation of Desonide loaded Niosomal Gel

2.2.1 Preparation of Desonide loaded Niosomes. Desonide loaded Niosomes were prepared using thin film hydration technique using a lipid mixture consisting of surfactant (Span 80) and Cholesterol (CHO), at different levels as given in Table 1. Surfactant, CHO and drug were dissolved in chloroform. The lipid mixture was then transferred to a round bottom flask, and the solvent was evaporated under reduced pressure at a temperature of 55°-65°C, using a rotary flash evaporator until the formation of a thin lipid film. The formed film was hydrated with pH 7.4 Phosphate buffer saline. The hydration was continued for 1 h, while the flask was kept rotating at 55°-65°C in the rotary evaporator.

The hydrated niosomes were sonicated at different levels as given in Table 1 using a bath sonicator to obtain niosomal dispersion containing both free and entrapped drug. This dispersion was washed several times using deionized water and final wash by pH 7.4 Phosphate buffer saline and niosomes with entrapped drug were collected. The DoE was performed using DesignExpert® software.

Table 1 *Variables and Responses for DoE study for Box- Behnken Design.*

Independent variables (X)	**Levels (coded values)**		
	Low (-1)	**Mid (0)**	**High (+1)**
Span 80 in mL (X1)	10	15	20
Cholesterol in mg (X2)	80	90	100
Sonication time in min (X3)	3	4	5
Responses (Y)	**Constraints**		
Vesicle size (Y1)	Minimize		
%Entrapment Efficiency (%EE) (Y2)	Maximize		
Polydispersivity Index (PDI) (Y3)	<1		

2.2.2 Preparation of Niosomal gel. The prepared Desonide loaded Niosomes were added to the dispersion of Carbomer 940 in purified water along with Glycerin (as humectant), Propylene glycol (as permeation enhancer) and preservative (Methylparaben). Then, neutralizing agent was (5N Sodium hydroxide solution) added to adjust pH to 7 to form a topical niosomal gel. Similarly, Desonide gel was prepared for comparison purposes.

2.3. Evaluation of Niosomes

2.3.1 Particle size and Polydispersive index (PDI). Particle size and polydispersity index of the prepared niosomal suspensions were measured by Photon Correlation Spectroscopy (PCS) using a Zetasizer Nano ZS-90 instrument (Malvern instruments, Worcestershire, UK). An aliquot of Niosomes was prepared by dispersing uniformly in deionized water. This sample was subjected to particle size and PDI analysis.

2.3.2 *Drug entrapment efficiency and drug loading.* 50 mg of drug loaded Niosomes were taken and added to diluent containing chloroform, Methanol and Water for extraction of Desonide. Lysis of niosomes was assisted by sonication followed by vortexing. The samples were analyzed at 248 nm using UV-visible detector.

2.3.2 In vitro permeation study. In vitro permeation study of Desonide Niosomes was performed using vertical diffusion cell or commonly known as

Franz diffusion cell (6 cell manual diffusion cell apparatus, Make: Hanson Research). Excised rat skin was used as testing membrane with pH 7.4 Phosphate buffer as receptor fluid at 32°±0.5°C. Finite dosing of 200 mg of drug product was employed for testing. Skin permeation levels were measured at the end of the study after 24 hours.

2.4. Results and Discussion

2.4.1 *Formulation of Desonide Loaded Niosomes.* A 3 level 3 factor design was adopted to optimize the formulation variables and a 17 runs were suggested by DesignExpert® software for Box-Behnken design. The selected independent variables were amount of surfactant i.e Span 80 (X1), Cholesterol (X2), and Sonication time (X3) (as shown in Table 1). The dependent variables were vesicle size (Y1) EE%(Y2) and PDI (Y3). The experimental results are presented in the Table 2.

Table 2 *Experimental runs and the responses.*

RUN	Span 80 (mL)	Cholesterol (mg)	Sonication time (min)	Particle size (nm)	EE%	PDI
	X1	X2	X3	Y1	Y2	Y3
1	20	100	4	187	83	0.43
2	15	90	4	186	77	0.38
3	15	90	4	175	76	0.36
4	15	90	4	181	79	0.39
5	15	100	5	216	88	0.44
6	10	90	3	181	75	0.36
7	15	90	4	177	78	0.35
8	15	100	3	191	85	0.45
9	15	90	4	158	78	0.37
10	20	90	5	168	78	0.38
11	15	80	5	119	72	0.31
12	10	80	4	120	75	0.32
13	20	80	4	114	73	0.33
14	15	80	3	191	73	0.35
15	10	100	4	189	82	0.42
16	20	90	3	189	75	0.36
17	10	90	5	175	76	0.35

ANOVA was performed and it was observed that the selected factors (i.e., Span 80 concentrations, Cholesterol concentrations and Sonication duration) had significant impact on the responses i.e., particle size, PDI and entrapment efficiency with p value <0.05 at 95% CI. It was observed that the lack of fit was also satisfactory. Optimization was run and observed prediction stated that Span at 10.76 mL, Cholesterol at 95.29 mg and Sonication time at 3.71 min would provide optimal Niosomal vesicles (Table 3).

Table 3 *Optimized levels for selected factors.*

Variable/Factor	**Optimized level**
Span 80 (X1)	10.76 mL
Cholesterol (X2)	95.29 mg
Sonication time (X3)	3.71in

2.4.2 Particle size and Polydispersive index (PDI). The particle size was found to be in the range of 114 nm to 216 nm. The optimized formulation was found to have 180 nm vs. the predicted 179.115 nm, indicating the optimization design is justified.

The PDI was found to be in the range of 0.31 to 0.45. The optimized formulation was found to have 0.42 vs. the predicted 0.39, indicating the optimization design is justified.

2.4.2 Drug entrapment efficiency and drug loading. The drug entrapment efficiency was found to be in the range of 72% to 88%. The optimized formulation was found to have 76% vs. the predicted 79.42%, indicating the optimization design is justified. The Table 4 provides the comparison of predicted mean and the obtained experimental value. The prediction was backed with the experimental results and found that the results were matching. See Figures 2-4.

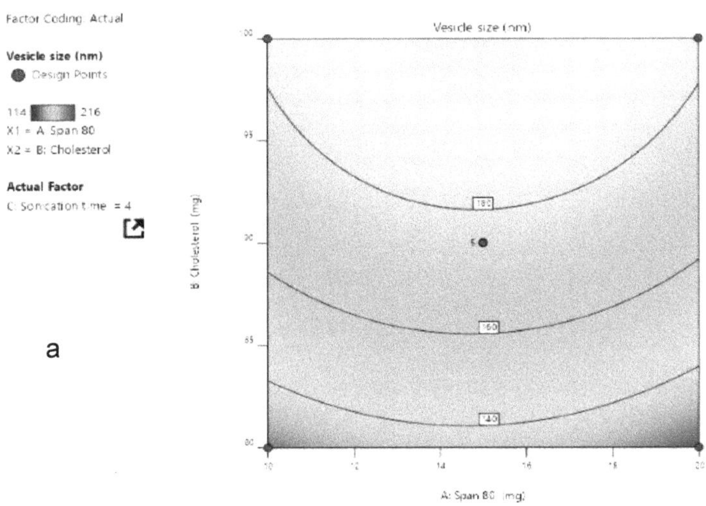

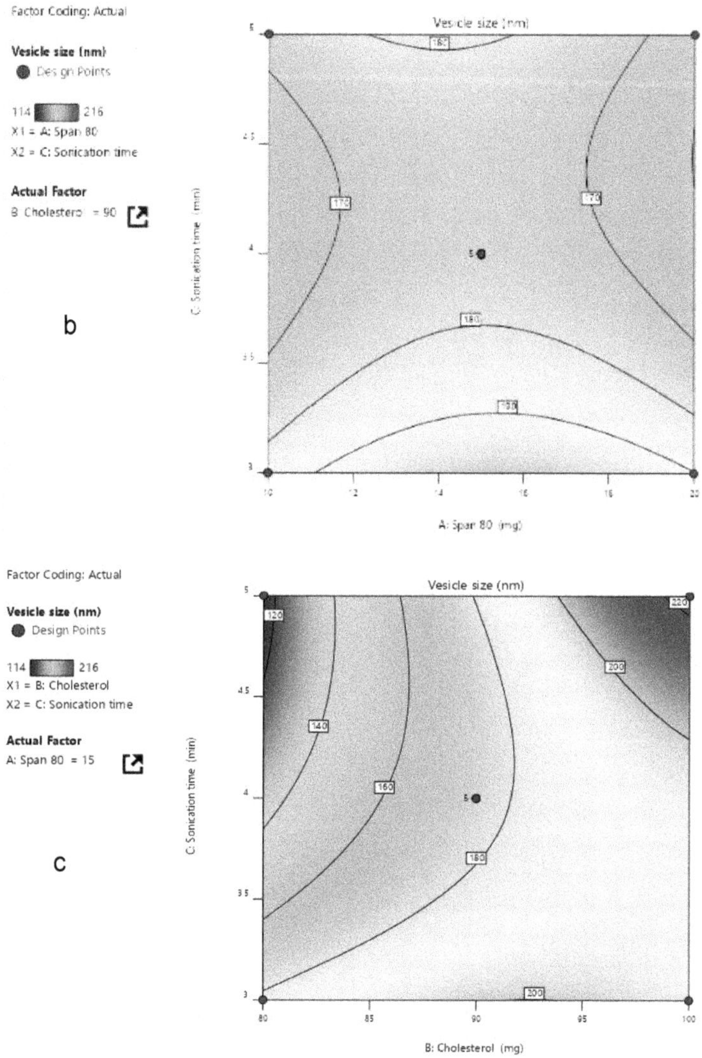

Figure 2 Counter plots for vesicle size of Desonide loaded Niosomes. a) Span 80 vs. Cholesterol. b) Span 80 vs. Sonication time. c) Cholesterol vs. Sonication time.

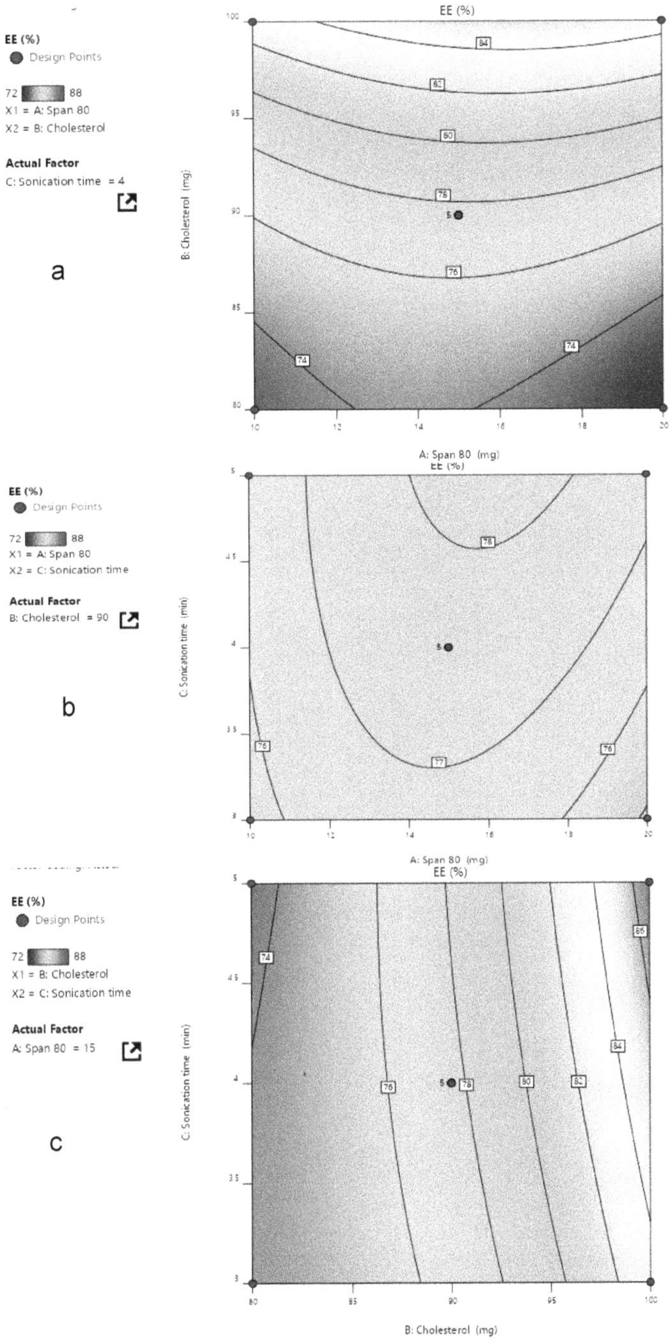

Figure 3 Counter plots for entrapment efficiency of Desonide loaded Niosomes. a) Span 80 vs. Cholesterol. b) Span 80 vs. Sonication time. c) Cholesterol vs. Sonication time.

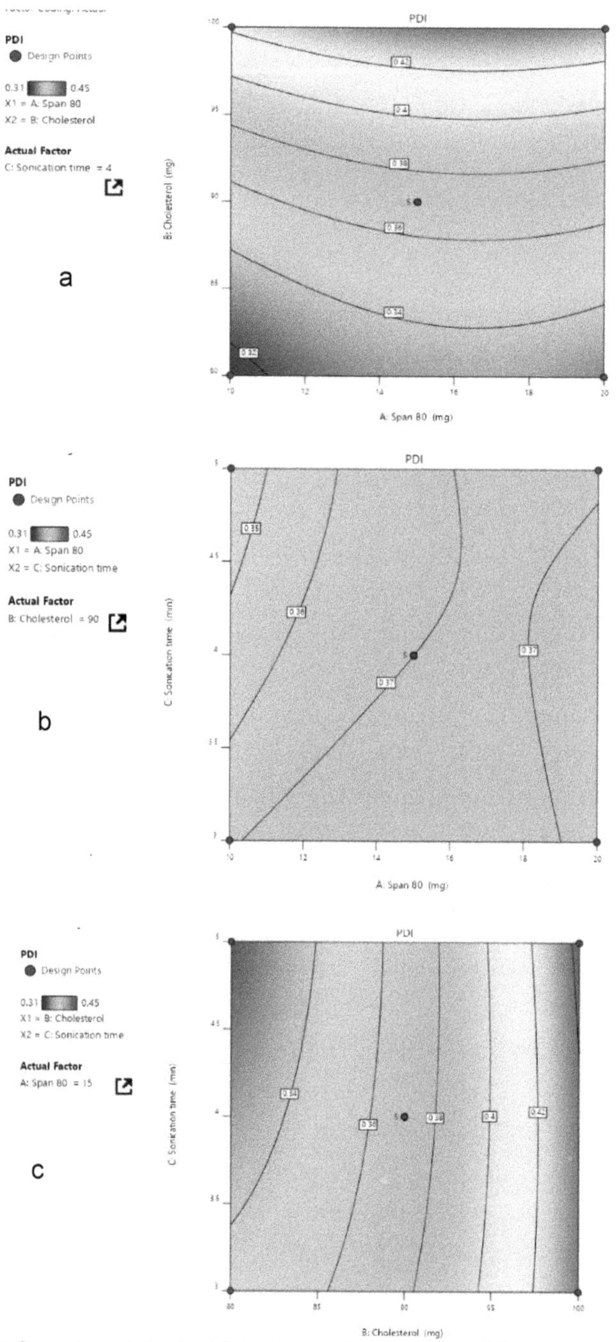

Figure 4 Counter plots for Polydispersivity index of Desonide loaded Niosomes. a) Span 80 vs. Cholesterol. b) Span 80 vs. Sonication time. c) Cholesterol vs. Sonication time.

Table 4 *Predicted mean vs. Experimental value.*

Response	**Predicted mean**	**Experimental value**
Particle size (nm)	179.115	180.23
Entrapment efficiency (%)	79.419	75.92
Polydispersive index	0.392	0.42

2.4.3 In vitro permeation study. The skin layers were precisely collected by skin stripping using cellophane tape and subjected for extraction, followed by analysis. It was observed that the Desonide gel and Niosomal gel were permeating through the skin at almost equal levels. However, the Niosomal gel was found to localize more within the skin layers (epidermis and dermis) as opposed to the Desonide gel. The Desonide Gel was found to traverse through the dermis and to the receptor fluid, indicating higher systemic exposure. The Niosomal Gel was found to be less available in the receptor fluid, thus indicating improved chances of localized drug delivery to skin and less systemic exposure. [8] See Table 5.

Table 5 *Amount of drug recovered in In vitro permeation study on excised rat skin after 24 h study period.*

Skin layer	**Desonide Gel**	**Desonide loaded Niosomal Gel**
Skin surface	0.08 mg	0.06 mg
Within skin (Epidermis & Dermis)	0.005 mg	0.033 mg
Receptor fluid	0.015 mg	0.009 mg

3. CONCLUSION

Box-Behnken design was employed using DesignExpert® software and found that the selected factors i.e., concentration of Span 80, concentration of Cholesterol and Sonication duration were having a significant impact on the responses – particle size, PDI and Entrapment efficiency. The point prediction (predicted mean) was found to be matching with that of the experimental value indicating the design accuracy. In vitro permeation testing was performed and it was observed that the skin permeation and retention was better with Desonide loaded Niosomal gel compared to the conventional gel dosage form of Desonide. This would conclude that Niosomes have the ability to permeate across Stratum corneum and localize within the skin for extending the local action at target site. This characteristic would help in reduction of adverse effects caused by corticosteroids during their long term use and also in dose or dosage regimen reduction. [1,3,7] Further animal studies are planned to establish the Niosomal carrier for topical delivery of Desonide.

4. ACKNOWLEDGEMENTS

The authors are thankful to KL College of Pharmacy (KL University), Vaddeswaram and Orbicular Pharmaceutical Technologies Pvt. Ltd., Hyderabad for providing the facilities to carry the work and for timely scientific insights.

5. REFERENCES

1. D. A. Seleci, M. Seleci, J. G. Walter, T. Scheper, *Journal of Nanomaterials,* 2016, Aricle ID: 7372306.
2. C. Cerqueira-Coutinho, E. P. dos Santos, C. R. E. Mansur, *Critical Reviews in Therapeutic Drug Carrier Systems*, 2016, 33, 195.
3. R. Rajera, K. Nagpal, S. K. Singh, D. N. Mishra, *Biological & Pharmaceutical Bulletin*, 2011, 34, 945.
4. D. G. Mercurio, L. S. Calixto, P. M. B. G. M. Campos, *Brazilian Journal of Pharmaceutical Sciences*, 2020, 56, e18502.
5. P. Rosa, J. dos Santos, T. F. Lehmen, J. Weber, F. C. Flores, C. de B. da Silva, S. M. Oliveira, I. Brusco, G. B. Milani, A. I. H. Adams, *Drug Development and Industrial Pharmacy*, 2015, DOI: 10.3109/03639045.2015.1022554.
6. Product label (Pack insert) of DesOwen® (Desonide Cream).
7. B. A. Witika, K. E. Bassey, P. H. Demana, X. Siwe-Noundou, M. S. Poka, *International Journal of Molecular Sciences*, 2022, 23, 9668.
8. M. Manian, P. Jain, D. Vora, A. K. Banga, *Pharmaceutics*, 2022, 14, 1892.

FORMULATION AND CHARACTERIZATION OF BILASTINE ORAL DISINTEGRATED TABLETS USING NATURAL AND SYNTHETIC SUPER DISINTEGRANTS

Sadhana Noothi[1,2], Narender Malothu[1], P. Vishnu[2]

[1]KL College of Pharmacy, Koneru Lakshmaiah Education Foundation, Vaddeswaram, AP, India;

[2]Department of Pharmaceutical Engineering, B V Raju Institute of Technology, Narsapur, Medak, Telangana, India;

Corresponding author: vishnu.p@bvrit.ac.in

1. INTRODUCTION

Oral drug delivery is a widely used for delivery of conventional and controlled drug release[1]. Traditional drug release systems like tablets, pills, powders, solutions, emulsions, suspensions, and aerosols have been used to treat acute or chronic diseases It is the most popular route of drug administration because of its versatility, easy administration, painlessness, and less manufacturing cost. In conventional drug delivery, one of the approaches is immediate drug delivery which releases the drug rapidly from the dosage form, has rapid absorption, and produces rapid onset of action[2].

Disintegration is a process of breakdown of tablets into smaller particles which leads to improvement of the dissolution rate[3] Disintegrating agents are the substances used for the disintegration of tablets. They promote moisture penetration onto the tablet matrix and initiate the breakdown of the tablet into smaller fragments, thereby increase in the surface area of the particles. Disintegration is a critical criterion for the rapid dissolution and absorption of the drug. Super disintegrating agents are substances, one type that is effective at low concentration and improves dissolution efficiency by swelling mechanism. Some examples are croscarmellose, cross povidone, sodium starch glycolate, primogel, etc. [4,5]

Histamine is a major cause of allergic inflammation. Histamine shows its biological activity by four different histamine receptors in that H1 plays a major role in allergic diseases[6] BLS is one of the second-generation antihistamine drugs that is used to relieve the symptoms of allergic rhino conjunctivitis and other forms of allergic rhinitis. It was more effective against both seasonal and perennial allergic rhinitis without sedation effect. In the present study an attempt was made to formulate BLS immediate-release tablets using various types of super disintegrants and evaluated for best release.

2. MATERIALS AND METHODS

BLS, Chitosan, Fenugreek mucilage, Sodium starch glycolate, Ludiflash, and Cross povidone were obtained as gift samples, poly (ethylene glycol) 4000, Microcrystalline cellulose, talc & magnesium stearate were purchased from SD fine chemicals, Mumbai.

2.2 Analytical method

The primary standard solution of BLS was prepared using ethanol. From this, after necessary dilution was made to achieve the concentration of 10 µg/mL A UV spectrophotometer then scanned the prepared stock solution at a wavelength ranging from 200nm to 400nm. The λmax for the BLS was determined and was found as 236 nm[7]. The stock solution was diluted using

0.1N HCl for pH 1.2 to prepare various concentrations of 2, 4, 6, 8, and 10 µg/ml. The absorbance was measured at 236 nm and drawn the calibration curve which was used to identify unknown concentrations observed in Figures 1 and 2.

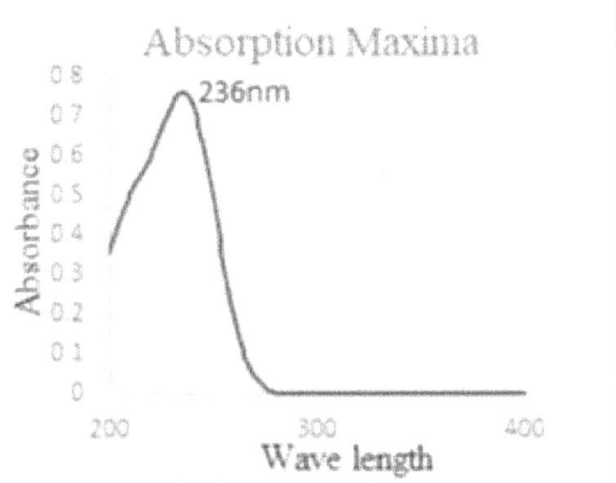

Figure 1. Absorption Maxima

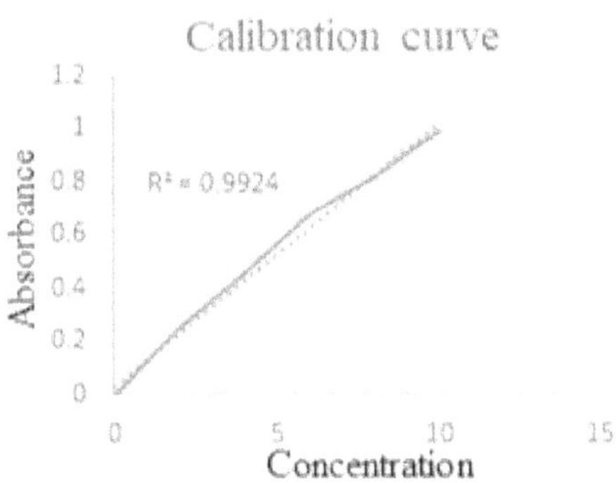

Figure 2. Calibration curve

2.2 Formulation of BLS tablets

The direct compression method was used to formulate BLS immediate-release tablets [8] The formulation makeup of various batches is shown in Table 1. The granules were all sieved through a 40-mesh sieve. The needed amount of BLS, super disintegrants, and fillers were carefully combined. As lubricants and glidants, magnesium stearate and talc were added. The dry blends were evaluated for a variety of pre-compression parameters and tablets were prepared on a 10-station rotary tablet punching machine, the evaluated powder mixture was compressed directly (circular concave-faced punches) BLS was present in 20 mg tablets and the tablets were kept in sealed containers for further research.

Table 1 *Formulation of BLS tablets.*

	BLS (mg)	Fenugreek seed powder (%)	Chitosan (%)	SSG (%)	Cross povidone (%)	Ludiflash (%)
BF1	20	2	-	-	-	-
BF2	20	-	2	-	-	-
BF3	20	-	-	2	-	-
BF4	20	-	-	-	2	-
BF5	20	-	-	-	-	2
BF6	20	4	-	-	-	-
BF7	20	-	4	-	-	-
BF8	20	-	-	4	-	-
BF9	20	-	-	-	4	-
BF10	20	-	-	-	-	4
BF11	20	6	-	-	-	-
BF12	20	-	6	-	-	-
BF13	20	-	-	6	-	-
BF14	20	-	-	-	6	-
BF15	20	-	-	-	-	6

Note: PEG: 10mg, MCC: qs, Talc: 1.5mg, Mg Stearate: 1.5mg

2.3 Evaluation of pre-compression parameters of dry powder blends BLS immediate release formulations

2.3.1 The angle of repose (θ). The funnel was fastened to a stand at a certain height, and the dry mix powders were allowed to flow through it (h). The height and radius of the newly created pile of powders were then measured to determine the angle of repose(i) [9].

$$\text{Angle of repose } \theta = \tan^{-1}\frac{h}{r} \qquad (1)$$

Where h is height and r is the radius of the powder heap, respectively, and θ was referred to as the angle of repose. The angle of repose value less than 25 indicates the excellent flow of a blend, whereas bad flow is indicated by an angle of repose value larger than 40.

2.3.2 Bulk density and tapped density. The following formula (I & ii) was used to figure out the bulk density and tapped density of all the ready-to-use BLS immediate-release dry blend powders.

$$\text{Bulk density} = \frac{\text{Wt. of powder}}{\text{Bulk volume}} \qquad (2)$$

$$\text{Tapped density} = \frac{\text{Wt. of powder}}{\text{Tapped volume}} \qquad (3)$$

2.3.3 Compressibility Index (Carr's index). Powder's ability to flow can be measured by comparing the bulk density (BD) of the powder to its tapped density (TD), as well as the pace at which it is packed down. This will give an idea of the powder's tapped density. The following equation was used to determine the compressibility index(iv) (also known as Carr's index) of the BLS immediate-release dry powders blend that had previously been developed.

$$\text{Carr's Index} = \frac{\text{TD} - \text{BD}}{\text{TD}} \times 100 \qquad (4)$$

According to the specification, 5-15 Carr's index values suggest excellent flow, while 12-16 indicate good flow. 18-21 is passable, 23-25 is terrible. 33-38 indicates very poorly, and above 40 indicates extremely poor flow

2.3.4 Hausner's ratio. BLS immediate release dry powders Hausner's ratios were determined by the following formula (v).

$$\text{Hausner's Ratio} = \frac{\text{Tapped density}}{\text{Bulk density}} \qquad (5)$$

Hausner's ratio < 1.25 implies good flow, whereas larger than 1.25 indicates bad flow. The flow rate improved by the addition of glidant between 1.25 and 1.5.

2.3.5 Drug and excipients compatibility studies. FTIR experiments were performed on BLS drugs and excipients from various batches to look for any potential physical and chemical interactions[10]. FTIR (Fourier Transform Infrared) spectroscopy was carried out to identify the peaks that indicate the presence of a specific functional group in the pure drug. FTIR tests were performed on the pure drug BLS as well as a physical mixture of the drug and all excipients (optimized formulation). The potassium bromide (KBr) pellet technique was used. The materials were triturated with KBr, and the pellet was formed by applying pressure for 2 min. The obtained pellet was examined for FTIR.

2.4 Post compression parameters of BLS immediate release formulations

2.4.1 Average thickness. Randomly chosen 10 tablets were selected from formulations (BLS) to identify gauge thickness. Digital Vernier Callipers were used to gauge each tablet's thickness. According to the specification, the tablet thickness shall be kept within a 5% range of the specified value.

2.4.2 Tablet Hardness. A Monsanto hardness tester was utilized to evaluate the degree of brittleness present in every formulation of an immediate-release tablet containing BLS. The crushing strength of 10 immediate-release tablets of known weight was measured in kg/cm2 for each formulation, and then the averages were calculated and shown together with standard deviations. It has been determined that the tablet meets the criteria for an acceptable limit according to the USP hardness guidelines, which are 3–4 kg for an instant release [11].

2.4.3 Friability. Ten BLS pills from each batch, which had previously been weighed, were placed in the Roche friabilator. tablets were found after one hundred revolutions cleaned of dust, and the total weight that remained was noted. This formula was used to determine friability(vi).

$$\%\text{Friability} = \frac{(\text{Wi} - \text{Wf})}{\text{Wi}} \times 100 \qquad (6)$$

2.4.4 Weight variation test. The USP standard states that the weight variation for uncoated tablets with an average weight of 130 mg or less is 10%, 7.5% for tablets with an average weight between 130 and 324 mg, and 5% for tablets with an average weight of 324 mg or more. The weight of the pill must not differ from the average weight by more than two tablets, and no tablet may deviate by more than 15%. BLS immediate-release tablet formulations were all examined for weight variation per the USP standard. Initially, twenty tablets from each batch were weighed individually and collectively using an electronic balance, and the average weight of each pill and its percent variance was then computed.

2.4.5 Content uniformity. Twenty tablets were crushed to make a powder so that the same amount of BLS could be found in all the different forms. One tablet's worth of powder was mixed with 100 ml of HCl buffer with a pH of 1.2 and heated at 37 °C for 15 to 20 min while stirring constantly. After cooling and filtering the solution, the BLS content was measured at 236 nm using a UV Spectrophotometer (LAB India). Each measurement was repeated three times, and each formulation's average drug amount was calculated [12].

2.4.6 Water absorption ratio. A straightforward method that involves weighing the tablet (Wb) before placing it on the Petri dish and again after recording the wetting time can be used to estimate the water absorption ratio (R). The wet tablet was taken out and weighed again (Wa). The following equation(vii) was used to calculate the water absorption ratio [13].

$$\text{R} = \frac{(\text{Wa} - \text{Wb})}{\text{Wb}} \times 100 \qquad (7)$$

2.4.7 In vitro disintegration time (Dt). The USP states that covering plastic discs is unnecessary when using the disintegration device for oral tablets. The allowable time limit for tablet disintegration meeting official standards is 2 min instead of < 2 min for immediate-release dosage forms. The tablet disintegration device was used for the test. 0.1N HCl buffer with a pH of 1.2 was placed in the disintegration apparatus and kept at 37°±0.5°C used to conduct the in-vitro disintegration test (n = 6). The time it took for each tablet to completely break down into smaller pieces after being placed in the basket was recorded [14].

2.4.8 In-vitro drug release (dissolution) study. Type II USP dissolution test apparatus was utilized for the in-vitro dissolution investigation, which was carried out for each of the formulations (Labtronics DS 8000). As the dissolution media, 900 ml of 0.1N HCl buffer with a PH of 1.2 was used, and the temperature was kept at 37°C +/- 0.5°C while the rotational speed was 50 rpm. At regular intervals, 5 ml of aliquots were removed, and the volume that was removed was replaced with an equal volume of freshly prepared dissolving medium. The Whatman filter paper was used to filter the samples after they were collected at five-minute intervals [15] To assess the amount of BLS that was released from immediate-release tablets, spectrophotometric analysis was performed at 236 nm on some samples.

2.4.8 Stability studies. The s stability studies of the optimized formulation of BLS immediate-release tablet were conducted. The optimized formulation was exposed to various conditions at 40 ° C ± 5 °C & 75% ± 5% RH for 90 days and the product was evaluated for friability, hardness, weight variation, thickness, drug content, and *in vitro* drug release study.

3. RESULTS AND DISCUSSION

FTIR research looking at BLS compatibility revealed that its spectra exhibit a peak owing to pure drug 935cm^{-1}, 1420 cm^{-1}, 1715 cm^{-1},2950cm^{-1} for OH bending, C-C stretching, and C-H stretching and 938 cm^{-1},1415 cm^{-1}, 1716 cm^{-1}, 2959 cm^{-1} for the physical mixture (BLS with all the Excipients). This means that the characteristic peaks seen in the spectra of pure pharmaceuticals can be seen exactly where they first appeared in the spectra of improved formulations. No clinically relevant drug-excipient interactions were observed with the BLS immediate-release tablet.

There was a wide variation in the bulk and tapped densities among all BLS instant release dry powder blend formulations, with the bulk densities ranging from 0.3260.25 to 0.4160.39 g/cm3 and the tapped densities ranging from 0.4030.19 to 0.4850.37 g/cm3, respectively. The high packing density of dry powder mixtures is demonstrated here. The density of dry powder blends was shown to be particle-packing-dependent and to fluctuate as the powder consolidated based on results from bulk density and tapped density measurements. All formulations had Carr's index values below 16%, which is indicative of favorable flow properties. The stability of power and granule columns can be assessed along with flow parameters using Hausner's ratio, a straightforward procedure. Hausner's ratio was found to have a little variation, suggesting a high degree of flow capability. As Hausner's ratios ranged from 1.08 to 1.23 across all formulations, it was clear that the dry powder blends exhibited favourable flow properties. Material with an angle of repose of less than 25° is considered free-flowing, while an angle of repose greater than 40° indicates a poor-flowing material. Dry powder mixes of BLS instant release tablet exhibited good flow characteristics and were suitable for direct compression, with angles of repose ranging from 23.46 to 22.74. For each formulation, the table below displays the pre-compression parameters that were determined. All formulations of BLS instant-release tablets were physically satisfactory, including hardness, average weight variation, average friability, and average thickness. Common tablet flaws like capping, chipping or picking we not observed. The pills had average thicknesses between 4.420mm and 4.25mm. Each lot had a consistent thickness that was within the specified tolerances. The weight differences across the various formulations ranged from 98±1.62 to 102±1.67mg. All formulations are within the allowable 10% range for average percentage variation in tablets weighing 200 mg. All the formulations were found to be within the acceptable range for weight uniformity.

All BLS immediate-release tablet formulations had an average hardness of 3.51±0.6 to 4.72±0.5 kg/cm2. The percentage friability of all formulations ranged from 0.32±0.38% to 0.54±0.05% and the percentage friability increased with increased super-disintegrant concentration. The % friability for all formulations in the current study was within the prescribed limits. Disintegration periods were determined for all formulations ranging from 50 to 128 sec, and it was discovered that raising the concentration of super-disintegrant decreases the disintegration time; however, increasing the concentration above 6% decreases the hardness value. The water absorption ratio of formulations BF1 to BF15 ranged from 9.36±0.44 to 39.41±0.69 The water absorption ratio increases with increasing super-disintegrant concentration, which could be attributed to an increase in formulation porosity with increasing super-disintegrant concentration. Table 2 shows the physicochemical characterizations of various batches of BLS immediate-release tablets.

Table 2. Post compression parameters of BLS tablets.

Formulation code	Average Hardness (kg/cm2)	Average Weight Variation (%)	Friability (% w/w)	Average thickness (mm)	Dt (Sec)	Water absorption ratio
BF1	4.31±0.5	102±1.67	0.34	4.82	128	18.37±0.2
BF2	3.84±0.6	99±2.54	0.41	4.67	112	20.46±0.24
BF3	3.89±0.5	98±2.71	0.36	4.78	107	19.62±0.35
BF4	3.83±0.6	102±1.52	0.38	4.80	102	23.45±0.28
BF5	4.72±0.7	98±1.62	0.32	4.64	96	9.36±0.27
BF6	3.73±0.5	102±1.45	0.42	4.59	100	21.44±0.34
BF7	3.67±0.8	102±2.06	0.48	4.73	87	28.51±0.3
BF8	3.71±0.9	101±1.67	0.46	4.63	81	27.67±0.42
BF9	3.68±0.4	100±1.85	0.43	4.45	76	30.78±0.35

BF10	4.15±0.8	101±1.92	0.38	4.83	74	17.81±0.27
BF11	4.01±0.3	101.46	0.48	4.79	72	29.45±0.39
BF12	3.52±0.4	99.34	0.54	4.69	58	33.46±0.42
BF13	3.58±0.5	99.68	0.51	4.42	63	34.37±0.32
BF14	3.51±0.6	101.56	0.47	4.56	54	39.41±0.46
BF15	4.23±0.8	100.67	0.43	4.68	50	26.45±0.23

3.1 In vitro drug release study

USP type-II (paddle type) dissolution apparatus was used to investigate the *in vitro* drug release characteristics of BLS instant release tablets in 0.1N HCl buffer PH 1.2 dissolution medium for 50 min. Getting the super disintegrant concentration up to the optimal 6% enhanced the dissolving rate. Within 40 min, 99.78±0.45% of the medication was released from the BF11 formulation containing 6% Fenugreek mucilage. The drug was released at a rate of 99.54±0.11% after 30 min from the BF12 formulation containing 6% chitosan, 99.68±0.16% after 40 min from the BF13 formulation containing 6% SSG, 99.76±0.48 % after 30 min from the BF14 formulation containing 6% cross povidone, and 99.97±0.46% after 20 min from the BF15 formulation containing 6% Ludiflash. Within 50 min, the medication was released from all super disintegrants. Since MCC (Avicel 101) has a higher disintegration capability than alternative diluents, it was employed as a diluent in all formulations. Each formulation's dissolving profile (from BF1 through BF15) is displayed in Table 3. Dissolution pattern were shown graphically in Figure 3.

Table 3. Dissolution of BLS tablets (n = 3).

	%Drug release at time (min) (±SD)						
Time(min)	0	5	10	20	30	40	50
BF1	0	8.45±0.34	15.45±0.19	34.45±0.88	58.45±0.45	76.67±0.36	96.54±0.45
BF2	0	9.67±0.25	19.68±0.28	35.54±0.23	60.32±0.64	82.68±0.57	-
BF3	0	10.56±0.36	19.34±0.37	39.56±0.63	61.45±0.45	83.52±0.45	-
BF4	0	11.23±0.57	21.46±0.57	41.38±0.52	68.56±0.63	87.46±0.75	-
BF5	0	14.45±0.78	28.65±0.46	62.56±0.61	84.35±0.33	-	-
BF6	0	11.24±0.65	20.14±0.58	42.34±0.14	67.58±0.42	87.53±0.48	-
BF7	0	12.34±0.35	25.47±0.84	50.43±0.38	72.68±0.51	98.68±0.76	-
BF8	0	14.38±0.54	29.45±0.24	61.23±0.93	89.75±0.63	-	-
BF9	0	17.46±0.45	35.23±0.31	69.45±0.76	99.54±0.56	-	-
BF10	0	20.45±0.67	42.68±0.42	76.45±0.46	99.98±0.82	-	-
BF11	0	15.34±0.76	29.34±0.53	57.34±0.4576	79.73±0.24	99.78±0.45	-
BF12	0	20.43±0.84	43.63±0.64	73.34±0.33	99.54±0.11	-	-
BF13	0	18.46±0.94	34.79±0.19	65.76±0.68	84.34±0.44	99.68±0.16	-

BF14	0	21.35±0.34	45.68±0.86	78.34±0.73	99.76±0.48	-	-
BF15	0	27.41±0.33	56.58±0.59	99.97±0.46	-	-	-

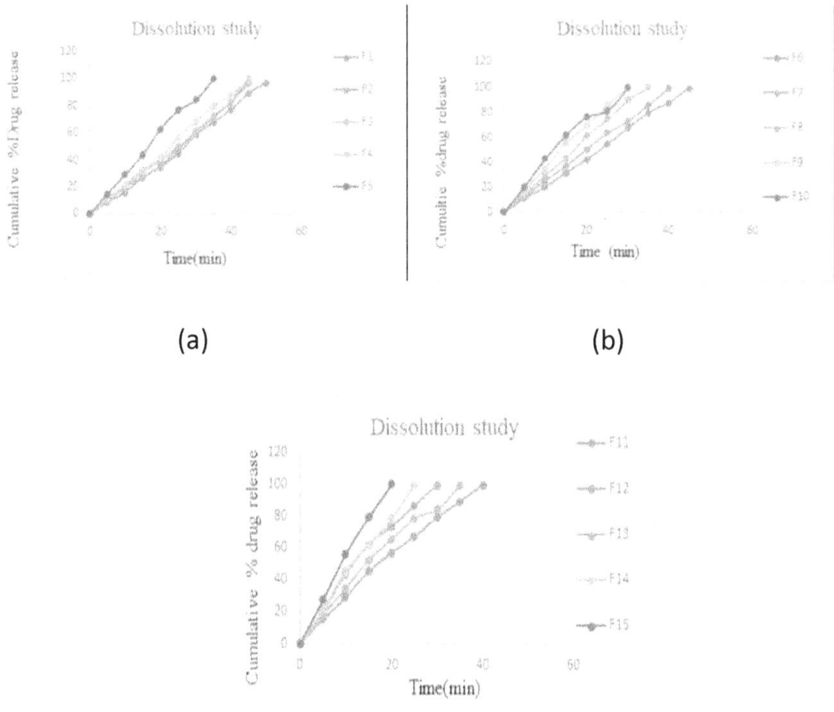

(a) (b) (c)

Figure 3: Drug release study for(a)(b)(c) BF1- BF15.

The stability study showed no significant difference in the evaluation tests for optimized formulation subjected to accelerated stress conditions at 40 °C ± 5 °C/ 75% ± 5% RH for 90 days shown in Table 4.

Table 4. Stability study of BLS tablets (n= 3).

Parameters	30 days 40 o C ± 5° C/ 75% ± 5% RH	60days 40 o C ± 5°C/ 75% ± 5% RH	90 days 40 o C ± 5° C/ 75% ± 5% RH
Hardness	3.54±0.34	3.58±0.12	3.34±32
Weight variation	100.67	100.32	99.89
Friability	0.43	0.41	0.47
Thickness	4.68	4.32	4.46
In vitro drug release study	99.97 in 20min	99.86 in 20min	99.94 in 20min

4. CONCLUSION

This study successfully developed BLS immediate-release pills. This study's main challenge was to compare synthetic, and natural Super disintegrants on BLS *in vitro* release rate. The instant-release medicine delivery technology was effective for acute disorders like urticaria. FTIR investigations showed medication and excipient compatibility and formulation. BLS immediate-release pills are compressed directly. All pre-compression metrics for the dry drug-excipient combination met acceptability standards, indicating excellent flow characteristics. All post-compression properties like thickness, hardness, friability, weight fluctuation, and disintegration are allowed. Formulation BF15, containing 6% Ludiflash, showed full drug release within 20 minutes (>99%). The optimum formulation contains Ludiflash, a super disintegrant, and a hydrophobic binder for flowability and minimal water absorption. It's hard, brittle, and stable. Thus, the current study suggests that BLS instant-release tablets can be utilized as an alternative to standard dosage forms because they release the medicine quickly and are effective for allergies and urticaria. Synthetic super disintegrants yield better medication release than natural super disintegrants, according to the study.

5. REFERENCES

1. S. Ahmed, M. Nazmi, I. Hasan, S. Sultana, S. Haldar and Reza M. S, *Bangladesh Pharmaceutical Journal*, 2013, **16(1)**, 1.
2. S.Bhanja, M. Sudhakar, *International Journal of Pharmaceutical Technology*, 2013, **5(3)**.
3. Ch, S., P, T, *Phytochemistry* 2021.
4. S.Dhanalakshmi, S.R. Baratam, *Journal of Applied Pharmaceutical Sciences,* 2018, **8(7)**, 72.
5. J. Dhiman, D. Dev and D. N. Prasad, *Journal of Drug Delivery and Therapeutics*. 2022, **12(1)**, 170.
6. J. N. R. Hazarika, P. Deb, *International Journal of Current Pharmaceutical research*, 2017, **9(3)**, 118.
7. H. Ijaz, J. Qureshi, Z. Danish, M. Zaman, M.Abdel-Daim and I. Bashir *Advances in Polymer Technology*, 2017, **36(2)**,152.
8. P. anathe, R. Jain, N. Jain and S. K. Jain, *Journal of Drug Delivery and Therapeutics*. 2021, **11(1)**, 42.
9. N. S.Krishna, , B. Jayanthi and A. Madhukar, *Interational Journal of Applied Pharmaceutics*, 2021, **13(4)**, 211.
10. P. S. Patil, A. K. More, S. Kadam, V. Vishwas rao and Y. Patel, *Journal of Drug Delivery and Therapeutics*,2013, **3(5)**,44.
11. A. M. Pethe, A. T. Patil, D. R. Telange and A. A. Tatode, *Asian Journal of Pharmaceutics*,2015, **9(4).**
12. A. Schröder, C. Petring, A. Damanaki, J. Jantsch, P. Proof and C. *J Orofacial Orthopedics* ,2021,**83(1)**,13.
13. V. Sekar, V. R Chellan, *Chemical and Pharmaceutical. Buletin*, 2008, 56**(4)**,575.
14. A. Singh Randhawa, N. Mohd Noor, Md Daud, and B. *Frontiers in Pharmacology*.2022, **12**.
15. G.Tiwari, R.Tiwar, S. Bannerjee, L. Bhati, S. Pandey, P. Pandey B Sriwastawa, *International Journal of Pharmaceutical Investigation*,2012, **2(1)**,2.

NANOFORMULATED EPIGALLOCATECHIN GALLATE FOR ITS POTENTIAL MANAGEMENT OF ORAL SQUAMMOUS CELL CARCINOMA

Surekha MKL[a], Sharon CF[a*], Sindhu Abraham[a], Shwetha K[a]

[a]Department of Pharmaceutics, Faculty of Pharmacy, M S Ramaiah University of Applied Sciences, Bangalore, Karnataka – 560 054, India

*Corresponding Author E-Mail ID: *sharoncaroline.ps.ph@msruas.ac.in*

1 INTRODUCTION

Oral squamous cell carcinoma (OSCC) is mostly preceded by oral premalignant lesions. In India, 90-95% of the cancer is OSCC and it is the most active cancer in oral cavity. It is defined by WHO (1997) as "a predominantly white lesion of the oral mucosa that cannot be characterized as any other definable lesion". It is the 11th most common cancer all over the world. The international agency for research on cancer predicts and increases in the incidence of cancer from 1 million in 2012 to more than 1.7 million in 2035. As a result, the death rate due to oral cancer is also expected to increase from 680000 to 1- 2 million. In India, 20 per 100000 populations are affected by oral cancer which accounts for about 30% of all types of cancer.[1] It is mainly diagnosed by biopsy and the treatment includes mainly by elimination of risk factors like tobacco, smoking and alcohol consumption.[2] Treatment includes both surgical and non-surgical methods. The surgical method is the preferred treatment modality. Non-surgical methods include chemoradiotherapy, photodynamic therapy, COX-2 inhibitors and EGFR inhibitors. The major drawback of these methods are side-effects associated with non-specific cell death and frequent relapse of lesions on cessation of medications.[3]

Nano formulations can minimize side effects due to non-specific cell death and also have the ability to increase the efficacy of therapeutic agents. Nanosponges are of microscopic particles with tiny mesh like structures having few nanometers wide cavities. A variety of substances both lipophilic and hydrophilic can be encapsulated in these cavities.[4]

Nanosponges have been used for targeting of therapeutic agents to the skin with an intention to achieve benefits such as dose reduction prolonged drug release, prolonged skin retention and avoidance of systemic absorption. Nanosponges can reduce drug irritation to a significant extent without compromising on therapeutic efficacy.[5] Nanosponge loaded topical preparations can deliver drugs to a localized area, prolong drug release and act as rate- limiting layer for modifying systemic absorption. Hence defeating the boundaries of topical preparation.[6]

Green tea extract [Camellia sinensisL. Ktze. (Theaceae)] contains 55.9% Epigallocatechin gallate (EGCG) other catechins such as 2.2% Epigallocatechin (EGC), 1.4% Epicatechin (EC), 12.6% Epicatechin gallate (ECG), 4.9% Gallocatechin gallate (GCG), 0.7% Catechins gallate (CG) and 0.5% Gallocatechin (GC). EGCG is the most biologically active catechin in green tea. The compound (–)-epigallocatechin-3-gallate (EGCG) has several activities such as antioxidant, anti-inflammatory, anticancer, inhibition of angiogenesis in oral cancer cell lines, helps in weight loss and protects the skin from ionizing radiations.[7]

2 METHODS AND RESULTS

2.1 Chemicals and Reagents Used

Epigallocatechin gallate was purchased from Hunan Insen Biotech Co., Ltd, China. Polyvinyl alcohol, Ethyl cellulose and HPMC K15M were obtained from Yarrow chem products, Mumbai. Ethanol was procured from Changshu Hongsheng Fine chemicals Co., Ltd. Xylitol was procured from Hi Media Laboratories Pvt. Ltd.,Mumbai. Sodium benzoate was obtained from Rankem RFCL Ltd., Gujarat.

2.2 Formulation of EGCG loaded nanosponges

Nanosponges were prepared by using emulsion solvent evaporation technique. Accurately weighed quantities of EGCG and ethyl cellulose (drug phase) was dispersed in ethanol with constant stirring using a mechanical stirrer at 600rpm for 10min. Aqueous phase was prepared by dissolving PVA in hot water for 20min. Drug phase was added drop by drop to the aqueous phase with continuous stirring for 2 hours at 1000rpm. The product was filtered, kept for drying in hot air oven for 24 hours and dried nanosponges were collected.[8] See Table 1.

Table 1 *Formulation of nanosponges.*

Ingredients	Formulation code			
	NS1	NS2	NS3	NS4
EGCG (mg)	500	500	500	500
Polyvinyl alcohol (%)	1	2	3	4
Ethyl cellulose (%)	1	1	1	1
Ethanol (ml)	25	25	25	25
Distilled water q.s to (ml)	100			

2.3 Characterization of nanosponges

2.3.1 Differential Scanning Colorimetry (DSC). EGCG, polymers and their physical mixture were subjected to DSC analysis using differential scanning colorimeter (Mettler, Star^e SW 12.10).

2.3.2 Particle size, polydispersibility index and zeta potential. Particle size determination and polydispersity index measurement was carried out using ZetaPals particle sizing software version 5.23 (Brookhaven instruments corp.). The liquid samples were prepared by diluting the sample in deionized water, filled into the cuvettes and analyzed. The samples were sonicated for 45 min before measurement.[9]

Zeta potential was measured by using additional electrode in Brookhaven instrument used for DLS

2.3.3 X-Ray Diffraction Studies. Powder XRD patterns of EGCG and nanosponges were recorded at ambient temperature using X-Ray Diffractometer (Rigaku®, Japan). The samples were analysed at 2θ angle in range of 5-50° for 1hour.[10]

2.3.4 Entrapment efficiency. To calculate the entrapment efficiency, nanosponge suspension was prepared by dispersing accurately weighed nanosponge (100mg) in 100ml of buffer pH 6.8. The dispersion was kept in bath sonicator for 15 min, filtered and the concentration of EGCG in the filtrate was determined spectrometrically at 272nm. Loading efficiency was calculated by using the formula:

$$\text{Loading efficiency} = \frac{\text{Actual drug content in nanosponges}}{\text{Theoretical drug content}} \text{ X } 100 \qquad (1)$$

2.3.5 Surface morphology using Scanning Electron Microscopy (SEM). A small quantity of nanosponges was mounted on to a slab and then coated with gold prior to examination. The acceleration voltage was 20 KV. The photomicrographs were taken at 5X, 15X and 20X magnifications.[9]

2.4 Formulation of nanosponge loaded mucoadhesive gel

The prepared nanosponges were incorporated into a mucoadhesive gel base for ease of application onto the oral mucosa. The polymer HPMC K15M was initially soaked in water for the gel for 2hours and dispersed by agitation to get smooth dispersion. Xylitol was added as an artificial sweetener. Sodium benzoate acts as preservative to prevent against microbial attack. Peppermint oil was added as a flavouring agent. The prepared and optimized nanosponges were dispersed in the gel. The composition of nanosponge loaded gels was shown below in Table 2.

Table 2 *Formulation of nanosponge loaded gel.*

Ingredients	Formulation code			
	NSG1	NSG2	NSG3	NSG4
Nanosponges (equivalent to 500mg EGCG)	NS1	NS2	NS3	NS4
HPMC K15 M (%)	1.25	1.25	1.25	1.25
Xylitol (%)	0.5	0.5	0.5	0.5
Sodium benzoate (%)	0.05	0.05	0.05	0.05
Peppermint flavor (ml)	0.1	0.1	0.1	0.1
Distilled water q.s to (g)	50			

2.5 Characterization of nanosponge loaded gel

2.5.1 Homogeneity. Gels were allowed to set in a container and then examined for appearance, colour, texture, homogeneity and presence of any aggregates by visual inspection.

2.5.2 Determination of pH. The pH of the mucoadhesive oral gels was determined using a digital pH meter. The electrode was dipped into gel formulation and the pH was noted after stabilization. The measurement of pH for each formulation was replicated three times.

2.5.3 Viscosity. Viscosity of gels was determined using Brookfield viscometer (Model DV-II + Viscometer), Spindle No. 4 at 20, 50 and 100 rpm. The spindle was dipped into the gel for about 5 minutes and then the readings were noted.

2.5.4 Spreadability. Spreadability of formulations was determined by a glass slide apparatus which was fabricated in the laboratory and used for study. Weight was added to the pan attached to the upper slide and the time and weight required for the upper slide to slide over the lower one completely was noted.

$$Spreadability\ (S) = \frac{Weight\ of\ upper\ slide\ (M)\ x\ Length\ (L)}{Time\ (T)} \qquad (2)$$

2.5.5 In vitro drug permeation studies. In vitro permeation studies for topical gel were performed using dialysis membrane-70 (Himedia) in a Franz diffusion cell. The membrane was soaked in 0.1N HCl for 24 hours. The receptor compartment was filled with 6.8pH phosphate buffer. The formulated gels equivalent to 100mg was applied evenly on surface of the membrane. The whole assembly was maintained at constant temperature (37°C). Aliquots of drug sample (2ml) was withdrawn at every 30min time interval and replaced with fresh buffer. The drug samples were analyzed by UV-spectrophotometer at 272nm and the percentage drug permeation was calculated. The permeation studies were carried out for 6hours.

2.5.6 In vitro cell line studies. The study was carried out on human oral squamous carcinoma cell line SCC-9. The inhibitory potential of the nanosponge gel formulation on these cells was determined by MTT assay. Percentage growth inhibition was calculated by using below formula and the concentration of test drug need to inhibit growth by 50% (IC_{50}) was extrapolated from dose response curves.

$$Percentage\ inhibition = \frac{Optical\ Density\ of\ Sample}{Optical\ Density\ of\ Control} \times 100 \qquad (3)$$

3 DISCUSSIONS

3.1 Characterization of nanosponges

3.1.1 Differential Scanning Colorimetry (DSC). The enthalpy of vaporization of EGCG determined by DSC was compared with reported enthalpy values. The thermogram of EGCG and physical mixture showed an endothermic peak at 130.18°C and 121.82°C Figure 1, corresponding to the vaporization enthalpy of EGCG, thus indicating the absence of interaction between them.

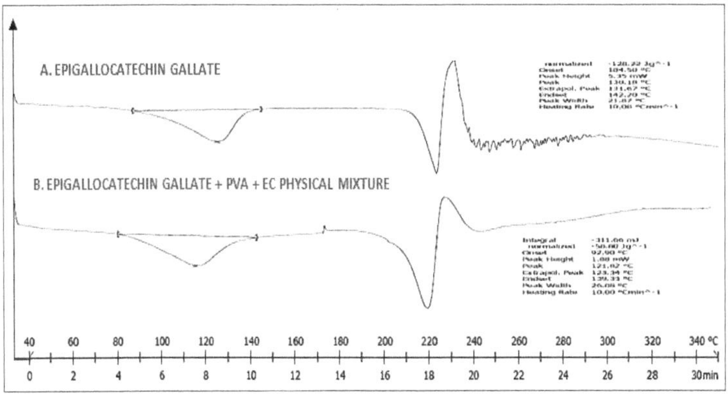

Figure 1 DSC Thermogram of EGCG and physical mixture of Nanosponges formulation

3.1.2 Particle size, polydispersibility index and zeta potential. The particle size analysis of the selected nanosponge formulations was measured by DLS. The particle size of NS1 and NS4 was found to be 372.7nm and 307.8nm respectively. The formulations displayed monodisperse size distribution with polydispersity index of 0.179 and 0.245 respectively.

The zeta potential of formulations NS1 and NS4 was -19.89 and -13.37 mV. These values indicate a stable colloidal suspension without tendency to agglomerate.

3.1.3 X-Ray Diffraction Studies. The powder X-ray diffraction patterns of pure drug and nanosponge formulation is shown in Figure 2. The diffraction patterns of pure drug show sharp and intense peaks indicating the crystalline nature of the sample. The PXRD of nanosponge show masking of all the crystalline peaks which suggest the successful encapsulation of the drug in the nanosponge.

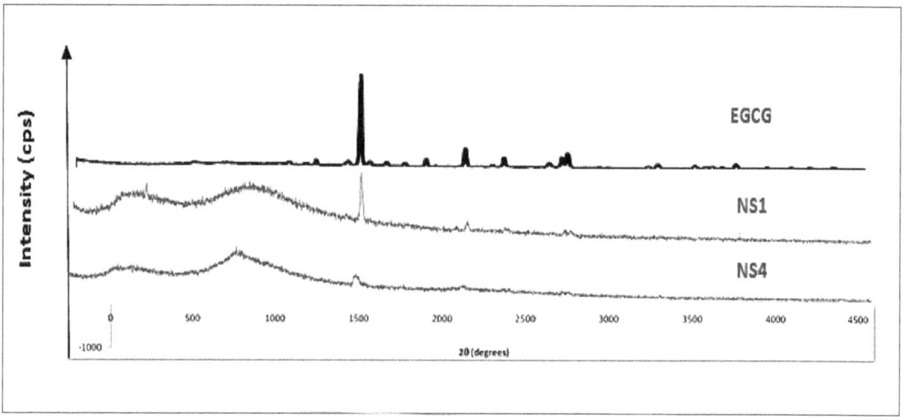

Figure 2 PXRD of EGCG, NS1 and NS4

3.1.4 Entrapment efficiency. The EGCG encapsulation in the nanosponges is shown in Table 3. The drug encapsulation is found to decrease with increasing PVA concentration. This finding is in agreement with particle size measurements. Increasing the concentration of PVA with respect to ethyl cellulose from 1:1 (NS1) to 4:1 (NS4) lead to a decrease in the particle size. The drug encapsulation is also found to decrease from 96.25 in NS1 to 76.85% in NS4

Table 3 *Percentage drug content of Nanosponges.*

SI.No	Formulation code	Percentage Drug content (%)
1	NS1	96.25 ± 0.0602
2	NS2	85.30 ± 0.0100
3	NS3	79.72 ± 0.0264
4	NS4	76.85 ± 0.0472

3.1.5 Surface morphology using Scanning Electron Microscopy (SEM). SEM of the selected nanosponge formulations is shown in Figure 3 (a and b). The images were captured in three different magnification 5KX, 10KX and 25KX. All the images show nanosponges with porous, spongy texture. Images with

25Kmagnifications show spherical morphology with a nanosize particle ranging from 100.1 to 150.1nm for NS1 and 48.32 to 133.6nm for NS4. These observations are in agreement with the particle size measurements using DLS. i.e., increasing the amount of PVA in the formulations decreased the particle of the nansponges.

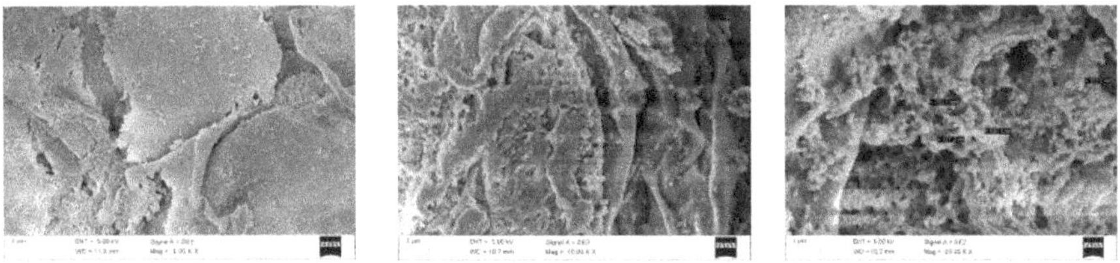

Figure 3(a) *SEM photomicrographs of NS1 at 5KX, 10KX and 25KX.*

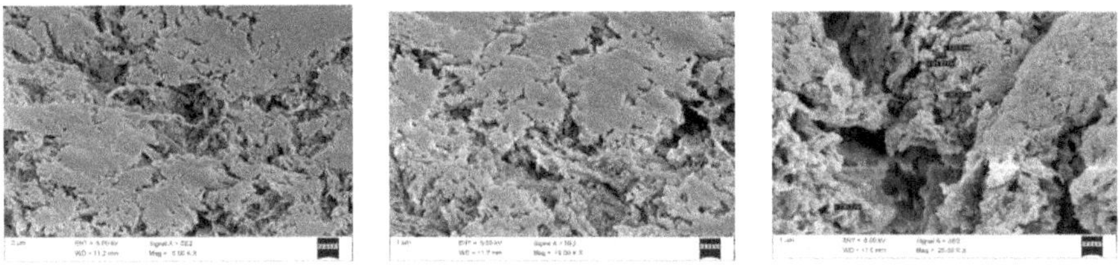

Figure 3(b) *SEM photomicrographs of NS4 at 5KX, 10KX and 25KX.*

3.2 Characterization of nanosponge loaded gels

3.2.1 Homogeneity, pH and drug content. All the sponge loaded gel formulations were found to be homogenous. The pH of the nanosponge loaded gels ranged from 6.80 to 7.11. The percentage encapsulation efficiency in nanosponge loaded gels ranged from 86.60 to 96.59 %.

3.2.2 Spreadability. The results of spreadability (gcm/sec) studies are shown in Figure 4. The values indicate highest spreadability for gels having the lowest carbopol percentage. As the percentage of carbopol in the formulation increased, the spreadability values decreased.

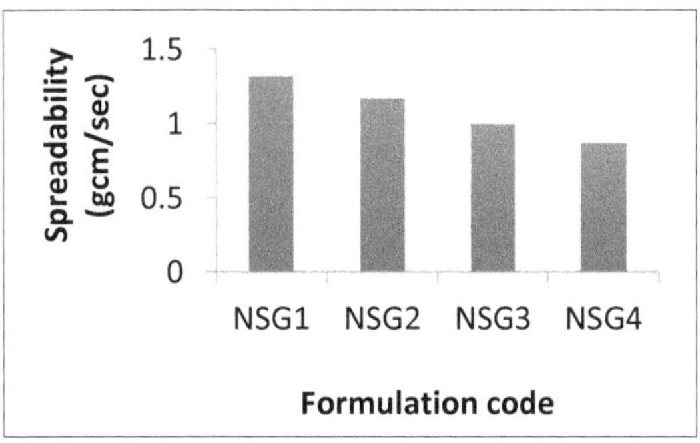

Figure 4 *Spreadability of nanosponge loaded gels.*

3.2.3 Viscosity. Viscosity of the nanosponge loaded gel formulations were determined and the results are shown in Figure 5 the viscosity readings indicate increased viscosity with increase in polymer concentration. The viscosity decreased with rpm indicating shear thinning behaviour of the gels

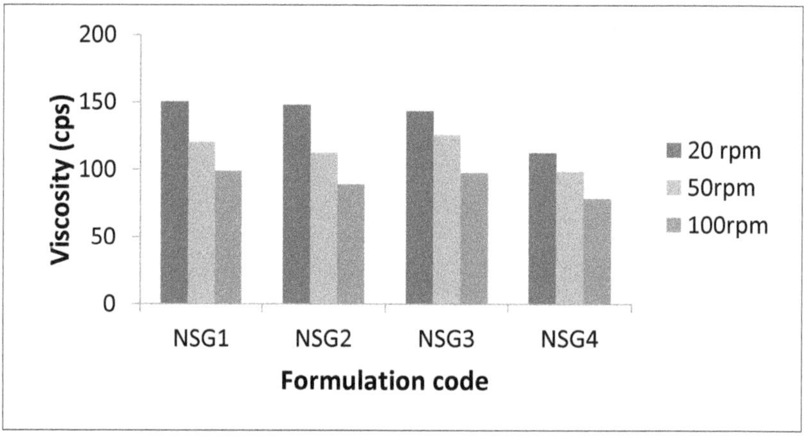

Figure 5 *Viscosity of nanosponge loaded gels.*

3.2.4 In vitro drug permeation studies. In vitro permeation study was performed for sponge loaded gel formulations for 6hours using dialysis membrane-70. The cumulative percentage drug release at various time intervals is shown in Figure 6.

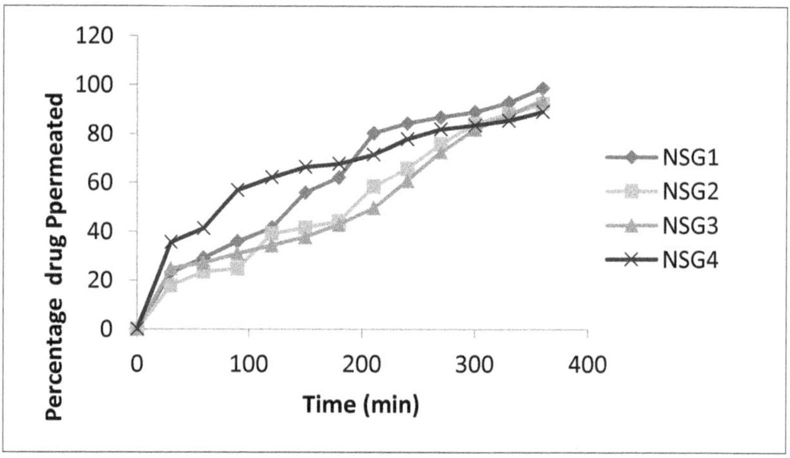

Figure 6 *Cumulative percentage drug permeated Vs Time profile for nanosponge loaded gels.*

The rate of drug release from NSG4 was the highest initially. This could be attributed to high PVA to Ethyl cellulose concentration in the incorporated nanosponges.

3.2.5 In vitro cell line studies. The results of in vitro cell line studies carried out on SSC-9 cell line are shown in Table 4. The IC50 value of EGCG nanosponge gels was found to be 60.3µM and that of standard doxorubicin was 21.38 µM. The photomicrographs of the cells is shown in Figure 7. The results indicate the inhibitory potential of EGCG on SSC-9 cells and thus support its potential use in the management of oral squamous cell carcinoma.

A B C

Figure 7 *Photomicrographs of SCC-9 cell line A) control B) 3.125µM sample C) 100µM sample.*

Table 4 *Cell culture studies for NSG1 (SCC-9 cell line).*

Compound name	Conc. µM	OD at 590nm	Percentage Inhibition	IC_{50} µM
Control	0	0.831	0.00	
Standard (Doxorubicin)	3.125	0.743	6.23	21.38
	6.25	0.686	13.38	
	12.5	0.533	32.67	
	25	0.376	52.56	
	50	0.283	64.29	
	100	0.136	82.86	
NSG1	3.125	0.765	7.92	60.3
	6.25	0.703	15.46	
	12.5	0.651	21.63	
	25	0.562	32.41	
	50	0.450	45.79	
	100	0.304	63.36	

4. CONCLUSION

Epigalocatechin gallate, derived from green tea extract has in recent years been intensively investigated for regulating or inhibiting the physiological changes during canceration. In the present work EGCG was formulated ino a nano sponge and incorporated into HPMC gel base. The gels were investigated for their anticancer pothential on oral squammous cell carcinome cell lines. The aforementioned results indicate that EGCG containing gels can restrain the growth and the proliferation of SSC-9 cells and can be viewed as an alternative option in the management of oral squammous cell carcinoma.

5. ACKNOWLEDGMENT

The authors are thankful to M S Ramaiah University of Applied Sciences, Bangalore for providing the facilities to carry out the work.

6. REFERENCES

1. V. Borse, A. Narayan Konwar and P. Buragohain, Sensors International, 2020, 1, 100046.
2. P. M. Williams, C. F. Poh, A. J. Hovan, S. Ng and M. P. Rosin, Journal (Canadian Dental Association), 2008, 74, 275–280.
3. S. A. Gharat, M. Momin and C. Bhavsar, Critical ReviewsTM in Therapeutic Drug Carrier Systems, 2016, 33, 363–400.
4. S. S, A. S, K. Krishnamoorthy and M. Rajappan, Journal of Pharmacy & Pharmaceutical Sciences, 2012, 15, 103.
5. P. Srinivas and A. Reddy, Pharmaceutical Nanotechnology, 2015, 3, 68–76.
6. H. Bhowmik, D. N. Venkatesh, A. Kuila and K. H. Kumar, International Journal of Applied Pharmaceutics, 2018, 10, 1.
7. D. G. Nagle, D. Ferreira and Y.-D. Zhou, Phytochemistry, 2006, 67, 1849–1855.
8. S. M. Badr-Eldin, H. Aldawsari, G. Labib and A. El-Kamel, International Journal of Nanomedicine, 2015, 10, 893.
9. K. A. Ansari, P. R. Vavia, F. Trotta and R. Cavalli, AAPS PharmSciTech, 2011, 12, 279–286.
10. G. Utzeri, P. M. C. Matias, D. Murtinho and A. J. M. Valente, Frontiers in chemistry, 2022, 10, 859406.

IN VIVO SKIN IRRITATION POTENTIAL OF A TOPICAL GEL LOADED WITH TEA TREE OIL NANOSPONGES

Amrita Rana[a], Sindhu Abraham[a], Sharon Furtado[a], Megha K[a] and Anbu J[b]

[a] Department of Pharmaceutics, [b] Department of Pharmacology, Faculty of Pharmacy, M S Ramaiah University of Applied Sciences, Gnanagangothri Campus, Bengaluru, INDIA

*Corresponding Author E-mail ID: sindhusa@gmail.com

1 INTRODUCTION

Tea tree oil (TTO), an essential oil extracted from the leaves of *Melaleuca alternifolia*, has been used as a traditional therapy for various ailments by the aborigines of Australia. The oil gained popularity around the world as an external remedy for acne, athlete's foot, fungal infections, nail infections, lice infestation, dandruff, vaginal yeast infections, and skin lesions.[1,2] TTO contains a mixture of almost 100 constituents, with Terpinen-4-ol (monocyclic terpene alcohol) being the major constituent to an extent of about 40%. TTO like other essential oils cannot be applied to the skin directly as it can cause erythema, photosensitization, dryness and hence should be used only after dilution with carrier oils. These oils are highly volatile and decompose on exposure to heat, humidity, light, and oxygen.[3] Nanosponges are porous, spherical, polymeric delivery systems for topical delivery of therapeutic agents to achieve benefits like enhanced safety, stability, and improved application.[4-5] Nano carriers in topical delivery systems are beneficial due to their increased penetrability and passive accumulation at the target site.[6,7] Encapsulation of essential oils aids in overcoming its instability and susceptibility to degradation. This technique is favorable for the formulation of cosmeceuticals, nutraceuticals, and personal care products. Encapsulation into nanosponges helps to prevent side effects, modulate their release, decrease their volatility, maintain stability, reduce toxicity, enhance their activity, and improve patient compliance.[8,9] Encapsulation of TTO in nanosponges can reduce its irritancy to a significant extent without compromising on therapeutic efficacy.[10-12]

The present study aimed to develop gels containing TTO loaded nanosponges that is devoid of skin irritation and other side effects.

2 METHODS AND RESULTS

2.1 Chemicals/Reagents Used

Tea tree oil and terpinen-4-ol were procured from Aramacs Ltd., India. Polyvinyl Alcohol (PVA), Ethyl Cellulose (EC), and Carbopol 934 were procured from Yarrow Chemicals, India. Sabouraud Dextrose Agar medium and Mueller-Hinton agar medium were purchased from Himedia Laboratories Ltd., India.

2.2 Analysis of components of tea tree oil by Gas Chromatography-Mass Spectrometry (GCMS)

Tea tree oil was analyzed by GC-MS (Agilent 7890A; MS: 5975C MSD) to determine its chemical constituents. A DB 5 MS column (30m L x 0.25mm ID x 0.25um thickness) was used for separation. The carrier gas Helium (99.999%) was introduced at a steady flow rate of 1.0 ml/min in splitless mode. 1µL of the sample was injected with the temperature maintained at 250°C. The quadrapole temperature was set at 180°C while the MS source was maintained at 250°C. Mass spectra was scanned from 30m/z to 700m/z and compared to identify the active constituents. The % peak area was measured and used for the quantitative analysis.

2.3 Estimation of terpinen-4-ol by High Performance Thin Layer Chromatography

High-performance thin-layer chromatography is commonly used for the quick detection of drugs and provides visual semi-quantitative data of the active constituents of drugs.[13,15] Terpinen-4-ol, the chief constituent of TTO was estimated by HPTLC in this study.

2.3.1 Instrumentation for HPTLC. Pre-coated Silica gel 60 F254 plates (10×10 cm, 200 μm thickness) was used as the stationary phase. The CAMAG HPTLC system included an automatic sample applicator (Linomat 5), 100 μL dosing syringe, twin-trough chamber (20×10 cm), and TLC Scanner equipped with winCATS software (V1.4.1).

2.3.2 Chromatographic conditions. Prewashed TLC plates were dried and activated at 105°C for 1h, before analysis. Samples were spotted at a constant rate of 150 nLs^{-1} as 6mm wide bands, placed 10 mm apart. The plates were developed in a chamber previously saturated with the mobile phase (toluene and ethyl acetate [93:7]) to a distance of 90 mm. The air-dried plates were scanned (slit dimension: 6.00 x 0.45 mm; scanning speed: 20 mm/s) at 366 nm in absorbance mode.

2.3.3 Preparation of standard stock solution of TTO. Standard stock solution of TTO was prepared by dissolving 100μl of TTO into 900μl of toluene. A series of dilutions were prepared for TTO over a range of 100 to 900 ng (n=3). Linear least square regression analysis was used to treat the data of area under the peak v/s drug concentration. Plates were treated with anisaldehyde-sulphuric acid reagent and heated in an oven at 105°C for 15 min to visualize the zones.

2.4 Preparation of tea tree oil loaded nanosponges

Nanosponges containing Tea tree oil were prepared by emulsion solvent evaporation technique. TTO and Ethyl Cellulose (EC) solubilized in ethanol was added slowly to Polyvinyl alcohol (PVA) solution and sonicated (QSONICA-CL-334, India) for 2 h to form nanosponges (Table 1).

Table 1 *Formulations of TTO loaded nanosponges.*

Ingredients	Formulations							
	N1	N2	N3	N4	N5	N6	N7	N8
Tea tree oil (ml)	2	2	2	2	2	2	2	2
Ethyl cellulose	1	1	1	1	2	2	2	2

(g)								
Polyvinyl alcohol (g)	1	2	3	4	1	2	3	4
Ethanol (ml)	20	20	20	20	20	20	20	20
Carbopol 934 (g)	0.5	0.5	0.5	0.5	0.5	0.5	0.5	0.5
Purified water (ml)	80	80	80	80	80	80	80	80

2.5 Characterization of nanosponges

2.5.1 Particle size Measurement. Phase Analysis Light Scattering technology (Zeta PALS, Brookhaven Instruments Corp., USA) with ZetaPALS Particle Sizing Software (Ver.5.23) was employed for the determination of particle size of TTO loaded nanosponges.

2.5.2 Scanning Electron Microscopy (SEM). The surface topography of nanosponges was analyzed by SEM (Ultra 55, Carl Zeiss, Germany). The samples were covered with a gold layer and affixed to aluminium stubs in an atmosphere of argon. Samples were observed at different magnifications with the acceleration voltage maintained at 15 kV.

2.5.3 TTO Entrapment Efficiency. The nanosponges were sonicated with methanol for 15 minutes to extract terpinen-4-ol completely and centrifuged. 6µl of supernatant along with 6µl of terpinen-4-ol standard were spotted on a single TLC plate, developed in the mobile phase to a distance of 90 mm, dried, and scanned at 366 nm to generate peaks. Rf and peak areas of TTO loaded nanosponges were compared with that of terpinen-4-ol standard. The zones were visualized by treating the spots with an anisaldehyde-sulphuric acid reagent and heating at 105°C for 15 min. The calculations were done based on a comparison of Area under curve (AUC) of test solution and standard.

% Drug entrapment = (Practical content/ Theoretical content) x 100

2.6 Loading of nanosponges into the gel

Carbopol 934 (0.5%) previously soaked in purified water was mixed with the nanosponge dispersion and left undisturbed for 15 min to remove air entrapment (Table 1). Eight such formulations were prepared, maintaining the concentration of TTO constant (2%) and varying the concentrations of PVA (1-4%) and EC (1-2%).

2.7 Evaluation of Nanosponge loaded gels

2.7.1 Viscosity determination. The viscosity of gels was determined with Brookfield viscometer DV II+ coupled with spindle no. 4 (Brookfield Engineering Laboratories Inc., USA). The spindle was rotated at 10 rpm. The average of three measurements was considered.

2.7.2 pH measurement. The pH of all the TTO loaded nanosponge gels was measured with a digital pH meter (Model ATC Chemlabs, India). The average of three readings was considered.

2.7.3 Spreadability test. The spreadability test apparatus consisted of two glass plates.[13] 0.5 g of the sample was placed between the plates and force was applied by adding increasing weight slowly at an interval of one minute into the pan connected to the upper plate. The time taken for the upper plate to slide over the lower plate until it separated was recorded. Each sample was evaluated in triplicate. Spreadability was then determined using the formula:

$$S = (ML) / T$$

Where, S= Spreadability; M = weight applied on the plate; L = length of the glass plate; T = time taken to separate the plates(sec)

2.7.4 Ex Vivo Skin permeation studies. These studies were performed using Franz diffusion cell having a receptor compartment of volume 60 ml and a surface area of 3.14 cm^2, using rat skin as the permeation membrane. The abdominal skin of sacrificed rats was excised, defatted to remove subcutaneous fat and cleaned thoroughly. 4 to 5 cm^2 (thickness) of the skin was mounted between the receptor and donor compartments with the dermal side of the skin facing the receptor compartment containing Phosphate buffer pH 7.4 maintained at 37±0.5 C. 1g of gel was evenly spread on the skin facing the donor compartment. The medium in the receptor compartment was stirred constantly at 100 rpm. At predetermined intervals of time (1, 2, 3, 4, 5, 6 h), 2 ml of sample was withdrawn from the receptor compartment and replaced with fresh buffer. The samples were suitably diluted and analyzed for terpinen-4-ol content.

2.7.5 Determination of antifungal and anti-bacterial activity of the TTO loaded nanosponge gels. In vitro antibacterial activity of TTO loaded nanosponges was determined by well diffusion method on Mueller-Hinton Agar (MHA) against bacterial strains *Staphylococcus aureus* and *Escherichia coli* while the antifungal activity was carried out on Sabouraud Dextrose Agar (SDA) medium against fungal strains *Aspergillus niger and Candida albicans*. TTO was used as the standard sample. All the formulations were placed in their respective wells (diameter 6 mm) and the diameter of inhibition zones were determined after 18 to 24 h of incubation at 37°C for bacteria and after 48 h of incubation at 28°C for fungi.

2.8 *In Vivo* Skin Irritation Test

Skin irritancy potential of nanosponge gel formulation was assessed by Draize patch test on healthy rabbits in accordance with OECD guideline 404 for testing of chemicals.[27] The study was carried out with prior approval from the Institutional Ethics committee (IAEC Ref No: XVI/MSRFPH/M-008/20.01.2016). 0.5 g of the gel was applied on the shaved skin of rabbits, covering an area of 4 cm^2 and kept in contact for 4 h. The skin was monitored after 1, 24, 48, and 72 h of application. Observations were graded on a scale of 0-4, with 0 denoting the absence of erythema/edema and 4 denoting severe erythema/edema.

3 DISCUSSION

Eight trials of TTO loaded nanosponges were prepared, each containing 2% of Tea tree oil and varying concentrations of polyvinyl alcohol and ethyl cellulose. The nanosponges were then incorporated into Carbopol 934 gels.

3.1 Analysis of components of tea tree oil by GCMS

In the present study, Gas chromatography conjugated with mass spectrometry was used for qualitative and quantitative analysis of Tea tree oil. Around 78 constituents of tea tree oil were identified out of which 13 major constituents were considered (Figure 1 and Table 2).

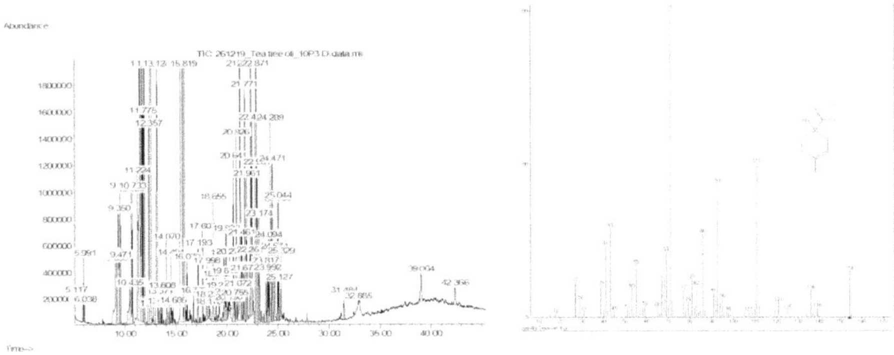

Figure 1 a) GC-MS Chromatogram of Tea tree oil b) Mass spectra of Terpinen-4-ol

Table 2 *Chemical constituents identified in Tea tree oil by GC-MS.*

No	CAS Number	Constituent	Peak RT (min)
1	000562-74-3	Terpinen-4-ol	15.6
2	000099-85-4	γ-Terpinene	11.22
3	000099-86-5	α-Terpinene	11.48
4	000470-82-6	1,8-Cineole	11.88
5	000586-62-9	Terpinolene	13.12
6	000098-55-5	ρ-Cymene	11.66
7	005989-27-5	α-Pinene	9.84
8	000099-87-6	α-Terpineol	15.82
9	000080-56-8	Aromadendrene	21.32
10	000489-39-4	Limonene	11.77
11	000489-41-8	Globulol	24.28
12	003387-41-5	Sabinene	9.486
13	000483-76-1	δ-Cadinene	21.959

3.2 Estimation of terpinen-4-ol by HPTLC method

A suitable HPTLC method was developed for the estimation of terpinen-4-ol in the formulations. Initial trials were performed to optimize the chromatographic parameters such as composition of mobile phase, detection wavelength, and concentrations of the standard solutions. The mobile phase comprising of toluene and ethyl acetate (93:7) could separate and resolve terpinen-4-ol peaks without interference from the other constituents. The peak areas were considered for the quantitative determination of terpinen-4-ol. The developed plates were scanned at 366nm using CAMAG HPTLC Scanner and the automated software generated a chromatogram of absorbance vs Rf. The peak of terpinen-4-ol was well resolved at Rf 0.68 ± 0.05 (Figure 2.a). The calibration curve generated by plotting peak area vs concentration of TTO (ng/spot) confirmed the linearity of the calibration curve over the range 100-800 ng/band with a high correlation coefficient (r^2 = 0.996). The 3-D chromatograms of the calibration concentrations are depicted in Figure 2b.

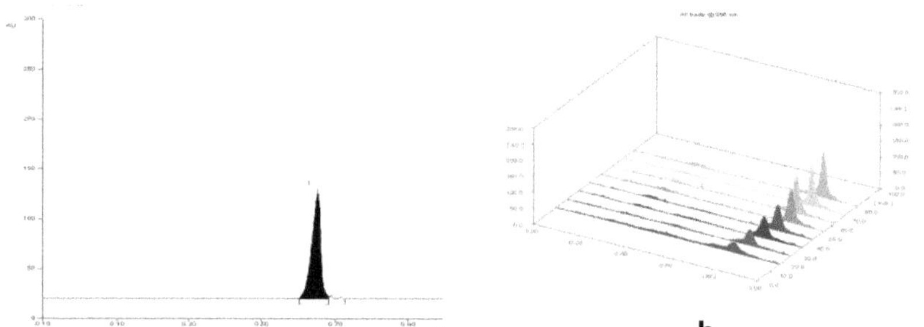

a b

Figure 2. a) Chromatogram of Standard Terpinen-4-ol peak of 400 ng (R_f =0.68) b) 3-D chromatogram indicating linearity of calibration curve.

3.3 Characterization of Nanosponges

3.3.1 Particle size Measurement. The mean particle size of all the formulations were found to be between 307.8 - 394.4nm. Nanosponges prepared with 1 % of ethyl cellulose exhibited particle sizes between 307.8 - 315.7nm (N1- N4), whereas formulations containing 2 % of ethyl cellulose (N5-N8) were in the higher size range (372.7 -394.4 nm) as shown in Table 3.3. N1 exhibited the least particle size of 307.8 nm and polydispersity index of 0.245.

3.3.2 Surface Morphology. SEM micrograph of the nanosponges at 21000X magnification demonstrated the presence of pleomorphic particles with a spongy surface (Figure 3 a, b).

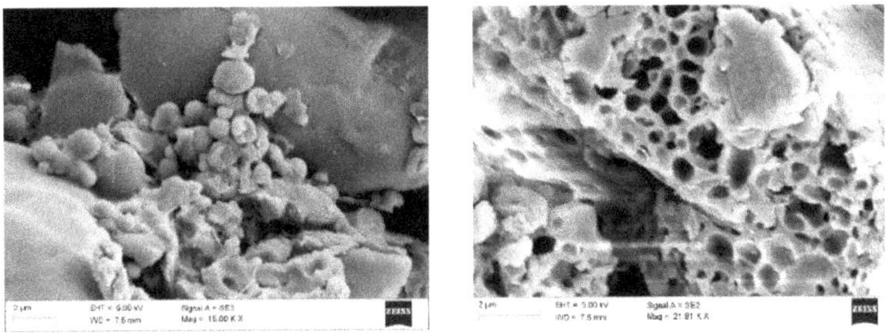

Figure 3 SEM micrographs showing **a)** TTO loaded nanosponges **b)** Spongy surface of nanosponges.

3.3.3 TTO Entrapment Efficiency. The entrapment of TTO within the nanosponges ranged from 62.4± 6.5 to 89.5 ± 4.43% (Table 3). Entrapment was significantly influenced by polymer concentration and formulation N1 showed the highest entrapment (89.5 ± 4.43 %).

3.4 Evaluation of Nanosponge loaded gels

3.4.1 Viscosity, pH, and Spreadability. The viscosities of all the formulations ranged from1265±2.0 to 3000±2.20 cp. pH evaluation of the topical formulation is important as variation in pH may causes irritation to the skin. The pH of the gels ranged from 6.37±0.11 to 6.45±0.01 (Table 3).
Spreadability determines the ease with which a semi-solid preparation can be applied to the skin. The values for all gels ranged between 3.60±0.15 to 4.28±0.11 g.cm/sec. Formulations with low viscosities showed better spreadability (Table 3). All the gels could be extruded uniformly and conveniently from the collapsible tubes.

Table 3 *Evaluation of TTO nanosponge loaded gels.*

Evaluation Parameters	***Formulation Code***							
	N1	**N2**	**N3**	**N4**	**N5**	**N6**	**N7**	**N8**
Particle Size (nm)	307.8	310.8	311.0	315.7	372.7	377.1	384.0	394.4
Entrapment Efficiency $^{\#}$ (%)	89.5 ±4.43 ^{ns}	88.6± 2.13	88.3 ±3.10^{ns}	85.75 ±4.50 ^{ns}	73.4± 4.60 ^{ns}	72.99 ±3.70 ^{ns}	64.1± 2.90 ^{ns}	62.4 ±6.50 ^{ns}
pH $^{\#}$	6.45 ±0.01*	6.38± 0.21*	6.43 ±0.21*	6.39± 0.04*	6.37± 0.11*	6.38± 0.12*	6.37± 0.23*	6.39 ±0.21*

Extrudability	Easy	Easy	Easy	Easy	Easy	Easy	Easy	Easy
Spreadability # (g.cm/sec)	4.28 ± 0.11 ns	3.85 ±0.22 ns	4.11 ±0.05 ns	3.40 ±0.21 ns	3.96 ±0.10 ns	3.60 ±0.15 ns	3.87 ±0.21 ns	3.83 ±0.05 ns
Viscosity # (cp)	1265 ± 2.0*	1295 ±2.50*	1287 ±2.60*	1265 ±2.89*	2290 ±2.50*	2295 ±2.00*	3000 ±2.20*	2997 ±2.50*

#Data expressed as mean ±S.D (n=3); Analysed by one way ANOVA followed by Tukey-Kramer test where ns (non-significant)$p>0.05$; *- (significant) $p<0.05$

3.4.2 Ex-Vivo skin permeation studies. TTO loaded nanosponge gels N1, N2, N3 and N4 were considered for permeation studies as they exhibited good entrapment efficiency. Permeation was assessed using Franz diffusion cell with rat skin as the permeation membrane. At the end of 6 h, formulation N1 showed the highest drug release (96.00 ± 4.09 %) and N4 showed the least drug release (85.00 ± 4.60 %). The permeation of TTO was rapid from formulation containing 1 % of ethyl cellulose and polyvinyl alcohol.

3.4.3 Determination of antifungal and antibacterial activity of the TTO loaded nanosponge gels. TTO and TTO loaded nanosponge gels (Formulations N1, N2 and N3) were evaluated for *in vitro* antifungal activity (against *Candida albicans, Aspergillus niger)* and antibacterial activity *(against Escherichia coli, and Staphylococcus aureus)* by quantitative assessment of presence or absence of inhibition zones. According to the results represented in Figure 4, TTO and TTO loaded nanosponge gel showed significant antifungal and antibacterial and activity against all tested microbial strains as evident by the inhibition zone in all the plates.

3.5 In Vivo Skin Irritation Test

The most important side effect associated with the use of pure TTO is erythema, which intensely restricts its effectiveness. The irritation potential of the TTO loaded nanosponges gel was compared with pure TTO by performing acute dermal irritation studies on rabbits. Pure TTO caused scarring of the skin whereas TTO loaded nanosponges gel showed no scarring or skin irritation as shown in Figure 5. The rabbits did not show any kind of discomfort with the nanosponge gel formulations.

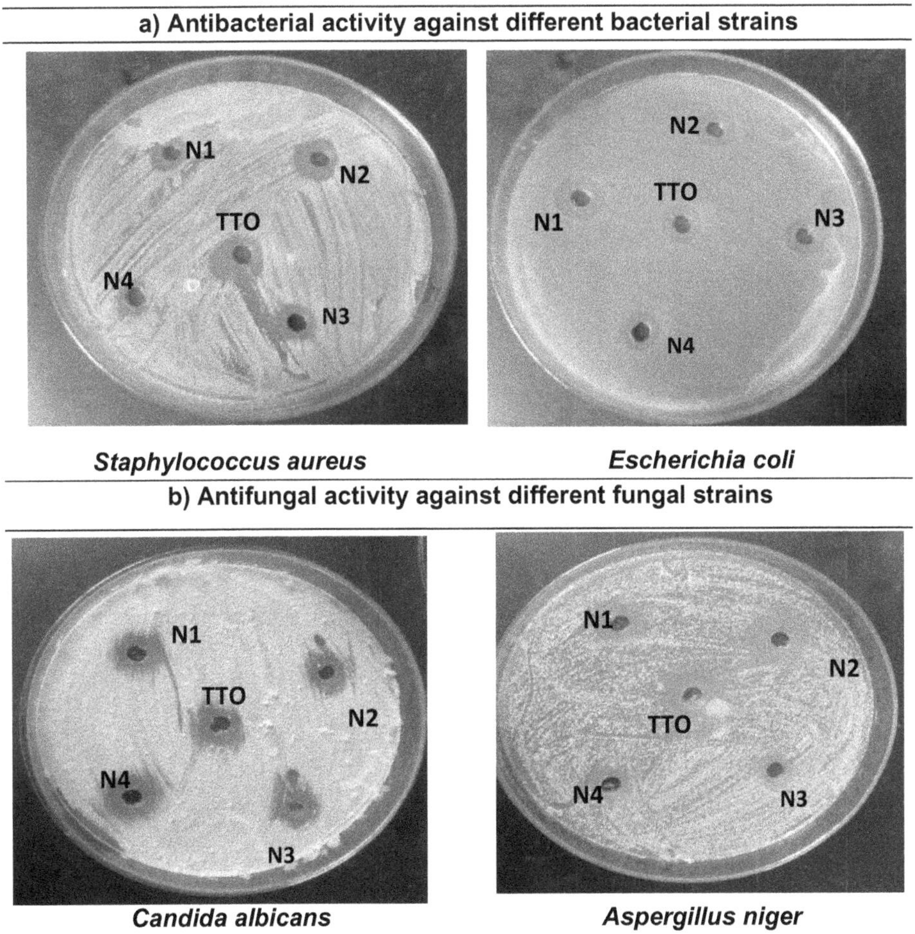

Figure 4 *Observations of antibacterial and antifungal activity of TTO and TTO loaded nanosponge gels.*

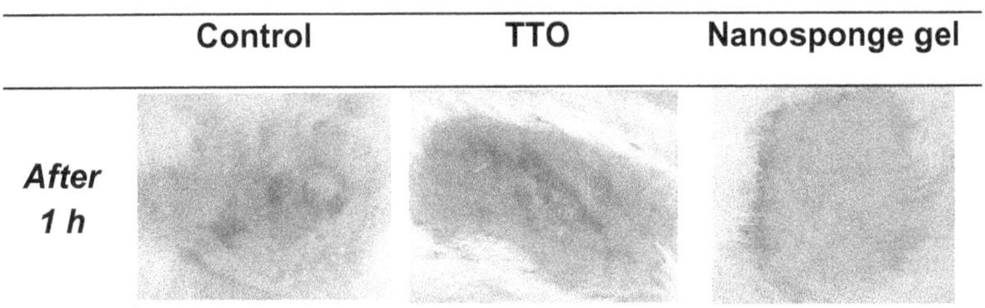

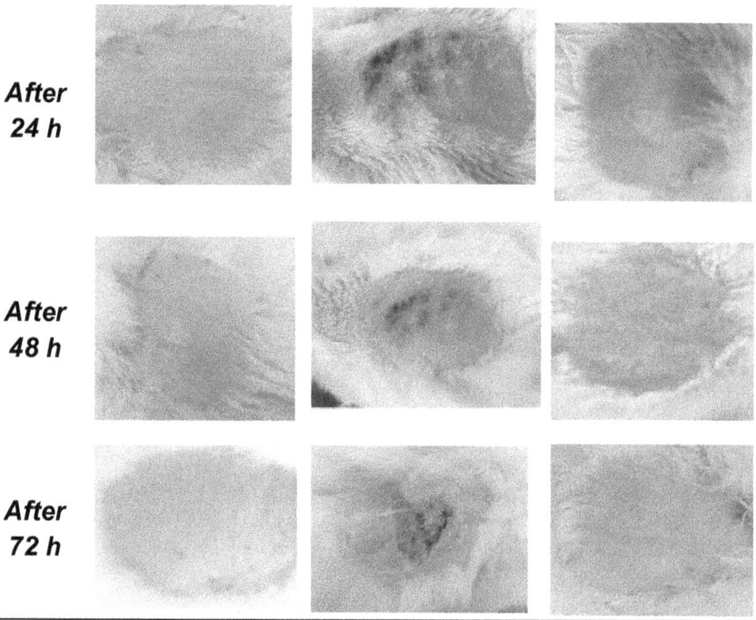

Figure 5 Photographs of skin irritation studies carried out on rabbits.

4 CONCLUSION

The study has proven that the encapsulation of TTO within nanosponges helps reduce its side effects. SEM studies showed the presence of pleomorphic, spongy nanosponges of sizes less than 384 nm. *In vitro* antimicrobial studies proved that the TTO loaded nanosponges had significant antifungal and antibacterial activity. The Draize patch test conducted on rabbits showed no skin reactions with the use of TTO loaded nanosponge gel. Thus, a dermatologically acceptable product has been formulated successfully.

5. ACKNOWLEDGMENT

The authors thank the Faculty of Pharmacy, MSRUAS for providing resources essential for carrying out this work. The authors wish to thank the staff of Micro and Nano Characterization Facility (MNCF), Centre for Nano Science and Engineering (CeNSE), IISc, Bengaluru for extending ZetaPALS and SEM facilities and the faculty of Pharmacognosy, MSRUAS for their valuable technical support on this project.

6. REFERENCES

1. S. Lahkar, M. Das, S. Bora, *Phytopharmacol Res.* 2013; **3**, 250.
2. F. Mondello, FD. Bernardis, A. Girolamo, G. Salvatore, and A. Cassone, *J Antimicrob Chemother*, 2003, **51**,1223.
3. Opinion on Tea tree oil. https://www.woundsource.com/blog/what-hydrocolloid-dressing (Accessed 4 December 2019).
4. S.Prathima , AR. Jahnavi, *Pharm Nanotechnol*, 2015, **3**, 68.
5. HM. Aldawsari, SM. Badr-Eldin, GS. Labib, AH. El-Kamel, *Int J Nanomedicine*, 2015, **10**:893.
6. MM. Ahmed, *Saudi Pharm J*, 2021, **29**, 467.
7. JS. Lee, H. Oh, S. Kim, JH. Lee, YC. Shin, WIA. Choi, *Pharmaceutics*, 2021, **13**, 1329.
8. AR. Bilia, C. Guccione, B. Isacchi, C. Righeschi, F. Firenzuoli, and C. Bergonzi, *Complement Alternat Med*, 2014, **651593**,1.
9. G. Jilsha G and V. Viswanad, *Int J Pharm*, 2015, **6**, 2781.
10. CD. Ferreira, and IL. Nunes, *Nanoscale Res Lett,* 2019, **14**, 9.
11. F. Donsi, M. Annunziata, M. Sessa, and G. Ferrari, *LWT - Food Sci. Technol*, 2011, **44**,1908.
12. C. Maes, S. Bouquillon, ML. Fauconnier, *Molecules*, 2019, **24**, 2539.
13. SS.Biju, A. Ahuja, MRM. Rafiullah, RK. Khar, *J Pharm Biomed Anal,* 2005, **38**, 41.
14. S. Abraham, R. Deveswaran, B. Srinivasan, S. Furtado, *Pharm. Methods*, 2015, **6**, 109.
15. HS. Marwa, GF. Metwally, *Trop J Pharm Res*, 2013, **12**, 877.

GABAPENTIN NOVEL CO-CRYSTALS: AN ATTEMPT TO ENHANCE PHYSICOCHEMICAL PARAMETERS

Ramu Samineni*[a] Jithendra Chimakurthy[a], Manikanta Billapati[b], Sai Bhaskar Bodduluru[b], Sarath Sai Kumar Kurra[b] and Venkata Guruvulu Madasu[b]

[a] Faculty of Pharmacy, Department of Pharmaceutical Sciences, Vignan's Foundation for Science Technology and Research, Vadlamudi, Guntur, Andhra Pradesh, 522213, India

[b] Department of Pharmaceutical Sciences, Vignan's Foundation for Science Technology and Research, Vadlamudi, Guntur, Andhra Pradesh, 522213, India

***Corresponding author email address:** saminenirmamu123@gmail.com

1 INTRODUCTION

Co-crystals are defined as multi-component systems, where one of the components is an active pharmaceutical ingredient and the other components should be substances or ingredients that are acceptable in the pharmaceutical industry.[1] Cocrystals are solids, neutral crystalline single phase materials made up of various types of molecular and ionic compounds.[2]

In terms of their mechanical characteristics, optical properties, release profiles, bioavailability, and therapeutic effects, pharmaceutical co-crystals are an efficient approach to improve their performance. Various factors affect stability, including air sensitivity, chemical structure, moisture content, and the impact of acids and bases.[3] The drug may be exposed to a variety of environments; for example, the pH in the mouth should be higher than 5.5 while the pH in the stomach ranges from 1.5 to 3.5. Because the co-crystal dissolves and there is a risk of the less soluble compound precipitating, adding surfactants to the drug can help avoid unneeded re-crystallization. Sensitivity to moisture plays an important role in preparation, packaging, and storage of goods.[4] Moisture might cause the API to go through unwanted phase changes. The efficiency of the medicine is also greatly influenced by its solubility kinetics; particularly slow dissolution rates may lead to less drug absorption. This calls for a higher compound dose, which might be hazardous. It can occasionally cause one or more of the co-crystals components to become super-saturated in the solution. Precipitation will then happen, blocking the effect of the API, and the solution becomes unstable.[5]

A variety of APIs can now be used in pharmacological therapy thanks to co-crystals. The most desirable characteristics of co-crystallized APIs are their enhanced solubility and the chemical and physical stability they provide. Due to their patentability, they are also effective from a financial and legal standpoint. Co-crystals are anticipated to play a significant part shortly after development.[6]

Gabapentin is an analgesic and antiepileptic drug used to treat neuropathic pain caused by the effect of the somatosensory system. It favours the pharmacokinetic profile and is absent in drug interaction and convenient dosing for patients with renal disease.[7] The changes in the solid-state structure of GBP through co-crystallization can have a marked influence on the physiological absorption characteristics, improving the drug's limited bioavailability. Gabapentin has a limited and variable bioavailability that may cause poor permeability and its dependence on low-capacity amino acid transporter expressed in a limited region of the upper small intestine.[8] The United States Food and Drug Administration authorised Gabapentin for the treatment of epilepsy and later post-herpetic neuralgia in 1993. It binds to the voltage-gated calcium channel's alpha-2-delta subunit. It was also used for anti-convulsive medication and antispasmodic medication.[9]

1.1 Co-crystal preparation methods

A good co-crystallization technique must still be chosen empirically, though. The two types of co-crystal formation techniques that are most frequently employed are generally referred to as solid-based techniques and solution-based techniques. To dissolve the co-crystal components in solution-based procedures, a lot of solvents are needed. Furthermore, since the solvent can alter the intermolecular interactions between API and co-former, the solvent choice has an impact on the co-crystallization outcomes. As opposed to this, solid-state techniques have the potential to reduce or perhaps completely remove the need for solvents in co-crystal synthesis.[10]

2 MATERIALS AND METHODS

2.1 Materials

Gabapentin received from A to Z Pharmaceutical from Chennai as a free sample. Acetylsalicylic acid and L-ascorbic acid were both procured from Sigma Aldrich chemicals. Double-distilled water is the solvent of choice for an analytical grade in contemporary research.

2.2 Preparation of co-crystal by co-grinding methods

Four novel Cocrystals of Gabapentin were synthesized by co-grinding method. Cocrystals are prepared by a molar ratio of 1:1 and 1:3 respectively. Weigh drug should be accurately 1gm of Gabapentin drug after which add 1gm or 3gm of co-formers like L-Ascorbic acid and Acetylsalicylic acid and transfer them right into a mortar. Then begin triturating the powder with the use of a pestle for as much as a time not less than an hour to lead them to uniform. Then weigh that one and allow the powder for drying for 30 minutes at 60°C. Collect the powder and weigh them. And take a look at how much moisture content of the material may be dried. This one will make the powder-free from moisture, so there may be no opportunity for powder degradation.[11]

3 EVALUATIONS

3.1 Standard Curve of Gabapentin

100 ml of 0.1N Hydrochloric acid are used to dissolve 100 mg of Gabapentin, and 1 ml is pipette out to form 10 ml. From there, 5-25 µg/ml solutions are generated. These solutions are then examined at 315 nm using a UV-visible spectrophotometer.[12]

3.2 Structural properties Evaluation

3.2.1 Fourier transforms infrared spectroscopy analysis (FTIR). FTIR spectra of pure drug, L-ASC, ASA, and synthesized Cocrystals like GABA-LASC, and GABA-ASA, is recorded using Micro labs Agilent technologies containing DLATGS detector with 2 cm^{-1} spectral resolution. Each spectrum was created from an average of three single scans taken in the range of 4000-400 cm, with 2-4 mg of each sample being put on the sample cell.[12]

3.2.2 Powder X-ray diffraction analysis (PXRD). Utilizing a PXRD diffractometer, the crystalline structural modifications of GBP and new Cocrystals were examined. The Philips Xpert MPD diffractometer with a Cu Target X-ray tube source was used to scan each sample of the pure drug, co-formers, and Cocrystals. This device was operated at 30 kV and 15 mA for two scan axis. Maintaining a 5–65° scan range with 0.02° of step width and 10.00°/min scan speed.[12]

3.2.3 Percentage yield. The dried crystals were weighed, and the prepared Cocrystals' % yield was determined using the formula below.[12]

$$\%\text{Yield} = PracticalYield\ (mg)/TheoreticalYield\ (mg) \times 100 \qquad (1)$$

3.2.4 Drug content:

For the determination of drug content, prepared supramolecular crystals (100mg) were dissolved in 100ml distilled water and the solution was analyzed spectrophotometric at 315 nm (Agilent technologies cary 60 UV Vis) for drug content, after sufficient dilution with distilled water.[13]

3.2.5 Melting point. Arrange the capillary tubes whose one side is closed on a table next to the samples. Inject the pinch of powder into the capillary tubes carefully. Then keep those tubes in the apparatus and turn on the switch. Keep Thermometer back side of capillary tubes which facilitates at which temp the powder will turn into liquid. Note down the temp where all samples of converted into liquids.[13]

3.2.6 SEM. Otsuka and Kaneniwa describe using SEM in the drug delivery systems for Preformulation stage. Utilizing SEM to define and distinguish the polymorphs and alterations in particle surface topography throughout processing, they investigated the polymorphic transformation of Gabapentin during the grinding process.[12]

3.3 Flow properties Evaluation

3.3.1 Angle of repose. The angle formed between peak heights of powder to that of a surface is called as the angle of repose. It will vary for different powders based on their nature. Crystalline powders show better flow properties whereas amorphous powders are not that much having.[12-13]

3.3.2 Bulk density. The ratio of blend weight and its volume occupied in the particular vessel is known as bulk density.[12]

$$\text{Bulk Density} = \frac{Blend\ weight}{Volume\ of\ blend} \qquad (2)$$

3.3.3 Tapped density. The ratio of the weight of blend to its volume of blend after tapping 100 times on the surface is known as tapped density[12].

$$\text{Tapped Density} = \frac{Weight\ of blend}{Tapped\ volume} \qquad (3)$$

3.3.4 Carrs index. It is the proportion of the difference between the tapped density and the bulk density[12]

$$\text{Carrs Index} = \frac{tapped\ density - bulk\ densit}{tapped\ density} * 100 \quad (4)$$

3.3.5 Hausners ratio (HR). It is the proportion of tapped density to bulk density[12]

$$\text{HR} = \frac{tapped\ density}{bulk\ density} \quad (5)$$

3.4 Intrinsic solubility

Intrinsic solubility analysis was conducted in triplicate; the equilibrium method[29] was utilized for determining the intrinsic solubility of pure drug GBP, Cocrystals GBP-L-ASC, GBP-ASA (1:1 and 1:3). Each sample was added in excess to vials with lids that held 10 ml of distilled water. Following a 15-minute sonication, these solutions were shaken in a water bath for 48 hours at a rotational speed of 100 rpm and a temperature of 30 ± 1 °C. Through 0.45µm filters, the supernatant solution was filtered. Methanol was used to dilute this filtrate, and a UV spectrophotometer was used to assess the drug concentration (Agilent technologies cary 60 UV VIS) at 315 nm.[12-13]

3.5 pH of solution at equilibrium

The vials holding 10 ml of distilled water and excess quantity of prepared Cocrystals was placed. For two hours at room temperature, the vials were shaken in an incubator at 100 revolutions per minute. The pH of prepared crystal formulations and pure drug are measured in digital pH meter.[14,15]

3.6 *In vitro* dissolution study:

The eight station USP type-2 (Paddle apparatus) was used to conduct the *in vitro* dissolution investigations in triplicate (Model Disso 2000, Lab India). Dissolution studies performed by using 900 ml of pH 1.2 (0.1N HCl) acidic buffer. The temperature was kept constant at 37 ± 0.5 °C throughout the study. The paddles were rotated at 100 rpm by maintaining 5 ml of aliquot was collected at 10, 20, 30, 40, 50, and 60 min by replacing with fresh dissolution medium. The collected aliquots were filtered through a 0.45µ, the filtrate was diluted and the drug concentration was calculated using a UV-Spectrophotometer (Agilent technologies cary 60 UV-VIS) at 270 nm.[15]

3.7 Release Kinetics

The matrix systems mimic the drug's diffusion release mechanism and peppas release rate. The gathered data was fitted into zero order, first order, higuchi matrix, and peppas to assess the mechanism for the dosage form release and release rate kinetics. The best-fit model in this was chosen by comparing the r-values that were obtained.[15]

Zero order kinetics

$$Qt = Qo + Ko\ t \qquad (6)$$

First order kinetics

$$Log\ Qt = \log Qo + K1t / 2.303 \qquad (7)$$

Higuchi model

$$Qt = KH \times t\ 1/2 \qquad (8)$$

Peppas release model

$$Mt / M\infty = K.tn \qquad (9)$$

4. RESULTS AND DISCUSSION

4.1 Pure Drug calibration by UV-Visible Spectrophotometer

Only a straight line relationship between two variables is explored in a basic linear regression study. Beer Lambert's rule was followed while creating a calibration curve for a concentration range of 5 -30 µg/ml in a pH 1.2 medium. The drug's linearity was between 5 - 30 µg/ml and its correlation value in 0.1 N HCl was 0.998.

4.2 FTIR

The FTIR results may be used to examine the distinctive peaks brought on by changes in intermolecular interactions relative to parent molecules. Figure 1 shows the FTIR spectra of pure drug and novel cocrystals forms. For GBP, the characteristic peaks corresponding to amine N-H Stretch and O-H Stretch are assigned at 2976.70, 2822.51 cm^{-1} and N-H bending at 1541.42 cm^{-1}. These are the primary functional groups involved in the formation of hydrogen bonds. FTIR analysis of GBP-ASA cocrystals forms showed the shift in the N-H, and O-H stretch peak position to 3350.07 cm^{-1}, 2855.01 cm^{-1} and N-H bending shifted to 1440.40 cm^{-1}. GBP-LASC cocrystals forms showed the shift in N-H, and O-H stretch peak position to 3584.21 cm^{-1},2850.49 cm^{-1} and N-H bending shifted to 1322.60 cm^{-1}.

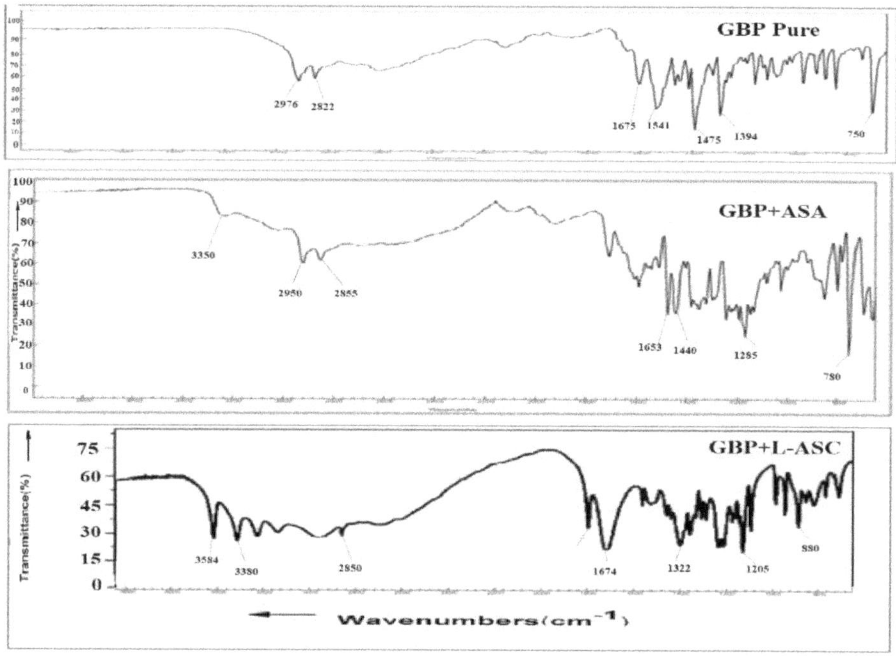

Figure 1 FTIR of GBP pure and prepared co-crystal forms.

4.3 PXRD

The Crystallinity of the drug molecule can be evaluated by powder X-ray diffraction (PXRD) analysis. PXRD spectra of the pure drug and the novel cocrystal forms were shown in Figure 2. Pure GBP has a distinctive peak of 2ɵ scattering angles at 19.68° with 100% relative intensity in its powder diffraction spectrum, which is distinct from produced cocrystal forms. GBP-ASA (1:1) cocrystal showed the major characteristic peak at 6.231° which were absent in the drug molecule and differ from ASA. In GBP-ASA (1:3) cocrystal showed the major characteristic peak of 2ɵ scattering at 27.300° which are new for GBP and differ from ASA. In case of GBP-LASC (1:1) cocrystal showed the major characteristic peak at 15.137° which are new for GBP and differ from L-ASC. In GBP-LASC (1:3) cocrystal showed the major characteristic peak at 28.199°. The degree of Crystallinity is more GBP-LASC cocrystal formulation when compared to GBP-ASA Cocrystals.

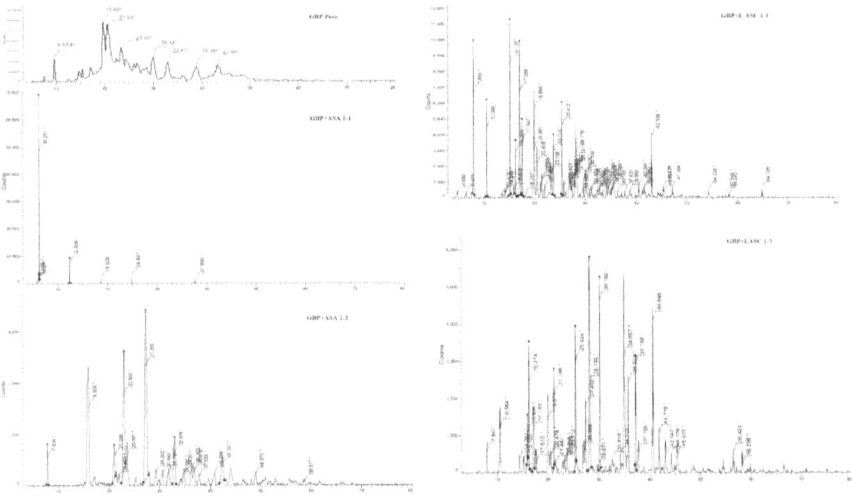

Figure 2 PXRD of pure drug and prepared co-crystal forms.

4.4 Percentage yield, Drug content and Melting point

For prepared Cocrystals formulations like GBP-ASA CF 4 (94.45%) and GBP-ASA CF 3 (92.82%) both are showing high percentage yield compared to other co-crystal formulations, represented in Figure 4. The drug content of the prepared multicomponent co-crystal formulations is assessed. in case of GBP-ASA CF 4 (99.89) and GBP-ASC CF 2 (99.24) both are showing high drug content compared to other cocrystals formulations. In case of GABA-ASA CF 3 and 4 showing low melting point while compared to other coformer Cocrystals and pure drug represented in Figure 3.

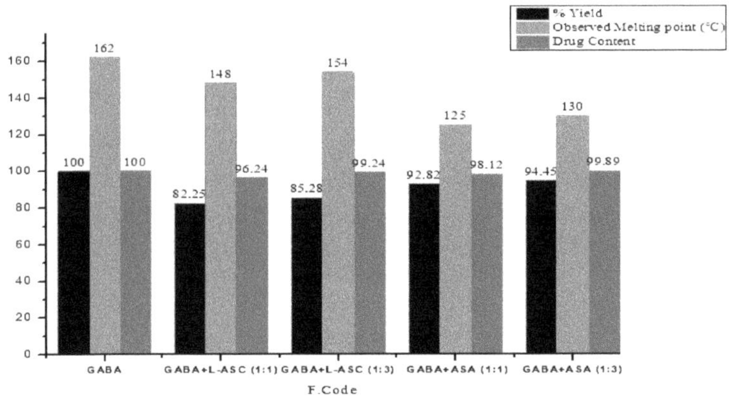

Figure 3 Percentage yield, drug content and melting point of pure drug and prepared Cocrystals formulations.

4.5 SEM

Surface topography, particle size and morphology were investigated with a SEM. Scanning electron microscopy (SEM) pictures are clearly indicating that in case of GBP-ASA and GBP-L-ASC prepared cocrystals in 1:3 ratio reductions in size and degree of crystalline more while compared to 1:1 ratio prepared cocrystal forms, GBP-L-ASC had shown flake like crystals are observed and GBP-ASA co-crystal had shown small uniform rod like crystals are represented in Figure 4.

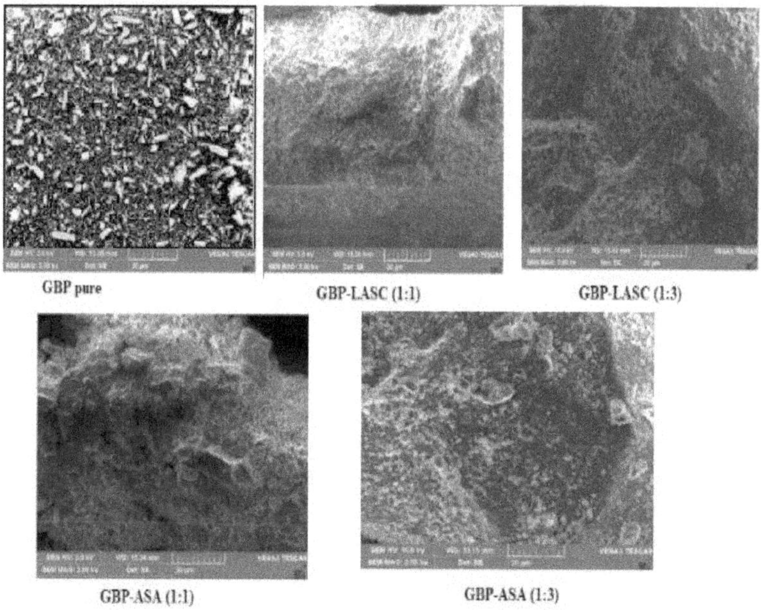

Figure 4 SEM images of pure drug and prepared novel Cocrystals formulation.

4.6 Pre-formulation

The prepared cocrystals formulations and pure Gabapentin angle of repose were measured using the fixed funnel technique. GBP-L-ASC CF-2 and GBP-ASA CF-4 prepared in 1:3 ratios had showing excellent flow properties while compared to other cocrystals formulations prepared in 1:2 ratios (GBP-L-ASC CF-1 and GBP+ASA CF-3) represented in Table 1.

Table 1 *Pure drug and prepared co-crystals flow properties. (N=3 ± S.D); GBP-Gabapentin, L-ASC-L-Ascorbic Acid, ASA-Acetyl Salicylic Acid*

F. Code	Bulk density (g/cc)	Tapped density (g/cc)	Compressibility index	Hausner's ratio	Angle of repose	Flow properties
GBP	0.52 ± 0.045	0.58 ± 0.04	13.56 ± 0.8	1.14 ± 0.09	33.69 ± 0.19	Passable
GBP+L-ASC CF-1 (1:1)	0.47 ± 0.065	0.53 ± 0.08	12.11± 0.7	1.13 ± 0.07	25.52 ± 0.17	Good
GBP+L-ASC CF-2 (1:3)	0.46 ± 0.035	0.49 ± 0.07	11.00 ± 0.6	1.12 ± 0.04	23.57 ± 0.15	Excellent
GBP+ASA CF-3 (1:1)	0.44 ± 0.054	0.52± 0.07	14.68 ± 0.6	1.17 ± 0.06	24.54 ± 0.14	Good
GBP+ASA CF-4 (1:3)	0.45 ± 0.044	0.51 ± 0.09	11.58 ± 0.8	1.12 ± 0.08	22.44 ± 0.11	Excellent

4.7 pH and Intrinsic Solubility

In comparison to pure GBP the prepared multicomponent Cocrystals showed high intrinsic solubility values of 59.12, and 44.63 two folds' solubility increases respectively, with GBP-ASA CF-4 (35.47 mg/ml) and GAB-L-ASC CF-2 (26.78 mg/ml) showing high solubility. The pH of pure Gabapentin solution at equilibrium showed at 3.89, while in case of GBP-L-ASC CF 1 and CF 2 solution showed at 2.54 and 1.84, GBP-ASA CF 3 and CF 4 solution showed at 2.85 and 3.89

4.8 Dissolution

GBP in pure form releases 88.9 % at the end of 360th minute. When compared to GBP-L-ASC CF-1 (94.45%) and GBP-ASA CF-3 (95.2%), the produced co-crystal formulations (1:3) GBP-ASA CF-4 and GABA-L-ASC CF-2 demonstrated high drug release at the 360th minute of 99.87% and 97.84%, respectively. as seen in Table 2.

Table 2 *Invitro Dissolution of pure drug and prepared Cocrystals.*

Time (Hours)	*GBP* pure	*GBP*-L-ASC CF-1 (1:1)	*GBP*-L-ASC CF-2 (1:3)	*GBP*-ASA CF-3 (1:1)	*GBP*-ASA CF-4 (1:3)
	%CDR				
0.5	7.8±1.17	10.6±1.10	12.4±1.15	11.1±1.16	15±1.11
1.0	10.8±1.20	22.8±1.22	24.4±1.23	23.6±1.29	25.7±1.43
1.5	16.8±1.22	28.2±1.21	30.2±1.16	29.6±1.23	31.6±1.12
2	25.4±1.25	34.5±1.23	41.8±1.21	36.6±1.24	42.4±1.23
2.5	32.8±1.22	41.2±1.25	52.4±1.14	42.9±1.26	56.6±1.28
3.0	40.2±1.26	54.24±1.28	66.8±1.32	56.6±1.28	68.2±1.27
3.5	50.4±1.26	72.6±1.26	74.8±1.36	73.8±1.27	75.4±1.26
4.0	58.2±1.25	78.86±1.28	80.6±1.29	79.8±1.30	82.2±1.30
4.5	71.4±1.24	85.12±1.34	88.2±1.54	86.2±1.25	90.3±1.33
5.0	78.9±1.26	89.25±1.13	91.2±1.24	90.6±1.12	91.8±1.21
5.5	84.5±1.25	90.54±1.24	93.6±1.32	92.3±1.23	94.6±1.23
6.0	88.9±1.22	94.45±1.22	97.84±1.13	95.2±1.24	99.87±1.28

(N=3 ± S.D); GBP-Gabapentin, L-ASC-L-Ascorbic Acid, ASA-Acetyl Salicylic Acid

4.9 Release Kinetics

For the prepared co-crystal formulations, the release kinetics was performed. In zero order, the regression value is high for GBP-ASA CF 4 (0.984). Peppas release kinetics of GBP-ASA CF 4 showed high regression value (0.998) and release exponent showed 0.890 and optimized formulation following non-fickian release mechanism. Release kinetics for prepared co-crystals and pure drug represented in Figure 5. GBP-ASA CF **4** is selected as optimized and used for *in vivo* study in future.

Figure 5 Release Kinetics plots of prepared Cocrystals formulations.

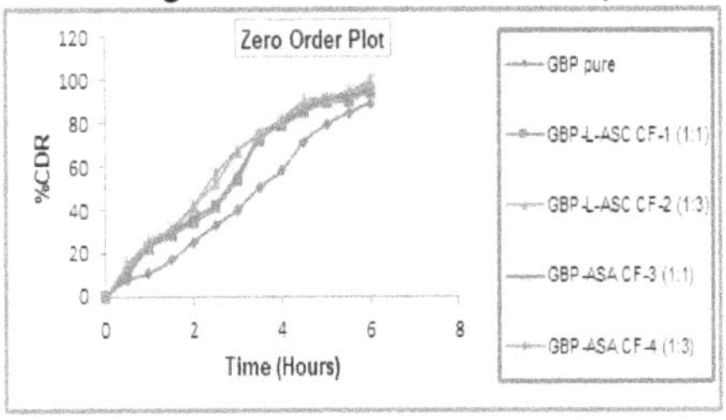

Zero order plot of prepared Cocrystals formulations

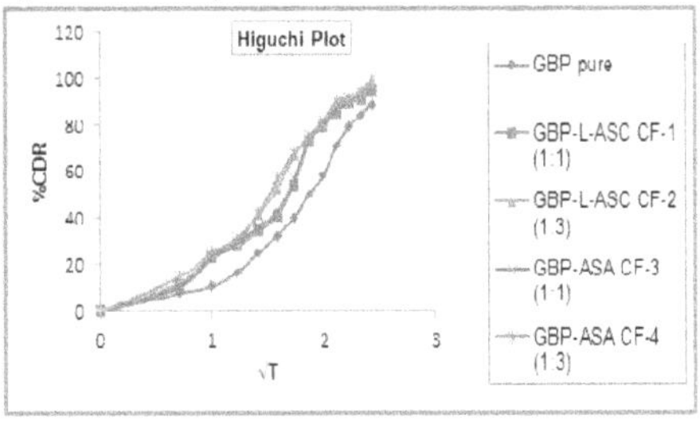

Higuchi plot of prepared Cocrystals formulations

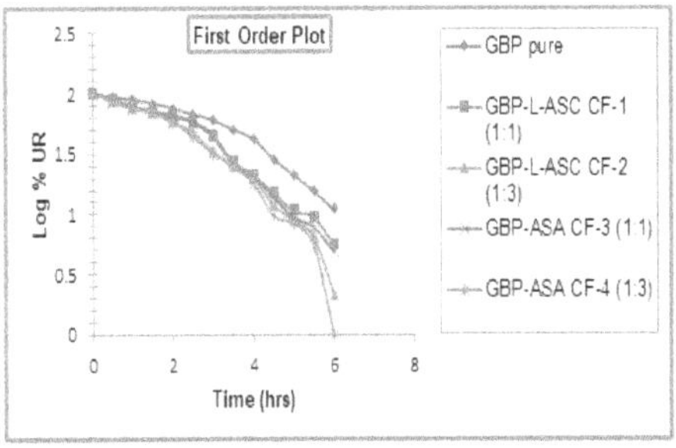

First order plot of prepared Cocrystals formulations

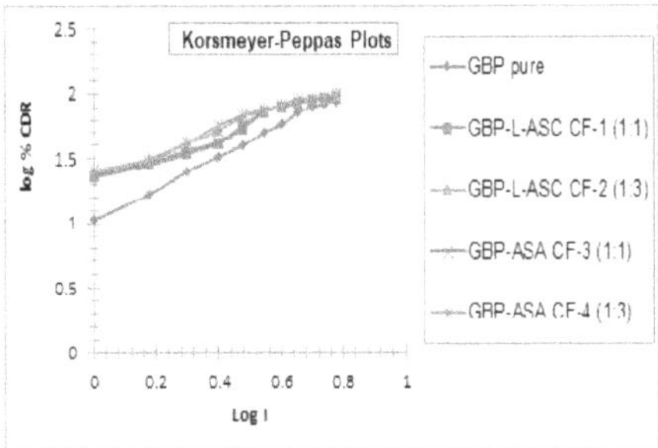

Korsemeyer Peppas plot of prepared Cocrystals formulations

5 CONCLUSION

The unique Gabapentin Cocrystals displayed a new FTIR and powder X-ray diffraction feature. These findings show that an intermolecular interaction has been established within the Cocrystals. It was shown that Gabapentin interacts via a hydrogen bond with complementary functional groups like acetyl salicylic acid and L-ascorbic acid in both of the new Cocrystals. Compared to the pure Gabapentin flow properties for prepared co-crystals (GBP-L-ASC CF2 and GBP-ASA CF 4) by using Cogrinding method are showing excellent flow properties. GBP-ASC CF 4, GBP-L-ASC CF 2 showed 59.12 folds and 44.63 folds, solubility improved respectively. The dissolution test showed that the GBP-ASA CF 4, GBP-LASC CF 2 higher dissolution rate than the pure Gabapentin in pH 1.2 medium respectively. Modification in the chemical environment, Intermolecular interactions established within the Cocrystals, improved flow properties, Enhanced intrinsic solubility and invitro dissolution rate than pure drug.

The effect of co-crystallization in improving physicochemical characteristics was proven in solubility tests. Additionally, the outcomes of the solubility analysis concur with the findings of the in vitro dissolution study. The pharmacokinetic characteristics of the co-crystals, such as Cmax and AUC, were not examined in the current experiment but will be in our upcoming paper. The results amply show that co-crystallization modifies the chemical structure, improves solubility, and improves dissolution of LTG.

AUTHOR CONTRIBUTIONS:

Ramu Samineni: Supervision, plan of investigation, literature review, formulation development, writing original draft preparation, visualization, and corrections.

Jithendra Chimakurthy: Plan of investigation and corrections.

Manikanta B, Sai Bhaskar B: Synthesis, Writing original draft preparation, and corrections.

Sarath Sai kumar K, and Venkata Guruvulu M*:* Synthesis, literature review, writing original draft preparation

6 ACKNOWLEDGMENTS

The author would like to thank Dr. Jithendra Chimakurthy, the principal, and the management of Vignan's Foundation for Science Technology and Research for their assistance in making the research study a success.

7 CONFLICTS OF INTEREST

There was no disclosed conflict of interest by the authors. The paper's writing and content are solely the authors' responsibility.

8 REFERENCES

1. Karki S, Friscic T, Jones W. Control and interconversion of cocrystal stoichiometry in grinding: stepwise mechanism for theformation of a hydrogen-bonded cocrystal. CrystEngComm 2009; 11:470-81. DOI:10.1039/B812531G
2. Kotak U, Prajapati V, Solanki H, Jani G, Jha P. Co-crystallization technique its rational and recent progress. World J Pharm Pharm Sci 2015;4(4);1484-1508.
3. Ramu Samineni, Jithendra Chimakurthy, Narahari Narayan Palei, Phani Krishna Yella, Sabareesh M, Guru Prasad M. Lamotrigine Novel Cocrystals: An Attempt To Enhance Physicochemical Parameters. Journal of Pharmaceutical Negative Results. 2022;13(5):622-633

4. Alhalaweh A, Velaga P. Formation of cocrystals from stoichiometric solutions of incongruently saturating systems by spray drying. Cryst Growth Des 2010;10(8):3302-3305. DOI: 10.1021/cg100451q
5. Grossjohann C, Serrano DR, Paluch KJ, O'connell P, Vella-zarb L, Manesiotis P, et al. Polymorphism in sulfadimidine/4- aminosalicylic acid cocrystals: solid-state characterization and physicochemical properties. J Pharm Sci 2015; 104:1385-98. DOI: 10.1002/jps.24345.
6. Abourahma H, Cocuzza DS, Melendez J, Urban JM. Pyrazinamide cocrystals and the search for polymorphs. CrystEngComm 2011; 13:1-22. DOI:10.1039/C1CE05598D
7. Bhatt PM, Azim Y, Thakur TS, Desiraju GR. Cocrystals of the antiHIV drugs lamivudine and zidovudine. Cryst Growth Des 2009;9(2):951-7. DOI: 10.1021/cg8007359.
8. Vaghela P, Tank HM, Jalpa P. Cocrystals: A novel approach to improve the physicochemical and mechanical properties. Indo Am J Pharm Res 2014;4(10):5055-65.
9. Ramu Samineni, P. Aravanan, Jithendra Chimakurthy, Y. Ratna Sindhu, B. Jeevan Kumar. Factorial Design and Optimization of Antiepileptic Drug: Lamotrigine Cocrystals Immediate Release Tablets. Journal of Positive School Psychology.2022,6(9); 4070-4077.
10. Cocrystals continuum: The influence of crystal structure on ionization state. Mol Pharm 2007:4;323-3819:1-11. DOI: 10.1021/mp0601345
11. Prabhakar Panzade,et,al, Pharmaceutical Cocrystal of Piroxicam: Design, Formulation and Evaluation, Journal of advanced pharmaceutical bulletin .2017,sep,v.3(7) 399–408. DOI: 10.15171/apb.2017.048
12. Pinki Rajbhar et al, Co-Crystals Formation of Clarithromycin with Urea: An Efficient Approach to Enhance the Solubility and Dissolution Rate. American Journal of Advanced Drug Delivery. 2016,3,13
13. Childs SL, Stahly GP, Park A. The salt-cocrystals continuum: The influence of crystal structure on ionization state. Mol Pharm 2007:4; 323-38. DOI: 10.1021/mp0601345.
14. Ramu Samineni et, al, Co-Crystals: A Review of Recent Trends in Co Crystallization of BCS Class II Drugs. Research J. Pharm. and Tech. 2019.12(7):3117-3124. DOI: 10.5958/0974-360X.2019.00527.4
15. Ramu Samineni, Jithendra Chimakurthy. Effect of Coformers on Novel Co-Crystals of Gabapentin: An in Vivo Approach. Journal of Pharmaceutical Sciences and Research.2020,12(5); 639-648

QUALITY BY DESIGN APPROACH FOR OPTIMIZATION AND DEVELOPMENT OF GASTRO-RETENTIVE HOLLOW MICROSPHERES FOR PRASUGREL

Grandhi Srikar[1*], Nagamma I[2]

[1]Department of Pharmaceutical Sciences, Vignan's Foundation for Science Technology and Research, Vadlamudi, Guntur, AP, India

[2]University College of Pharmaceutical Sciences, Acharya Nagarjuna University, Nagarjuna nagar, Guntur, AP, India

Corresponding Author E-Mail ID: srikar.grandhi@gmail.com

1. INTRODUCTION

Rate-controlled oral drug-delivery systems are gaining importance in the recent years and technological advancements are evolving rapidly in this area. But in oral route, the controlled drug delivery system has to pass through different parts of GIT of varying environment like acidic in upper GIT and alkaline environment in lower GIT, which may alter the uniformity of drug absorption. Besides pH, other inconstant physiological conditions across the GIT including unpredictable gastric emptying time cause ebb and flow in drug absorption which finally result in fluctuations in plasma drug concentrations. However, to achieve predictable and constant drug absorption, the above uncertainties should be overcome.[1] These contemplations have led to the development of oral controlled release dosage forms which retain in the stomach for a longer and more predictable time, called gastroretentive systems. Specifically, these kinds of drug delivery systems are more appropriate for drugs with an absorption window in the stomach or the duodenum.

Several techniques, [2,3] including but not limited to bioadhesive drug delivery systems and altered density systems have been adopted for this purpose.

Bioadhesive systems adhere to the gastric membrane and release the total drug at a particular site, which may cause local tissue irritation. Another promising approach for enhancing gastric residence time is floating drug delivery systems which have less density than gastric fluids. Floating drug delivery systems can be developed into single unit and multiple unit systems. The single unit floating systems are extensively studied and these are having the problem of a release all-or-nothing emptying process that leads to high variability in gastric transit time. However, the multiple unit floating systems don not have this problem and distribute widely on the surface of gastric fluids. These systems can reduce the does dumping as well as inter-subject variability in gastric residence time and further in absorption. Generally, most of the multiple-unit systems are effervescent matrices prepared with swellable polymers and effervescent materials. The disadvantage of these systems is high floating lag time as gas generation takes some time. This problem can be overcome by designing a novel dosage form, hollow microspheres or microballoons, because of their characteristic inner hollowness, they can float immediately when they are placed in aqueous medium.

Prasugrel hydrochloride (PGH) is a thienopyridine derivatives and anti-platelet category. The short half-life (approximately 7.4 hrs)[4] and moderate dose (10mg) make this drug a suitable candidate for developing extended release product. PGH is a drug with alkaline pKa of 5.48 indicating that it is a weak basic drug. It is practically insoluble at pH 6 to 7.5 and the solubility is decreased upon increase in pH, [5] which indicated that this cannot be sufficiently dissolved in lower GIT and cannot be sufficiently absorbed. These characteristics suggest that an extended release drug delivery system that resides and release the drug only in the stomach is the best dosage form into which can be developed. Hence, development of floating hollow microspheres for dipyridamole is the aim of this work.

Quality by design approach using Design Expert software was employed to optimize and develop hollow microspheres for PGH. Different formulation and process parameters viz. concentration of polymer, concentration of methanol in the solvent mixture, volume of internal phase and mixing speed were taken as the independent factors. Particle size, entrapment efficiency and drug release rate constant of the microspheres were taken as the response variables. Design of experiments (DoE) was employed to elucidate the effects of the factors on the response. Later, optimization by desirability functions approach was performed to identify the best combination of the levels of the factors in order to get the hollow microspheres with desired characteristics.

2. MATERIALS AND METHODS

2.1 Materials

Prasugrel Hydrochloride was provided by Aurobindo Pharmaceuticals, Hyderabad; Ethyl cellulose 100 cps was procured from Capricorn Pharma,

Hyderabad. Diethyl ether, methanol, petroleum ether, liquid paraffin and Span 80 were purchased from Merk Specialties Pvt. Ltd, Mumbai. All other chemicals of reagent grade were used.

Table 1 *Combinations of the selected factors according to the selected BBD.*

Formulation	Standard order	Run order	Polymer (% w/w)	Methanol (% v/v)	Volume of solvent (ml)	Speed (rpm)
F_1	1	22	50.00	20.00	7.50	550.00
F_2	17	10	50.00	40.00	5.00	550.00
F_3	9	9	50.00	40.00	7.50	400.00
F_4	11	7	50.00	40.00	7.50	700.00
F_5	19	24	50.00	40.00	10.00	550.00
F_6	3	25	50.00	60.00	7.50	550.00
F_7	13	23	62.50	20.00	5.00	550.00
F_8	21	5	62.50	20.00	7.50	400.00
F_9	23	20	62.50	20.00	7.50	700.00
F_{10}	15	21	62.50	20.00	10.00	550.00
F_{11}	5	6	62.50	40.00	5.00	400.00
F_{12}	7	11	62.50	40.00	5.00	700.00
F_{13}	25	2	62.50	40.00	7.50	550.00
F_{14}	6	19	62.50	40.00	10.00	400.00
F_{15}	8	14	62.50	40.00	10.00	700.00
F_{16}	14	4	62.50	60.00	5.00	550.00
F_{17}	22	13	62.50	60.00	7.50	400.00
F_{18}	24	17	62.50	60.00	7.50	700.00
F_{19}	16	8	62.50	60.00	10.00	550.00
F_{20}	2	1	75.00	20.00	7.50	550.00
F_{21}	18	18	75.00	40.00	5.00	550.00
F_{22}	10	16	75.00	40.00	7.50	400.00
F_{23}	12	15	75.00	40.00	7.50	700.00
F_{24}	20	3	75.00	40.00	10.00	550.00
F_{25}	20	3	75.00	40.00	10.00	550.00
F_{26}	4	12	75.00	60.00	7.50	550.00

2.2 Development of PGH hollow microspheres

2.2.1 Experimental Design. Box-Behnken design (BBD) under Response surface model (RSM) was selected to optimize the influence of various formulation and process variables (independent variables) on the characteristics (responses) of hollow microspheres or microballoons. The microballoons were prepared by emulsion solvent evaporation/diffusion method. Here, four different independent variables were taken viz. polymer concentration (factor A); concentration of methanol in the internal phase (factor b); volume of internal phase (factor C) and speed of rotation (factor D).

All the four factors were taken at three levels each. The selected responses were particle size (response R1); entrapment efficiency (response R1) and drug release rate constant (response R3) of the microballoons. The combinations of various factors given by the software according to the BBD were shown in Table 1.

2.2.2 Preparation of PGH Hollow Microspheres. Emulsion evaporation-diffusion method was employed to prepare hollow microspheres. Methanol and diethyl ether were taken as per the factorial combinations in which first Ethyl cellulose 100 cps was dissolved and then 1g of PGH was also dissolved to form the organic phase (the quantities of the polymer and the solvents were taken as per the Table 1). In another beaker, 80 mL of liquid paraffin was taken to which 0.25% v/v of Span 80 was added to form the oil phase that was immiscible with the above organic phase. A mechanical stirrer was adjusted into the oil phase and speed of the stirrer was set as 400/550/700 rpm as per the factorial combination. The organic phase was added slowly into the oil phase to form emulsion. This was kept for continuous stirring for about 4-5h until complete evaporation of both methanol and diethyl ether which then result in formation of solidified microspheres. The obtained microspheres were separated by filtration and rinsed to remove any liquid paraffin with petroleum ether and then dried.

2.3 Characterization studies on the prepared PGH hollow microspheres

2.3.1 Morphological study using. Surface morphology of the prepared hollow microspheres was studied using scanning electron microscopy (SEM). Sample preparation and analysis were performed as reported by SD. Vanshiv *et. al.* [6]

2.3.2 Percentage yield. Percentage yield of the prepared microballoons was calculated by dividing actual weight of the microballoons obtained to the total weight of all non-volatile components that were included in the preparation.

$$\text{Percentage yield} = \frac{\text{Actual weight of the microballoons obtained}}{\text{Total weight of drug and excipients}} \; x\; 100 \qquad (1)$$

2.3.3 Floating time. 100 microballoons were added into in a 250ml beaker containing water, then allowed for standing for 24hrs. Observed the number of microballoons remain floated after 24hrs. [6]

2.3.4 Percentage floating (buoyancy). 100 floating micro balloons were taken in a 100ml beaker contain 70ml of water, then allow standing for 24hrs. Observe the number of floating microballoons on surface after 24hrs. [6]

$$\text{Percentage floating} = \frac{\text{No.of microballoons floating after 24hrs}}{\text{No.of microballoons taken}} \; x\; 100 \qquad (2)$$

2.3.5 Particle size and size distribution. The particle size of the floating microballoons was measured using an optical microscopic method and mean micro balloons size was calculated by measuring 100 particles with the help of a calibrated ocular micrometer.

2.3.6 Drug entrapment efficiency. The microballoons were crushed and 10 mg drug equivalent crushed powder was transferred into 100ml beaker contain 50ml of 0.1N HCl. The beaker was placed on magnetic stirrer and set suitable rpm, until stirring was continued after 3hrs samples was withdrawn and the absorbance is measured at 204nm by UV spectrophotometer. The percentage drug entrapment was calculated as follows:

$$\text{Drug entrapment efficiency (\%)} = \frac{\text{Amount of drug present in the microballoons}}{\text{Amount of drug taken}} \, x \, 100 \qquad (3)$$

2.3.7 Drug release studies. In vitro drug release studies were performed for the prepared floating balloons using USP II Dissolution test apparatus at 100 rpm. 900 mL of 0.1M HCl was taken as the medium. At regular time intervals 5 mL of sample was withdrawn and subjected to analysis using spectrophotometer at maximum wavelength of 204nm. The obtained data was subjected to zero-order and first-order kinetic models to find the drug release kinetics and to calculate the release rate constant.

2.4 Design validation and optimization

The regression model for every response with the factors was identified by sequential model sum of squares analysis. The identified model was validated by analysis of variance (ANOVA) and normal plot of residuals. Upon successful validation of the design and the model, optimization was performed to identify the best combination of the factors for achieving the hollow microspheres with desired quality by setting the desirability criteria for the responses.

3. RESULTS AND DISCUSSION

The developed hollow microspheres or microballoons were subjected to various characterization studies. Immediately after preparation, the microballoons were tested for their floating nature by adding into 0.1M HCl and observed that they just remain floated on the surface only without any floating lag time. This could be due to the presence of inner hollowness. This observation confirmed that the method employed and solvents used were effective in producing microspheres with inner hollowness. Diethyl ether due to its high volatility could evaporate rapidly and made the outer surface of the polymer solution droplet rigid. But the methanol still present in inside of the droplet might diffuse slowly due to its less volatility. This slow diffusion of methanol form inside of the microspheres could allow some protrusions on the outer surface and hollowness inside. This was evidenced from the SEM images which were displayed in Figure 1. The dents on the surface were also observed. These dents could form due to applied external pressure on the microspheres only when they had inner hollowness. [7] Hence these results indicated that microspheres with inner hollowness were effectively in this work.

Table 2 *Results of physical characterization of microballoons of all the formulations $F_1 - F_{25}$.*

Formulation	Yield (%)	Buoyancy (%)	Particle size (µm)	Entrapment efficiency (%)	Drug release rate constant (hr^{-1})
F_1	94.0 ± 3.5	96.0 ±1.00	522.5 ± 24.9	84.1 ± 2.7	0.198 ± 0.05
F_2	82.2 ± 4.8	83.3 ± 3.7	699.4 ± 41.8	85.2 ± 3.6	0.156 ± 0.02
F_3	88.5 ± 2.6	96.7 ± 2.3	711.3 ± 26.9	87.8 ± 1.9	0.168 ± 0.03
F_4	82.6 ± 1.9	93.3 ± 1.2	413.9 ± 30.4	75.2 ± 4.2	0.235 ± 0.08
F_5	80.3 ± 4.3	98.7 ± 0.7	452.5 ± 37.2	77.3 ± 6.6	0.261 ± 0.06
F_6	86.5 ± 2.7	95.3 ± 2.1	567.2 ± 18.6	78.6 ± 2.7	0.178 ± 0.03
F_7	88.9 ± 5.4	94.0 ± 3.0	769.4 ± 51.2	94.1 ± 4.5	0.152 ± 0.04
F_8	92.5 ± 3.8	93.3 ± 1.6	904.6 ± 24.3	89.6 ± 3.1	0.147 ± 0.02
F_9	92.8 ± 4.1	83.3 ± 2.1	522.3 ± 33.6	85.7 ± 1.5	0.232 ± 0.11
F_{10}	94.6 ± 3.4	96.0 ± 2.0	584.6 ± 42.7	84.9 ± 5.4	0.209 ± 0.06
F_{11}	91.3 ± 5.1	91.7 ± 3.5	1018.1 ± 68.5	93.2 ± 2.9	0.132 ± 0.03
F_{12}	95.5 ± 2.9	98.3 ± 1.3	719.4 ± 29.8	76.4 ± 5.4	0.214 ± 0.09
F_{13}	92.4 ± 1.5	96.7 ± 2.6	723.9 ± 52.6	82.9 ± 3.2	0.193 ± 0.10
F_{14}	91.2 ± 2.4	98.3 ± 1.1	682.6 ± 21.6	78.3 ± 1.7	0.171 ± 0.05
F_{15}	92.5 ± 3.3	97.7 ± 2.9	359.2 ± 34.1	78.9 ± 4.6	0.242 ± 0.06
F_{16}	93.1 ± 2.6	98.0 ± 1.0	883.5 ± 40.7	87.9 ± 7.2	0.116 ± 0.01
F_{17}	91.0 ± 1.4	86.7 ± 4.2	946.6 ± 61.4	83.5 ± 1.8	0.103 ± 0.02
F_{18}	92.4 ± 2.3	95.7 ± 0.6	597.7 ± 18.5	81.6 ± 3.5	0.187 ± 0.07
F_{19}	90.7 ± 1.6	98.3 ± 0.6	674.5 ± 31.7	78.2 ± 4.4	0.164 ± 0.03
F_{20}	85.9 ± 2.2	93.3 ± 3.1	526.2 ± 40.6	91.8 ± 5.6	0.102 ± 0.02
F_{21}	91.3 ± 1.4	99.0 ± 1.0	917.1 ± 55.9	94.7 ± 1.2	0.109 ± 0.03
F_{22}	94.6 ± 3.1	96.7 ± 1.2	1040.8 ± 71.2	92.4 ± 3.4	0.087 ± 0.01
F_{23}	92.9 ± 1.1	98.0 ± 1.0	889.6 ± 43.8	82.5 ± 4.7	0.102 ± 0.02
F_{24}	93.2 ± 2.9	96.7 ± 1.1	641.6 ± 36.2	80.2 ± 1.1	0.151 ± 0.03
F_{25}	92.7 ± 2.1	97.3 ± 2.6	836.3 ± 44.1	90.1 ± 3.5	0.106 ± 0.01

Percentage yield of the microballoons of all the formulations was found to be ranging from 80.3 – 95.5 (showed in Table 2). These high yield values explain that the processing conditions and the formulation compositions were effective in preparing microballoons with minimum wastage. Floating time and percentage floating (buoyancy) were studied by adding 100 microballoons in a 250 mL beaker containing water. The microballoons from almost all the formulations remained float for 24 h. after 24 h, the number of microballoons remained float were counted and percentage floating was calculated (the results were showed in Table 2). The floating percentage or buoyancy of all the formulations was found to be in the range of 83.3 – 99.0%. These results indicated that the prepared microballoons have predominant floating characteristics.

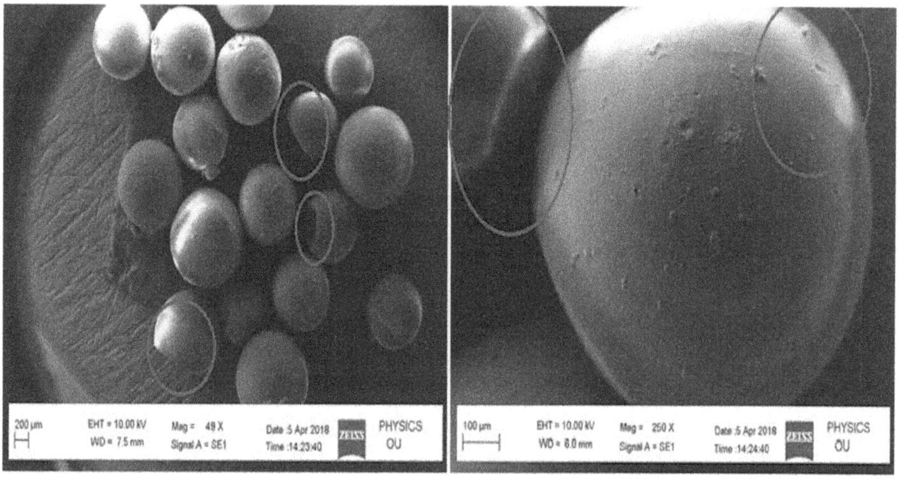

Figure 1 Scanning electron micrograph of F_{23} formulation. The encircled portions indicate dents on the surface.

3.1 Design of experiments (DoE) analysis

The influence of the selected four factors on each of the three responses has to be elucidated in order to proceed for optimization which was done by using Design Expert software. Sequential sum of squares analysis was performed for every response. It was observed that every factor was found to be in linear regression with the factors. Hence, linear regression model was chosen to elucidate the effect of the factors on every response. The regression equations were found to be as following

R1: Particle size $= +704.19 + 123.73*A + 56.35*B - 134.33*C - 150.16*D$

R2: Entrapment efficiency $= +84.60 + 3.63*A - 2.52*B - 4.48*C - 3.71*D$

R3: Release rate constant $= +0.16 - 0.045*A - 0.016*B + 0.027*C + 0.034*D$

The above equations indicate either positive or negative effect of every factor on every response. The effects of these factors on the responses were analyzed by ANOVA and were found that all the factors were found to have significant effect on all the three responses (showed in Table 3).

Table 3: *Results of ANOVA for response surface linear model for the response variables.*

Response	Source	SS[a]	Df[b]	MSS[c]	F value	p-Value[d]
Particle size (µm)	Model	7.089×10^5	4	1.77×10^5	23.60	< 0.0001
	A[e]	1.837×10^5	1	1.837×10^5	24.46	< 0.0001
	B[f]	38103.87	1	38103.87	5.07	0.0357
	C[g]	2.165×10^5	1	2.165×10^5	28.83	< 0.0001
	D[h]	2.706×10^5	1	2.706×10^5	36.03	< 0.0001
	Residual	1.50×10^5	20	7510.11	--	--
	Cor total	8.591×10^5	24	--	--	--
EE[i] (%)	Model	639.52	4	159.88	15.04	< 0.0001
	A	157.69	1	157.69	14.84	0.0010
	B	76.51	1	76.51	7.20	0.0143
	C	240.31	1	240.31	22.61	0.0001
	D	165.02	1	165.02	15.53	0.0008
	Residual	212.57	20	10.63	--	--
	Cor total	852.09	24	--	--	--
K[j] (h^{-1})	Model	0.049	4	0.012	22.34	< 0.0001
	A	0.024	1	0.024	43.99	< 0.0001
	B	2.883×10^3	1	2.883×10^3	5.24	0.0331
	C	8.48×10^3	1	8.48×10^3	15.41	0.0008
	D	0.014	1	0.014	24.71	< 0.0001
	Residual	0.011	20	5.504×10^4	--	--
	Cor total	0.060	24	--	--	--

Note: [a] Sum of Squares; [b] Degrees of Freedom; [c] Mean Sum of Squares; [d] p-Value less than 0.05 indicates model terms are significant; [e] Polymer concentration (%w/w); [f] Methanol concentration in the solvent mixture (%v/v); [g] Volume of solvent mixture (mL); [h] Speed (rpm); [i] Entrapment Efficiency; [j] Release rate constant.

3.1.1 Particle size. The particle size results of all the formulation are shown in Table 2. The effects of the factors on the particle size are illustrated in Figure 2(a) and 2(b). Upon increase in polymer concentration (factor A), the particle size was found to be increased. At high amounts of polymer, the viscosity of the dispersion might be increased which inversely affect the fine globule formation in the emulsion, which lead to the increase in the size of the microballoons. [8] Upon increase in concentration of methanol in the internal phase (factor B), particle size was found to be increased. This might be because the rate of evaporation might be decreased at higher methanol concentrations due to its higher boiling point than that of diethyl ether. The decreased rate of evaporation might cause aggregation of globules thus resulted in microspheres with increased particle size. [9] Volume of internal phase (factor C) was found to have negative effect on the particle size. This could be due to the lesser viscosity at higher volumes of the internal phase that resulted in the formation smaller globules in the emulsion, which upon solidification resulted in microballoons with smaller size. Speed of rotation (factor D) was also found to have negative effect on the particle size. As the speed of rotation increase, more energy could transfer into the emulsion that could break the globules of the emulsion to result in smaller size and thus lead to decrease in in the size of the obtained microballoons. [8]

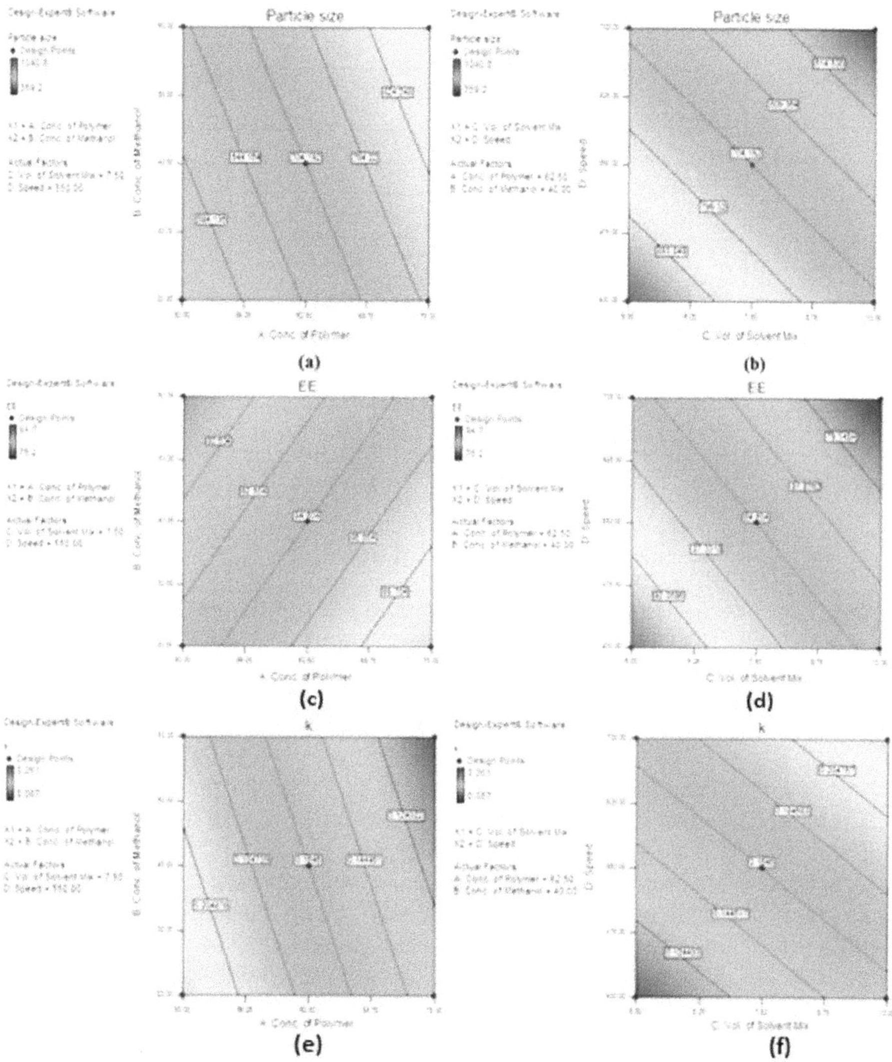

Figure 2 Contour plots illustrating the influences of (a) the factors A & B and (b) the factors C & D on the particle size; (c) the factors A & B and (d) the factors C & D on the entrapment efficiency; (e) the factors A & B; and (f) the factors C & D on the drug release rate constant.

3.1.2 Entrapment efficiency. The results of EE of all the formulations are showed in the Table 2 and were found to be ranging from 75.2 – 94.7%. The effects of the factors on the particle size are illustrated in Figure 2(c) and 2(d). Upon increase in the polymer concentration (factor A), entrapment efficiency was observed to be increased. High amount of polymer result in strong polymer network that can entrap more amount of drug tightly and reduce leakage of drug upon evaporation of the solvent. [8,10] Upon increase in the concentration of methanol in the internal phase (factor B), entrapment efficiency was found to be decreased. This could be attributed to the freely soluble nature of PGH in methanol and insolubility in diethyl ether. As the methanol concentration is increased the drug's diffusion along with methanol out of the microparticles could also be increased and thus resulted in decreased entrapment efficiency. This might be also because of slow evaporation for longer times. [9,11] When the evaporation occurs for longer time, more amount of drug can diffuse out of the microspheres thus cause decrease in entrapment efficiency. Upon increase in the volume of internal phase (factor C), entrapment efficiency was found to be decreased. This might be because of the high diffusion rate of the solvent at lower viscosities upon increasing volume of internal phase. [7,8] The rapidly moving solvent might bring the drug out of the microspheres which lead to decrease in entrapment efficiency. Upon increase in speed of rotation (factor D), entrapment efficiency was found to be decreased. This might be because of the high diffusion rate of the solvent due to higher evaporation rate at higher speeds. High diffusion rate of the solvent might bring the drug out of the microspheres which lead to decrease in entrapment efficiency. [12]

3.1.3 Drug release rate. The data obtained from the *in vitro* drug release studies was subjected to zero-order and first-order kinetics. From the kinetics analysis, it was observed that the drug release from all the formulations followed first-order kinetics. The first-order rate constants were calculated for all the formulations and were presented in Table 2. The effects of the factors on the particle size are illustrated in Figure 2(e) and 2(f).

Polymer concentration (factor A) was found to have negative effect on the drug release rate. This could be because of the increased path length for diffusion of the entrapped drug at high polymer concentrations (due to bigger particle size), that lead to lessening in drug release rate. [13,14] Concentration of methanol (factor B) was also observed to have negative effect on the drug release rate. Slow evaporation rate at high concentration on methanol might result in controlled formation of microspheres with tight surface which lead to decrease in drug release rate. Upon increase in the volume of internal phase (factor C), the release rate was found to be increased. This might be because of the decreased particle size, upon increase in the volume of internal phase, which could enhance the release of the drug because of higher surface area. [15] Upon increase in speed of rotation (factor D), drug release rate was observed to be increased. At higher speeds, smaller microballoons were obtained[8,15] which had more contact with the dissolution medium owing to their high surface area. This might cause release of the drug at relatively high rate.

3.2 Design validation and optimization

The selected BBD and the linear models for the effects of the factors on the responses were validated by ANOVA and normal plots of residuals for every response. The results of ANOVA (presented in Table 3) indicated that the selected models for all the three responses were significant and even the influences of all the four factors on the three selected responses were significant at $p < 0.05$. The adjusted and predicted regression coefficients were found to be 0.790 & 0.722; 0.701 & 0.605; and 0.780 & 0.716 for the particle size, entrapment efficiency and drug release rate constant respectively. For any response, the difference between these predicted and adjusted regression coefficient was found to be less than 0.2. This demonstrated that the selected models for the responses were significant. The normal plots of residuals displayed in Figure 3(a), 3(b) and 3(c) for the responses particle size, entrapment efficiency and drug release rate constant respectively. In all the three cases, the plot resembled like almost a straight line but not any sigmoid shape. Hence, these plots illustrated that the selected models were significant enough and can be navigated for optimization.

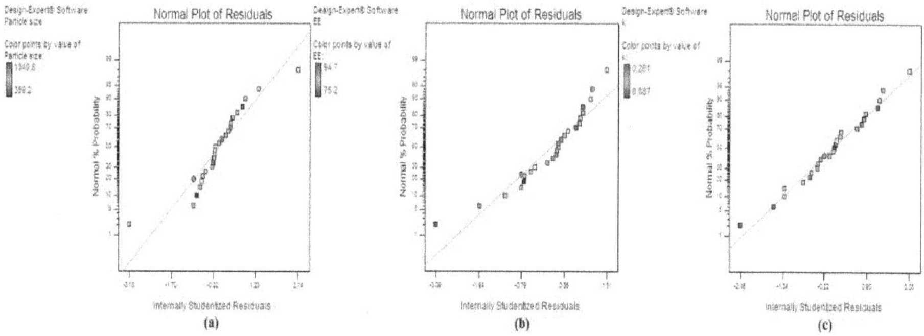

Figure 3 Normal plot of residuals for the responses (a) Particle size; (b) Entrapment efficiency; and (c) Drug release rate constant.

Optimization was performed by desirability functions approach. The desirability constraints for the responses were taken as minimizing the particle size with upper limit of 600 μm, maximizing the entrapment efficiency with lower limit of 85%, and keeping the drug release rate constant within 0.18 – 0.21 h^{-1} (corresponding to extending the drug release for around 24h). The overlay plot obtained with the set desirability was depicted in Figure 4. The yellow color region indicates the design space within which any combination of the factors will produce the microballoons with the desired response values. One such best combination was identified by the software (showed in the Table 4) was taken and the microballoons were again prepared at this combination. Then they were evaluated for the characterization tests including the responses. The observed response values for this optimized combination were found to be 546.2 μm of size; 87.4% of entrapment efficiency and 0.2 h^{-1} of drug release rate constant. These observed values of all the three responses were found to be well correlated with the predicted values by lying in the 95% confidence intervals (CI) range of the predicted values. Hence, these results indicated that optimization was successful for the development of hollow microspheres with the selected factors, their levels and the design.

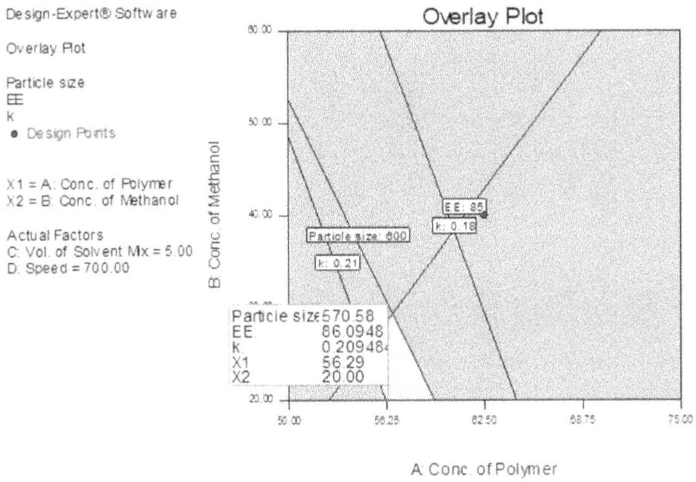

Figure 4 Overlay plot after optimization indicating the design space.

Table 4 *Optimized combination of the factors with the Predicted and observed/obtained values of the responses of the hollow microspheres prepared with the optimized combination of the factors.*

Optimized factors combination	Responses	Predicted values	95% CI low	95% CI High	Observed responses
A: Conc. of Polymer (56.29% w/w) B: Conc. of Methanol (20% v/v) C: Vol. of Solvent Mix (5 mL) D: Speed (700 mL)	R1: Particle size (μm)	570.6	469.8	671.3	**546.2**
	R2: Entrapment efficiency (%)	86.1	82.3	89.9	**87.4**
	R3: Drug release rate constant (h^{-1})	0.21	0.18	0.24	**0.20**

4. CONCLUSION

Hollow microspheres of microballoons have the advantage of inherent floating nature without any lag time when they added to water. Hence they prove as the best alternative for gastro-retentive drug delivery systems. Prasugrel HCl is water insoluble in the pH of 6.0 and above. Hence it should retain in upper GIT only for continuous and extended absorption to achieve greater therapeutic advantage. In the current work, emulsion solvent evaporation-diffusion method was used to develop hollow microspheres. For this purpose, QbD approach was adopted and BBD was selected as the experimental design to elucidate the influence of several formulation and process factors on the quality characteristics of the microballoons. The developed microballoons exhibited floating for more than 24h. The design validation indicated that all the factors had statistically significant linear effect on the responses and also these linear models were significant at $p < 0.05$. Later, optimization was performed and the results indicated that microballoons prepared at 56.29% w/w of polymer; 20% v/v of methanol in the internal phase of 5 mL with 700 rpm as the speed of rotation exhibited the desired quality characteristics of minimum particle size, maximum entrapment efficiency and extended drug release for 24h.

5. ACKNOWLEDGEMENT

The authors are acknowledged to the authority of Acharya Nagarjuna University, Guntur and Vignan's Foundation for Science Technology and Research, Vadlamudi, Guntur for providing necessary facilities to carry out this research work.

6. REFERENCES

1. AH. El-Kamel, MS. Sokar, SS. Al Gamal, VF. Naggar, *Int J Pharm*. 2001, **220**, 13.
2. HS. Chang, H. Park, P. Kelly, JR. Robinson. *J Pharm Sci*. 1985, **74**, 399.
3. BN. Singh, KH. Kim, *J Control Rel*, 2000, **63**, 235.
4. PP. Dobesh, *Pharmacotherapy*, 2009, **29**, 1089
5. C. Rigobello, M. Steppe. *J Chem Pharm Res*, 2016, **8**, 285.
6. SD. Vanshiv, HP. Joshi, AB. Aware, Indian J Pharm Sci, 2018, **80**, 181.
7. A. Porwal, G. Swami, SA. Saraf. *Daru*. 2011, **19**, 193.
8. A. Yadav, DK. Jain, *Trop J Pharm Res*, 2012, **11**, 561.
9. S. Ravi, KK. Peh, Y. Darwis, BK. Murthy, TRR. Singh, C. Mallikarjun, Indian J Pharm Sci, 2008, **70**, 303
10. A. Patel, R. Shah, P. Prajapati, A. Jadhav. *World J Pharm Pharm Sci*, 2015, 3, 335.
11. D. Sharma, D. Maheshwari, G. Philip, R. Rana, S. Bhatia, M. Singh, *et.al.*, *BioMed Res Int*, 2014, 156010.
12. AA. Mahmouda, GS. El-Fekya, R. Kamel, GEA. Awad, *Int J Pharm*, 2011, **413**, 229.
13. Y. Krishnamachari, P. Madan, S. Lin, *Int J Pharm*, 2007, **338**, 238.
14. X. Fu, Q. Ping, Y. Gao. *J Microencapsul*, 2005, **22**, 705.
15. R. Kumar, S. Kamboj, A. Chandra, PK. Gautam, VK. Sharma, *UK J Pharm Biosci*, 2016, **4**, 30.

FORMULATION OF CANDISARTAN MUCOADHESIVE BUCCAL TABLETS AFTER IMPROVING THE SOLUBILITY BY SOLID DISPERSION AND INCLUSION COMPLEX METHODS.

Narendra Pentu[a*] Manikanta Murahari[b], Narayana Murthy G[b]

[a] Department of Pharmaceutics, CMR College of Pharmacy, Kandlakoya(V), Medchal, Hyderabad-501401.,

[b] Department of Pharmacy, Koneru Lakshmaiah Education Foundation, Vaddeswaram, AP, India.
*narendrapentu@gmail.com

1 INTRODUCTION

In the previous two decades, has been a lot of research interest in drug delivery through the buccal mucosa. It was created with the goal in order to administer different medicinal ingredients through the buccal route.[1] There has been a resurgence of interest in since the early 1980s in using bio adhesive polymers to extend contact period in the various mucosal drug delivery methods. Using oral medications has been the the most used method for delivering systemic drugs. Its ineffectiveness of certain drugs as a result of Pre-systemic elimination, irregular, inconsistent absorption, reduced bioavailability, and GI discomfort another feasible route of administration. Recent advancements in drug delivery have escalated interest in mucosal drug delivery, including pulmonary, nasal, ocular, buccal, and oral routes, among others.[2]

A potential location for medication distribution to the systemic circulation is the buccal mucosa. An oral medication bypasses first-pass hepatic metabolism and negative gastrointestinal effects and enters the systemic circulation immediately.[3] The primary issue with oral dose formulations is their low bioavailability. Aqueous solubility, drug permeability, dissolving rate, first-pass metabolism, presystemic metabolism, and susceptibility to efflux mechanisms are some characteristics that affect oral bioavailability. Limited oral bioavailability is frequently caused by poor solubility and low permeability.

To achieve the required drug concentration in the systemic circulation and the desired pharmacological response, solubility is one of the most important considerations. The main problem with the formulation creation of novel chemical entities and generic development is low water solubility. Any drug that is meant to be absorbed must be present as an aqueous solution at the location of absorption.[4] The pharmaceutical business produces more than 40% of novel chemical compounds, that are essentially water insoluble. The lack of and inconsistent bioavailability of these medications is brought on by their poor water solubility and their sluggish drug absorption, as well as gastro-intestinal mucosal toxicity. Solubility is the most important rate-limiting factor when utilizing oral drugs since it determines whether the drug will reach the required concentration in systemic circulation for a pharmacological response.[5]

2 MATERIALS AND METHODS

2.1 Materials

Candesartan (CAND) purchased from Allied Fabrichem Pvt Lmt. Hyderabad. Dr. Reddy's Laboratories provided Poloxamer 188 as a free sample (Hyderabad). Signet Chemical Corporation Pvt Ltd supplied the -cyclodextrin (Mumbai, India). Laboratory of Dr. Reddy's (Hyderabad), SD Fine Chemicals (Mumbai, India), provided the Sodium CMC, Sodium alginate polymers, and HPMC K4M, respectively.

2.2 Studies on Phase Solubility

Phase solubility investigations were carried out using the Higuchi and Connors method. (Higuchi T and Connors, 1965). In summary, buffer solutions with different amounts of Cycldextrin were combined with excess medicines (1-10mM). The samples were filtered using a 0.45 m membrane filter after the suspensions were given a vigorous shaking in a mechanical shaker. A UV spectrophotometer with a maximum 212 nm setting was used to measure the concentration of Candesartan. Phase solubility diagrams can be used to derive the apparent stability constant (Kst) by using the equation.

Apparent stability constant (K_{st}) = $\frac{\text{Slope}}{\text{So}\ (1-\text{Slope})}$

Slope is calculated from the graph using the formula So = Intercept (Candesartan solubility without -Cyclodextrin).

2.3 Preparation of Binary Systems

Various ratios of Poloxamer 188 and -Cyclodextrin were used in two separate preparation techniques, melt agglomeration and kneading, to create candesartan and carrier binary systems.

2.4 Preparation of Solid Dispersions

The melting agglomeration (MA) method was used to create solid dispersions of Candesartan. This procedure involved combining the needed amount of the medication with Poloxamer 188 in various ratios of 1:0.5, 1:1, and 1:2 in a china dish, the melting point of the carrier plus a little bit more, and completely combining it. The mixture was broken up and allowed to warm up before being run through sieve number 40.[10]

2.5 Preparation of Complexes For Inclusion

Candesartan and -Cyclodextrin were combined in a mortar at various molar ratios (1:0.5, 1:0.75, and 1:1), and then water (in proportions) was added to make a consistent paste. Then after, the combination was milled for 30 minutes. During this process, the mixture contained the proper amount of water to maintain a satisfactory consistency. The paste was dried in a 400°C oven. The dried complex was powdered up after drying[9].

2.6 Bioadhesive Tablets Preparation

Buccal tablets were created by directly compressing a variety of mucoadhesive polymers with a medication that was either gratis or present with the help of inclusion and solid-state complexes.

According to the batch formula, every component, including the medication, polymers, and excipients, was weighed accurately. All of the materials were blended for ten minutes in a blown-up polyethylene bag, with the exception of the lubricants, they were combined in ascending weights. The lubricant and glidant were added after the components had been thoroughly mixed, and they were combined for a further two minutes. Each formulation's prepared blend was compressed on a rotary tablet punching machine with an 8mm punch[6].

2.7 Candesartan Buccal Tablets *In-Vitro* Evaluation

2.7.1 Total Tablet Swelling Index Determination. Separately, weighted tablets (W1) were put through Petri dishes with 20 milliliters of distilled water. At room temperature, the plates were stored. At predetermined intervals, the discs were removed, and using filter paper, the surplus water on their surface was carefully removed[11]. The swollen discs were reweighed (W2), and the formula below was used to calculate the swelling index.

Swelling index (SI) = $\frac{W2-W1}{W1} x100$

The weights of the tablet before and after swelling are, respectively, W1 and W2.

2.7.2 Measurement of Surface P^h. For this, a merged glass electrode was used. The surface pH of the mucosa pills' microenvironment was measured to look at potential adverse consequences in vivo. Because an acidic or alkaline pH may irritate the buccal mucosa, the surface pH was maintained as neutrally as possible. The tablet was allowed to swell for two hours at room temperature by submerging it in distilled water, 5 mL (pH 6.5 ± 0.05). By placing the electrode on the tablets' surface and letting it equalise for a minute, the pH was determined[12].

2.7.3 Mucoadhesion Time Measurement. Using newly cut pig buccal mucosa, after applying the buccal patch, and the *ex-vivo* mucoadhesion time was monitored. Fresh pig buccal mucosa was secured to an optical slide, and the central portion of each patch mucoadhesive was moistened with one drop of buffer (pH 6.8) and applied for 30 seconds with a fingertip. An optical slide was then placed in the beaker, it was held at 37°± 0.50°C while the equipment was running, and contained 800 ml of phosphate buffer. The duration of the tablet's mucoadhesion to the buccal mucosa was measured.

2.7.4 Measurement of Mucoadhesive Strength. The weight in g. required to remove the device from the porcine buccal mucosa used to calculate bio adhesive strength. The weight increase stopped when the pill separated from the porcine buccal mucosa. The *ex-vivo* mucoadhesive strength was calculated as the weight required to remove the pill from the buccal mucosa testing for mucoadhesive strength was carried out in triplicate, and the average Bio adhesive strength was calculated. The measurements were all taken at 37±1^{0}C[11].

2.7.5 Content of Drugs. The contents of 20 pills were broken down and pulverised. A precise weight of 30 mg CAND was dissolved 20 milliliters of methanol, rattled for 5 minutes, and filtered. Methanol was employed to dilute the filtrate in a 25 ml measuring flask. A portion of the diluted drug solution was treated exactly as specified. The found tablet content was calculated.

2.7.6 In-Vitro Dissolution Studies. Using the rotating paddle method described in the United States Pharmacopoeia (USP) XXIV, the drug release of buccal tablets was investigated. 500 milliliters of buffer served as the medium for dissolving (pH 6.8). The release was carried out at 37 ± 0.5°C and a spindle speed of 50 revolutions per minute. Five milliliter samples were taken out and fresh medium was added at specified intervals. CAND was detected using a UV spectrophotometer at max 212nm after the samples were filtered by Whitman filter paper No. 40. The percentage of drug release was calculated using the medication's calibration curve in buffer pH 6.8.

2.7.7 Studies on Ex-Vivo Permeation. This test was performed in a typical Franz diffusion cell with a 4.09 cm^2 a diffusion zone and a 25 ml acceptor compartment volume. Fresh swine buccal mucosa from the neighbourhood butcher was used to clamp the donor and acceptor compartments together. A magnetic stirrer was used to stir the 6.8 pH phosphate buffer continuously at 600 rpm in the receptor compartment. In the donor compartment, the tablet was put, and 1 cc of phosphate buffer was added. By removing aliquots from the receptor compartment and substituting the equivalent amount of buffer therein, the amount of drug that pierced the membrane could be measured. A UV spectrophotometer set to max 212nm was used to analyse the samples. The following equation was used to determine the CAND flux through the membrane

$$J = \frac{\mathrm{dQ}}{\mathrm{A\,dt}}$$

the steady-state flow J and the diffusion region A.

2.7.8 Statistical Analysis. To better understand the drug release kinetics and mechanism from produced tablets, in vitro drug release data was studied using kinetic models including zero-order, first order, Higuchi, and Korsmeyer-Peppas. The table displays the kinetic model-derived regression coefficient (R2) values. The Korsmeyer-Peppas model's kinetic data can be utilized to verify the drug release procedure. This model's "n" release exponent explains how medications are released from made-to-order tablets. Assume that there is Fickian diffusion, or n=0.45. Additionally, the n value ranges from 0.45 to 0.89, suggesting a non-Fickian transport mechanism. In the event when "n" is more than 0.89, the case II transport technique is used.

3 RESULTS AND DISCUSSION

3.1 Cand Binary Systems

The CAND-CYCLODEXTRIN complex's phase solubility diagram had an AL-type system as its form. It was found that Ks, the apparent stability constant, is 1.128 x 103 M-1. An equimolecular CAND-CYCLODEXTRIN solid system was created since the phase solubility studies predicted that a CAND-CYCLODEXTRIN inclusion complex with a 1:1 stoichiometry would arise. A solid amorphous product was created by freeze drying CAND that had been dissolved in cyclodextrin solution.

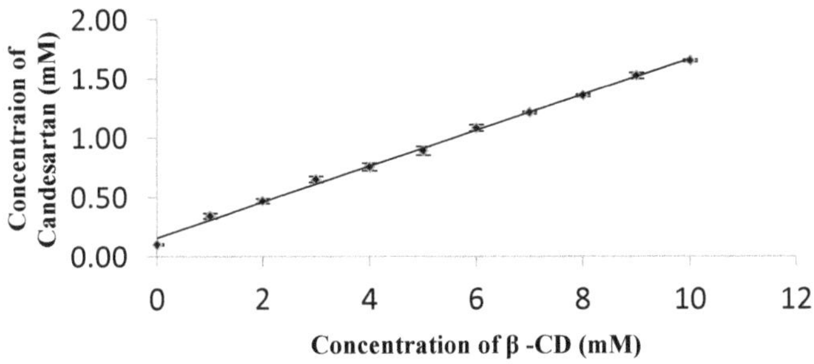

Figure 1 CAND-CYCLODEXTRIN complex phase solubility diagram.

According to the graph (See Figure 1), as the concentration of CYCLODEXTRIN increases, so does the concentration of Candesartan increases. Thus, Candesartan—CD Inclusion complexes exhibit the ALL types of Phase solubility graphs and form complexes with a stoichiometric ratio of m: 1.

The 1:2 ratio of Candesartan-Poloximer 188 and the 1:1 ratio of Candesartan—CD have higher solubility, according the Phase solubility diagram and solubility tests. These proportions are used to make the buccal tablets.

Additional proof of complex creation was supplied by X-ray diffractometry and differential scanning calorimetry. See Tables 1 and 2.

Table 1 *Composition of buccal tablets containing the drug loaded with Poloxamer 188.*

Ingredients(mg)	**Formulation Code**						
	F1	**F2**	**F3**	**F4**	**F5**	**F6**	**F7**
Candesartan	16	16	16	16	16	16	16
Poloxamer 188	-	32	32	32	32	32	32
HPMC K4M	-	50	-	-	25	25	-
Sodium CMC	-	-	50	-	25	-	25
Sodium alginate	-	-	-	50	-	25	25
MCC	123.5	41.5	41.5	41.5	41.5	41.5	41.5
PVP K30	7.5	7.5	7.5	7.5	7.5	7.5	7.5
Magnesium stearate	1.5	1.5	1.5	1.5	1.5	1.5	1.5
Talc	1.5	1.5	1.5	1.5	1.5	1.5	1.5

Table 2 *Composition of the Formlation in buccal tablets filled with cyclodextrin.*

Ingredients(mg)	**Formulation Code**						
	F1	**F8**	**F9**	**F10**	**F11**	**F12**	**F13**
Candesartan	16	16	16	16	16	16	16
β-CD	-	40	40	40	40	40	40
HPMC K4M	-	50	-	-	25	25	-
Sodium CMC	-	-	50	-	25	-	25
Sodium alginate	-	-	-	50	-	25	25
MCC	123.5	33.5	33.5	33.5	33.5	33.5	33.5

PVP K30	7.5	7.5	7.5	7.5	7.5	7.5	7.5
Magnesium stearate	1.5	1.5	1.5	1.5	1.5	1.5	1.5
Talc	1.5	1.5	1.5	1.5	1.5	1.5	1.5

3.2 Diffractometry by X-Ray

Studies using X-ray diffractometry proved that CAN is crystalline and that CD is amorphous. The physical mixture's X-ray diffraction pattern can be interpreted as a superposition of CAN and amorphous -CD. The medicine had stronger peaks than those that could be seen. a medication's diffractogram The CD complex had weaker crystal signals, indicating the substance was amorphous. (See Figures 2-4).

X-ray Diffractometry (X-RD)

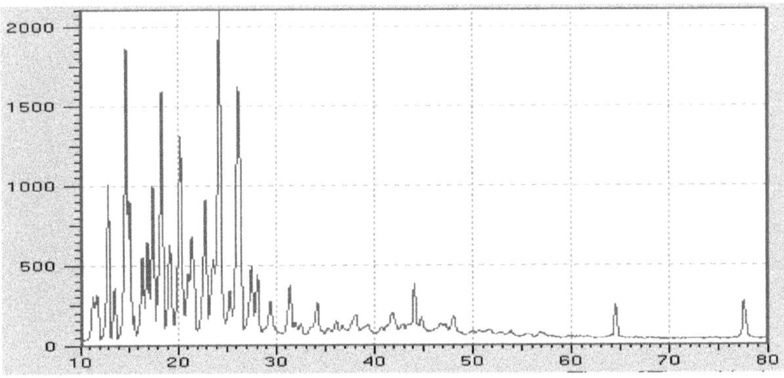

Figure 2 Candesartan drug X-ray Diffraction report.

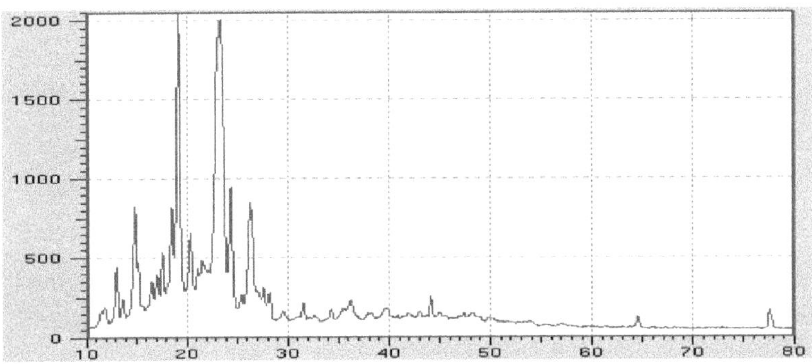

Figure 3 Poloxamer 188 (1:2) X-ray Diffraction study on candesartan.

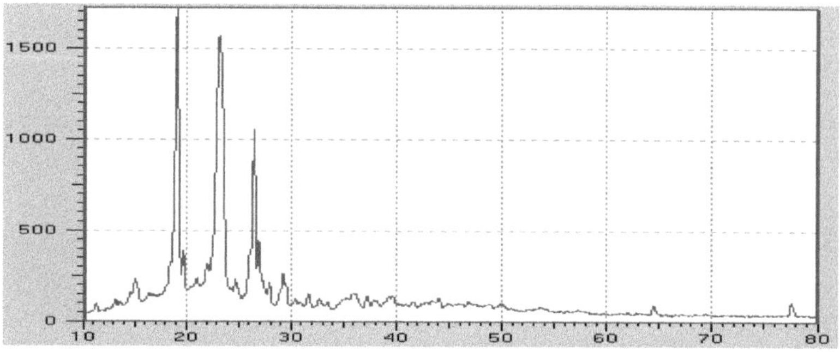

Figure 4 CAND: X-ray Diffraction report with CD (1:1).

3.3 Evaluation of Buccoadhesive Tablets

3.3.1 Surface Ph Study. All of the tablets' surfaces had pH values ranging from 6.33 to 6.84. These results demonstrated that there was no danger of mucosal irritation or injury when these formulations were administered to the buccal mucosa.

3.3.2 Ex Vivo Mucoadhesion Time. For all of the prepared pills, the ex vivo mucoadhesive time ranges from 11.1 to 12 hours. The F6 with HPMC was used in the medication formulations created for Poloxamer 188. Comparing other formulations, the combination of sodium alginate shows the longest mucoadhesion time.

In the medication compositions, the F12 with HPMC was used. Comparing different formulations, the combination of sodium alginate displays the longest mucoadhesion time.

The F5 and F11 formulations exhibit the shortest mucoadhesion times because to Sodium CMC's poor mucoadhesion properties. As a result, when comparison to competing formulations, formulations containing Na CMC have the shortest mucoadhesion time.

Due to the flexibility of sodium alginate, which easily diffuses and interpenetrates into the mucin and becomes entangled with that of mucin, it has better mucoadhesion. In comparison to HPMC and Sodium CMC, sodium alginate has the largest mucoadhesion effect on mucins.

3.3.3 Mucoadhesive Strength. The buccal pills' bioadhesive forces were impacted by the properties of Plymers. In order to enable bioadhesion, polymer swelling helped mucus and polymer interact.

In formulations made using the medication Poloxamer 188, formulation F6 with sodium alginate: HPMC K4M combination showed the strongest adhesion strength, followed by formulation F7. They demonstrate maximum mucoadhesion durations of 120.34 and 120.22 seconds, respectively. Sodium alginate has a stronger mucoadhesion effect on mucins than HPMC and Sodium CMC. This might be because sodium alginate is flexible and can easily infiltrate into mucin, interpenetrate with it, and entangle with it. The least amount of bioadhesion was found in the F3 formulation made with Na CMC. In the medication formulation made using the polymer: -CD, the formulation F12 containing sodium alginate showed the strongest adhesion strength: Combination HPMC K4M, then F13 formulas. The mucoadhesion times are displayed as up to 120.37 and 120.24, respectively. Compared to HPMC and Na CMC, sodium alginate has the strongest mucin's mucoadhesion effect. This might be as a result of the flexibility of sodium alginate, which readily diffuses and interpenetrates into mucin and combines with it. Additionally, F9 formulations containing Na CMC alone showed the least bioadhesion. In comparison to the single polymer, the polymer mixtures exhibit stronger mucoadhesion.

See Tables 3 & 4..

Table 3 *Formulations of the medication that have Poloxamer 188 added.*

Formulation Code	Surface pH Measurement	Swelling Index	Mucoadhesion time	Mucoadhesive Strength	% Drug Content
F1	6.56±0.22	10.32±0.88	0.15±0.41	2.64±0.14	98.56±0.16
F2	6.54±0.38	67.435±0.38	12.5±0.21	19.05±0.14	98.4±0.11
F3	6.64± 0.04	97.16±0.75	11±0.51	15.24±0.26	98.2±0.21
F4	6.81± 0.07	45.56±0.22	11.3±0.32	21.51±0.81	98±0.81
F5	6.84±0.15	80±0.45	11.5±0.21	19.41±0.41	98.45±0.51
F6	6.75±0.12	41.01±1.02	12±0.31	26.15±0.21	98.74±0.81
F7	6.68±0.22	53.81±0.71	12±0.11	21±0.89	97.1±0.11

Every value is expressed as the mean ± SD. (n=3)

Table 4 *Formulations containing the drug-loaded with β-Cyclodextrin.*

Formulation Code	Surface pH measurement	Swelling Index	Mucoadhesion time	Mucoadhesive Strength	% Drug Content
F1	6.67±0.21	11.11±0.81	0.15±0.41	2.65±0.15	97.64±41
F8	6.5±0.14	68.89±0.38	11.8±0.14	20.21±1.85	97.21±0.25
F9	6.57±0.11	98.56±0.65	11.2±0.21	16±0.51	98.57±0.87
F10	6.51±0.21	47±0.21	11.4±0.51	22±0.21	98.51±0.45
F11	6.61±0.14	82.25±0.41	11.3±0.21	20±0.31	98.58±0.21
F12	6.7±0.21	45±0.51	12±0.31	26.5±0.51	97.24±0.58
F13	6.21±0.41	59±2.75	12±0.21	20.14±0.41	98.51±0.51

Every value is expressed as the mean ± SD. (n=3)

3.3.4 Swelling Studies of Buccal Tablets The swelling behaviour of the tablets influences the bioadhesion and drug release profile. The swelling index was computed over time (See Figure 5). The swelling index increased according to the rate of hydration as the weight gain of the tablets increased. The rate of water spread was determined by the ratio of two polymers used. See tables 5 and 6.

The swelling index was 97.16±0.75 in the formulation F3 containing sodium alginate and HPMC K4M. The swelling index for formulation F6, which contained a combination of HPMC K4M and sodium alginate, was 41.01±1.02. The F9 formulation with Sodium alginate and HPMC K4M had a higher swelling index of 98.56±0.65. In comparison, the swelling index of the formulation F12 containing a combination of HPMC K4M and Sodium alginate was 45±0.51.

Table 5 *Formulations containing the drug-loaded with Poloxamer 188.*

Time (Hr)	**SWELLING INDEX**						
	F1	**F2**	**F3**	**F4**	**F5**	**F6**	**F7**
0	0	0	0	0	0	0	0
0.5	1.83	24.51	25.10	11	24.21	20.11	6.55
1	3.05	28.21	39.15	25.21	35.32	25.25	20.15
2	7.60	42	55.43	31.25	45.0	29.75	33.95

3	8.71	52.21	70	37.45	61.25	31.25	41.35
4	10	64.21	86.45	39.52	67.11	38.25	45.25
6	10.2	67.10	96.25	41.02	78.48	40.85	51.45
8	10.21	67.32	96.05	44.11	79.11	41.89	54.12
12	10.21	67.77	97.19	44.56	79.58	42.85	54.85

Every value is expressed as the mean ± SD (n = 3)

Table 6 *Formulations containing drug-loaded with β-Cyclodextrin.*

Time (hr)	**SWELLING INDEX**						
	F1	**F8**	**F9**	**F10**	**F11**	**F12**	**F13**
0	0	0	0	0	0	0	0
0.5	1.83±0.23	24.85±0.32	27.90±0.32	7.26±0.59	27.64±0.06	33.30±1.64	10.03±0.31
1	3.04±0.94	35.48±0.51	41.14±0.20	26.81±1.57	37.40±0.10	39.54±0.50	23.68±0.52
2	7.61±0.74	45.40±0.10	58.66±1.14	35.69±0.99	51.12±1.03	43.26±2.19	38.16±1.63
3	8.72±082	57.77±0.83	72.61±0.21	41.23±1.39	68.29±1.79	44.67±1.63	44.88±1.72
4	10.24±0.63	66.16±1.07	89.10±1.23	45.36±1.41	76.32±0.72	44.87±0.62	50.26±0.45
6	10.34±1.53	68.10±0.77	98.88±0.62	46.98±1.23	81.50±1.26	44.97±0.51	57.50±1.39
8	10.34±1.02	69.53±0.81	99.19±1.18	47.48±1.15	82.32±1.14	45.07±1.78	58.63±1.23
12	10.34±0.89	69.53±1.39	99.69±1.65	46.99±2.06	82.42±0.44	45.07±0.50	58.84±2.75

Every value is expressed as the mean ± SD (n = 3)

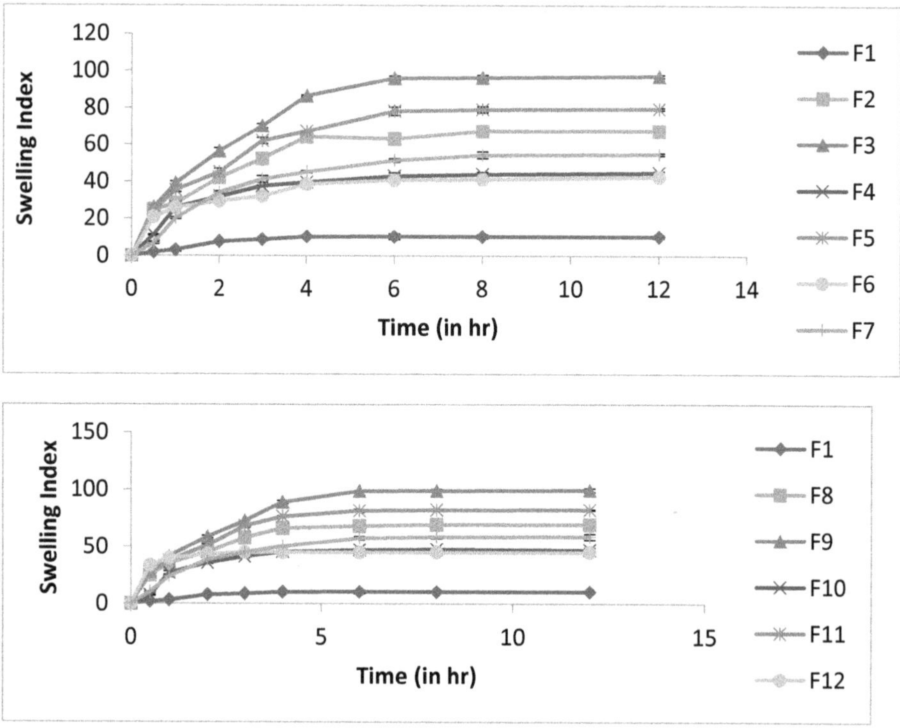

Figure 5 Swelling studies.

3.3.5 In-Vitro Studies on Drugs' Release. Dissolution of the different formulations having a 1:2 ratio of solid dispersion and complexes having a 1:1 ratio are shown in Table. The formulations contacting β-CD complexes gave the best results than those contacting the solid dispersions.

At the conclusion of 12 hours, only 66.54% of the pure candesartan had dissolved. The Candesartan dissolving was better in all other samples. Dissolution of the different formulations having 1:2 ratios of solid dispersion and complexities with 1:1 ratios are shown in Table. The formulations contacting β-CD complexes gave the best results than those contacting the solid dispersions.
After 12 hours, pure candesartan had dissolved to a maximum of 61.80%. The Candesartan dissolving was better in all other samples.

Drug release is improved in F12 formulations using the sodium alginate: HPMC polymer combination up to 95.33 percent. The formulations F3 with HPMC: Sodium alginate polymer combination exhibit the lowest drug release, up to 80.47%.

F12 formulations containing the Sodium alginate: HPMC polymer combination show a better drug release of 96.07%. And the formulation F9 with polymer combination containing the HPMC: Sodium alginate shows the lowest drug release, up to 80.91%.

Candesartan dissolution was slower in Poloxamer 188-based tablets than in tablets containing the cyclodextrin complex. In a hydrated polymeric environment, the Candesartan:-CD complex dissolves easily, causing a greater driving force for dissemination and a quicker the release of medication only a small portion of CAND can dissipate within the hydrated polymeric matrix due to its low water solubility. The matrix's -CD addition enhanced the drug's solubility and rate of dissolution. Its solubility was increased by the combination that CAND and the dissolved -CD in the gel matrix created. The main reason for increased CAND release from -CD containing polymeric matrices was solubilization caused by in situ complex formation.

3.4 RELATIONSHIP BETWEEN THE SWELLING INDEX AND THE DRUGS RELEASED IN THE FORM OF BUCCAL TABLETS CONTAINING POLOXAMER 188-LOADED DRUGS

Most formulations' drug releases decreased as the swelling index of the formulations increased. The drug release rates for the percentages of Sodium CMC in formulations F3 and F5 are 80.47 % and 82.97±0.14, respectively, respectively. These formulations have swelling indexes of 97.19 and 79.550.62, respectively. As a result of the viscous solution surrounding the tablet, drug release was prolonged.

Formulation F6 exhibits erosion prior to full swelling as well as low swelling up to 42.64, leading to 95.33 percent faster drug release.

3.5 Buccal Tablets Including the Medicine Loaded With B-Cyclodextrin Were Released, As Was The Medicine Itself, And The Swelling Index

As the formulations' swelling indices rose, the medication releases from the majority of them decreased. The drug release rates for formulations F9 and F11 containing Sodium CMC are 80.91 percent and 83.301.94, respectively. These preparations have highest swelling values of 99.68 and 82.41±0.44, respectively. As a result of the viscous solution surrounding the tablet, drug release was prolonged.

Formulations F12 exhibit erosion before to full swelling mild swelling up to 45.07, which led to 96.07 percent quicker drug release. All formulations have undergone swelling trials, and as the polymer's viscosity gets higher, so does the swelling percentage. Higher viscosity polymers have a larger capacity to absorb water, which causes rapid swelling and the creation of robust, homogeneous structures.

3.6 In Vitro Drug Release Kinetics of Buccal Tablets

To further characterise the method by which CAND is released from buccoadhesive tablets, the dissolution data were submitted to a variety of models, incorporating the Higuchi diffusion model, the first-order, zero-order, and Korsmeyer-Peppas models (See Figure 6 and Tables 7 and 8). The R2 values for zero-order release are higher in all formulations, and all formulations exhibit zero-order drug release kinetics. R2 values were in the range of 0.978 to 0.996.

The kinetic information from the Korsmeyer-Peppas model can be used to confirm the drug release process. The release exponent "n" in this model Describes the mechanism behind medication release from manufactured tablets. Fickian diffusion is used if n is 0.45. Similar to that, a non-Fickian transport The mechanism is suggested by n value, which varies from 0.44 to 0.90. Case II transport is employed if 'n' is higher than 0.89.

Table 7 *Studies on the drug's in-vitro dissolution in Buccal tablets with Poloxamer 188.*
Every value is expressed as the mean ± SD (n =3).

Time (hr)	% DRUG RELEASE						
	F1	F2	F3	F4	F5	F6	F7
0	0	0	0	0	0	0	0
0.5	1.89±1.09	8.37±1.27	7.05±1.38	14.24±1.24	7.92±1.89	15.07±0.79	9.29±1.39
1	2.91±1.24	12.44±1.43	8.85±.54	19.65±1.77	10.61±1.66	20.64±1.42	14.50±1.21
2	5.54±1.60	18.43±1.67	13.65±1.24	23.94±1.55	15.70±1.59	24.69±1.95	19.47±1.82
3	9.82±1.67	24.10±1.05	18.01±1.97	34.19±1.58	22.20±1.29	36.87±1.92	27.23±1.25
4	18.56±1.97	31.92±1.27	27.23±1.58	39.93±1.79	29.15±1.10	41.72±1.44	33.65±0.98
5	25.64±1.38	40.05±1.64	35.44±1.91	51.62±1.86	38.45±1.19	53.05±1.50	40.54±1.43
6	32.28±1.82	49.11±2.95	46.07±1.23	58.83±1.80	48.51±1.41	63.26±0.96	50.58±1.27
7	41.54±1.57	58.92±3.24	53.02±1.94	69.96±1.40	56.44±1.29	76.81±1.24	61.68±1.19
8	52.38±1.95	67.60±4.46	61.47±1.74	78.99±1.39	67.59±0.82	79.26±2.62	69.24±1.28
10	60.90±1.41	80.62±2.12	75.20±1.70	91.60±1.30	78.65±0.52	92.70±1.12	81.20±1.67
12	61.80±1.30	85.90±1.06	80.47±1.68	94.04±1.89	82.97±0.14	95.33±1.31	86.80±1.35

Table 8 *Studies on the drug's in vitro dissolution in buccal tablets with - Cyclodextrin.*

Time (hr)	% DRUG RELEASE						
	F1	F8	F9	F10	F11	F12	F13
0	0	0	0	0	0	0	0
0.5	1.89±1.09	8.90±1.59	7.21±1.31	14.76±1.27	8.07±1.27	15.61±1.10	10.25±1.45
1	2.91±1.24	13.01±1.57	9.64±0.52	20.20±1.07	11.82±1.60	21.55±0.48	15.22±0.52
2	5.54±1.60	18.61±0.99	14.75±1.69	24.24±1.46	16.55±1.29	26.77±1.23	19.79±1.37
3	9.82±1.67	25.20±1.39	19.87±1.72	35.92±0.70	23.26±1.47	38.20±0.97	28.62±1.13
4	18.56±1.97	32.74±1.41	27.82±0.45	40.29±1.97	30.86±1.38	42.60±1.11	34.37±1.99
5	25.64±1.38	40.43±1.24	36.05±1.39	52.15±1.65	39.74±1.71	453.39±1.06	41.78±1.64
6	32.28±1.82	50.08±1.15	47.24±1.23	61.73±0.56	48.97±0.93	64.03±0.10	51.83±0.50
7	41.54±1.57	59.64±2.06	53.81±2.75	71.24±1.81	57.52±1.86	76.83±1.03	62.19±2.19
8	52.38±1.95	68.48±0.15	62.15±1.51	79.40±1.81	66.50±1.18	80.59±1.79	71.27±1.63
10	60.90±1.41	80.79±1.93	76.60±0.43	91.94±0.95	79.66±1.86	95.58±1.26	88.21±0.51
12	61.80±1.30	86.46±0.82	80.91±1.37	94.25±1.73	83.30±1.94	96.07±0.44	87.52±0.50

Every value is expressed as the mean ± SD (n = 3)

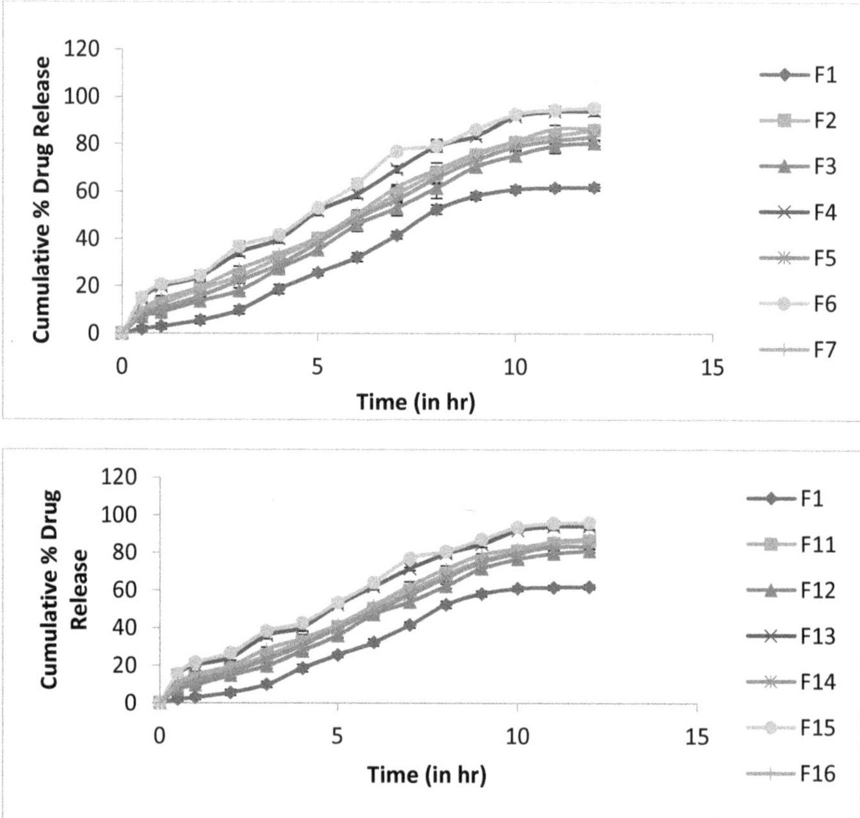

Figure 6 Drug release profile of different formulation.

Table 9 *Buccal tablet In Vitro Drug Release Kinetics.*

Formulation Code	Zero		Higchi		Peppas			First		Hixon Crowell	
	R^2	k	R^2	k	R^2	k	n	R^2	k	R^2	k
F1	0.978	6.616	0.854	21.385	0.962	15.286	1.184	0.948	0.015	0.961	0.133
F2	0.996	7.985	0.926	26.646	0.982	4.966	0.696	0.950	0.018	0.975	0.189
F3	0.992	7.614	0.903	25.121	0.941	5.649	0.752	0.952	0.018	0.972	0.170
F4	0.99	8.672	0.953	29.437	0.963	3.656	0.563	0.923	0.025	0.969	0.237
F5	0.995	7.878	0.916	26.153	0.962	5.272	0.722	0.952	0.018	0.975	0.182

F6	0.984	8.903	0.957	30.378	0.957	3.656	0.563	0.934	0.025	0.975	0.251
F7	0.994	7.988	0.935	26.809	0.984	4.529	0.656	0.955	0.020	0.978	0.192
F8	0.996	7.991	0.930	26.727	0.980	4.775	0.679	0.953	0.018	0.977	0.190
F9	0.994	7.678	0.909	25.396	0.958	5.534	0.743	0.950	0.018	0.972	0.174
F10	0.988	8.746	0.957	29.768	0.957	3.639	0.561	0.931	0.025	0.974	0.242
F11	0.996	7.95	0.922	26.466	0.971	5.152	0.712	0.951	0.018	0.975	0.186
F12	0.984	8.909	0.962	30.469	0.969	3.540	0.549	0.928	0.025	0.973	0.256
F13	0.993	8.124	0.936	27.285	0.978	4.276	0.631	0.951	0.021	0.975	0.199

3.7 Ex Vivo Permeation Study

The F1 formulation with pure CAND exhibits the lowest permeability up to 7.845 mg (See Table 10). The amount penetrated (Q) in the F6 formulation, which contains a mixture of sodium alginate and HPMC polymer, increased to 13.12 mg over the course of 12 hours. See Table 11.

The F1 formulation with pure CAND exhibits the lowest permeability up to 7.845 mg. The amount penetrated (Q) in the F12 formulation, which contains a mixture of sodium alginate and HPMC polymer, increased to 13.29 mg over the course of 12 hours. See Table 12.

Table 10 *Ex vivo permeation studies of CAND Buccal tablets for F1, F6, and F12.*

Time (hr)	**Cumulative amount Q (mg)**		
	Pure Candesartan	**Candesartan: Poloxamer 188**	**Candesartan: β-CD**
	F1	**F6**	**F12**
0	0	0	0
0.5	0.20±0.01	1.09±0.01	1.11±0.01
1	0.70±0.21	1.84±0.12	1.87±0.16
2	0.79±0.54	2.76±0.69	2.81±0.13

3	1.37±0.71	3.75±0.72	3.81±0.42
4	2.09±0.65	4.61±0.45	4.66±0.69
5	3.15±0.21	5.80±0.39	6.00±0.35
6	4.22±0.92	7.51±0.23	7.64±0.14
7	5.31±0.44	9.29±0.75	9.42±0.32
8	6.85±0.25	10.49±0.51	10.84±0.14
9	7.08±0.95	11.66±0.87	11.86±0.18
10	7.24±0.86	12.34±0.43	12.45±0.17
11	7.37±0.98	12.87±0.92	13.11±0.68
12	7.85±0.55	13.12±0.37	13.29±0.20

Every value is expressed as the mean± SD (n = 3)

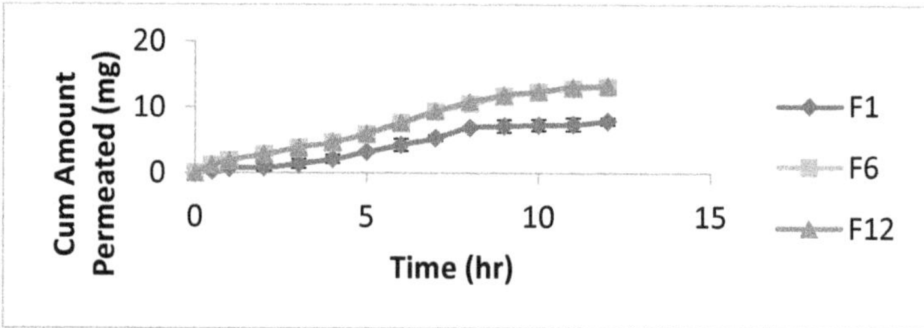

Figure 7 Ex vivo permeation studies of CAND for F1, F6, and F12 formulations.

Table 11 *A formulation containing the Drug: Poloxamer 188 with Sodium alginate and HPMC K4M combination.*

Formulation code	**Cumulative amount Permeated (Q) mg**	**Flux (J) µg/cm²/hr**
F1	7.85	10.881
F6	13.12	18.559

Table 12 *Formulation containing the drug: β-Cyclodextrin with Sodium alginate and HPMC K4M combination.*

Formulation code	**Cumulative amount Permeated (Q) mg**	**Flux (J) µg/cm²/hr**
F1	7.85	10.881
F12	13.29	18.875

4. CONCLUSION

The outcomes of the XRD studies revealed No interactions occurred between the medication and the carriers. When the medication is changed from its crystalline form to an amorphous one, its solubility increases. The findings demonstrate the considerable potential for buccal medication delivery of formulations including the medication complexed with -cyclodextrins and a solid binary system. The qualities of mucoadhesion, dissolution, and penetration were determined to be well-balanced.

Two drug-carrier combinations were chosen as optimised medication loading formulations for mucoadhesive formulations based on the saturation solubility data: Poloxamer (MA 1:2) and -Cyclodextrin (1:1 M). The F6 and F12 formulations showed better ex vivo mucoadhesion strength (26.151.23 g) and mucoadhesion time (120.34, 120.37 hr), as well as lower swelling,increased release of drugs *in-vitro*.

5. ACKNOWLEDGEMENT

My sincere thanks to P. Mahender Chaitanya College of Pharmacy, Warangal, to gave inspiration to write and publish this work.

6. REFERENCES

1. Patel VF, Liu F, Brown MB. Advances in oral transmucosal drug delivery. J Controll Rel. 2011; 153:106-116.
2. Jain NK. Controlled and Novel Drug Delivery. New Delhi: CBS Publishers & Distributors; 1997.
3. Singh B, Ahuja N. Development of controlled-release buccoadhesive hydrophilic matrices of diltiazem hydrochloride: optimization of bioadhesion, dissolution, and diffusion parameters. Drug Develop. Ind Pharm. 2002; 28(4):431-442.
4. Vemula VR, Lagishetty V, Lingala S. Solubility enhancement techniques. Int J Pharm Sci Rev Res. 2010; 5(1):41-51.
5. Sharma D, Soni M, Kumar S, Gupta GD. Solubility enhancement-eminent role in poorly soluble drugs. Research J Pharm and Tech. 2009; 2(2):220-224.
6. Mario Jug, Mira Becirevic-Lacan. Influence of hydroxypropyl-β-cyclodextrin complexation on piroxicam release from buccoadhesive tablets. European Journal of Pharmaceutical Sciences 2004; 21:251-260.

7. Brunella Cappello, Giuseppe De Rosa, Maria Immacolata La Rotonda, Giuseppe Mensitieri, Roberto Russo. Cyclodextrin-containing poly(ethylene oxide) tablets for the delivery of poorly soluble drugs: Potential as a buccal delivery system. 2006; 319:63-70.

8. Higuchi T, Connors KA. Phase solubility techniques. Adv Anal Chem Instrum. 1965; 4:117-212.

9. Al Omari AA, Al Omari MM, Badwan AA, Al-Sou'od KA. Effect of cyclodextrins on the solubility and stability of candesartan cilexetil in solution and solid state. J Pharm Biomed Anal. 2011; 54(3):503-9.

10. Madhuri Newa, Krishna Hari Bhandari, Dong Xun Li, Tae-Hyub Kwon, Bong Kyu Yoo, Jong Soo Woo, Won Seok Lyoo, Chul Soon Yong, Jung Ae Kim, Han Gon Choi. Preparation, characterization and in vivo evaluation of ibuprofen binary solid dispersions with poloxamer 188. International Journal of Pharmaceutics 2007; 343:228-237.

11. Harikrishna Boyapally, Ravi Kumar Nukala, Pranav Bhujbal, Dennis Douroumis. Controlled release from directly compressible theophylline buccal tablets. Colloids and Surfaces B: Biointerfaces 2010; 77:227–233.

12. Varshosaz J, Dehghan Z. Development and characterization of buccoadhesive nifedipine tablets. European Journal of Pharmaceutics and Biopharmaceutics 2002; 54:135–141.

13. Shrikant Charde, Madri Mudgal, Lajwinder Kumar, Ranendra Saha. Development and Evaluation of Buccoadhesive Controlled Relea

PREPARATION & EVALUATION OF ACETYLATED JACK FRUIT SEED STARCH AS A CARRIER FOR CONTROLLED RELEASE DRUG DELIVERY OF ZIDOVUDINE TABLETS

P. Shailaja*, N. Sridevi and CH. Dinesh Kumar
Department of Pharmaceutical Technology, AU College of Pharmaceutical Sciences [Andhra University], Visakhapatnam, Andhra Pradesh- 530003

Corresponding Author E-Mail ID: dr.pshailaja.andhrauniversity.edu.in

1.INTRODUCTION

Zidovudine, an anti-HIV compound approved by the U.S in 1987, is commonly used to treat AIDS and illnesses that are associated with AIDS, either by itself or in combination with other antiretrovirals. therapeutic effectiveness of this drug is limited by key factors like poor bioavailability, liver toxicity which is dose dependent, rapid first-pass metabolism, and very short $t_{½}$.[1] It is quickly absorbed from the digestive tract after oral administration, reaching C_{max} (1.2g/ml) in 0.8h. Additionally, zidovudine is quickly converted to an inactive glucuronide compound, with a 1 h $t_{1/2}$. The concentration of the drug in the systemic circulation must be maintained within the therapeutic efficacy range throughout the treatment period, requiring repeated administration of high doses (200 mg every 4 hours). The drug should be administered in a controlled manner in order to avoid this.

Starches have been used in pharmaceutical dosage forms as excipients for a variety of purposes over the years,[2] as disintegrants,[3] bio-adhesive excipients,[4] binders in wet granulation,[5] in dressings of wounds,[6] in oral disintegrating tablets (ORD),[7] in colon targeting of drugs,[8] and as matrix forming agents (hydrophilic) in oral control release tablets,[9] to name a few. Starches are readily available from a variety of plant sources, renewable, biodegradable, and biocompatible.[10] As a reserve carbohydrate, starch is extensively distributed in a variety of plant species and is highly common in cereal grains, legumes, tubers, and young fruits. Jack fruit seeds (JFS) range in length from 2-4 cm and can make up 8-15% of the fruit's total weight because each fruit can contain 100-500 seeds. All JFS powder variants showed moisture values ranging from 6.28-9.16%, proteins 9.19-11.34%, fats 1.18-1.40%, ash 1.53-2.66%, amylose[11] 49%-30.21%, and starch contents ranging from 81.05%-82.52%[12]. The molecular composition of amylose and amylopectin, as well as their arrangement in starch granules, determine the functional property of starch. The amylose and amylopectin composition (arrangement and structure) plays a major role in determining the starch paste consistency, which depends on the degree of gelatinization and swelling power of the starch granule.

The goal of the present research is to extract starch from the JFS and modify the chemical properties by acetylation, the use of modified extracted starch as a rate-controlling polymer for the release of zidovudine, and drug release kinetics of the formulations are studied.

2. MATERIALS AND METHODS

2.1 Materials

Zidovudine was generously given as a gift sample from the Mylan laboratories, PVPK 30 and spray dried lactose were obtained from Lee Pharma laboratories, JFSs (JFS) are obtained from the local market of Visakhapatnam, ingredients like magnesium stearate, talc, sodium hydroxide, acetic anhydride, pyridine and other chemicals are purchased from an authorized vendor.

2.2. Extraction of starch from JFSs[12]

An alkali extraction procedure was used to extract the starch from the JFSs. Weighing 5g of JFS powder, we added it to a 100ml beaker of 0.1N sodium hydroxide solution. For 6 to 8 hours at room temperature, the contents of the beaker were constantly stirred. The slurry was filtered, and the sediment was washed three times thoroughly with distilled water. At 4°C, the filtrate was precipitated overnight. The supernatant was thrown away, and the precipitate was washed three times with distilled water. Using a tray dryer, the produced precipitate (starch cake) dried for 24h at 40°C. The acquired starch powder was then crushed with a mortar and pestle and kept at room temperature in sealed containers for later usage.

2.3 Phytochemical tests for JFS starch

The JFS starch extract was examined for the presence of phytochemicals like carbohydrates, saponins, alkaloids, proteins, glycosides, and steroids by conducting a number of tests like Mayer's test, biuret test, iodine test, foam test, ferric chloride test, Molisch's test, etc.,

2.4 Evaluation of physicochemical properties of JFS starch

2.4.1 Swelling index (SI). Accurately weigh 1 g of starch extract, it was then transferred to a 100 ml measuring cylinder which has a stopper. After measuring the initial volume, filled with the powder, distilled water was used to adjust the volume to 100 ml. After a gentle shaking, the cylinder was kept undisturbed for 24 hours. After 24 hours, the starch sediment volume was measured. The equation below was used to calculate the % swelling index (SI).

$$SI\ (\%) = [V_t - V_0/V_0] \times 100 \tag{1}$$

V_0 = Initial volume, V_t = Final volume of JFS starch after 24 h

2.5 Preparation of acetylated JFS starch (AJFSS)[13]

25g of JFS starch was weighed and sieved through sieve number # 60. To this 100g of acetic anhydride was added and 200 g of pyridine was used as a medium, this reaction was performed by varying temperatures from 50°C to 150°C for 6h. The final product was precipitated with ethanol, filtered, vacuum dried, passed through sieve no #60, and stored.

2.6 Characterization of AJFSS

2.6.1 Swelling index (SI). This test was performed as mentioned above in section 2.4.1 with AJFSS.

2.6.2 FTIR studies. The FTIR of the AJFSS was determined by using the α FTIR Bruker by KBr pressed pellet method. The sample was scanned for checking interactions and impurities over a wave number ranging from 400 to 4000cm^{-1}. The peaks obtained in the graph were studied with standard values.

2.7 Flow properties of AJFSS

2.7.1 Bulk density (BD). BD is the mass of the powder divided by the volume taken up by loose powder bed[14]. The particle volume, the inter and intra particulate space is considered as the bulk volume. The three-tap method was used to calculate the bulk density. Without compacting, the appropriate quantity of powder was precisely weighed, taken into measuring cylinder (25 ml) and total volume filled by the poured powder was measured. The equation below was used to estimate bulk density. Three replicates of each experiment were carried out.

$$\text{Bulk density} = \frac{\text{mass ofthe powder}}{\text{untapped volume}} \tag{2}$$

2.7.2 Tapped density (TD). TD is the ratio of powder's mass to its volume after a significant number of taps[14]. It is an accurate representation of its dense packing. A measuring cylinder of 25ml capacity was filled with the accurately weighed starch powder, and is kept on the tapped density tester, and tested using the USP II technique, which involves 250 taps/min keeping a height of 3±0.3 mm for 250 tapings. After 500 tapings, the powder bed's volume (V_t) was measured. An additional 750 times tapping was performed and the total volume is expressed as (V_b). Supposing that, the difference observed between V_b and V_t (final and initial) is less than 2%; else, 1250 more taps are performed. The final volume was noted and the following equation was used to determine the tapped density. Triplicates of each experiment were carried out.

$$\text{Tapped density} = \frac{\text{mass of the powder}}{\text{tapped volume}} \tag{3}$$

2.7.3 Angle of repose(AR) (θ). AR is the angle formed between the powder pile and the flat surface[14]. The flow property is improved with a lower AR. The fixed funnel and free-standing cone approach were used to make the determination. A funnel was fixed to the stand at a specific height (h) from the graph paper that was kept on the flat surface, with the end of the stem perpendicular to its axis of symmetry. The powder is gently transferred into the funnel until the edge of the powder pile comes in contact with the funnel's tip. By using the below equation angle of repose can be calculated.

$\tan \theta = h/r$ Or

$\theta = \tan^{-1}(h/r)$ (4)

Where, h = height of the pile, r = radius of the pile, and θ = AR

2.7.4 Compressibility index (CI). This compressibility index[15] can be used for measuring the powder's strength that could potentially form in the arch of a hopper as well as how easily such an arch could be broken. The below equation mentioned is used to measure compressibility index also knows as carrs index.

$$\text{CI}\ (\%) = \frac{\text{Tapped density} - \text{Bulk density}}{\text{Tapped density}} \times 100 \tag{5}$$

*2.7.5 Hausner's ratio(HR).*As a result of its connection to interparticle friction, the HR[15] can be utilized to estimate the characteristics of powder flow. In contrast to more cohesive, less free-flowing powders like flakes, which have Hausner ratios larger than 1.6, coarse spheres and other powders having low interparticle friction show ratios of about 1.2. Given below is the relationship between the Hausner ratio and flow behavior. The formula below is used to calculate it.

$$\text{Hausner's ratio} = \frac{\text{Tapped density}}{\text{Bulk density}} \tag{6}$$

2.8 Construction of calibration curve for zidovudine[16]

2.8.1 Calibration curve of zidovudine with pH 1.2 acid buffer. Stock I: Accurately weighed 25 mg of zidovudine was taken into a volumetric flask (25 ml). To this 5 ml of ethanol was added and dilutions were made with pH 1.2 acid buffer up to the required mark, to produce 1000 µg/ml solution.
Stock II: From stock I, take 10 ml of solution and transfer it into volumetric flask of 100ml, volume was made up to the mark with pH 1.2 acid buffer to produce 100 µg/ml solution.

From stock II, serial dilutions were made by taking 0.2-1 ml and diluting up to the mark in 10 ml volumetric flask with pH 1.2 acid buffer to produce 2 µg/mL-10 µg/ml. Using UV double beam spectrophotometer, the absorbance for each concentration was observed at 267 nm against pH 1.2 buffer as the blank. The experiment was conducted for three times and the mean values were noted. the obtained absorbances are used to construct the calibration curve.

2.8.2 Calibration curve of Zidovudine in pH 6.8 phosphate buffer. Instead of pH 1.2 buffer, pH 6.8 phosphate buffer is used to make stock I, stock II and the serial dilutions. The method was mentioned in above section is used for the construction of calibration curve with pH 6.8 phosphate buffer.

2.9 Formulation of Controlled Release Tablets of Zidovudine

Controlled release tablets (CRT's) of zidovudine were formulated, using AJFSS as a carrier for controlled release. The CRTs are made by using direct compression method. The formulae for preparation of CRT's of Zidovudine was presented in Table 1.

Table 1 *Composition of controlled release tablet of zidovudine.*

Ingredients	**Purpose**	**Formulations**			
		F1(mg)	**F2(mg)**	**F3(mg)**	**F4(mg)**
Zidovudine	Active pharmaceutical ingredient	300	300	300	300
AJFSS	Rate controlling polymer	10	20	30	40
PVPK30	Dry binder	25	25	25	25
Spay dried lactose	Diluent	159	149	139	129
Magnesium stearate	Lubricant	2	2	2	2
Talc	Glidant	4	4	4	4
Total weight		500	500	500	500

2.9.1 Preparation of powdered blend. Ingredients required to make a batch size of 60 tablets based on formulations in Table 2 were passed through

B.S.S # 60 mesh (250mm) the drug and excipients (AJFSS, PVPK-30, spray dried lactose) were mixed by using the geometric dilution method. The lubrication of the blend was done by adding magnesium stearate and talc. The powder blend was evaluated for pre-compression characteristics for their flow characteristics.

2.10 Flow properties of prepared powder blend of formulation

The powder blend formulation was evaluated for their flow properties using the procedure described in section 2.7.

2.11 Compression of CRTs

500 mg powder blend was compressed using a tablet compression machine round of 16 stations, with flat faced punches of 10mm diameter punches to obtain hardness 4-8 kg/cm2. Then the tablets were evaluated for physicochemical properties like general appearance, weight variation, friability, hardness, drug content and dissolution profile.

2.12 Evaluation of CRTs of Zidovudine

General appearance: The appearance, visual identification, and overall elegance of a tablet are essential for acceptance from a customer. This includes tablet size, shape, color & thickness, etc.

2.12.1 Weight variation. In accordance with the Indian Pharmacopoeia, 20 tablets were randomly chosen and weighed both individually and together. The average weight (avg wt), mean(m), and standard deviation (s.d) were calculated.[17]

2.12.2 Friability. The dual chamber friabilator was used to conduct this test (Electro lab, India). 6.5 g of weighed tablets (W_1) were placed in a rotating drum chamber. Then, these tablets were made to fall from 6 inches for 100 times (25 rpm for 4 min). Following this step, the tablets were weighed(W_2) and dusted using a brush. The formula shown below was used to compute the percentage decrease in weight or friability.

$$\text{Friability } (\%) = \frac{W1-W2}{W1} \times 100 \qquad (7)$$

Where, W_1 = Initial weight of Tablets (Before Tumbling) & W_2 = Final Weight of Tablets (After Tumbling or friability)

2.12.3 Thickness. The compressibility of the powder and Hausner's ratio affect the tablet's thickness. The tablet's disintegration and thus, the release, vary depending on its thickness. Therefore, tablet thickness is quite important.

5 tablets were picked at random and average thickness is measured using Vernier callipers.

2.12.4 Hardness. Tablets need to be firm enough to overcome breakage during handling and transportation yet sufficiently soft to break down and disintegrate into granules after swallowing. A Monsanto hardness tester was used to test each tablet's hardness, five tablets were arbitrarily chosen and checked for hardness and the average hardness (kg/cm^2) is calculated.

2.12.5 Drug content. 10 tablets were selected at random from each batch and were pounded in a glass mortar until a fine powder is formed, and then from the resulting powder take equivalent powder of 50 mg zidovudine and add it to a volumetric flask (50 mL). The drug extraction was done using methanol (25ml). The drug and methanol solution is kept on a mechanical shaker with continuous and vigorous shaking for 1 hour, then passed through a 0.45 μm nylon filter disc (Millipore)into volumetric flask (50 mL), and the resultant filtrate was diluted to the desired strength with methanol. Using a UV visible spectrophotometer, additional suitable dilutions were made with pH 6.8 buffer, and the absorbance at 267 nm was evaluated against a blank prepared under the same conditions without the drug.

2.12.6. In vitro Dissolution studies. A reliable *in vitro* dissolution method is a crucial quality control tool for assessing batch to batch release performance and ensuring the physiological availability of the drug. The drug release was verified using a single point dissolution study. For the *in vitro* investigations, the dissolution conditions like stirrer depth of 25mm, RPM of 50 and temperature of 37.0±0.5°C. The dissolution of tablet was done up to 2h in HCl buffer and later up to 24h it was continued in 6.8 pH buffer in basket type dissolution apparatus. For each test, one tablet containing 300 mg of zidovudine was utilised. At varying time intervals, samples of the dissolution fluid (5mL) were taken and passed through a 0.45-micron filter before being assayed for zidovudine using a UV spectrophotometer at 267 nm. Fresh dissolution media was used to replace the sample that was obtained at each sampling interval.

3. RESULTS

3.1 characterization of the JFS starch

The extracted starch from the JFS is characterized by phytochemical and physical properties of the starch.

3.1.1 Phytochemical properties. The characterization of JFS starch is evaluated by performing different phytochemical tests and the chemical composition of JFS starch is determined. Various phytochemical tests on JFSS were performed and results were presented in Table 3.

Table 3 *Phytochemical test results of the JFS starch.*

Test	Observation	Inference
Carbohydrates		
Molisch's test	Violet color appears	Carbohydrates are present
Benedicts test	Sample turns yellow	Reducing sugars are present
Polysaccharides		
Iodine test	Sample turns blue color	Starch is present
Proteins		
Biuret test	Pink color was not observed	Proteins are absent
Alkaloids		
Mayer's test	Creamy precipitate was not absorbed	Alkaloids are absent
Steroids		
Salkowiesk test	Purple or brown color was not observed	Steroids are absent
Liebermann–Burchard test	Green color was not observed	Steroids are absent
Glycosides		
Keller–Killani test	Reddish brown color was not observed	Glycosides are absent
Flavonoids		
Ferric chloride test	Blackish red color was not observed	Flavonoids are absent
Lead acetate test	Yellow color was not observed	Flavonoids are absent
Saponins		
Foam test	Foam was not observed	Saponins are absent

The phytochemical tests confirmed that the alkali extraction of starch powder contains only polysaccharides and there is no presence of proteins, alkaloids, steroids, glycosides, flavonoids, and saponins.

3.1.2 Swelling index. The JFS starch used in the present investigation is evaluated for swelling index. The result obtained is 0% which indicates that the obtained starch doesn't show any swelling property.

3.2 Characterization of AJFSS

The AJFSS was evaluated by using swelling index and FTIR studies

3.2.1 Swelling index. The AJFSS used in present investigation is evaluated for swelling index. The result obtained is 128 which indicates it shows swelling property.

3.2.2 FTIR study. FTIR spectra for AJFSS were obtained and various functional groups were detected. The major peaks observed in AJFSS FTIR spectra and normal starch FTIR spectra were compared. See Figure 1.

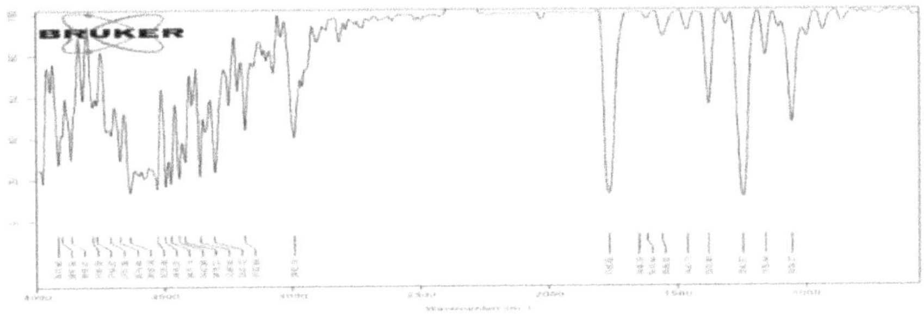

Figure 1 FTIR of the AJFFS.

The IR spectra of AJFSS showed the acetyl carbonyl stretching at 1765 cm^{-1} which was absent in the IR spectrum of normal starch. Carbonyl stretching frequencies, where C-O stretching occurs, appear as two or more bands in the 1300-1000 cm-1 range it indicating the acetylation of the normal starch The hydroxyl group (OH) in normal starch at 3000cm^{-1} was replaced by acetyl carbonyl C=O stretching frequency at 1765cm^{-1} and the C-O stretch appears as two or more bands in the region from 1300 to 1000 cm^{-1}

3.2.3: Evaluation of flow properties of the AJFSS. The BD and TD of AJFSS was found to be around 0.66 ± 0.01 and 0.74 ± 0.01 respectively. The modified extracted starch shows angle of repose value of 27 ± 0.23, carrs index value of 09 ± 0.32% and hausners ratio of 1.10 ± 0.11 which indicates the powder has excellent flow properties. The tests were done in triplicates and the results were given.

3.3 Construction of calibration curve of zidovudine

The calibration curve (Figure 2) of zidovudine was plotted in pH 1.2 buffer and pH 6.8 buffer using UV-spectrophotometry by observing the absorbance at 267 nm. From the concentration ranges of 2-10 µg/mL, the absorbance vs. concentration obeyed Beer's law. There is a good correlation between absorbance and concentration and the coefficient of determination R^2 was found to be 0.999 for both buffers. The absorbance values in different buffers were given in Table 4.

Table 4 Absorbance *of zidovudine in 1.2 HCl buffer and 6.8 phosphate buffer.*

Concentration (µg/mL) (conc)	Absorbance(abs)	
	1.2 HCl buffer	**6.8 phosphate buffer**
2	0.089 ± 0.002	0.085 ± 0.002
4	0.169 ± 0.005	0.170 ± 0.005
6	0.245 ± 0.003	0.249 ± 0.003
8	0.323 ± 0.010	0.330 ± 0.010
10	0.398 ± 0.007	0.400 ± 0.007

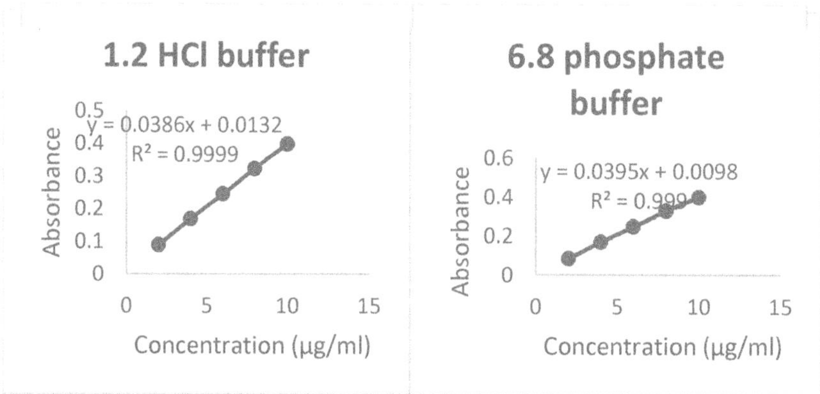

2(a) 2(b)

Figure 2 Calibration curve of zidovudine.

3.4 Evaluation of tablets

3.4.1 Evaluation of flow properties of the powder blend. The powder blend of F1 to F4 was evaluated for BD, TD, AR(θ), % CI and HR. The tests were done in triplicates. All formulation powder blends show AR values in the range of 25±0.12 to 28±0.09, CI value range from10±0.09% to 13±0.11%, and HR value ranges from 1.18±0.02 to 1.2±0.02 which indicates the powder has good flow properties.

3.4.2 Evaluation of controlled release tablet of zidovudine. General appearance of all the formulations was examined, the shape was round, flat beveled edge, surface was smooth, and color was found to be white to off white. The thickness was in the range of 2.48 ± 0.02 to 2.50 ± 0.02 for all the formulations.

The weight variation, hardness, content uniformity, and friability were evaluated. All the tests were done in triplicates and reported in Table 5.

Table 5 *Weight variation, Hardness, Friability, and %Drug content of zidovudine-CRTs.*

Formulation	Weight variation	Hardness	Friability	%Drug content
F1	500.02 ± 0.01	7.02 ± 0.05	0.15 ± 0.01	100.01 ± 0.12
F2	500.03 ± 0.01	7.16 ± 0.04	0.07 ± 0.03	100.03 ± 0.11
F3	500.01 ± 0.01	7.22 ± 0.06	0.05 ± 0.05	100.02 ± 0.08
F4	500.02 ± 0.01	7.18 ± 0.08	0.07 ± 0.05	100.06 ± 0.14

3.4.5 In-vitro drug release studies. The *in-vitro* drug release profiles of all the prepared CR tablets of zidovudine using AJFSS were shown in the Table 6 for all the formulations the study was done in triplicates. See also Figure 3.

Table 6 *% Drug release of prepared zidovudine-CRTs.*

Time (h)	F1	F2	F3	F4
Dissolution media is 1.2 pH HCl buffer				
1	12.16 ± 0.01	10.28 ± 0.21	9.03 ± 0.01	8.9 ± 0.09
2	14.62 ± 0.23	13.65 ± 0.12	12.63 ± 0.03	11.61 ± 0.28
Dissolution media is 6.8 pH phosphate buffer				
3	20.79 ± 0.23	18.76 ± 0.09	16.5 ± 0.06	15.45 ± 0.12
4	26.41 ± 0.11	24.20 ± 0.32	22.4 ±0.09	20.2 ± 0.23
6	32.6 ± 0.03	27.26 ± 0.13	25.33 ± 0.18	24.21 ± 0.22
8	47.68 ± 0.07	32.05 ± 0.17	29.27 ± 0.17	27.42 ± 0.19
10	60.04 ± 0.34	55.7 ± 0.29	45.33 ± 0.11	40.43 ± 0.32
12	68.66 ± 0.29	65.05 ± 0.11	54.6 ± 0.29	46.24 ± 0.19
14	79.8 ± 0.11	72.98 ± 0.25	62.68 ± 0.09	54.37 ± 0.23
16	88.56 ± 0.09	85.87 ± 0.19	70.55 ± 0.17	64.28 ± 0.43
20	98.5 ± 0.13	95.5 ± 0.17	88.5 ± 0.37	78.65 ± 0.11
24			98.76 ± 0.22	88.95 ± 0.09

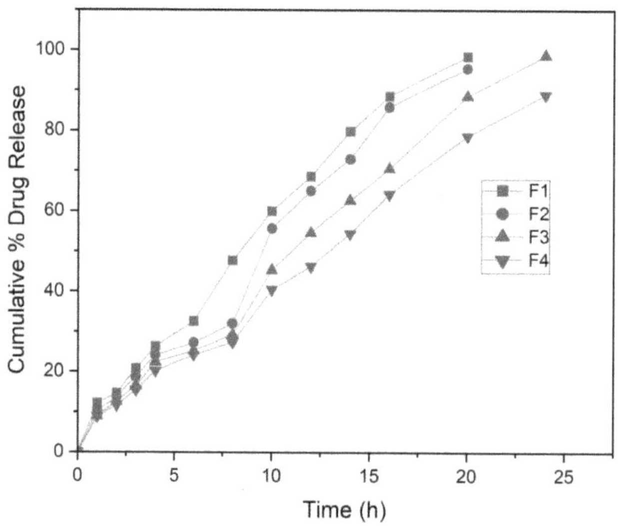

Figure 3 Comparative dissolution profile of zidovudine-CRTs.

The information related to the kinetics of the drug release is given below in graphical representation (Figure 4). All the formulations followed zero order kinetics and non fickian diffusion mechanism. The R^2 value of all the release mechanism of all the formulations are given in the Table 7.

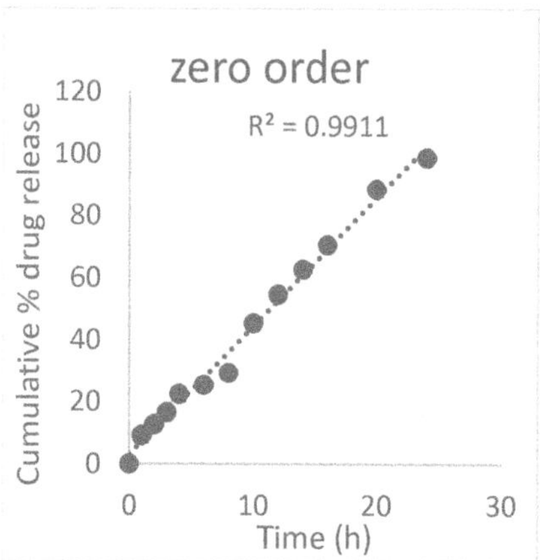

(a)

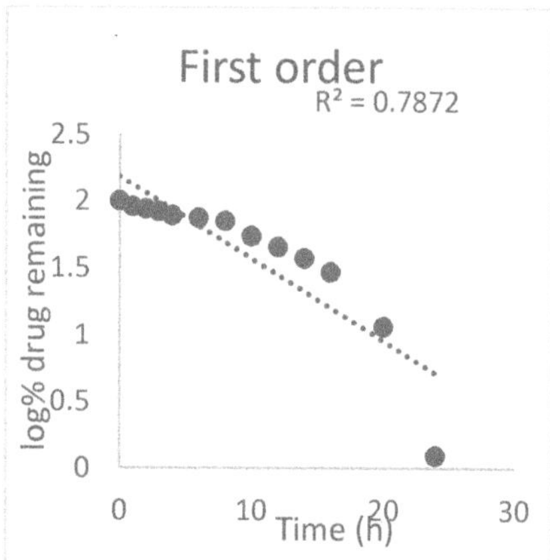

(b)

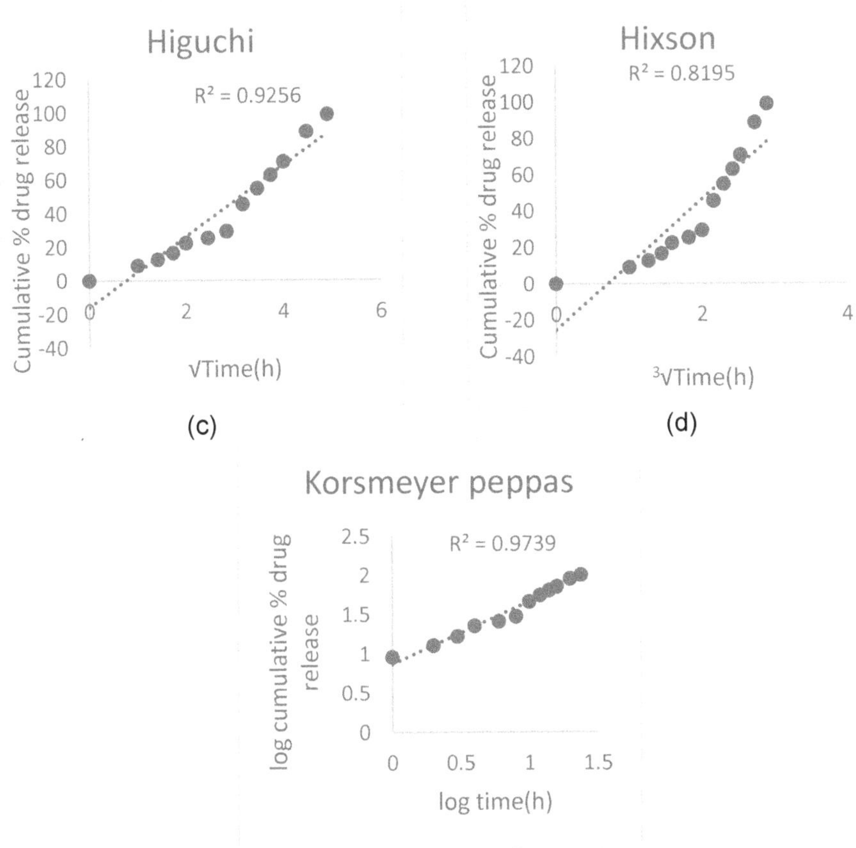

(c) (d)

(e)

Figure 4 Drug release kinetics of zidovudine tablets of F3 formulation (a-e).

Table 7 *Mechanism of Drug Release Studies on Zidovudine CRTs.*

Tablet	Zero-order		First-order		Higuchi	Hixon	Korsmeyer peppas	
	R^2	K_o(%h)	R^2	K_1(hr^{-1})	R^2	R^2	R^2	n
F1	0.987	5.057	0.845	0.172	0.9492	0.850	0.977	0.772
F2	0.980	4.898	0.884	0.135	0.9129	0.805	0.959	0.795
F3	**0.991**	**4.134**	**0.787**	**0.141**	**0.9256**	**0.819**	**0.973**	**0.789**
F4	0.992	3.662	0.922	0.081	0.9267	0.821	0.973	0.760

All formulations show zero order drug release. F1 and F2 shows drug release up to 20 hr, F3 formulation shows 100% drug release up to 24 h and F4 formulation up to 24 h 88% drug only released.

4. CONCLUSION

The starch extracted from jackfruit was stable and free-flowing in nature. In order to develop zidovudine CRTs, the starches derived from JFS were chosen as a rate-controlling polymer. The powder blend has good flow properties, according to the evaluated precompression parameters. All manufactured matrix tablets underwent *in vitro* dissolution tests in USP I (basket) which consists of 900mL of HCl buffer(1.2 pH) for 2 h, followed by 6.8 pH phosphate buffer for up to 24 h. The type of starch used as a rate-controlling agent and the amount of starch was found to have a prominent effect on the dissolution parameters of different formulations, according to the results obtained in dissolution studies. Jackfruit starches that have been modified could be used as a rate-controlling agent. In the F1, F2, F3, and F4 samples, the starch concentrations were 2%, 4%, 6%, and 8%, respectively. All of the formulations' drug release mechanisms were discovered to be non-fickian diffusion processes. The time required for the complete release of drug from the tablets was shown to increase as the starch content in the tablets increased. It was determined that the AJFSS can be efficiently used as a controlled release polymer in the formulation of zidovudine CR tablets based on the current experiment and the acquired results. The drug polymer ratio and the matrix's composition show effect on the drug release. F1 and F2 formulations failed to regulate the release for up to 24 hours. It is due to the controlled release of the Zidovudine for up to 24 hours, formulation F3 was selected as optimized formulation. The optimized formulation was tested using the Peppas model, the Higuchi matrix, the Zero order, and the first order. The results of the kinetic experiments showed that the formulation followed zero order, which means that the zidovudine release is independent of concentration. Goal of the current work was to develop zidovudine 24-h continuous release tablets (CRTs) using AJFSS as the matrix-forming hydrophobic polymer. The tablets regulated release the drug over a 24-h period at higher AJFSS concentrations. The release mechanism revealed anomalous non-fickian diffusion that combined the erosion and diffusion mechanisms.

5. REFERENCES

1. Kieburtz KD, Seidlin M Lambert, JS DR. Extended follow-up of peripheral neuropathy in patients with AIDS and AIDS-related complex treated with dideoxyinosine. J Acquir Immune Defic Syndr. 1992;41:407.
2. S. F. Chin, A. N. B. Romainor, S. C. Pang, and S. Lihan, Antimicrobial starch-citrate hydrogel for potential applications as drug delivery carriers, J. Drug Deliv. Sci. Technol. 54 (2019) 101239. doi: 10.1016/j.jddst.2019.101239.
3. W. Lowenthal and R. A. Burruss, Mechanism of Action of Starch as a Tablet Disintegrant IV: Effect of Medicaments and Disintegrants on Mean Pore Diameter and Porosity, J. Pharm. Sci. 60 (1971) 1325–1332. doi: 10.1002/jps.2600600907.

4. M. Jelkmann, C. Leichner, C. Menzel, V. Kreb, and A. Bernkop-Schnürch, Cationic starch derivatives as mucoadhesive and soluble excipients in drug delivery, Int. J. Pharm. 570 (2019) 118664. doi: 10.1016/j.ijpharm.2019.118664.
5. L. Vandevivere et al., Native starch as in situ binder for continuous twin screw wet granulation, Int. J. Pharm., 571 (2019) 118760. doi: 10.1016/j.ijpharm.2019.118760.
6. S. Batool, Z. Hussain, M. B. K. Niazi, U. Liaqat, and M. Afzal, Biogenic synthesis of silver nanoparticles and evaluation of physical and antimicrobial properties of Ag/PVA/starch nanocomposites hydrogel membranes for wound dressing application, J. Drug Deliv. Sci. Technol. 52 (2019) 403–414. doi: 10.1016/j.jddst.2019.05.016.
7. A. Alalaiwe et al., Application of design of experiment approach for investigating the effect of partially pre-gelatinized starch on critical quality attributes of rapid orally disintegrating tablets, J. Drug Deliv. Sci. Technol. 49 (2019) 227–234. doi: 10.1016/j.jddst.2018.11.018.
8. F. Varum, A. C. Freire, H. M. Fadda, R. Bravo, and A. W. Basit, A dual pH and microbiota-triggered coating (PhloralTM) for fail-safe colonic drug release, Int. J. Pharm. 583 (2020) 119379 doi: 10.1016/j.ijpharm.2020.119379.
9. P. van Aerde and J. P. Remon, In vitro evaluation of modified starches as matrices for sustained release dosage forms, Int. J. Pharm. 45 (1988) 145–152. doi: 10.1016/0378-5173(88)90043-9.
10. A. O. Ashogbon and E. T. Akintayo, Recent trend in the physical and chemical modification of starches from different botanical sources: A review, Starch-Stärke. 66 (2014) 41–57. doi: 10.1002/star.201300106.
11. N. Charoenthai, A. Wickramanayaka, S. Sungthongjeen, and S. Puttipipatkhachorn, Use of cassava starch nanocrystals to make a robust rupturable pulsatile release pellet, J. Drug Deliv. Sci. Technol. 47 (2018) 283–290. doi: 10.1016/j.jddst.2018.07.026.
12. Noor F, Rahman MJ, Mahomud MS, Akter MS, Talukder MA, Ahmed M. Physicochemical properties of flour and extraction of starch from jackfruit seed. International Journal of Nutrition and Food Sciences. 2014;3(4):347.
13. Singh AV, Nath LK. Synthesis and evaluation of physicochemical properties of cross-linked sago starch. International Journal of Biological Macromolecules. 2012 Jan 1;50(1):14-8.
14. United States Pharmacopeia-National Formulary. The official compendia of standards Volume 1. The United States Pharmacopoeial Publication. 2007; p.242, 643.
15. Carr RL. Evaluating flow properties of solids. Chemical Engineering Journal, 1965;72(3):163-168.
16. Analytical profiles of Drug Substances and Excipients- Volume 20, 1991.
17. Indian Pharmacopoeia 2018. Vol. I. The Indian Pharmacopeia Commission; 2018. p.308, 309.

PREPARATION AND CHARACTERIZATION OF MONTELUKAST SODIUM LOADED CUBOSOMAL VESICLES FOR ORAL DELIVERY USING CENTRAL COMPOSITE DESIGN

P.Shailaja, K.Pavani, K.Swathi
Department of Pharmaceutical Technology, A.U College of Pharmaceutical Sciences
Andhra University, Visakhapatnam, 530003, India.
Corresponding author email id: dr.pshailaja@andhrauniversity.edu.in

1.INTRODUCTION

Montelukast sodium is a leukotriene receptor antagonist used to treat asthma and chronic asthma attacks, as well as to relieve the symptoms of seasonal allergies. It is a BCS Class II drug that has low solubility and high permeability. Oral bioavailability is moderate accounting for its low solubility and extensive first pass metabolism.[1]

A drug delivery system is usually a formulation or a device that helps with the inception of a drug or therapeutic agent into the body which increases the drug's efficacy and safety profile by managing the parameters such as rate of drug release or, time taken and site of drug release in the body. In order to achieve controlled release, vesicular systems such as liposomes, cubosomes, and transferosomes are formulated.

Cubosomes are distinct, sub-micron, nanoparticles consisting of a cubic liquid crystalline phase that is bicontinuous and has two individual hydrophilic regions that are separated by the bilayer which are arranged in three-dimensional honeycombed structures that are further divided into two internal aqueous channels. These channels can accommodate drugs, peptides, proteins, and other bioactive substances. Cubosomes can deal with many challenges offered by oral delivery such as poor aqueous solubility, absorption, and large molecular size.

Another application of cubosomes is to encapsulate large proteins that show local action in the gastrointestinal tract. Combined functions of targeting and control release can be achieved by Liquid crystalline nanoparticle technology.

These particles are designed to get released at a controlled pace, which accredits for an effective *in vivo* drug distribution. Drugs, with a narrow absorption window, can be released at different absorption sites, like in the upper or lower intestine for example with the help of cubosomes.[2,3,4]

They are formed by the mixture of phospholipid (glyceryl monooleate) and nonionic surfactant in aqueous media by applying a high-energy dispersion mechanism such as sonication and homogenization.[5,6,7]

2.MATERIALS AND METHODS

Montelukast sodium was obtained as a gift sample from Mylan laboratories, Hyderabad, Glyceryl monooleate was obtained from STEPAN, MAYWOOD, USA. Pluronic F 127 was obtained from BASF, USA.

2.1 Pre formulation studies

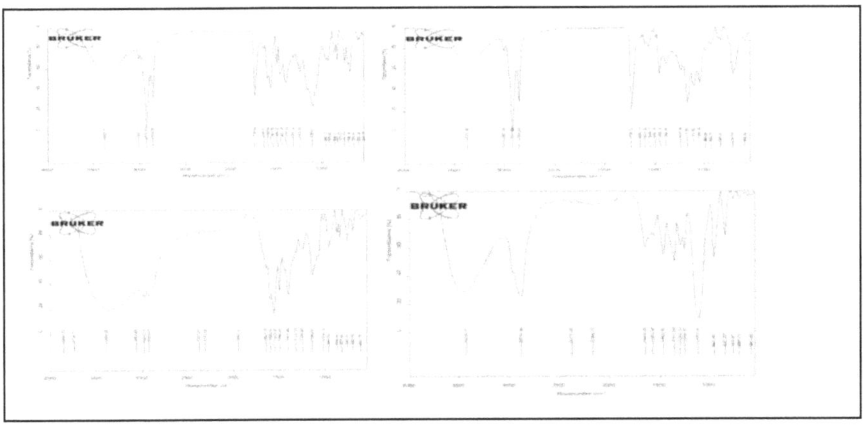

Figure 1 FT-IR spectra of Montelukast sodium, GMO and drug mixture (1:1), PLF 127 and drug mixture (1:1), drug, GMO and PLF 127 Mixture (1:1).

2.1.1 Drug-polymer compatibility studies FTIR spectroscopy.

For the FTIR studies, the samples in this study were crushed tenderly with anhydrous KBr and compacted to form a pellet. 400-4000 cm^{-1} was the scanning range. It was found that there was no interaction between the drug and excipients. See Figure 1.

2.1.2 Preparation of stock for standard curve in phosphate buffer pH 6.8.

Stock SI is prepared by taking10 mg of the drug and made up to the volume of 10 ml with pH 6.8 phosphate buffer. Stock II is prepared by taking 5 ml of stock

I solution and made up to the volume of 50 ml with buffer. From this 0.2 ml, 0.4 ml, 0.6ml, 0.8 ml, 1.0 ml, and 1.2 ml solutions was obtained and made up to 10 ml with same buffer to yield concentrations of 2 µg/ml, 4 µg/ml, 6 µg/ml, 8 µg/ml, 10 µg/ml and 12 µg/ml respectively. The absorbance was measured at 287 nm in UV spectrophotometer.[8] A calibration curve was plotted for the absorbance against the concentration of Montelukast sodium. A regression equation was derived from the plot, which was used for the estimation of Montelukast sodium

2.1.3 Preparation of stock for standard calibration curve in phosphate buffer pH 7.4. Stock I is prepared by taking 10 mg of the drug and made up to 10 ml with phosphate buffer of pH 7.4. Stock II is prepared by taking 5 ml of stock I solution and made up to the volume of 50 ml with buffer. From this 0.2 ml, 0.4 ml, 0.6 ml, 0.8 ml, 1.0 ml, 1.2ml, 1.4 ml solutions were taken and made up to 10 ml with the same buffer to obtain concentrations of 2µg/ml, 4 µg/ml, 6 µg/ml, 8µg/ml, 10 µg/ml, 12 µg/ml, 14µg/ml respectively. The absorbance was noted at 287 nm in UV spectrophotometer. See Figure 2.

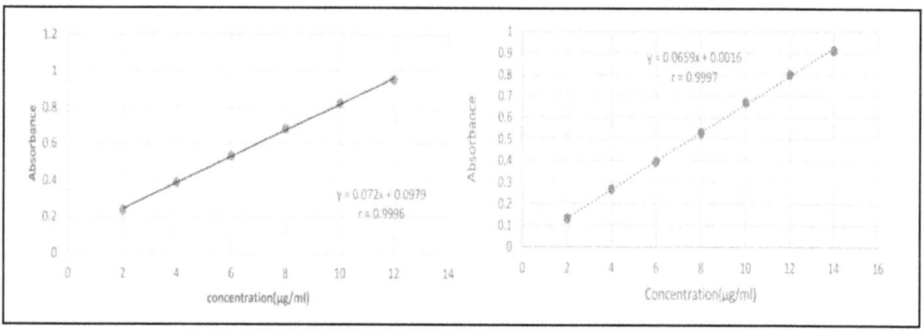

Figure 2 Calibration curve for the estimation of montelukast sodium in Phosphate buffer (pH 7.4 and pH6.8 respectively).

2.2 Experimental design

A Central Composite Design using Design expert software® version 11.0 was used for the development and optimization of cubosomes of Montelukast sodium. Two factors were evaluated at 5 levels (+α, +1, 0, -1, -α) and experimental trials were conducted for the 13 combinations. One face centered value was repeated five times.

Quantity of GMO (X1 or A) = independent variable

PLF 127 (X2 or B) =independent variable

% Drug content (Y1), % Transmittance (Y2), % Entrapment Efficiency (Y3) =Dependant variables

The above variables were taken to optimize the response data.

2.3 PREPARATION OF CUBOSOMES[9,10]

The appropriate amounts of the lipid (GMO) and the stabilizer (PLF 127) were weighed into a crucible. This lipid mixture was melted up to the phase transition temperature of the lipid (70°C). To this molten mixture, the fixed amount of drug Montelukast sodium (10 mg) was added. The heating was continued until a clear solution was formed. This molten mixture was added dropwise with aid of a syringe to a beaker of 10 ml distilled water which was kept for magnetic stirring at RPM of 450-500.

After transferring it, the mixture was left for 10 more minutes for stirring at an RPM of 750-800. Later it was subjected to bath sonication for 10 minutes. Formed cubosomes were subjected to probe sonication for 4 minutes at 2 minutes cycle with a break time of 1 minute. The formed cubosomes were left for 48 hours for stabilization and evaluations were done. The cubosomal suspension batch size of 10 ml was prepared at a time by using the above method.

Table 1 *Formulae and characterization of different cubosomal formulations.*

Formulation code	**Drug (mg)**	**GMO (ml)**	**PLF 127 (mg)**	**Water (up to ml)**	**Drug content* (%)**	**Entrapment efficiency* (%)**	**Transmittance* (%)**
F1	100	27.50	3.00	100	97.33±0.56	93.41±1.55	89.48±0.01
F2	100	27.50	3.00	100	97.18±0.27	**94.17±1.00**	88.72±0.01
F3	100	50.00	3.00	100	98.69±0.09	54.72±1.04	81.95±0.02
F4	100	27.50	1.00	100	93.69±0.49	85.83±0.22	90.45±0.02
F5	100	27.50	5.00	100	97.78±0.66	93.12±1.05	93.56±0.01
F6	100	43.41	4.41	100	**99.30±1.26**	70.15±0,15	84.53±0.01
F7	100	11.59	1.59	100	92.02±0.74	71.87±0.96	**98.26±0.01**
F8	100	43.41	1.59	100	94.45±0.55	75.82±0.60	82.25±0.01
F9	100	27.50	3.00	100	97.63±0.92	92.36±0.29	88.79±0.01
F10	100	27.50	3.00	100	96.27±1.50	87.65±0.98	90.20±0.02
F11	100	27.50	3.00	100	97.48±0.21	93.72±0.77	87.65±0.01
F12	100	11.59	4.41	100	94.75±0.87	78.70±0.20	94.25±0.01
F13	100	5.00	3.00	100	91.11±0.77	51.39±0.53	93.99±0.00

* mean ± s.d(n=3)

2.4 Characterization of formulations[11,12]

The formulation obtained from the Design Expert® Software version11.0 prepared and characterized for the responses drug content (%), entrapment efficiency (%), and transmittance (%).

2.4.1 Zeta potential. Laser Doppler electrophoretic mobility technique was employed in estimating the zeta potential of the cubosomes. For this purpose, Zetasizer 300 HSA (Malvern Instruments) is used at a temperature of 25°C.

2.4.2 Particle size and Poly dispersibility index. Photon Correlation Spectroscopy (PCS) Delsa Nano C (Beckman Coulter Counter, USA) particle size analyzer was used to analyse particle size and poly dispersibility index of cubosomes. Dilution of the samples was done by the suitable aqueous phase. Polystyrene cuvettes were used to analyse the samples, and values were noted down at a fixed degree. Poly dispersity was determined using the following equation:

$$\text{PDI} = \frac{(\text{D90\%} - \text{D10\%})}{\text{D50\%}} \tag{1}$$

Where D90% =particle size just above the sizes of 90% of the nanoparticles,
D50% =particle size just above the sizes of 50% of the nanoparticles,
D10% =to particle size just above the sizes of 10% of the nanoparticles.

2.4.3 Drug content. Take a volumetric flask, and add 1 ml of cubosomal suspension and 7 ml of methanol in order to aid the lysis. It was thoroughly shaken for 10min using a cyclomixer and subjected to magnetic stirring at 600 RPM for 1 hr. Then, filtration was done using micro filters of pore size 22µ. From this, 1 ml of filtrate was taken and made up to the volume of 10 ml with pH 6.8 phosphate buffer. Suitable dilutions were made and absorbance was noted in UV Spectrophotometer at 287 nm.

2.4.4 Entrapment efficiency. Cubosomal suspension of 1 ml was taken in an ependorff tube and was cooling centrifuged (4°c) for 45 min at 7000 RPM. The supernatant was taken and methanol was added to it to aid the lysis of vesicles and estimation of the drug. Then the necessary dilutions were made with phosphate buffer pH 6.8 and filtered using micro filters 22µ. The absorbance was checked in UV Spectrophotometer at 287 nm.

$$EE\ (\%) = \frac{D_T D_E}{D_T} X\ 100 \quad (2)$$

Where EE (%) =percentage entrapment efficiency,
D_T =total amount of drug
D_E =amount of drug entrapped.

2.5 STATISTICAL ANALYSIS:

The above obtained data were subjected to multiple regression analysis using Design Expert software® version 11.0 and were fitted in the equation:

$$Y= ß_0+ ß_1X_1+ ß_2X_2+ ß_{12}\ X_1X_2+ ß_{11}X_1X_1+ ß_{22}X_2X_2$$

Table 2 *Variables.*

Y	Dependent variable
$ß_0$	Arithmetic mean response of the 13 runs
$ß_1$ and $ß_2$	Estimated coefficients for the factors x_1 and x_2
Polynomial terms (X_1X_2 and X_2X_2)	Investigate nonlinearity
The main effect of $X_1\,X_2$	Changing one factor at a time from low to high value yields the best overall result.
Interaction effect of $X_1\,X_2$	Shows how the response changes when two factors are modified at the same time.

The effect shown by the variables was explained by reviewing the absolute magnitude of correlation and the positive or negative sign it carried.

The competency of the obtained fitted model was checked by analysis of variance (ANOVA). Design Expert software® version 11.0 gives response surface plots after feeding the results of the responses which helps to know about the main and interaction effects of the independent variables.

From the above statistical analysis, two formulations were optimized with specified constraints. The formula for optimized suspensions for a batch size of 100 ml was presented in Table 2.

Table 3 *Formulae of optimized cubosomes.*

FC	D	G	P	W	EE			T			DC			REM
					E	Z	R	E	Z	R	E	Z	R	
OF1	100	22.3	2.85	100	88	90	2	88	90	2	93	96	3	
OF2	100	27.6	4.19	100	92	92	0	89	90	1	97	98	1	S

FC- **Formulation code;** D- **Drug(mg);** G-**GMO (ml);** P- **PLF 127(mg);** W- **Water(up to ml);** EE- **Entrapment efficiency (%);** T- **Transmittance (%);** DC- **Drug content (%);** RE- **Relative error (%);** REM-**Remarks**

The optimized formulations of cubosomal suspensions were prepared based on the given formulae and the evaluation tests such as % drug content, % Entrapment Efficiency, and % transmittance was repeated for all the optimized formulations and one formulation was finally optimized by taking into consideration the % relative error of the predicted and the experimental values. The formulation having the least % relative error was taken into consideration.

2.5.4 Overlay plots of optimized cubosomes.

The surface overlay plots of the optimised cubosomes are shown in the below Fig 3.

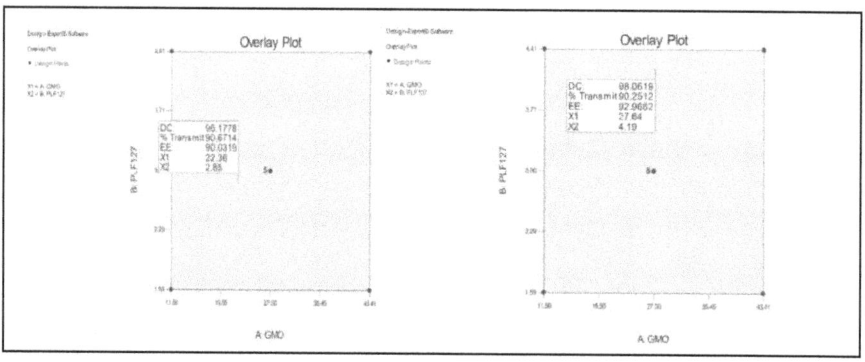

Figure 3 Overlay plots of OF1 and OF2.

2.6 Evaluation of optimized cubosomes

2.6.1 %Drug content. The results of drug content are given in Table 1. OF2 was found to have the maximum drug content of 98.0624.

2.6.2 Entrapment efficiency (%). The results of Entrapment efficiency are given in Table 1. OF2 was found to have the highest entrapment efficiency of 92.956.

2.6.3 Transmittance (%). Its results are tabulated in Table 1. OF2 was found to have the highest transmittance% of 90.2503.

2.6.4 Particle size and poly dispersity index. The observed particle size of the cubosomal suspension (OF2) was within a satisfactory range of 155.45nm. Polydispersibility index values are studied. The PDI of the cubosomes was observed to be 0.204.

2.6.5 Zeta potential. Zeta potential values for the prepared cubosomes were observed to be -25.00mV..

2.6.6 Morphology. The morphology of optimized cubosomal suspension is studied using SEM. Particles showed spherical shape with smooth surfaces. See Figure 4.

Figure 4 SEM image of OF2.

2.6.7 Relative error. By comparing the obtained responses with that of the desired responses, the prediction error was determined by the below equation, and it was less than 5% as shown in Table 3. These lower values of the relative error indicate that experimental values complied with the predicted values proving the predictability and validity of the model and determining the effects of concentration of lipid and poloxamer on responses.

2.6.8 In vitro diffusion studies. (See Table 4) Dialysis bag method was employed to study release of the drug from cubosomes. In order to remove the preservative, 12 hours before using, dialysis bags (HIMEDIA) were soaked in distilled water at room temperature followed by rinsing thoroughly with distilled water. One ml of cubosomal suspension was taken in the dialysis bag such that it gets completely submerged in a 1000 ml beaker containing 900 ml phosphate buffer pH 7.4. By using a twine thread, the bag was closed on both ends and was fully immersed. The beaker was kept on the magnetic stirrer and the working conditions were 100 RPM and 37 ± 0.2°C. At various time intervals, aliquots of the media were removed and replaced with new medium. Release studies were done for pure drug suspension and optimized formulation for a period of 24 hours. The samples were diluted as necessary and absorbance was measured at 287 nm.

Table 4 *In vitro diffusion studies.*

Time (hours)	% Drug release	
	Pure drug	**Cubosomal Formulation**
0.00	0.000±0.000	0.000±0.000
0.08	0.410±0.174	6.009±0.001
0.16	1.639±0.078	7.374±0.005
0.25	2.258±0.058	10.106±0.002
0.33	3.256±0.222	11.471±0.064
0.50	3.874±0.203	15.569±0.112
1	5.657±0.122	21.031±0.568
2	8.894±0.097	25.128±0.405
4	12.421±0.277	30.591±0.245
8	21.214±0.172	46.980±0.006
12	24.897±0.156	66.100±0.062
20	45.287±0.065	92.048±0.096
24	49.623±0.284	98.877±0.035

* mean ± s.d(n=3)

The maximum percentage of drug diffused from the cubosomal suspension in over 24 hrs was found to be which released about 98.877±0.035 and for drug suspension, it was found to be 49.623±0.284 for 24 hrs, High amount of drug diffusion from cubosomal suspension was in coincidence with the high entrapment efficiency of drug in cubic nanoparticles, denoting the importance of the vesicular systems in the diffusion of drug molecules. It was observed that diffusion was enhanced in cubosomal suspension with optimum concentrations of GMO and optimum concentrations of PLF 127 compared to drug suspension, The Correlation coefficient and kinetic model fitting of *in vitro* drug release studies is tabulated in Table 6.

2.6.9 Ex vivo studies. The *ex vivo* studies of the optimized Montelukast sodium loaded cubosomal suspension were performed in a 1000 ml beaker with a receptor cell capacity of 900 ml utilising the pouch method at 37±1°C and it was kept spinning on a magnetic stirrer at 100 RPM. The pouch made by the chicken's intestine was mounted in the beaker containing phosphate buffer of pH 7.4. Cubosomal suspension of 1 ml was placed inside the pouch. Samples were withdrawn from the buffer containing beaker at different time points for 24 hrs and tested for drug content using a UV spectrophotometer. An equal volume of fresh buffer is added to the receptor phase to replace the withdrawn sample. See Table 5.

Table 5 *Ex vivo* diffusion studies.

Time (hours)	% Drug release	
	Pure drug	**Cubosomal Formulation**
0.00	0.000±0.000	0.000±0.000
0.08	0.236±0.022	3.277±0.047
0.16	01.098±0.024	4.643±0.330
0.25	1.215±0.049	7.374±0.057
0.33	3.158±0.124	10.106±0.004
0.50	4.125±0.049	12.837±0.043
1	4.957±0.036	18.300±0.057
2	5.098±0.015	22.397±0.240
4	11.364±0.285	27.860±0.003
8	13.778±0.132	44.248±0.003
12	24.654±0.105	64.734±0.999
20	38.852±0.083	89.317±0.904
24	42.587±0.099	96.145±0.103

* mean ± s.d(n=3)

The maximum percentage of drug diffused from the cubosomal suspension in over 24hrs was found to be which released about 96.146±0.103 and for drug suspension, it was found to be 42.587±0.026 for 24hrs, High amount of drug diffusion from cubosomal suspension was in coincidence with the high entrapment efficiency of drug in cubic nanoparticles, showing the efficiency of the vesicular systems in the diffusion of drug molecules. It was observed that diffusion was enhanced in cubosomal suspension with optimum concentrations of GMO and optimum concentrations of PLF 127 compared to the drug suspension.

2.6.10 Kinetic model fitting. In vitro Kinetic plots of optimized formulation (OF2).

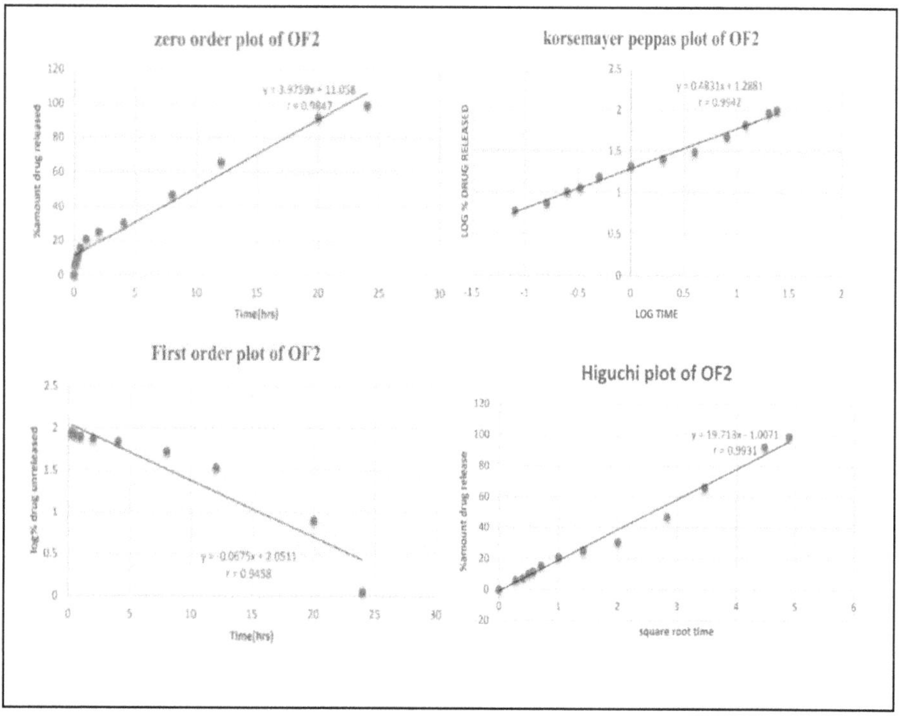

Figure 5 Zero order, First order, Higuchi release, Korsmeyer-Peppas kinetic plots of OF2.

The cubosomal suspension followed zero order kinetics for up to 24 hours. Cubosomal suspension showed controlled release and it follows non fickian diffusion. Cubosomal suspension showed controlled release and from Peppas and Korsmeyer plots the n value was observed to be 0.4831, which indicates that cubosomal suspension follows non fickian diffusion. From plots, Korsmeyer and Peppas plot showed higher r (0.994) which indicates that polymeric system drug release followed both diffusion and erosion-controlled mechanisms. The release of the vesicular system may be controlled by the polymer coating. See Figure 5.

EX VIVO Kinetic plots of optimized formulation (OF2):

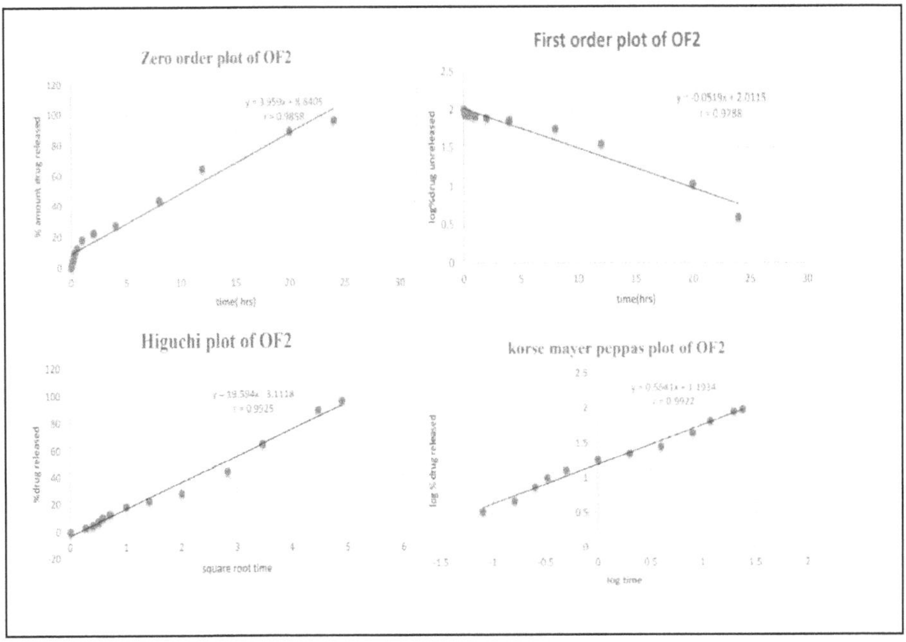

Figure 6 Zero order, First order, Higuchi release, Korsmeyer-Peppas kinetic plots of OF2.

The maximum percentage of drug diffused from the cubosomal suspension in over 24hrs was found to be which released about 96.146±0.103 and for drug suspension it was found to be 42.587±0.026 for 24hrs, High amount of drug diffusion from cubosomal suspension was in coincidence with the high entrapment efficiency of drug in cubic Nano particles, indicating the prominence of the vesicular systems in the diffusion of drug molecules. It was observed that diffusion was enhanced in cubosomal suspension with optimum concentrations of GMO and optimum concentrations of PLF 127 compared to drug suspension. See Figure 6.

Table 6 *Correlation coefficient and kinetic model fitting of in vitro and ex vivo drug release studies.*

IN VITRO						
Zero order		First order		Higuchi	Korsmeyer and Peppas	
r	K_0	r	K_1	r	r	n
0.9847	0.045	0.9458	0.037	0.9931	**0.9942**	0.4831
EX VIVO						
0.9858	0.045	0.9788	0.0281	0.9925	0.9922	0.5641

2.6.11 Calculation of flux. For calculating the *in vitro* flux of Montelukast sodium containing cubosomes by plotting a graph of time versus active compound permeation through the skin and dividing the resultant slope by the area of the skin surface through which diffusion occurred the flux ranging from 0.5µg/cm ^{2} h^{-1} indicating a noticeable permeation of drug. The flux of cubosomal suspensions was found to be 7.2125±0.220 µg/cm ^{2} h^{-1}.

3.DISCUSSION

The present work is focused on development of montelukast sodium loaded cubosomal suspension which serves the purpose of increasing solubility by the use of GMO and PLF 127. The lipidic delivery systems which help in lymphatic absorption also support the fact that of surpassing first pass metabolism. To the compatibility between the drugs and various components of formulation, FT-IR studies were performed. There was no shift in the major peaks of montelukast sodium confirming the drug-excipient compatibility. Using Design Expert Software® version 11.0 various proportions of the GMO and PLF 127 was selected. Montelukast sodium loaded cubosomes were formulated and evaluated. They were evaluated for % transmittance, drug content and entrapment efficiency and analyzed by Design Expert Software® version 11.0. Finally, optimized cubosomal suspension formulation was subjected to in vitro diffusion studies, kinetic fitting model and ex vivo diffusion studies.

The cubosomal suspension consisting of GMO 27.64mL and PLF 127 4.19mg are the optimal concentrations required in formulating an effective cubosomal formulation with desired responses. Thus, we could conclude that cubosomes can be used as efficient means of formulation to increase the solubility and permeability of the BCS II drug, thereby enhancing its bioavailability and controlling the release of drug for 24 hrs.

4.CONCLUSION

Cubosomes can be used as efficient means of formulation to increase the solubility and permeability of the BCS II drug, thereby enhancing its bioavailability and controlling the release of the drug for 24 hrs.

As the suspension releases the drug for up to 24 hrs, repeated administration can be avoided and patient compliance can be enhanced. The solubility and permeation of the drug are being enhanced. Thus, in this research work, the formulation technique used was effective and reproducible.

The developed cubosomal formulations of the anti-asthmatic drug, montelukast sodium were safe and effective for controlling the release of the drug.

5.CONFLICTS OF INTEREST

The authors declared no conflict of interest.

6 REFERENCES

1. https://go.drugbank.com/drugs/DB00471
2. Tilekar K, Khade P, Kakade S, Kotwal S, Patil R. CUBOSOMES-A DRUG DELIVERY SYSTEM. International Journal of Pharmaceutical, Chemical & Biological Sciences. 2014 Oct 1;4(4). 6. Gopal Garg, Shailendra Saraf; Cubosomes: An Overview; Biological and Pharmaceutical Bulletin; 2007; 30; 350-353.
3. Madhurilatha Thadanki: Overview of Cubosomes: A Nano particle; international Journal of Research in Pharmacy and chemistry; 2011; 1(3).
4. Karami Z, Hamidi M. Cubosomes: remarkable drug delivery potential. Drug discovery today. 2016 May 1;21(5):789-801.
5. Yaghmur A, Laggner P, Almgren M, Rappolt M. Self-assembly in monoelaidin aqueous dispersions: direct vesicles to cubosomes transition. PLoS One. 2008 Nov 18;3(11):e3747.
6. Jyoti Kailas Bachhav, Bushan A. Bhairav; Cubosomal Technique: A Review; International Journal of Pharmacy and Technology; 2015; 7 (2); 3263-3272.
7. Spicer PT, Hayden KL, Lynch ML, Ofori-Boateng A, Burns JL. Novel process for producing cubic liquid crystalline nanoparticles (cubosomes). Langmuir. 2001 Sep 18;17(19):5748-56.
8. Lai J, Chen J, Lu Y, Sun J, Hu F, Yin Z, Wu W. Glyceryl monooleate/Poloxamer 407 cubic nanoparticles as oral drug delivery systems: I. In vitro evaluation and enhanced oral bioavailability of the poorly water-soluble drug simvastatin. AAPS Pharm SciTech. 2009 Sep 1;10(3):960.

9. Elnaggar YS, Etman SM, Abdelmonsif DA, Abdallah OY. Novel piperine-loaded Tweenintegrated monoolein cubosomes as brain-targeted oral nanomedicine in Alzheimer's disease: pharmaceutical, biological, and toxicological studies. International journal of nanomedicine. 2015;10:5459.
10. Sharma R, Kaur G, Kapoor DN. Fluconazole loaded cubosomal vesicles for topical delivery. Int J Drug Dev Res. 2015;7(1):32-41.
11. Shaikh HK, Kshirsagar RV, Patil SG. Mathematical models for drug release characterization: a review. World J Pharm Pharm Sci. 2015 Jan 26;4(4):324-38.
12. Siepmann J, Peppas NA. Higuchi equation: derivation, applications, use and misuse. International journal of pharmaceutics. 2011 Oct 10;418(1):6-12.

DEVELOPMENT AND OPTIMIZATION OF NIOSOMES FOR BRAIN TARGETING USING RESPONSE SURFACE METHODOLOGY

A.Veenadevi [a], P. Srinivasa Babu [a], Pallavi Kurra [a*]

Vignan Pharmacy College, Vadlamudi, Guntur, AP, India

[*]Corresponding author email id: pallavi1203phd@gmail.com

1. INTRODUCTION

The brain delivery of a drug is a versatile task which was tremendously challenging due to the existence of several barriers in the brain. Blood Brain Barrier (BBB) is a structural & biological barrier that guard the brain from harmful substances that enter into the brain. Endothelial cells existing in the brain are categorized by the manifestation of fenestrations, widespread tight junctions & these junctions limit Paracellular transport of hydrophilic drug molecule across barrier with spare pinocytic vesicular transport was occur within the brain cells.[1] Abundant scientific methodologies was employed to disrupt the these tight junction like BBB disruption, cell penetrating peptides, receptor mediated transport, and targeted delivery using prodrugs approach, have been considered for brain targeting. Considering the exclusive connecting the nose and CNS, the intranasal route has gained popularity in transporting therapeutic agents along the olfactory nerve cells through nose to the brain avoiding the BBB. Absorption of drug through the olfactory region of the nose delivers a distinctive feature and superior option for target of drugs to brain. Additionally, intranasal administration is a noninvasive technique of drug delivery that permits therapeutic materials directly to CNS.

Treatment of various pathological diseases such as glioblastoma and neurodegenerative diseases nanostructures was consider depending upon the biomaterial and morphology of the drug-delivery system. Nanoparticles can be prepared from polymeric substances, metals, some other formulation include nanogel, colloidal systems, and particular or vesicular systems was employed for targeting of drugs into brain. Vesicular drug delivery systems comprise of shielding liposomes, ethosomes, transfersomes, bilosomes, and niosomes. Among those vesicular system niosomes was significantly deliver the drug molecule passing through tissues via blood vessels. [3] Niosomes was fabricated by using various non-ionic surfactants with bilayered structures which can entrap both hydrophilic as well as lipophilic drugs molecule with in their structure and retains the drug concentration for longer time period for directing their therapeutic action at the target site.[4] Capecitabine is a chemotherapeutic agent with antineoplastic action consumed orally for the treatment of colorectal & metastatic breast cancer. It has a clinically proven efficacy against brain metastases. However, the role of capecitabine in treating CNS disorders is unexplored. Hence, an attempt was made to develop a niosomal formulation which can target brain effectively when administered through intranasal route.[5]

2 MATERALS AND EXPERIMENTAL METHODS

2.1 Material

Capecitabine was obtained from the Hetero pharmaceuticals, Hyderabad as a gift source. Chloroform and Isopropyl alcohol were purchased from Loba chem pvt Ltd Mumbai. Cholesterol were from nice chemicals pvt ltd Kochi. Spans, Brij, Tweens, Diacetyl phosphate was obtained from NR chemicals Mumbai. All chemicals and reagents available were analytical grade used in the research work.

2.2 Method

2.2.1 Optimization employing experimental design Niosomes was prepared by the Thin Film Hydration technique by using Design Expert® software version 11 (Stat Ease Inc, USA). In the present study was performed by using Central Composite Design which a part of surface response method to formulate and optimize the niosomal formulation. In this method employing two independent variables like Span-80 & Brij-78 concentration as two factors and surfactant concentration was measured at two levels at low, high concentration ranges and 13 runs were selected for optimization study. Two dependent variables like Drug content and In-vitro drug release as two responses parameters as shown in the Table 1.

Table 1 *CCD Design that representing the input variables used for the formulation of niosomes.*

Name	**Units**	**Type**	**Low**	**High**
Span-80	Mg	Factor	15	300
Brij-78	Mg	Factor	15	300
Drug content	Mg	Response		
In-vitro drug release	%	Response		

2.2.2 Preparation of drug-loaded Niosomes. The accurate amount of Non-ionic surfactant cholesterol & drug into cleaned Round Bottomed Flask (RBF) according to the design as mention in Table 2, & then add 5ml of chloroform to dissolve the contents. The chloroform evaporated was by using a rotary evaporator at 60°C. Thin-film was developed inside the RBF & further hydrate the film with phosphate buffer of pH 7.4 for the time period of 30 minutes and rest for 24 hours at room temperature conditions. The developed niosomal suspension was sonicated for 10 minutes in Ultrasonic Homogenizer (Probe sonicator) for particle size reduction. Niosomes were stored at 4° C for further evaluation. [5, 6]

Table 2 *Formulation table for Niosomes by using design expert software V11.*

Runs	**Factor -1 (span-80) (mg)**	**Factor-2 (Brij -78) (mg)**	**Capecitabine (mg)**	**Cholesterol (mg)**	**Diacetyl phosphate (mg)**
1	157.5	359.025	150	50	10
2	15	15	150	50	10
3	157.5	157.5	150	50	10
4	157.5	157.5	150	50	10
5	15	300	150	50	10
6	300	300	150	50	10
7	0	157.5	150	50	10

8	157.5	0	150	50	10
9	157.5	157.5	150	50	10
10	300	15	150	50	10
11	157.5	157.5	150	50	10
12	157.5	157.5	150	50	10
13	359.025	157.5	150	50	10

2.3 Characterization of niosomes

2.3.1 Microscopic examination. Freshly formulated niosomes was placed on the glass slide and the normal size of vesicles was calculated using optical microscope at a magnifying lens of 45x.

2.3.2 Particle size & particle size distribution. Niosomal preparation was diluted to 100 times to get a transparent solution by using pH 7.4 phosphate buffer for determination of particle size using particle size analyzer (HORIBA Scientific SZ-100) version 2.20.

2.3.3 Zeta-potential measurement. Zeta potential also plays a significant role in vesicle stability as well as the charge on the vesicle also plays a key role in the prediction of in-vivo performance. Further the Zeta potential designates the repulsion forces among the neighboring vesicles nor the charged particles within the niosomal suspension.[7] The diluted niosomal samples was analyzed by using HORIBA Scientific SZ-100) version 2.20 analyzer for the charged species that are developed on the surface of niosomal vesicles.

2.3.4 Drug entrapment efficiency. Entrapment efficiency of drug with in the niosomes was governed by using centrifugation method for the determination of unentrapped drug. 1ml of Niosomal preparation was centrifuge at 1400 rpm for 60 minutes and measure the absorbance of unentrapped drug with in the supernatant was measured at a wavelength of 240 nm by using UV - visible spectrophotometry (ELICO® Double beam SL210).[8]

$$\%\, Drug\ entrapment = \frac{Total\ drug - drug\ in\ supernatent\ liquid}{Total\ drug} * 100 \qquad (1)$$

2.3.5 Drug content. The amount of drug with in the niosomal formulation was governed by the breakdown of vesicle using 50% Isopropyl alcohol. 1ml of formulation was pipette out & further add isopropyl alcohol and shake well for lysis of Niosomes. Further dilute with pH 7.4 phosphate buffer & measure the absorbance at the wavelength of 240 nm using UV - visible spectrophotometry.[9]

2.3.6 In-Vitro drug release. Dissolution studies was carried by using the USP Type II dissolution apparatus (ELECTROLAB TDT-08L) & 1ml of Niosomal preparation was placed inside the cellophane membrane. This membrane was tied to the paddle which is rotated at the speed of 50 rpm, the temperature was maintained at 37°C ± 1 °C & pH 7.4 phosphate buffer as a dissolution media at the volume of 900ml. At regular time intervals, 5ml samples were collected up to 12 hours & replaced with a 5ml of fresh buffer. The sample absorbance was measured by using UV - visible spectrophotometry at 240 nm.

2.3.7 Kinetics studies. Kinetic models were predominantly termed as the drug discharge from the dosage form. The various mathematical model analyzed were:

$$Zero\ order: C = C_0 - Kt \quad (2)$$

$$First\ order: Log\ C = LogC_0 - \frac{Kt}{2.303} \quad (3)$$

$$Higuchi\ model: C = Kt^{1/2} \quad (4)$$

$$Peppas\ model: C = Kt^n \quad (5)$$

The terms mostly used in the kinetic studies in which C is the concentration at a time interval t, C_0 will be the initial concentration of the drug, K will be the rate constant and n might be exponential drug release.

3 RESULTS & DISCUSSION

3.1 Optimization of Niosomal formulation

Optimization using central composite design for two factors, two levels offer less offer fewer experimental runs of (13 run). Experimental runs and their responses observed were given in Table 3.

Table 3 *Optimization table of Niosomes by using design expert software V11.*

Runs	**Factor-1 (span-80)**	**Factor-2 (Brij -78)**	**Response-1 (Drug content)**	**Response-2 (In-vitro drug release) after 12 hours**
1	157.5	359.025	8.98	67.4

2	15	15	6.87	83.025
3	157.5	157.5	8.84	72.11
4	157.5	157.5	8.82	71.79
5	15	300	7.62	76.49
6	300	300	9.6	62.56
7	0	157.5	7.29	81.56
8	157.5	0	8.38	74.05
9	157.5	157.5	8.79	71.74
10	300	15	9.2	57.01
11	157.5	157.5	8.74	70.79
12	157.5	157.5	8.69	70.64
13	359.025	157.5	9.89	64.15

3.1.1 Effect of surfactant on the drug content of the formulation. The ANOVA results may revealed that the Drug content followed the Quadratic model represent in Table 4 indicate that the model was significant. The response parameters selected based on the input variables concentration influence was estimated using ANOVA methods. The **Model F-value** of 250.08 suggests that the model was significant. As per literature, now only 0.01% chance occur whereas F-value might be large due to noise. **P-values** less than 0.0500 designate that the model terms are significant. In this case, A, B, A^2, B^2 are significant model terms as illustrated in Table 4. Values greater than 0.1000 specified that the model terms are not significant. The **Lack of Fit F-value** of 3.17 for response factor i.e. drug content suggests that the Lack of Fit was not significant proportional to the pure error representing the positive results for fit statistics. The Lack of Fit F-value of 14.07% chance & this might be large due to noise.

Table 4 *ANOVA for Quadratic model representing drug content response variable.*

Source	**Sum of Squares**	**Df**	**Mean Square**	**F-value**	**p-value**	
Model	9.01	5	1.80	250.08	< 0.0001	Significant
A-span-80	8.30	1	8.30	1151.86	< 0.0001	
B-Brij - 78	0.5711	1	0.5711	79.30	< 0.0001	
AB	0.0306	1	0.0306	4.25	0.0781	
A^2	0.3746	1	0.3746	52.02	0.0002	
B^2	0.1007	1	0.1007	13.98	0.0073	
Residual	0.0504	7	0.0072			
Lack of Fit	0.0355	3	0.0118	3.17	0.1470	Not significant
Pure Error	0.0149	4	0.0037			
Cor Total	9.06	12				

Polynomial equation related to response parameter drug content was given below:

Drug content = + 8.79 + 1.10 A + 0.2878 B - 0.0875 AB - 0.2674 A^2 - 0.1386 B^2 (6)

Surface response method gives the relationship between surfactant concentration & drug content was represented by graphical methods was illustrated in (Figure 1)

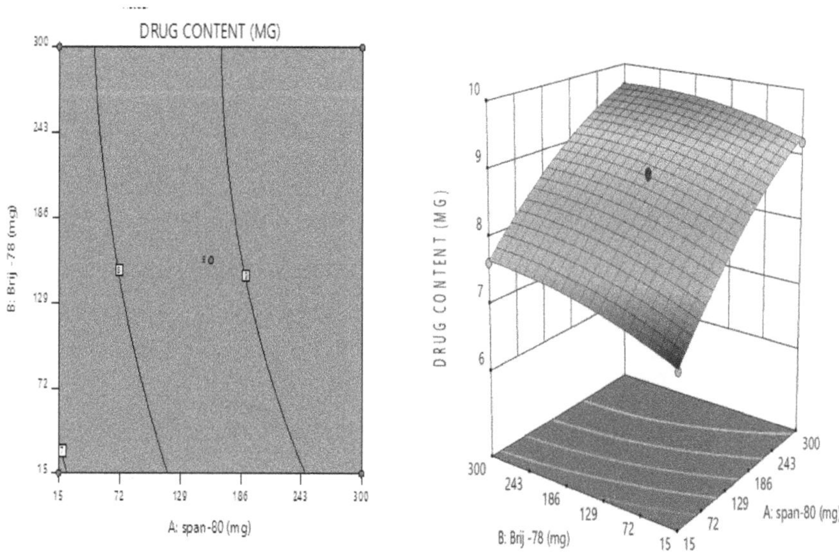

(a) (b)

Figure 1 Representing the (a) Contour plots (b) 3D plot of drug content variable on surfactant concentration.

3.1.2 Effect of surfactant on the in-vitro drug release of the formulation. The results of ANOVA for the In-vitro drug release followed 2FI model represent in Table 5 designate that the model was significant. The **Model F-value** of 236.43 suggested that the model was significant. As per literature, there is only a 0.01% chance that an F-value this large might be due to noise. The **P-values** less than 0.0500 indicate that the model terms are significant. In this case, A, B, AB are significant model terms designated in the Table 5. Further the values greater than 0.1000 suggested that the model terms are not significant. The **Lack of Fit F-value** of 2.84 denoted that the Lack of Fit is not significant comparative to the pure error. Lack of Fit F-value was large i.e. 16.68% chance that must be large due to noise.

Table 5 *ANOVA for 2FI model representing In-vitro drug release response parameter.*

Source	Sum of Squares	Df	Mean Square	F-value	p-value	
Model	617.13	3	205.71	236.43	< 0.0001	Significant
A-span-80	572.85	1	572.85	658.40	< 0.0001	
B-Brij - 78	38.47	1	38.47	44.22	< 0.0001	
AB	6.11	1	6.11	7.03	0.0264	
Residual	7.83	9	0.8701			
Lack of Fit	6.11	5	1.22	2.84	0.1668	Not significant
Pure Error	1.72	4	0.4301			
Cor Total	624.96	12				

Polynomial equation related to response parameter In-vitro drug release was given below:

In-vitro drug release = + 71.29 – 8.91 A – 2.31 B +1.24 AB (7)

Surface response method gives the relationship between surfactant concentration & In-vitro drug release was represented by graphical methods was illustrated in (Figure 2)

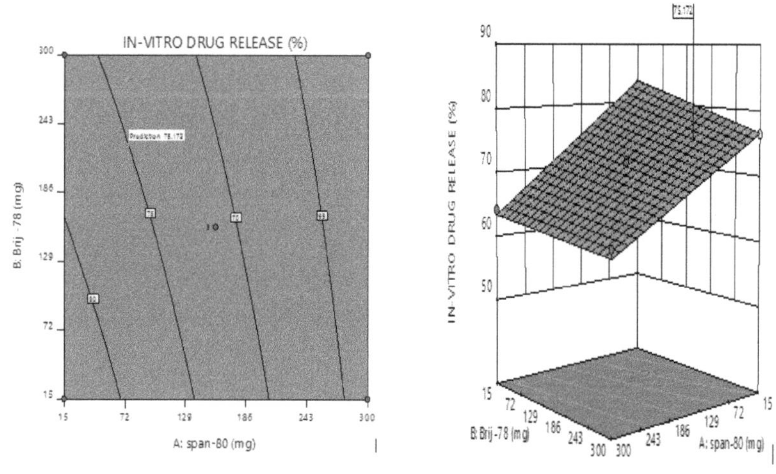

(a) (b)

Figure 2 representing the (a) Contour plots (b) 3D plot of In-vitro drug release variable on surfactant concentration.

3.2 Characterization of niosomes

3.2.1 Microscopic examination. Microscopic examination of the niosomal formulation F5 & F7 was observed under the magnifying lens of 10X microscope vesicle formation was clustery with fused vesicle formation. F2 formulation vesicle formation was not observed. F1, F3, F4, F6, F8, F9, F10, F11, F12 niosomal formulation vesicles were formed with a round shape with smooth surface & discrete structure in larger amounts without any aggregation or agglomeration.

3.2.2 Particle size & particle size distribution. Particle size ranges of niosomes were varied from 0.5976 to 7.0897 µm. F13 formulation has large vesicle of 7.08 µm whereas optimized formulation with vesicle size of 4.07 µm as shown in Table 6. HLB value having major influence in the vesicular size and their distribution due to the following reason i.e. The HLB value moves towards the hydrophilic region then the water uptake into the surfactant increases due to increased surface energy & these consequence lead to larger size of vesicles.[10]

3.2.3 Zeta-potential measurement. Zeta potential values were varied from the -0.1 to -4.2 mV. Highest Zeta- potential value obtained was – 4.2 mV and optimized formulation has -2.1 Mv as presented in Table 6. Negative charge established on the surface of the vesicle properly due to accumulation of diacetyl phosphate with in the formulation. Negative values increases due to subsequent reason i.e. HLB values moves towards the hydrophilic region tend to increase the surface energy may leads to an increase in the values of zeta potential towards negative.[11, 12]

3.2.4 Drug entrapment efficiency. HLB value was low for span-80 and high value for Brij-78 also influence the drug entrapment efficiency. Drug entrapment efficiency values were varied from 63.63 % to 99.85% as displayed in Table 6 Optimized formulation having drug entrapment efficiency of 94.82%. Drug entrapment efficiency reduced due following reason i.e. Span-80 and Brij-78 which is a non-ionic surfactant concentration increases, % drug entrapment also increases. Surfactant chain length with alkyl functional group has noticeable effect on permeability whereas surfactant chain length increases % EE also rises similarly chain length reduces % EE also diminished.[13] Although Bayindir create a direct relationship between HLB value and drug entrapment efficiency in which low HLB value of non - ionic surfactant yield higher drug entrapment efficiency. [14] It was not the case span 80 non-ionic surfactant due to the existence of unsaturated double bound in span-80 doesn't permit the rotation of the alkyl chain and further leads to permeability of thin film increases causes reduction of % EE. As a result, it prevents the tight packing of the noisome membrane and decreases its capacity to entrap a higher amount of drug. Also, a low degree of hydration that was detected with span -80 noisome has a direct influence to their low entrapment efficiency. [15, 16]

Table 6 *Evaluation parameter of formulated Niosomes.*

Formulation code	**Particle size (Mean vesicular diameter (μm) ± S.D**	**Zeta potential (mV) ± S.D**	**Drug entrapment (%) ± S.D**
F1	1.0419 ± 0.13	- 4.2 ± 0.45	99.85 ± 0.12
F2	0.5978 ± 0.25	- 0.1 ± 0.31	70.13 ± 0.21
F3	1.0262 ± 0.54	- 0.70 ± 0.74	87.28 ± 0.31
F4	1.0325 ± 0.21	-0.62 ± 0.52	83.33 ± 0.45
F5	1.9875 ± 0.21	-3.5 ± 0.14	63.63 ± 0.25
F6	2.7779 ± 0.31	- 1.7 ± 0.56	86.37 ± 0.31
F7	1.2356 ± 0.25	-2.8 ± 0.25	93.69 ± 0.45
F8	3.8808 ± 0.34	- 0.9 ± 0.25	90.96 ± 0.54

F9	1.0897 ± 0.24	-0.59 ± 0.45	86.37 ± 0.65
F10	4.0799 ± 0.54	- 2.1 ± 0.54	94.82 ± 0.61
F11	1.07965 ± 0.26	-0.67 ± 0.24	85.67 ± 0.56
F12	1.0523 ± 0.85	-0.72 ± 0.45	84.92 ± 0.95
F13	7.0897 ± 0.45	-0.4 ± 0.23	93.01 ± 0.15

3.2.5 Drug content. Drug content values varied from 6.87 to 9.87. As the surfactant concentration increases drug content values also increase from the results obtained from the optimization of 3D graph, fit statistics and various polynomial equations.

3.2.6 In-vitro drug release. In-vitro drug release values ranging from 57.01% to 83.025% at a period of 12 hours. As the surfactant concentration increases in-vitro drug release decreases. The order of In-vitro drug release from the niosomal formulation was F2> F7> F5> F8> F3> F4 >F9 > F11> F12 > F1 > F13 > F6> F10. F10 formulation was best formulation which was more stable compared other formulation showing low drug from niosomal vesicles was 57.01% for a period of 12 hours' time intervals. From the 3D plot span-80 has major influence than brij-78 non-ionic surfactant on in-vitro drug release due the following reasons. Eventhough the span-80 has major influence which shows increased drug release from niosomal formulation due to following reason i.e., Span -80 having HLB value of 4.3 and (C_{18}) alkyl chain length with double bond in their structure causes drug leaking out of the niosomal vesicle increases & hence release increases. The low transition temperature of span-80 increased drug release. [17] Brij-78 niosome with the highest HLB value of 15.3 which is hydrophilic in nature shows better drug release from the formulation eventhough it has low drug entrapment efficiency.[18] Comparative graph of all formulation was given below in the Graph 1.

Graph 1 *Comparative In-vitro dissolution profile of niosomal formulation.*

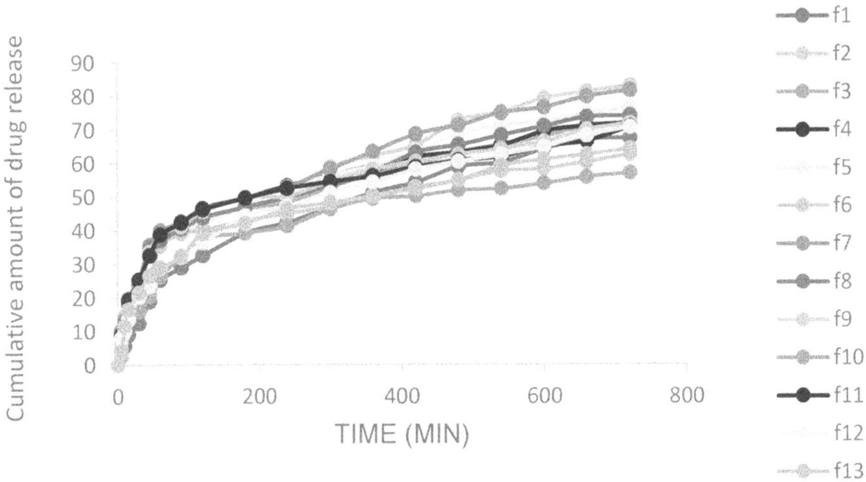

3.2.7 Kinetics studies of drug release data. Kinetic of optimized niosomal formulation was represented in the Table 7. The drug release data was fitted into various mathematical model term for the estimation of R^2 value i.e. regression co-efficient with highest value was selected for getting better results. The optimized F10 with R^2 value of 0.9803 was highest compared to other models follows first order kinetics. First order kinetics was dose dependent process in which drug release from the dosage form depends on the surfactant concentration. Korsmeyer Peppas model with release exponent (n) value of 0.0564 for the optimized Niosomal formulation follows a Non-Fickian type of drug transport mechanism or anomalous diffusion process.

Table 7 *Curve fitting results of optimized Niosomes loaded with Capecitabine.*

Optimized Formulation code	**R^2 value of optimized formulation**				**Release exponent 'n' value**
	Zero order	***First order***	***Higuchi model***	***Peppas model***	
F 10	0.9456	0.9803	0.9517	0.728	0.0564

3.2.8 Simulation studies. Simulation studies were performed using PK Sim® with the construction of physiological based pharmacokinetic models of both population & human individuals as well as large animal species.[21] In the current study, building blocks were generated using standard European humans as a selected individual. Input in terms of physicochemical properties i.e. molecular weight of 359.35 g/mole, the solubility of water 26 mg/ml, fraction unbound to albumin of 65%, Pka value of 8.3, log P value of 0.83 present in the literature for Capecitabine were given to simulation software. Further, generation of formulation & administration protocol for simple molecule into simulation study of individual humans yielded a comparative profile of simulated & experimental data as illustrated in (Figure 3). The results conclude thatthe simulated PK parameter values were in correlation with bibliographic values. The obtained PK parameter values which were within the acceptable bibliographicranges indicate that the simulation model was significant.

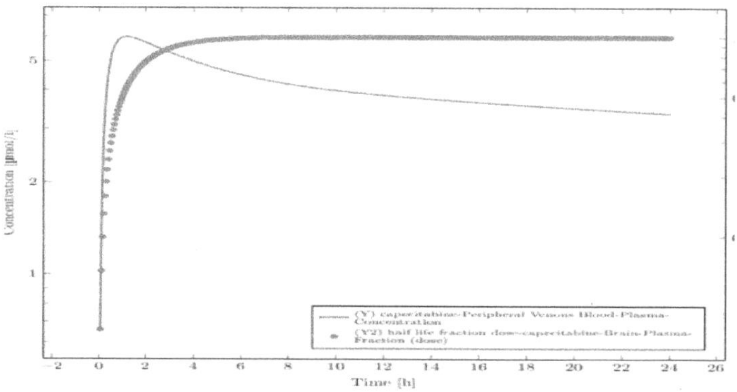

Figure 3 Comparative plasma concentration time profile of experimental and simulated data.

4. CONCLUSION

The current study achieved optimized niosomal formulation which is anticipated to target brain by crossing BBB when administered through intranasal route. The influence of varying concentration of non-ionic surfactants analyzed using Central Composite Design have reported a significant effect on drug content and invitro drug release. The optimized formulation exhibited a vesicle size of 4.07 µm, zeta potential of -2.1 mV and a drug entrapment efficiency of 94.8% indicating the increased efficacy of the formulation to target and exhibit sustained release in the brain region. The study also portrays the scope of exploring the potential of optimized formulation in various CNS disorders. Simulation studies performed using PBPK modeling of PK Sim software aided in extrapolating the *invitro* drug release data to *invivo* healthy human volunteer data and the obtained PK parameters were in simulation with the bibliographic values. Hence, the study finds the scope of the optimized formulation in delivering the drug incorporated at the brain region when incorporated intranasally. The study also portrays the scope of exploring the potential of optimized formulation in treating various CNS disorders.

5. REFERENCES

1. Z. Richard Alan Mitteer, k. Yonggao, Y. Fan, Multimodality Targeting of Glioma Cells, In glioblastoma; Elsevier: Amsterdam, Netherlands, 1, 2016.
2. K. Vyas, A. Shahiwala, S. Marathe and M. Ambikanandan, Intranasal Drug Delivery for Brain Targeting, 2005, 2, 165-175.
3. M. Gharbavi, J. Amani, H. Kheiri-Manjili, H. Danafar, and A. Sharaf, Niosome: A Promising Nanocarrier for Natural Drug Delivery through Blood-Brain Barrier, 2018, 1, 1-15.
4. B. Anbarasan, S. Rekha, K. Elango, B. Shriya and S. Ramaprabhu, Optimization of the formulation and in-vitro evaluation of capecitabine niosomes for the treatment of colon cancer, 2013,4,1504-1513.
5. R. Daniel, A. Budman, Capecitabine, 2000, 18, 355–363.
6. R. Nazari-Vanani, K. Karimian, N. Azarpira & H. Heli, Capecitabine-loaded Nanoniosomes and evaluation of anticancer efficacy, 2019,47, 420–426.
7. S. Kamboj, V. Saini, and K. Suman Bala, Formulation and Characterization of Drug Loaded Nonionic Surfactant Vesicles (Niosomes) for Oral Bioavailability Enhancement, 2014,2,1-8.
8. K. Srikanth, Y. Anand Kumar, C. Mallikarjuna Setty, Design and Optimization of Capecitabine Proniosomes, 2018, 6, 2717-22.
9. S. Nahar, S. M. Ashraful Islam, K. Swarnali Islam, W. Nasreen, Formulation and Evaluation of Metoprolol Tartrate Loaded Niosomes Using 2^3 Factorial Design, 2018, 22, 1-17.
10. N. Mishra*, V. Srivastava, A. Kaushik, K. Vivek Chauhan, Gunjan Srivastava, Formulation and in-vitro evaluation of Niosomes of Aceclofenac, 2014, 3, 337-341.
11. A. Manosroi, P. Wongtrakul, J. Manosroi et al, Characterization of vesicles prepared with various non-ionic surfactants mixed with cholesterol, 2003, 30, 129–138.
12. P. Balakrishnana, S. Shanmugama, W. S. Leeaetal, Formulation and Invitro assessment of minoxidil niosomes for enhances skin delivery, 2009, 377, 2, 1–8.
13. A. Fathi Azarbayjani, E. H. Tan, Y.W. Chan, and S. Y. Chan. Transdermal delivery of haloperidol by proniosomal formulations with non-ionic surfactants. 2009, 32, 1453–1458.
14. H. A. Pawar and V. B. Attarde, Study of Effect of Formulation Variables on Bifonazole Niosomes by Applying Plackett Burman Design, 2017, 4,

1-5.C.T. Lo, A. Jahn, Locascio, L. E. Vreeland, Controlled self-assembly of monodisperse niosomes by microfluidic hydrodynamic focusing, 2010, 26, 8559-8566.

15. L. F. Uchegbu, R. Duncan, Niosomes containing N-(2- Hydroxypropyl) methacrylamide co-polymer- doxorubicin (PK1): effect of method of preparation and choice of surfactants on niosome characteristics and a preliminary study of body distribution, 1997, 1, 155, 1-5.
16. K. Venkata ramana reddy, R. Sravani., S. Vaishnavi., U. Srilatha, R. Srilatha., M. Poornima Dr. B. Venkateswara reddy, Factors affecting in formation of niosomes, 2014, 4,2087- 2091.
17. Z. Sezgin Bayindir, N. Yuksel, Characterization of Niosomes prepared with various Non-ionic surfactants for Paclitaxel oral delivery, 2010, 99, 2049-2060.
18. S. Dash, P. Narasimha Murthy, K. Lilakanta Nath, P. Chowdhury. Kinetic modeling on drug release from controlled drug delivery systems, 2010, 67, 217-223.
19. G. Sandra, J. Parojcic, and Z. Djuric, Computer-aided biopharmaceutical characterization: gastrointestinal absorption simulation, Woodhead publishing limited, Sawston, 5, 2013.
20. M. Moawia A. Tabakha and J. Muaed Alomar, In Vitro Dissolution and in Silico Modeling Shortcuts in Bioequivalence Testing. 2020, 12, 1-16.

MICROWAVE ASSISTED NANOCOMPOSITE LOADED CINNARIZINE FOR IMPROVING ABSORPTION IN GASTROINTESTINAL TRACT

Sourav Mohanto[a,b*], Mohammed Gulzar Ahmed[a], Adrija Bhunia[a], and Mithun Bhowmick[b]

[a]Department of Pharmaceutics, Yenepoya Pharmacy College & Research Centre, Yenepoya (Deemed to be University), Mangalore, Karnataka-575018.
[b]Department of Pharmaceutics, Bengal College of Pharmaceutical Sciences and Research, Durgapur, West Bengal-713212.
Corresponding Author E-mail address: souravmohanto@yenepoya.edu.in

1 INTRODUCTION

Among several other routes of drug administration, conventional oral route is considered as the most ingested and convenient route from patient's outlook.[1] Although, several biopharmaceutical limitations associated with oral drug administration, i.e., poor solubility in the aqueous medium, lead to narrow absorption and poor bioavailability, which can be encountered via physicochemical properties modification or offering a new delivery approach. [2,3] Motion illness is a prosaic and complicated syndrome that looks in reaction to an absolute or perceived movement, particularly for the duration of traveling. Its expression includes diverse reactions such as gastrointestinal, autonomic and nervous symptoms. The exposure to movement illness is a massive personality variation, as few people might also go through minimum provocation, Whereas in others, it can be hard to elicit symptoms. [4] As per the latest data, 1 out of 3 people is sensitive to motion sickness, especially during travel. Even though 80% of the general population prone to insomnia has been seen as more susceptible to motion sickness in the Asian region. Several other factors, i.e., age, body weight, emotional balance, weather, diet, eating behaviors, etc., may impact the possibilities of motion sickness during traveling and also has been witnessed in females compared to men. [5,6]

Cinnarizine, a popular antihistaminic or calcium channel blocker, belongs to BCS Class type II of drugs, which affects the solubility and permeability in aqueous medium developing a major concern to these class of drug's that further affects the absorption pattern in gastrointestinal (GI) tract. Several investigations reported the pH-dependent behavior of Cinnarizine in GIT and hence bioavailability of the drug is restricted. [7] Microwave assisted/ induced nanocomposite preparation plays a significant role in reducing the particle size of the materials via break down of intermolecular bonding between drug and excipients/polymers. [8] Thus, effective surface area may improve the rate of absorption and further affects the bioavailability of poor aqueous soluble drugs. [8,9]

In this current study, an effort has been applied to formulate and characterize the Cinnarizine loaded Bio-nanocomposites consisting of xanthan and guar gum for enhanced solubility of Cinnarizine and significant drug release from nanocomposite, which further devise the permeability and bioavailability compared to the marketed formulations.

2 MATERIALS AND METHODS

2.1 Chemicals and Reagents used

Standard and pure sample of Cinnarizine was provided as a gift sample. The Guar gum and Xanthan gum were obtained from Fisher Scientific India Pvt. Ltd. to prepare composite system. Acetone, ethanol, chloroform and other chemicals were also obtained from Fisher Scientific India Pvt. Ltd. With best industry quality and specifications.

2.2 Pre-formulation studies of Cinnarizine

2.2.1 UV spectrum of Cinnarizine. The spectrophotometer operates on the exposition of light rays in the ultraviolet/visible range leading to the absorption of the light from the different molecules in a solution, having a similar wavelength based on the transition of electrons associated with absorption. The spectrum graph is recorded by plotting the absorbance against wavelength. In the current study, UV-spectrophotometer (Model no: Shimadzu, UV-1800, Japan) was used to determine the λ_{max} of Cinnarizine. A solution of Cinnarizine of 10 µg/ml was prepared and determined in the range of 200-800 nm. [10,11]

2.2.2 Estimation of Cinnarizine by UV-visible spectrophotometer

2.2.2.1 Preparation of Stock Solution. The stock solution of 1000 µg/ml was prepared by adding 10mg Cinnarizine in 10 ml methanol. From this stock solution, 10 ml was withdrawn and further diluted up to 100 ml with methanol as solvent, producing a concentration of 100 µg/ml. [10,11]

2.2.2.2 Preparation of Working Solution. The concentration of 100 µg/ml of Cinnarizine in methanol as standard stock solution was further utilized to prepare various working solutions. The standard stock solution was used to obtain various dilutions (1-10 µg/ml). Working solutions were diluted and scanned at 252 nm against methanol as blank using a UV-visible spectrophotometer. The straight line equation, slope and correlation coefficient were accessed from standard curve plot against concentration and was further reported. [10,11]

2.2.3 Determination of solubility of Cinnarizine. Various solvents were taken i.e., Methanol, Ethanol, Acetone, Chloroform, 0.1N Hcl, water, and PBS pH 6.8 in test tubes with an excess amount of drug to determine the solubility of Cinnarizine. The drug was thoroughly mixed in culture tubes followed by water bath shaker at room temperature for 24 hrs. Each sample were further centrifuged at 15,000 rpm, and supernatant was withdrawn. The filtrates from centrifugation was further diluted and determine spectrophotometrically at 252 nm. [12]

2.2.4 Determination of Partition Coefficient of Cinnarizine. The partition coefficient determination of Cinnarizine was done using shake flask method where it was identified and confirmed that the nature of the drug molecule was hydrophilic/ lipophilic in nature. To determine the same an excess amount of the drug was dissolved for 24 hrs. in n-octanol: Water at a ratio of 1:1. After 24 hrs, centrifugation was done for 15 mins at 15,000 rpm to separate the two layers and the absorbance was recorded via UV visible spectrophotometer at the 252 nm. [13,14]

2.3 Characterization of polymers for composite preparation

2.3.1 Determination of Swelling index (SI). The swelling index was determined with guar gum and Xanthan gums to understand optimize composite capacity and the changes in drug release behavior. In a 100 ml of measuring cylinder, 10 g of Guar gum and Xanthan gum was added and noted the initial volume of the powder. 100 ml of milli Q water was further added to both the cylinders and left aside for about 24 hrs. Final volume of the polymers were further noted and swelling index was calculated via the following equation:

$$\%\,Swelling = \frac{Xt - Xo}{Xo} \times 100 \tag{1}$$

Where, X_0 is the powder's initial height in the graduated cylinder and X_t is the occupied height of the swollen gum after 24 hrs. [15]

2.3.2 Determination of Foaming index (FI). To understand the properties of the surfactant in the polymeric matrix, 1 gm of Xanthan gum and Guar gum were added in individual 250 ml of measuring cylinder and diluted with 10 ml of distilled water for proper dispersion. The dispersion was further shaken vigorously for 2 mins. [16] The following equation calculates the foaming index of each polymer:

$$Foaming\ index = H_f - H_i \tag{2}$$

Where, H_f and H_i denotes height of gum solution after shaking and height of gum solution before shaking.

2.4 Preparation of Bio-nanocomposites loaded Cinnarizine

The Bio-nanocomposites (BNCs) loaded Cinnarizine developed via Microwave induced technique (MIND) at optimized radiation (see Figure 1). The BNCs were prepared by accurately weighing the quantity of Cinnarizine, and carriers (i.e., Xanthan gum, and guar gum) which was taken in 1:0.1, 1:0.2, 1:0.3, and up to 1:0.5 w/w proportions. A homogeneous mixture was formulated of drug and carrier using mortar-pestle. Further the slurry was prepared by mixing 10 ml of milli Q water to each gram of the physical mixture of drug and carrier. Further, 6 gm of slurry was kept on a glass beaker and radiated with microwave radiations at optimized power of 100 W. [17]

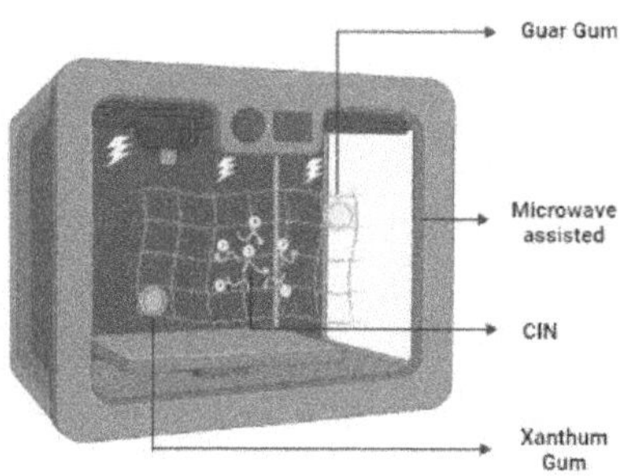

Figure 1 Demonstrated graphical representation of Microwave induced technique (MIND) associated Bio-nanocomposite loaded Cinnarizine.

2.5 Characterization of bio nanocomposites loaded Cinnarizine

2.5.1 Determination of percentage (%) of Yield. Firstly, the bio nanocomposites prepared, were collected and weighed. Further, the obtained weight was divided by the overall weight of all the drug and excipients. Hence, by using the below mentioned formula [18] the % yield was calculated:

$$Percentage\ yield = \frac{Practical\ mass\ of\ Bio\ nanocomposite}{Theoretical\ mass\ (polymer+Drug)} X100$$
(3)

2.5.2 Solubility study of Bio-nanocomposites (BNCs) of Cinnarizine. The determination of solubility was carried out via adding an appropriate amount of BNCs in culture flask containing 10 ml of water. The orbital shaker incubator was used to stir the mixture at 25°C temperature for about 24 h. The membrane filter of 0.2 μ was used to filtrate the supernatant and further analyzed via UV-Visible spectrophotometer at 252 nm. The solubility study was optimized using the drug: carrier ratio in the nanocomposite. [18]

2.5.3 Determination of Percentage (%) Drug Content from bio nanocomposite. The concentration of Cinnarizine in BNCs was further confirmed via dissolving in 25 ml of water. The solution was further passed through 0.2μ membrane filter and finally analyzed by UV-visible spectrophotometer against blank as methanol at 252 nm. [17]

2.5.4 Determination of Particle Size & Zeta Potential. The Zeta Potential/ Particle Sizer analyzer is used for the determination of the particle size diameter, vesicle properties and zeta potential at room temperature. The formulation was ultra-sonicated for 30 mins. [19]

2.5.5 Transmission Electron Microscopy (TEM). The TEM was utilized to understand the morphology and structure of BNC at 200 kV with magnifications order of 7,50,000X (H–7500, Hitachi, Japan). The water diluted BNC in the ratio of 1:10 was deposited on the carbon-coated copper grid (200 mesh) and was finally observed after drying. The morphology and size of the BNC was further determined via adjusting diffraction modes and magnification at bright field imaging. [20,21]

2.5.6 Determination of percentage (%) of drug release. The % drug release of BNC and pure drug (30 mg) was carried out via USP dissolution apparatus type II (Paddle). The pH 6.8 phosphate buffer saline solution was used in dissolution media and rate of stirring was adjusted up to 100 ± 2 rpm throughout the process. The 05 ml of samples were withdrawn at different intervals (5, 10, 15, 20, 30, 45, 60, 90, 120 mins), and same amount of fresh PBS was added to maintain the equilibrium. The samples were further passed through 0.45 mm Millipore filter, diluted, and finally analyzed by a UV-visible spectrophotometer at 252 nm respectively. [16]

3 RESULTS

3.1 Determination of absorption maxima in methanol

The maximum absorption of Cinnarizine in methanol was recorded at 252 nm and showed similar to the literature [22] (shown in Figure 2).

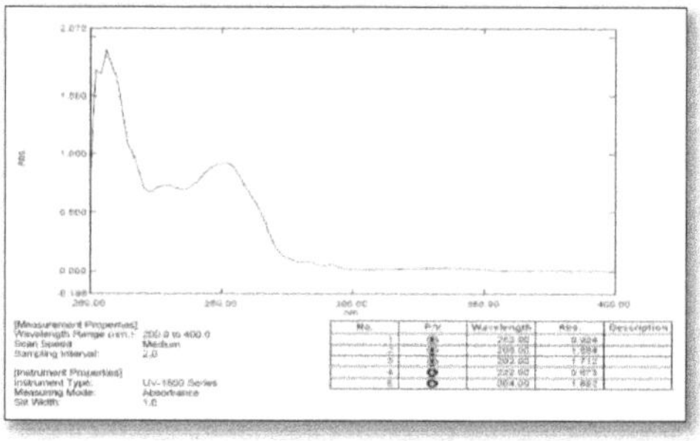

Figure 2 Maximum wavelength of Cinnarizine in methanol recorded via UV spectrophotometer.

3.1.1 Calibration/ standard curve of Cinnarizine in methanol for linearity. The maximum absorbance of Cinnarizine in methanol was recorded at 252 nm at various concentrations (1-10 µg/ml) via a UV spectrophotometer to determine the standard curve and R2 value for further consideration. The linearity at standard curve equation Y=0.0896x + 0.0366 showed an R2 value of 0.9992, which further confirmed the good linearity and purity of the drug sample [22], (see Figure 3).

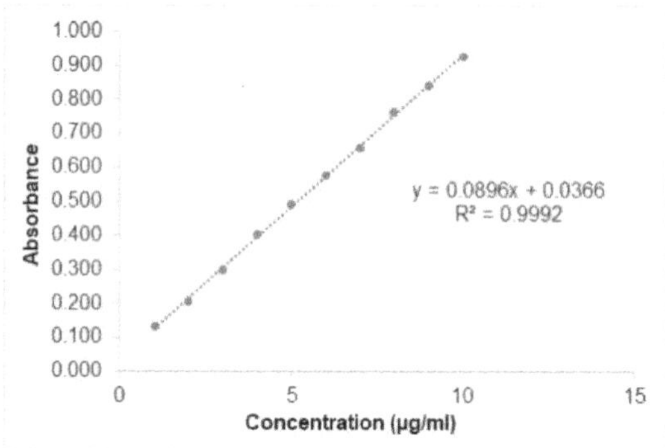

Figure 3 Graph showed linearity of Cinnarizine at various concentration in methanol.

3.2 Determination of solubility of Cinnarizine

The determination of solubility in various polar and non-polar solvents was carried out to comprehend the physicochemical characteristics of Cinnarizine. The quantitative solubility analysis was recorded at 252 nm via UV spectrophotometer and was found to be highly soluble in ethanol, followed by methanol, chloroform, and acetone, as summarized in Table 1.

Table 1 *Summarized the quantitative determination of solubility of Cinnarizine in various solvents.*

Name of Solvents	Solubility (mg/ml)	Inference
Water	0.003±0.0003	Insoluble
Phosphate buffer saline pH 7.4	1.031±0.0010	Slightly soluble
Acetone	22.347±0.0224	Slightly soluble
Ethanol	28.000±0.665	Sparingly soluble
Methanol	27.714±0.0332	Sparingly soluble
Chloroform	25.230±0.0331	Sparingly soluble
0.1 N Hcl	2.491±0.0025	Slightly soluble

*Each value is mean of three independent determinations.

3.3 Determination of Partition coefficient of Cinnarizine

The partition coefficient (Log P) determination of Cinnarizine was analyzed via the conventional shake flask method, and the obtained values were referred to the pharmacopoeia to understand the physical properties of the Cinnarizine. A Log P value greater than one indicates the drug is lipophilic, whereas those with a Log P value less than one indicate a hydrophilic drug, (see Table 2). The partition coefficient of Cinnarizine was found to be 5.519±0.037039, which indicated lipophilicity when referred to the literature.

Table 2 Referred the Log P value of Cinnarizine observed in n-octanol: water phase.

Partition coefficient of drug	Solvent system	Reported Log P Values	Observed Log P Values
Cinnarizine	n-octanol: water	5.77	5.519±0.037039

*Value is expressed as mean ± SD; n = 3

3.4 Characterization of Polymers for Bio-nanocomposite preparation

The conventional/ most common plant-derived polymers (Guar gum and Xanthan Gum) were subjected to screen out to develop the optimized composite preparation for immediate drug release in the GIT. The swelling and foaming index of the polymers/ matrix carrier is needed to understand the capability of solubility enhancement and dissolution characteristics of the Cinnarizine.

*3.4.1 Determination of Swelling Index (SI). (*See Figure 4), it was seen that the Percentage Swelling Index of carriers was found to be in a range of 63.41±0.591 to 76.56±0.170, with optimum swelling observed in xanthan gum, which was found to be maximum and best fit for bio-nanocomposite formation due to high swelling index, playing a crucial role in solubility enhancement.

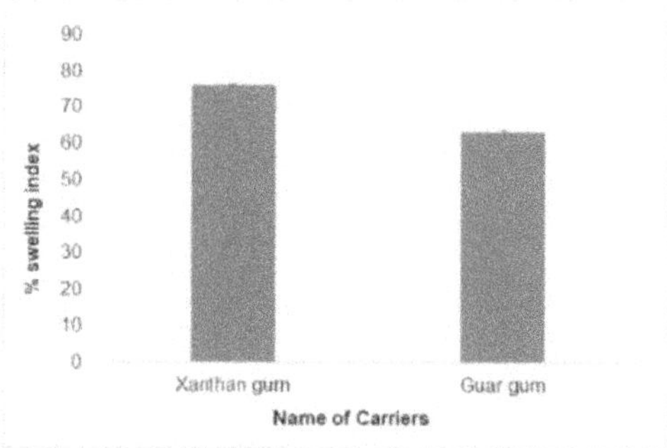

Figure 4 Referred % Swelling Index of Xanthan gum and Guar gum (Carriers) for optimized composite preparation.

3.4.2 Determination of Foaming index (FI). (See Figure 5) shows that the Foaming Index of all carriers was found to be in a range of 14±0.173 to 17±0.577, with the optimum foaming index observed in xanthan gum to be maximum and best fit for bio-nanocomposite formation due to improvising the solubility and dissolution enhancement.

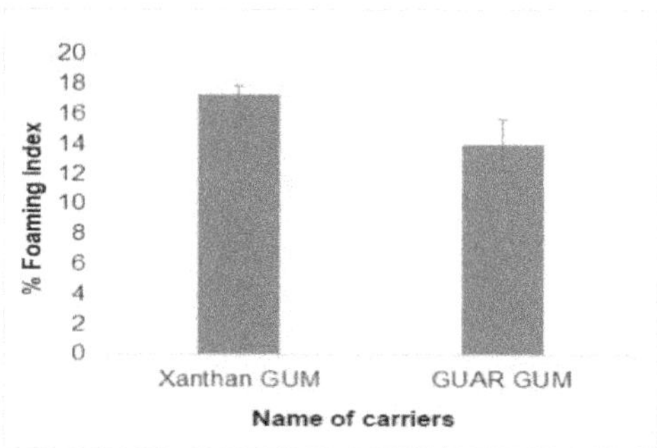

Figure 5 Referred the % Foaming Index of Xanthan gum and Guar gum (Carriers) for optimized composite preparation.

3.5 Preparation of Bio-nanocomposites loaded Cinnarizine

In the present investigation, the bio-nanocomposite loaded Cinnarizine with the capability of providing immediate release was developed using different carriers such as Xanthan gum and guar gum. The technique for developing bio-nanocomposite was reported as simple without any changes in the physicochemical characteristics of the drug molecule or matrix systems.

3.6 Characterization of Bio nanocomposites loaded Cinnarizine

3.6.1 Percentage Yield of bio nanocomposites. Figure 6 shows that the yield of bio-nanocomposites was found to be in the range of 58.97±0.634 to 80.80±0.8%. The result showed that the higher carrier concentrations in the bio-nanocomposite development slightly affected the percentage of yield. Hence, F4 was shown as an improved percentage of yield compared to other formulations, as mentioned in Table 3.

Table 3 *Referred Percentage yield of bio nanocomposites loaded Cinnarizine.*

Formulation Code	% of yield
F1	58.97±0.634
F2	62.69±0.529
F3	68.48±0.447
F4	74.14±0.925
F5	62.25±1.391
F6	66.67±0.539
F7	78.41±0.549
F8	80.80±0.8

*Value is expressed as mean ± SD; n=3

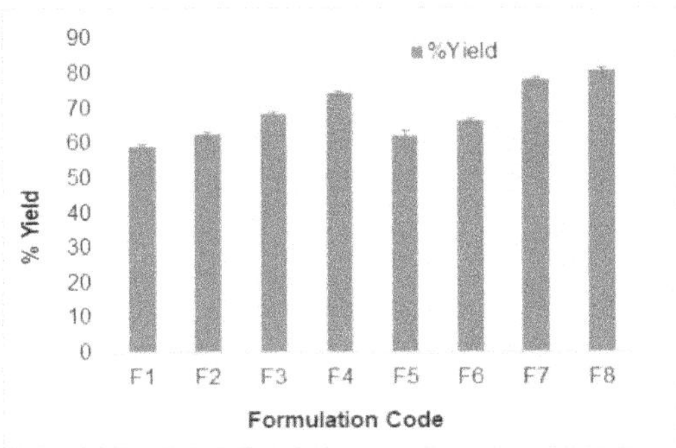

Figure 6 *Showed Percent of yield of bio-nanocomposites loaded Cinnarizine.*

3.6.2 Solubility determination of Bio-nanocomposite (BNCs) of Cinnarizine. The solubility study of BNCs was carried out to understand whether Cinnarizine was released from the bio nanocomposites, further providing inputs regarding absorption pattern. The improved solubility of the prepared Bio-nanocomposites was calculated in distilled water compared with pure drug, (see Figure 6 and Table 4). It was found that there was a maximum 830 folds increase in solubility when xanthan gum was used as a carrier in the ratio of 1:0.4 (drug: carrier).

Table 4 *Referred the quantitative solubility determinations of Cinnarizine loaded Bio-nanocomposites.*

Sl. No.	Formulation Code	Solubility (mg/ml)
1	Pure Drug	0.001±0.0001
2	F1	0.265±0.0111
3	F2	0.666±0.005
4	F3	0.493±0.014
5	F4	0.830±0.008
6	F5	0.231±0.003
7	F6	0.004±0.003
8	F7	0.028±0.008
9	F8	0.105±0.006

*Value is expressed as mean ± SD; n = 3

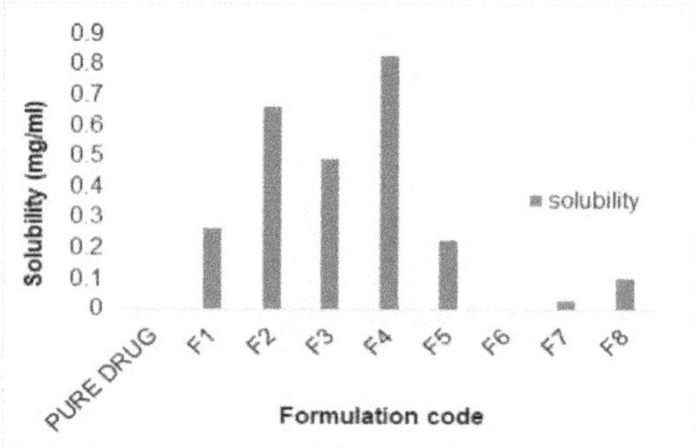

Figure 7 *Showed Solubility Profile of Bio-nanocomposites loaded Cinnarizine.*

3.6.3 Determination of Percentage (%) of Drug Content from Bio nanocomposites. The cinnarizine-loaded Bio-nanocomposites were further subjected to the determination of % drug content in aqueous media. See Figure 8 and Table 5, it was concluded that % drug content of the bio-nanocomposites varied from 75.740±0.153 to 89.398±0.088%, which further indicated the optimized characteristics of Xanthan gum in the preparation of nanocomposite and immediate release of the drug.

Table 5 *Percentage (%) of Drug Content of Cinnarizine loaded Bio-nanocomposites.*

Sl. No.	Formulation Code	Percentage (%) Drug Content
1	F1	75.740±0.153
2	F2	86.850±0.318
3	F3	89.551±0.233
4	F4	92.608±1.300
5	F5	70.083±0.306
6	F6	74.262±0.491
7	F7	87.257±0.722
8	F8	89.398±0.088

*Value is expressed as mean ± SD; n = 3

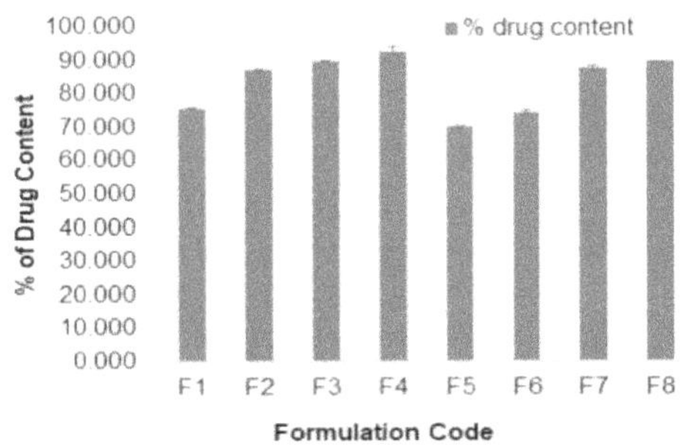

Figure 8 Showed Percentage drug content of Bio-nanocomposites loaded Cinnarizine.

3.6.4 Determination of Particle Size and zeta potential of F4 formulation. The percentage of drug content, solubility study of the nanocomposites, and % of yield may further conclude that the F4 formulation result was higher than other formulations. Figure 9 demonstrated hydrodynamic diameter or average particle size of the F4 formulation was 169.36 nm with PDI 23.6%. Figure 10 demonstrated zeta potential of the F4 formulation was 34.4 mV, representing higher stability of the F4 nanocomposites.

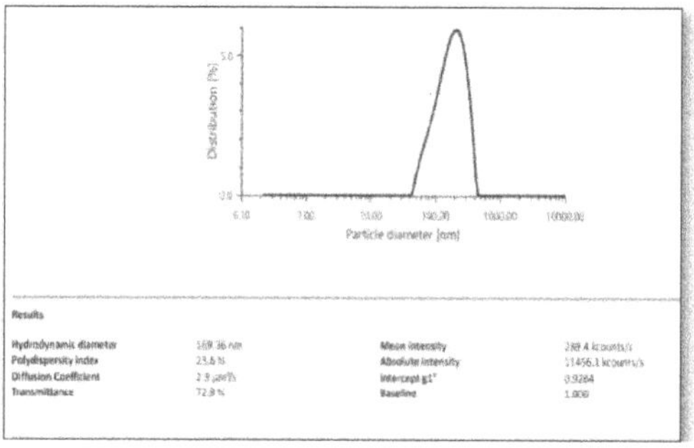

Figure 9 Demonstrated the average particle size determination of F4 nanocomposites via Dynamic light scattering.

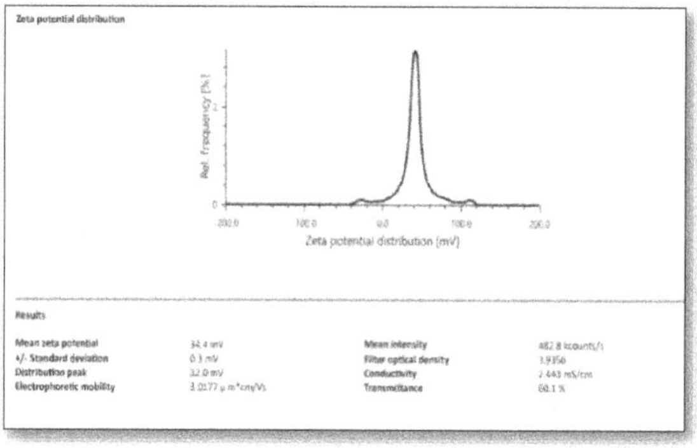

Figure 10 Showed improved zeta potential of F4 nanocomposite.

3.6.5 Transmission Scanning Microscopy (TEM). The TEM study was oriented to understand/ determine the surface morphology and particle size of the F4 nanocomposites, which further gives an idea to load the maximum amount of Cinnarizine in the composite. Figure 11 demonstrated that the oval shape of the crystalline Cinnarizine was transformed entirely in the BNC showing embedded Cinnarizine crystals in the matrix.

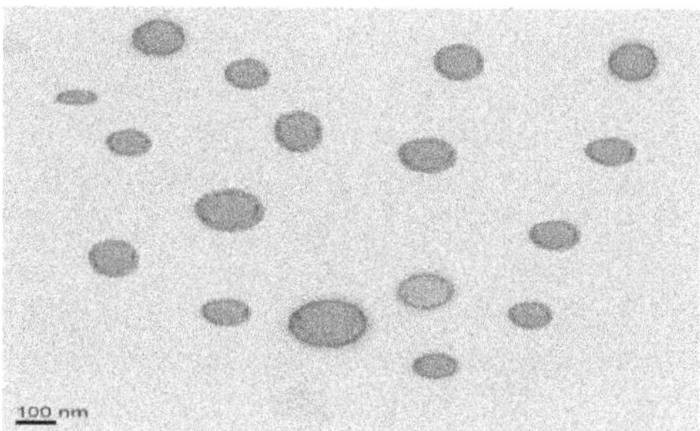

Figure 11 Demonstrated TEM Image of Formulation F4 nanocomposite.

3.6.6 In vitro Drug release study. The *In vitro* drug release study via USP types II apparatus was initiated to attain further the objective of immediate cinnarizine release in the GIT environment from the bio-nanocomposite. Table

6 and Figure 12 showed that *In vitro* release of drug from the Bio-nanocomposite was higher than pure drug suspension, which showed the effect of the carrier in drug release property. Table 6 indicated that *In vitro* release of Cinnarizine loaded Bio-nanocomposite showed 94.77±0.193% released within 2 hours. The release profiles of the Bio-nanocomposite of Cinnarizine employed yielded an immediate Cinnarizine release in the GIT medium. The polymer matrix enhanced the solubility of the F4 nanocomposite system in the PBS environment, and hence Cinnarizine showed immediate release from the nanocomposite compared with pure nanosuspension of Cinnarizine [23].

Table 6 *Demonstrated % drug release of F4 nanocomposite and pure drug in the suspension form.*

Time (Mins)	% Drug release of F4 Formulation	% Drug release of Pure Drug Suspension
0	0	0
5	80.60±0.193	0.404±0.084
10	81.49±0.334	1.051±0.120
15	82.61±0.193	0.594±0.008
20	83.61±0.386	0.939±0.005
30	85.62±0.511	1.133±0.005
45	87.07±0.386	1.801±0.003
60	89.75±1.353	2.904±0.107
90	92.87±0.334	3.161±0.033
120	94.77±0.193	3.272±0.019

*Value is expressed as mean ± SD; n = 3

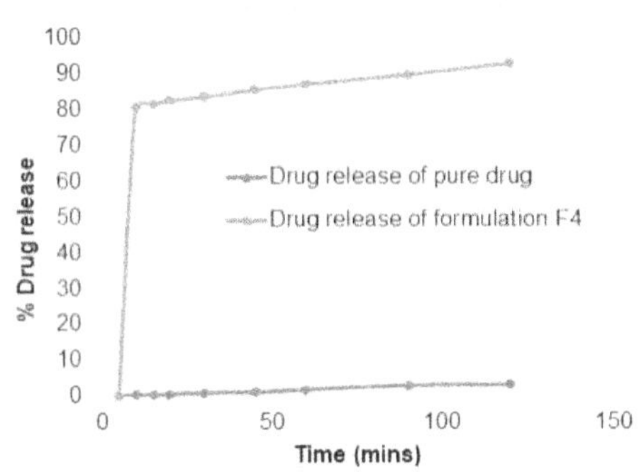

Figure 12 In vitro Drug release profile of Bio-nanocomposite loaded Cinnarizine and Pure Drug.

3.6.6.1 In vitro drug release kinetic. The release kinetics model *In vitro* was introduced to understand the release mechanism and comparison of release profiles based on models, i.e., zero order, first order, Higuchi and Korsmeyer-Peppas. The mathematical models were tested based on log percent drug remaining vs. time (first order), % drug release vs. time (zero order), log of log % drug release vs. log time (Korsmeyer-Peppas Exponential Equation), and log percent drug release vs. square root of time (Higuchi plot) and further reported the R2 value in Table 7. Considering the determination coefficients, the First Order model was found (R2=0.997) to be best fitted and confirmed the immediate release of Cinnarizine in the PBS environment.

Table 7 *Reported the R^2 value from various In vitro drug release kinetics model.*

Formulation name	**Zero order**		**First order**		**Higuchi**		**Korsmeyer-**	
	R_2	K_0	R_2	K_0	R_2	K_0	R_2	K_0
F4	0.97	0.126	0.9978	-0.005	0.9945	1.707	0.9386	0.0533

4. CONCLUSION

The oral drug administration required the absorption of the drug in GIT to attain maximum bioavailability. The 1/3rd of the drug molecules in the market are lipophilic or poor aqueous soluble; hence, enhancing the solubility and absorption rate in the GIT is always crucial if targeted to deliver via the oral route. Several highly yielded cutting-edge technologies involved enhancing solubility (at the molecular and/or nanoscale level) and absorption of poorly aqueous soluble drugs in the GIT. Microwave-induced/ assisted synthesis is recently introduced in pharmaceutical formulation development, i.e., microbead formation, solid dispersions, nanospheres, microspheres, composite preparation, etc., due to higher yield and less time-consuming process. The Microwave induced technique conferred the transfer of energy inside the materials, which reaches quick and homogenous heating to the synthesis process even for low-heat conductive materials, i.e., polymers. Since many excipients employed in pharmaceutical formulations are polymers, this technique is essential. This method is frequently used to create polymeric cross-linkages, interact drugs with polymers and dry materials, and alter the structural properties of drug crystals through heating. The solubility is a physical characteristic of the drug molecule, which further improvising leads to higher dissolution kinetics and rate of absorption in nanocomposite preparation via Microwave induced technique keeping the integrity of the polymers in the system.

The UV spectrophotometric analysis in methanol was carried out to understand the preformulation prerequisite of Cinnarizine and reported 262 nm, similar to the literature. The solubility of the Cinnarizine was reported to be maximum in methanol, followed by ethanol, chloroform, and acetone. The partition coefficient value of Cinnarizine further confirmed the lipophilic nature of the molecule, and hence solubility remains a concern, especially in the GIT environment. Thus, an attempt was constructed to prepare nanocomposite-loaded Cinnarizine using different carriers (Xanthan gum, guar gum) to enhance the solubility and immediate release in the GIT microenvironment. The optimized carriers were selected based on the polymers' swelling and foaming capacity, which can improvise the solubility and dissolution rate in the PBS medium. The quantitative solubility analysis of Cinnarizine loaded nanocomposite formulation further supports the solubility enhancement objective and found a maximum 830-fold increase in solubility when Xanthan gum was incorporated as a carrier in the ratio of 1:4 (drug: carrier). The *In vitro* drug release study further confirms the immediate release of the Cinnarizine (94.77±0.193%) within 2 hrs in the PBS microenvironment, which also showed a similar pattern as First order kinetics model (R2= 0.9997) when fitted into various mathematical kinetics models. Hence from all aspects, we concluded that the release of the drug Cinnarizine from bio-nanocomposite was a primary mechanism. This could be achieved by adequately designing the formulation and selecting a suitable preparation strategy.

Therefore, the study concluded that the development of bio nanocomposite was simple, reproducible, and provided good solubility and drug content. The *In vitro* data obtained for Cinnarizine loaded bio-nanocomposite showed increased solubility in distilled water. The development of nanocomposite was mainly responsible for enhancing the solubility of the drug compared pure form of the drug in the suspension. Thus, the conventional polymer, i.e., Xanthan gum could be considered an effective carrier system for preparing nanocomposite via crystallization of the Cinnarizine with suitable microstructural properties. The present further opens the chances of In vivo investigation of the bio-nanocomposite loaded Cinnarizine in the animal model for the near future.

5. ACKNOWLEDGEMENTS

The authors are thankful to Bengal College of Pharmaceutical Sciences & Research, West Bengal for providing the facilities to carry out the work.

6. REFERENCES

1. M. Sugawara, S. Kadomura, X. He, Y. Takekuma, N. Kohri and K. Miyazaki, The use of an in vitro dissolution and absorption system to evaluate oral absorption of two weak bases in pH-independent controlled-release formulations., *Eur. J. Pharm. Sci. Off. J. Eur. Fed. Pharm. Sci.*, 2005, **26**, 1–8.
2. A. Streubel, J. Siepmann and R. Bodmeier, Drug delivery to the upper small intestine window using gastroretentive technologies., *Curr. Opin. Pharmacol.*, 2006, **6**, 501–508.
3. R. Malviya, P. Srivastava, M. Bansal and P. Sharma, IMPROVEMENT OF DISSOLUTION BEHAVIOR OF PARACETAMOL USING SOLID DISPERSION TECHNIQUE.
4. L.-L. Zhang, J.-Q. Wang, R.-R. Qi, L.-L. Pan, M. Li and Y.-L. Cai, Motion Sickness: Current Knowledge and Recent Advance., *CNS Neurosci. Ther.*, 2016, **22**, 15–24.
5. B. S. Hromatka, J. Y. Tung, A. K. Kiefer, C. B. Do, D. A. Hinds and N.

Eriksson, Genetic variants associated with motion sickness point to roles for inner ear development, neurological processes and glucose homeostasis, *Hum. Mol. Genet.*, 2015, **24**, 2700–2708.

6. V. Takov and P. Tadi, Treasure Island (FL), 2022.
7. S. Raghuvanshi and K. Pathak, Recent advances in delivery systems and therapeutics of cinnarizine: a poorly water soluble drug with absorption window in stomach., *J. Drug Deliv.*, 2014, **2014**, 479246.
8. T. P. Naik, I. Singh and A. K. Sharma, Processing of polymer matrix composites using microwave energy: A review, *Compos. Part A Appl. Sci. Manuf.*, 2022, **156**, 106870.
9. K. Yu, Y. Liu and J. Leng, Shape memory polymer/CNT composites and their microwave induced shape memory behaviors, *RSC Adv.*, 2014, **4**, 2961–2968.
10. H. Abdine, F. Belal and N. Zoman, Simple spectrophotometric determination of cinnarizine in its dosage forms., *Farmaco*, 2002, **57**, 267–271.
11. N. T. Lamie, Comparative study of spectrophotometric methods manipulating ratio spectra: an application on pharmaceutical binary mixture of cinnarizine and dimenhydrinate., *Spectrochim. Acta. A. Mol. Biomol. Spectrosc.*, 2015, **141**, 193–201.
12. T. S. Tambawala, P. J. R. Shah and S. Sha, 2015.
13. M. Francois, E. Snoeckx, P. Putteman, F. Wouters, E. De Proost, U. Delaet, J. Peeters and M. E. Brewster, A mucoadhesive, cyclodextrin-based vaginal cream formulation of itraconazole., *AAPS PharmSci*, 2003, **5**, E5.
14. F. Shakeel, M. Kazi, F. K. Alanazi and P. Alam, Solubility of Cinnarizine in (Transcutol + Water) Mixtures: Determination, Hansen Solubility Parameters, Correlation, and Thermodynamics, *Molecules*, , DOI:10.3390/molecules26227052.
15. G. L. Amidon, H. Lennernäs, V. P. Shah and J. R. Crison, A theoretical basis for a biopharmaceutic drug classification: the correlation of in vitro drug product dissolution and in vivo bioavailability., *Pharm. Res.*, 1995, **12**, 413–420.
16. M. Bhat, S. Sharma, G. K. Derkar, R. M. Chimkode and S. Payghan, Microwave-generated bionanocomposites for solubility enhancement of nifedipine, *Asian J. Pharm.*, 2016, **10**, S741–S749.
17. M. Bhat, S. Sharma, G. Derkar and S. Mansharamanic, OPTIMIZATION BIONANOCOMPOSITES OF FENOFIBRATE FOR ENHANCEMENT OF SOLUBILITY AND DISSOLUTION USING MICROWAVE INDUCED DIFFUSION TECHNIQUE.
18. S. S. Kushare and S. G. Gattani, Microwave-generated bionanocomposites for solubility and dissolution enhancement of poorly water-soluble drug glipizide: in-vitro and in-vivo studies., *J. Pharm. Pharmacol.*, 2013, **65**, 79–93.
19. B. Mishra, J. Sahoo and P. Dixit, Enhanced bioavailability of cinnarizine nanosuspensions by particle size engineering: Optimization and physicochemical investigations, *Mater. Sci. Eng. C*, ,

DOI:10.1016/j.msec.2016.02.046.

20. S. Saber-Samandari and S. Saber-Samandari, Biocompatible nanocomposite scaffolds based on copolymer-grafted chitosan for bone tissue engineering with drug delivery capability., *Mater. Sci. Eng. C. Mater. Biol. Appl.*, 2017, **75**, 721–732.
21. Y. Patel, A. Poddar and K. Sawant, Improved Oral Bioavailability of Cefuroxime Axetil Utilizing Nanosuspensions Developed by Media Milling Technique, *Pharm. Nanotechnol.*, , DOI:10.2174/2211738502666140804233343.
22. L. K. Yeo, T. O. B. Olusanya, C. S. Chaw and A. A. Elkordy, Brief Effect of a Small Hydrophobic Drug (Cinnarizine) on the Physicochemical Characterisation of Niosomes Produced by Thin-Film Hydration and Microfluidic Methods., *Pharmaceutics*, , DOI:10.3390/pharmaceutics10040185.
23. P. Joyce, R. Yasmin, A. Bhatt, B. J. Boyd, A. Pham and C. A. Prestidge, Comparison across Three Hybrid Lipid-Based Drug Delivery Systems for Improving the Oral Absorption of the Poorly Water-Soluble Weak Base Cinnarizine., *Mol. Pharm.*, 2017, **14**, 4008–4018.

FORMULATION, PREPARATION, AND EVALUATION OF NEBIVOLOL FAST-DISSOLVING TABLETS BY USING 3^2 FACTORIAL DESIGN

V. Rupa Sree Devi*, M. Sukanya, K. Ravi Shankar, G. Ramana Reddy, and KNV. Chenchu Lakshmi

KVSR Siddhartha College of Pharmaceutical Sciences, Vijayawada

*vemurirupasreedevi@gmail.com

1 INTRODUCTION

The drug administration through oral route is the most reliable drug delivery system because of its comfort and easy way of consumption. A solid drug dosage is more robust, stable; and also exhibits additional advantages such as easy to handle and popular way of medicine consumption. Hence, a better subject compliance and drug treatment can be observed with oral route of medications than any other administration routes of dosage forms. Nebivolol hydrochloride is chemically named as "α,α/{iminobis [methylene] }bis {6-flouro3,4-dihydro-2H-1-benzopyran-2-methanol}HCL". It is a third-generation lipophilic molecule which is highly selective β1- adrenergic receptor blocking agent selectively blocks the β1-adrenoceptor with N-O coordinated vascular dilatory action and significant action on the function of vascular endothelial cells. Nebivolol is preferred in the treatment of hypertension, decrease rate of heart, myocardial contractility rate and also reduces systolic pressure, when diastolic pause increases. Based on the permeability of drug, Nebivolol comes under biopharmaceutical classification system of class II which distinguish by low dissolution rate (due to its less aqueous solubility) and high permeability through membrane. However, Nebivolol has poor solubility in GI medium, and high metabolism which results in its poor oral bioavailability of 12%. So, the extent and rate of Nebivolol absorption tends to less oral bioavailability. The Enhancement of rate of dissolution and bioavailability of nebivolol is therefore a prominent step to achieve therapeutic action. The usual dose of Nebivolol is available in 5 mg and 10 mg with a preliminary 2.5 mg dose which is used to treat in case of older and for patients of nephropathy. Many methods are executed to develop characteristics like dissolution rate and aqueous solubility of less aqueous-soluble drugs, including strong electrolyte salt formation, micronization, nanosuspensions, polymorphs, complexation, solid dispersion (SD), prodrugs, buffers and by inclusion of solvent or surfactants. [1] Solid dispersion has been the most active technique from above methods, which was simple, economic, and promising marked improvement in the bioavailability and dissolving rate of insoluble drugs. Solid dispersion is another term for the distribution of active substances in crystalline, amorphous, and molecular forms within an inert carrier, with certain benefits such as flavour masking strips.[2]

The objective of this study was to procure the techniques that to enhancing the nebivolol solubility and dissolution rate by executing solid dispersion techniques with using like as PEG4000, Poloxamer, β -Cyclodextrin, Starch 1500 polymers by kneading method. The Optimized solid dispersion was used to develop formulations by using 3^2 Factorial Design which were characterized for *in vitro* dissolution studies. [3,4] The characterization of optimized formulation was performed through "Differential scanning calorimetry (DSC), X-ray powder diffractometry (XRD), Fourier transform infrared (FTIR)", In addition, *in vitro* dissolution study was carry out to compare drug with SD.

2 MATERIALS AND METHODS

2.1 Preparation of Solid Dispersion by Kneading Method

The predetermined quantity of Nebivolol and carriers of PEG4000, β - Cyclodextrin, Poloxamer, Starch1500 in different drug: carrier ratios were weighed and triturated using small quantity of organic solvent usually methanol The mix is kneaded for 30-40 minutes and allow to dry at 45ºC through hot air oven until the fraction was dried. The fraction was ground and sieved through 80 no sieve and was bagged for further studies.[5]

2.2 Formulation of Tablet

Optimised Solid dispersion β-CD (1:3M) which 5mg equivalent dose of Nebivolol obtained by kneading method were combined with diluent, superdisintegrants and excipients in a container as per the composition given in Table-2. Powder mixes were compacted to a hardness of 4-6 kg/cm2 using 8mm-round-shaped and 9mm- flat punches in a multi station tablet compression machine.

2.3 FT-IR

The FT-IR spectrum of pure nebivolol and optimized solid dispersion was recorded by FTIR-8700 Shimadzu spectrophotometer. For this, KBR disc technique was used. The sample was combined with potassium bromide dry powder and was compacted into a transparent disc which placed in IR spectrophotometer using specialised dies at high pressure. The spectrum read over a frequency range 4000-400 cm^{-1} to evaluate polymer drug interaction studies.

2.4 Differential Scanning Calorimetry (Dsc)

The probability of any interaction between the drug and the carriers throughout various techniques was examined using DSC to conduct investigation for pure drug, optimised solid dispersion, and optimised formulation. Evaluation of samples was accessed by using DSC 60 Shimadzu-Thermal Analyzer (Japan). Samples were exposed to heat at the intensity of 10°C/min in a disclosed pan which over a temperature of 100°C-250°C in along with a nitrogen atmosphere.

2.5 X-Ray Powder Diffractometry

To evaluate the sample characteristics, the drug and optimised solid dispersion β-cd(1:3M) was impelled to X-ray powder diffractometry (Shimadzu) under copper as a target with expelling voltage of 40 Kv and 15 mA current. With detector SC-70 XRD scan and were recorded with scanning range 3.0000-90.0000deg.

3 CHARACTERIZATION OF TABLET

The formulations were evaluated for tablet hardness, tablet friability, tablet disintegration time, tablet drug content and *in vitro* dissolution.

3.1 Hardness

The tablet's hardness was measured by positioning it longitudinally between both tester's plungers, and the result was expressed in kilogrammes per square centimetre (kg/sq.cm).

3.2 Friability Test

The forces that break tablets are friction and shock. The friability test, which measures a tablet's resistance to abrasion during handling, packaging, and transportation, which related to tablet hardness. Typically, it is assessed with a Roche friabilator. Tablets were weighted and subjected to abrasion as they fell 6 inches every time the device turned. The weight of the pills was measured after 100 spins and contrasted with the starting weight. Weight loss serves as a proxy for tablet friability, which is reported as a percentage. The common consensus is that a weight loss of no more than 1% of total weight is acceptable.

3.3 Drug Content

Solid dispersions which 5mg equivalent dose of Nebivolol was placed into a volumetric flask measuring 100 ml, diluted in methanol, and appropriately made upto the mark with 1.2 pH Hydrochloric acid buffer. The content was filtered after subjected to sonication for 10minutes was determined for Nebivolol concentration spectrophotometrically using UV-visible spectrophotometer (UV 1800 shimandzu spectrophotometer) at 280nm.[6]

3.4 Disintegration test

The U.S.P device which is used to evaluate disintegration comprises of 6 glass tubes with mesh screens at the bottom. One tablet is positioned in each tube which arranged perforated discs on its top were placed in a 1-L water beaker that mimics gastric fluid and intestinal fluid at 37±2C and moves up and down movement over 5–6 cm at a rate of 28–32 cycles per minute by the help of motor. According to the test the time that tablet requires to disintegrated and passed through the mesh screen is considered as Disintegration time.
3.5 In Vitro Dissolution. This study was executed using Type II USP dissolution test apparatus. Solid dispersions which 5mg equivalent dose of Nebivolol was placed into the dissolution medium holder which hold 900 mL of 0.1N HCL acid buffer at 37°C± 0.5°C and fix at 75rpm. Aliquotes of 5ml sample collected through 0.45µm microfilter periodically at predetermined intervals of time and refill with 5ml of new dissolution medium. The Nebivolol concentration was evaluated spectrophotometrically by UV spectrophotometer (UV 1800 shimandzu spectrophotometer) at λ max 280 nm.

4 DISCUSSION OF RESULTS

The motive of the current work is to prepare the fast-dissolving tablets of Nebivolol by using solid dispersions of PEG 4000, Starch 1500, Poloxamer 188 and β-cyclodextrin (β-CD). Nebivolol tablets with fast dissolution characteristics employing βCD were prepared using 3^2 factorial study employing by direct compression technique.

An UV-visible spectrophotometer used for the measurement of Nebivolol concentration at absorbance 280nm in 0.1N Hydrochloric acid buffer. A calibration curve was constructed, and approach was verified for linearity, accuracy, precision, and interference, which followed Beer's rule at concentrations between 10-50µg/ml. Low RSD values (<1.21%) ensured the reproducibility of the method.

The compatibility of drug with the excipients used namely β-cyclodextrin, PEG4000, Starch1500and Poloxamer was evaluated by FTIR studies. The FT-IR spectrum of Optimized solid dispersion was shown in Figure 1.The above IR peaks were also observed in the spectra of nebivolol and evident that preparations have no interaction between nebivolol and the excipients used in the work.

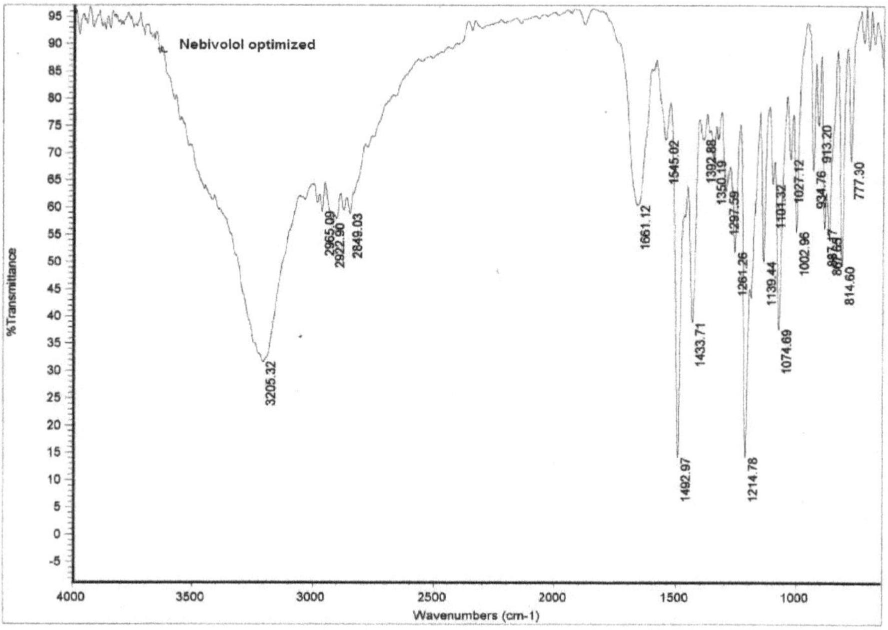

Figure 1 FT-IR Peaks of Optimized solid dispersion.

Solid dispersions of Nebivolol were prepared to enhance the dissolution by kneading method employing different carriers like PEG4000, Poloxamer in various drug: polymer ratios of 10% and 20% and β-Cyclodextrin, Starch1500 with drug:polymer ratios of 1:1M,1:2M,1:3M and Dissolution investigations were conducted on each of the 10 formulations and identified that solid dispersion with β-CD (1:3M) showed higher dissolution rate and was selected for optimization. The obtained Parameters of dissolution study are given in Table 1.

Table 1 *Dissolution study Parameters of Solid dispersions.*

Release Kinetics	**K values**	**DE_{15}**	**PD_{10}**
Polaxomer10%	0.044	49.78±0.99	60.74 ± 1.21
Polaxomer20%	0.048	53.27±1.06	63.88 ± 1.27
β-CD (1:1M)	0.0983	53.64±1.05	72.44 ± 1.44
β-CD (1:2M)	0.107	63.79±1.22	74.24 ± 1.48
β-CD (1:3M)	0.346	65.47±1.25	100 ± 2

PEG10%	0.047	56.50±1.14	68.61 ± 1.37
PEG20%	0.102	57.72±1.15	70.58 ± 1.41
Starch1500 (1:1M)	0.035	31.72±0.60	38.49 ± 0.76
Starch1500 (1:2M)	0.044	42.01±0.81	52.13 ± 1.04
Starch1500 (1:3M)	0.052	51.69±1.01	61.60 ± 1.23

Preparation and characterization of Nebivolol Tablets with βCD by using 3^2 factorial study using Primojel and Crospovidone was adopted.

Nebivolol containing 5 mg in each tablet were formulated with βCD using Primojel and Crospovidone as per the composition given in Table 2. In each case the tablets were formulated by direct compression technique. The blends of tablet ingredients prepared in each case were evaluated for flow characteristics before compression by determining angle of repose and compressibility index. The values of angle of repose were in the limit of 15^0-22^0 and values of compressibility index (%) were in the limit of 10-12 % with various formulations. These findings concluded that the tablet powder mix in each case provide excellent flow characteristics recommended for direct compression.

Table 2 *Formulae of Nebivolol Tablet Formulations.*

Ingredients (mg)	**F.1**	**F.2**	**F.3**	**F.4**	**F.5**	**F.6**	**F.7**	**F.8**	**F.9**	**OPT1**	**OPT2**
Nebivolol SD	47	47	47	47	47	47	47	47	47	47	47
Acacia	5	5	5	5	5	5	5	5	5	5	5
Primojel	7.5	10	12.5	7.5	10	12.5	7.5	10	12.5	10	11
Cross Povidone	7.5	7.5	7.5	10	10	10	12.5	12.5	12.5	11.75	12.5
Pearletol	QS	QS	QS	QS	QS	QS	QS	QS	QS	QS	QS
Magnesium Stearate	2.5	2.5	2.5	2.5	2.5	2.5	2.5	2.5	2.5	2.5	2.5
Talc	2.5	2.5	2.5	2.5	2.5	2.5	2.5	2.5	2.5	2.5	2.5
Total weight	250	250	250	250	250	250	250	250	250	250	250

All the tablets were characterized for hardness, friability, disintegration time, drug content, and dissolution rate. The tablets hardness were in between 4.0-6.0 kg/cm^2. Friability test provides weight loss of less than 1 % in each case. Nebivolol drug content of all tablet preparations were in limit 100±3 %. The disintegration times were within15-30 sec. All tablets prepared satisfied the I.P disintegration time recommendations of noncoated tablets.

The rate of dissolution of various Nebivolol tablets formulated were investigated in 0. 1NHClbuffer.In each case the data of dissolution were examined using zero order and first order kinetics. The values of correlation coefficient (r) were more in the first order model with regard to zero order model specified that the Nebivolol rate of dissolution followed first order kinetics. The values of correlation coefficient (r) in the first order model were in between of 0.918-0.998. The various Nebivolol dissolution parameters are tabulated in Table 3 and Figure 2.

Table 3 *Dissolution Parameters of Nebivolol Tablets.*

Release Kinetics	K values	DE_{15}	PD10
F1	0.114	58.14±1.11	78.03±1.51
F2	0.123	68.46±1.32	82.5±1.68
F3	0.302	80.18±1.59	100±1.95
F4	0.176	62.82±1.20	68.7±1.33
F5	0.364	77.19±1.50	90.2±1.82
F6	0.943	91.30±1.80	100±1.96
F7	0.299	72.99±1.42	68.1±1.38
F8	0.893	86.69±1.71	84.6±1.67
F9	2.134	93.14±1.82	100±1.96

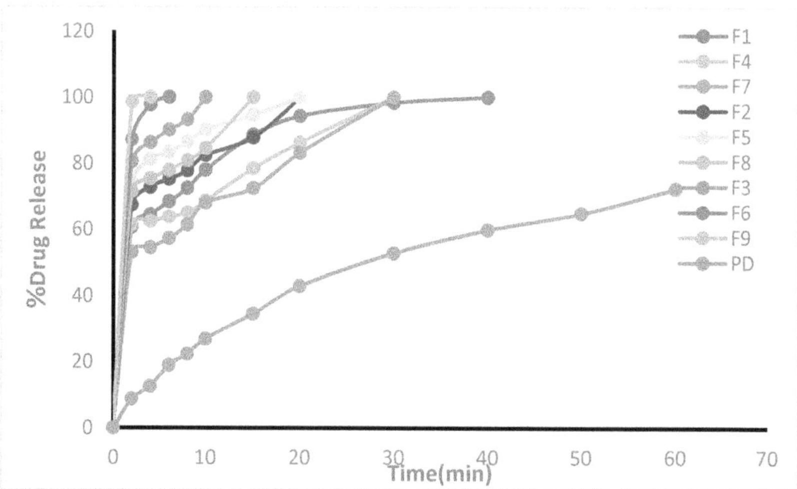

Figure 2 *In* vitro dissolution Profile of formulated inclusion complex tablets by using Crospovidone and SSG as super disintegrants.

The percentage that obtained for independent variables "Crospovidone" and 'Primojel' were utilised in the preparation of Nebivolol Fast Dissolving solid dosage form were described by a chosen three level, two factor plan (3^2 factorial study). As dependent variables Disintegration time (DT), percent drug dissolved in 6 min and percent drug dissolved in 10 min were chosen. For the final equations, significance parameters with a 95% confidence interval (p0.05) were adopted and the polynomial equations were constructed for disintegration time (DT), percentile of drug dissolved in 6 min., and percentile of drug dissolved in 10 min.

The factorial design was used in formulation of nebivolol tablets in which all three levels of factor X1 (Primojel) and factor X2 (Crospovidone) are of concentration 3%, 4%, 5% (% calculated based on total Tablet weight, i.e250 mg) were chosen as the hypothesis for the execution of the rapid dissolution of Nebivolol formulations. Overall nine nebivolol fast Dissolving tablet formulations were obtained by implementing factors (X_1, X_2**)** as of 3^2 Factorial study[7] and analysed to choose the best composition needed to acquire the required release of drug and to determine the relevance of the combined effects of X_1, X_2. All three levels of Crospovidone and primojel were opted and coded as -1= 3%, 0=4%, +1=5%.

PCP Disso software was used to derive Polynomial equations for Disintegration time (DT), percent drug dissolved in 6 min PD_6 and percent drug dissolved in 10min (PD_{10}) and SIGMAPLOT V13 software was utilised for constructing response surface plots which were depicted in **Figures 3-4** for disintegrating time, PD6 , and PD10 using both the X1 and X2 factors on either side of the axes respectively.

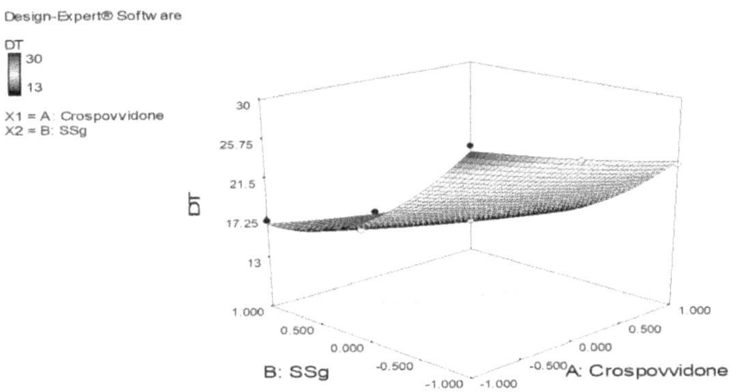

Figure 3 3D Surface Plots for Disintegration Time.

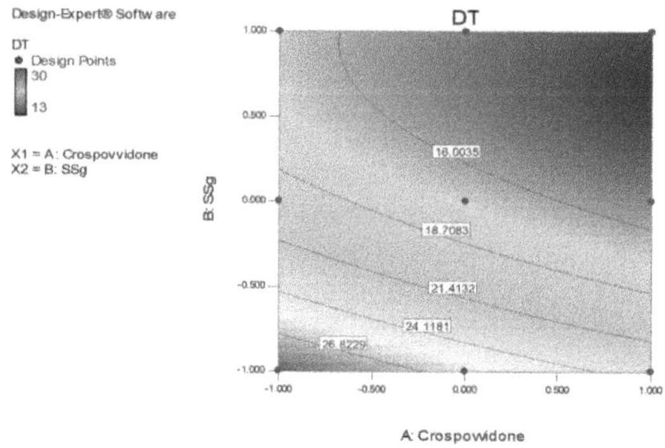

Figure 4 Contour Plot for disintegration Time.

The given Equation is a polynomial expression for 3^2 Factorial study

$$Y= b_0+b_1 X1+b_2 X2+b_{12} X1X2+b_{11} X1^2+b_{22} X2^2... \quad (1)$$

Where Y constitute dependent factor, b_0 represents the response of arithmetic mean across the F1-F9 batches, and b1 represents the predicted value for X1 factor[8,9]. The primary impacts (X_1 and X_2) show the report that obtain when you gradually increase each factor from a low value to the high. The reaction changes when two factors are altered at once, as shown by the interaction term (X_1X_2). To study non-linearity, the polynomial terms ($X1^2$ and $X2^2$) are inserted in equation. By creating two intermediate concentration check point formulations, the validity of the resulting equations was confirmed (C1, C2). The expression for Disintegration time (DT), percent drug dissolved in 6 min (PD_6) and percent drug dissolved in 10 min (PD_{10}) developed as follows.

$$Y_1= 17.11 - 2.33 X_1 -5.67 X_2+0.75 X_1X_2 +0.33 X_1^2 +3.33 X_2^2 \textbf{(DT)} \quad (2)$$

$$Y_2= 81.70 + 0.234 X_1 + 16.75 X_2 + 5.28 X_1X_2 - 4.2 X_1^2 + 1.05X_2^2 \textbf{(PD}_{\textbf{6m}}\textbf{)} \quad (3)$$

$$Y_3= 86.27 - 1.31 X_1 +14.2 X_2 +2.48 X_1X_2 - 0.762 X_1^2 + 0.038 X_2^2 \textbf{(PD}_{\textbf{10m}}\textbf{)} \quad (4)$$

The negative sign for factor X1 in Y1 equation represents that when the concentration of Crospovidone decreases, disintegrating time increases. The data show that the amounts of X1 (crospovidone) and X2 (primojel) both influence the length of time needed to release drug.[10]
From the report, it is evident that an increase in the quantity of Superdisintegrants tends to decline in the time required for the tablet to disintegrate and increase in drug dissolution.

The dissolution parameters that were observed from experimental data and those anticipated from the polynomial equations are compared. The response of X1 and X2 on Disintegration time (DT), percent of drug dissolved in 6 minutes (PD6min), and percent of drug dissolved in 10 minutes

(PD10min) were depicted using response surface plots. Optimized formulations represent the validity of derived expression for dependent variables namely Disintegration time (DT), Amount of % drug dissolved in 6 min (PD_{6}) and Amount of % drug dissolved in 10 min (PD_{10}).

The probability of any interaction between the drug with carriers throughout various techniques and the sample characteristics, was examined using DSC and XR-D studies respectively which shown in Figures 5-6.

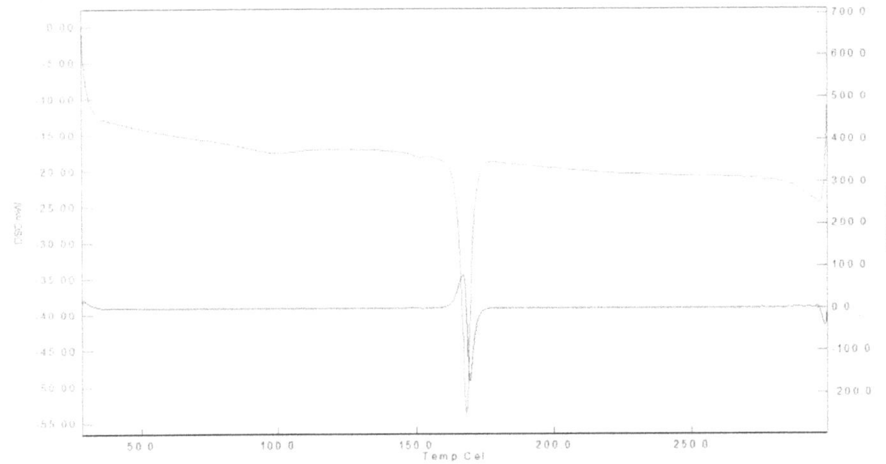

Figure 5 DSC spectra of optimized formulation.

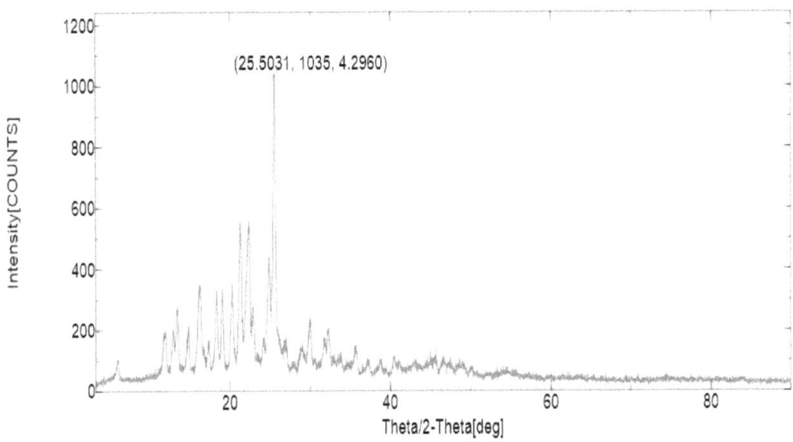

Figure 6 XRD spectra of optimized formulation.

5 CONCLUSION

The motive of the current study is to prepare the fast-dissolving tablets of Nebivolol employing complexation with β-cyclodextrin (1:3M) and by using solid dispersions of PEG 4000, Starch 1500, and Poloxamer 188. Nebivolol tablets with fast dissolution characteristics employing βCD were prepared using 3^2 factorial study and were also employed by direct compression technique.

1. The FT-IR spectrum of nebivolol was evident that has no interaction between nebivolol and the excipients used in the work.
2. Solid inclusion complexation with β CD – Nebivolol were employed by kneading method using various drug: β CD ratios namely 1:1M, 1:2M, and 1:3 M.
3. Dissolution of Nebivolol from the β CD complexes was rapid and higher when compared to that of Nebivolol pure drug. As the proportion of β CD was increased the dissolution rate also increased.
4. Solid dispersions of Nebivolol in Poloxamer 188, PEG 4000 and Starch 1500 was rapid and higher when compared to that of Nebivolol pure drug.
5. When compared with different carriers, βCD (1:3 M) ratio gave the highest dissolution rate when compared with other carriers like Poloxamer, PEG 4000 and Starch 1500.
6. Hence βCD (1:3 M) ratio was selected for the preparation of tablets by 3^2 factorial study.
7. Nebivolol containing 5 mg Tablets were formulated using βCD (1:3 M) employing primojel and Crospovidone by direct compression technique.
8. The tablet powder mix in each case possess excellent flow characteristics suitable for direct compression.
9. All tablets prepared were fulfilled the I.P disintegration time recommendation of noncoated tablets.

10. The values of correlation coefficient (r) were more in the first order model with regard to zero order model specified that the Nebivolol rate of dissolution followed first order kinetics.

11. The factorial study was used in formulation of nebivolol tablets in which all three levels of factor X1 (Primojel) and factor X2 (Crospovidone) are of concentration 3%, 4%, 5% (% calculated based on total Tablet weight , i.e250 mg) were chosen as the hypothesis for the execution of the rapid dissolution of Nebivolol formulations
12. The equations for Disintegration time (DT), percent drug dissolved in 6 min (PD_6) and percent drug dissolved in 10 min (PD_{10}) developed are as in equations (2,3,4.)
13. From the overview, it is evident that an enhancement in the portion of Superdisintegrant tends to decline in the time taken for disintegration of the tablet and increase in drug dissolution.
14. The dissolution parameters for those actually observed from experimental data and those anticipated from the polynomial equations are compared and identified as they are very close which indicating the validity of the equation.

Dissolution of Nebivolol from the β CD complexes was rapid and higher when compared to that of Nebivolol pure drug when compared with different carriers, like Poloxamer, PEG 4000 and Starch 1500. Hence βCD (1:3 M) ratio was selected for the preparation of tablets by 3^2 factorial study. Nebivolol equivalent to 5mg tablets were prepared by immersing βCD (1:3 M) using Primojel and Crospovidone as super disintegrants through direct compression technique. The tablet powder blends in each case possess excellent flow characteristics suitable for direct compression.All tablets prepared were satisified the I.P disintegration time recommendation of noncoated tablets. The factorial study was used in formulation of nebivolol tablets in which all three levels of factor X1 (Primojel) and factor X2 (Crospovidone) are of concentration 3%, 4%, 5% (% calculated based on total Tablet weight , i.e250 mg) were chosen as the hypothesis for the execution of the rapid dissolution of Nebivolol formulations

From the overview, it is evident that an enhancement in the portion of Superdisintegrant tends to decline in the time taken for disintegration of the tablet and increase in drug dissolution.

6 REFERENCES

1. K Kavitha, KumuthaSubramaniam, BoeyJiaHui, K. Santhi, SA Dhanaraj, and M Rupesh Kumar. Research Journal of Pharmaceutical, Potential Drug Candidates for Fast Dissolving Drug Delivery - A Review. Biological and Chemical Sciences, 2013; **4(4)**: 1510-1526.

2. Sehgal P, Gupta R, Umesh Kumar S, Chaturvedi A, Gulati A, Sharma M. Am. J. Pharm Tech Res, Fast Dissolving Tablets: A New Venture in Drug Delivery, 2012; **2(4)**: 252-279.

3. Raghavendra Kumar Gunda., International J of PharmTech Research, Formulation development and evaluation of rosiglitazone maleate sustained release tablets using 32 factorial design, 2015; **8(4)**: 713-724.

4. K.P.R.chowdary, Journal of Global Trends in Pharmaceutical Sciences, Optimization of valsartan tablet formulation by 23 factorial design; 2014,Volume **5(1),**1374-1379 .

5. Notari RE. Biopharmaceutics and clinical pharmacokinetics. New York: Marcel Dekker Inc; 4th ed. 1987; 6-21.

6. Higuchi T, J Pharm Sci.Mechanism of sustained-action medication. Theoretical analysis of rate of release of solid drugs dispersed in solid matrices,1963; 51:1145-9.

7. Peppas NA. Pharm ActaHelv1, Analysis of fickian and non-fickian drug release from polymers **985**; 60:110-1.

8. Schwartz BJ, Connor RE. J Drugs and Pharm Sciin Modern Pharmaceutics, Optimization technique in pharmaceutical formulations and processing. 1996; **72(3):**727-54.

9. A. A. Kharia, s. N. Hiremath, a. K. Singhai, K. Omray and S. K. Jain, Indian J. Pharm. Sci.Design and optimization of floating drug delivery system of acyclovir . 2010; **72 (5)**: 599-606.

10. Raghavendra Kumar Gunda, J. N. Suresh Kumar, Ch Ajay Babu and M. V. Anjaneyulu, International Journal of Pharmaceutical Sciences and Research.Formulation development and evaluation of lamotrigine sustained release tablets using 32 factorial design., 2015; **6(4)**: 1746-1752.

OPTIMIZATION OF VANILLIN CROSSLINKED CHITOSAN/POLYCARBOPHIL SUPERPOROUS HYDROGEL HYBRIDS CONTAINING TABLETS FOR REGULATED DRUG DELIVERY

Shwetha K [a], Basavaraj. B.V [a]

[a] Department of Pharmaceutics, Faculty of Pharmacy, Ramaiah University of Applied Sciences, Bangalore-560054, India.

Email: shwetha.ps.ph@msruas.ac.in

1 INTRODUCTION

Substantial research has been made to develop novel superporous hydrogel hybrids using natural polysaccharides and crosslinking agents.[1] Polysaccharides are recognized as most sustainable building blocks with improved characteristics for formulation of superporous hydrogels.[2] Chitosan is a most abundantly available marine based polysaccharide possessing extensive industrial applications due to their distinctive properties such as hydrophilicity, mechanical stability, and flexible functionality. [3,4] Many approaches have been attempted to formulate swellable tablets using hydrophilic and pH dependent polymers. Also, swellable systems need large quantity of fluid to swell. To overcome these hindrances, incorporation of superporous hydrogel hybrids made up of combination of pH sensitive polymers into tablet formulation will aid in good swelling with minimum fluid and at wide pH range.[5-7] Present research work emphasis on development and optimization of tablets comprising vanillin crosslinked chitosan/polycarbophil (CVP) superporous hydrogel hybrids for regulated drug delivery.

2 METHODS AND RESULTS

2.1 Preparation of CVP superporous hydrogel hybrids[8, 9]

The formulation of CVP superporous hydrogel hybrids were prepared using different concentrations of chitosan (polymer), vanillin (crosslinking agent) and polycarbophil (hybrid agent). These hybrids were prepared using microwave irradiation technique. 2% chitosan solution in 0.8%w/v lactic acid was prepared. To this, 2% vanillin solution was added and microwaved at power 100 for 60 seconds. Mixture was allowed to stand for 15mins and 4% polycarbophil solution was added and microwaved for another 60 seconds. Resultant formulation was subjected for characterization techniques.

2.2 Characterization of CVP superporous hydrogel hybrids

The CVP superporous hydrogel hybrids were characterized by Fourier Transform Infrared Spectroscopy (FT-IR), Differential Scanning Calorimetry (DSC), Thermal Gravimetric Analysis (TGA) and Nuclear Magnetic Resonance (NMR) characterization. Surface morphology of CVP hybrids and its powdered form were studied using scanning electron microscopy (SEM).

2.2.1 Fourier Transform Infrared Spectroscopy (FT-IR). The IR spectra of chitosan, polycarbophil, CVP hybrids and CVP tablets (shown in Fig.1). The IR spectroscopic analysis of chitosan shows typical characteristic board peaks at 3546cm^{-1} , 3470cm^{-1}and 3414 cm^{-1} (–OH stretching), 2918 cm^{-1} and 2880 cm^{-1}(-CH_2 stretching), 1638 cm^{-1}(N-H deformation vibrations), 1617 cm^{-1} (C=O stretching), 1153 cm^{-1} (C-O-C stretching) and 1092 cm^{-1} broad peak (skeletal vibrations of C-O and C-N stretching). N-H stretching also occurs in the 3315-3215 cm^{-1} region but OH stretching from carbohydrate ring is very strong and hence overlaps N-H stretching in this region. Peaks at 620cm^{-1} and 522cm^{-1} signifies polymeric backbone of chitosan. The IR spectroscopic analysis of polycarbophil shows typical characteristic –OH stretching peak at 3414 cm^{-1}, CH and CH_2 stretching at 3116 and 2957cm^{-1}. Polycarbophil exhibits intense C = O stretching peak at 1713 cm^{-1}, weak band at 1415 cm^{-1}indicates symmetrical stretching of COO- (carboxylate group), peaks at 1245 cm^{-1} and 1175 cm^{-1} signifies C-O stretching. Peaks at 803 cm^{-1},625 cm^{-1}and 522 cm^{-1}signifies polymeric backbone of polycarbophil. The IR spectroscopic analysis of chitosan – polycarbophil superporous hydrogels showed intense –OH stretching peak at 3413 cm^{-1}, sharp CH_2 stretching at 2919 cm^{-1} and CH (aldehyde) stretching at 2851cm^{-1}. Intense C = O stretching peak at 1713 cm^{-1}of polycarbophil is not seen and new peak at 1616 cm-1 is observed indicating interactions of polycarbophil and chitosan. Chitosan – polycarbophil superporous hydrogels exhibited strong and new peaks at 1561 cm^{-1}indicating interaction between amine group (-NH_2) of chitosan with carboxylate group (-COO^-) polycarbophil. Many sharp new peaks are noticed in 1161 cm^{-1},1113 cm^{-1},1053 cm^{-1} and 1023 cm^{-1}. The IR spectroscopic analysis of chitosan – polycarbophil superporous hydrogel powder with tablet excipients reveals all characteristics peaks of superporous hydrogels –OH stretching peak at 3414cm^{-1}, CH and CH_2 stretching at 2942 cm^{-1}, C=O from 1615 cm^{-1} similar to superporous hydrogels formulation, -CH deformation vibration is observed at 1428 cm^{-1}. Along with typical characteristic peaks many other additional and intense peaks showed the presence of excipients along with formulated superporous hydrogels.

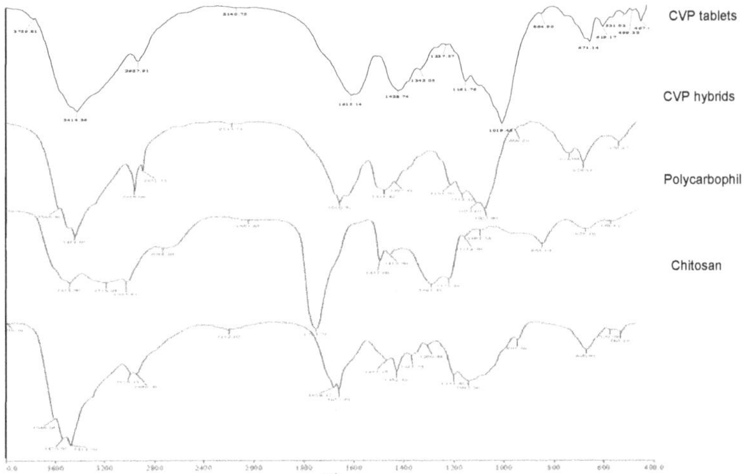

Figure 1 FTIR spectra of chitosan, polycarbophil, CVP hybrids and CVP tablets.

2.2.2. Differential Scanning Calorimetry (DSC). DSC thermogram (Fig 2) of chitosan showed an endothermic peak at 80°C due to evaporation of bound water molecules, a peak at 240°C due to glass transition and an exothermic peak at may be due to polymer degradation. Polycarbophil showed endothermic peak at 80°C due to evaporation of bound water molecules and peak between 230 – 245°C due to polymer degradation. CVP superporous hydrogels showed peak at 53°C may be due to evaporation of bound water molecules, broad exothermic peak at 100°C may be due to cleavage of polymeric rings as a result of electrostatic interactions between oppositely charged polymers and a broad peak at 240°C showed distinctive thermal characteristics of CVP superporous hydrogels.

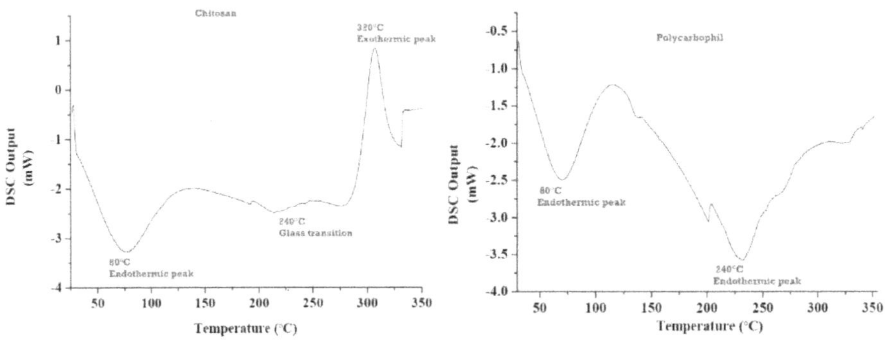

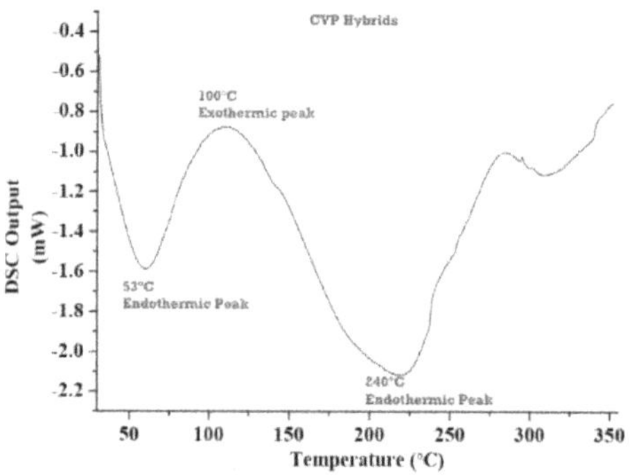

Figure 2 DSC thermogram of chitosan, polycarbophil and CVP hybrids.

2.2.3 Thermal Gravimetric Analysis (TGA). TGA thermograms helps to assess the thermal stability of polymer and hybrids. These thermograms (Fig 3) revealed single large peak for chitosan, polycarbophil and resultant hybrids indicating single degradation mechanism. TGA thermogram of chitosan revealed degradation process started at 279.98°C with midpoint of 292.94°C and ended at 309.64°C with a weight loss of 39.052%. polycarbophil thermogram revealed degradation process started at 202.42°C, with midpoint of 246.64°C. Degradation ended at 266.82°C with 24.738% weight loss of sample. Thermogram of CVP hybrids showed degradation process started at 30.92°C and ends at 38.70°C with 24.516% weight loss of sample. This thermogram reveals that at body temperature resultant formulation swells and forms gel like thicker consistency. This characteristic has wide application in drug delivery approaches.

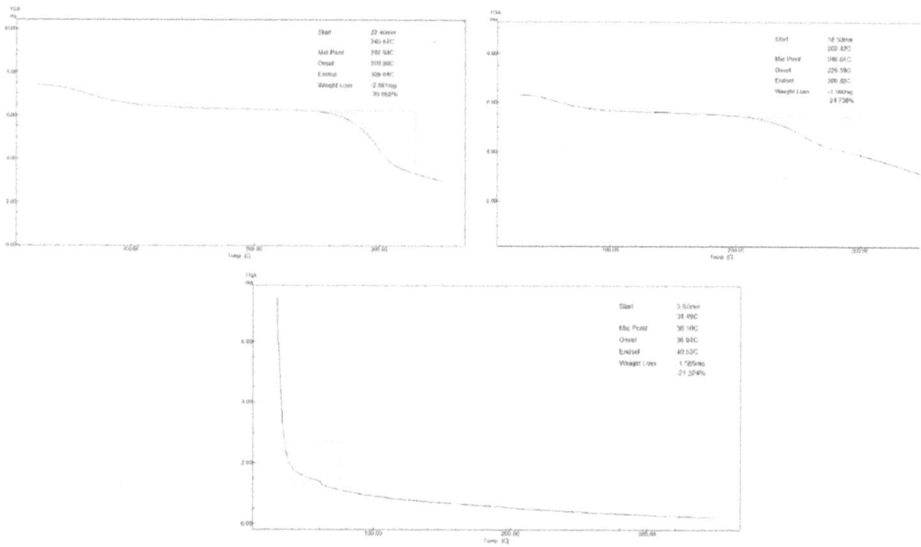

Figure 3 TGA thermogram of chitosan, polycarbophil and CVP hybrids.

2.2.4 Nuclear Magnetic Resonance (NMR) characterization. Carbon-13 NMR is a significant tool to predict the characteristics of biopolymers. The spectra of pure chitosan, polycarbophil and CVP hybrids (are given in Fig 4). The set of peaks complies with the literature data available. Chitosan shows peak at 75ppm symbolizing deacetylated chitosan. C=O, C_1, C_2, C_3, C_4, C_6 peaks are observed at 174.22ppm, 105ppm, 57.57ppm, 75.45ppm, 85ppm and 60ppm respectively. Sharp peak at 57.57ppm represents C-NH_2 group. Similarly, polycarbophil shows overlapping peaks at 182.52 and 179.78, which represents carboxyl group (-COOH) and carbonyl group (R-C=OR'), respectively. Sharp intense peaks are seen at 40.869 and 25.042 denotes the presence of aliphatic carbon chain (CH) and CH_2 carbon, respectively. The set of peaks for CVP hybrids revealed some shifts from the peak observed in chitosan and polycarbophil and some new peaks were also seen denoting formation of newer bonds. In comparison to chitosan and polycarbophil, intense sharp peak is observed at 180.179 indicating interactions with carbonyl group of polycarbophil, 1-2ppm shifts are observed in C_2 and C_3 carbon atoms of chitosan. This shift may be due to interactions with free amino group. Peak at 149.30ppm indicates hydroxyl group of vanillin (C-OH) used for crosslinking, peak at 129.18ppm indicates formation of amide bond (–$CONH_2$) and mild peak at 170ppm due to imine group (C=N) formation. Intense peak at 42.52ppm denotes RCH_2NH_2 group and peak at 23.47ppm represents methyl CH_3 carbon atom.

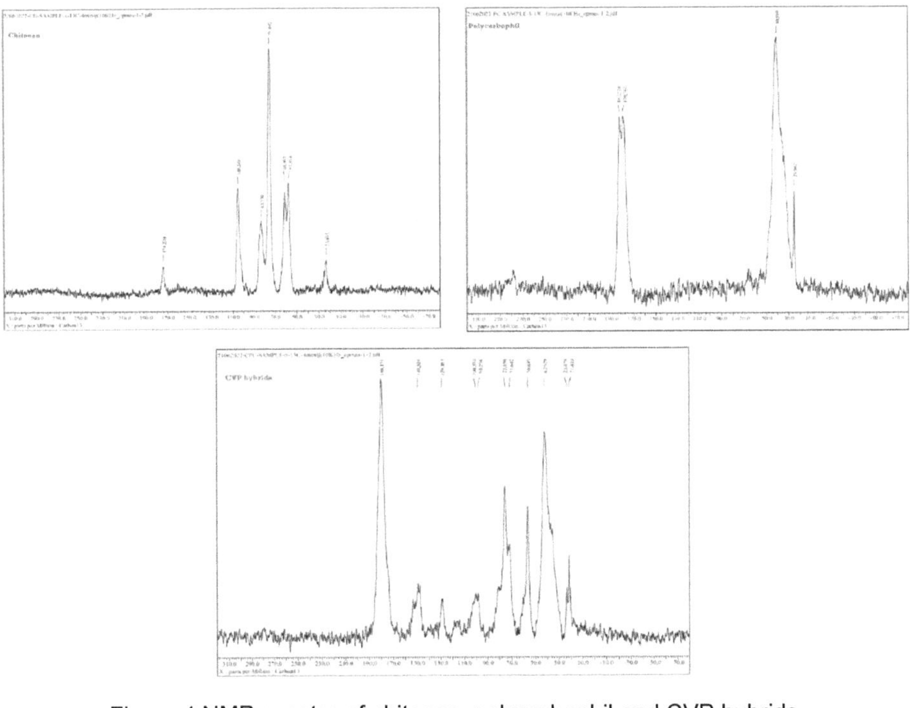

Figure 4 NMR spectra of chitosan, polycarbophil and CVP hybrids.

2.2.5 Surface morphology studies. SEM images (Fig. 5a) of optimized superporous hydrogel formulations revealed highly porous structure with unevenly distributed pores indicating abundant of interconnected pores with smaller pores at the terminal side of hydrogels and larger pores at the centre. These types of arrangement of pores are mainly due to freeze drying technique. The porous structure also confirmed better penetration of water molecules through the pores causing rapid swelling of product. The SEM images (Fig. 5b) of powdered CVP hybrids indicated the presence of porous structure in spite of powdering the formed hydrogels. This structural explanation using SEM images justifies the higher swelling ratio of the tablets.

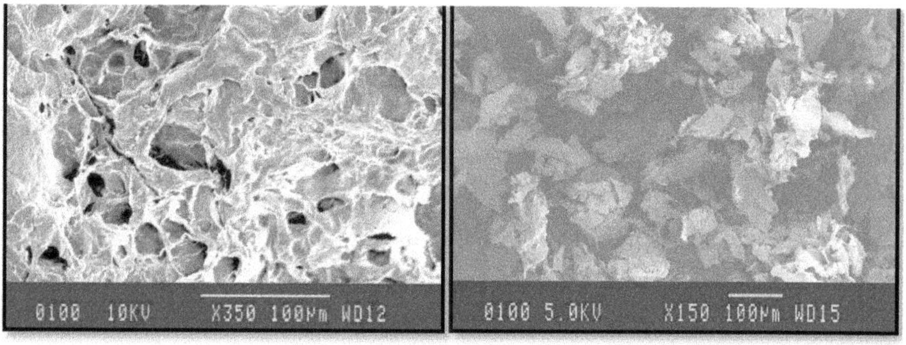

Figure 5 SEM images of CVP hybrids (a) and powdered CVP hybrids (b).

2.3 Preparation of tablets using CVP hybrids

Prepared CVP hybrids were powdered and blended with other excipients to formulate CVP tablets. Wet granulation technique was adopted for tablet preparation. CVP hybrids were blended with croscarmellose, sodium starch glycolate, lactose monohydrate. Using 5% of HPMC K100 in alcohol solution as binder, granules were prepared and dried at temperature less than 54°C until complete drying was achieved. These granules were punched using tablet punching machine (Rimek, RSB-4, MINI PRESS) and were evaluated for its characteristics.

2.4 Optimization of tablets by Box Behnken design[10]

These tablets were formulated and systematically optimized using Box Behnken design. This design was selected to determine higher order response surface using fewer required trials. Box Behnken design was used with three independent and dependent variables. Concentration of powdered CVP hybrids (A), croscarmellose (B) and sodium starch glycolate (C) were considered as independent variables whereas tablet hardness, swelling time and swelling ratio were taken as responses. The independent variables were studied at three different levels such as low (-1), intermediate (0) and high (+1) and a total of 17 experimental runs were evaluated to optimize the formulation. The obtained hardness ranges from 3-5 kg/N, swelling time ranges from 90 seconds to 180seconds, swelling ratio ranges from 15.5303 - 21.7224. 2D contour plots and 3D response surface plots were developed to predict the effect of independent variables on responses.

Based on analysis of hardness on the tablet, fit summary suggested linear model. Effect of individual factors on hardness of tablet was given mathematically as Hardness = 1.75 + 0.015 * A + 0.005 * B + -3.92523e-19 * C and is graphically represented in Figure 6 2D and 3D contour plots. Presence of straight lines indicates that interactions between the independent variables does not influence hardness. ANOVA for this model was significant and lack of fit was insignificant as desired.

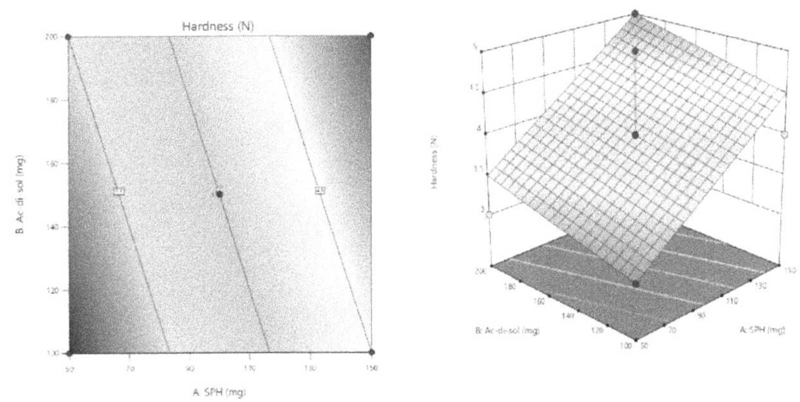

Figure 6 2D contour plot and 3D response surface plot for hardness of tablet

Based on analysis of swelling time, fit summary suggested quadratic model. Effect of each individual factors on swelling time was mathematically given as Swelling time = -0.0625 + 0.01125 * A + 0.04 * B + -0.01375 * C + 5e-05 * AB + -0.0001 * AC + -5e-05 * BC + -1.70681e-19 * A^2 + -0.00015 * B^2 + 0.0001 * C^2 and is graphically represented in Figure 7 2D and 3D contour plots. Presence of slightly curved lines in graph revealed influence of few interactions on swelling time of the tablet. ANOVA for this model was significant and lack of fit was insignificant as desired.

Based on analysis of swelling ratio, fit summary suggested 2 factor interaction model. Effect of each individual factors on swelling ratio was mathematically given as Swelling ratio = 39.1236 + -.00550394 * A + -0.14741 * B + -0.136666 * C + 0.000132695 * AB + 0.0001348 * AC + 0.000817539 * BC and are graphically represented in Figure 8 2D and 3D contour plots. Presence of slightly curved lines in graph revealed influence of few interactions on swelling ratio of the tablet. ANOVA for this model was significant and lack of fit was insignificant as desired.

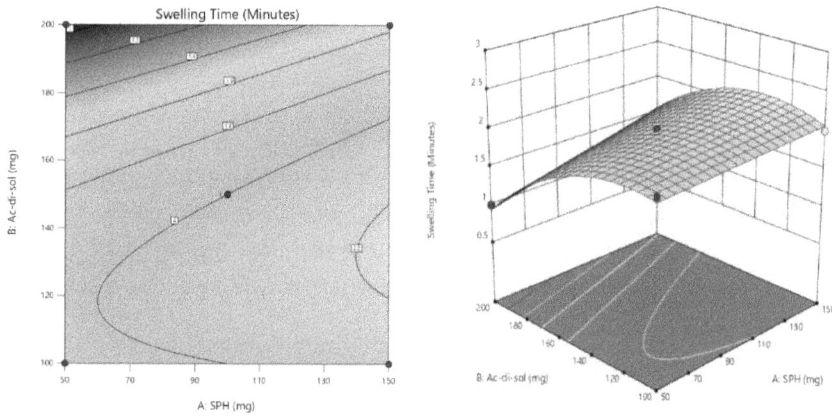

Figure 7 2D contour plot and 3D response surface plot for swelling time of tablet.

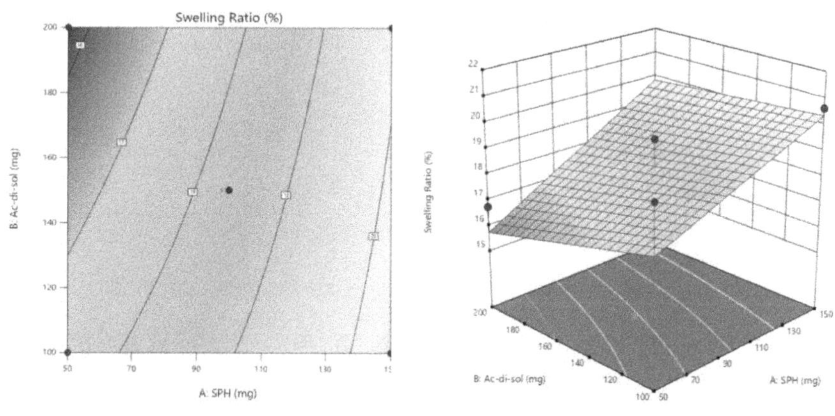

Figure 8 2D contour plot and 3D response surface plot for swelling ratio of tablet.

Based on the constraints given, software generated 100 solutions with desirability of 1, out of which one was selected, and experiment was repeated. Experimental values were compared with predicted values and it was found that difference between them was minimum and acceptable as shown in following Table 1. Hence solution two was considered as optimized formulation (Table 2) for further studies.

Table 1 Point *prediction of Box Behnken Design.*

Solution 2 of 100 Response	Predicted Mean	Predicted Median	Observed	Std Dev	n	SE Pred	95 % PI low	Data Mean	95 % PI high
Hardness	4	4	4	0.480384	3	0.300828	3.3501	4	4.6499
Swelling Time	2	2	2	0.211289	3	0.154303	1.63513	2	2.36487
Swelling Ratio	18.3689	18.3689	18.6929	1.46274	3	0.916006	16.3279	18.6929	20.4098

Table 2 Optimized *formulation.*

Ingredients	Quantity (mg)
CVP hybrids	100
Ac-di-sol	150
Sodium starch glycolate	150
Magnesium stearate	2.5
Talc	2.5
Lactose monohydrate	q.s to 700
HPMC 5 % in Isopropyl alcohol	q.s

2.5 Evaluation of optimized tablets

Optimized CVP tablets shown in Fig. 9 were evaluated for hardness, friability, weight variation swelling time and swelling ratio. Hardness of the tablet was found to be 4kg/cm^2, friability was 0.72 %-0.84 %, average weight of the tablets was between 688mg to 705mg. Optimized CVP tablets complied with pharmacopeial specifications for hardness, friability and weight variation. Swelling time was less than 2 minutes. Optimized CVP tablets revealed potential swelling characteristics, chitosan aided swelling at pH 1 to 4 whereas polycarbophil revealed good swelling from pH 4 to 8 with swelling ratio of 17.6531±0.1134 and 21.7031± 0.0791 respectively.

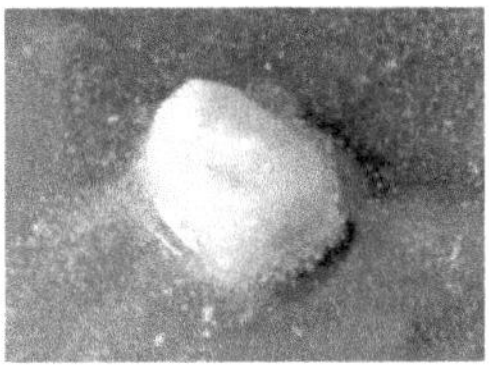

Figure 9 Images of CVP hybrid tablet in dried state and swollen state.

3 CONCLUSION

The physical and chemical properties of CVP hybrids were determined successfully to understand interactions between polymers and crosslinking agent and their influence on swelling characteristics. FTIR showed shifts in existing peak indicating formation of newer bonds. DSC thermograms revealed thermal stability of the CVP hybrids. TGA thermograms indicated formulation alters at body temperature. NMR studies confirmed formation of bonds that aid in improved swelling behaviour. SEM images of CVP-hybrids and its powdered form revealed highly interconnected porous structure. These SEM images indicated the presence of porous structure despite size reduction of the formed hybrids. This structural explanation using SEM images justifies the higher swelling ratio of the tablets. Optimized CVP tablets complied with pharmacopeial specifications for hardness, friability, and weight variation. Optimized CVP tablets revealed potential swelling characteristics from pH 1 to 8. These results suggested that vanillin crosslinked chitosan polycarbophil superporous hydrogel hybrids can be successfully formulated as tablets by wet granulation technique and can be explored for regulated delivery of drugs to alimentary canal for both local and systemic effects.

4 ACKNOWLEDGEMENTS

The authors are thankful to Ramaiah University of Applied Sciences for providing all facilities to conduct the research. The authors are also grateful to Lubrizol, USA for providing gift sample of polycarbophil.

5 REFERENCES

1. D. J. Mastropietro, H. Omidian and K. Park, *Expert Opinion on Drug Delivery*, 2011, **9**, 71–89.
2. G. Auriemma, P. Russo, P. Del Gaudio, C. A. García-González, M. Landín and R. P. Aquino, *Molecules*, 2020, **25**, 3156.
3. T. G. Barclay, C. M. Day, N. Petrovsky and S. Garg, *Carbohydrate Polymers*, 2019, **221**, 94–112.
4. Q. Chai, Y. Jiao and X. Yu, *Gels*, 2017, **3**, 6.
5. R. Anil, K. Arza, C. Sekhara, R. Gonugunta and P. Reddy Veerareddy, *AAPS PharmSciTech,* DOI:10.1208/s12249-009-9200-y.
6. Ali Demir Sezer, *Smart drug delivery system*, Intech, Rijeka, Croatia, 2016.
7. K.-M. Hwang, T.-T. Nguyen, S. Seok, H.-I. Jo, C.-H. Cho, K.-M. Hwang, J.-Y. Kim, H. Seok, H-I. Jo, C-H. Cho, K-M. Hwang, J-Y. Kim, C-W. Park and Y-S. Rhee, *International Journal of Pharmaceutics-in vivo evaluation*, DOI:10.1016/j.ijpharm.2019.118783.
8. S. Abraham, D. Rajamanickam and B. Srinivasan, *Science International*, 2018, **6**, 18–30.
9. T. N. Le, J. Her, T. Sim, C. E. Jung, J. K. Kang and K. T. Oh, *AAPS PharmSciTech*, DOI:10.1208/s12249-020-01851-5.
10. O. A. Attallah and W. Mamdouh, *International Journal of Environmental Science and Technology*, 2020, **18**, 131–140.

OPTIMIZATION OF TELMISARTAN FAST DISSOLVING TABLETS BY 3^2 FACTORIAL DESIGN

M. Sukanya*, V.Rupa Sree Devi, K.Ravi Shankar, G.Ramana Reddy and R.Anusha
KVSR Siddhartha College of Pharmaceutical Sciences,Vijayawada
*munnangisukanya@gmail.com

1 INTRODUCTION

Since most prospective therapeutic medicines (APIs) have low water solubility at physiological pH and consequently slow dissolution rates, they provide difficult issues in the creation of pharmaceutical products because of their low solubility and high permeability, which place them in BCS class II classification. Micronization, cyclodextrin complexation, use of surfactants and solubilizers, solid dispersion in water soluble and dispersible carriers, use of salts, prodrugs and polymorphs that exhibit high solubility, micro emulsions, and self-emulsifying micro and nano disperse systems are a few techniques that have been used to enhance the solubility, dissolution rate, and bioavailability of low soluble drugs. Out of the approaches mentioned above, cyclodextrin complexation, solid dispersion technique has proven to be the most effective one.[1] They represent a straightforward, affordable, and promising methods to significantly enhance bioavailability, dissolution rate, and the bioavailability of medications with low solubility. The widely used antihypertensive drug telmisartan, also known as [4-[[4-methyl-6-(1-methylbenzimidazol-2-yl)-2-propylbenzimidazol-1-yl]methyl] 1,1 biphenyl]- 2 carboxylic acid, is classified as class-II drug under the BCS classification and has a low and variable oral bioavailability because of its poor aqueous solubility.[2] To achieve maximal therapeutic efficacy, it therefore requires improvement in the solubility and dissolution rate in the formulation development. The typical dose of telmisartan, which is used in patients and the elderly, ranges from 20 mg to 80 mg.

The major goal of this research is to increase Telmisartan's solubility and rate of dissolution by making solid dispersions with certain polymers using the kneading method using PEG4000, Mannitol, β-Cyclodextrin, Starch 1500, Microcrystalline Cellulose, Primogel. [2,3,4] Utilizing a 3^2-factorial design, the optimized solid dispersion was used to create tablet formulations that were tested for physical parameters and In vitro dissolution rates. Fourier transform infrared (FTIR), X-ray powder diffractometry (XRD), and Differential scanning calorimetry (DSC) are used to distinguish the optimized formulation.

2 MATERIALS AND METHODS

2.1 Solid Dispersion Preparation by Kneading Method[5]

The calculated amount of Telmisartan and used carriers of PEG4000, β-Cyclodextrin, Primogel, Starch 1500, MCC, and Mannitol in various drug:carrier ratios were weighed and triturated using a small amount of organic solvent, often methanol. Until the dried mass was obtained, the slurry was kneaded for 30 to 40 minutes before drying at 45 °c in a hot air oven. The mass that was dried was ground and passed through sieve number 80, and the fraction was collected, kept in a desiccator, and used in further investigations.[6]

2.2 Formulation of Tablet

As prepared by the kneading method, solid dispersion of Telmisartan- β-CD (1:3M) equivalent to 20 mg of drug was taken and mixed with pliable diluent, superdisintegrants, and other excipients in plastic container. Table 1 lists the ingredients in the tablet formulation. In a multi-station tablet compression machine (Cadmach, Ahmedabad, India), powder blends were directly compressed using an 8mm flat punch with a round form to a hardness of 4-6 kg/cm^2.

2.3 FT-IR Studies

Shimadzu FTIR-8700 spectrophotometer was used to record the FT-IR spectrum of the pure drug and the optimized solid dispersion. Potassium bromide disc method was used for this study. Dry potassium bromide powder was added to the powdered sample, which was subsequently compacted into a transparent disc under intense pressure using specialized dies. To assess polymer drug interaction investigations, the disc was placed in an IR spectrophotometer with sample holder, and the spectrum was scanned across the frequency range of 4000-400 cm^{-1}.

2.4 Differential Scanning Calorimetry (DSC)

The probability of any interactions between drug and excipients was evaluated using DSC to perform drug thermal analysis. Samples equivalent to 1 to 5 mg, were used in the Shimadzu-Thermal Analyzer DSC 60 (Japan) was used to evaluate. In presence of nitrogen environment, samples were heated in a disclosed pan at a rate of 10°C/min over a temperature range of 100°C to 250°C.

2.5 X-Ray Powder Diffraction Studies

An X-ray powder diffraction system (Shimadzu, Japan) is used to conduct XRD studies of drug and optimised solid dispersion (β-CD1:3M) employing copper target, at a voltage of 40 Kv, and a current of 15 mA. Scanning range for the SC-70 detector's XRD scan was between 3.000 and 90.0000 degrees.

3 CHARACTERIZATION OF TABLETS

The tablets' hardness, friability, drug content analysis, disintegration time, and in vitro dissolution were all evaluated.

3.1 Hardness

The tablet's hardness was measured by placing it longitudinally between the tester's two plungers, and the result was expressed in kilogrammes per square centimetre (kg/sq.cm).

3.2 Test of Friability

Tablets crack due to forces like friction and shock. The friability test, which measures a tablet's resistance to abrasion during handling, packaging, and transportation, is related to tablet hardness. Usually, it is assessed with a Roche friabilator. A number of tablets were weighted and put inside the device, where they were subjected to abrasion as they fell 6 inches everytime the device turned. The weight of the pills was measured after 100 spins and compared with the initial weight. Weight loss serves as a measure of tablet friability, which is expressed as a percentage. The common consensus is that a weight loss of no more than 1% of one's body weight is acceptable.

3.3 Estimation Of Drug Content[3]:-An optimized solid dispersion (β-CD 1:3M) which is equivalent to 20 mg of telmisartan is placed in a volumetric flask of 100 ml, which was then made up with 7.4 pH phosphate buffer. After sonicating for 15 minutes, the solution was then filtered using Whatman filter paper. Using UV-visible spectrophotometer at 296 nm, the amount of telmisartan (from the above solution) was measured against an appropriate blank (1601, Shimadzu, Kyoto, Japan).

3.4 Disintegration test

The U.S.P device to evaluate disintegration comprises 6 glass tubes with mesh screens at the bottom. One tablet is positioned in each tube which arranged perforated discs on its top were placed in a 1-L water beaker that mimics gastric fluid and intestinal fluid at 37±2C and moves up and down movement over 5–6 cm at a rate of 28–32 cycles per minute by the help of motor. According to the test the time that tablet requires to disintegrated and passed through the mesh screen is considered as Disintegration time.

3.5 In-Vitro Dissolution

Utilizing USP dissolution test type II equipment, the in-vitro dissolution investigation was carried at 37°C±0.5°C and 75 rpm. Solid dispersion samples holding 20 mg of drug /Tablet holding 20 mg of drug were added to a dissolution container holding 900 mL of pH 7.4 phosphate buffer. 5 ml samples that were periodically taken out at set intervals of time were filtered and are replaced with 5 ml of fresh dissolution medium. Samples filtered via 0.45 m microfilter were measured for the concentration spectrophotometrically by UV spectrophotometer (UV 1800 shimandzu spectrophotometer) at a λmax of 296.

4 DISCUSSION OF RESULTS

Telmisartan was estimated using UV Spectrophotometric approach based on the readings of absorbance at 296 nm in pH 7.4 phosphate buffer. The method is verified for linearity, accuracy, precision, and interference after creating a calibration curve. The procedure followed Beer's rule at concentrations between 2 and 10 g/ml.

By using FTIR tests, the compatibility of the drug with the excipients were evaluated. Figures 1 and 2 depict the FT-IR spectra of the powdered form of pure Telmisartan and optimized formulation respectively. There is no interaction between Telmisartan and the excipients utilized in the investigation, as evidenced by the observation of the IR peaks in the spectra of Telmisartan and preparations.

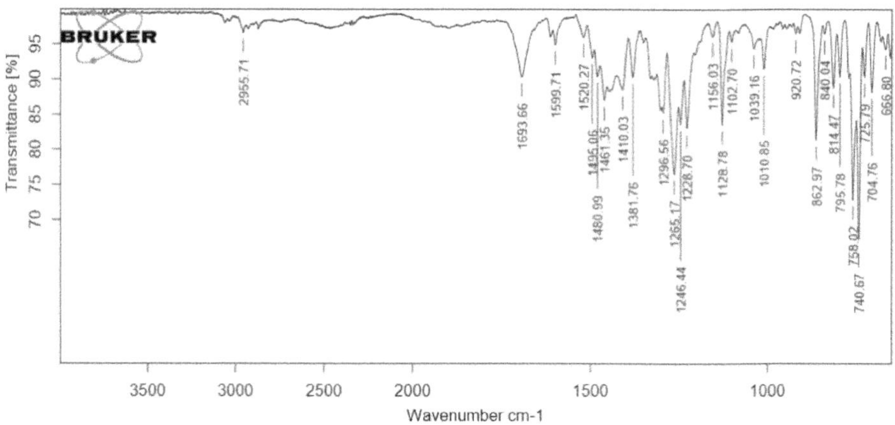

Figure 1 FT-IR spectra of Pure Telmisartan.

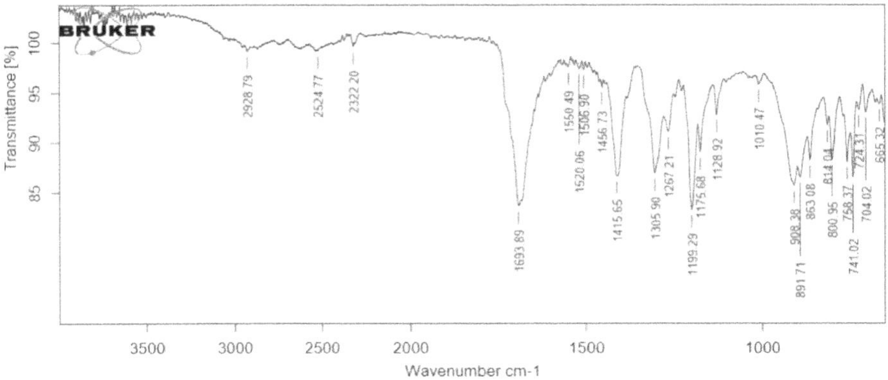

Figure 2 FT-IR spectra of Compressed tablet.

To increase Telmisartan's solubility, it was attempted to complex the drug with β cyclodextrin (βCD) in the ratios of 1:1, 1:2, and 1:3 M using kneading method which were undergone for Phase solubility studies to examine the impact of βCD and its concentration on Telmisartan's solubility. Comparing the rate of Telmisartan dissolution from the βCD complexes to that of the pure drug, it was quicker and higher. Other carriers like Primogel, PEG 4000, Mannitol, MCC and Starch 1500 were used to create similar dispersions.[7] The βCD (1:3 M) ratio gave the highest dissolving rate when compared to other carriers, hence βCD (1:3 M) ratio was chosen for tablet manufacture using 3^2 factorial study. The dissolution parameters of various solid dispersions were given in Table 1

Table 1: *Dissolution Parameters of Telmisartan Inclusion complexes and Solid Dispersions.*

Release Kinetics	R^2 values zero order	R^2 values first order	K values	DE_{30} values
β-CD (1:1M)	0.951	0.982	0.060	55.39
β-CD (1:2M)	0.914	0.973	0.124	71.46
β-CD (1:3M)	0.912	0.951	0.162	83.67
Mannitol(1:1)	0.942	0.986	0.047	51.315
Mannitol(1:2)	0.884	0.981	0.080	62.06

Mannitol(1:3)	0.872	0.992	0.106	68.91
PEG10%	0.954	0.980	0.081	57.116
PEG20%	0.962	0.997	0.074	62.71
Primojel(1:1)	0.859	0.977	0.074	46.67
Primojel(1:2)	0.875	0.983	0.051	48.33
Primojel(1:3)	0.897	0.961	0.114	52.03
Starch1500 (1:1)	0.951	0.992	0.138	60.70
Starch1500 (1:2)	0.942	0.991	0.039	64.29
Starch1500 (1:3)	0.938	0.990	0.040	67.01
MCC(1:1)	0.958	0.986	0.044	45.25
MCC(1:2)	0.957	0.972	0.052	45.72
MCC(1:3)	0.955	0.995	0.053	50.38

Preparation and characterization of Telmisartan tablet formulation and evaluation utilizing a 3^2 factorial design using Sodium Starch Glycolate and Crospovidone.was adopted

Tablets containing 20mg of Telmisartan each were formulated using the formula given in Table 2, together with sodium starch glycolate and crospovidone. The tablets were made using the direct compression technique. By evaluating the compressibility index and angle of repose, the formulations of tablet ingredients prepared in each case were assessed for their flow characteristics prior to compression. With varied formulations, compressibility index (%) values are in between of 10-12% and the angle of repose (θ) values ranged from 150° to 220°. These results showed that each case tablet powder blends had good flow properties suited for direct compression.

Table 2 *Formulae of Telmisartan Tablets Prepared.*

Ingredients (mg)	**F1**	**F2**	**F3**	**F4**	**F5**	**F6**	**F7**	**F8**	**F9**	**OPT1**
Telmisartan SD	132	132	132	132	132	132	132	132	132	132
Acacia	5	5	5	5	5	5	5	5	5	5
Sodium Starch Glycolate	7.5	10	12.5	7.5	10	12.5	7.5	10	12.5	12.5
Cross Povidone	7.5	7.5	7.5	10	10	10	12.5	12.5	12.5	12.5

Starch 1500	qs	qs	qs	qs	qs	qs	qs	qs	qs	qs
Magnesium Stearate	2.5	2.5	2.5	2.5	2.5	2.5	2.5	2.5	2.5	2.5
Talc	2.5	2.5	2.5	2.5	2.5	2.5	2.5	2.5	2.5	2.5
Total weight (mg)	250	250	250	250	250	250	250	250	250	250SS

Drug content, hardness, friability, disintegration time, and dissolution rate parameters for each manufactured tablet were assessed. The tablet's hardness ranged from 4.0 to 6.0 kg/cm^2. In every case, the weight lost during the friability test is less than 1%. The drug content is within 3%. The disintegration time ranged from 15 to 30 seconds. All of the manufactured tablets met the standard uncoated tablet disintegration time criterion.

In 7.4 phosphate buffer, the rate of dissolution of various Telmisartan tablets were investigated. Fig.3 display the dissolution profiles of several tablets that have been manufactured.

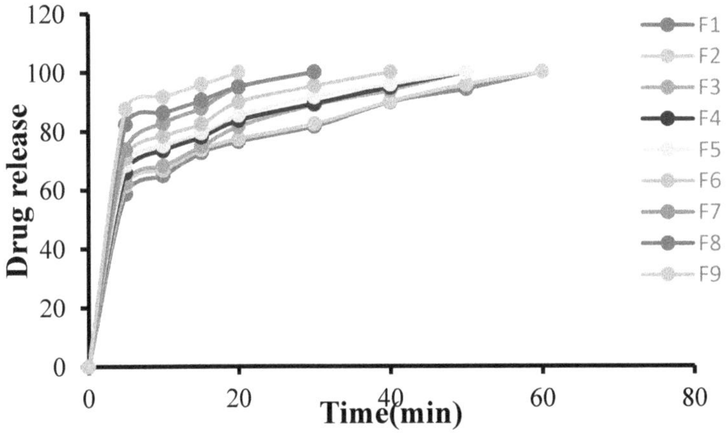

Figure 3 *In* vitro dissolution Profile of Telmisartan tablets prepared by using Crospovidone and SSG as super disintegrants.

In each case, the dissolution data was examined using zero order and first order kinetics. First order model had correlation coefficient(r) values that were higher than those in the zero-order model, showing that first order kinetics underlay the dissolution of Telmisartan. The first order model's correlation coefficient (r) values were discovered to be between 0.918 and 0.998. Table 3 provides a summary of the Telmisartan tablet dissolution parameters and Table 4 provides the actual and predicted values of optimized telmisartan tablet formulation.

Table 3 *Dissolution Parameters of Telmisartan Tablets.*

Release Kinetics	R^2 values zero order	R^2 values first order	K values	DE_{30} values
F1	0.804	0.970	0.048	65.45
F2	0.791	0.966	0.052	72.07
F3	0.795	0.971	0.061	81.15
F4	0.761	0.961	0.063	66.8
F5	0.746	0.967	0.069	73.78
F6	0.753	0.965	0.092	83.65
F7	0.777	0.970	0.134	69.46
F8	0.727	0.934	0.13	76.99
F9	0.780	0.938	0.199	87.53
Telmikind	0.680	0.943	0.157	68.42
OPT1	0.793	0.987	0.070	73.56

Table 4 *Actual and Predicted values of Optimized Telmisartan Tablet Formulation.*

FORMULATION CODE	PREDICTED VALUE		ACTUAL OBSERVED VALUE	
	DT (Sec)	PD 20 (%)	DT (Sec)	PD 20 (%)
OPT1	17	84.99	16.85	84.25

The percentage in which independent variables Crosspovidone and Sodium Starch Glycolate were utilized in the formulation of Telmisartan Fast Dissolving Tablets is described by 3^2 factorial design. The dependent variables chosen were disintegration time(DT), percent drug dissolved in 10min and percent drug dissolved in 20min. For the final equations, significance terms with a 95% confidence interval (p0.05) were chosen. Polynomial equations were developed for disintegration time (DT), percent of drug dissolved in 10min and percent of drug dissolved in 20min.

The 3 levels of factor X1 (Crospovidone) at 3%, 4%, and 5%, as well as the 3 levels of factor X2 (Sodium starch glycolate) at 3%, 4%, and 5% (% calculated based on total weight of the tablet, i.e.,250mg) were used to design telmisartan fast dissolving tablet formulations. Nine different telmisartan rapid-dissolving tablet formulations were developed using various combinations of the two factors, X1 and X2, according to the 3^2-factorial design. These formulations were then evaluated to determine the importance of the combined effects of factors X1, X2, which is the best combination. The factor X1(Crospovidone) and X2(Primogel) concentrations were divided into three levels and assigned the codes as -1=3%, 0=4%, and +1=5%.

Using Design Expert 7 software, polynomial equations were generated for the disintegration time(DT), percent of drug dissolved in 10 min(PD10), and percent of drug dissolved in 20min(PD20). Figures 4-9 display the response surface plots for disintegrating time(DT), PD10, and PD20 using the X1, X2 , respectively.

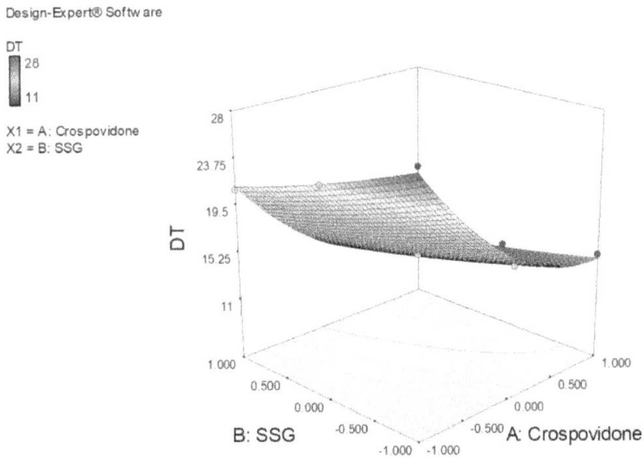

Figure 4 3D Response surface Plot for Disintegration Time (DT).

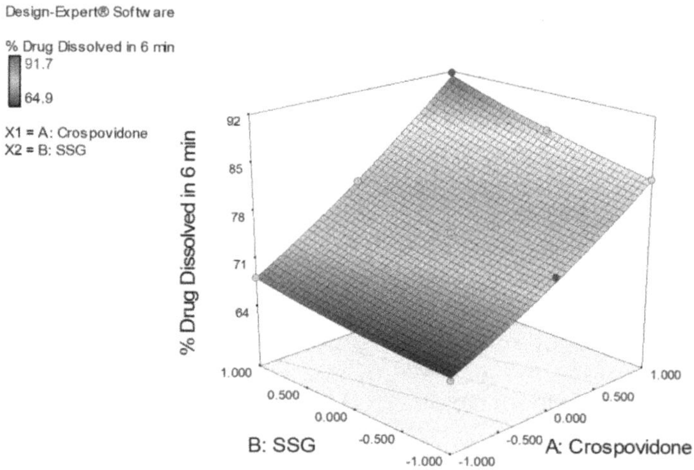

Figure 5 3D Response surface Plot for drug dissolved in 10min(PD10).

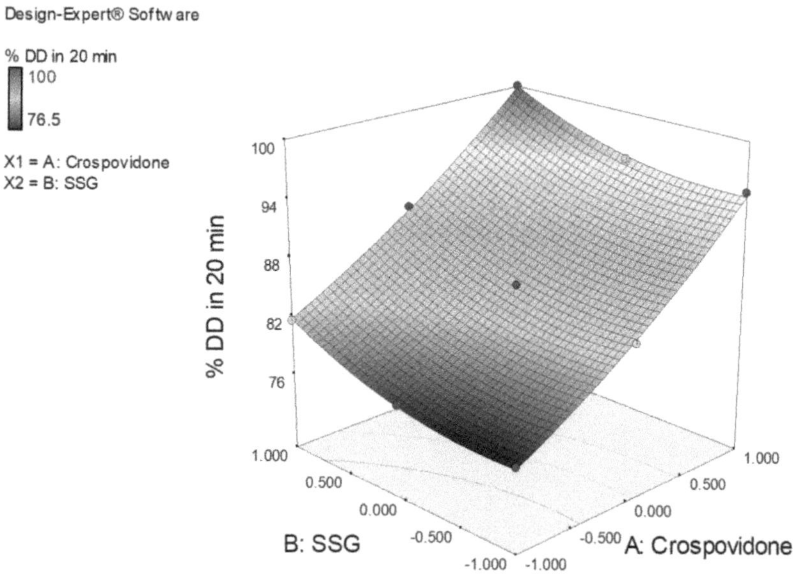

Figure 6 3D Response surface Plot for drug dissolved in 20min(PD20).

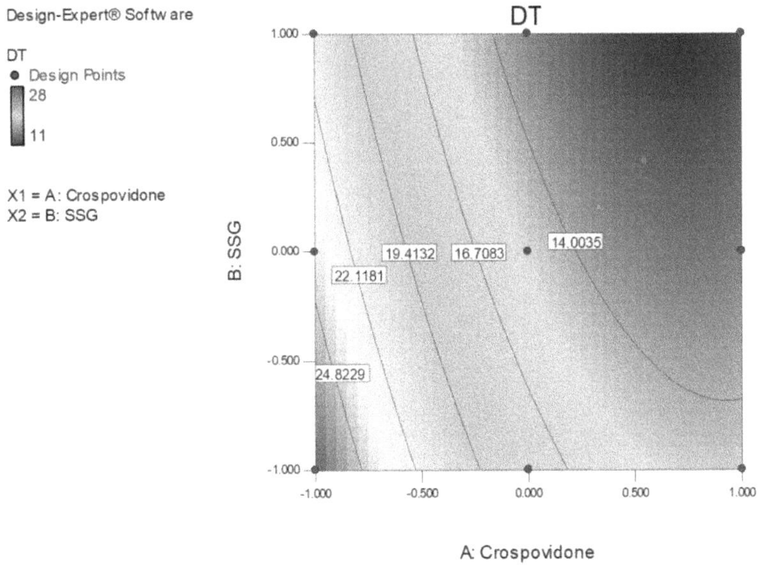

Figure 7 Contour Plot for Disintegration Time (DT).

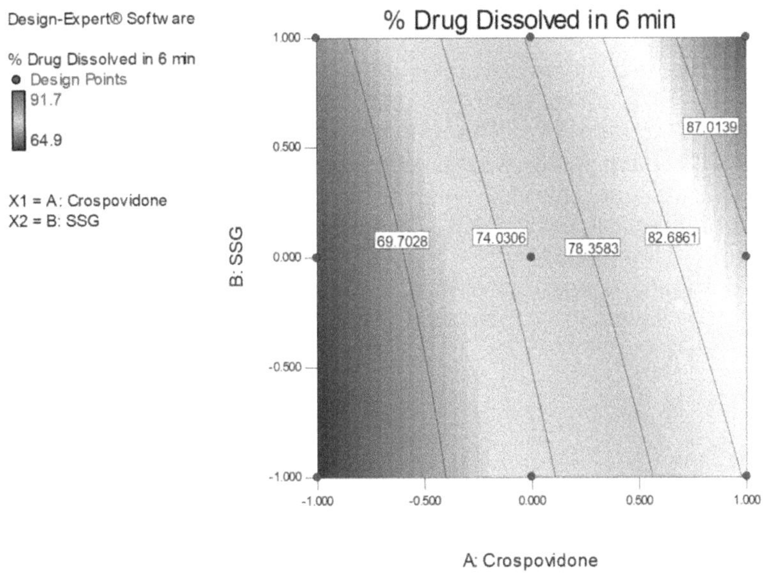

Figure 8 Contour Plot for drug dissolved in 10min(PD 10).

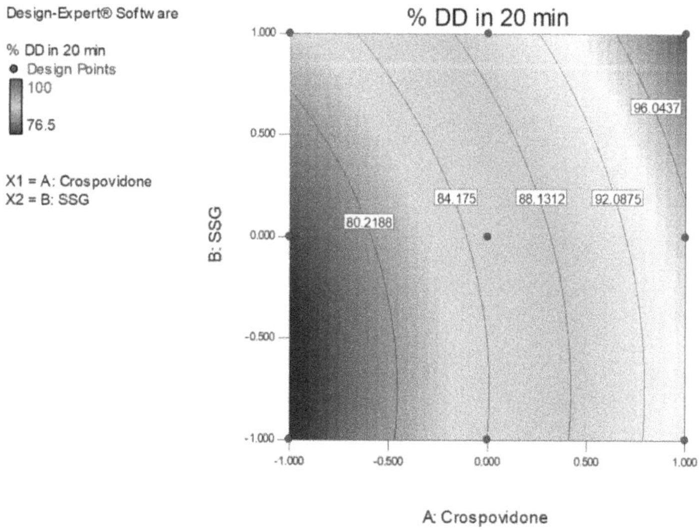

Figure 9 Contour Plot for drug dissolved in 20min(PD20).

The 3^2 full factorial design polynomial equation is given by

$$Y=b_0+b_1 X_1+b_2 X_2+b_{12} X_1X_2 X_1^2+b_{22} X_2^2... \quad (1)$$

Here, Y is the dependent variable, b_0 represents the arithmetic mean response across the nine batches, and b1 represents the calculated coefficient for component X_1. The primary impacts (X_1, X_2) show what happens when an each factor is gradually increased from a low value to a high value. The reaction changes when two factors are altered at once, as shown by the interaction term (X_1X_2). To study non-linearity, the polynomial terms (X_1^2 and X_2^2) are added.

The following equations were created to represent disintegration time (DT), percent of drug dissolved in 10 minutes (PD10), and percent of drug dissolved in 20 minutes (PD20).

$$Y_1= 15.11 - 5.67 X_1 -2.33 X_2+0.75 X_1X_2 +3.33 X_1^2 +0.33 X_2^2 \textbf{(DT)} \quad (2)$$
$$Y_2= 75.33 + 10.17 X_1 + 2.82 X_2 + 1.37 X_1X_2 - 1.10 X_1^2 + 0.55 X_2^2 \textbf{(PD}_{\textbf{10min}}\textbf{)} \quad (3)$$
$$Y_3= 85.13 - 9.08 X_1 +2.73 X_2 +0.000 X_1X_2 - 1.25 X_1^2 + 1.80 X_2^2 \textbf{(PD}_{\textbf{20min}}\textbf{)} \quad (4)$$

The co-efficient of X_1 in the Y_1 equation has a negative sign, which means that as Crospovidone concentration decreases, a delay in disintegration time is observed. In other words, the statistics show that the amounts of X1 (crospovidone) and X2 (sodium starch glycolate) both influence the drug release.

The findings suggest that increasing the Super disintegrant concentration reduces the dosage form's disintegration time and the release pattern of the drug can be altered by choosing right X_1, X_2 levels. [8,9,10].Dissolution parameters for those actually observed from experimental data and those anticipated from the polynomial equations are compared. The effects of X1 and X2 on Disintegration time (DT), percentage of drug dissolved in 10 minutes (PD10min), and percentage of drug dissolved in 20 minutes (PD20min) were displayed using response surface plots.The rationality of the developed equations for dependent variables disintegration time (DT), percent of the medication dissolved in ten minutes (PD10), and percent of the drug dissolved in twenty minutes (PD20) is shown using optimized formulations.

The probability of any interaction between the drug with carriers throughout various techniques and the sample characteristics, was examined using DSC and XR-D studies respectively which shown in figures 10-13.

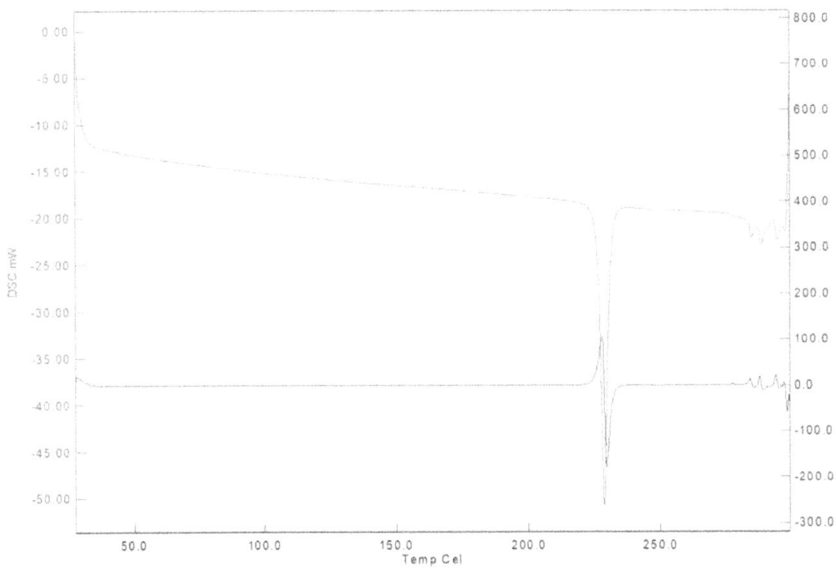

Figure 10 DSC spectra of Pure drug Telmisartan.

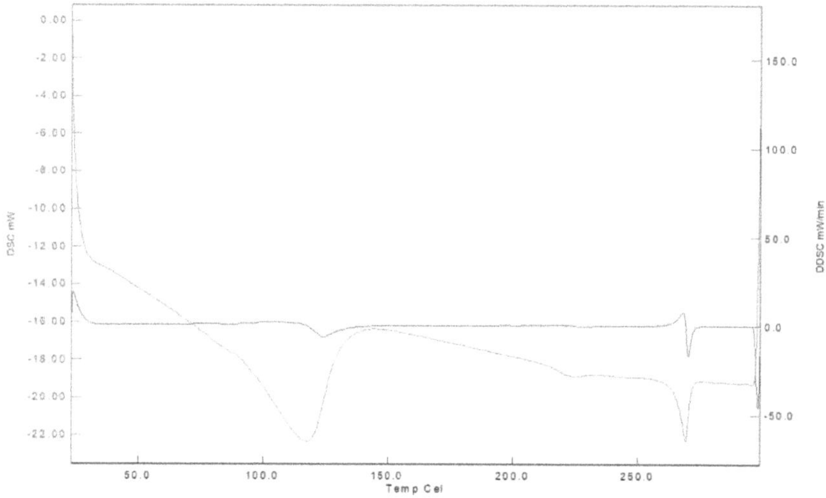

Figure 11 DSC spectra of Formulated and Optimized formulation.

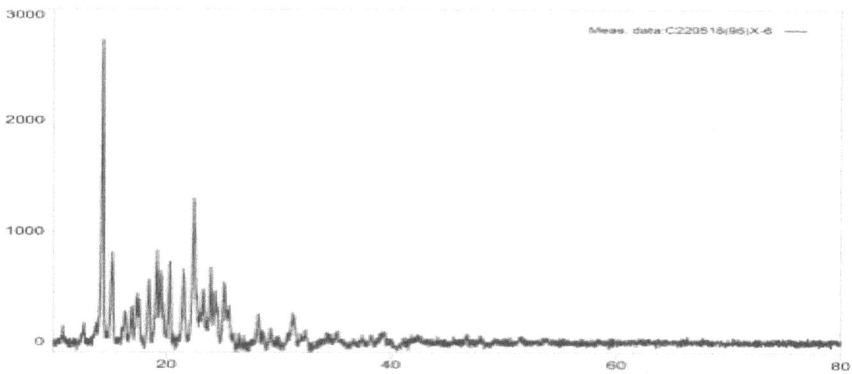

Figure 12 XRD spectra of Pure drug Telmisartan.

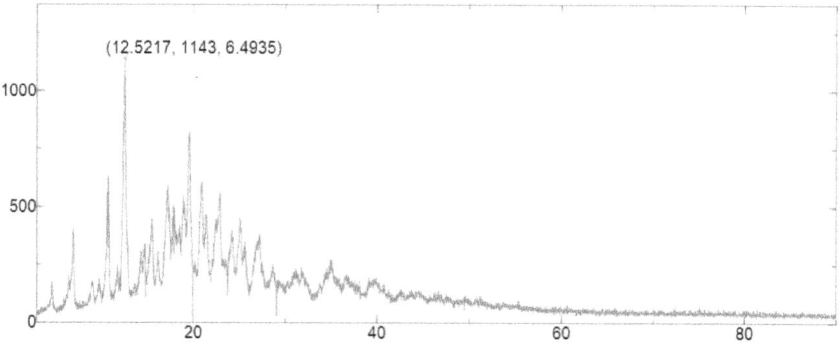

Figure 13 XRD spectra of Solid dispersions with β -CD.

5 CONCLUSION

The goal of the current work is to formulate telmisartan tablets with a rapid dissolution rate by complexing it with β-cyclodextrin and employing solid dispersions of PEG 4000, Starch 1500, MCC, Mannitol and Primojel. Telmisartan fast dissolving tablets were prepared with βCD using 3^2 factorial design and compressed using direct compression method.

1. Telmisartan and the excipients used in the study had no interaction, according to the FT-IR spectra.
2. The kneading method was used to develop solid inclusion complexes of Telmisartan-βCD in the ratios of 1:1, 1:2, and 1:3M.
3. Comparing the rate of Telmisartan dissolution from the βCD complexes to that of the pure drug, it was faster and higher. The dissolving rate increased as the proportion of βCD increased.
4. In comparison to the pure drug, Telmisartan's dissolution from solid dispersions in Primojel, PEG 4000, Mannitol, MCC and Starch 1500 were more fast and effective.
5. When compared to other carriers like Primogel, PEG 4000, Mannitol, MCC and Starch 1500, βCD (1:3 M) ratio exhibited the highest rate of dissolution.
6. As a result, the 3^2 factorial study chose the βCD (1:3 M) ratio for tablet formulation.
7. Telmisartan 20mg tablets were produced using the direct compression method using βCD (1:3 M) formulation with sodium starch glycolate and crospovidone.
8. In each case, the tablet powder blends have excellent flow properties suited for direct compression. All the tablets met the I.P requirements.
9. The first order model had correlation coefficient(r) values that were higher than those in the zero-order model, showing that first order kinetics follow the dissolution of Telmisartan.
10. In order to create the Telmisartan tablets, 3^2 factorial design was employed. The concentrations for the three levels of factor X1 (Crospovidone) and the three levels of factor X2 (sodium starch glycolate) are 3%, 4%, and 5%, respectively.
11. The formulation of the telmisartan Fast Dissolving tablet was developed using the following justification:(% calculated based on total weight of the tablet, i.e.,250mg)
12. The equations for disintegration time (DT), percent of drug dissolved in 10 minutes (PD10), and percent of drug dissolved in 20 minutes (PD20) were developed as in equations (2),(3),(4).
13. The results suggest that increasing the Super disintegrant concentration reduces the dosage form's disintegration time and the release pattern of the drug can be varied choosing the right X1 and X2 levels.
14. The validity of the developed equations for the dependent variables disintegration time (DT), percent of the medication dissolved in ten minutes (PD10), and percent of the drug dissolved in twenty minutes (PD20) is was confirmed using optimized formulations (PD10).

6 REFERENCES

1. K Kavitha, KumuthaSubramaniam, BoeyJiaHui, K. Santhi, SA Dhanaraj, and M Rupesh Kumar. *Potential Drug Candidates for Fast Dissolving Drug Delivery - A Review.* Research Journal of Pharmaceutical, Biological and Chemical Sciences, 2013; **4(4)**: 1510-1526.

2. Hetal Patel, Hiral Patel, Mukesh Gohe, Sanjay Tiwari,, *Dissolution rate improvement of telmisartan through modified MCC pellets using 3^2 full factorial design*, Saudi Pharmaceutical Journal , 2016; **24(5)** 579-587

3. Raghavendra Kumar Gunda. *Formulation development and evaluation of rosiglitazone maleate sustained release tablets using 3^2 factorial design.* International J of PharmTech Research, 2015; **8(4):** 713-724.

4. K.P.R.chowdary, *Optimization of valsartan tablet formulation by 2^3 factorial design* Journal of Global Trends in Pharmaceutical Sciences, 2014; Volume **5(1),**1374-1379 .

5. Notari RE. *Biopharmaceutics and clinical pharmacokinetics.* 4th ed. New York: Marcel Dekker Inc; 1987; 6-21.

6. Higuchi T,*Mechanism of sustained-action medication. Theoretical analysis of rate of release of solid drugs dispersed in solid matrices*. J Pharm Sci.1963; 51:1145-9.

7. Peppas NA. *Analysis of fickian and non-fickian drug release from polymers* Pharm ActaHelv1985; 60:110-1.

8. Schwartz BJ, Connor RE. *Optimization technique in pharmaceutical formulations and processing.* J Drugs and Pharm Sciin Modern Pharmaceutics 1996; **72(3):**727-54.

9. A. A. Kharia, s. N. Hiremath, a. K. Singhai, K. Omray and S. K. Jain, *Design and optimization of floating drug delivery system of acyclovir* . Indian J. Pharm. Sci.2010; **72 (5)**: 599-606.

10. Raghavendra Kumar Gunda, J. N. Suresh Kumar, Ch Ajay Babu and M. V. Anjaneyulu, *Formulation development and evaluation of lamotrigine sustained release tablets using 3^2 factorial design.* International Journal of Pharmaceutical Sciences and Research, 2015; **6(4):** 1746-1752.

STATISTICAL OPTIMIZATION AND EVALUATION OF EXTENDED RELEASE TABLETS FOR DIVALPROEX SODIUM

Ramineni Sunitha*[a], Praveen Sivadasu[a], Raghavendra Kumar Gunda[b]

*[a] KL College of Pharmacy, Koneru Lakshmaiah Education Foundation, Vaddeswaram, Andhra Pradesh, India.

[b] Department of Pharmaceutics, Narasaraopeta Institute of Pharmaceutical Sciences, Yellamanda, Narasaraopeta, Andhra Pradesh, India.

*Corresponding author email address: raminenisunitha@gmail.com

1 INTRODUCTION

Extended release formulations prolong effective plasma concentrations for longer periods of time while simultaneously reducing the frequency of dosage in a two-fold decline pattern. They raise patient adherence. Additionally, they give enhanced *in-vivo* clinical results.[1-2]

XL; LA (Long acting); XR were popularly used symbols for extended release dosage forms.[3] Many problems were experienced by formulation scientist while developing novel formulations for obtaining good absorption and enhanced bioavailability with sustained or prolonged release medications.

Divalproex is popularly used anticonvulsant agent for the effective management of bipolar disorders, epilepsy. It also used for the prophylactic measure in case of migrane. It acts by the prolongation of sodium channel blockade or inactivation. It also shows significant effect on GABA levels in brain. It inhibits the GABA degradation thereby increases its levels in brain; exerts anticonvulsant property. Due to its lower elimination half-life (5±1 hr), for achieving good clinical outcome need to administer 3-4 time a day. It is a difficult task. It was found that also exhibits first pass effect. By considering all these issue; we made an attempt to design and develop an extended release tablet formulation for divalproex for the effective management of epilepsy.[4]

Several tools were available to the formulation scientist for optimising the developed formulations with statistical significance. Among all available statistical tools response surface methodology was widely used technique in Industries as well as academia. Popular methods in the above mentioned category were factorial approach, central composite approach, Box-behnkenapproach etc.[5]

Direct Compression method was considered to be widest method of manufacture to produce tablets was seen in many cases.[6]

The present investigation focus on the development of extended release tablet formulations for divalproexsodium to reduce the dosing frequency and so as to enhance the patient compliance. ER formulations for divalproex were prepared with the help of polymers EudragitL/100-55 (partially neutralized pH dependent polymer) along with HPMC-K-100M (pH independent polymer) as per 3^2 factorial approach to know the relative effect of combination polymers on the drug release.[7]

2 MATERIALS AND METHODS

2.1 Materials

Divalproex sodium procured from GulanPharma, India as a complementary sample. EudragitL/100-55,HPMC-K-100Mwere purchased from commercial sources. All other excipients were obtained from Amanscientifics, Vijayawada, India.

2.2 Design and Development of gastro retentive Floating tablets for Divalproex sodium

The quantities needed to formulate divalproex sodium extended release tablets (EudragitL/100-55, HPMC-K-100M) were designated as independent variables (X_1,X_2). Drug release times were designated as dependent variables ($t_{10\%}$,$t_{50\%}$,$t_{75\%}$&$t_{90\%}$). Using PCP Disso software, polynomial equations were created for dependent variables.[8]

The three X_1 values (EudragitL/100-55) were 3.75, 6.25, and 8.75 percent. The X_2 (HPMC-K-100M) levels were 3.75, 6.25, and 8.75 percent. (% based on the active ingredient's weight). As part of a 3^2 factorial design, nine different divalproex sodium extended release tablet formulations were created. In Table 1, the design layout was severed.

Table 1. *Experimental design layout*

Formulation Code	X_1	X_2
SOD_1	1	1
SOD_2	1	0
SOD_3	1	-1
SOD_4	0	1
SOD_5	0	0
SOD_6	0	-1
SOD_7	-1	1
SOD_8	-1	0
SOD_9	-1	-1
CD_1	-0.5	-0.5
CD_2	+0.5	+0.5

2.3 Preparation of divalproex sodium extended release formulations

523 milligram of divalproex sodium equivalent to 500 mg of divalproex was taken as dose and tablets were prepared by direct compression technique. Table 2 a and b represents the formulation table for the preparation of Tablets. All ingredients were weighed accurately and subjected to sieving to ensure uniform size and to prevent segregation. Mixing operation was carried out in polybag for 10 min to obtain uniform blend. Powder mix was compressed to obtain tablets by using tablet punching machine. Obtained tablets were subjected to pharmaceutical product performance tests. [9]

Table 2 (a). *Formulae for Divalproate Sodium Extended-Release Tablets*

Name of Ingredients	Quantity of Ingredients per each Tablet (mg)				
	SOD_1	SOD_2	SOD_3	SOD_4	SOD_5
Divalproate sodium	523	523	523	523	523
Dicalcium phosphate	18.5	25	31	25	31
Starch	18	24	30.5	24	30.5
Eudragit L 100-55	43.75	43.75	43.75	31.25	31.25
HPMCK100M	43.75	31.25	18.75	43.75	31.25
Magnesium stearate	7	7	7	7	7
Talc	6	6	6	6	6
Total Weight	660	660	660	660	660

Table 2(b). *Formulae for Divalproate Sodium Extended-Release Tablets*

Name of Ingredients	**Quantity of Ingredients per each Tablet (mg)**			
	SOD_6	**SOD_7**	**SOD_8**	**SOD_9**
Divalproate sodium	523	523	523	523
Dicalcium phosphate	37	31	37	43.5
Starch	37	30.5	37	43
Eudragit L 100-55	31.25	18.75	18.75	18.75
HPMCK100M	18.75	43.75	31.25	18.75
Magnesium stearate	7	7	7	7
Talc	6	6	6	6
Total Weight	660	660	660	660

2.4 Characterization of ER Tablets of divalproex

Hardness

Hardness of tablets was measured as per diametric break down when operated with Pfizer tablet hardness tester.[10]

Friability

This test is carried out with the help of friabilator. Twenty tablets selected randomly and their weight was recorded as W_0. The tablets were placed in the drum of apparatus and subjected to 100 freefalls and weight of tablets again taken and recorded as W_1. Percentage weight loss was measured as follows

$$\text{Friability} = (W_0\text{-}W_1/W_0)100$$

Estimation of Drug Content

Twenty tablets were selected randomly and pulverized them to fine powder. A powder equivalent to 100 mg of Divalproex was taken into volumetric flask and then add 100 mL of pH 1.2 buffer and get dissolved. The resultant solution was subjected to estimation of drug content by measuring the absorbance using spectrophotometer at 210 nm. [11]

Thickness

On the basis of object longitudinal measurement theory, it was obtained using vernier calipers. [12]

Dissolution Test

This test was carried out with the help of Tablet dissolution test apparatus (USP-23) containing Paddle as rotating mechanism. It simulates the physiological circumstances like 900 mL of pH 1.2 buffer as SGF and is maintained for first 2 h and phosphate buffer for subsequent time intervals up to end of the current study. Temperature maintained throughout the study period is constant (37±0.5°C) and paddle was operated at 50 rpm speed. samples were obtained using standard procedures. The samples were analyzed for the estimation of drug content using spectrophotometer at 210 nm. The same was repeated to get triplicate result. [13-14]

Data obtained from the study was fitted to kinetic modeling to find out the mechanism of release of drug from tablet. [15]

3 RESULTS AND DISCUSSION

Extended release tablets of divalproex sodium were formulated according to 3^2 factorial approach. In Table 1, the formulation design was displayed. Time needed for drug release was designated as the dependent variable ($t_{10\%}$, $t_{50\%}$, $t_{75\%}$, and $t_{90\%}$), while the quantity of EudragitL/100-55 and HPMC-K-100M was designated as independent variables (X_1 and X_2, respectively). Table 2a and b represents about formulae for the preparation of ER Tablets for divalproex. All trials have Divalproex sodium 523 mg equivalent to 500 mg Divalproex; as an extended-release tablet dosage form prepared using Direct compression method.

Prepared tablets were subjected to evaluation tests. Results were summarized in Table 3a and b. Every formulation has sufficient mechanical strength. All formulations were discovered to be less brittle. All formulations passed the weight variation test as well drug content. drug release rate study was performed as per IP. To identify the drug release mechanism, kinetic analysis was applied to the data from the drug release investigation. Results were shown in Figure 1. After analysing the findings, it was evident that there was a direct correlation between the amounts of polymers combined and the rate of drug release (both were inversely proportional to each other). Predicted With the right dosages of EudragitL/100-55 and HPMC-K-100M, divalproex was released for a longer period of time. Table 4 provides a summary of the dissolution parameters. Response surface morphology (RSM)plots, shown in Figure 2, were used to examine the combined impact of different polymer ratios on the drug delivery of divalproex sodium. Sigma plot V13 was used to create RSM graphs.

SOD_5 is regarded as the ideal formulation out of all batches (on the basis of desirability factor). SOD_5 displayed promising dissolution properties, which helped in achieving the goal of the research by extending the time of drug release (optimum delivery of the drug from the dosage form). SOD_5 was constituted of both EudragitL/100-55 and HPMC-K-100M in equal amounts, or 31.25 mg each.

The following are the polynomial equations that were created to determine the projected drug release parameter:

$Y_1= 0.811+0.462X_1+0.037X_2-0.023X_1X_2+0.23 X_1^2+0.04X_2^2$ **($t_{10\%}$)**

$Y_2= 2.215+1.24X_1+0.085X_2-0.07 X_1X_2+0.64 X_1^2+0.11 X_2^2$ **($t_{25\%}$)**

$Y_3= 5.29+3.03X_1+0.22X_2-0.16 X_1X_2+1.53 X_1^2+0.24 X_2^2$ **($t_{50\%}$)**

$Y_4= 10.66+6.14X_1+0.42X_2-0.33 X_1X_2+3.05 X_1^2+0.501 X_2^2$ **($t_{75\%}$)**

$Y_5 = 17.702+10.09X_1+0.68 X_2-0.518 X_1X_2+5.05 X_1^2+0.77 X_2^2$ **($t_{90\%}$)**

Comparative results for both original dissolution parameters as well as predicted parameters were shown in Table 5a and b. Closeness was observed between the original and theoretical responses. It confirms, the developed equation was valid. SOD_5 has showed greater similarity with marketed product DIVALEX {f_2=85.91,f_1=2.25}.

Table 3 (a): *Post-Compression Parameters.*

Batch Code	**Hardness (kg/cm^2)**	**Thickness (mm)**	**Friability (%)**
SOD_1	8.46±0.26	4.05±0.08	0.10±0.001
SOD_2	8.19±0.29	3.99±0.085	0.11±0.001
SOD_3	7.92±0.27	3.91±0.08	0.09±0.001
SOD_4	8.51±0.41	4.12±0.06	0.06±0.001
SOD_5	8.12±0.42	4.06±0.06	0.07±0.001
SOD_6	7.69±0.40	3.98±0.05	0.07±0.001
SOD_7	8.34±0.43	4.19±0.05	0.05±0.001
SOD_8	7.90±0.41	4.06±0.06	0.04±0.001
SOD_9	7.5±0.40	4.01±0.06	0.05±0.001

Table 3 (b): *Post-Compression Parameters.*

Batch Code	**Average Weight (mg)**	**Drug Content (%)**
SOD_1	661.09±0.01	99.94±0.49
SOD_2	661.11±0.01	99.45±0.50
SOD_3	661.10±0.01	99.11±0.51
SOD_4	661.14±0.02	99.74±0.32
SOD_5	661.2±0.02	99.95±0.33
SOD_6	661.31±0.02	99.11±0.34
SOD_7	660.66±0.02	99.70±0.43
SOD_8	661.2±0.01	99.23±0.47
SOD_9	660.65±0.01	98.77±0.35

Table 4: *Dissolution parameters.*

Formulation Code	Dissolution Parameters				
	$t_{10\%}$ (h)	$t_{25\%}$ (h)	$t_{1/2}$ (h)	$t_{75\%}$ (h)	$t_{90\%}$ (h)
SOD_1	1.362	3.719	8.962	17.923	29.779
SOD_2	1.348	3.679	8.865	17.731	29.460
SOD_3	1.333	3.639	8.769	17.537	29.139
SOD_4	0.701	1.913	4.611	9.221	15.321
SOD_5	0.596	1.627	3.921	7.843	13.030
SOD_6	0.669	1.827	4.401	8.803	14.626
SOD_7	0.493	1.345	3.242	6.484	10.773
SOD_8	0.415	1.132	2.728	5.456	9.065
SOD_9	0.369	1.007	2.426	4.852	8.061
MP	1.362	1.447	3.487	17.923	11.589

Table 5(a): *Dissolution parameters for check point formulations*

Formulation Code	Predicted value				
	$t_{10\%}$ (h)	$t_{25\%}$ (h)	$t_{50\%}$ (h)	$t_{75\%}$ (h)	$t_{90\%}$ (h)
CD_1	0.625	1.71	4.11	8.22	13.65
CD_2	1.117	3.05	7.35	14.69	24.401

Table 5(b): *Dissolution parameters for check point formulations*

Formulation Code	Actual observed value				
	$t_{10\%}$ (h)	$t_{25\%}$ (h)	$t_{50\%}$ (h)	$t_{75\%}$ (h)	$t_{90\%}$ (h)
CD_1	0.63	1.75	4.25	8.25	13.72
CD_2	1.119	3.07	7.45	14.75	24.64

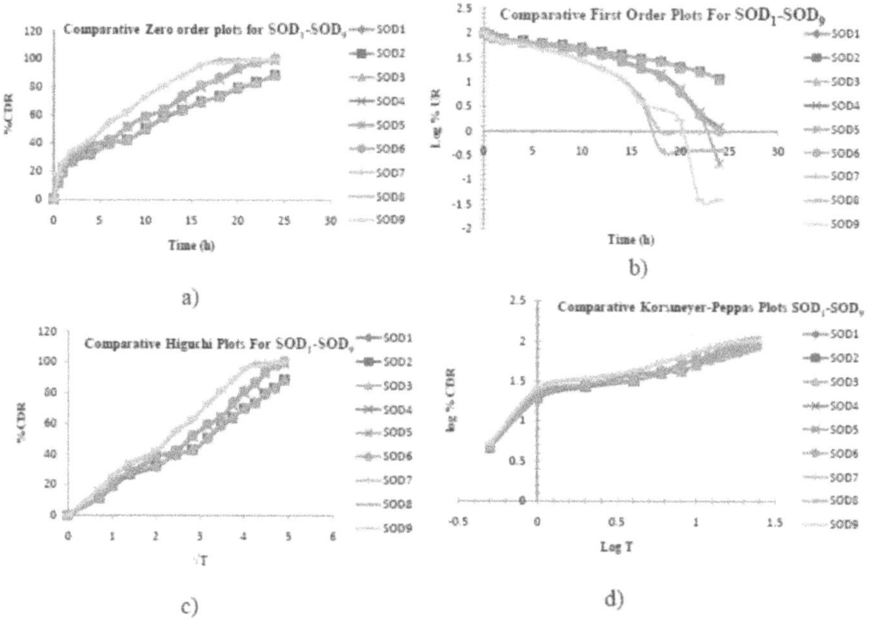

Figure 1 Comparative Kinetic Plots.

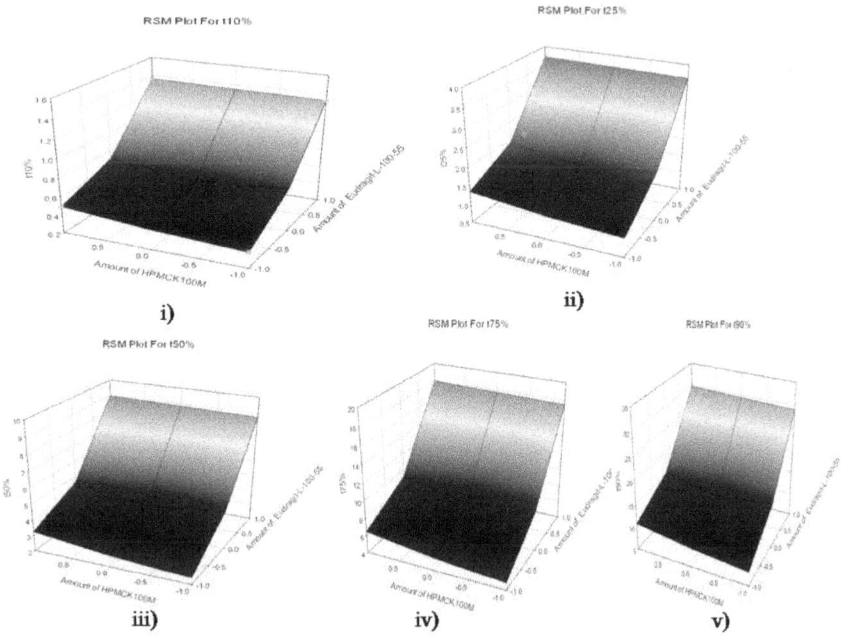

Figure 2 RSM Plots for Dependent Variables.

4 CONCLUSION

Based on the results of the current investigation, combining large molecules (polymers) offered advantages for maintaining the formulation's integrity and extending drug release. The necessary extended drug release will result from the proper proportional mix of partially neutralised pH dependent polymer and pH independent polymer, which finally results in a 2-fold decrease in the dose frequency of divalproex sodium. The Divalproex sodium is prepared utilising a 3^2 factorial design approach along with additional excipients and combination polymers (EudragitL/100-55, HPMC-K-100M). The formulation (SOD_5) was thought to be the best formulation among the several ER formulations investigated since it achieved the best results across all objective metrics. optimal formulation SOD_5 uses Non-Fickian Diffusion and Zero Order Release. By lowering the dose frequency by two or more times, it may increase patient compliance and, as a result, enhance therapeutic response.

5 REFERENCES

1. Gunda R.K. and Vijayalakshmi A, Thai J Pharm Sci. 2019,43(3),138.
2. Murthy TEGK., Bala Vishnu Priya Mukkala. and Suresh Babu VV, Asi J Pharm, 2011,5(3),162.
3. Jawed Akhtar, PrashantRana. and Bharat Jawaher, J Sci, 2011, 1(1), 21.
4. Rangasamy Manivannan, Parthiban. and Rajiv Gandhi, Int J Pharm Pharm Sci, 2010, 2(3), 83.
5. Farnaz Monajjemzadeh and Hadi Valizadeh, AAPS Pharm Sci Tech, 2013, 14(1), 245.
6. Qiu Y, Cheskin HS, Engh KR and Poska RP, J Pharm Sci, 2003, 92,1166.
7. Dutta S and Reed RC, Clin Drug Investig, 2004, 24,495.
8. Centorrino F, Kelleher JP, Berry JM, Salvatore P. and Baldessarini RJ, Am J Psychiatry, 2003, 160,1348.
9. Phaechamud T, Mueannoom W, Tuntarawongsa S. and Chitrattha S, Ind J Pharm Sci, 2010, 72,173.
10. Dutta S, Zhang Y, Selness DS, Lee LL, Williams LA, And Sommerville KW, Epilepsy Res, 2002, 49,1.
11. Ravi Chaudhari and Mohammed Gulzar Ahmed, Der Pharm Let, 2015, 7(5), 87.
12. Vamsy krishna, Srinath, Palanisamy, and Vijayasankar, Int J Res Pharm Bio Sc, 2011, 2(2),1.
13. Chitra Latha A, Jeeven Mani Babu D. and Nagendra Babu B, Ind J Res Pharm Bio, 2017, 5(1), 41.
14. Arunachalam A, Karthikeyan, Prathap, and Sabarikumar, J Pharm Res, 2012, 5(2),1195.
15. Raghavendra Kumar Gunda, Int J Pharm Tech Res, 2015, 8, 713

IRBESARTAN SOLID LIPID NANOPARTICLES FORMULATION AND EVALUATION USING NON-IONIC SURFACTANTS

Medapati Saritha[a*], Ajitha Palli[a],Bandaru Divya[a],N.Lalitha[a],Dr.Y.SrinivasaRao[a]

[a]Department of Pharmaceutics, Vignan Institute of Pharmaceutical Technology, Duvvada, Visakhapatnam
*Corresponding Author E-Mail: chsaritha1975@gmail.com

1 INTRODUCTION

An alternative to the regular colloidal system carriers suchas liposomes emulsions and polymeric microparticles and nanoparticles are the solid lipid nanoparticles.Sub-micron colloidal particles, or SLNs, ranging between 50 and 1000 nm are usually composed of a solid lipid core which are spherical in shape and are stabilized by an emulsifier [surfactant] .Large surface area,Small size, and high drug loading are some the unique properties of solid lipid nanoparticles.[1-2] Drug targeting, biocompatibility, nontoxicity, drug release regulation, and small-scale manufacture are some of the benefits of SLNs[3].One of the newest areas of nanotechnology is the development of SLNs, which has many potential uses in research, clinical treatment, and medication delivery.[4-5] The Drug is protected from a variety of chemical degradation by the solid lipid matrix of the SLN, increasing drug stability. Lowering the fluctuation in the therapeutic zone can prolong medication release and produce a regulated drug release profile with little harm. The methods for preparing of solid lipid nanoparticles are solvent injection, high pressure homogenization, and ultra-sonication.[6]

Angiotensin-2 receptor antagonist irbesartan is used to treat hypertension. The medication is practically insoluble in water. It exhibits 60–70% bioavailability and 90% plasma protein binding.[7] Irbesartan, because of its extremely low water solubility [0.0084mg/mL] and first-pass metabolism, has limited therapeutic utility and efficacy as an oral dosage form.[8-9] In order to avoid hepatic first-pass metabolism and to improve oral bioavailability SLNs can be used These solid lipid nanoparticles increased bioavailability and improve lymphatic transport. [10-12]

2 MATERIALS AND METHODS

Irbesartan was obtainedfrom Hetero drugs limited, Stearic acid (lipid) from Oxford laboratory reagent, PVA (stabilizer) from Rankem, and Span 80, Span 20, Tween 80, Tween 20 (surfactants) were obtained from Oxford LAB FINE CHEM LLP.

2.1 Method of preparation of solid lipid nanoparticles

A modified solvent injection methodwas employed to prepare Irbesartan SLN'S 50mg of Irbesartan and 100mg of stearic acid were weighed and dissolved in 3ml volume of organic solvent of methanol & 2 ml dichloromethane. This organic phase consisting of the drug-polymer mixture ofpolyvinyl alcohol [PVA]was added into 20ml of water (aqueous phase) containing Tween 80, Tween20, Span80, and Span20 with a 2,000 rpm continuously stirring with (Magnetic stirrer). The organic phase could be introduced to the aqueous stabiliser at a rate of 0.5 ml/min using a syringe equipped with a needle.The dispersion was stirred continuously for half an hour at 2,000 rpm. The resultant dispersion was then held overnight to completely evaporate the solvent (Table 1).

Table 1 *Formulation of Irbesartan Solid Lipid Nanoparticles.*

Ingredients	**F1 (1:5)**	**F2 (1:3)**	**F3 (1:5)**	**F4 (1:3)**	**F5 (1:5)**	**F6 (1:3)**	**F7 (1:5)**	**F8 (1:3)**
Irbesartan (mg)	50	50	50	50	50	50	50	50
Stearic acid (mg)	100	100	100	100	100	100	100	10
Polyvinylalcohol (mg)	500	500	500	500	500	500	500	500
Tween-80 (mg)	500	300	-	-	-	-	-	-
Tween-20 (mg)	-	-	500	300	-	-	-	-
Span-80 (mg)	-	-	-	-	500	300	-	-
Span-20 (mg)	-	-	-	-	-	-	500	300
Methanol (ml)	3	3	3	3	3	3	3	3
Dichloromethane (ml)	2	2	2	2	2	2	2	2

2.2 Characterization of The Prepared Solid Lipid Nanoparticles

2.2.1 FT-IR Studies. FT-IR spectroscopy is used to determine drug-excipient compatibility.

2.2.2 Particle size. Scanning electron microscopy was used to determine the particle size of all the solid lipid nanoparticles in all of the formulations (SEM).

2.2.3 Zeta Potential and Poly Dispersity Index (PDI). The Zeta potential of formulated NPs was measured by a zeta potential analyzer (Zeta sizer Ver 6.20 with serial number: MAL1004428) by placing the sample in zeta dip cell to check the size distribution of particles. Polydispersity is used to measure the heterogeneity of a sample based on the size of the particle. Zeta sizer was used to measure polydispersity index was by placing the sample in a disposable sizing cuvette, at a temperature of 25°C.

2.3 Drug Entrapment Efficiency.

The Entrapment efficiency of SLN loaded with irbesartan was estimated through the described method with modification. In brief, prepared SLN separation was performed by centrifugation. at 6000 RPM for 30 mins then the solid lipid nanoparticles, and supernatant were separated out. The compounds from SLNs were analysed using this technique using adequately diluted supernatant in water as a blank at 235nm using UV-visible spectrophotometer.The qualification is taken in triplicate and based on percentage entrapment, the best one is selected for further evaluation.The following equation was used to calculate entrapment efficiency.

EE % = W(Added drug) −W(free drug) / W(Added drug) x 100

Where, W(added drug) is the quantity of drug added during preparation of SLN, W(free drug) is the quantity of free drug measured in supernatant after centrifugation

2.4 In-vitro drug release

Drug release studies from SLNs was carried out using the dialysis bag diffusion method .4-5 cm dialysis bag was made into sac by blindly off one end with thread. It was filled with 7.4pHbuffer and inspected for leaks. The buffer was discarded from the sac and it filled with the 4ml SLN'S dispersion and is suspended in glass beaker receptor contains 0.1 HCL for 2hrs latter on the 0.1N HCL is replaced by ph 7.4 phosphate buffer. At specific intervals 5ml of the sample was withdrawn from the beaker and its absorbance is noted using UV spectrophotometer. The sample is replaced by blank. The Study was carried out up to 8hrs.

2.5 Stability studies

The Optimized preparation was evaluated for accelerated stability studies at room temperature [40°C] and at [60°C] for 45 days and specific time interval's the sample was withdrawn at that is at 15 days,30 days, and 45 days and evaluated for entrapment efficiency and particle size.

3. RESULTS AND DISCUSSION

3.1 FT-IR Studies

FT-IR spectroscopy was used to study the compatibilitynature between thepolymers and the drug.FT-IR studies show there is no compatibility between the drug and the excipients.

3.2 Particle size

Scanning electron microscopy was used to measure the SLN'sparticle size. Different surfactants concentrations for Tween_20, Tween_80, Span_80, and Span_20 were used to disperse SLN's, which revealed different particle sizes. The particles' sizes grew larger as the surfactant concentration decreased. Comparing the size of the SLN's made byusing Tween-80 and Span-80 with those made with Tween-20 and Span-20, the earlier showed smaller particle size. This could be a result of Tween-80 and Span-80 having a higher molecular weight than Tween-20 and Span-20, as well as having a higher HLB value. Every particle is contained within the predetermined range. The Figure 1 represented Particle size image of SLN's by Scanning Electron Microscopy.

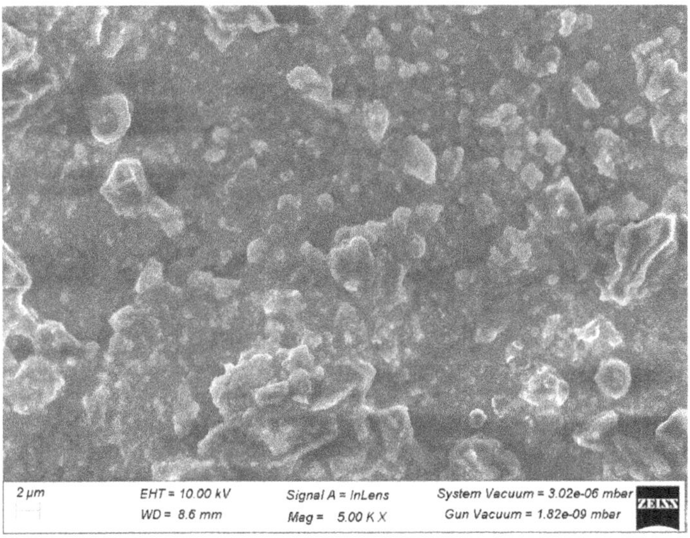

Figure 1 Particle Size image of SLN's by Scanning Electron Microscopy.

3.3 Zeta-Potential

The Zeta-Potential of all the SLN's dispersions was found to be in the range of -15.00 to -24.00 mv showing that all dispersions' prepared are stable.

3.4 Polydispersity Index

It is used to determine the size distribution of SLN's particle, it ranged from 0.012 to 0.7, polydispersity index decreased the concentration of the surfactant increased. The results obtained are of acceptable range.

3.5 Drug Entrapment Efficiency

Drug Entrapment efficiency is important for characterizing solid lipid nanoparticles. Various factors affect the encapsulation efficiency like the time, Lipid concentration and surfactant used. The entrapment efficiency of all the solid lipid nanoparticle dispersions is found to be in the range of 38.58 to 86.40 %. The Entrapment efficiency of the prepared SLN's. After 8 hours, the samples were taken out. The modified formulation was tested for invitro release in buffer made of phosphateat pH 7.4 and distilled waterwith the drug solution using the same technique as shown above. This improved as the surfactant concentration increased, this may be due to increase in the solubility of the drug in the lipid as the concentration of surfactant increased there by increasing the entrapment efficiency. The Span-80 formulations had more entrapment efficiency than the Tween-80formulations and the entrapment efficiency of the Span-20formulations is found to be more than entrapment efficiency of nanoparticles prepared by using tween 20 this may be because of the more HLB value of tween's compared to that of the span's. Substances having less HLB value are lipophilic in nature. **Table 2** represents the values of Zeta potential, Polydispersity index, % Entrapment efficiency of Irbesartan SLN's.

Table 2 *Zeta potential, Polydispersity index, % Entrapment efficiency of IrbesartanSLN's.*

Formulation code	Particle Size [nm]	Zeta Potential [mv]	Polydispersity index	%Encapsulation efficiency
F1	145	-21.69	0.180	54.57%
F2	160	-19.88	0.240	48.49%
F3	104	-14.84	0.179	45.85%
F4	125	-22.15	0.198	38.58%
F5	325	-21.50	0.188	86.40%
F6	350	-16.20	0.238	80.23%
F7	250	-14.80	0.185	79.80%
F8	270	-20.78	0.204	57.16%

3.6 In-vitro Drug Release

In-vitro Drug release of irbesartan SLN's was carried out by dialysis method by using 0.1 HCl for the first 2 hours and followed by 7.4 phosphate buffer up to 8 hrs. the drug release in the first 2 hours in 0.1 HCL showed nearly 10% ofthe formulations containing tween has surfactant and formulations containing span released almost a maximum of 15% of the drug by the end of 2 hours.

By the end of 8 hours, the formulation containing tween has surfactants F1,F2,F3, and F4 showed 30.9%, 25.00%, 29.89%, and 24.56% while the irbesartan SLN's dispersion containing span as the surfactants F5,F6,F7,F8 released 73.00%,68.13%,54.45%, and 50.00% shows that SLN's prepared by using span have the surfactant showed better release of the drug than the formulations prepared by using tween.This may lead due to the tweenbeing hydrophilic and spansbeing hydrophobic in nature.

Among the formulations prepared by using span-80 that is F-6 is far better than the formulation containing the surfactants span-80 having less HLB value than span-20 as a result the solubility of lipophilic drug irbesartan is increased presence of span-80.

Compared to the 2 formulations prepared by using different ratios of lipid: span-80 i.e 1:5 and 1:3 i.e F5 and F6. F5 formulation containing span-80 and the lipid: surfactant ratio of 1;5 showed the highest cumulative % drug release of about 73.00% compared to F6 containing lipid: span:80 ratio of 1:3 this may be because of the higher amount of span-80 in F5 formulation leading to increase the solubility of the drug and thereby release F5 formulation is taken from the formulation as the optimized formulation compared to all other formulations release. The Figure 2 represented the Invitro drug release of formulation codes F1 to F4. The Figure 3 represented the Invitro drug release of formulation codes F5 to F8.

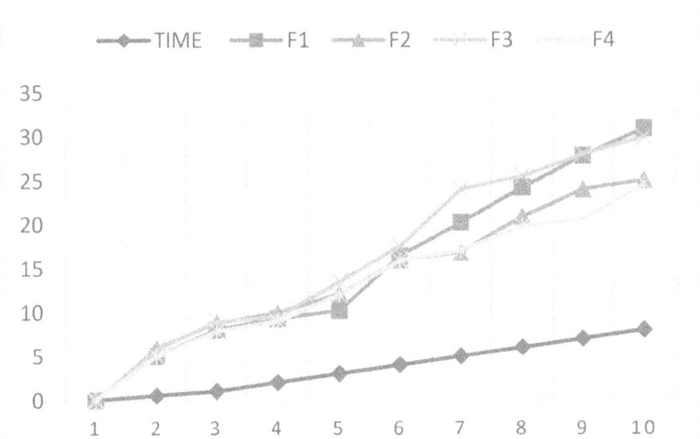

Figure 2 Drug release of Tween- 80 & Tween -20.

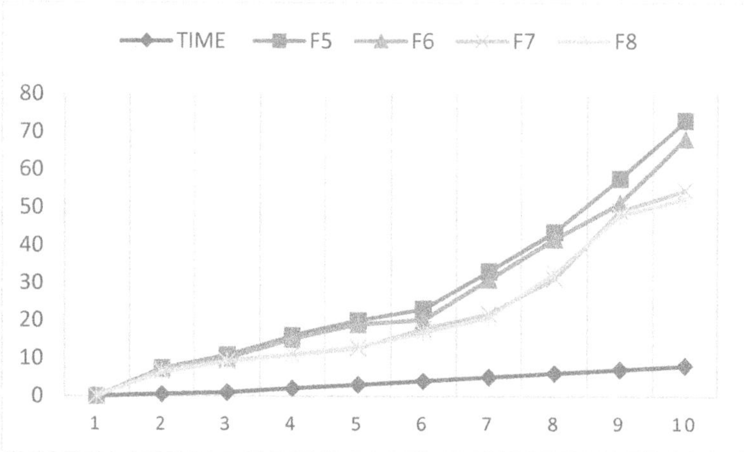

Figure 3 Drug release of Span -80 & Span-20.

3.7 Stability Studies.

The samples were analysed for size of a particle and entrapment efficiency during 45-day stability trials of the optimized formulation at a room temperature of 40°C and an evaluation temperature of 60°C. After 45 days, significant change was not observed in the particle size and entrapment efficiencyof the solid lipid nanoparticles.

4 CONCLUSION

Irbesartan is a poorly dissoluble in water, used for the treatment of hypertension. In the present study, Irbesartan-loaded SLN's were prepared by solventinjection method in order to improve the solubility and the bioavailability of the drug.

The irbesartan SLN'sprepared by solvent injection method containingspan as the surfactantshowed better drug releasecompared to the pure drug which is practically insoluble in nature. The in-vitro release studies also showed a sustained release from the prepared solid lipid nanoparticles Thus by preparing irbesartan into SLN'S the solubility of the drug is increased and thereby the bioavailability is improved as a result the therapeutic efficacy of the drug can be improved

5 REFERENCES

1.K.H. Ramteke, S.A. Joshi, S.N. Dhole, IOSR, Journal of pharmacy,2012,2, 34.
2.P. Ekambaram, A.H. Sathali, Priyanka, Sci Revs ChemCommun 2012,2, 80.
3. RH. Müller, M. Radtke, S.A.Wissing, Adv Drug Deliv Rev, 2002,54,131.
4. M. Trotta, F. Debernardi, O. Capato, Int J Pharm 2003,257,153.
5. R. Hirlekar and V. Kadam, AAPS PharmSciTech 2009,10(1),276
6. R.H. Muller, K. Mader, S. Gohla., Eur. J. Pharm. Biopharm.,2000,50,161.

7. B.R. Ganapuram, M. Alle, R. Dadigala, G.M. Kotu, V. Guttena, J. Pharm. Res, 2013,7,472.
8. K.F. Croom, G.L. Plosker, Drugs, 2008, 68(11), 1543.
9. R.R. Kamisetti, B. Vinitha, S. Bennuru, Sr. Cheeli, R.M. Gupta Vankadari, Asian J Pharm Res. 2020,10(1),01.
10. D. Narendar, V. Kishan, Drug Dev Ind Pharm. 2015; 41 (12),1968.
11. D.Thulasi Ram, Subhashis Debnath, M. Niranjan Babu, T. ChakradharNath, B. Thejeswi, Research J Pharm and Tech. 2012, 5(11): 1359.
12. G. Usha, D. Narendar, V. Kishan, Int J Pharma Sci Nanotech. 2015, 8 (4), 2995.

GLOBAL PERSPECTIVE ON SOFTWARE AS A MEDICAL DEVICE (SAMD): DEVELOPMENT PROCESS, ASSOCIATED RISKS, AND LEGISLATIVE ASPECTS

Koushik Yetukuri[a,b], M.S. Umashankar[a*]

[a]Deptartment of Pharmaceutics, SRM Institute of Science and Technology, Kattankulathur, Tamil Nadu - 603 203.
[b]Department of Regulatory Affairs, Chalapathi Institute of Pharmaceutical Sciences, Guntur, Andhra Pradesh - 522 034.

*Corresponding Author E-Mail ID: umashans@srmist.edu.in

1 INTRODUCTION

To keep up with the growing need for innovative medical technology, a substantial number of medical inventions are developed each year, millions of dollars are put into their development. According to research, medical technology is incredibly important to the successful adoption of healthcare services globally. The Organization for Economic Co-operation and Development (OECD) has released statistics on the use of healthcare technologies, which show a steady increase in the number of medical technologies.[1] It is therefore a demand for higher healthcare technologies.[2] It is becoming more common to employ software as a medical device. It can be used for a broad range of technology platforms, including platforms for virtual networks, commercial "off-the-shelf" systems, and platforms for medical devices, to cite a few. Such software was alluded to as "standalone software," "medical device software," and/or "health software" by the business, international bodies, and healthcare providers.[3]

To meet the requirements of contemporary pharmacological therapies, new medical device development procedures are currently being researched and promoted. Companies frequently emphasize regulatory clearance goals without a comprehensive strategy that accounts for developing a less expensive and more sustainable approach.[4] Well-planned and well-considered marketing plans that immediately recognize inefficiencies and reduced overall costs are vital for ensuring an effective product development cycle and achieving stakeholder objectives. The process for the development of a product is an exceptionally difficult procedure.[5] It significantly relies on the application of exact techniques that enable programmers to efficiently advance development, experimentation, validating, and market release.

2 METHODS

2.1 Research Strategy

Focusing on PRISMA principles, the study was performed systematically. Aside from the official websites of international regulatory authorities, databases such as Scopus and Web of Science were also searched. Additionally, websites for medical firms that are involved with the phases of software as a medical device (SaMD) development and legislative records on the research topic were examined. The terms "medical device development" and "software as a medical device" were used as keywords in Web of Science and Scopus. While using the more focused keyword categories, a limited number of studies were retrieved. The search process involved a Boolean operator technique.[6]

2.2 Data Collection

There were 3 phases to the article selection process, and when integrated with analysis, as shown in (see Figure 1). We began looking for articles in Scopus between 2010 and 2021 in the first phase, and between 2010 and 2021 in Web of Science (WOS) in the second segment; additionally, in the third section, we took a peek through the official websites of IMDRF, ISO, the US FDA, and the EU for any recent publications and current regulatory guidelines (2022). Considering how risk affects the development process, we chose publications, and 21 articles were evaluated against the inclusion and exclusion criteria for our study. In the final analysis, 09 studies were chosen based on the criteria.

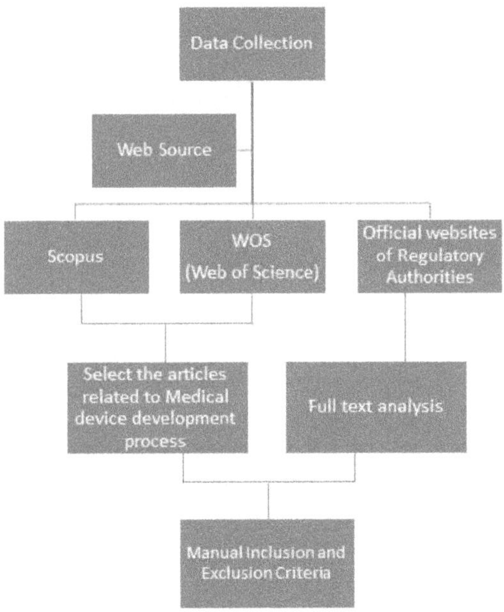

Figure 1 Illustration of the search strategy.

2.3 Text Analysis

The text analysis was carried out throughout the review process to support the reviewer's choice. Before identifying the keywords to employ in our search, we analysed the correlation between the co-occurrence of variables.

2.4 SaMD Clinical Evaluation Process

In order to confirm SaMD's performance, safety, and efficacy, the International Medical Device Regulators Forum (IMDRF) intends to provide a standardised and harmonised method to clinical evaluation. (see Figure 2) provides an illustration.[7] A convergent strategy should be used to establish the clinical evaluation process for SaMDs to assurance:

• There is a legitimate clinical correlation between a SaMD's output and the intended clinical condition, along with any pathological changes or condition; and the SaMD delivers the desired technical and clinical information.

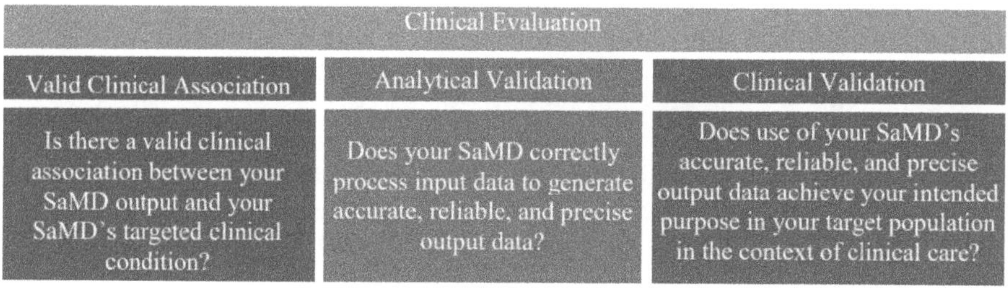

Figure 2 Clinical Evaluation Process (IMDRF/SaMD WG/N41FINAL:201).

SaMD's output is increasingly used to inform healthcare decisions, which can have an impact on patient care and clinical results. Global regulators, therefore, demand that performance indicators for a SaMD exhibit assurance of safety, effectiveness, and performance at a standard of integrity consistent with risk and consequences of the SaMD.[8]

2.4.1 SaMD Clinical Evaluation Process: General Principles and Context. The best route to describe a SaMD is as software that implements an algorithm (reasoning, logic, or modelling) to the data entry (electronic data information) to produce an output that is meant for medical uses as stated by the SaMD developer, as shown in (see Figure 3). The chance that the SaMD would produce faulty or incorrect data, which could affect a patient's clinical supervision, is a significant issue in the drawbacks and advantages connected with SaMD outputs.[9]

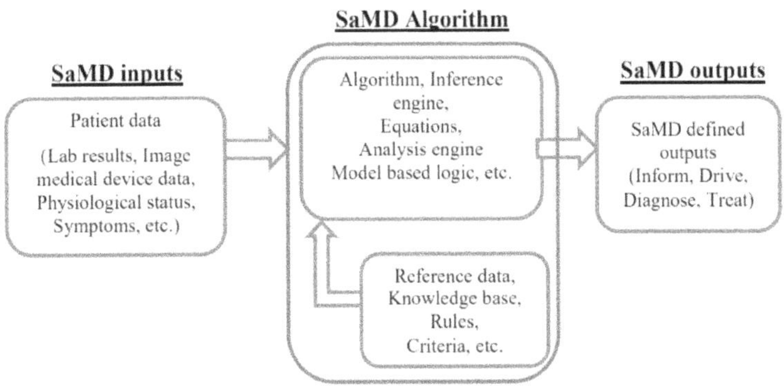

Figure 3 *SaMD* Basic Programming Model (IMDRF/SaMD WG/N41FINAL:201).

2.5 Approach to Continuous Learning Using Performance Data from the Real World

In order to continuously assess the device's performance and effectiveness, SaMD may utilize connectivity between individuals, objects, and both. A SaMD developer may have an idea about the functioning and intended usage of a SaMD based on continuing analysis and data collection on SaMD use in a post-market scenario. SaMD functionality and desired use can alter after being first delivered to the market with the aid of tracking actual performance data.[10] This information could include safety data, performance study results, ongoing clinical scientific data generation for medical devices, brand-new research publications or findings that strengthen or assist the clinical correlation of the SaMD output to a clinical situation, or direct end-user feedback that can help the SaMD manufacturer fully understand the SaMD's current outcomes. The SaMD defining statement might change if this is supported by the clinical evidence generated by clinical review utilising actual performance data from regular inspection, as shown in (see Figure 4).[9]

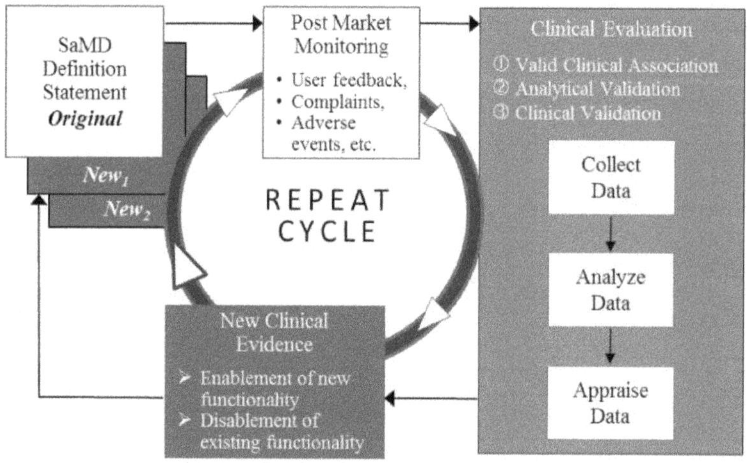

Figure 4 Pathway for Continuous Learning - Use of Real World SaMD Performance Data in Ongoing SaMD Clinical Evaluations (IMDRF/SaMD WG/N41FINAL:201).

Learning may change the initial classification of a SaMD in the following ways:

• Performance improvement data may demonstrate that somehow a SaMD's analytical or clinical accuracy exceeds the performance measures previously assessed by the SaMD manufacturer; or

• Performance data from application scenarios may demonstrate that a SaMD's analytical or clinical accuracy falls short of the metrics initially determined by the SaMD manufacturer.

2.5.1 Examples of Software as a Medical Device. Software that allows a smartphone to view images obtained from a magnetic resonance imaging (MRI) medical device for clinical diagnosis is referred to as "software as a medical device." This category also includes computer-aided detection (CAD) software that processes image data after they have been captured by the device.[11]

Other examples include

• Software as a Medical Devices (SaMDs) might interact with other mechanical medical devices, broad sense software, other SaMDs, and other medical devices. software that acts as a medical device by providing input variables for other hardware or other software. Software that offers information for a linear accelerator, for instance, is an example of software used as a medical device.

- A software component is classified as medical device if it performs a medical function and is run on a general-purpose computing system or a platform without a dedicated medical function. Software that diagnoses a condition using the tri-axial accelerometer and core processor of a retail digital camera is an example of software that qualifies as a medical device.

- Software that is connected to a physical medical device but is not required for that device to perform medically seems to be a diagnostic tool in its own right and is not that real medical device's accessory.

- Systems made for general-purpose (non-medical) computing can support software as a medical device. On these summary computing platforms, a hardware medical device may include software that performs the same operations as a medical device.

3 CHALLENGES WITH SOFTWARE AS A MEDICAL DEVICE

To enable all stakeholders, including regulators, to support safe innovation and protect patient safety, regulators from around the world acknowledged the need to collaborate on a common framework and principles for Software as a Medical Device (SaMD). This was because software as a medical device has unique features that go above and beyond what is typically seen in medical devices. IMDRF produces documentation that is generally regarded on a variety of medical device-related topics. The Software as a Medical Device Working Group was created by the IMDRF in order to promote innovation and speedy access to reliable Software as a Medical Device across the globe (WG).

4 RISK ASPECTS

The risks that are present at each stage of the SaMD process are directly tied to the different stages of development. For example, developing new medical equipment is an expensive and dangerous process; the project's success strongly depends on the application of exact methods. Intense competition forces product designers and developers to investigate the sources of risks inside the MDD process, which could threaten the MDD process in terms of cost, speed, and quality. Despite their best efforts to reduce such concerns, these potential remains. [12,13,14,15]

The three primary categories of user-related, internal, and third-party-related sources of hazards were used to characterize the most important risks. User-related hazards include, but are not limited to, inadequate user requirements translation and unmet needs/requirements. The absence of appropriate standards for evaluating device performance leads to internal risks, from a failure to take into account how packaging and Labeling may affect a product, or from a breakdown in communication between marketers, product developers, and end users. Lack of end-user training, inaccurate or subpar reviewers' assessments of progress, and inadequate planning for clinical and regulatory clearances and tests are only a few examples of risk factors associated with third parties. According to their results, unsatisfied user demands are the main cause of risks, hence to successfully market any technology, user needs must be satisfied. [16,17]

In order to prevent many bad effects of end users utilising medical devices, hasten the design and development of the medical device, and reduce costs, it is essential for the SaMD process to identify problems and their causes. The SaMD procedure should be thoroughly designed and carried out as a result to lessen the number and severity of risks.[18,19]

4.1 Potential IMDRF Framework for Software as a Medical Device Risk Categorization

The paradigm for categorising risks in software as a medical device has 4 types (I, II, III, and IV). These divisions are based on how seriously a medical device's software affects a patient's or the general public's health. For the software to effectively treat, diagnose, assist, or inform clinical treatment and avert death, permanent disability, or other serious health problems, accurate information must be delivered. Levels I through IV are used to categorise software as medical devices, with Level I possessing the slightest effect on patient or public health.[20]

5 LEGISLATIVE ASPECTS

Beginning in the 1980s, the pharmaceutical sector set the stage for a global harmonisation of medical device regulation. Regulatory agencies in the US, EU, and Japan determined in 1989 that it would be better for the industry to adopt worldwide standards for all parts of medication development, manufacture, and pharmacovigilance in order to increase the safety of the pharmaceutical manufacturing process. The International Council for Harmonization of Technical Requirements for Pharmaceuticals for Human Use was chosen to be created (ICH). Since then, almost all large pharmaceutical markets have implemented laws based on ICH criteria. The adoption of innovative rules for the security of pharmaceuticals depends on ICH.[21,22,23]

The Global Harmonization Task Force (GHTF), which was founded in 1992, was replaced by the International Medical Device Regulators Forum (IMDRF), which was established in 2011. If medical device safety is to be satisfied in a market that is continuously changing globally, harmonisation must be given priority. In the 1980s, there were basically no laws governing medical devices. There have been a few restrictions put in place during the 1990s, mostly in the US, the EU, and East Asia, particularly in Japan and Taiwan. Since the turn of the millennium, the GHTF's efforts have resulted in convergence in the regulation of the medical device sector. But the modern world demands global strategies. This is why the IMDRF was established.[24] The two largest markets for medical equipment are in North America and Europe. The laws on both sides of the Atlantic are similar, but until the new medical devices rule was recently approved by the European Parliament, American laws were stricter than those in Europe. Regulations governing medical devices in the EU had previously been much more flexible. Despite being comprehensive, it still falls short of the United States' good manufacturing practises regulations included in Title 21, Parts 812 and 820 of the Code of Federal Regulations (cGMP).[25]

In the EU, the so-called "accredited notified bodies," which are for-profit companies with private ownership, are extremely important. The entire system is in grave jeopardy because of their character. Due to the small number of such bodies (which is steadily falling) and the mandate that all medical devices licenced for the EU common market undergo required re-evaluation, there will be a scarcity of available capacity for re-evaluation. While tiny enterprises often produce a limited variety of medical devices and are less equipped to adapt to new requirements, notified bodies would probably be rather eager to inspect them.[26] Because huge companies with broad portfolios and well-organized paperwork will be far more profitable to inspect. Another barrier is the relative abundance of these groups in comparison to the US, where all inspectors are employed by the Food and Drug Administration and, to a lesser extent, the Center for Devices and Radiological Health (CDRH).[27]

It seems plausible that organisations based in various EU regions will undergo various levels of examination. This flaw might endanger the crucial aspect of the suggested rule as a result. Medical device post-marketing surveillance is covered by the medical device regulations. The new law mandates post-marketing follow-up for any devices promoted in the EU.[28] There are four different danger levels in Japan: Class I, Class II, Class III, and Class IV. The highest risk levels are found in Class III and Class IV, which are highly regulated medical devices. These classes also call for approval from the minister and review by the Pharmaceutical and Medical Devices Agency (PMDA).[29]

6 CONCLUSION

The basic model utilised in the SaMD process should follow the tried-and-true linear stage-gate method, which is technically sound and manageable from the standpoint of innovation. This is because it is consistent with the phases involved in SaMD as well as the associated rules and risk factors. A strategy for integrated customer interaction, for instance, should be established, which is one of the most recent strategies that have been advised. The most critical element of healthcare safety is the constancy of elements over a long production period. Good manufacturing methods need to become ingrained in the production process. These objectives are pursued by CGMP and medical device regulations. Future resolutions must focus on maintaining the integrity of the certification processes and the uniformity of device classification.

7 ACKNOWLEDGMENT

The facilities needed to complete the experiment were provided by Chalapathi Institute of Pharmaceutical Sciences (Autonomous), Guntur, Andhra Pradesh, for which the authors are grateful.

8 REFERENCES

1. Health - OECD Statistics Data online, https://www.oecd.org/els/health-systems/health-data.htm, (Accessed July 2022).
2. Y. Song kajorn, N. Thawesaengskulthai, Medical device innovation development process, Int J Innov Technol Manag, 2014, **11,** 24.
3. Software as a medical device – U.S. FDA Online, https://www.fda.gov/medical-devices/digital-health-center excellence/software-medical-device-samd, (Accessed August 2022).
4. S.J. Goldenberg, J. Gravagna. A real-world perspective: building and executing an integrated customer engagement roadmap that bridges the gaps in traditional medical device development processes. J Med Market, 2018, **16**, 41.
5. A. Sharma, S. Jha. Innovation from emerging market firms: what happens when market ambitions meet technology challenges, J Busin Indust Market, 2016, **31,** 507.
6. J.B. Pietzsch, L.A. Shluzas, M.E. Pate Cornell, P.G. Yock, J.H. Linehan, Stage-gate process for the development of medical devices, J Med Dev, 2009, **3,** 14.
7. FDA Digital Health Program: Software as a Medical Device (SAMD): Clinical Evaluation, Guidance for Industry and Food and Drug Administration Staff Document online, https://www.fda.gov/files/medical%20devices/published/Software-as-a-Medical-Device-%28SAMD%29--Clinical-Evaluation---Guidance-for-Industry-and-Food-and-Drug-Administration-Staff.pdf, (Accessed December 2021).
8. SaMD N10: Software as a Medical Device (SaMD): Key Definitions online, http://www.imdrf.org/docs/imdrf/final/technical/imdrf-tech-131209-samd-key-definitions-140901.pdf, (Accessed July 2022).
9. SaMD N12: Software as a Medical Device (SaMD): Possible Framework for Risk Categorization and Corresponding Considerations online, http://www.imdrf.org/docs/imdrf/final/technical/imdrf-tech-140918-samd-framework-risk-categorization-141013.pdf, (Accessed March 2022).
10. SaMD N23: Software as a Medical Device (SaMD): Application of Quality Management System online, http://www.imdrf.org/docs/imdrf/final/technical/imdrf-tech-151002-samd-qms.pdf, (Accessed June 2022).

11. FDA Digital Health Care Guidelines for SaMD Examples online, https://www.fda.gov/medical-devices/software-medical-device-samd/what-are-examples-software-medical-device, (Accessed August 2022).
12. C. Gilmartin, E.H. Arbe-Barnes, M. Diamond, S. Fretwell, E. McGivern, M. Vlazaki et al., Varsity medical ethics debate 2018: constant health monitoring - the advance of technology into healthcare, Phil Ethics Hum Med, 2018, **13,** 12.
13. A. Ho M, A. Saha, K.K. McCleary, B. Levitan, S. Christopher, K. Zandlo et al., A framework for incorporating patient preferences regarding benefits and risks into regulatory assessment of medical technologies, Value Health, 2016, **19,**746.
14. I. Ivlev, P. Kneppo, M. Bartak, Multicriteria decision analysis: a multifaceted approach to medical equipment management, Technol Econ Dev Econ, 2014, **20**, 576.
15. M.R. Millson, D. Wilemon, Impact of new product development (NPD) proficiency and NPD entry strategies on product quality and risk, R&D Manag, 2008, **38**, 491.
16. M.B. Privitera, M. Evans, D. Southee, Human factors in the design of medical devices – Approaches to meeting international standards in the European Union and USA, Appl Ergonom, 2017, **59**, 251.
17. C. Gerber, K. Goevert, S. Schweigert-Recksiek, U. Lindemann, Agile development of physical products—A case study of medical device product development, Smart Innov Syst Technol, 2019, **135**, 823.
18. S. Niimi, Practice of regulatory science (Development of medical devices), Yakugaku Zasshi, 2017, **137**, 431.
19. S.B. Rane, M.S. Kirkire, Interpretive structural modelling of risk sources in medical device development process, Int J Syst Assur Eng Manag, 2017, **8**, 451.
20. GHTF Pre-market: GHTF Study Group 1 documents online, http://www.imdrf.org/documents/doc-ghtf-sg1.asp, (Accessed August 2022).
21. GHTF Post-market: GHTF Study Group 2 documents online, http://www.imdrf.org/documents/doc-ghtf-sg2.asp, (Accessed August 2022).
22. GHTF SG3 N18:2010: Quality management system –Medical Devices – Guidance on corrective action and preventive action and related QMS processes online, http://www.imdrf.org/docs/ghtf/final/sg3/technical-docs/ghtf-sg3-n18-2010-qms-guidance-on-corrective-preventative-action-101104.pdf, (Accessed August 2022).
23. ISO 14155:2011: Clinical investigation of medical devices for human subjects - Good clinical practice online, http://www.iso.org/iso/home/store/catalogue_ics/catalogue_detail_ics.htm?csnumber=45557, (Accessed August 2022).
24. ISO 14971:2007: Application of risk management to medical devices online, http://www.iso.org/iso/en/CatalogueDetailPage.CatalogueDetail?CSNUMBER=31550&ICS 1=11&ICS2=40&ICS3=1, (Accessed August 2022).

25. ISO 62304/A1: 2015: Medical device software – Software life-cycle processes online, https://www.iso.org/standard/64686.html, (Accessed August 2022).
26. IEC 62366-1:2015: Medical Devices – Part 1: Application of usability engineering to medical devices online, https://www.iso.org/standard/63179.html , (Accessed August 2022).
27. IEC 80002-1:2009: Medical device software -- Part 1: Guidance on the application of ISO 14971 to medical device software online, http://www.iso.org/iso/iso_catalogue/catalogue_tc/catalogue_detail.htm?csnumber=54146, (Accessed June 2022).
28. ISO 82304-1:2016: Health software – Part 1: General requirements for product safety online, https://www.iso.org/standard/59543.html, (Accessed July 2022).
29. Japan Medical Device Regulations online, https://www.pmda.go.jp/english/index.html, (Accessed August 2022).

SOLUBILITY ENCHANCEMENT AND *IN-VITRO* DISSOLUTION STUDIES OF GLIPIZIDE USING RECRYSTALLIZATION TECHNIQUE

K. Jyothsna, K Anil Kumar, TP Rao

Department of pharmaceutics, K.V.S.R Siddhartha College of Pharmaceutical Sciences, Vijayawada-520010, AP, INDIA

*Corresponding Author: tprao2020@gmail.com

1 INTRODUCTION

The challenging aspects of drug development is improving oral bioavailability of poorly water-soluble drugs. The solubility and permeability of the drug are crucial factors in determining its oral bioavailability. There have always been drugs whose solubility poses a barrier for the development of an oral formulation. Crystallization is the formation of crystals from a uniform (or) homogeneous solution, either naturally or artificially. [1,2] It is a solid-liquid separation in which crystals of various shapes and sizes are formed depending on the downstream processing (or) final product requirements. Cubic, tetragonal, orthorhombic hexagonal, monoclinic, triclinic, and trigonal crystal forms are among the many available. The concept of crystallization is of paramount interest as the changes in properties of a solid substance (EX: polymorphism) may affect the pharmacokinetic parameters (like dissolution rate etc.) and the product's compliance which can be controlled by monitoring the formation & isolation of solid crystals.[3] Crystals can be transformed by recrystallizing the drug substance in various ways, which affects its physicochemical characteristics such as melting point, solubility, true density, flow ability, tabletability, and dissolution profile. GPZ is a sulfonylurea-class anti-diabetic drug used to treat type 2 diabetes and absorbs rapidly through

GIT. It has a quick onset of action when taken orally in combination with a diabetic diet and physical activity. It is not recommended for usage alone in patients with type 1 diabetes.[4] However, GPZ is insoluble in water, very slightly soluble in acetone & methylene chloride. It dissolves in dilute solutions of alkalihydroxide.

The literature review states that, all the research work was published with cocrystals to achieve enhanced solubility and dissolution. No reports were reported on the recrystallization methods to improve the pharmaceutical properties like solubility and drug (GPZ) *in-vitro* dissolution characteristics.[5,6]

2 METHODS AND MATERIALS

2.1 Chemicals

Aurobindo Pharma Ltd., (India) supplied Glipizide (GPZ). All the other chemicals like sodium hydroxide, potassium dihydrogen phosphate, methanol, isopropyl alcohol (IPA), acetonitrile, acetone, ethyl acetate, ethanol, dichloromethane were procured from Loba Chemie Pvt, Ltd,, (Mumbai)., Jiangsu Huaxi International Trade Co.Ltd, (China)., Merck Specialities Pvt, Ltd,, (Mumbai).

2.2 Preparation of GPZ Recrystallization Products

In a 25 mL beaker, 500 mg GPZ was added to 20 mL specified pure organic solvent and the mixture was gently heated to 45°C. A supersaturated solution is produced was then cooled to room temperature and dried at 40°C for one hour.[7] The dried powder is sieved using #80 mesh and packaged in glass container, kept in desiccator until analysis.

2.3 Analytical Validation Procedures

In this analytical study, UV-VIS spectrophotometric (UV spectrophotometer Single Beam, ELICO 159) at absorbance 276nm were used.

2.4 Physicochemical Characterization of prepared Recrystallized Products

2.4.1 Scanning Electron Microscopy (SEM). It was used to examine the crystal surface morphology of GPZ and its recrystallized derivatives.[8] Prior to observation, the sample were mounted on metal stubs with adhesive tape that was coated with gold under vacuum.

2.4.2 Differential Scanning Calorimetry (DSC). The Shimadzu DSC 60 device was utilized for the thermal analysis that was performed. GPZ that had been accurately weighed as well as its samples were placed in aluminum pans, sealed, and then heated at 10 degrees Celsius per minute at a temperature ranging from 30 to 300 degrees Celsius.[9]

2.4.3 Powder X-ray Diffraction (X-RD). The study of GPZ and its samples (recrystallized) were recorded on a Bruker D8 Advance diffractometer using

Cu-K X-radiation at 40 kV and 30mA power.[10] Diffraction patterns were gathered throughout a temperature 5-50 degrees Celsius at a scan 1 degree Celsius per minute.

2.4.4 Fourier Transform Infrared Spectroscopy (FTIR). The Shimadzu FTIR spectrophotometer, Model 8033, from Japan was used to obtain FTIR spectrum measurements of GPZ and recrystallized GPZ at room temperature. In KBr powder, the samples were first scattered, and then pellets were created by pressing them under pressure. Following that, FTIR spectrums were acquired with the application of powder diffuse reflectance.[11]

2.4.5 In-vitro Dissolution Studies. The tests of GPZ and its recrystallized derivatives were performed using a USP type 2 (paddle technique) Dissolution Rate Test Apparatus (DS 8000, Lab India) in 900mL of phosphate buffer with a pH of 7.4.[12] The samples were disseminated in a medium. In each test, 5 mg of GPZ-equivalent samples were utilized at speed of 50 rpm and a temperature 37±1°C. At various time intervals, 5mL of liquid was taken, filtered and replaced with 5mL of new dissolving media.[13] If required, the filtered samples were appropriately diluted. Percentage of GPZ dissolved at various intervals of time.

3 RESULTS AND DISCUSSION

3.1 GPZ Recrystallized Products preparation

Different organic solvent, a solvent was chosen with low boiling points for GPZ recrystallized. since the faster the drying process, the lesser the solvent's boiling point. Recrystallization of GPZ in various solvents required varying amounts of time to create dry product, although samples were dried under identical temperature and humidity factors. Even at ambient temperature, solvents like methanol, ethanol, isopropanol (IPA), acetone, ethyl acetate, dichloromethane, and acetonitrile dried the product in a few minutes.

3.2 SEM Studies

Crystal morphology and its habit is an essential aspect that can affect a variety of qualities, including powder flow, compaction, stability, and dissolving rate. The SEM analysis were conducted to determine the impact of recrystallization on the crystal form and size of GPZ. Figure 1 depicts SEM images of untreated GPZ, methanol, and ethanol products. GPZ and its recrystallized derivatives (ethanol, methanol) are tetragonal; the crystal structure and size do not vary significantly.

Figure 1 SEM images of GPZ recrystallized products.

3.3 DSC Studies

One of the methods used to characterize the crystals was differential scanning calorimetry (DSC). DSC analyses were conducted to examine changes in crystal nature. DSC thermograms were also utilized to infer the physico-chemical parameters of the recrystallized GPZ products MET, ETH, IPA, DCM, ACE, EA, and AN, as well as pure GPZ. At 214.51^{0}C, the untreated GPZ displays a sharp endothermic peak. For the seven recrystallized products MET, ETH, ACE, IPA, DCM, AN, and EA, broad endothermic peaks were observed at 214.34^{0}C, 213.54^{0}C, 211.59^{0}C, 209.03^{0}C, 213.17^{0}C, 212.52^{0}C, 214.18^{0}C respectively indicates that melting temperature has not changed much.

3.4 Powder X-RD studies

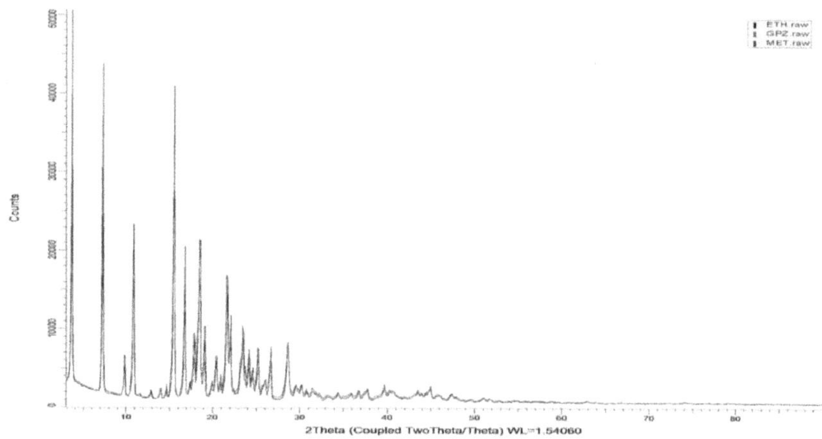

Figure 2 X-RD overlay of GPZ and its recrystallized products.

To obtain insight into the crystallinity variations, X-ray diffraction patterns of GPZ and its recrystallized derivatives MET, ETH, ACE, DCM, IPA, EA, and AN were also examined. Figure 2 demonstrates the powder X-ray diffractograms.

3.5 FTIR Studies

GPZ FTIR spectra and its recrystallized products were visualized in the Figure 3. The peak at 3352.66, 3224.68 cm^{-1} shows N-H functional group. The band 2943.08, 2914.18, 2853.95 cm^{-1} demonstrates the C-H functional group. 1688.20,1649.67,1525.9,1332.15 cm^{-1} indicates the presence of C=O, O=C-NH, C=N, S=O stretching. The band 1290.88 cm^{-1} is because of C-N group. 1032.78 cm^{-1} is due to =C-H Bond Bending. This all revealed that there were no significant structural changes when compared to pure GPZ.

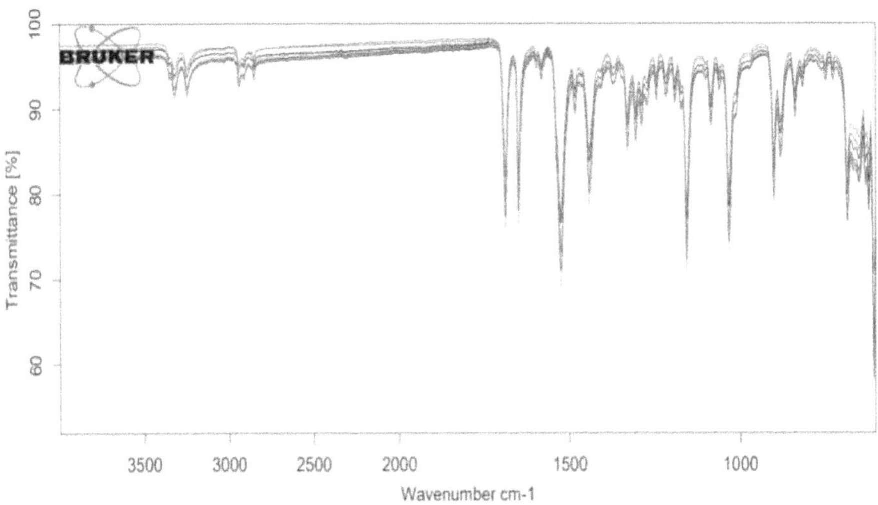

Figure 3 FTIR overlay of GPZ and its recrystallized products.

3.6 *In-vitro* Studies

This was conducted to determine the solubility properties of the recrystallized GPZ samples and to compare the results with those of the GPZ untreated. All samples were subjected to *in vitro* dissolution studies using phosphate buffer of pH 7.4 to evaluate dissolution parameters, including % drug (GPZ) released at 10 to 60 minutes (DP10 and DP60), dissolution efficiency at 20 minutes (DE20), time to release 50% of GPZ, and first order and Hixon-Crowell rate constants. The profiles are depicted in Figures 4, and 5.

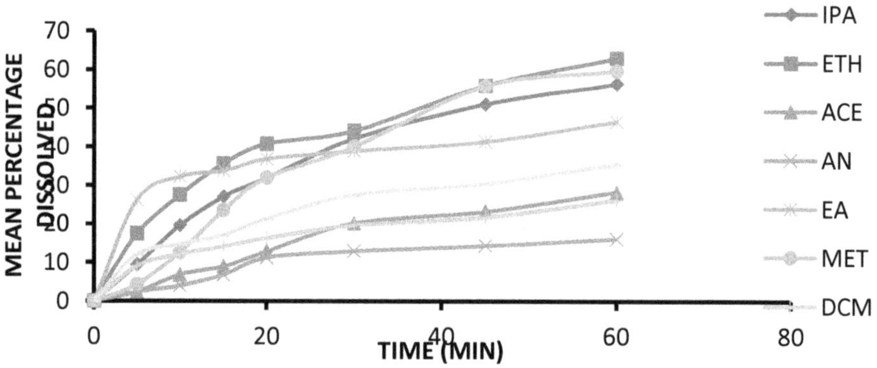

Figure 4 Comparative GPZ *In-vitro* profiles with its recrystallized derivatives.

The $t_{50\%}$ for GPZ and its samples (recrystallized) of MET, ETH, ACE, IPA, DCM, EA, and AN were >60, 39, 37, >60, 42, >60, >60, >60 min respectively. The DP_{10} for GPZ and its samples (recrystallized) of MET, ETH, ACE, IPA, DCM, EA, and AN are 15.02, 12.70, 27.78, 19.62, 12.18, 32.30,4.00. Whereas DP_{60} values are for GPZ, and its recrystallized products MET, ETH, ACE, IPA, DCM, EA, and AN are 35.55, 60.11, 63.03, 28.33, 56.40, 26.27, 46.5, 16.15. At DE_{20} the values are 13.63, 14.09, 25.21, 6.13, 17.9, 10.8, 4.68 and 25.8. See Table 1.

Table 1 *GPZ In-Vitro Dissolution with its recrystallized products.*

Sample	DP_{10}	DP_{60}	DE_{20}	$t_{50\%}$
GPZ	15.02	35.55	13.63	115
MET	12.70	60.11	14.09	39
ETH	27.78	63.03	25.21	37
ACE	7.17	28.33	6.13	-
IPA	19.62	56.40	17.9	42
DCM	12.18	26.27	10.8	-
EA	32.30	46.5	4.68	-
AN	4.00	16.15	25.8	-

The t50% values for MET, ETH, ACE, IPA, DCM, EA, AN and GPZ were significantly in the order ETH> MET> IPA> GPZ. The DP_{10} values order of EA > ETH > MET > DCM > GPZ > IPA > ACE > AN. Whereas, for DP_{60} the values are of ETH > IPA > MET > EA > GPZ > ACE > DCM > AN. the variation in the values may be because of the wettability of recrystallized samples. The DE_{20} values for MET, ETH, ACE, IPA, DCM, EA, AN and PURE are in the order of AN > ACE > DCM > GPZ > MET > IPA > ETH > EA.

Table 2 *GPZ Release kinetics data with its samples (recrystallized).*

SAMPLE	First order		Hixson-Crowell	
	$K^{(-min)}$	r^2	$K_{HC}^{(-min)}$	r^2
GPZ	0.00345	0.9198	0.02141	0.9084
MET	0.01635	0.9779	0.0499	0.9693
ETH	0.0138	0.9511	0.0455	0.928
ACE	0.00552	0.9714	0.01911	0.9677
IPA	0.01312	0.9681	0.04260	0.9536
DCM	0.00368	0.8862	0.0145	0.8772
EA	0.00506	0.6874	0.0239	0.6577
AN	0.00276	0.8812	0.01013	0.8783

The % drug release data was fitted to release kinetics models to get release rate constant that can be comparable with the disso data. The information was presented in Table 2. The first order (k^{-min}, 0-60min) for MET, ACE, IPA, ETH, DCM, AN, EA and GPZ order MET > ETH > IPA > ACE > EA > DCM > GPZ > AN. The Hixson-Crowell release (K_{HC}^{-min}) for MET, ETH, ACE, IPA, DCM, AN, EA and GPZ in order of MET > ETH > IPA > EA > GPZ > ACE > DCM > AN.

However, the R^2 values for first order kinetics better than the ones from Hixson-Crowell Kinetic model.

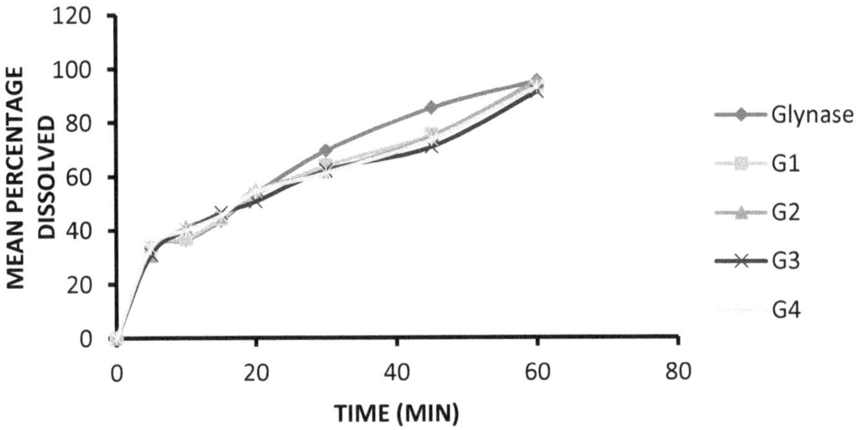

Figure 5 Comparative In-vitro profiles of GLYNASE (marketed) and formulated tablets.

The $t_{50\%}$ values for GLYNASE and formulated tablets G1, G2, G3, G4 were 18, 18, 19, 20, 19 min respectively. The DP_{10} values for GLYNASE and formulated tablets G1, G2, G3, G4 were 36.31, 36.70, 41.24, 40.05, 39.98 mins respectively. Whereas, DP_{60} values for GLYNASE and formulated tablets G1, G2, G3, G4 are 95.30, 92.95, 94.85, 91.05, 92.83 mins respectively. The DE_{20} values for GLYNASE and formulated tablets G1, G2, G3, G4 are 32.16, 32.78, 36.12, 35.68, 36.47.

The t50% values for GLYNASE and formulated tablets were significantly in order G3 > G2 > G4 > GLYNASE > G1. The DP_{10} values are in the order of G2 > G3 > G4 > G1 > GLYNASE. Whereas, for DP_{60} the values are in the order of GLYNASE > G2 > G1 > G4 > G3. the variation in the values may be because of the wettability of recrystallized products in the dissolution medium. The DE_{20} values for marketed and formulated tablets are in the order of G4 > G2 > G3 > G1 > GLYNASE. See Table 3.

Table 3 *In-vitro dissolution marketed and formulated tablets.*

Sample	DP_{10}	DP_{60}	DE_{20}	$t_{50\%}$
Glynase	36.31	95.30	32.16	18
G1	36.7	92.95	32.78	18
G2	41.24	94.85	36.12	19
G3	40.05	91.05	35.68	20
G4	39.98	92.83	36.47	19

As described in earlier, the data was kept into release kinetics models. Table 4 showed the information. The first order (k^{-min} , 0-60min) for GLYNASE and formulated tablets G1, G2, G3, G4 in the order GLYNASE > G2 > G1 > G4 > G3. The Hixson-Crowell release (K_{HC}^{-min}) for GLYNASE and formulated tablets G1, G2, G3, G4 in the order GLYNASE > G2 > G1 > G4 > G3.

However, the R^2 values for first order kinetics better than the ones from Hixson-Crowell Kinetic model.

Table 4: Release kinetics of marketed and formulated tablets.

SAMPLE	First order release kinetics		Hixson-Crowell release kinetics	
	$K^{(-min)}$	r^2	$K_{HC}^{(-min)}$	r^2
Glynase	0.04744	0.977	0.1060	0.9894
G1	0.03822	0.9361	0.09073	0.9638
G2	0.04168	0.8989	0.09511	0.9475
G3	0.03408	0.9253	0.08313	0.9479
G4	0.03707	0.9184	0.08820	0.9493

4 CONCLUSION

As a result, by utilizing the technique of recrystallization, GPZ may be transformed into either amorphous phases or dispersions, as demonstrated by the current DSC and X-ray diffraction. These amorphous phases and dispersions have enhanced water solubility and dissolving capabilities, as well as enhanced therapeutic effectiveness.

5 ACKNOWLEDGMENT

The author's want to thank the Siddhartha Academy of General and Technical Education, Vijayawada, AP, India, for giving them the tools they needed to do this research.

6 REFERENCES

1. Giovanna Bruni,Ines Ghione et al, 2021 Jun; **26**(11): 3142.
2. Shivarani Eesam, Jaswanth S. Bhandaru et al, 7,124 (2021).
3. Narendra Kumar pandey et al, AAPS. 2017 Oct;18(7):2454-2465.
4. Dimpy Rani, Sarb Sukhmani Kaur Dargan, Parnika Goyal and Renu Chadha, ISSN:2641-2020.
5. Ali Samie, Gautam R. Desiraju et al,2017,17,5,2406.
6. Developing solid oral dosage forms: pharmaceutical theory and practice., 2009, Elsevier Inc.,
7. Hlaeblian, J., Mccrone, W., (1969), pharmaceutical application of polymorphism. J. pharm. Sci. 58 (8) 911.
8. Rasenack, N., Hartenhauer, H., Muller, B.W. (2003), Microcrystal for dissolution rate enhancement of poorly water-soluble drugs. Int. J. Pharm. **254**, 137.
9. Chiou, W.L., Riegelman, S., (1971), pharmaceutical application of solid dispersion system . J. pharm. Sci. **60**, 1281.
10. Huttenrauch, R. Modification of Starting Materials to Improve Tableting Properties. Pharm. Ind. 1983, **4**,435.
11. York, P. Solid-State Properties of Powders in the Formulation and Processing of Solid Dosage Forms.Int. J. pharm. 1983,**14**,1.
12. Marshall, P.V.; York, p. The Compaction Properties of Nitrofurantonin Samples crystalized from Different Solvents. Int. J. Pharm. 1991,**67**,59.
13. PHARMACEUTICAL PREFORMULATION AND FORMULATION A Partical Guide from Candidate Drug Selection to Commerical Dosage Form by *Mark Gibson.*

STATISTICAL OPTIMIZATION AND IN VITRO EVALUATION OF NIZATIDINE FLOATING TABLETS

S. Madhavi Latha[a*], P. Bharghava Bhushan[b]
[a]Department of Pharmaceutics, Aditya Pharmacy College, Surampalem, Andhra Pradesh, India.
[b]A. M. Reddy Memorial College of Pharmacy, Narsaraopeta, Andhra Pradesh, India.

Corresponding Author E-Mail ID: drbharghav84@gmail.com

1 INTRODUCTION

Floating drug delivery can be used to increase the permeability and bioavailability of drugs by increasing the gastric residence time.[1] H2 receptor antagonists are used for the management of various disorders like Gastric Esophageal Reflux Disorder (GERD), Peptic Ulcer Disease (PUD), gastric and duodenal ulcer.[2] Nizatidine is chosen as a model drug for this work because of its need for local action, high solubility, low permeability (class III of Biopharmaceutical Classification System (BCS), oral bioavailability of 70% and half-life of 1-2 hours.[3-5] Non-efferevescent floating matrix release technique can be employed for Nizatidine to overcome these demerits of the drug where buoyancy improves the permeability of the drug while matrix release sustains the action of the drug for prolonged period of time avoiding the necessity of frequent drug administration.[6-7] As the buoyancy is an intricate parameter based upon the root level of the drug, the process of granulation should be carefully controlled so as to maintain the density of final formulation lower than that of the gastric medium.[8] Hot melt granulation technique was utilized in the study and the evaluation of post compression parameters were designed to evaluate the effect of granulation of final formulations. The drug release retardant effect of Glyceryl Laurate (GL) was explored in the study and its potential for sustaining the drug action is evaluated with the help of response surface methodology.

2 METHODS AND RESULTS

2.1 Materials used

Nizatidine was received as a gift sample from Aurobindo Pharma (Hyderabad, India). Glyceryl Laurate was kindly provided by Dr Reddy's Laboratories (Hyderabad, India). Lactose, Aerosil, Magnesium stearate was provided as gift sample from Loba Chemie Pvt Ltd (Mumbai, India). All other chemicals and solvents were of analytical grade or highest quality and were used as such as obtained.

2.2 Methods

2.2.1 Factorial Design. A 3^2 full factorial design was used for development and optimization of Nizatidine floating tablets, 2 factors were evaluated each at 3 levels and experimental trials were performed using all 9 possible combinations. Formulation design, optimization and other investigations were done using Design Expert software from Stat-Ease Inc, Minneapolis, trial version 12. The experimental design used in the present study is shown in Table 1.

Table 1 *Levels of independent variables in 3^2 Factorial design.*

	Level		
	Low	Middle	High
X1 (amount of GL, mg/tablet) *	-1(42.5)	0(85)	+1(127.5)
X2 (percentage of Lactose) **	0	5	10

* *1:0.5,1:1,1:1.5 ratio of drug to polymer **%w/w to total weight of drug and polymer*

2.2.2 Preparation of NEGFMT by compression of melt granules. Melt granules were prepared by hot fusion technique. Glyceryl Laurate was melted with continuous stirring in a porcelain dish on a water bath maintained at 70° C. The required amount of drug (Nizatidine) were added to the molten GL with proper mixing and cooled to room temperature. The solidified mass was ground and passed through #30 mesh (600 µm) to obtain uniform sized granules. The melt granules obtained from hot fusion technique were geometrically mixed with other excipients and compressed into tablets on a 16 station rotary tablet machine (M/s. Cadmach Machinery Co. Pvt. Ltd., India) using 7 mm round, flat plain punches with a sufficient compression force to obtain hardness of 4 to 5 Kg/cm2 containing the Nizatidine equivalent to a dose of 85 mg.[9] Formulation details were shown in Table 2.

Table 2 *The Composition of 9 formulations as per 3^2 Factorial design.*

Ingredients (mg)	**F1**	**F2**	**F3**	**F4**	**F5**	**F6**	**F7**	**F8**	**F9**
Nizatidine	85	85	85	85	85	85	85	85	85
Glyceryl Laurate	42.5	42.5	42.5	85	85	85	127.5	127.5	127.5
Lactose	0	6.4	12.8	0	8.5	17	0	10.6	21.3
Aerosil	3	3	3	3	3	3	3	3	3
Mag.Stearate	2	2	2	2	2	2	2	2	2
Total Wt (mg)	132.5	138.9	145.3	175	183.5	192	217.5	228.1	238.8

2.2.3 Evaluation of tablets. The formulated tablets were assessed for in vitro buoyancy properties, consistency of weight, hardness, friability, consistency of active ingredient and in vitro drug release concentrates according to the systems described.[10] The disintegration information was studied by model dependent approach.

2.2.4 Data analysis, optimization of model. The measurable assessment of dependent factors, forecast of improved formulations and cross-approval of model were preceded as described [10]. Optimized formulations were set up with the ideal qualities and assessed for consistency of weight, consistency of content, hardness, friability, in vitro buoyancy and in vitro dissolution. The test values of dependent factors (%DR1 and T100) were resolved from the in vitro active ingredient information of the optimized formulas.

2.2.5 *Drug-GL interaction studies.* Drug- carrier interaction studies were performed by FTIR spectra of samples using Perkin Elmer 2000 FTIR system (Perkin–Elmer, Norwalk, CT) by KBr disk method (2 mg sample in 200 mg KBr). The scanning range was 450–4000 cm-1 and the resolution was 1 cm-1. DSC was performed using a Differential scanning calorimeter (DSC 220C, Seiko, Japan) at a heating rate of 100 ºC/min from 30 to 3000 C in nitrogen atmosphere.

3 DISCUSSIONS

NEGFMT were set up with various proportions of active ingredient- GL (1:0.5, 1:1 and 1:1.5) to look at the impact of GL on the arrival of Nizatidine. GL being a hydrophobic polymer, the impact of channelling specialist was concentrated on the drug delivery and lactose was choice being dissolvable in nature. Henceforth the amount of lactose was selected as an independent variable alongside with GL to evaluate its impact on drug discharge and studied at three levels viz. 0, 5 and 10%.

3.1 Flow properties

The progression of medication (θ = 52.1°) was upgraded with the inclusion of GL. The angle of repose for all the molten granules arranged with drug GL blends likewise displayed great flow qualities demonstrating their suitability for compression (appeared in Table 3).

Table 3 *Flow properties of pure drug and melt granules with GL*(mean, n=3).

DRUG:GL	Angle of Repose(θ)	Inference
1:0 (Nizatidine)	52.1^0	Poor
1:0.5	39.0^0	Passable
1:1	32.6^0	Passable
1:1.5	27.9^0	Good

3.2 Evaluation of Nizatidine non effervescent GFMT

The consequences of *in vitro* floating attributes, consistency of weight, hardness, friability and consistency of substance are appeared in Table 4. All the prepared formulations floated immediately after placing into the beaker and the floating was maintained more than 24 hrs. This may be because of low thickness of the tablet with the availability of low dense waxy material (GL), which created a vertical movement of the dosage form and showed its floating. The outcomes show that the nizatidine NEGFMT arranged with GL were discovered to be of acceptable quality satisfying every one of the official and different necessities of tablets.

Table 4 *In vitro floating properties and tabletting characteristics of Nizatidine NEGFMT.*

NEGFMT	FLT (sec)	TFT (hrs)	Uniformity of weight[a] (mg)	Uniformity of Content[b] (mg)	Hardness[c] (Kg/cm^2)	Friability (%)
F1	0	>24	133.5±0.36	99.0±0.25	4-5	0.02
F2	0	>24	137.9±0.21	99.9±0.37	4-5	0.04
F3	0	>24	143.3±0.24	99.9±0.68	4-5	0.08
F4	0	>24	177±0.37	99.9±0.31	4-5	0.09
F5	0	>24	182.5±0.41	99.9±0.52	4-5	0.07
F6	0	>24	190±1.25	99.9±0.36	4-5	0.05
F7	0	>24	215.5±0.39	99.9±0.24	4-5	0.08
F8	0	>24	227.1±0.52	99.9±0.39	4-5	0.01
F9	0	>24	236.8±0.44	99.9±0.33	4-5	0.06

a; mean±% deviation, n=20, b; mean±s.d., n=10, c; mean, n=5

3.2.1 *In vitro drug release studies*

USP XXIV dissolution rate test apparatus (LABINDIA) with 900 ml of 0.1 N HCl as dissolution medium kept up at a temperature of 37± 0.5 ºC utilizing paddle stirrer at 50 rpm was utilized for in vitro drug release considers. The samples withdrawn at defined time stretches were reasonably diluted and absorbance was estimated at 325 nm utilizing UV spectrophotometer (UV-1800, Shimadzu, Japan). The mean drug release profiles of the formulations were appeared in Figure 1. The entirety of the tablets showed an impressive retardation in the arrival of the drug. All the formulated NEGFMT released total medication (~100%) after 12 hrs other than F3 which delivered 100% of the medication within 12 hrs. Dissolution information showed that the arrival of the medication from the formulated NEGFMT rely upon the substance of glyceryl laurate (GL) and lactose. It was studied that as measure of GL expanded, release of medication from matrix diminished. It could be expected to more slow passage of medium into the hydrophobic frameworks shaped by GL. Improved content of GL caused moderately more hindrance in drug discharge because of expansion in the way length for the release of medication. Availability of lactose in the formulations upgraded the medication release from the hydrophilic lattice. Expanding the content of lactose from 0 to

10% expanded the active ingredient discharge from NEGFMT. This conduct might be because of the quick and high diffusion of lactose in water allowing the arrangement of pores in the matrix facilitating with the diffusion of medication particles through the pores.

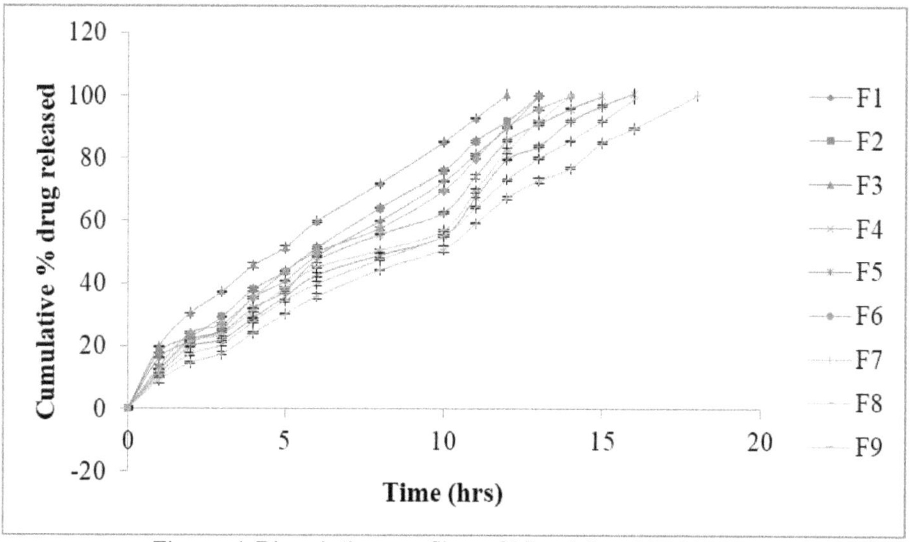

Figure 1 Dissolution profiles of Nizatidine NEGFMT.

3.2.1.1 *Drug release kinetics and mechanism.* The connection coefficient (r) values of zero and first order of all the NEGFMT are presented in Table 5. It was seen that all the NEGFMT followed zero order release. The medication discharge component was controlled by fitting the dissolution information to Higuchi, Hixon-Crowell and Korsemeyer-Peppas equations and the outcomes are presented in the Table 5. It was observed that all the formulated NEGFMT followed diffusion mechanism instead of erosion. The 'r' values of these matrices were discovered to be 0.9891 to 0.9983 demonstrating that the release likewise followed Korsmeyer–Peppas model. The delivery rate values (n) were in the scope of 0.54 to 0.86 affirming the non Fickian diffusion release mechanism.

Table 5 *Correlation coefficient (r) and Release mechanisms of NEGFMT.*

NEGFMT	Zero order	First order	Higuchi	Hixson Crowell	Korsmeyer-Peppas	
	r	r	r	R	r	n
F1	0.9965	0.8438	0.9682	0.8992	0.9874	0.74
F2	0.9963	0.8718	0.9862	0.9836	0.9900	0.71
F3	0.9900	0.8929	0.9887	0.9882	0.9978	0.65
F4	0.9945	0.8770	0.9638	0.9111	0.9924	0.80
F5	0.9956	0.9210	0.9708	0.9349	0.9942	0.78

F6	0.9953	0.9177	0.9747	0.9360	0.9953	0.76
F7	0.9982	0.9510	0.9649	0.9769	0.9951	0.86
F8	0.9968	0.9397	0.9665	0.9390	0.9953	0.83
F9	0.9878	0.9064	0.9550	0.8859	0.9891	0.78

3.2.2 *Data analysis, optimization and cross-validation of model*

3.2.2.1 *Data analysis.* The outline of insights alongside with comparative R2, adjusted R2, predicted R2, PRESS, s.d., F-values and p-values are presented in Table 6. An appropriate model for depicting the information was chosen dependent on coefficient of determination (R2) and PRESS values. The responses Y1 and Y2 were found to follow quadratic model. These models showed higher R2 and F-values and lower PRESS and p-values. Subsequently these models were chosen for additional optimization. Consequences of ANOVA for estimated reactions (%DR1 and T100) are appeared in Table 7. Higher F-values and lower p-values ($p<0.05$) for every one of the reactions demonstrated the meaning of the models.

Table 6 *Summary of Model statistics for Responses Y1 and Y2 for Nizatidine.*

Model	R^2	Adjusted R^2	Predicted R^2	PRESS	s.d.	F-value	p-value	Remarks
Response Y_1 (%DR1)								
Linear	0.9141	0.8896	0.8188	19.96	1.16	37.25	0.0002	
Interactive	0.9163	0.8744	0.6318	39.67	1.24	0.1563	0.7063	
Quadratic	0.9973	0.9940	0.9718	3.11	0.2706	60.96	0.0010	Suggested
Response Y_2 (T_{100})								
Linear	0.8744	0.8385	0.6500	9.94	0.7138	24.37	0.0007	
Interactive	0.9536	0.9305	0.7650	6.67	0.4684	10.25	0.0186	
Quadratic	0.9929	0.9840	0.9266	2.09	0.2249	33.21	0.0002	Suggested

A decent relationship was seen between the independent and dependent as shown by the R2 results of %DR1 and T100 responses of soften granule formulations (0.9973 and 0.9929). The model was found to be significant with result to adjusted coefficient of determination (Adj R2 > 0.9000). In every one of the cases 'Predicted R2' values were in sensible agreement with the 'Adj R2'values.

Table 7 *ANOVA for the responses of NEGFMT.*

Source	SS	Df	MS	F-value	P-value
Response Y_1 (DR_1): Quadratic model					
Model	109.84	5	21.97	300.02	< 0.0001*
X_1:Glyceryl Laurate	87.17	1	87.17	1190.52	< 0.0001*
X_2:Lactose	13.50	1	13.50	184.37	0.0002*
X_1^2	8.42	1	8.42	114.95	0.0004*
X_2^2	0.0486	1	0.0486	0.6634	0.4611
X_1X_2	0.2401	1	0.2401	3.28	0.1444
Residual	0.2929	4	0.0732		
Total	110.13	9			
R^2:0.9973, Adjusted R^2:0.9940, Predicted R^2: 0.9718					
Response Y_2(T_{100}): Quadratic model					
Model	28.20	5	5.64	111.46	0.0002*
X_1:Glyceryl Laurate	16.67	1	16.67	329.41	< 0.0001*
X_2:Lactose	8.17	1	8.17	161.41	0.0002*
X_1^2	0.9643	1	0.9643	19.06	0.0120*
X_2^2	0.0476	1	0.0476	0.9412	0.3869
X_1X_2	2.25	1	2.25	44.47	0.0026*
Residual	0.2024	4	0.0506		
Total	28.40	9			
R^2: 0.9929, Adjusted R^2: 0.9840, Predicted R^2: 0.9266					

* *Significant (p<0.05), SS: Sum of squares; MS: Mean sum of squares*

The application of response surface methodology yielded the following regression equations which are an empirical relationship between the logarithm values of %DR1 and T100.

$$\%DR1=12.13-3.81\,X1+1.50X2-0.2450X1X2+1.90X1X1+0.1443X2X2$$

$$T100=15.07+1.67X1-1.17X2-0.7500{*}X1X2-0.6429X1X1-0.1429X2X2$$

Where, X1 and X2 are the coded values of the test variables of the glyceryl laurate quantity and % w/w of lactose to the total weight of drug and glyceryl laurate respectively.

The point by point rundown of results by multiple regression investigation of dependant factors are appeared in Table 8. Main impacts of the selected independent factors like glyceryl laurate amount and % w/w of lactose are exceptionally huge (p<0.05). Opposing and added substance impact was noticed for glyceryl laurate (X1) separately for %DR1 and T100 and converse circumstances were noticed for lactose (X2). That implies expanding the glyceryl laurate diminished %DR1 and expanded T100 results for example controlled the arrival of the medication and the other way around noticed for lactose. The VIF esteems for every one of the models were discovered to be one, showing acceptable assessment of coefficient.

Table 8 *Model coefficients of NEGFMT for Nizatidine.*

Factor	Co efficient of Estimate	p-value	SE	95% CI Low	95% CI High	VIF
Response Y_1 (%DR1): Quadratic model						
Intercept	12.13(β_0)	1	0.1617	11.68	12.58	
X_1:Glyceryl Laurate	-3.81(β_1)	< 0.0001*	0.1105	-4.12	-3.50	1.000
X_2:Lactose	1.50(β_2)	0.0002*	0.1105	1.19	1.81	1.000
X_1X_2	-0.2450(β_{12})	0.1444**	0.1353	-0.6206	0.1306	1.000
X_1X_1	1.90(β_{11})	0.0004*	0.1771	1.41	2.39	1.03
X_2X_2	0.1443(β_{22})	0.4611**	0.1771	-0.3476	0.6361	1.03
Response Y_2 (T100): Quadratic model						
Intercept	15.07(β_0)	0.001*	0.1344	14.70	15.44	
X_1:Glyceryl Laurate	1.67(β_1)	< 0.0001*	0.0918	1.41	1.92	1.0000
X_2:Lactose	-1.17(β_2)	0.0002*	0.0918	-1.42	-0.9117	1.0000
X_1X_2	-0.7500(β_{12})	0.0026*	0.1125	-1.06	-0.4377	1.000
X_1X_1	-0.6429(β_{11})	0.0120*	0.1473	-1.05	-0.2340	1.03
X_2X_2	-0.1429(β_{22})	0.3869**	0.1473	-0.5517	0.2660	1.03

%DR1: % drug released in 1 hr; T100: Time to release 100% of drug; SE: Standard error; CI: Confidence interval; VIF: variance of inflation factor
** Significant (p<0.05) ** Not significant (p>0.05)*

Contour and response surface plots were generated for %DR1 and T100 to exhibit graphically the impact of GL and lactose and are appeared in Figure 2, 3 and 4, 5. The discoveries propose that the measure of wax material (GL) and channeling agent (lactose) have a relationship for accomplishing a formula with better controlled medication discharge.

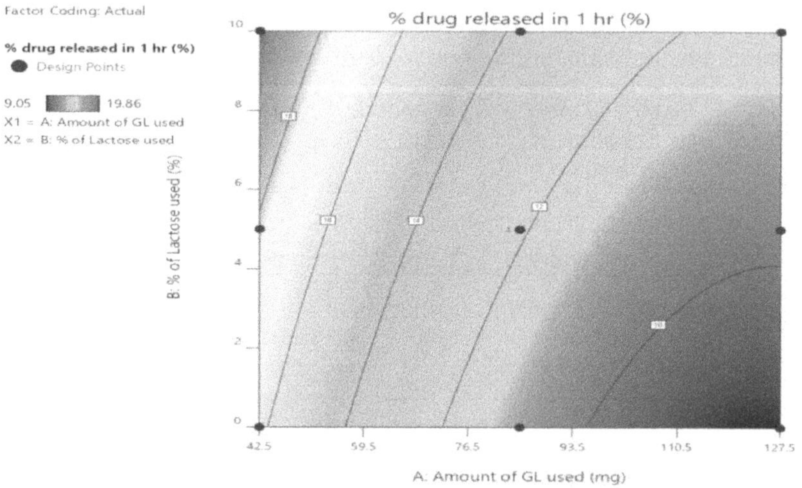

Figure 2 Contour plots for %DR1 of Nizatidine NEGFMT.

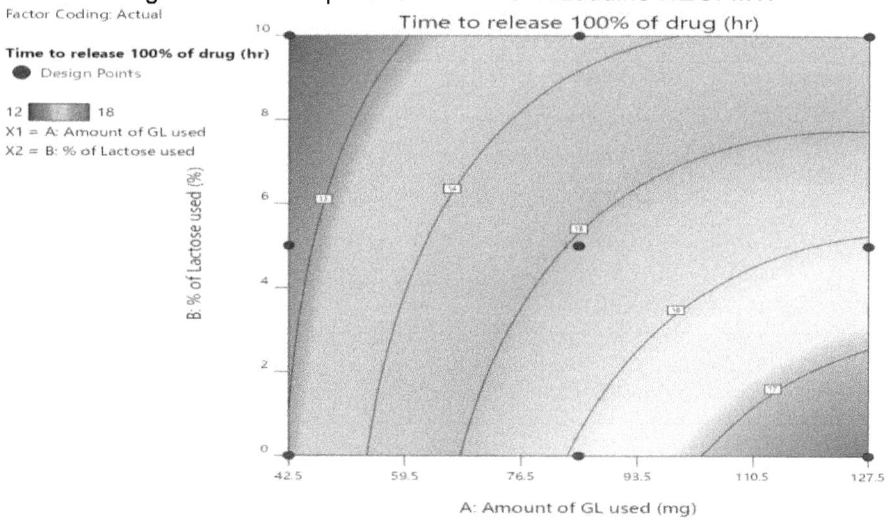

Figure 3 Contour plots for T100 of Nizatidine NEGFMT.

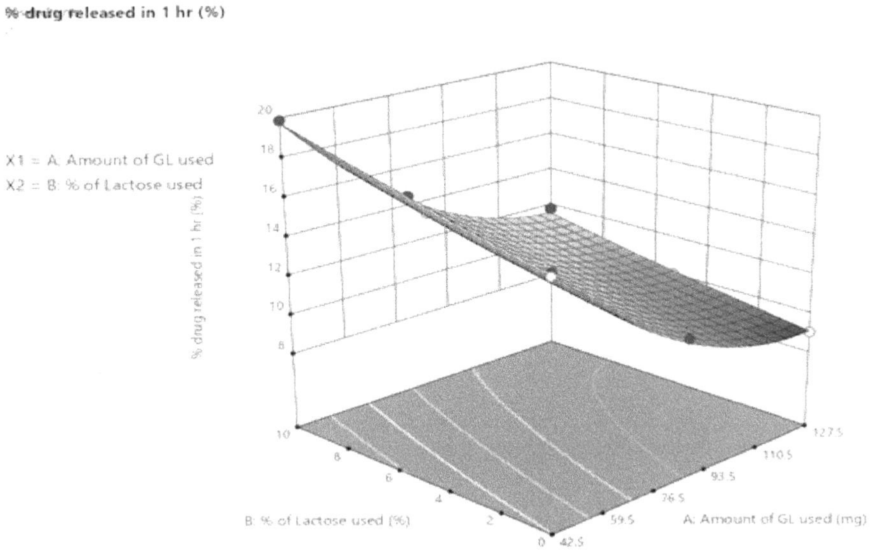

Figure 4 *Response surface plot for %DR1 of NEGFMT.*

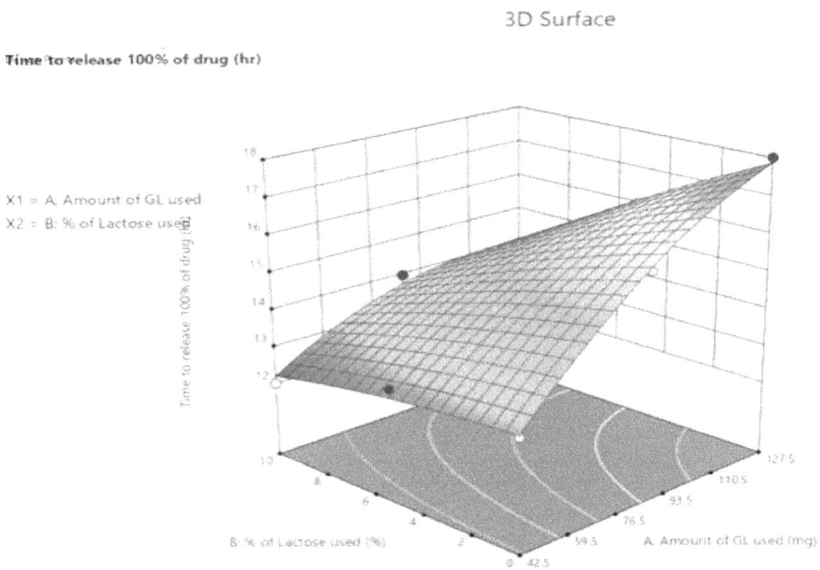

Figure 5 Response surface plot for T100 of NEGFMT.

3.2.2.2 *Optimization.* Improvement was done by both numerical enhancement utilizing attractive quality capacity and graphical streamlining utilizing overlay plots. The desirability and overlay plots are shown individually in Figures 6 and 7. The attractive quality capacity was discovered to be higher (close to 1) for the advanced equation showing the appropriateness of the formulations.

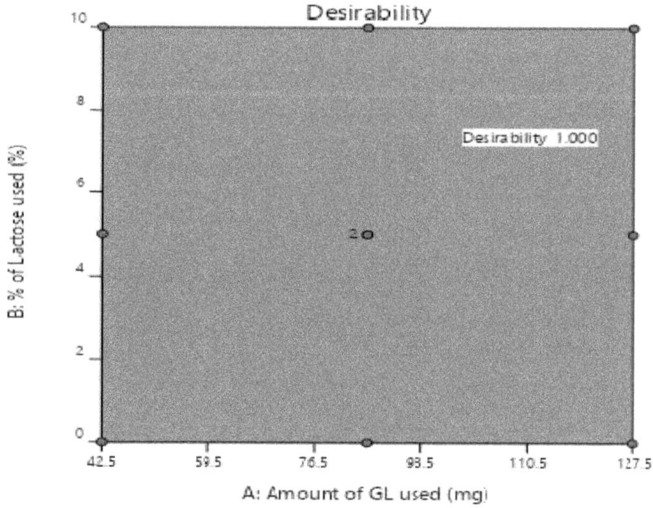

Figure 6 Desirability plot for NEGFMT.

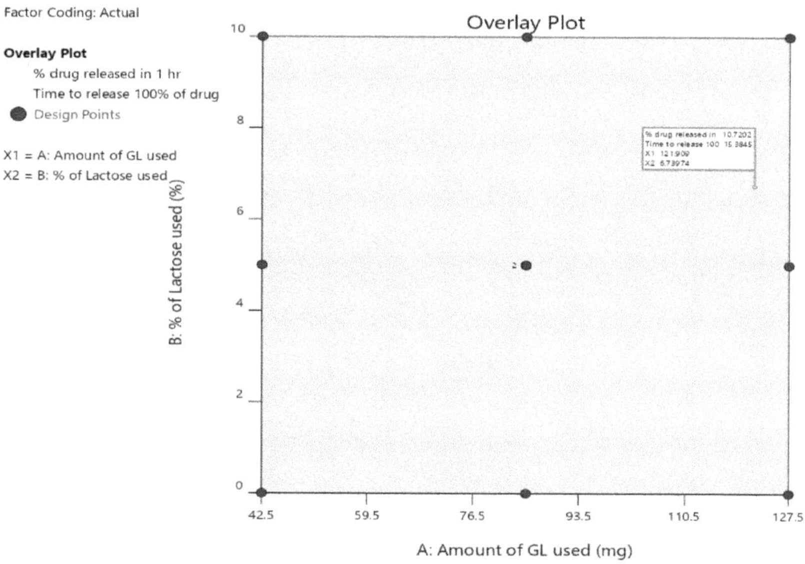

Figure 7 Overlay plot for NEGFMT.

The ideal values of individual test factors are appeared in Table 9. The optimized formulation of molten granules contained 61.41 mg of GL and 12.79 mg (8.74%) of lactose.

Table 9 *Formula of optimized NEGFMT and its in vitro characteristics.*

Ingredients(mg/Tab)	Optimized formulation
Nizatidine	85
Esomeprazole	-
Glyceryl Laurate	61.41
Lactose	12.79 (8.74%)
Aerosil	3
Magnesium Stearate	2
Total	164.2
Characteristics	
FLT(sec)	0
TFT(hrs)	>24
Uniformity of weight[a](mg)	164.2±0.47
Uniformity of Content[b] (%)	99.9±0.52
Hardness[c]	4.6
Friability[d]	0.09

3.2.2.3 *Cross-validation of model.* The prepared optimized formulations were found to be of good quality fulfilling all the official and other requirements of tablet (tabletting characteristics for both drugs are shown in Table 9). No lag time observed for floating of the prepared formulations and the floating was remained more than 24 hrs. Photograph was taken during the in vitro floating of optimized formulation and shown in Figure 8.

Figure 8 in vitro floating of optimized formulation.

The comparative dissolution profile is shown in Figure 9 along with theoretical profile.

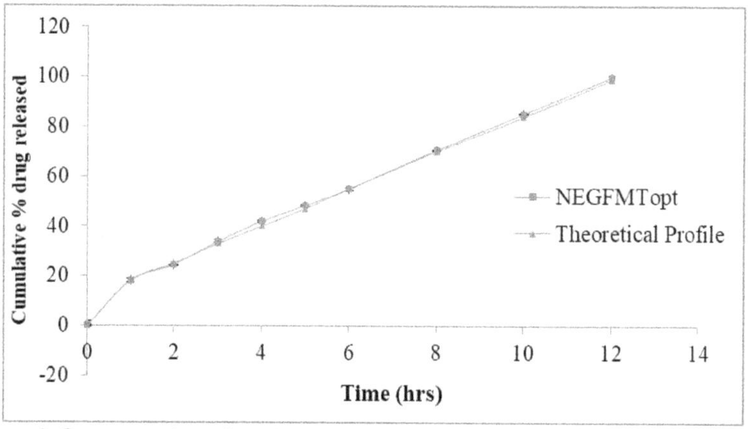

Figure 9 Comparative dissolution profiles of NEGFMTopt and theoretical release.

NEGFMT followed zero order release kinetics with anomalous (non-Fickian) diffusion mechanism. Correlation coefficient values were shown in Table 10.

Table 10 Correlation coefficients values of optimized NEGFMT.

Correlation value	NEGFMTopt
Zero order 'r' value	0.9957
First order 'r' value	0.9579
Higuchi order 'r' value	0.9787
Hixson crowell 'r' value	0.9305
Korsmeyer Peppas 'r' value	0.9943
Korsmeyer Peppas 'n' value	0.71

The experimental values, predicted values and percentage relative error of optimized formulation responses (%DR1 and T100) are represented in Table 11. A reasonable agreement between predicted and experimental values was observed as indicated by low values (<5%) of the relative error. This proved the validity of model and ascertained the effects of GL and the amount of lactose on drug release.

Table 11 *Cross-validation of model obtained using experimental and predicted results of both optimized NEGFMT.*

Optimized formulation	Response	Predicted value	Experimental value	%prediction error#
NEGFMTopt	% DR1	18	18.23	-1.28
	T100	12.128	12.00	1.06

Percent Error was calculated using the formula: ([Predicted value-Experimental value]/Predicted value) x 100

3.3 Drug-excipients interaction studies

3.3.1 *FTIR analysis*. The FTIR spectra of pure drug Nizatidine, GL, NEGFMTopt were shown in Figure 10.

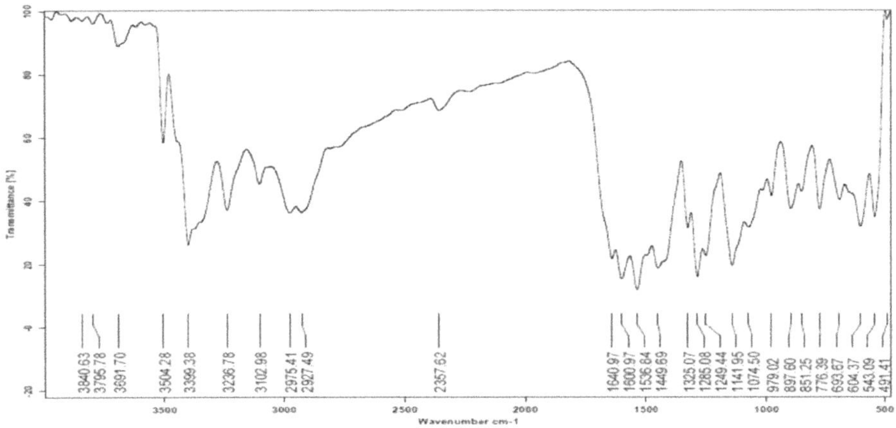

(a)

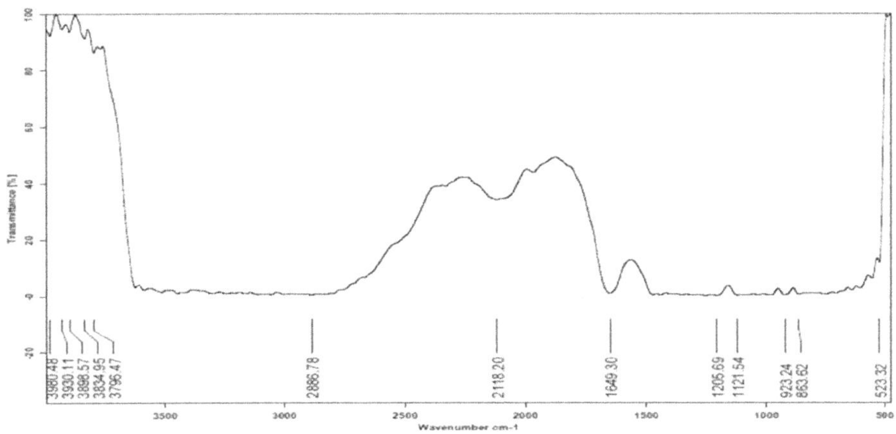

(b)

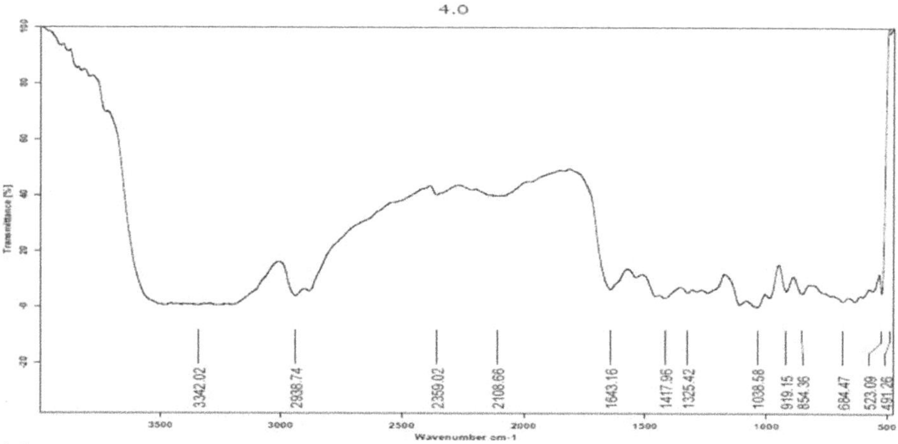

(c)

Figure 10 FTIR spectra of a) Nizatidine, b) Glyceryl laurate, c) NEGFMTopt

3.3.2 DSC analysis.

Thermograms of pure drug Nizatidine, GL and NEGFMTopt were shown in Figure 11. A single sharp endothermic peak at 132° C was obtained for Nizatidine, 132° C corresponding to its melting point. GL showed broad endothermic peaks at 60.3° C. Characteristic peak of Nizatidine was slightly shifted to left for NEGFMTopt (142.1° C).

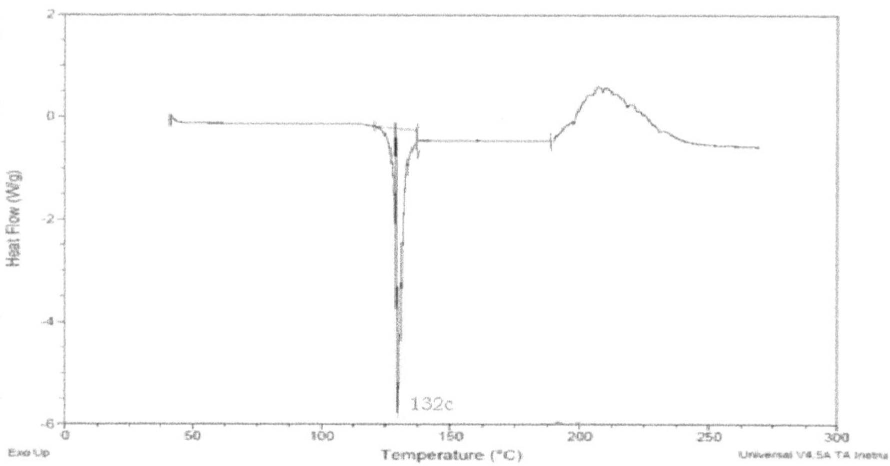

(a)

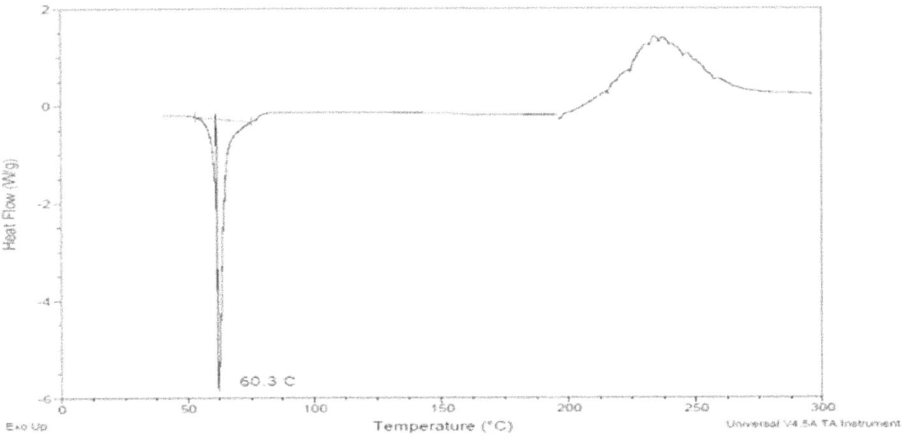

(b)

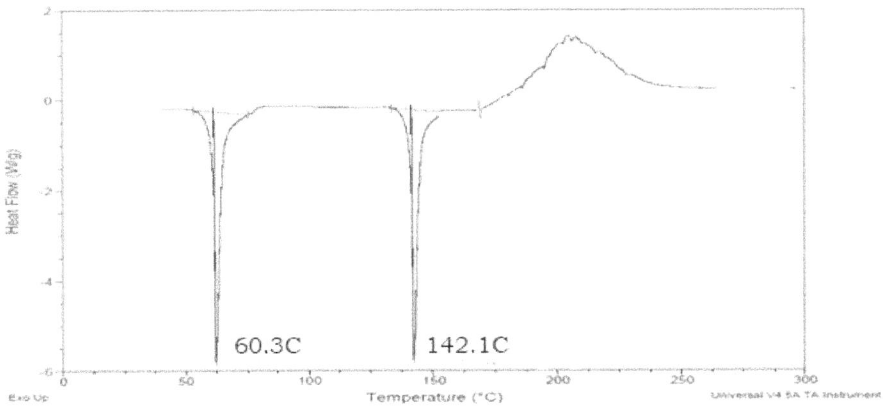

(c)

Figure 11 DSC of a) Nizatidine, b) Glyceryl Laurate, c)NEGFMTopt.

FTIR data indicated the absence of any chemical interactions between the drug and GL.

In the DSC thermograms, the characteristic peak for optimized formulation was not altered even after the physical interaction between the drug and wax material. In addition, the studied GL was hydrophobic in nature with melting point less than that of Nizatidine. The DSC data confirmed that the shift in the melting peaks of the drug in the selected formulation may be due to physical interaction between the drug and GL without any chemical interaction or complexation.

4 CONCLUSIONS

Non effervescent gastric floating technique was successfully employed in increasing the permeability of Nizatidine and enhanced its therapeutic action by releasing the drug in gastric medium for prolonged period of time. Melt granulation technique is found to be more superior in improving the gastric floating time and floating lag time when compared to conventional direct granulation technique. Glyceryl Laurate (GL) and lactose used in the formulations had a significant effect on drug release profile as evident from the desirability and response surface methodology plots. Statistical optimization can further be explored in the development of non-effervescent gastric floating tablets to achieve pilot level formulations with more desirability.

5. ACKNOWLEDGMENT

The authors are thankful to Aditya Pharmacy College, Surampalem for providing the facilities to carry the work.

6. REFERENCES

1. S. Arora, J. Ali, A. Ahuja, R.K. Khar, S. Baboota, Floating drug delivery systems: A review, AAPS PharmSciTech. (2005). https://doi.org/10.1208/pt060347.
2. L. Kuna, J. Jakab, R. Smolic, N. Raguz-Lucic, A. Vcev, M. Smolic, Peptic Ulcer Disease: A Brief Review of Conventional Therapy and Herbal Treatment Options, J. Clin. Med. (2019). https://doi.org/10.3390/jcm8020179.
3. R. Jewell, Nizatidine, in: XPharm Compr. Pharmacol. Ref., 2007. https://doi.org/10.1016/B978-008055232-3.62292-5.
4. J. Koskenpato, J.M. Punkkinen, K. Kairemo, M. Färkkilä, Nizatidine and gastric emptying in functional dyspepsia, Dig. Dis. Sci. (2008). https://doi.org/10.1007/s10620-007-9898-x.
5. S.R. Orenstein, D.A. Gremse, C.D. Pantaleon, D.F. Kling, K.S. Rotenberg, Nizatidine for the treatment of pediatric gastroesophageal reflux symptoms: An open-label, multiple-dose, randomized, multicenter clinical trial in 210 children, Clin.Ther.(2005).https://doi.org/10.1016/j.clinthera.2005.04.008
6. A. Streubel, J. Siepmann, R. Bodmeier, Floating matrix tablets based on low density foam powder: Effects of formulation and processing parameters on drug release, Eur. J. Pharm. Sci. (2003). https://doi.org/10.1016/S0928-0987(02)00223-3.
7. T.O. Oh, J.Y. Kim, J.M. Ha, S.C. Chi, Y.S. Rhee, C.W. Park, E.S. Park, Preparation of highly porous gastroretentive metformin tablets using a sublimation method, Eur. J. Pharm. Biopharm. (2013). https://doi.org/10.1016/j.ejpb.2012.11.009.
8. S. Kim, K.M. Hwang, Y.S. Park, T.T. Nguyen, E.S. Park, Preparation and evaluation of non-effervescent gastroretentive tablets containing pregabalin for once-daily administration and dose proportional

pharmacokinetics, Int. J. Pharm. (2018). https://doi.org/10.1016/j.ijpharm.2018.08.038.

9. Ankit U. Patel, Dipti V. Chaudary, Pranav J. Shah, Shailesh A. Shah, Hot melt granulation method for preparation of floating tablets of Tolperison hydrochloride, Futur. J. Pharm. Sci (2018). https://doi.org/10.1016/j.fjps.2017.12.002.
10. Samala, M.L., Janga, R.B. Design, statistical optimization of Nizatidine floating tablets using natural polymer. Futur J Pharm Sci 7, 2 (2021). https://doi.org/10.1186/s43094-020-00140-z

OPHTHALMIC QUERCETIN FILMS PREPARATION AND EVALUATION USING EUPHORBIA HIRTA

K.Srikala[a], Dr.CH K V L S N Anjana Male[a], Sr. G.Nirmala Jyothi[a], K.Snehitha*[a], P.Mary Asha[a], N.Anusha[a]
[a]Nirmala college of Pharmacy, Atmakuru, Mangalgiri-522503
Corresponding author:*snehithakasu@gmail.com

1 INTRODUCTION

Cataracts cause the lens to become cloudy, which blurs the image. The arrangement of the protein and water in the lens keeps it clear and allows light to pass through. Age-related protein clumping may cause a cataract, which lowers the amount of light reaching the retina, by blocking a small part of the lens.[1] The cataract may gradually enlarge and cover more of the lens over time, obstructing vision.[11] There are three primary cataract types: The most prevalent type of cataracts are called nuclear cataracts because they develop in the lens's middle. Cortical cataracts begin in the cortex (periphery) of the lens and gradually move toward the lens's core. Opacities that are concentrated below or inside the lens capsule are known as sub capsular cataracts. Aging includes the development of cataracts. Cataracts cannot be prevented even though some people may experience a slower rate of development as a result of lifestyle changes. Surgery, however, can treat the issue.[2]

1.1 Plant Profile

There are numerous Euphorbia species that are utilised in conventional medicine. With over 1600 species, Euphorbia is the biggest genus in the Euphorbiaceae family. It is distinguished by the presence of a more or less poisonous white milky latex. Among those *Euphorbia hirta* is the special plant with lot of pharmacological activities like antibacterial, anthelmintic, anti asthamatic, sedative, antispasmodic, antifertility, antifungal, and antimalarial activities due to the presence of phyto constituents like flavonoids, triterpenoids, phenols, essential oils and alkaloids.

2 MATERIALS AND METHODS

The components were weighed according to the instructions in Table No. 1 for F-1, F-2, F-3, (PVPK-30), F-4, F-5, and F-6, HPMC, to make the quercetin ocular inserts (K-100). Then, these components were completely blended, and water was added to make the volume 100 ml. The materials listed above were ultrasonically treated at an 80 MHZ speed for 20 minutes. After the formation of dispersion 5 ml of propylene glycol(30% w/v) were added as a plasticizer. A thin layer of glycerin was pre-coated on top of the prepared petridish before the prepared drug and polymeric solution were mixed. Place the petridish in hot air oven at a temperature of 35 to 200^{0}C. After 24 hours the petridish was removed, and the dried films were collected and then cut to the desired size. Table No 1: Optimization of the formulation of ocular inserts containing 100 concentration.
Quercetin ocular films were produced by using a solvent evaporation method using glycerin as a lubricant and were assessed for physical and chemical characteristics, in-vitro drug release experiments, and in vivo release studies. Each polymer's formulations were assessed for number of factors, including thickness, surface pH, moisture absorption rate, moisture loss rate, and quercetin drug release, *in vitro* dissolution studies.[3]

2.1 *In-vitro* Dissolution Studies of Occular Film

Using Franz diffusion cells and dialysis membranes, the in-vitro drug release from various ocular inserts was investigated.[4] The dialysis membrane mimics corneal epithelium. Fresh artificial tear fluid was placed in the receptor compartment. At pH 7.4 the receptor film was kept at 37°C with continuous stirring using a magnetic stirrer, 1.5 cm ^{2} portion of the ocular film was placed on the dialysis membrane.[5] The donor compartment's aperture was then sealed with a glass coverslip. At various times up to 8 hours, 1 ml of material was taken out of the receptor compartment and subjected to spectrophotometric analysis between 256 and 369 nm. Withdrawn samples were replaced with an equal volume of synthetic tear fluid.

2.2 In-Vivo Animal Studies

Naphthalene was fed to rats as part of an experimental protocol to cause cataract genesis 0.5gm/kg/day for the first three days, then 1 gm/kg/day for the following 28 days. [6]

2.3 Examination of The Eyes

An ophthalmoscope was used to look for morphological changes in the lens of the rats' eyes. The examination was carried out after the pupil was dilated with a 1% tropicamide solution.[6] Based on the several stages, cataract formation was evaluated.[7] Stage 1 is characterized by a clear, normal lens; Stage 2 by vesicles; Stage 3 by vesicles with cortical opacities; Stage 4 by diffuse, central opacities; and Stage 5 by opacity enclosing the entire lens (Mature cataract)[8] Cataract development was completed when the red fundus reflex was no longer visible through any area of the lens and the lens appeared dull white to the naked eyes (stage 5).[9]

3 RESULTS AND DISCUSSION

Quercetin ocular films were produced by means of the solvent evaporation method various formulas used in the various formulations were represented in the table:1 and they were evaluated for their in-vitro drug dissolution studies the results were mentioned in the Table 2. *In vivo* by animal studies the results were represented in the Table 3 and physico-chemical properties were given in the Table 4 and 5, *in vitro* drug release is given in the Table 6.

Table 1 *Optimization of the ocular inserts containing 200mg of the concentration (test 2).*

Formula	F1	F2	F3	F4	F5	F6
Drug(Quercetin)	200 mg	200 mg	200 mg	200 mg	200 mg	200 mg
Poly vinyl Pyrrolidine (k30)	200 mg	300mg	400 mg	-	-	-
HPMC (K100)	-	-	-	200	300	400
Propylene glycol	5 ml	5 ml	5ml	5ml	5ml	5ml
Distilled Water	Quantity sufficient					

Table 2 *In vitro drug release profile of ocular inserts consisting of 200mg of the drug (test 2).*

S.No	Time (hrs)	PVP(k30)			HPMC(K-100)		
Formulation code		F1	F2	F3	F4	F5	F6
1	0	0	0	0	0	0	0
2	0.5	4.3	3	1.1	6.07	4.8	2.69
3	1	6.9	6.3	2.8	9.6	6.8	3.4
4	2	13.1	9.2	5.6	17.9	11.8	6.9
5	3	18	11.6	8.4	9	18	9.8
6	4	21.3	14.7	11.3	34.1	22.1	13.1
7	5	26.2	18.7	14	43.8	26.6	17.6
8	6	30.5	21	16.6	53.8	32.4	21.7
9	7	35.8	23.1	18.1	60	36.8	24
10	8	38.8	24.9	19.6	64.6	40	27

The following formula was used of calculation for the cataract incidence rate:[10]

$$\text{Opacity index} = \frac{\text{Number of eyes in each stage}}{\text{Sum of all the eyes}} \qquad (1)$$

The naphthalene solution was made by dissolving 10% naphthalene in warm maize oil at 60°C for 30 minutes.[12]

Table 3 *Effect of the e. hirta loaded film % incidence on 28th day.*

S.no	Drug treatment	Percentage Incidence of Cataract				
		1st Stage	2nd Stage	3rd Stage	4th Stage	5th Stage
1	Normal Control	100	0	0	0	0
2	Cataract Control	0	0	0	83.3	66.7
3	Standard (Isotine)	33.3	50	16.7	0	0
4	Test 1 (100mg)	0	50	33.3	33.3	16.7
5	Test 2 (200 mg)	0	66.7	50	33.3	0

Among all six formulations, F4 formulation drug release rate was high, with 78.9%, at 200 mg concentration. Rats pre-treated with naphthalene for 28 days shows profound decrease in opacity index. Hence may be ophthalmic quercetin films are promising options for the treatment of cataract in upcoming days.

Table 4 *Evaluation of different formulations of drug concentration of 100 mg.*

Formulation Code	Weight (mg)	Surface pH	% Moisture absorption	Uniformity of drug Content	Thickness
F1	65±0.3	6.8	5.45±0.3	1.99±0.2	0.41±0.01
F2	68±0.4	7.0	6.30±0.5	2.09±0.4	0.53±0.04
F3	70±0.7	7.1	5.93±0.1	2.01±0.1	0.36±0.05
F4	75±0.8	7.3	4.91±0.2	2.19±0.1	0.36±0.06
F5	78±0.4	6.9	5.53±0.1	2.13±0.8	0.39±0.01
F6	76±0.2	7.0	5.21±0.3	2.19±0.3	0.42±0.02

Table 5 *Evaluation of different formulation of drug concentration of 200 mg.*

Formultion code	Weight (mg)	Surface pH	% Moisture absorption	Uniformity of drug content	Thickness
F1	73±0.2	7.0	6.54±0.4	2.0±0.1	0.51±0.01
F2	62±0.3	7.4	4.76±0.5	1.96±0.2	0.62±0.03
F3	75±0.1	6.9	5.89±0.2	1.86±0.4	0.66±0.04
F4	78±0.6	7.2	4.73±0.5	2.13±0.6	0.42±0.04
F5	69±0.9	6.8	5.81±0.6	1.94±0.3	0.42±0.04
F6	78±0.3	7.3	0.73±0.8	1.03±0.1	0.61±0.2

Table 6 *In-vitro drug release profile of Ocular inserts consisting of 100 mg of the active pharmaceutical ingredient.*

S.No	**Time (hrs)**	**PVP(k30)**			**HPMC(K-100)**		
Formulation code		**F1**	**F2**	**F3**	**F4**	**F5**	**F6**
1	0	0	0	0	0	0	0
2	0.5	4.3	3	1.1	6.07	4.8	2.69
3	1	6.9	6.3	2.8	9.6	6.8	3.4
4	2	13.1	9.2	5.6	17.9	11.8	6.9
5	3	18	11.6	8.4	9	18	9.8
6	4	21.3	14.7	11.3	34.1	22.1	13.1
7	5	26.2	18.7	14	43.8	26.6	17.6
8	6	30.5	21	16.6	53.8	32.4	21.7
9	7	35.8	23.1	18.1	60	36.8	24
10	8	38.8	24.9	19.6	64.6	40	27

4 CONCLUSION

The group that underwent tests 1 and 2 demonstrated a decrease in opacity index, with values of results, when compared to the naphthalene control. It was discovered that the usual drug's opacity index had significantly decreased to 1.8. Each polymer's formulations were assessed for a number of factors, including thickness, surface pH, moisture absorption rate, moisture loss rate, and quercetin drug release. Among all six formulations,F4 formulation drug release rate is high, with 78.9%, at 200 mg concentration. Rats pre-treated with naphthalene for 28 days shows profound decrease in opacity index. Hence may be ophthalmic quercetin films are promising options for the treatment of cataract in upcoming days.

5 REFERENCES

1. BE.Klein, R.Klein, and KL. Linton. *Ophthalmology*. 1992, **99**, 546.
2. FW. Newell. National Eye Institute. *American Journal of Ophthalmology*. 1969, **68**, 726.
3. A S. Mundada, BK Shrikhande. *Drug development and industrial pharmacy*. 2006 ,**32**, 443.
4. M. Pattnaik, P. Pandey, GJ. Martin, NH. Mishra and M. Ashok kumar . *Foods*. 2021.**10**.279.
5. P. Nayak and A. Thirunavoukkarasu . *J. Phytopharmacol*. 2016,**5**,83
6. PK. Kulabhusan, S. Agrawal, J. Jeevanandam and MK. Danquah. *Poisonous Plants and Phytochemicals in Drug Discovery*. 2020,**12**,269.
7. SR.Ahmed, MM.Al-Sanea, EM.Mostafa, S.Qasim, N. Abelyan, and FA. Mokhtar. *ACS omega*. 2022 .
8. JC.Ikewuchi, EN. Onyeike, AA. Uwakwe, and CC. Ikewuchi. *Journal of Ethnopharmacology*. 2011,**137**,1415.
9. S. Perumal and R. Mahmud. *BMC complementary and alternative medicine*. 2013,**13**,1.
10. RS. Konya, LF.Leelee and KO.Chukwu . *Animal Research International*. 2013,**10**,1747.
11. L.Tona, NP. Ngimbi, M. Tsakala, K. Mesia, K. Cimanga, S. Apers, T. De Bruyne, L. Pieters, J. Totte and J. Vlietinck . *Journal of ethnopharmacology*. 1999,**68,**193.
12. M.Umamaheswari, K.Asokkuma, V.Lalitha, T.Sivashanmugam and V. Subhadradevi. *Journal of Pharmacy Research*. 2011,**4**,680.

PREPARATION OF RILPIVIRINE NANOSPONGES USING DIFFERENT METHODOLOGIES

KVNR Aishwarya [a], Praveen Sivadasu* [b]

[a, b] Department of Pharmaceutics, KL College of Pharmacy, Koneru Lakshmaiah Education Foundation, Vaddeswaram, AP, India, *praveen.sivadasu399@gmail.com

1 INTRODUCTION

Rilpivirine (RLP) is BCS class II medication utilized for treatment of HIV contamination. The medication has low fluid solvency (0.0166 mg/ml) and disintegration rate prompting low bioavailability (32%). Point of this work was to upgrade dissolvability and disintegration of rilpivirine utilizing regular polymer-based nanosponges. These nanosponges are biocompatible nanoporous particles having high stacking ability to shape supramolecular consideration and exclusion buildings with hydrophilic and lipophilic medications for dissolvability upgrade.[1] The goal of any drug delivery system is to provide a therapeutic amount of drug to the proper site of action in the body and also to achieve and to maintain the valuable plasma concentration of the drug for a particular period of time. The development of new and complex molecules called Nanosponges have the potential to achieve this goal.

The most popular method of administering drugs is orally because it is simple and painless. The bulk of novel active pharmaceutical compounds have poor oral bioavailability due to their slow dissolving rate and restricted absorption. Although there has been an increase in the use of lipid-based drug delivery systems and other formulation techniques such as pH adjustment, solid dispersions, co crystal formation, and complexion changes, drug absorption has improved noticeably. Long-standing goals include the development of an effective drug delivery system, targeted drug delivery, and bioavailability augmentation of medicines that are poorly water soluble. The development of effective medicine delivery systems using nanotechnology has gained importance recently.

Nanosponge is a name used to describe nanoparticles with porous structures. Nanosponges are extremely small sponges, with an average diameter of less than 1 m and the size of a virus. They can improve the bioavailability of poorly soluble medications by binding them within the matrix due to their small size and porous nature. This is done by altering the pharmacokinetic characteristics of the actives. Nanosponges are three-dimensional, solid, porous, biocompatible drug delivery systems that may entrap both hydrophilic and hydrophobic medications and solve the issues of drug toxicity and poor bioavailability.[2] For the delivery of medications for oral administration, topical administration, and parental administration, nano sponge technology has received extensive study. Vaccines, antibodies, proteins, and enzymes can all be effectively transported via nanosponges.

Biodegradable Polymers which are synthetic or natural are used in the preparation of Nanosponges to improve the solubility and bioavailability of the drug. Synthetic polymers such as Ethyl cellulose, Beta cyclodextrins, poly valerol acetone, Eudragit RS 100 etc. have been used in the formulation of Nanosponges. Various natural polymers which can also be used includes kondagogu gum, olibanum gum, gum Arabica, Xanthan gum, locust bean gum, etc. in the formulation of Nanosponges. Polymers of natural origin are preferred over synthetic ones because they are comparatively cheaper, abundantly available, non-irritating and nontoxic in nature. Rilpivirine is in a class of medications called non-nucleoside reverse transcriptase inhibitors (NNRTIs) Rilpivirine inhibits the replication of HIV-1 by binding in a non-competitive manner directly to the reverse transcriptase (RT) enzyme. [3]

2 MATERIALS AND METHODS

2.1 Standard calibration curve of Rilpivirine

10 mg of Pure RLP was weighed accurately on digital balance and it was separately dissolved in each 10 ml of methanol which was added in a 100ml volumetric flask respectively. The above solution was sonicated for 15 mins. After that volume was adjusted up to 100ml with phosphate buffer pH 6.8. This stock solution of 100µg/ml was then used for further dilutions with phosphate buffer pH 6.8 to obtain standard solutions for pure RLP of concentrations of 2 to 20 µg/ml. The absorbances of three replicates of the standard solutions of each concentration were measured at the selected analytical wavelength (245 nm).

2.2 FT-IR

The dry sample of RLP was combined in a 100:1 ratio with IR grade KBr. By using a hydraulic press with 10 tonnes of pressure, this mixture was compacted into a pellet. The pellets were scanned using an FTIR instrument (Perkin Elmer, Spectrum BX, USA) throughout a wave number range of 4000 to 400 cm^{-1}.

2.3 Preparation of Nanosponges (NSG) by Emulsion solvent diffusion (ESD) method.

By using varying amounts of various polymers, stabilisers, and cross-linkers that were tested with varied proportions and ratios, nanosponges can be created using this process as depicted in Table 1. The dispersed phase, which contained the medication and polymer, was dissolved in the proper amount of cross-linkers before being gradually added to the predetermined quantity of stabilizers in the aqueous continuous phase. For two hours, the reaction mixture was agitated at various rpms. The resulting nanosponges were filtered out and dried for 24 hours at 400°C in an oven. Vacuum desiccators were used to preserve the dried NSGs to ensure that any remaining solvent was removed. [3,7]

Table 1 *Different formulations prepared by emulsion solvent diffusion method.*

Ingredients	F1	F2	F3	F4	F5	F6	F7	F8	F9
Weight of drug (mg)-RLP	25	25	25	25	25	25	25	25	25
PVA: EC (%w/w)	1:1	-	-	1:2	-	-	1:3	-	-
KGG:EC (%w/w)	-	1:2	-	-	1:3	-	-	1:1	-
OLG:EC (%w/w)	-	-	1:3	-	-	1:1	-	-	1:2
DCM (ml)	20	20	20	-	-	-	-	-	-
GTAD (ml)	-	-	-	20	20	20	-	-	-
DMC (ml)	-	-	-	-	-	-	20	20	20
DW (ml)	150	150	150	150	150	150	150	150	150

2.4 Preparation of Nanosponges (NSG) by Ultrasound-Assisted synthesis.

NSGs can be obtained by responding polymers with cross-linkers in the absence of dissolvable and under sonication. The accomplished NSGs will be circle molded, steady in size and more modest than 5 microns. In this technique DMC (or) pyromellitic anhydride is utilized as cross-linker. A measure of anhydrous Cd was put to respond in softened DMC at 90 °C for 5 hrs. Then, the strong was ground in a mortar and soxhlet separated with ethanol to eliminate either contamination (or) unreacted dimethyl carbonate. After purging nanosponges were put away at 25 °C until additional utilization. Further, Complexes of Drug NS were prepared at two different weight ratios of 1:5 and 1:10 (Drug, BNS w/w) .2 Using a magnetic stirrer, precisely weighed amounts of NS 1:1, 1:2, 1:3, 1:4, and 1:5 was suspended in 20 mL of distilled water as depicted in Table 2. The calculated amount of drug was added and the mixture was then sonicated for 10 min and maintained for 24 hrs while being stirred. The un-complexed medication was separated from the suspensions as a residue below the colloidal supernatant by centrifuging them at 2,000 rpm for 10 min. A desiccator was used to preserve the dried powder. [2]

2.5 Preparation of Nanosponges (NSG) by hyper cross-linked β- CD

To complete entire disintegration, 100 mL of anhydrous dimethyl formamide and 17.42 g of anhydrous cadmium (15.34 mmol) were added to a round base carafe. After adding 9.96 g of dimethyl carbonate (carbonyl di-imidazole) (61.42 mmol), the system was allowed to react for 4 hours at 100 °C. The simple block of hyper-cross-connected -Disc was normally ground when the build up polymerization was complete, and deionized water was added to remove Dimethyl formamide. Further, ethanol-based Soxhlet extraction was used to completely eliminate any residual by-products or unreacted chemicals. The resulting white powder was baked in a 60°C oven overnight before being pulverised in a mortar. Furthermore, correctly weighed amounts of NS 1:2 or 1:4 were accurately suspended in 20 mL of distilled water using a magnetic stirrer, followed by the addition of the determined amount of drug, sonicated it for 10 min, and storage for 24 h under stirring. The un-complexed medication was separated from the suspensions as a residue below the colloidal supernatant by centrifuging them at 2,000 rpm for 10 min. At an operating pressure of 13.33 mbar and a temperature of 20 °C, the supernatant was lyophilized. A desiccator was used to preserve the dried powder.[1] All formulations were allocated 1.0 g of the lot volume as depicted in Table 3.

Table 2 *Different formulations prepared by using Ultrasound-Assisted synthesis method.*

Step I: Synthesis of β-CD Nanosponges					
Ingredients	**F1**	**F2**	**F3**	**F4**	**F5**
β-CD:DMC (mmol or gm)	1:1	1:2	1:3	1:4	1:5
β-CD (gm)	17.42	17.42	17.42	17.42	17.42
DMC (gm)	9.96	19.92	29.88	39.84	49.8
Ethanol (ml)	100	100	100	100	100
Step II: Preparation of RLP-Loaded NS					
RLP: β-CD (w/w)	1:1	1:2	1:3	1:4	1:5
Drug	25	25	25	25	25
β-CD NSGs	40	80	120	160	200
Dist. water (ml)	20	20	20	20	20
Centrifugation (RPM)	2000	2000	2000	2000	2000

Table 3 *Different formulations prepared from hypercross-linked β-cyclodextrins.*

Step I: Synthesis of β-CD Nanosponges			
Ingredients	F1	F2	F3
DMF	100	100	100
β-CD : DMC (mmol or gm)	1:2	1:4	1:8
β-CD (gm)	17.42	17.42	17.42
β-CD (mmol)	15.34	15.34	15.34
DMC (gm)	19.92	39.84	79.68
DMC (mmol)	30.68	61.36	122.72
Ethanol (ml)	100	100	100
Dist. water (ml)	100	100	100
Step II: Preparation of RLP-Loaded NS			
RLP: β-CD (w/w)	1:5	1:10	1:15
Drug(mg)	25	25	25
β-CD NSGs	200	400	600
Dist. water (ml)	20	20	20
Centrifugation (RPM)	2000	2000	2000

2.6 Preparation of Nanosponges (NSG) by Solvent Method

A typical procedure involves combining the polymer with a polar aprotic solvent, such as dimethylformamide (DMF), dimethyl sulfoxide, or another appropriate solvent (DMSO). Add the excess cross-linker and the medication combination next, preferably in a cross-linker/polymer molar ratio of 4 to 16. Perform the reaction for a duration of 1 to 48 hours at a temperature range from 10°C to the solvent's reflux temperature. Carbonyl molecules, such as dimethyl carbonate and carbonyl diimidazole, are preferred cross linkers. Post completion of the reaction, the solution was allowed to cool. Further, the product was filtered out by using large quantities of distilled water and cleaned by soxhlet extraction with ethanol as shown in Table 4.

Table 4 *Different formulations of nanosponges prepared by solvent method.*

Ingredients	**F1**	**F2**	**F3**	**F4**	**F5**	**F6**	**F7**	**F8**	**F9**
Weight of drug (mg)	50	50	50	50	50	50	50	50	50
Dimethyl sulfoxide (ml)	10	10	10	10	10	10	10	10	10
β-CD (mg)	500	500	500	1000	1000	1000	1500	1500	1500
Ethyl cellulose (mg)	500	500	500	1000	1000	1000	1500	1500	1500
Kondagogu gum (mg)	100	100	100	100	100	100	100	100	100
Dimethyl carbonate (ml)	23.15	46.3	69.4	23.1	46.3	69.4	23.1	46.3	69.4
Distilled water (ml)	150	150	150	150	150	150	150	150	150
Ethanol (ml)	100	100	100	100	100	100	100	100	100

2.7 Characterization of formulated nanosponges by using FESEM

Microphotographs were recorded for the formulated nanosponges by various methods and the interpretations were reported.[4, 5]

3 DISCUSSION OF RESULTS

An UV-visible spectrophotometer used for the measurement of Rilpivirine concentration at absorbance 245nm in 6.8 pH phosphate buffer. A calibration curve was constructed, and approach was verified for linearity, accuracy, precision, and interference, which followed Beer's rule at concentrations between 2-20µg/ml. The Standard curve as shown in Figure 1 was found to be linear in the concentration range between 2-20 µg/ml having coefficient of regression value R^2=0.998.

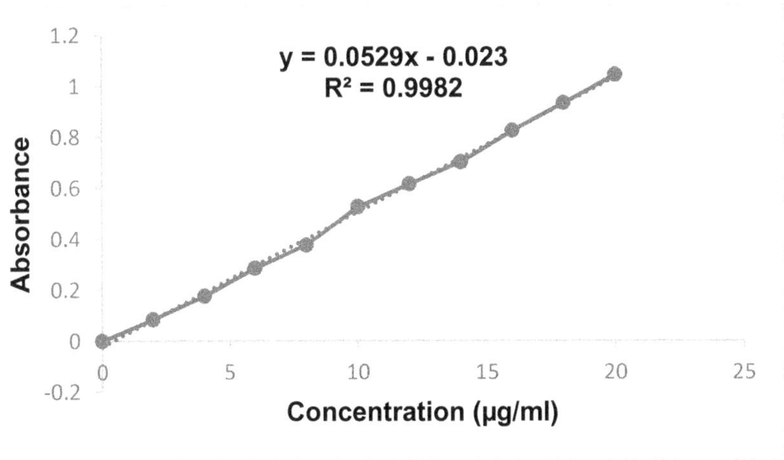

Figure 1 Standard calibration curve of Rilpivirine.

The compatibility of drug with the excipients used was evaluated by FTIR studies. The FT-IR spectrum of Rilpivirine powder was shown in Figures 2-3 from the above FT-IR images it was evident that all the functional groups of pure drug and other excipients were present in the spectrum. Therefore, it can be inferred that the selected drug and excipients were compatible and do not possess any interactions.

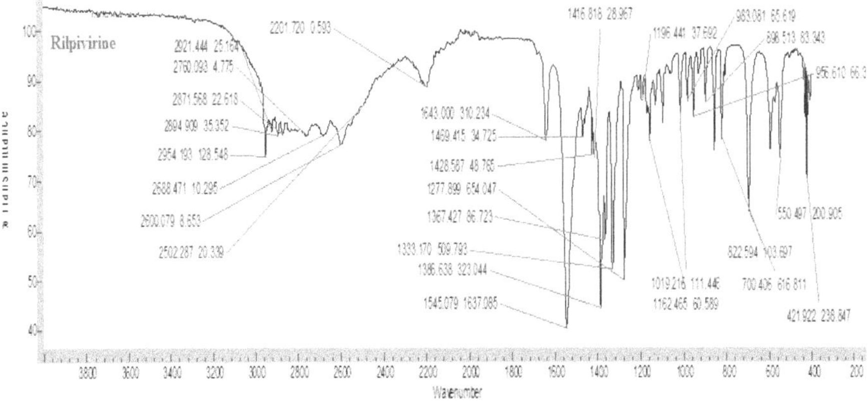

Figure 2 FT-IR Peaks of Rilpivirine Pure.

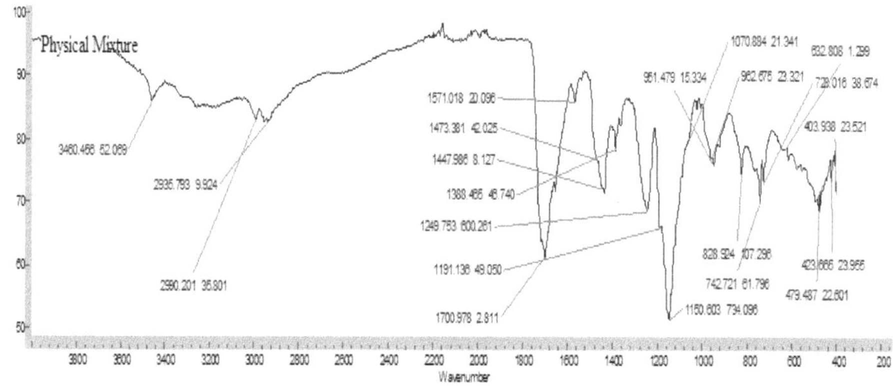

Figure 3 FT-IR Peaks of physical mixture.

FESEM for nanosponges formulated by Emulsion solvent diffusion method

In this method, various ratios of PVA: EC (%w/w), KGG: EC (%w/w), OLG: EC (%w/w) in the proportion of 1:1, 1:2 and 1:3 (%w/w) were used respectively.[4,9] Whereas dichloromethane (DCM), glutaraldehyde (GTAD), dimethyl carbonate (DMC) were used as cross-linkers in different batches. Total nine batches were prepared and on the basis of observation and obtained result batch F2 has shown optimal results and it is considered for further development. From the Figure 4 it was observed that formulation (F2) was found to be irregular in shape as well as dry mass with no cavity was observed.

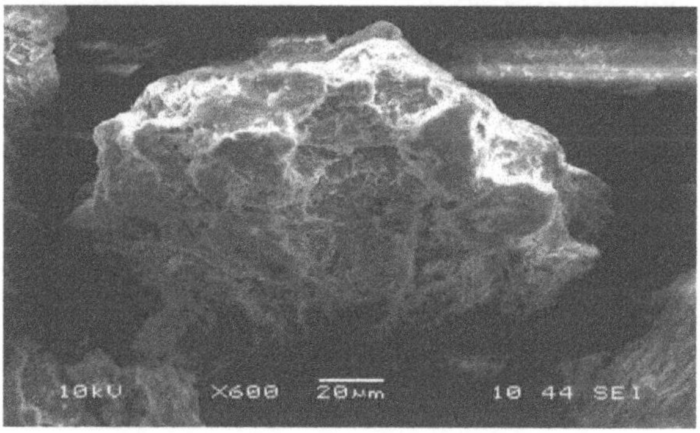

Figure 4 FESEM image nanosponges prepared by Emulsion solvent diffusion (ESD) method (F2).

FESEM for nanosponges formulated by Ultrasound-Assisted synthesis

In this method, various ratios of β-CD: DMC (1:1, 1:2, 1:3, 1:4 and 1:5) was used as polymer: cross-linker ratio. Total five batches were formulated.[8,9] From the Figure 5 it was evident that the formulated nanosponges surface was found to be irregular in shape and few porous cavities was also seen.

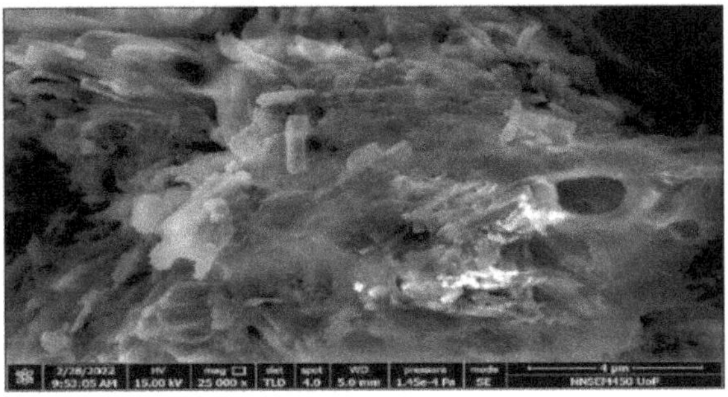

Figure 5 FESEM image nanosponges prepared by Ultrasound-Assisted synthesis (F3).

FESEM for nanosponges formulated by hypercross-linked β-cyclodextrins

A typical procedure involved in formulating nanosponges using various ratios of β-CD: DMC (1:1, 1:2 and 1:3) was used as polymer: cross- linker ratio and dimethyl formamide (DMF) was used as a solvent. Total three batches were formulated. From the Figure 6 it was evident that the formulated nanosponges surface was found to be irregular and crystal in shape; dry mass with no cavity formation was also observed.[6]

Figure 6 FESEM image nanosponges prepared by hypercross-linked β-cyclodextrins (F2).

FESEM for nanosponges formulated by Solvent Method

A typical procedure involved in formulating nanosponges using various ratios of β-CD: DMC (%w/w), EC: DMC (%w/w) in various proportions were used. Whereas dimethyl sulfoxide (DMSO), was used as a solvent and KGG was used as stabilizers for different batches. FESEM image Figure 7 revealed that the surface was found to be smooth and porous surface with cavity formation was observed. Hence, this method was considered for further development of batches with further scale up.[5]

Figure 7 FESEM image nanosponges prepared by Solvent Method (F8).

4 CONCLUSION

In contrast to the stable nanosponges made with kondagogum gum, those made with olibanum gum in the emulsion solvent method did not hold their shape and were thus unstable. In order to prepare nanosponges, kondagogum is utilised instead of olibanum gum. The FESEM analysis revealed that the surface of the nanosponges made using the solvent approach was both smooth and porous, while the surface of the nanosponges made using the other methods was rough and non-porous. Thus, the solvent approach was considered for further development of batches with larger scale up. Based on this finding, we continued our formulation development, creating many batches using various polymer (EC): cross-linker (DMC), solvent (DMSO), and stabiliser (KGG) combinations.

5 REFERENCES

1. Lala R, Thorat A, Gargote C, Int J Res Ayur Pharm, 2011, 2, 1520.
2. Renuka Sharma, Roderick BW, Kamla Pathak, Ind J Pham Edu Res, 2011, 45, 25.
3. Ansari KA, Torne SJ, Pradeep RV, Trotta F, Cavalli R, Curr Drug Deliv, 2011, 8(2), 194.
4. Rana Z Ahmed, Zahid zaheer, Jaiprakash, Drug development and industrial pharmacy, 2017, 43 (12),1.
5. Vyas SP, Khar RK, Molecular Basis of Targeted Drug Delivery. In: Targeted and Controlled Drug Delivery, CBS Publishers and Distributors, New Delhi, 2008.
6. Idawsari HM, Badr-Eldin SM, Labib GS, El-Kamel AH, Int J Nanomedicine. 2015, 10, 893.
7. Moin A, Deb TK, Osmani RA, Bhosale RR, Hani U, J Basic Clin Pharm. 2016, 7, 39.
8. El-Housiny S, Shams Eldeen MA, El-Attar YA, Salem HA, Attia D, Bendas ER, et al, Drug Deliv. 2018, 25, 78.
9. Ansari KA, Vavia PR, Trotta F, Cavalli R, AAPS PharmSciTech. 2011, 12, 279.

FORMULATION AND EVALUATION OF MINOCYCLINE USING MULTIPLE POLYMERS

S. Nigama Chandra*, A. Anka Rao, Amareswarapu.V Surendra

[a]Department of Pharmaceutics, KL College of Pharmacy, Koneru Lakshmaiah Education Foundation, Vaddeswaram, AP, India.

Corresponding Author E-Mail ID: nigamac111@gmail.com

1. INTRODUCTION

Minocycline is a semi-synthetic tetracycline anti-biotic. It is used to treat many bacterial infections like acne, urinary tract infections, gonorrhea, tick fever etc. and hence act both as bactericidal and bacteriostatic. Minocycline exhibits bactericidal activity by combining with A-site of 30S ribosomal subunit and prevents extension of peptide chain which inhibit protein synthesis process in bacteria. As the drug is freely soluble in water, the excipients must be selected prudently to achieve constant *in vivo* drug release. [1] The medication level in the blood cannot be sustained for a lengthy amount of time with conventional dosing forms. The usual dosage form's inability to regulate temporal delivery is the cause of the action's brief duration. Technical developments have resulted in the creation of controlled drug delivery systems, which have the potential to change the practise of medication and offer a variety of therapeutic benefits, in order to overcome the limitations of conventional drug delivery systems. [2] The simplest strategy to control release dosage forms is to use matrix tablets, which are made of a medication and a polymer that retards absorption of the drug. These matrix-controlled preparations can exhibit slow controlled release of active compound in blood and thus bring greater compliance in long term and oral controlled dosage forms.[3]

Matrix controlled tablets are more effective because of their ease in preparation, cost effective and the drug release is reckoned flawlessly. In this study, hydrophilic polymers are selected (Xanthun gum, guar gum and HPMCK100) as they have high gelling capacity which retards the release of drug on contacting the aqueous medium where they hydrate at solid-liquid interphase by forming a viscous layer.

The present work is aimed at studying the role of various hydrophilic polymers in controlled-release matrix tablets of Minocycline. The drug release mechanism is calculated by fitting the kinetic data in mathematical equations such as zero order and first order release, Higuichi and Peppas models. The prepared formulation with different polymers is compared and a suitable polymer is optimized which exhibit high dissolution rates. [4]

2 MATERIALS AND METHODS

2.1. Chemicals and Reagents

Minocycline HCl and HPMC(K4M) were obtained from Aurobindo pharma. Ltd. (Hyderabad, India). MCC was provided by Loba Chemie Pvt. Ltd (Mumbai, India). Talc and magnesium stearate were provided by QualiChem's Fine Chemicals Pvt. Ltd (New Delhi, India). All the excipients and other ingredients used of analytical grade.

2.2. Preparation of Controlled release Matrix Tablets

Different formulations were prepared by direct compression method. Different ratios of formulations were prepared with different polymers. Accurately weighed quantities of pre sieved drug along with release retardants (polymer matrix) were mixed thoroughly using tumbling method. The prepared granules were sieved and fines were separated to obtain uniform granules. The use of talc and magnesium stearate served as the final anti-frictional additive. Before proceeding to compression, granules were evaluated for pre-compression parameters. The uniform granules were compressed using rotary press compressing machine (Acura, Mumbai). All of the matrix tablets contain 100 mg of minocycline, and it has been determined how varying polymer ratios affect how much of the medication is released.

2.3. Evaluation of Matrix Tablets

Table 1 *Composition of Matrix Tablets Containing HPMC K100, guar gum, xanthum gum.*

Ingredients (mg/tab)	Formulation code								
	F1	F2	F3	F4	F5	F6	F7	F8	F9
Minocycline	100	100	100	100	100	100	100	100	100
HPMC K100	50	100	150	-	-	-	-	-	-
Guar gum	-	-	-	50	100	150	-	-	-

Xanthun gum	-	-	-	-	-	-	50	100	150
MCC	245	195	145	245	195	145	245	195	145
Magnesium stearate	2	2	2	2	2	2	2	2	2
Talc	3	3	3	3	3	3	3	3	3
Total	400	400	400	400	400	400	400	400	400

Bulk weight of tablet is 400mg containing 100mg of minocycline HCl F-1 to F-9 represents formulations of minocycline HCl. formulation F-1 to F-3 was prepared using ratio of HPMCK100; formulation F-3 To F-6 was prepared using ration of guar gum; F-6 to F-9 contain ratio of Xanthun gum. See Table 1.

2.4. Evaluation of granules

The bulk density and tapped density of granules were calculated in order to assess their flow characteristics (See Table 2). In order to measure the volume (V0) and properly weigh the quantity (W), it was carefully transferred into a graduated cylinder. The graduated cylinder was then covered with a lid and placed into a tap density tester (USP) to measure the density. Final volume (Vf) was taken after the density apparatus had been set up for 500 tabs, and readings were kept until two consecutive readings were equal (Lachman et al.,1987). The BD and TD were calculated by the formula,

Bulk density = W/V_0
Tapped density = W/V_f

Carr's compressibility index, which equals [(TBD-LBD) *100]/TBD, was used to calculate the compressibility index. [5] By placing the precisely weighed granules in the funnel, the angle of repose was ascertained (FUNNNEL METHOD). From the 4-inch surface where the funnel is stored, the granules were permitted to freely flow through the funnel and onto the surface. The formula Tan =h/r was used to determine the angle of repose and estimate the cone's diameter.

Angle of repose is, powder height is h in cm, and powder cone radius is r in cm. To determine Karl Fischer instrument, a Karl Fischer instrument was issued. The granules were weighed, added to the instrument's reagent solution, mixed, and the tare weight was fed into the apparatus. After a predetermined amount of time, the monitor displayed the moisture content as %w/w.

Table 2 *Pre compression parameters of CR release minocycline HCl granules.*

FORMULATION CODE	BULK DENSITY(g/ml)	TAPPED DENSITY(g/ml)	CARR'S INDEX (%)	HAUSNER RATIO	ANGLE OF REPOSE (°)
F1	0.55± 0.02	0.65± 0.02	1.25± 0.47	0.25± 0.01	23.45 ± 0.88
F2	0.54± 0.01	0.62± 0.03	1.19± 0.59	0.16± 0.01	19.65 ± 0.44
F3	0.56± 0.02	0.64± 0.01	1.23± 0.43	0.18± 0.01	22.35 ± 0.85
F4	0.54± 0.02	0.63± 0.02	1.12± 0.45	0.11± 0.01	20.69 ± 0.45
F5	0.50± 0.01	0.67± 0.02	1.24± 0.84	0.22± 0.01	20.82 ± 0.89
F6	0.53± 0.02	0.64± 0.02	1.23± 0.48	0.18± 0.01	20.72 ± 0.48
F7	0.51± 0.01	0.67± 0.01	1.24± 0.89	0.19± 0.01	20.89 ± 0.34
F8	0.52± 0.02	0.69± 0.02	1.3± 0.68	0.23± 0.01	20.78± 0.58
F9	0.56± 0.01	0.68± 0.03	1.21± 0.45	0.17± 0.01	22.6± 0.89
Standards	-	-	12-16	<1.25	20-30

2.5. Evaluation of Matrix Tablets

The weight fluctuation, friability, thickness, hardness, diameter, and drug content of the prepared matrix tablets were assessed (See Table 3). Monsanto hardness tester (Campbell electronics, Mumbai) was used to determine the hardness of tablets by using 10 tablets. [6] The weight variation test was performed according to the official (I.P) method with an electronic balance (Presario 310M, Mumbai); evaluated for 20 tablets, the thickness and hardness were determined in conjunction with Vernier callipers (Mitutoyo, Japan) for 10 tablets. Roche friabilator (Yamto, USA) was used to determine the friability for 20 tablets for 4min at 25RPM.

Make as solution of 100 microgram/ml in 6.8 phosphate buffer and absorbance was measured against blank at 283nm using UV/vis spectrophotometer (Labman, West Bengal) and the calibration curve approach, constructed with standard minocycline HCl in the same medium, was used to compare the content.

Table 3 *Physico-chemical parameters of CR release minocycline HCl tablets.*

Formulation Code	Thickness (mm)	Weight Variation(mg)	Friability (%)	Hardness	%Drug Content
F1	3.41± 0.04	399.6± 1.02	0.16± 0.00	5.4± 0.02	96.19± 0.36
F2	3.45± 0.02	398.75± 1.02	0.18± 0.00	5.5± 0.03	99.69± 0.42
F3	3.43± 0.02	401.67± 1.62	0.17± 0.00	5.3± 0.02	99.77± 0.33
F4	3.35± 0.04	396.40± 1.12	0.25± 0.00	5.6± 0.04	100.38± 0.63
F5	3.54± 0.04	402.56± 1.02	0.22± 0.00	5.3± 0.04	99.38± 0.36
F6	3.60± 0.02	399.67± 1.52	0.3± 0.00	6.0± 0.02	96.5± 0.03
F7	3.63± 0.02	397.40± 1.03	0.48± 0.00	5.6± 0.03	99.49± 0.43
F8	3.72± 0.03	400.89± 1.02	0.25± 0.00	5.5± 0.02	98.17± 0.43
F9	3.46± 0.02	401.7± 0.52	0.42± 0.00	5.0± 0.03	99.38± 0.53

2.6. *Invitro* Release Rate Studies

By using a USP type II dissolve device (paddle method) at 50 RPM in 900ml of pH 6.8 phosphate buffer for 24 hours while keeping the temperature at 37°c 0. 5°c, in vitro drug release was evaluated. At specified intervals, an aliquot (5 ml) was removed and replaced with an equal volume of freshly prepared medium that had been preheated to 37 0.5 ^{oC}.The samples were taken out, filtered through membrane filters with a pore size of 0.4 microns, and their drug concentration was measured using a UV-visible spectrophotometer at 283 nm. All dissolution investigations used a blank canvas and were repeated at least three times. The calibration curve, which was created using standard minocycline HCl, was used to compute the actual amount.

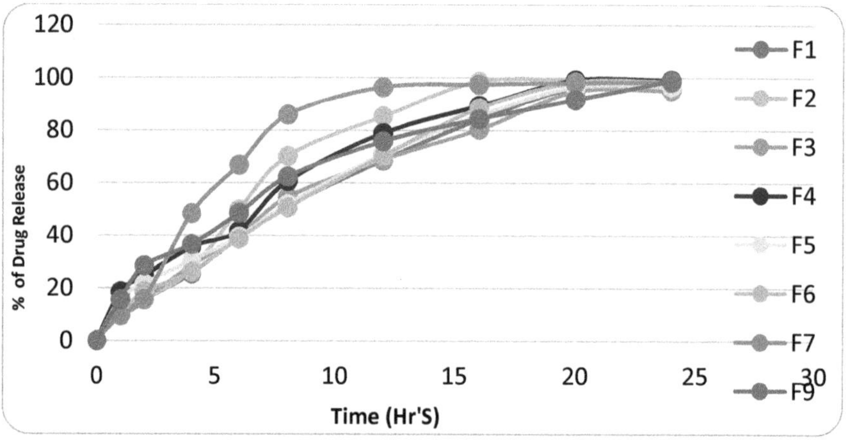

Figure 1 Graphical Representation of cumulative percent Dissolution Profiles of Formulations in vitro release formulation (F1 to F9).

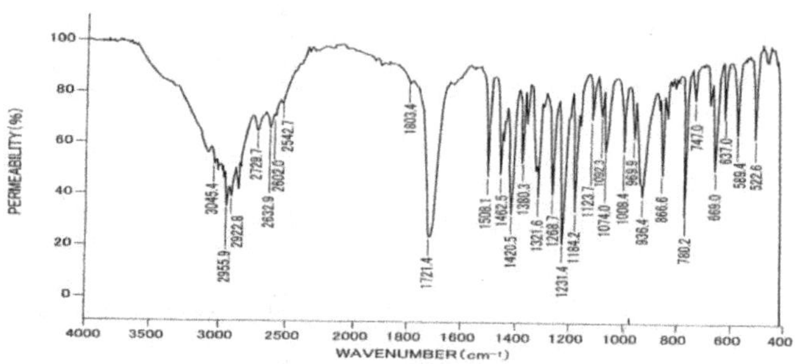

Figure 2 IR spectral data of minocycline HCl.

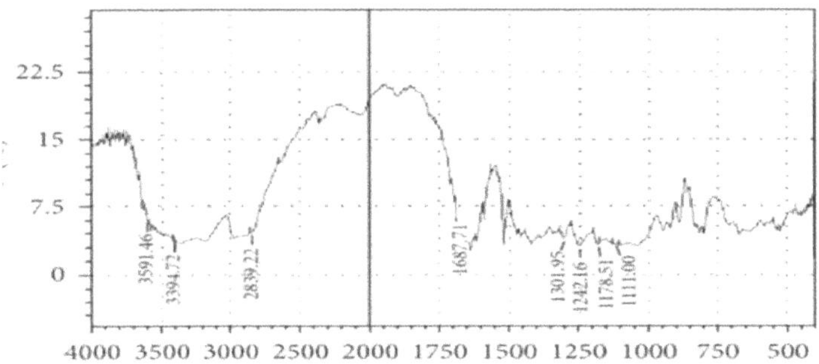

Figure 3 IR spectral data of minocycline HCl and HPMK100 polymer.

2.7. Swelling and Erosion Studies

A technique comparable to that described by Avachat and Vikram was used to determine swelling and eroding behaviour.[7] There were time stamps of 0.5, 1, 2, 3, 4, 6, and 12 hours on the dissolution jars. Each dissolution jar held one tablet and 900 cc of phosphate buffer pH 6.8 at 37 °C 0.5 °C. The paddle-driven equipment was run at 50 rpm. After the above-mentioned stipulated time period had passed, the tablets were retrieved and weighed after the excess water at the surface had been wiped away with filter paper. After being wet, the samples were dried in a 40 °C oven until they reached a consistent weight. The weight of the liquid absorption is reflected in the increased weight of the tablet. It was calculated using the equation below: (Table.5)

$Q = 100(Ww - Wi) / Wi$

where Q is the percentage swelling,

Ww and Wi, respectively, stand for the masses of the hydrated samples prior to drying and the first beginning dry weight.[8] The following equation was used to calculate the degree of erosion (stated as percentage erosion of the polymer content, E).

$E = 100(Wi - W_f) / s$

Where, The identical dried and partially degraded sample's final mass is called Wf.

3. RESULTS AND DISCUSSION

The formulation of granules is crucial in the creation of controlled release tablet dosage form. Physical qualities including hardness, area, size, and surface properties can have a significant impact on how quickly medications dissolve in the body. Choosing a direct compression process with cost-effective production in mind. [9] To get the necessary tablet weight, this excipient was utilised as a bulking agent. Magnesium stearate served as a glidant and talc served as a lubricant.[10]

TBD, compressibility index, LBD, angle of repose, and moisture content are all assessed for every formulation. Granules often fall within the LBD and TBD ranges of 0.52-0.56g/cc and 0.62-0.69g/cc, respectively. Given that the Hausner's ratio ranged from 1.16 to 1.25 (or 1.3), all batches of powder mixes had reasonable compressibility. Angles of repose values between 19.65 and 25.8 were discovered, indicating that the powder mass blend was well-flowing.[9] All of the aforementioned findings imply that the formed granules have adequate flow characteristics. All formulations' moisture content was deemed to be satisfactory. [11] The developed matrix tablets have hardness and friability that range from 5.0 kg/cm2 to 6 kg/cm2 and 0.33 to 0.52 correspondingly. All of the generated batches of the matrix formulations contain minocycline HCl in concentrations between 96.5% and 1.13% and 100.38% and 0.63%. Every batch of manufactured tablets is, thus, of high quality in terms of hardness, friability, and medication content.[11] The diameter and thickness of the formulated tablet range from 10.0 mm to 10.1 mm and 3.4 mm to 3.72 mm, respectively. Therefore, it suggests that the thickness and diameter of all formulas were consistent. The weight variation of matrix tablets yields values between 498.7 mg and 501.3 mg and 1.20 mg. For weight variation tests involving tablets containing more than 250 mg, the pharmacopeial limit for percentage deviation of tablets is 5%. As a result, it was determined that all pill formulations' percentage deviation was within the suggested limit and in accordance with established criteria. The effect of different concentrations of HPMCK100—12.5% (F-1), 25% (F-2) and 37.5% w/w of tablet (F-3)—on the percent release of minocycline HCl (68.6, 85.4, and 68.9 during 12 hours of the dissolving study, respectively—is shown in Fig. 1.The effect of different guargum concentrations—12.5% (F-4), 25% (F-5) and 37.5% w/w of tablet (F-6) on the percentage release of minocycline HCl—78.9, 70.9, and 70.4 within 12 hours of the dissolution study—is shown in Fig. 2. Guar gum formulations (F-4 to F-6) were found to have a higher percentage of extended drug release than formulations using HPMCK100 (F-1to F-3). The drug release was found to be decreased with increase in the polymer ratio.

The ratio of drug to polymer affects drug release in addition to matrix type. As the amount of polymer rises, the drug's kinetics are reduced. This might be the result of the hydrophilic HPMC polymer being reorganised. When HPMC is in a hydrophilic environment, it quickly hydrates, experiences chain relaxation, and forms a thick gel layer. As a result of its failure to produce a coherent and homogenous gel, the drug release is quick. Higher hardness tablets are also known to have a dense mass of polymer, which causes a slower release of the medication.

Considering the above two factors, the next three batches were formulated with Xanthun gum (F-7 to F-9). Each set has three different drug and polymer ratios of Xanthun gum polymer. The effect of three different Xanthun gum concentrations—12.5% (F-7), 25% (F-8) and 37.5% w/w of tablet (F-9)—on the percentage release of minocycline HCl is depicted in Fig. 3. The drug release was found to be satisfied with Xanthun gum. F-9 formulation showed highest drug release rate of about 98±0.98% at the end of 20h (Fig.3). Because this F9 formulation was proven to be superior to other batch formulations, it is therefore deemed to be an optimised formulation.

Various equations, such as zero-order, which characterises the system where the release rate is independent of the concentration of the dissolved particles, were used to describe the kinetics of the drug release process in all formulations (See Table 4). The discharge from systems where the rate of dissolution is influenced by the concentration of dissolving particles is described by the Higuichi square root equation. The Korsmeyer-Peppas equation can be used to analyse the release of pharmacological polymeric dosage forms when the release mechanism is unknown or when multiple types of release mechanisms are involved.

When the formulations were plotted using a zero-order equation, they displayed a high degree of linearity, and the regression coefficient was 0.98 (Figure 4). First order and Higuchi's equation of the F-9 formulation were found to have correlation coefficient values of 0.73 and 0.86, respectively (Figures 5 and 6). The results were fitted into Korsmeyer-equation[12], Peppa's and correlation was found to be 0.98, confirming the super case II transport mechanism (Figure 7).

Table 4 *Release rates of different formulations.*

Formulation code	Zero order	First order	Higuchi	Peppas	
	R2	R2	R2	R2	N
F9	0.89	0.7	0.87	0.97	1.03

Figure 4 Graphical representation of Zero Order.

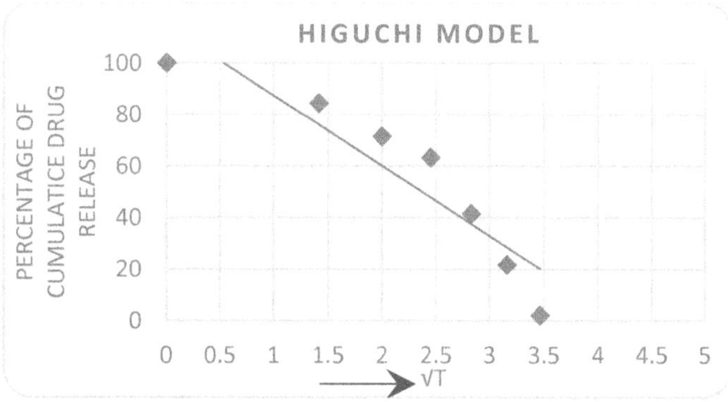

Figure 5 Graphical representation of Higuchi Model.

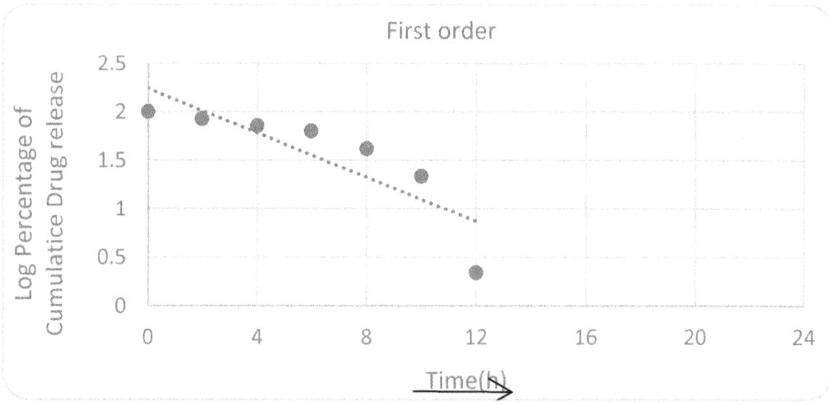

Figure 6 Graphical representation of First Order.

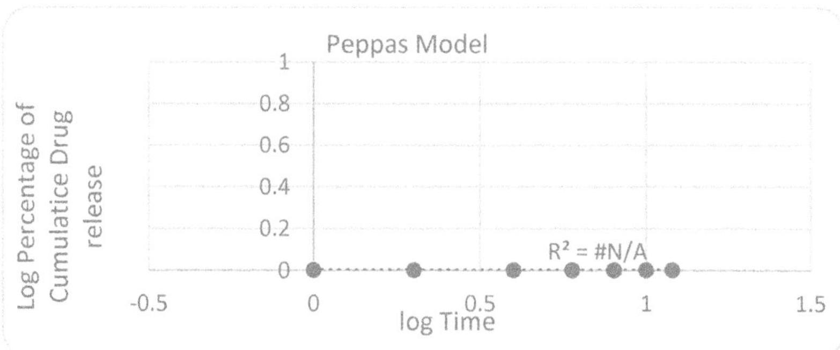

Figure 7 Graphical representation of Peppas Model.

Swelling and Erosion Studies

The heterogeneous erosion study explains the drug release from spheres, slabs, and infinite cylinders. In this paper, the applicability of each of these equations was evaluated. See Table 5.

Table 5 *Swelling index of different formulations.*

Formulation code	**Swelling index ±SD**
F1	22.41
F2	24.071
F3	23.68
F4	28.6
F5	29.76
F6	31.8
F7	38.7
F8	29.4
F9	30.1

5. CONCLUSION

In present study was done in order to optimize the controlled drug release up to 24h. Tablets containing minocycline HCl were created and tested. The concentrations of HPMCK100, guar gum, and Xanthun gum used to make the minocycline HCl were adjusted through a series of trials. There is no drug-excipient interaction in the improved formulation, according to IR spectroscopic measurements. The optimized formulation F9 can be considered as a promising Controlled drug delivery system of Minocycline providing nearly zero order drug release over a period of 24 hrs. Success of the In-vitro drug release studies recommends the product for further in vivo studies, which may improve patient compliance.

6. ACKNOWLEDGMENT

The authors are thankful to KL College of Pharmacy for providing the facilities to carry the work.

7. REFERENCES

1. Yie W. Chien, Novel drug delivery systems,1992,12 edition,12-24.
2. Plott RT, Wortzman MS. Key bioavailability features of a new extended release formulation of minocycline hydrochloride tablets. Cutis. 2006;78:6–10
3. R. Baker (Ed.), Controlled Release of Biologically Active Agents, vol. XV, Wiley-Interscience Publication, New York (1987), pp. 1-18

4. C. Olbrich, O. Kayser, R.H. Muller . Lipase degradation of Dynasan 114 and 116 solid lipid nanoparticles (SLN)—effect of surfactants, storage time and crystallinity. Int. J. Pharm., 237 (2002), pp. 119-128
5. Dollery C. Therapeutic Drugs. Vol. 2. Edinburgh: Churchill Livingstone; 1991. pp. M203–6
6. Chambers HF. Protein Synthesis Inhibitors and Miscellaneous Antibacterial agents. In: Brunton LL, Lazo KS, Parker KL, editors. The Pharmacological Basis of Therapeutics. 11th ed. New York: McGraw-Hill Professional; 2005. pp. 1173–87
7. Avachat A, Kotwal V. Design and evaluation of matrix-based controlled release tablets of diclofenac sodium and chondroitin sulphate. AAPS PharmSciTech. 2007 Oct 19;8(4):E88.
8. Lopes, Paulo & Grewal, Daisy & Pepper, Jessica & Gall, Michelle & Salovey, Peter. (2006). Evidence that emotional intelligence is related to job performance and affect and attitudes at work. Psicothema. 18 Suppl. 132-8.
9. Nath BS, Venkatesh, Hiremath D. Formulation and evaluation of sustained release dosage form of theophylline using a combined hydrophobic and hydrophilic matrix. Indian J Pharm Sci. 2000;62:33–6
10. Hajare AA, More HN, Dsouza JI. Design and evaluation of sustained release tablets of diltiazem hydrochloride. Indian Drugs. 2004;41:175–6
11. Behl AK, Dhake AS. Formulation and release characteristics of sustained release ofloxacin Tablets. Indian Drugs. 2005;42:316–8
12. P. Kallinteri, S.G. Antimisiaris, D. Karnabatidis, C. Kalogeropoulou, I. Tsota, D. Siablis. Dexamethasone incorporating liposomes: an in vitro study of their applicability as a slow releasing delivery system of dexamethasone from covered metallic stents. Biomaterials, 23 (2002), pp. 4819-4826

PART 4

PHYTOCHEMICAL SCREENING AND BIOLOGICAL EVALUATION OF INDIAN MEDICINAL PLANTS

HERBAL FORMULATION AND QUALITY CONTROL EVALUATION OF *MORINGA OLEIFERA* LEAF EXTRACT

V.Swathi*[1], Dr. Sailendra Kumar Mahanta [2], Dr.J.N.Suresh Kumar[1], Sd. Salma[1], M.Rani[1], Sk.Azam Aslam[1], K.Tipsy[1].

[1]Narasaraopet Institute of Pharmaceutical Sciences, Yellamanda, Narasaraopet.

[2]Department of Pharmacology, The Assam Kaziranga University, Assam.

Corresponding Author Email ID: swathi.pharmabud@gmail.com

1. INTRODUCTION

A series of events that occurs in response to noxious stimuli, infection, or trauma is known as an inflammatory phase.[1] Local redness, swelling, discomfort, heat, and loss of function are classic inflammation signs. Kinins, eicosanoids, complement proteins, histamine, and monokines are among the chemical mediators that cause and control the inflammation that underpins these manifestations. [2] The property of a drug or medication that decreases inflammation or swelling is known as an anti-inflammatory. Herbal medicines function in a way that relies on an orchestral approach, unlike modern allopathic drugs, which are single active components that target one particular pathway. A plant contains several molecules that target particular elements of the complex cellular pathway .[3] For millennia, medicinal plants have been a source of a wide range of biologically active compounds, used widely as crude material or as pure compounds to treat several diseases .[4]

Medicinal plants are a valuable natural resource whose valuation necessitates a detailed understanding of the resources to be valued.[5] In preventive medicine, the use of plant extracts containing bioactive components has become increasingly significant.[6] Many of the anti-inflammatory medications currently on the market have well-known side and adverse effects.[7] Purified natural compounds from plants can be used as a starting point for producing new anti-inflammatory drugs with lower toxicity and greater therapeutic utility.[8] Inflammation can be handled with medicinal plants, including Aegle marmelos, Bryophyllum pinnatum, Turmeric, Green Tea, White Willow Bark, Chili Peppers, Black Pepper, Resveratrol, Rosemary, Cloves, Ginger, Cinnamon, and Moringa Oleifera· Chili peppers and Moringa Oleifera have predominantly anti-inflammatory properties. On the other hand, turmeric is used to treat digestive disorders, wounds, and diseases and minimize inflammation. Green tea can be a powerful anti-inflammatory, particularly when used to treat arthritis. It may also help to reduce gastrointestinal inflammation. Many of these herbs have antioxidant and antibiotic properties and the ability to minimize pain perception.

Cinnamon has antioxidant, anti-diabetic, anti-microbial and anti-inflammatory properties, and it's also used to treat neurological disorders. As a result, the widespread use of herbal remedies and healthcare preparations, such as those mentioned in standard books like Vedas and the Bible, made from widely used traditional herbs and medicinal plants, has been related to the occurrence of natural products with medicinal properties.[9] With the world's largest repository of medicinal plants, India may continue to play a key role for raw materials production, either directly for crude drugs or as bioactive compounds in pharmaceutical and cosmetic formulations.[10]

Ayurveda is one of India's oldest healing systems, focused on lifestyle, diet and herbs. It means "science of life and longevity" in ancient Sanskrit.[11,12] Anti-inflammatory natural products have long been used in folk medicine to treat inflammatory disorders like fevers, pain, migraines, and arthritis.

This study aimed to establish an M. Oleifera ointment and test its anti-inflammatory properties. Moringa oleifera, also known as the horseradish tree or drumstick tree, is one of the most commonly grown and well-known thirteen Moringaceae plants.[13] Moringa oleifera is considered to as a "miracle tree" since virtually every aspect of it is beneficial to humans. This tree is found to have high nutritional value, it is often referred to as "natural nutrition for the tropics".[14] The anti-inflammatory property for moringa leaf has been proven in albino mice using aqueous and alcoholic extract by carrageenan-induced paw edema model done by Rao .. *et al.* 1999 in Ancient Science of Life.

1.1 Introduction to plant

Moringa oleifera is a fast-growing, drought-resistant moringaceae tree that is native to tropical and subtropical region of Asia content. its commonly known as Drum stick. Propagation is done by seeds, the leaves are eaten as vegetables and used in traditional herbal medicine. The bark is a whitish-grey color with dense cork covering it. Purplish or greenish-white, hairy bark protects young shoots. The tree's crown is open, with drooping, delicate branches, and the leaves form feathery foliage of tripinnate leaves.

1.1.1 Taxonomy. The species was called Balanus myrepsica by French botanist François Alexandre Pierre de Garsault, but his names are not valid because he did not always give his definitions binomial names.

The species was first described in 1785 by French naturalist Jean-Baptiste Lamarck. Moringa comes from the Tamil word murungai, which means "twisted pod" and refers to the young fruit. The name derives from oleum, which means "oil," and ferre, which means "to bear." It is most closely related to M. concanensis, according to morphology and DNA analysis, and their shared ancestor diverged from the ancestry of M. peregrina.

2. EXPERIMENTAL WORK

2.1 Chemicals Used

Methanol, distilled water, Propyl alcohol, Stearyl alcohol, Liquid paraffin, Methylparaben, Propylparaben, Paraffin wax. All the chemicals and reagents used were of analytical grade

2.2 Plant Material and Extraction

Leaves of selected plant were collected and dried in the shade for two weeks. Using a mixture, the shade dried leaves are ground into a fine powder. Five hundred grams of *Moringa Oleifera* were weighed and extracted where hydro-alcohol (1:1) was used as a solvent. Continuous hot percolation is used for extraction. Filtration was used to absorb the extract, which was then evaporated in a water bath to eliminate the solvent. The extract was allowed to cool to room temperature before being placed in an airtight jar.

2.3 Phytochemical Screening Test

Phytochemical screening was performed on the extract to assess the presence of phytochemical constituents. With the aid of reagents, a small amount of extract was taken and analyzed for the presence of various phytochemical constituents. Gallic acid, tannins, catechol tannins, coumarins, glycosides, and flavonoids are all screened in the extract. The result is mentioned in Table 1.

Table 1 *Table showing phytochemical screening results.*

Name of the Phytocomponent	Result
Gallic tannins	+
Catechol tannins	++
Coumarins	-
Steroids and triterpenoids	++
Saponins	+
Anthraquinones	++

Alkaloids	-
Reducing sugars	++
Glycosides	++
Flavonoids	++

+: present at low concentrations **++:** present at moderate concentrations **+++:** present at high concentrations

The results proved the presence of Catechol tannins, Steroids and Triterpenoids, Anthraquinones, Reducing sugars, Glycosides and flavonoids in major. To a lesser extent, gallic tannins and saponins are present. The result also proved the absence of coumarins and alkaloids in the extract.

2.4 Method of Preparation

The formula for the preparation of cream is presented below:

R.x.

(1) Moringa Oleifera -50gm

(2) Propylene glycol-10gm

(3) Stearyl alcohol-20gm

(4) White petroleum-20gm

(5) Methylparaben- 0.1ml

(6) Propylparaben- 0.1 ml

(7) Sodium benzoate- Q.S

(8) NaOH-Q.S

(9) Water -Q.S.

(10) Ethanol- 500ml

2.5 Method of Dispensing

2.5.1 Step-1. The leaf powder was accurately weighed and subjected to a continuous hot percolation process with a hydro alcoholic solvent in 1:5. It is a continuous process for seven days where the temperature is maintained.

The above extract was filtered with the muslin cloth; the extract was made solvent-free using a water bath. Obtained extract was stored in an airtight container until further process.

2.5.2 Step- 2. Melt stearyl alcohol and white petroleum on the water bath by heating the mixture at temperature 70^{0}c. The aqueous soluble gradients are accurately weighed and heated at 70^{0}c in another beaker until they completely dissolve.

Now the extract was weighed and dissolved in an aqueous phase with continuous stirring. If needed, trituration is done for proper mixing.

The oily phase and aqueous phases are mixed in hot conditions with continuous stirring. Mixing is made even by using motor and pestle.

2.5.3 Step-3. The above Formulation was transferred into a clean airtight container and stored in a place free from contamination.

3. EVALUATION OF FORMULATION

3.1 Physicochemical parameters.[15]

Preliminary evaluation of Formulation was carried out for colour, odour and appearance. The results are subjected in Table 2.

Table 2 *Table Showing the result of physicochemical parameters.*

S.NO	Property	Formulation
1.	Colour	Pale green
2.	Odour	Characteristic
3.	Appearance	Semisolid

3.2 pH

pH refers to the potentiality of hydrogen atoms that signifies the activity of hydrogen atoms in an aqueous solution. PH of the skin ranges from 4 to 6, with an average of 4.7. The formulations prepared should have pH at optimum to skin.

A solution was prepared taking one gram of Formulation dissolved in 100 ml of water and is kept aside for two hours. A digital pH meter is used to determine the pH of the Formulation. 2-3 readings with altered time intervals are noted, and an average value is optimized.

The pH of the formulation is found to be 5.45.

3.3 Thermal stability

Thermal stability is defined as "the ability of a fluid to resist breaking down under heat stress."

The thermal stability of the formulations is determined by the humidity chamber controlled at 60- 70% R.H. and 37 ± 10°C.

After subjecting the Formulation in a humidity chamber at 60- 70% R.H. and 37 ± 10°C, the Formulation was found to be stable with no oil separation. This shows that the Formulation is stable at altered temperature and humidity conditions.

3.4 Spreadability

It is a term as expressed "to denote the extent of the area to which the ointments readily spreads on application to the skin or affected part." The spreadability was expressed in "terms of times in seconds taken by two slides to slip off from ointment and placed in between the slides under the direction of certain load."

The cream (0.5gm) was weighed accurately and placed between two slides. The time taken for the separation of cream between two slides was noted.

The time taken for spreading the Formulation is of 15sec. It gives a clear idea that the Formulation can be spread easily on hand.

3.5 Accelerated stability studies[4]

Accelerated stability studies of the Formulation were determined by placing the Formulation at different conditions for different duration time intervals to explore the effect of altered situations instability of Formulation.

The Formulation is placed at room temperature for one week. after that, the Formulation is kept at temperature 37°C ± 1°C for one month. The Formulation is observed for change 0th, 5th, 10th , 15th , 20th and 30th day for physical appearance, texture, colour, odor and thermal stability. The results are mentioned in Table 3:

Table 3 *Accelerated stability studies for Formulation.*

Test	**Initial month**	**After one month**
Physical appearance	Semisolid	Semisolid
Texture	Palatable	Palatable
Colour	Pale green	Pale green
Odor	Characteristic	Characteristic
Thermal stability	Thermally stable	Thermally stable
Degradation of product	No degradation	No degradation

The accelerated stability studies for the Formulation resulted in no change in physical appearance, texture, color, odour, thermal stability, and product degradation. This indicates clearly that the product is safe when exposed at different duration in altered temperature and humidity conditions.

3.6 Microbiological test[6,7]

The microbial test is performed for the Formulation to test the presence of microorganisms. The viable count is determined by taking one gram of sample dissolved in buffer solution. The sample is inoculated into a conical flask using fluid thioglycollate medium by pour plate technique. The inoculated flask is incubated at 37 ^{0}C for 24 hours. After incubation, the samples are evaluated for the presence of viable microorganisms. It was observed that no growth of viable microorganisms is observed, which concludes that the Formulation is free from microbial contamination and prepared in aseptic condition.

3.7 Patch test or Irritancy test.[5]

The Formulation was applied to a specified area of about one sq. cm at the left-hand dorsal surface or skin behind the human ears. For up to 24hrs duration of time at regular time intervals, the Formulation is checked for causing any Irritancy, Erythema, Oedema and Patches.

The irritancy results revealed that no chances for irritancy, erythema, oedema and patches. This gives a clear idea that the Formulation is safe to use.

4. CONCLUSION

Enormous therapeutic approaches for herbal drugs have been explored, making the use of herbs in Formulation. Particle size and lipid solubility make it advantageous to utilize the herbs for formulating topical dosage forms. The topical route of administration becomes advantageous in concentrating the drug at the desired place, and rubbing pressure facilitates more absorption of the drug through the skin and yielding a good pharmacological response. Literature support reveals the use of leaves of moringa for inflammation and its associated pain. The herb is collected, extracted and subjected to phytochemical screening. The phytochemical screening results revealed Catechol tannins, Steroids and Triterpenoids, Anthraquinones, Reducing sugars, Glycosides and flavonoids, gallic tannins and saponins with the absence of coumarins and alkaloids.

A topical cream is formulated where 50% of the cream contains moringa extract. The Formulation is subjected to evaluation tests like Physico-chemical properties, pH, Thermal stability, spreadability, accelerated stability studies, microbiological test, patch tests or irritancy. The evaluation test proved that the Formulation is safer for skin with pH 5.45, thermally stable with good spreadability properties, and no irritant side effects. No growth in microorganisms is observed and also the Formulation is found to be stable at altered temperature and duration conditions.

Prospects: planning to do clinical trials with permission from the hospital ethical committee.

5. ACKNOWLEDGEMENTS

We acknowledge the management of Narasaraopeta Institute of Pharmaceutical Sciences for their constant support and encouragement.

6. REFERENCES

1. Calixto JB, Campos MM, Otuki MF, Santos AR. Anti-inflammatory compounds of plant origin. Part II. Modulation of pro-inflammatory cytokines, chemokines and adhesion molecules. Planta medica. 2004 Feb;70(02):93-103.
2. Vikrant A, Arya ML. A review on anti-inflammatory plant barks. International Journal of PharmTech Research. 2011;3(2):899-908.MElm, X. D., Yan-Feng, C. A. O., Yan-Yun, C. H. E., Jing, L. I., Shang, Z. P., Wen-Jing, Z. H. A. O., ... & Zhang, J. Y. (2019). Danshen: a phytochemical and pharmacological overview. Chinese journal of natural medicines, 17(1), 59-80.
3. Kim C, Hwang JK. Flavonoids: nutraceutical potential for counteracting muscle atrophy. Food science and biotechnology. 2020 Dec;29(12):1619-40.Parashar Bharat1, Sharma Pankaj2*, Kabra Atul3, Sharma Pankaj on "Formulation and evaluation of polyherbal fairness cream" in INTERNATIONALE PHARMACEUTICA SCIENCE | July-September 2013 | Vol. 3 | Issue 3 |pg:63-68.
4. Barkin RL. The pharmacology of topical analgesics. Postgraduate medicine. 2013 Jul 2;125(sup1):7-18.Seiwerth, J., Tasiopoulou, G., Hoffmann, J., Wölfle, U., Schwabe, K., Quirin, K. W., & Schempp, C. M. (2019). The anti-inflammatory effect of a novel topical herbal composition (VEL-091604) consists of gentian, licorice, and willow bark extract. Planta medica, 85(07), 608-614.
5. Jadhav ST, Gaikwad NA, Mane IV, Kengar MD, Pujari AS, Devkar SB. Formulation and Evaluation of Ointment and Liniment using Moringa oleifera Seed Extract. Research Journal of Pharmacognosy and Phytochemistry. 2019 Apr 1;11(2):45-8.Anil K.M. Ethnomedicinal plants as anti-inflammatory and analgesic agents. Ethnomedicine 2010; 10:267-3.
6. Hoareau L, DaSilva EJ. Medicinal plants: a re-emerging health aid. Electronic Journal of biotechnology. 1999 Aug;2(2):3-4.Tiwari S. "Plants: A Rich Source of Herbal Medicine." Journal of natural products 2008; 1:27-35.
7. Shah VK, Zalavadiya VI, Sheth NR, Raval MK, Dudhrejiya AV, Golwala DK. Standardization and assessment of preformulation parameters of rasayana Tablet. International Journal of Pharmaceutical Sciences and Drug Research. 2010;2(1):58-62.Gupta M, Shaw BP. "Uses of medicinal

plants in Panchkarma Ayurvedic therapy." Indian Journal of traditional knowledge 2009;8(3), 372-378.

8. Ramachandran C, Peter KV, Gopalakrishnan PK. Drumstick (Moringa oleifera): a multipurpose Indian vegetable. Economic botany. 1980 Jul 1:276-83.Fahey JW. Moringa O. A Review of the Medical Evidence for Its Nutritional, Therapeutic, and Prophylactic Properties. Part 1. Trees for life journal. 2005;1:5.
9. Anilkumar M. 10. Ethnomedicinal plants as anti-inflammatory and analgesic agents. Ethnomedicine: A source of complementary therapeutics. 2010:267-93.Dasilva E.J." Medicinal plants: a re-emerging health aid." Electronic Journal of Biotechnology 1999; 2(2):57-70.
10. Anilkumar M. 10. Ethnomedicinal plants as anti-inflammatory and analgesic agents. Ethnomedicine: A source of complementary therapeutics. 2010:267-93.Shah V K, Zalavadiya VI, Sheth NR, Raval MK, Dudherjiya AV, Golwala DK. "Standardisation and Assessment of Preformulation Parameters of Rasayana Tablet." International Journal of Pharmaceutical Sciences and Drug Research 2010; 2(1):58-62.
11. Gupta M, Shaw BP. Uses of medicinal plants in Panchakarma Ayurvedic therapy.Ramachandran C, Peter KV, Gopalakrishnan P.K. Drumstick (Moringa oleifera): a multipurpose Indian vegetable. Econ Bot. 1980; 34:276-83.
12. Fa Fahey JW. Moringa oleifera: a review of the medical evidence for its nutritional, therapeutic, and prophylactic properties. Part 1. Trees for life Journal. 2005 Dec 1;1(5)
13. hey JW. Moringa O. A Review of the Medical Evidence for Its Nutritional, Therapeutic, and Prophylactic Properties. Part 1. Trees for life journal. 2005;1:5.
14. Ramachandran C, Peter KV, Gopalakrishnan PK. Drumstick (Moringa oleifera): a multipurpose Indian vegetable. Economic botany. 1980 Jul 1:276-83.
15. Fahey JW. Moringa oleifera: a review of the medical evidence for its nutritional, therapeutic, and prophylactic properties. Part 1. Trees for life Journal. 2005 Dec 1;1(5):1-5.

PROTECTIVE EFFECT OF *Psidium guajava* L. STEM BARK ON ASPIRIN PLUS PYLORUS LIGATION-INDUCED GASTRIC ULCERS IN RATS

A. Anka Rao[a], G. Jemimah[b*], M. Sreekanth[c]

[a,b]Department of Pharmacy, Koneru Lakshmaiah Education Foundation, Vaddeswaram, AP, India.

[c]Department of Pharmacology, Sarojini Naidu Vanita Pharmacy Maha Vidyalaya, Secunderabad, Telangana, India.

* Corresponding author details: Email ID gerajemimah7@gmail.com

1. INTRODUCTION

Up to 10% of people get peptic ulcers and other symptoms of acidity severe enough to require medical intervention. Ulcers and gastroesophageal illnesses are the more serious medical issues that call for treatment. There are around 4 million peptic ulcers across the United States (duodenal and stomach varieties), and each year, 350 000 new cases are diagnosed, 180 000 people are admitted to hospitals, treated with medications, and about 5,000 of these patients pass away due to their ulcerative situations. For American males, the likelihood of having a peptic ulcer is approximately 10%, while it is just 4% for women.[1]

The population of younger to older adults is most frequently impacted by peptic ulcers, which are wounds or lesions. After a few days to several months of the disease's active phase, they frequently manifest without any outward symptoms or indications and may get better with or without medication. H. pylori infection is the primary cause.[2]

Alcohol, coffee, tea, aspirin, and related medications (nonsteroidal anti-inflammatory medicines) can all hinder the healing of peptic ulcers. Smoking may also slow the recovery from ulcers.[3-5] Additionally, it has been noted that sucrose raises stomach pH. Salt may irritate the intestines and stomach. A higher risk of stomach ulcers has been linked to high salt intake.

The cells lining the stomach and intestine use glutamine, one of the amino acids, as a substantial source of energy.

2. METHODS AND MATERIALS

2.1 Plant material collection, authenticity, and extraction

The stem bark of *Psidium guajava*L. used in the present study was collected from Joginpally BR Pharmacy College medicinal garden and authenticated by Dr. P.V.Prasanna, Scientist'F', Botanical Survey of India, Hyderabad (Ref. no: BSI/DRC/2018-19/Tech./821). The dried stem bark was coarsely powdered, packed into soxhlet thimble and extracted with 99.9% (v/v) ethanol for 18 h at 60°C.

2.2 Chemicals

Analytical quality chemicals and reagents were employed throughout the study, and they were all purchased from reputable Indian producers.

2.3 Pharmacological screening

2.3.1 Experimental protocol. Ensuing regulations of CPCSEA-Committee For The Purpose Of Control &Supervision of Experiments on Animals (Reg. no: 1758/PO/ac/14/CPCSEA), New Delhi, India, this research protocol was authorized by the Institutional Animal Ethics Committee -IAEC, Bhaskar Medical College. Male adult Wistar rats weighing between 150 and 200 g were bought from "Mahaveera Enterprises" Central Animal House facility in Hyderabad. Std. laboratory procedures with 12 hrs of both light and darkness, were applied to the animal's housing in polypropylene cages, and they were given unlimited access to a commercial pellet food and tap water. In an animal home that had CPCSEA, BMC, Hyderabad approval, the rats were kept in accordance with normal procedures. A 200 mg/kg dosage of aspirin was given orally to non-fasted rats once daily for six days while it was suspended in a 0.5% CMC solution. Omeprazole and an ethanol extract of the stem bark of *Psidium guajava* L. were given orally to the corresponding treatment groups 30 minutes before each aspirin dose, while the control group merely received the vehicle (2 mL/kg of normal saline). Rats that had fasted for 48 hours. On the seventh day, they underwent pylorus ligation while being sedated with ether. After each rat's pylorus was tied on the seventh day, water intake was restricted while gastric juice was allowed to build for 19 hours. After that, the rats' stomachs were removed and they were euthanised by an ether overdose. Following that, the esophagus was used to gather the gastric contents so that the volume could be assessed. For 20 minutes, they were centrifuged at 3000 rpm. The p^H, total amount of gastric juice, and titrable acidity of the supernatant were all measured. Gastric lesions were seen after the stomachs were cleaned and opened along the larger curvature.[6–8]

2.4 Evaluation of ulcer parameters

2.4.1 Assessment of total and free acidity. Pipette 1ml clear lysated liquid and diluted with10ml distilled H_2O. Using phenolphthalein as an indicator, Topfer's reagent (which employs dimethyl amino azobenzene) was used for titrating the solution towards 0.01N NaOH. The products sol. transition to orange, which represents the end point and the free acidity, was observed. In order to assess overall acidity, titration was maintained until the solution once again turned pink. Using the following formula, acidity was determined:

Acidity = NaOH Vol. x Nx 100/0.1 mEq/Lt/100g (1)

2.4.2 Ulcer severity score. A stereo-microscope was used to count the number and severity of ulcers on the stomach after it had been excised and mounted on a cork plate.

Score of severity

0 indicates a neutral stomach color
0.5 indicates coloration in red
1 indicates spot ulceration
1.5 indicates Bloody streaks
2 indicates ulcers more than 3 but less than 5
3 indicates ulcers more than 5

2.4.3 Index of ulcers. The formula below is used to calculate the ulcer index;

$UI=UN+US+UPX10^{-1}$ (2)

Where,

UI stands for Ulcer Index

UN stands for "avg. no. of ulcers per animal."

US is the avg. intensity range

UP is the proportion with affected animals who had ulcers

2.4.4 Gastric juice volume: A small incision was made along the stomach's larger curvature to collect its contents, which were then put in a calibrated centrifuge tube after the stomach was removed. The juice volume was determined.

2.4.5 pH. After being centrifuged for 10 minutes at 3000 rpm with the stomach contents, the samplings are drained & subjected through a pH test by using Digital pH meter, Type DPH-100 Data equipments.

2.4 H and E staining protocol

Hematoxylin and Eosin (H&E) staining is the most widely used staining technique in histopathology. After the slides were deparaffinised, they were hydrated and stained with H & E stains. The slides were mounted and analyzed.

2.5 Statistical evaluation

Software called Graph Pad Prism 8.0 employed for data analysis. One-way ANOVA was employed to examine data, and Graph Pad Instat3 was used for Dunnett's multiple comparative analysis. Ranges were presented with Mean ± SEM. The p values <0.05 were considered statistically significant, $p<0.01$, as high significant, $p<0.001$, as very high significant, and $p>0.05$ was not significant.

3. RESULTS

Alkaloids, saponins, tannins, steroids, cardiac glycosides, reducing sugars, and flavonoids were discovered through phytochemical analysis[9]. The effect of ethanolic extract of *Psidium guajava* L. stem bark (EEPGSB) on Aspirin plus pylorus ligation induced ulceration. Aspirin plus pylorus ligation caused the build-up of gastric secretions of 10.26 ± 0.06 mL with pH 2.10 ± 0.14 in aspirin plus ligation control group. The full acidity of the gastric secretions was found to be 198.13 ± 1.12 mEq/L (see figure 2). Pre-treatment with the *Psidium guajava* L. stem bark extract, significantly ($P<0.01$) reduced the amount of gastric secretions 7.60 ± 0.09 and 4.30 ± 0.15mL at the doses of 200 and 400 mg/kg, respectively (see figure: 1). The pH of the gastric secretion was significantly ($P<0.01$) elevated to 4.43 ± 0.12 and 6.39 ± 0.10 at a dose of 200 and 400 mg/kg/day of *Psidium guajava* L. stem bark extract respectively (see figure 3). Additionally, total acidity was also reduced significantly ($P<0.01$) as compared to the aspirin plus ligation control group. In aspirin plus ligation ulcer model, the EEPGSB treatment at 200 and 400 mg/kg (no ulcer number, ulcer index, gastric lesion and ulcer severity score) significantly ($P < 0.01$) inhibited the ulcer number, ulcer index, gastric lesion and ulcer severity score to 2.00 ± 0.36, 0.40 ± 0.07, 6.16 ± 0.40 and 0.50 ± 0.12 when compared with aspirin plus pylorus ligation control rats (see figure 4-7). The proportion % protection (ulcer index) and inhibition (ulcer number) of the ulcer was 60%, 100% and 49%, 100% in the groups treated with 200 and 400 mg/kg of EEPGSB, and 0% and 100 % in the aspirin plus pylorus ligation control (see figure: 8,9). Histopathological evaluation of aspirin plus pyloric ligation induced ulcer models showed perforated ulcer, deep ulceration of granular epithelium and loss of the histological structure, almost reducing the submucosa. EEPGSB at 200 mg/kg showed partial healing of ulcer with few inflammatory cells and at 400 mg/kg showed no ulcers (see figure: 10, 11).

Further, it was observed that aspirin plus pyloric ligation caused gastric ulcerations, and pre-treatment with *Psidium guajava* L. stem bark extract had reduced them significantly in a dose-dependent manner.

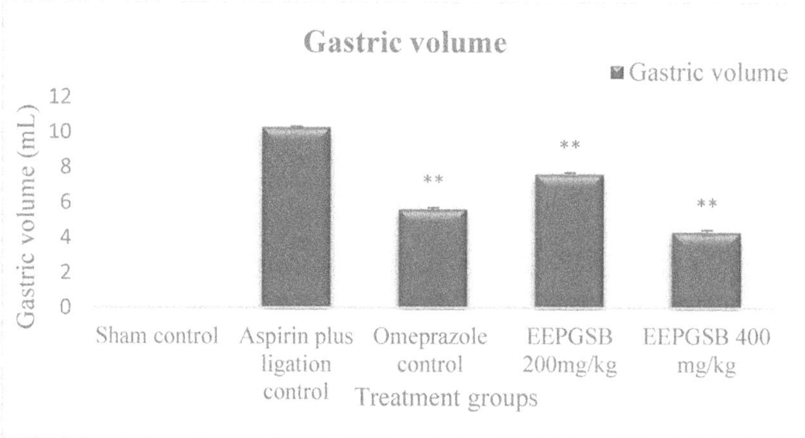

Figure 1: Effect with EEPGSB on gastric volume of aspirin & pyloric ligation ulcerated rats

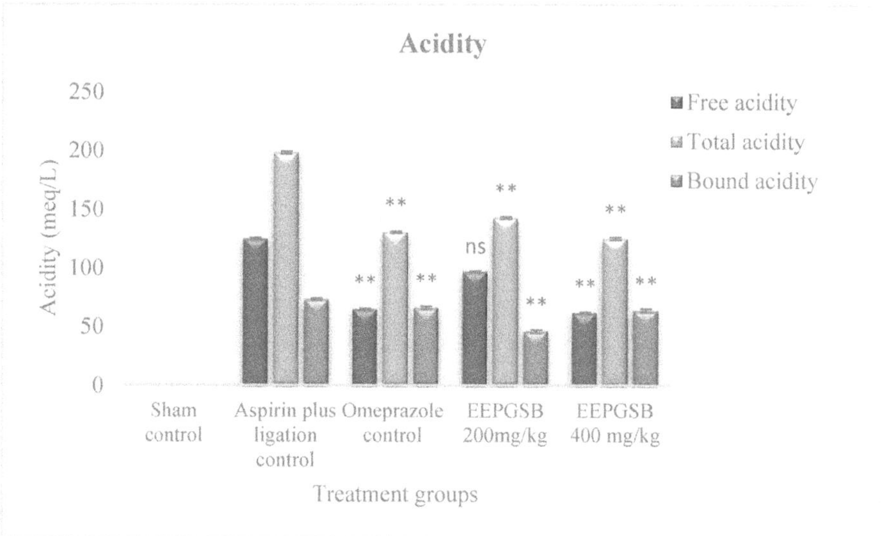

Figure 2: Impact with EEPGSB on acidity of aspirin and pyloric ligation ulcerated rats

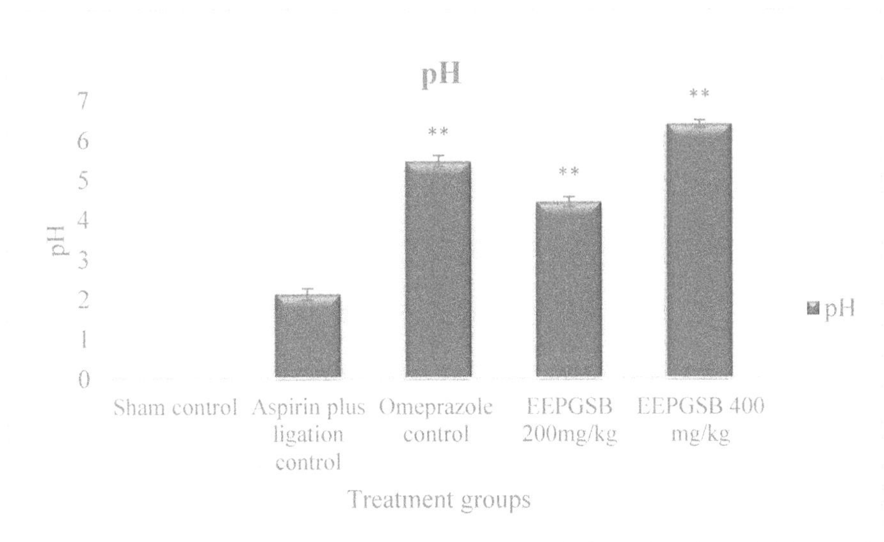

Figure 3: Impact with EEPGSB with pH of aspirin and pyloric ligation ulcerated rats

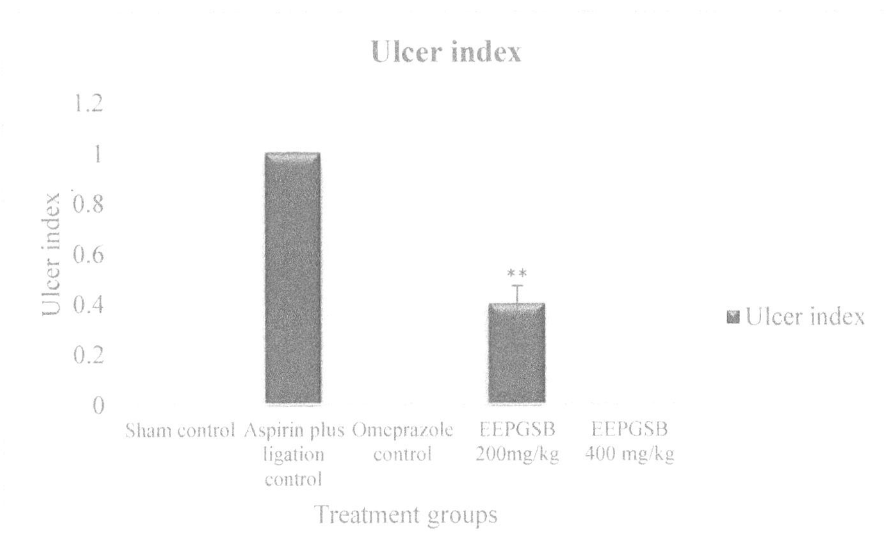

Figure 4: Impact with EEPGSB with ulcer index of aspirin and pyloric ligation ulcerated rats

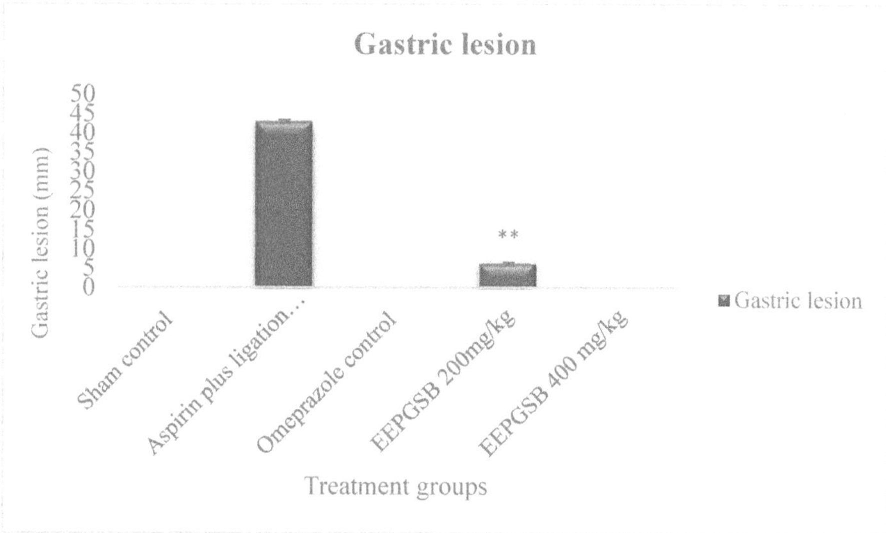

Figure 5: Impact with EEPGSB on gastric lesion (mm) of aspirin and pyloric ligation ulcerated rats

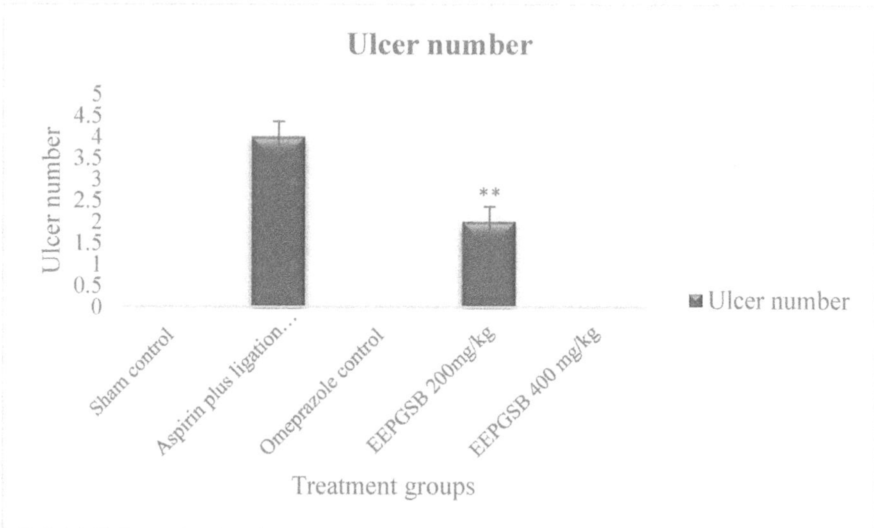

Figure 6: Impact with EEPGSB on ulcer number of aspirin and pyloric ligation ulcerated rats

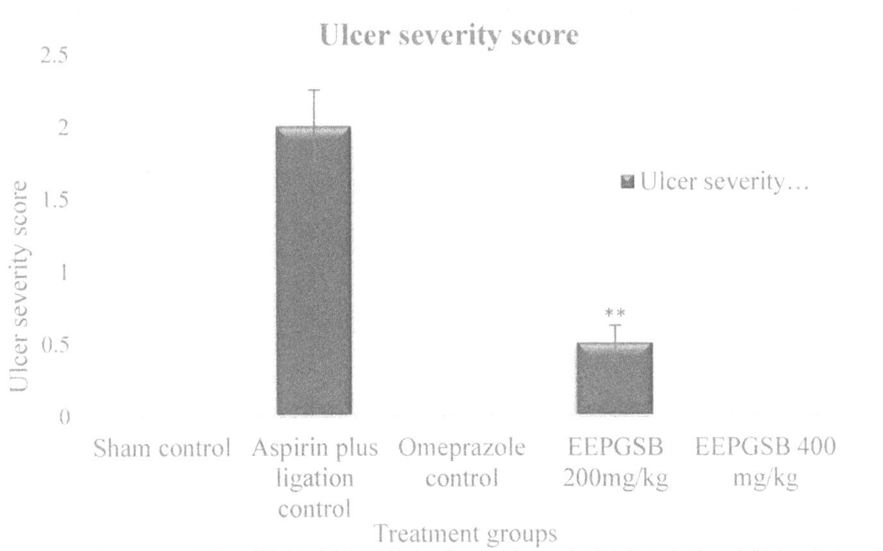

Figure 7: Impact with EEPGSB on ulcer severity score of aspirin and pyloric ligation ulcerated rats

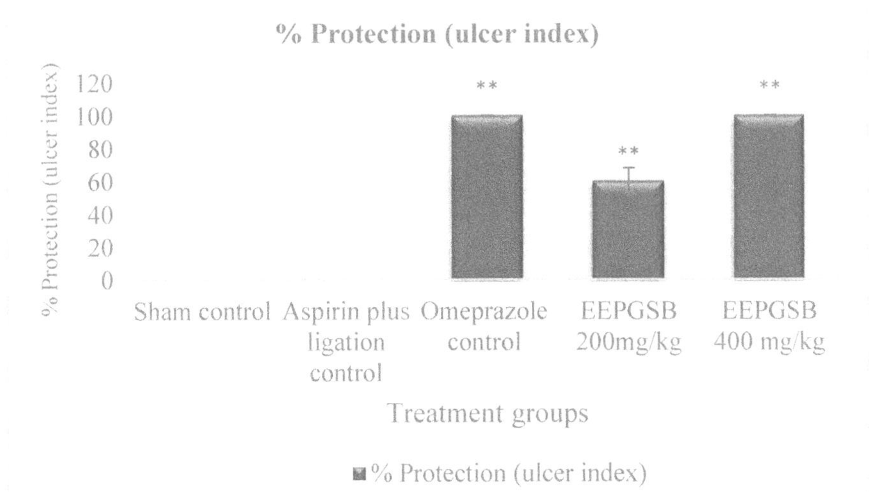

Figure 8: Effect of EEPGSB on percentage protection (ulcer index) of aspirin and pyloric ligation ulcerated rat

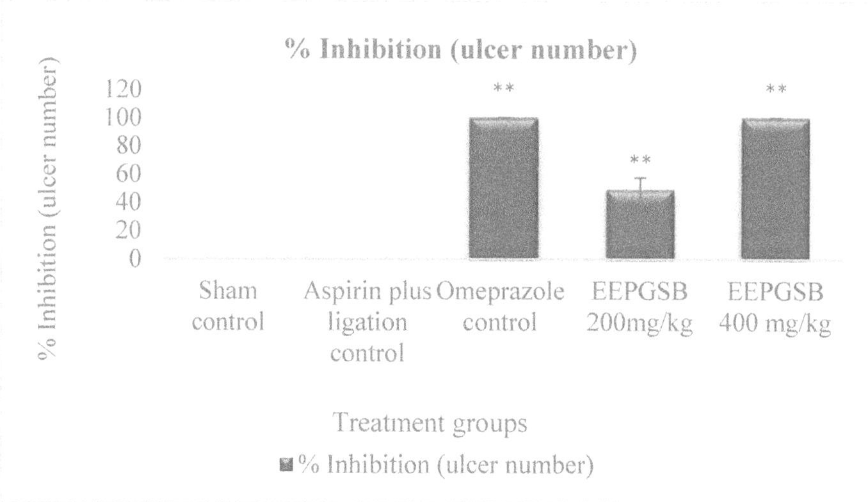

Figure 9: Impact with EEPGSB on percentage inhibition (ulcer number) of aspirin and pyloric ligation ulcerated rats

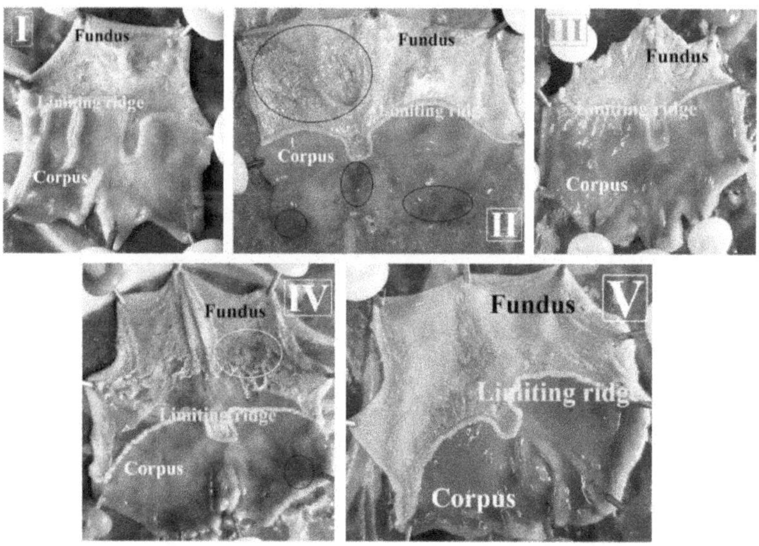

Figure 10: Effect of *Psidium guajava* L. stem bark extract on aspirin & pylorus ligation-induced gastric ulcer model in rats. I = Sham control, II = Aspirin + ligation control, III = Omeprazole control 20mg/ kg, IV = EEPGSB 200mg/ kg, V = EEPGSB 400mg/ kg. In this figure, we can see that red coloration (indicated by black circle) in fundus (non-glandular stomach) and deep ulcers (indicated by black circles) in corpus (glandular stomach) of Aspirin + ligation control, abrasion (indicated by yellow circle) in fundus and red coloration (indicated by black circle) in corpus were observed in EEPGSB 200mg/ kg and normal structure in remaining groups.

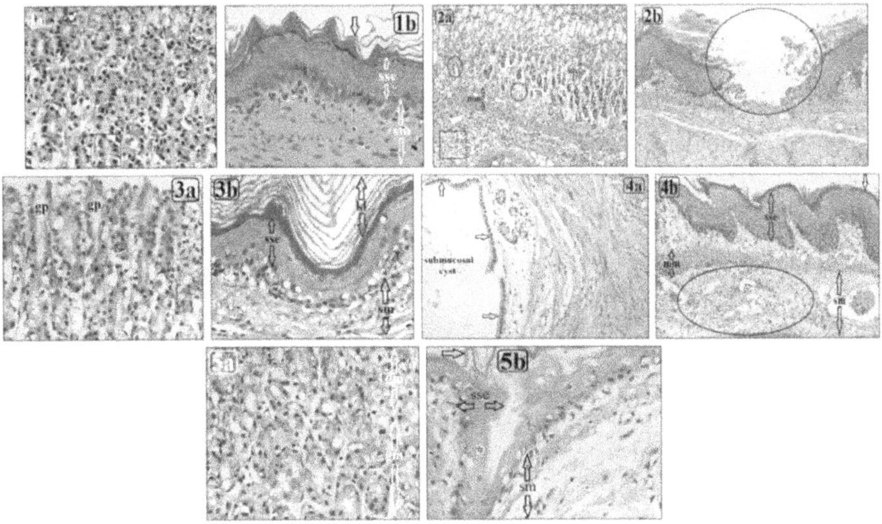

Figure 11: Photomicrographs of stomach sections at 50x magnifications i.e., 1a: Sham control–Normal glandular mucosa, circles- Parietal cells, square- Chief cells, 1b: Sham control– Normal non-glandular mucosa, black arrow- Keratin layer, sse- Stratified squamous epithelium, sm- Submucosa, 2a: Aspirin+ ligation control– Glandular stomach,circles- Mild submucosal mononuclear cell infiltration, mm- Muscularis mucosae, square- Submucosa with inflammatory cell infiltration, 2b: Aspirin+ ligation control–Non-glandular mucosa, circle- Loss and discontinuation of mucosa, 3a: Omeprazole 20 mg/kg– Normal glandular mucosa, gp- Gastric pits, 3b: Omeprazole 20 mg/kg– Normal non-glandular mucosa, kl: Keratin layer, sse- Stratified squamous epithelium, sm- Submucosa, 4a: EEPGSB 200 mg/kg– Glandular stomach, arrows- Lining epithelium of cyst., 4b: EEPGSB 200 mg/kg– Non-glandular stomach, arrow- Keratin layer, sse- Stratified squamous epithelium, sm- Submucosa, mm- Muscularis mucosae, 5a: EEPGSB 400 mg/kg– Normal glandular mucosa, gm – Glandular mucosa, 5b: EEPGSB 400 mg/kg– Normal non-glandular mucosa, arrow- Keratin layer, sse – Stratified squamous epithelium, sm – Submucosa.

4. DISCUSSION

The findings demonstrated that administration of the Psidium guajava L. stem bark extract on aspirin plus pyloric ligation-induced ulcer in rats resulted in a considerable reduction in gastric acidity and stomach secretion, as well as a significant rise in pH and mucus content. Furthermore, there was a dose-dependent reduction in ulcer development in aspirin plus pyloric ligation and generated ulcer models.

The aspirin plus pyloric ligation ulcer model is used to assess the anti-secretory and gastro protective effects of experimental drugs. The closure of the stomach's pyloric end creates a build-up of gastric acid, which causes ulcers owing to auto digestion of the mucosa[10,11]. Aspirin is a nonsteroidal anti-inflammatory medicine that causes ulcers by decreasing prostaglandin production in the stomach via the inhibition of cyclooxygenase enzymes. Nonsteroidal anti-inflammatory medications elicit an inflammatory response in the stomach mucosa, raising reactive oxygen species[12,13]. EEPGSB reduced stomach volume, free acidity, and total acidity significantly as compared to the control group, suggesting an anti-secretory action. The suppression of the H^+-K^+-ATPase enzyme[12] might explain the anti-secretory action. The ulcer index (UI) and percentage inhibition (PI) are two measures widely employed to assess an experimental agent's gastro protective impact. When compared to the control group, EEPGSB at all dosages resulted in a substantial reduction in UI and an improvement in PI, suggesting a gastro protective effect. The percentage protection of ulcer genesis by pylorus ligation was found to be 60% and 100% at EEPGSB doses of 200 and 400 mg/kg, respectively, and the gastro protection provided by EEPGSB at 400 mg/kg was comparable to the standard (100%). This finding suggests that EEPGSB has a dose-dependent efficacy in preventing stomach ulcers.

5. CONCLUSION

India has a large population of the South American native plant *Psidium guajava* L. Researchers have looked into the anti-ulcer properties of the plant's leaves, fruits, and seeds, but they haven't looked into the anti-ulcer properties of the stem bark.

The results of this study suggests that ethanol extract of *Psidium guajava* L. stem bark, which contains active chemical components such quercetin, gallic acid, tannic acid, and guaijavarin, has a strong protective effect against stomach ulcers. In comparison to a lower dose (200 mg/kg), a larger dose (400 mg/kg) of the extract shows a stronger protective effect. Longer-term, lower-dose research may reveal these finding's characteristics more clearly.

6. ACKNOWLEDGEMENTS

The authors wish to express their deepest gratitude to the Management, Dr. JVC Sharma, M. Pharm., Ph.D., Principal, and AVSSS Gupta Atyam, M. Pharm., (Ph.D.) Vice Principal of Joginpally B.R. Pharmacy College, Hyderabad, Telangana, India, for *bestowing* all the necessary laboratory needs of the research and their constant support.

DECLARATIONS: Nil

Financial support and sponsorship: Nil

Ethical approval: The Institutional Animal Ethics Committee has given its approval.

CONFLICTS OF INTEREST: Nil

7. BIBLIOGRAPHY

1. Lalchand D. Devhare, Niharika Gokhale, Antioxidant and Antiulcer property of different solvent extracts of *Cassia tora* Linn., *Res. J. Pharm. Technol* 15(3) (2022) 1109-3.
2. Alka Zade, Jugalkishore Vyas, Vivek Paithankar, Anjali Wankhede, Evaluation of antiulcer activity of herbal drugs on experimental animal: *Myrica nagi* (Myricaceae), *GSC biol. pharm. sci.* 20(02) (2022) 138–144.
3. Habtalem Adane, Seyfe Asrade Atnafie, Zemene Demelash Kifle, Digambar Ambikar, Evaluation of In Vivo Antiulcer Activity of Hydro-Methanol Extract and Solvent Fractions of the Stem Bark of *Ficus thonningii* (Moraceae) on Rodent Models, Biomed Res. Int (2021) 1-10.
4. Woretaw Sisay Zewdu, Tezera Jemere Aragaw, Evaluation of the Anti-Ulcer Activity of Hydromethanolic Crude Extract and Solvent Fractions of the Root of Rumex nepalensis in Rats, J. Exp. Pharmacol (12) (2020) 325-337.
5. Bikram Tewari1, Mayanglambam Medhabati, Anti-Ulcer Effect of Aqueous Extract of *Gynura cusimbua* (D.Don) S.Moore on Swimming Stress Ulcer Model in Albino Rats, J. Evolution Med. Dent. 8 (33) (2019) 2602-2606.
6. Damayanthi Dalu, Anti Ulcer Activity of *Leea Indica* in Wistar Albino Rats, Arch Nano Op Acc J 1(1) (2018) 11-16.
7. Ehile Ehilé Hervé, Goze Nomane Bernard, Kouakou Kouakou Léandre, Yapo Angoué Paul, Ehile Ehouan Etienne, Acute toxicity and gastric anti-ulcer activity of an aqueous extract of the leaves of Macaranga barteri Müll. Arg (Euphorbiaceae) on rat models, J. Med. Plant Res 12(9) (2018) 96-105.
8. Abebaw M, Mishra B, Gelayee DA, Evaluation of anti-ulcer activity of the leaf extract of *Osyris quadripartita* Decne. (Santalaceae) in rats, J. Exp. Pharmacol, (9) (2017) 1–11.
9. Bhajoni P.S., Meshram G.G., Lahkar M., Evaluation of the Antiulcer Activity of the Leaves of *Azadirachta indica:* An Experimental Study, Integr Med Int (3) (2016) 10-16.
10. Dev Prakash, Nishant Singh Katiyar, Amrit Pal Singh, A study on anti-ulcer activity of stem extracts of *Cuscuta reflexa* (Roxb) against pylorus ligation induced gasrtic ulcer in rats World J. Pharm. Res 5 (3) (2016) 1461-1470.
11. Dharma Swathi, K. V. S. R. G. Prasad, Antioxidant and antiulcer potential of ethanolic extract of bark of *Myrica esculenta* in pyloric ligation ulcer model, Int J Pharm Pharm Sci, 7 (10) (2015) 195-198.
12. Rathnakumar K., Jaikumar S., Sengottuvelu S., Antiulcer Activity of *Euphorbia hirta* against Experimentally Induced Ulcer in Rats, Asian J. Pharm. Res. 3(3) (2013) 121-124.
13. Nitin Mahurkar, Sayeed ul hasan S.M, Antiulcer Activity of *Commicarpus chinensis* in Ethanol and Aspirin Induced Ulcers, Asian J. Pharm. Res. 4 (3) (2014) 119-122.

PHYTOCHEMICAL STUDY, IN VITRO ANTIOXIDANT AND ANTIDIABETIC ACTIVITY OF METHONALIC LEAF EXTRACT OF LINDERINA PARVIFLORA

Malik Yahia Abdul Resool Hussein, Narender Malothu*, Haritha Lanka

Department of Pharmacy, Koneru Lakshmaiah Education Foundation, Vaddeswaram, Guntur, AP, India

*Corresponding author: narendermalothu@gmail.com

1 INTRODUCTION

Diabetes mellitus (DM), a chronic condition caused by severe or partial insulin release failure together with or without varying degrees of insulin resistance,[1] impacts the metabolism of carbohydrates, proteins, and fats. It also increases the risk of vascular issues, such as cardiovascular disease, and contributes to poor glucose metabolism, which results in ketoacidosis. The environment or genetic variables may have an impact on how DM develops. Though many abnormalities, including insulin resistance, fatty liver, brain atrophy, and anorexia nervosa, have been demonstrated,[2] the insulin therapy is by far the most successful medication used to combat DM.

A well-known medicinal plant known as "Small-Flowered Lindernia" and a synonym of Ilysanthes parviflora (Roxb.) Benth) is termed *L. parviflora* (Roxb.) Haines (*Scrophulariaceae*) (Figure 1). Particularly in the continents of Asia and Africa, *Lindernia* species are found throughout moist tropical lands. These species have been found in a number of Indian states, including West Bengal, Assam, Bihar, Karnataka, Kerala, Madhya Pradesh, Maharashtra, Orissa, Punjab, Tamil Nadu, and Uttar Pradesh.[3-4] *L. Parviflora* has therapeutic value for treating conditions like blood dysentery and livestock ephemeral fever. An effort was undertaken to investigate the anti-diabetic and antioxidant activities of leaf extracts since leaves are a plentiful source of various phytoconstituents, also because traditional claims and discoveries in literature studies exist.

Figure 1. Picture of *L. parviflora* herb.

2 MATERIAL AND METHODS

Materials: Acarbose API was obtained as a gift sample from Hetero laboratories, Hyderabad. In addition to this, other materials and reagents were bought from SD Fine chemicals in Mumbai.

Collection and extraction: The plant is widespread in India's southern region. Dr. K. Madhava Shetty, a taxonomist at Sri Venkateswara University, Tirupati, Andhra Pradesh, acquired the entire plant from the Tirumala Hills, Chittoor. For future utilization, a voucher specimen copy was retained. Initially, the leaves were dried out at room temperature (in the air). Approximately 250 g of powdered dried leaf material was used for solvent extraction. Using methanol (1 lit) as the extraction solvent, the powder was macerated for between 24 and 72 h. The final leaf extract (MELP) was concentrated with a rotary evaporator (*in vacuo*), and then it was stored in cold temperatures.

Phytochemical investigation: In order to detect whether flavonoids, phenols, glycosides, carbohydrates, alkaloids, tannins, sterols, and saponins were present in the extract (MELP), qualitative phytochemical analyses were performed (Trease and Evans, 1989).[5-7]

Estimation of total flavonoid content: Flavonoids are admixtures of polyphenols that exhibit a variety of pharmacological effects. The test was carried out by soaking roughly 2.5 mL of leaf extract (MELP) in 90% ethanol and 2.5 mL of $AlCl_3$ for 40 min. By using a UV- visible spectrophotometer (Effendorf), the mixture's absorbance was measured at 415 nm. The tests were carried out in triplicate using quercetin as the reference standard and ethanol as the control. The calculated total flavonoid content was expressed in dry extract as g/QE/mg.[8]

Estimation of total phenolic content: Plants' secondary metabolites are a rich source of phenols, which often have a variety of pharmacological effects. Gallic acid was used as the reference substance for the test employing FC reagent. Approximately 0.2 mL of the MELP was treated with 10% w/v FC (1 mL) reagent in 7.5% w/v sodium carbonate (0.8 mL). After incubation (1h), the resulting mixture's phenolic content was estimated using the absorbance measured at 760 nm. The amount of total phenolic content in dry extract was given as mg per g of gallic acid equivalents (mg GAE/g).

GC-MS analysis: GC-MS is a reliable chemical analytical technique for identifying phytoconstituents. It will provide an approximate spectral output of all the likely constituents that are separated from the sample. The process is initiated with the sample being injected into the GC-MS device's port, where it is vaporised before being separated using an analyzer. Each component was generating a perfect, distinct peak, which was electronically recorded on a paper chart. At SAIF, IIT-Madras, the GC-MS analysis of extract was carried out. The data was collected using an integrated gas chromatograph and mass spectrometer (Agilent GC-MS5977B), equipped with an HP-5 MS fused silica column (5% phenyl methyl siloxane 30.0 m 250 M, film thickness 0.25 M), and interfaced with a detector.

Antioxidant activity: DPPH radical scavenging assay: A stock solution containing 0.5 mg/mL of MELP was prepared. About 4 mL of DPPH reagent was added to 1 mL of test samples (MELP) with varied concentrations. Similarly the control was made without the test sample. For blanks, ethanol was used in the place of DPPH reagent. For about 30 min, the reaction was allowed to run its course in complete darkness. Then, at 517 nm, the absorbance was determined using vitamin C as standard. The formula [((Control-Test)/Control) x 100] was used to get the percentage (%) scavenging. By displaying concentration *versus* % inhibition on a graph, a linear regression equation was drawn. Utilizing a linear regression analysis, the sample concentration (IC_{50}) needed to achieve a 50% decrease in absorbance was determined. The average of three replicated readings was calculated.

In vitro antidiabetic activity: Inhibition of α-amylase: Potato starch (0.125 g) was stirred for 15 min in a boiling water bath with 25 mL of 20 mM sodium phosphate buffer and 6.7 mM sodium chloride (pH 6.9) to produce the starch solution (0.5% w/v). In the same buffer, 1 U/mL of α-amylase was combined to make the α-amylase solution. A solution of sodium potassium tartrate tetrahydrate and a solution of 96 mM 3, 5-dinitro salicylic acid (DNS) were mixed in equal parts to form the colorimetric reagent. Increasing concentrations of an enzyme inhibitor, such as test substance (10, 20, 40, 60, 80, and 100 µg/mL) or acarbose (1, 2, 4, 6, 8, and 10 µg/mL), were added to the starch solution (1000 µL), and this mixture was then combined with 1mL of α-amylase solution and incubated at 25°C for 3 min to produce the desired results.[9-12]

Inhibition of α-Glucosidase: The ability of the test substance (100, 200, 400, 600, 800, and 1000 μg/mL) or acarbose (0.1, 0.2, 0.4, 0.6, 0.8, and 1 μg/mL) to inhibit the activity of the α-glucosidase enzyme was assessed by incubating 100 μL of α-glucosidase enzyme (1 U/mL) solution with 100 μL of phosphate buffer. The above reaction mixture was placed in boiling water for 2 min before cooling to inhibit the action of α-glucosidase on the maltose. In order to determine how much glucose was released by the activity of the enzyme α-glucosidase, 2 mL of glucose reagent was added to the mixture, and its absorbance was measured at 540 nm. Calculations were made to determine the IC_{50} value and the percentage (%) of inhibition.[9-12] The % inhibition of α-glucosidase was calculated as per the formula given under α-amylase inhibition assay.

Statistical Analysis: The results were presented statistically using Graph Pad Prism (Version 5.0) software. Through one-way analysis of variance and Tukey's multiple comparison tests, the statistical divergence was investigated. The acquired data were reported as mean standard error mean, and $p \leq 0.05$ was used to determine the degree of significance.

3 RESULTS AND DISCUSSION

Collection and extraction: The extraction of *L. parviflora* leaves was successfully performed by maceration technique with the help of methanol as a solvent. The concentrated macerated methanolic extract of leaves constituted 4.3% w/w crude MELP which was further utilized for testing phytoconstituents and assessing in vitro antioxidant and antidiabetic studies.

Phytochemical investigation: In this evaluation MELP showed the existence of carbohydrates, alkaloids, flavonoid, tannins and steroids. Hence the results explored its crucial importance as they can show the specific chemical nature of the plant that useful for medicinal purposes.

Total phenolic content in MELP: The concentrated methanolic extract was constituted about 12.97 mg GAE/g of total phenols.

Total flavonoid content in MELP: The concentrated macerated methanolic extract of the leaves comprised a total flavonoid concentration of 6.42 mg/QE/mg.

GC-MS analysis of MELP: GC-MS analysis of MELP demonstrated the presence of seven different components in total (Figure 2). The compounds' varied peak areas and retention durations indicate the presence of chemically diverse plant components. Table 1 shows the components that were eluted, retention time (Rt), peak area (A), and percentage area (%A).

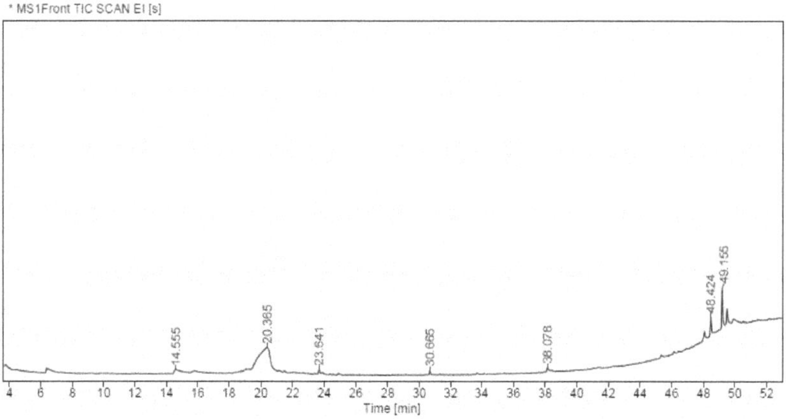

Figure 2: GC-MS chromatogram of MELP.

Table 1 *Chromatographic parameters of GC-MS data.*

S. No.	Chemical Name	Retention time (min)	Peak area	Percentage (%)
1	Guanosine	14.552	674828.75	3.22
2	3-O-Methyl-d-glucose	20.365	12507845.34	59.71
3	3,7,11,15-Tetramethyl-2-hexadecen-1-ol	23.641	283100.167	1.35
4	Phytol	30.667	297628.587	1.42
5	Hexadecanoic acid, 2-hydroxy-1-(hydroxymethyl)ethyl ester	38.075	645659.258	3.08
6	Stigmasterol	48.420	1277141.554	6.10
7	γ-Sitosterol	49.152	5262925.7	25.12

The phyto-constituents in GC-MS analysis were recognized by taking into consideration the % elution, then interpreting the mass spectrum with the relevant hits (NIST library). Table 2 displays the recognized compounds' molecular weight (MW), molecular formula (MF), and structure. Only a few number of these components were proven to be biologically active, and they are now known to act as antioxidants (phytol), anti-inflammatory agents (Hexadecanoic acid, 2-hydroxy-1-(hydroxymethyl)ethyl ester), and anti-cancer agents (Stigmasterol and -Sitosterol) (Table 2).

Table 2 *Chemical composition in MELP by GC-MS.*

S. No	Chemical Name	Mass (g/mol)	Molecular formula	Structure
1	Guanosine	283.0	$C_5H_5N_5O$	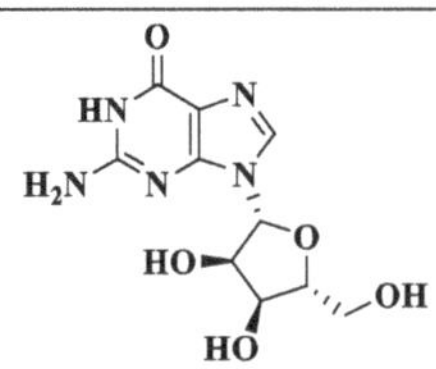
2	3-O-Methyl-d-glucose	194.18	$C_7H_{14}O_6$	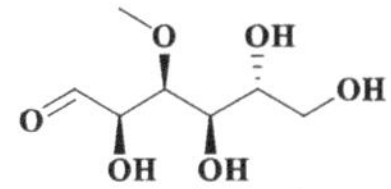
3	3,7,11,15-Tetramethyl-2-hexadecen-1-ol	296.5	$C_{20}H_4O$	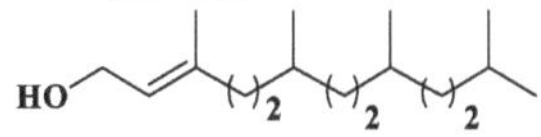
4	Phytol	296.5	$C_{20}H_4O$	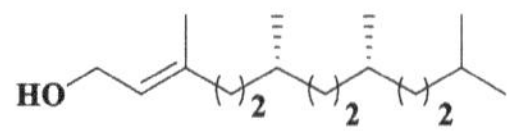
5	Hexadecanoic acid, 2-hydroxy-1-(hydroxyl methyl)ethyl ester	146.11	$C_{19}H_{38}O_4$	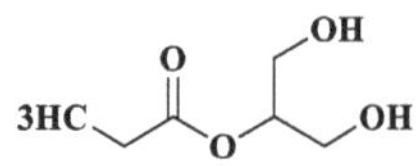
6	Stigmasterol	412.7	$C_{29}H_{48}O$	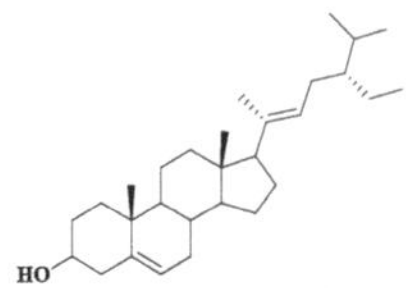
7	γ-Sitosterol	414.7	$C_{29}H_5O$	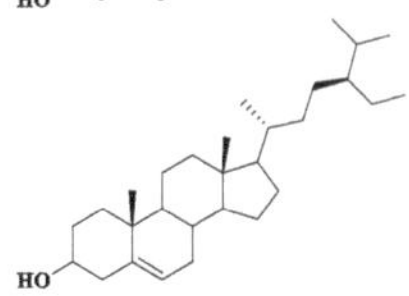

The occurrence of several plant components was discovered through analysis of the chromatogram and mass spectral data of the hit compounds. Among this 3-O-methyl-D-glucose (59.71%), γ-Sitosterol (25.12%) was comprised majorly. The mass spectrums of the eluted compounds were shown in Figure 3.

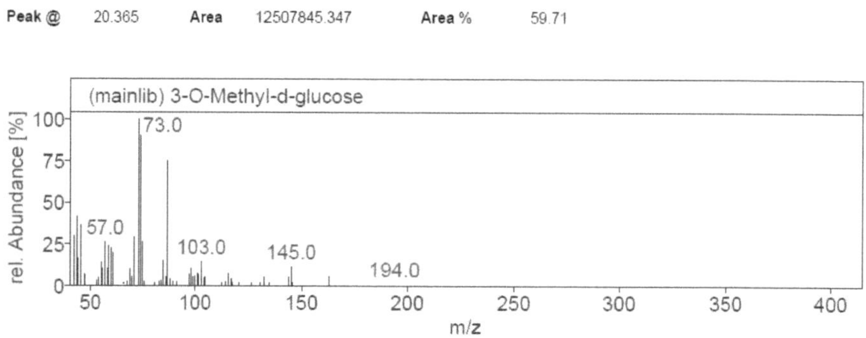

Figure 3: Mass spectrum of eluted compounds, 3-O-Methyl-d-glucose.

Antioxidant activity: DPPH radical scavenging assay: By employing the DPPH assay, the antioxidant activity of MELP was quantified. It was discovered that the MELP dose of 40 µg/mL closely resembled the standard's dose of 10 µg/mL in terms of % inhibition. The reduction in DPPH scavenging activity was indicated by the IC_{50} value of MELP, and the results were determined to be statistically significant (Table 3).

Table 3 *DPPH radical scavenging activity of MELP.*

Test compound	**Percentage (%) DPPH scavenging activity in Mean±SEM**					**IC_{50} (µg/mL)**
	2.5 µg/mL	**5 µg/mL**	**10 µg/mL**	**20 µg/mL**	**40 µg/mL**	
Vit. C	26.38±1.42	42.14±3.18	67.15±8.2	77.72±9.32	94.62±8.43	7.8±0.51
MELP	9.43±0.72	22.07±1.27	31.23±2.14	52.45±1.28	67.38±3.20	26.48±0.41

In vitro anti-diabetic activity: Inhibition of α-amylase: *In vitro* α-amylase inhibitory effect was assessed for MELP at the dose of of 10, 20, 40, 60, 800 and 100 µg. The IC_{50} value was found at 14.56±0.23 µg/mL whereas Acarbose standard was appeared at 3.72±0.22. The results were summarised in Table 4.

Table 4 *% Inhibition of αamylase enzyme activity.*

Sample	**Concentration (µg/mL)**	**% Inhibition of α-amylase enzyme activity**	**IC_{50} value (µg/mL)**
Acarbose	1	12.15 ± 0.14	
	2	43.99 ±0.23	3.72±0.22
	4	52.41±0.34	
	6	61.92± 0.11	
	8	73.61± 0.31	
	10	84.3± 0.33	

	10	43.33± 0.45	
	20	54.22±0.01	14.56±0.23
MELP	40	64.21± 0.33	
	60	84.31±0.34	
	80	93.21± 0.23	
	100	109.3±0.26	

Note: Results are represented in Mean±SD (n=3).

Inhibition of α-glycosidase: In vitro α-glycosidase inhibitory effect was assessed for MELP at the dose of of 100, 200, 400, 600, 800 and 1000 µg. The IC_{50} value was found at 359.5± 0.12 µg/mL whereas acarbose standard was appeared at 0.75± 0.22. The results were summarised in Table 5.

Table 5 *% Inhibition of α-glucosidase enzyme activity.*

Sample	Concentration (µg/ml)	%inhibition of α-glucosidase enzyme activity	IC_{50} value (µg/mL)
	0.1	6.19±0.41	
	0.2	15.04±0.33	0.75± 0.22
Acarbose	0.4	22.54±0.24	
	0.6	48.39±0.23	
	0.8	71.62±0.33	
	1	86.32±0.14	
	100	21.45±0.22	
	200	42.74± 0.25	359.5± 0.12
MELP	400	52.18± 0.45	
	600	61.23±0.11	
	800	73.22±0.22	
	1000	92.34± 0.45	

Note: Results were represented in mean±SD (n=3).

4 DISCUSSION

Newer pharmacological drugs were developed owing to recent advancements in basic understanding of the function of intestinal enzymes. By preventing α-glucosidase from functioning at the intestinal brush border, α-glucosidase inhibitors decrease the intestinal absorption of starch, dextrin, and disaccharides. This enzyme's inhibition lowers the GI tract's absorption of carbohydrates and slows the rate at which postprandial glucose rises. This delaying of starch digestion and breakdown may help diabetics with insulin resistance and glycemic index control.[13]

Acarbose is an inhibitor of the enzyme α-glucosidase that slows digestion of complex carbohydrates and their absorption from the gut. These medications may help lower blood sugar levels by increasing the circulation of the glucoregulatory hormone glucagon-like peptide-1. However, they potentially result in adverse effects such malabsorption, discomfort in the abdomen, gas, and diarrhoea, which would increase the possibility that patients would discontinue utilizing them. Patients with renal impairment shouldn't be administered acarbose. Patients with hepatic disorders should use acarbose cautiously since it may cause reversible increase of liver enzymes.

Experimental results showed that the MELP significantly inhibited the α-amylase activity over α-glucosidase enzyme. The antioxidant profile suggests the existence of potential plant constituents which interfere with oxidative stress. The GC-MS reports revealed the existence of several constituents among them few are medicinally useful.

5 CONCLUSION

Herbal medicines are known for their safety and efficacy in treating diseases effectively without producing any untoward effects. The present research work is being planned to determine the antioxidant and antidiabetic activity of *L. parviflora* which could be serve as a potent conventional therapy. The plant material was procured and conducted the solvent extraction. The phytochemical tests and GC-MS characterization was carried out which showed occurrence of different phytoconstituents. In DPPH radical scavenging assay, the antioxidant activity found IC_{50} value at 25.47±0.51 μg/mL. The *in vitro* anti diabetic activity of leaf extract exhibited IC_{50} value of 14.56±0.23 for α-amylase inhibition, where as it was showed IC_{50} value of 359.5± 0.12 α-Glucosidase inhibitions. The results suggest that the plant possess abundant potential phytoconstituents, further optimization may give new scope in development of new antidiabetic agents.

6 ACKNOWLEDGMENTS

The authors thank the authorities of KL College of Pharmacy, Koneru Lakshmaiah Education Foundation, Guntur, Andhra Pradesh, India for providing infrastructure support to conduct the research and to the complete present work.

7 REFERENCES

1. R. Cressey, W. Kumsaiyai and A. Mangklabruks. *Indian J Exp Biol.* 2014, **52**, 1173.
2. T. Ghosh, M.T. Kumar, P. Sengupta, D. K. Dash and A. Bose. *Iran J Pharm Res.* 2008, **7**, 61.
3. A. Gupta and V. N. Pandey. *Int. J Pharma Bio Sci.* 2014, **5(1)**, 300.
4. S. A. Umakrithika. *J. Pharmacogn. Phytochem.* 2021, **10(5)**, 42.
5. S. K Gugulothu, M Narender and DSNBK Prasanth. *Pharmacogn. Res.* 2022, **14(1)**, 53.
6. N. S. Bembde, P. V. Meshram, M. K. Patil and J. Shaikh. *Res. J. Pharmacogn. Phytochem.* 2020, **12(1)**, 1.
7. G. E. Trease and W. C. Evans. Introduction and General Methods in Pharmacognosy. United Kingdom: Cambridge University Press, 13th Edition, 1989.

8. R. B. Rathor, D. R. Rao and P. Rao. *Indian J. Res. Pharm. Biotechnol.* 2013, **1(4)**, 575.
9. S. S. Momina and V. S. Swaroopa Rani. *J. Young Pharm*, 2020, **12(2),** Suppl: s72.
10. S. Nallapaty, M Narender, R. Pasupula, A. Ketha. *J. Pharm. Pharmacog. Res.* 2022, **10 (3),** 496.
11. N. K. Sachan, Y. Kumar, S. Pushkar, R. N. Thakur, S. S. Gangwar. *Int J Pharm Sci Res.* 2009, **1**, 24.
12. N. Uddin, M. R. Hasan, M. M. Hossain, A. Sarker, A. H. Hasan, A. F. Islam, *et al. Asian Pac J Trop Biomed.* 2014, **4(6)**, 473.
13. P. N. Bennett and M.J. Brown. Clinical Pharmacology. Edinburgh: Churchill Livingstone Elsevier Publisher, 11th edition, 2012, 580.

EXTRACTION, ISOLATION, PHYTOCHEMICAL SCREENING AND EVALUATION OF ANTI OXIDANT AND HYPOGLYCEMIC ACTIVITY OF METHANOLIC EXTRACT OF TAGETES PATULA LINN FLOWERS

Vijayalakshmi Nandikatti[a], M. Bujji Babu[a], K.J.Vergelia Jenifer[a], Sk. Abida[a]

[a] Department of Chemistry, Hindu College of Pharmacy, Guntur

* Corresponding author details E-mail.I.D: nandikatti2014@gmail.com

1. INTRODUCTION

Free radicals are unstable molecules involved in oxidative reactions, produced by cells continuously by enzymatic and nonenzymatic reactions in our body, and will produce harmful or beneficial effects. As these are having oxygen (nascent-with uneven electrons) in their structure, they get involve in number of chemical reactions readily with other molecules. If there is an imbalance between free radicals produced, they get and accumulated in the biological system leading to oxidative stress. The oxidative stress weakens the immune system of our body, which leads to susceptibility towards pathogens like bacterial, fungal and viral infections and also causes disorders in different biological systems like GIT, CVS, neurological, renal systems and also carcinogenesis. To overcome these effects, body itself produces antioxidants, which makes free radicals to stable molecule by donating an electron. [1,2,5]

Diabetes mellitus is the one of most important disorder which makes immune system weak and reduces the life span of patient. [2,5]
*Tagetes patula Linn*are (TPL), which is commonly known as Dwarf or French marigold, belongs to the family of Asteraceae and traditionally used for ornamental purpose. It is also used as remedy in folk medicine, because of the presence of potent phytochemicals in different plants like flowers, leaves and roots. Particularly flowers are composed of essential oils (monoterpenes and sesquiterpenes), Flavonoids (like Quercetin, Patulin, Lutein), flavonoid glycosides (Quercetin -3,7 –diglucoside), and lutein. [1]

The flowers possess medicinal properties like anti pyretic, aromatic, carminative, digestive, diuretic, sedative and also used in the treatment of cough, constipation, renal and hepatic disorders, pneumonia, rheumatism and as an eye wash. [1]

There are number of plants with potent medicinal values. In the present work we used the methanolic extract of *Tagetus patula Linn* flowers for the evaluation of anti-oxidant and anti-diabetic activity. [1,2]

2. MATERIALS AND METHODS

2.1. Collection, identification and authentication of the TPL

The fresh flowers of *Tagetes patula Linn*are collected from the farm, nearby Guntur, A.P., India, and was authenticated and certified by K. Madhava Chetty, taxonomist, Department of Botany, Sri Venkateswara University, Tirupathi, A.P.

Maceration process:

The fresh flowers are collected, cleaned and macerated in methanol for 2 days. The extract was concentrated with rota vaporizer. The crude extract was collected and dried. The yield was 2.8%. The phytochemical screening was done by following standard tests and performed TLC. The column chromatography was performed to the isolated active ingredients by using different solvents petroleum ether, hexane, ethanol, chloroform, ethyl acetate, methanol and water based on polarity the fractions were tested for their phytochemicals. [3, 4]

The fractions which are collected from the column were subjected for both phytochemical screening and TLC. The ethanolic fractions of column have flavonoids particularly flavanones and flavanols (are confirmed with specific tests for flavonoids). So, these fractions were used for biological evaluation, i.e., fraction number from 16-20. This is because flavonoids have good antioxidant activity.

2.2 Antioxidant activity

The anti-oxidant activity was evaluated for both crude and the collected column fractions by using following methods.

2.2.1. Hydrogen peroxide (H_2O_2) scavenging assay. Hydrogen peroxide solution having the concentration 20mM was prepared by dissolving it in pH7.4 phosphate buffer solution. Serial dilutions were prepared from extract, fractions and standard having the concentration range from 50-250 µg/mL in ethanol. In to the test tube 1ml of each dilution was taken and 2ml of H_2O_2 buffer solution was added and kept a side for 10minutes. The absorbance reading at 230nm using phosphate buffer solution with hydrogen peroxide as control was recorded. [6]

2.2.2. Nitric oxide scavenging method. Into a series of test tube add 0.5ml of phosphate buffer solution having a pH of 7.4, each ml of sample dilutions ranging from 50-250 µg/mL and 2ml of sodium nitroprusside solution having 10mM concentration was added and incubate the mixture for 2.5hrs. To 0.5ml of incubated mixture, 0.5ml of Griess reagent was added and again incubates it for 30minutes and the absorbance reading was taken at 546nm. [6]

2.3. Anti-diabetic activity

The anti-diabetic activity was evaluated by using Inhibition of α-Amylase enzyme method. In 10ml volumetric flask, 1ml of α-amylase (0.5 mg/ml) solution and the sample at various concentrations (50,100,150,200 and 250µg/ml) were placed. To this mixture a 1% solution of starch and 0.1ml of pH-6.9 phosphate buffer having the concentration 0.2mM was added and kept a side for 5minute at 37°C. Now the mixture is heated for 15 min by adding 2 ml of 3,5-dinitrosalicylic acid and make up the volume to 10ml with distilled water. Measured the intensity of color at 540 nm using spectrophotometer.

For all the activities percentage inhibition was calculated by using following formula.

$$\% \text{ inhibitory activity} = (Ac-As)/Ac \times 100.$$

Absorbance and the calculated percentage inhibition of all standards, extracts and the fractions were reported in Tables 1,2 and 3 respectively. [7]

3. RESULTS AND DISCUSSION

The schematic procedure used for identification of phyto-constituents showed the presence of flavonoids, glycosides, terpenoids, Saponins, amino acids, proteins, alkaloids and phenols in crude extract of flowers.

3.1. Effects of treatment on Anti-oxidant activity

Table 1: Absorbance and Percent inhibition of crude, F17, F18 and Ascorbic acid by Hydrogen peroxide (H_2O_2) scavenging assay.

S.No	Conc. (µg/ml)	Hydrogen peroxide (H_2O_2) scavenging assay							
		Crude		F_{17}		F_{18}		Standard	
		Abs	% Inh	Abs	% Inh	Abs	% Inh	Abs	% Inh
1	50	0.801	2.31	0.726	11.4	0.703	14.2	1.046	14.2
2	100	0.749	8.65	0.646	21.2	0.626	23.6	0.916	24.9
3	150	0.678	17.31	0.563	31.3	0.553	32.5	0.786	35.5
4	200	0.596	27.31	0.493	39.8	0.473	42.3	0.633	48.1
5	250	0.453	44.75	0.413	49.6	0.406	50.4	0.496	59.3
IC_{50} value		50.67µg/ml		8.95 µg/ml		26.49 µg/ml		8.53 µg/ml	

The anti-oxidant of crude extract and column fractions (F_{16}-F_{20}) were evaluated by Hydrogen peroxide (H_2O_2) and Nitric oxide scavenging methods.

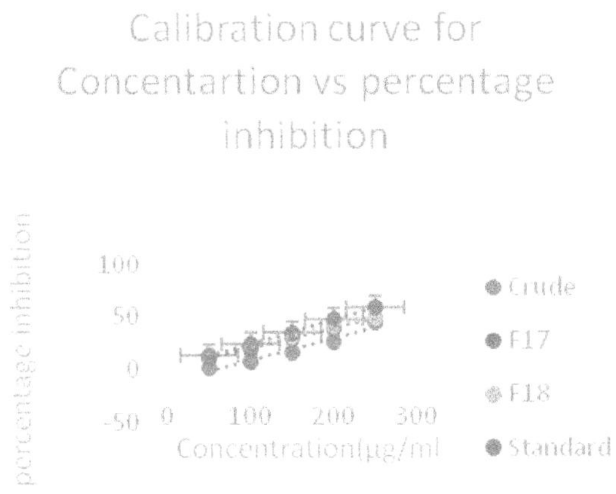

Figure 1 Calibration graph for % inhibition by H_2O_2 scavenging assay.

Table 2 *Absorbance and Percent inhibition of crude, F17, F18 and Ascorbic acid by Nitric oxide scavenging method.*

S.No	Conc. (µg/ml)	Nitric oxide scavenging method							
		Crude		F_{17}		F_{18}		Standard	
		Abs	% Inh	Abs	% Inh	Abs	% Inh	Abs	% Inh
1	50	0.912	17.0	0.896	18.5	0.983	10.6	1.043	17.2
2	100	0.874	20.5	0.783	28.8	0.92	16.3	0.956	24.1
3	150	0.756	31.2	0.703	36.0	0.836	24	0.836	33.6
4	200	0.687	37.5	0.646	44.3	0.763	30.6	0.723	42.6
5	250	0.593	46.0	0.576	52.6	0.716	34.9	0.646	48.7
IC_{50} value		50.04 µg/ml		62.63 µg/ml		31.68 µg/ml		29.38 µg/ml	

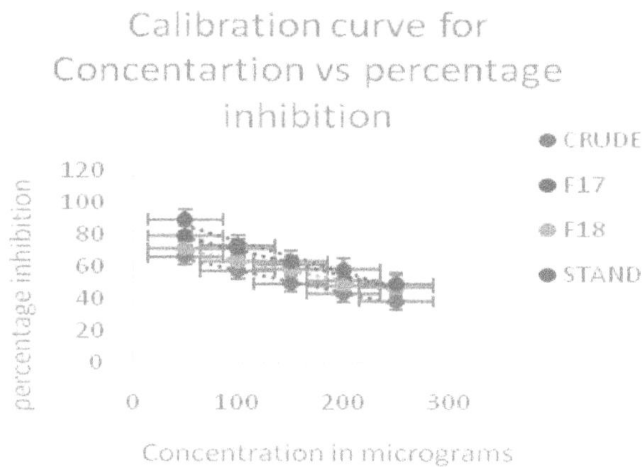

Figure 2 Calibration graph for % inhibition by Nitric oxide scavenging method.

The absorbance values shows that the crude extract and the fractions of Tagetus have good anti-oxidant activity. Among them fraction no 17 has significant activity than that of crude and other fractions by hydrogen peroxide scavenging assay (Figure 1) and the fraction no 18 having significant activity than that of crude and other fractions by Nitric oxide scavenging method (Figure 2). When comparing with standard the fractions showed potent activity.

The linearity of calibration graph indicates that as the concentration of sample increases there is an increase in the activity. IC_{50} values were calculated from the calibration graph plotted between percent inhibition and concentration. The IC_{50} values for fraction no 17 and fraction no 18 are equal to that of standard i.e. 8.53µg/ml and 29.38µg/ml.

3.2. Effects of treatment on Anti-diabetic activity

Table 3: Absorbance and % inhibition of crude, F_{17}, F_{18} and standard of by α-amylase inhibition method.

S.No	Conc. (µg/ml)	α-Amylase Inhibition method							
		Crude		F_{17}		F_{18}		Standard	
		Abs	% Inh	Abs	% Inh	Abs	%Inh	Abs	%Inh
1	50	0.66	66	0.79	79	0.71	71	0.89	89
2	100	0.57	57	0.71	71	0.63	63	0.736	73
3	150	0.49	49	0.61	61	0.58	58	0.636	63
4	200	0.43	43	0.51	51	0.48	48	0.583	58
5	250	0.38	38	0.49	49	0.47	47	0.49	49
IC_{50} value		**159 µg/ml**		**245 µg/ml**		**258 µg/ml**		**295.38 µg/ml**	

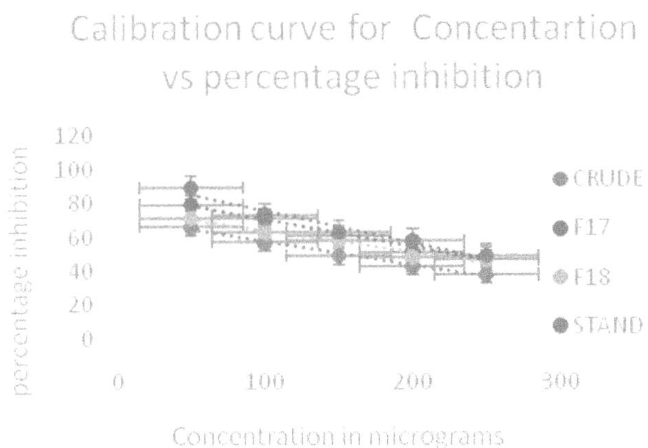

Figure 3 Calibration graph for % inhibition by Nitric oxide scavenging method.

From the results it was noted that the methanolic extract and the column fractions F_{17}& F_{18} of Tagetus showed significant inhibitory effect on α-amylase activity. From the calibration graph, it shows that as the concentration of extract increases, the activity of α-amylase enzyme decreased (Figure 3). From this study we report that the ethanolic fraction of methanolic extract of TPL having anti-diabetic activity, which is also dose dependent. This was also supported by percentage inhibition values along with IC_{50} value.

4. CONCLUSION

We conclude that the methanolic extract of *Tagetes patula Linn* flowers shown the presence of phyto constituents like Flavonoids, Phenolic compounds, Saponins, Alkaloids, Proteins, Amino acids, Terpenoids. The crude extract and some column fractions were evaluated for their biological activities like antioxidant activity by Hydrogen peroxide and Nitric oxide scavenging assay methods and Anti diabetic activity by α- amylase method. Both crude and isolated column fraction (F_{17}) of *Tagetes patula* flowers showed significant activity when compared with standard.

5. ACKNOWLEDGMENT

The authors are thankful to HCOP,GNT, for providing the facilities to carry the work.

6. REFERENCES

1. Ausama A Safar, Anwar O Ghafoor, Dara Dastan,*Journal of reports in pharmaceutical sciences*, 2020, **Volume 9**, 59.

2. Muhammad Kashif, Samina Bano, Sabira Naqvi, Shaheen Faizi, *Pharmaceutical Biology*, 2014, **Volume 5**, 1.

Deepshikha Kushwaha, Yashodhara Verma, *Annual Research & Review in Biology*, 2017, **Volume 13**, 1.

4. C. Eswari and A. John merina, *Indian journal of applied research*, 2015, **Volume 5**, 274.

5. Cicevan R, Al Hassan M, Sestras AF, Prohens J, Vicente O, Sestras RE, Boscaiu M., *PeerJ*, 2016, **Volume 4**, 21.

EVALUATION OF NEUROPROTECTIVE ACTIVITY OF *SIDA CORDIFOLIA* AGAINST β-AMYLOID PROTEIN INDUCED LOSS OF COGNITIVE FUNCTIONS IN SWISS ALBINO MICE

Annie Mande [a, b], Narender Malothu[a], Tata Prasanna Kumari[a]

[a] K L College of Pharmacy, Koneru Lakshmaiah Education Foundation, Vaddeswaram, Guntur, AP, India.
[b] Department of Pharmacology, Bapatla College of Pharmacy, Bapatla, Andhra Pradesh, 522101.

Corresponding Author: anniemande@gmail.com

1. INTRODUCTION

Neurodegeneration is the word which is used to denote the cumulative deprivation of both anatomical and functional capacity of neuronal cells[1.] It is caused due to oxidative stress or due to the natural apoptotic pathways.[2] No permanent cure was reported for any diseases like Parkinson's disease, Alzheimer's disease (AD) and celebral palsy etc...where there is degeneration in neuronal plasticity. Alzheimer's disease is most prevalent in elder people and symptoms includes loss of memory, loss of proper judgment and loss of cognitive functions of brain. Current drugs have the capacity only to reduce the symptoms and to retard the progressive damage of neurons. The mechanisms underlying to delay the process of neurodegeneration includes free radical scavenging, brain neurotransmitter modulation, and hormonal effect. Ayurvedic medicine is the traditional system having the history over 5000 years in which most of the herbal medicines were proved to have profitable effects in the prevention of various diseases, rejuvenation of different body systems, and in the extension of life span. Plants like *Rauwolfia serpentina, Bacopa monneria, Centella asiatica, Ocimum santum, and Mucuna pruriens* etc. were proved to have beneficial effects on neurodegenerative disorders in improving cognition of the patients.[3] β-Amyloid protein is a small fragment of a large membrane protein called as Amyloid precursor protein (APP). It can penetrate through the neuronal membrane and it is more crucial for the growth of the neuron's viability, and quick injury repair in neurodegenerative disease like AD. An unknown cause divides this APP into small fragments by enzymatic breakdown. Among these one of the filaments gives rise to fibrils of β-Amyloid which forms into a clot and deposits outside of the neurons as dense bodies which are called as senile plaques.[4]

Sida cardifolia is a perennial sub-shrub which belongs to the family of Malvaceae. The vernacular names of this plant are country mallow in English and in Telugu ayurveda it is called as Bala. This drug can be categorized into low toxic chemicals based on in its toxicity studies done by Franco et al.[5] This plant was proved to have antioxidant activity. According to a study in1999, ethyl acetate extract of *S. cordifolia* was proved to show both analgesic and hypoglycaemic activities. An aqueous solution of S. cordifolia was found to stimulate liver regeneration in hepatectomy rats and possess hypotensive activities.[6] The important chemical components which are proved to present in this plant aerial parts are palmitic acid, stearic acid, hexacosanoic acid and sitosterol. Based on the literature review and chemicals present in this plant this is the first ever approach to screen the neuroprotective potential of the plant extract of *S. cordifolia* in mice. The objective of the present study is screening of the neuroprotective potential of the plant *S. Cordifolia*.

2. MATERIALS AND METHODS

2.1 Chemicals and Reagents

The chemicals and reagents used in the current study are of analytical grade. The chemicals like phosphate buffer, ethylenediamine tetra acetic acid (EDTA), Dithiobis nitro benzoic acid, acetylthiocholine iodide, serotonin, butyl acetate, pyrogallol sodium azide, reduced glutathione, dinitro phenyl hydrazine, copper sulphate and ascorbic acid were purchased from SD fine chemicals, India.

2.2 Plant collection and authentication

S. Cordifolia was identified and collected from Western Ghats, in Nilgiris district of Tamil Nadu and it was authenticated by Dr.S. Rajan, field botanist, Department of Survey of medicinal plants and collection unit, Indira Nagar, Emerald post, Nilgiris District, Tamil Nadu.

2.3 Drying and size reduction of the plant material

The whole plant of *S. cordifolia* was collected and shade dried for about two months. The dried plant material was crushed to powder, sieved (sieve no.22) and packed in an airtight container.

2.4 Extraction of the plant Material

The coarse powder of plant material (500 g) was packed in Soxhlet apparatus. The plant material was extracted with ethanol for 16-18 hr. Then the solvent with extract was filtered and concentrated by the process of evaporation. The residue was concentrated and final extract was stored in the desiccator.

2.5 Preliminary phytochemical analysis

The preliminary phytochemical screening was performed based on the procedures explained by Khandelwal *et al*., 2006. The evaluation tests were carried to know the existence of secondary metabolites like alkaloids, carbohydrates, proteins, flavonoids and saponins etc.

2.6 Acute Toxicity Studies

Albino mice of adult age which weighs in between 20-25 g were selected for this work. The initial dose level of methanol extract of *S. Cordifolia* was supposed to be 200 mg/kg body weight (p.o.) because large number of natural extracts were proved to have LD_{50} value larger than 2000 mg/kg, (p.o.). These compounds can be considered as safest natural compounds. Therefore, the initial dose proposed for the use was 200 mg/kg, (p.o.). The volume given to the animal was 1 mL/100 gm body weight. The mice were kept fasting over night with water *ad libitum*. Food is supposed to withhold for about 4 hr following administration (p.o.) of drugs and should be examined to identify any sign of lethal effects.[7]

Mice should be weighed before and post administration of crude extracts. Thereby the mice were observed and recorded for any changes in skin, eyes and mucus membranes including circulatory, respiratory, central, and autonomic nervous systems, locomotors activity and behavioural pattern. Specific signs like sleep, convulsions, salivation, diarrhoea, lethargy, sedation, sleep, and unconsciousness should also be noted. The onset and duration of toxic reactions are also recorded.

2.7 Experimental Design and animal groupings

The present study was approved by the IAEC of Delve Labs Pvt. Limited with the approval number 1675/RO/c/12/CPCSEA/040/14. Mice were grouped into four with six animals in each.

Group I: Animals which receive saline with phosphate buffer,

Group II: Animals which receive β-amyloid peptide,

Group III: Animals which receive β-amyloid protein, and 200 mg/kg (p.o) AESC,

Group IV: Animals which receive β-amyloid protein and 400 mg/kg (p.o) AESC (Alcoholic extract of *S. cordifolia).*

The required quantity of sample was weighed and dissolved in 0.5% Carboxy Methyl Cellulose (CMC) and the final volume was adjusted according to the dose that should be given. Both extracts were administered through oral route by using oral gavage needles.

2.8 Evaluation of Neuroprotective activity by using behavioural tests

2.8.1 By using Y-Maze. In general, it is being used to assess spatial working memory in mice. Whole apparatus is manufactured with black painted wood or grey plastic. Each arm measures about 40 cm in length, 13 cm in height, 3 cm in width at the bottom, and top width 10 cm, with convergence at same angle. Each mouse dropped at the end of one arm and permitted to explore for about 8 mins throughout the maze. Mice opt to move through the maze sequentially every arm in return. The capacity to enter possible arm each time needs cognition to remember previously visited one. The actual entries into different arms, alterations, and feasible re-entries into the same arm are counted. Alternation can be explained as the entries into all arms, on triple sets which are overlapping. The % alternations exhibited by mice is 60-70%, and exhibits 25-35 arm entries within 8 min session. The percentage (%) of alternations is calculated as per the given formula (i)
% alternation = (Actual alternation)/ (Total number of arm entries -2) x100[8]- (i)

2.8.2 By using Open field habituation test. An open field with dimensions 40 cm x 50 cm x 60 cm is floored with brown linoleum and is partitioned into 16 squares of same size separated by white lines. The mice were allowed twice to move in this area within a 24-hr interval to control the possible effects on locomotors activity. In two sessions, mice were kept in the left back square which is remained to inspect for 5 min. During this the number of intersectional lines in opposite directions were counted.

2.8.3 Step Down Inhibitory Avoidance. This contains an acrylic box with measurements 50 cm x 25 cm x 25 cm whose floor consists of a platform at the centre. During the training session the animal has given a shock of 0.4 mA at foot for 2 sec immediately after stepping down. In test session foot shock was not given and step-down latency was calculated as a measure of retention of memory up to a maximum time of 300 sec. To evaluate the short-term memory, retention time was measured up to 90 mins and long-term memory was evaluated by the same measuring up to 7 days.[9]

2.8.4 By using Water Maze Test. This was done by using the instrument called Morris Water Maze. This apparatus consists of a water tank which is circular in shape with a diameter of 100cm and height of about 35 cm. It contains water at 28°C temperature and should be maintained down to the bottom of 15 cm and made non-transparent by mixing milk in powdered form.

This consists of a submerged platform 0.5 cm beneath the water surface and kept at the middle point of one quadrant. On day 1 the animals are allowed to explore the maze completely. From day 2-7 three trails were conducted per day and escape latencies were recorded.

2.9 Biochemical Estimation

The animals were sacrificed and brains were isolated and the samples were assayed for the estimation of biochemical parameters like Acetylcholine esterase (AchE), Vitamin-C, Mono amino oxidase-A (MAO-A), Glutathione peroxidase, Tissue Protein, Superoxide Dismutase (SOD), and Glutathione reductase.

2.9.1 Estimation of AchE. The principle involved in the estimation of AchE is the development of yellow colour as a result of reaction between thiocholine and thiobisnitrobenzoate (DTNB). The isolated brain tissue was subjected to homogenisation by using a homogeniser by using solvent as Phosphate buffer (pH 8, 0.1M). 0.5 mL of this brain homogenate was taken out and poured into 25 mL volumetric flask and diluted with fresh DTNB solution. From this solution two portions each of 4 mL were taken out into two test tubes.1 mL of substrate solution (75 mg of Acetylcholine iodide per 50 mL distilled water) was poured in both test tubes and incubated for 10 min at 30°C, to these 2 drops of serine solution was added. The solution with serine solution was used for zeroing the value in colorimeter and absorbance of the sample was recorded at 420 nm.

The quantity of AchE was calculated by using the below formula (ii),

R = (δO. D X Volume of Assay (3 ml)/ (E X mg of Protein) ---- (ii)

Where, R = rate of enzyme activity in n mole of Ach iodide hydrolysed/min/mg of protein; δO. D = Change in absorbance/min; E = Extinction coefficient = 13600/M/cm.

The enzyme activity is expressed as μmoles/min/mg of brain tissue.

2.9.2 Evaluation of Superoxide Dismutase. For this Marklund and Marklund method is used which is based on the ability of enzyme to reduce the auto oxidation of Pyrogallol. 1 mL of Pyrogallol- Tris-DETPA, was combined with 0.2 mL of brain homogenate and 0.8 mL of water. The rate of Pyrogallol autooxidation is measured as changes in absorbance value at 420 nm. The quantity of SOD has been stated as Units/mins/mg of protein.

2.9.3 Estimation of Monoamine Oxidases. For the estimation of MAO, the brain tissue was cleaned and homogenised with the help of 0.25 m Sucrose 0-1 M Tris 0-02 M EDTA. The resultant mixture comprises of 4 mM of Serotonin or 2 mM of β-phenyl ethyl amine as specific substrates for MAO-A and MAO-B respectively, 250 ml of brain homogenate, and 100 mM sodium phosphate buffer (pH 7.4) to make final volume of 1 mL. The mixture was stand to react for 10 mins and then 1 M HCl (200 μL) was added and the resultant solution was extracted with 5 mL of butyl acetate for MAO A and 5 mL of cyclohexane for MAO B, respectively. The absorbance of the mixture was calculated at 242 nm and 1M HCl was served as blank.

2.9.4 Estimation of Total Protein content. It was measured by Lowry method. To 0.1 mL of diluted solution of brain tissue 0.9 mL of water and 4.5 mL of Alkaline copper reagent were mixed and stand for about 10 min at room temperature. To it 0.5 mL of Folins reagent was mixed after that the resultant colour was recorded after 20 min at 640 nm.

2.9.5 Estimation of Glutathione Peroxidase. It was evaluated by Lawrence & Bruke method. A mixture of 0.2 mL of EDTA, Sodium azide, H_2O_2, 0.4 mL of phosphate buffer, and 0.1 mL of diluted tissue was incubated at 37°C. To the above mixture TCA (0.5 mL) was added and centrifuged at 2000 rpm. To 0.5 mL 0f homogenate 4 mL of disodium hydrogen phosphate and 0.5 mL of DTNB were mixed and absorbance was measured immediately at 420 nm.

2.9.6 Estimation of Reduced Glutathione. For this estimation a mixture of 0.5 mL of oxidized glutathione,1 mL of Phosphate buffer, and 0.2 mL of NADPH were mixed and finally it was made up to 3 mL with water. To this resultant mixture 0.1 mL of diluted brain tissue homogenate was added. The optical density variations were measured at 340 nm.

The level of GR was expressed as n moles of NADPH oxidized/min/mg Protein.

2.9.7 Estimation of Ascorbic acid. The Ascorbic acid levels were estimated by the method described by Omaye. It was performed by mixing 1mL of TCA, 0.5 mL of water to 1 mL of supernatant 0.2 mL of DTC reagent was mixed and subjected to incubation at 37^0 C for about 3 hr. To this mixture 1.5 mL of sulfuric acid was added and permitted to kept at room temperature for about 30 min. The absorbance was measured at 520 nm. The quantity of Ascorbic acid was represented as μg/ mg of protein.

2.10 Statistical analysis

The results were denoted as Mean ± SEM and statistical significance between different groups were analysed by using one way ANOVA.

3. RESULTS

3.1 List of phytochemicals present in *S. cordifolia*.

The plant extract was proved to have phytochemical constituents like alkaloids, carbohydrates, proteins, steroids, flavonoids and saponins.

3.2 Evaluation of Neuroprotective activity

3.2.1 Results of Step-down inhibitory Avoidance Test, Water maze Test, and Y-maze Test. There was considerable variation among the Group-I and Group-II STM and LTM values in the step-down latency evaluation test. The group which has taken AESC showed considerable tendency to enhance STM and LTM retention. Both the low dose ($p < 0.05$) and high dose ($p < 0.001$) of AESC has shown significant values.

There was an increase in escape latency in negative compared with the control group animals. The group treated with 200 and 400 mg/kg treated showed the significant decrease in escape latency ($p < 0.05$) respectively. The effect of AESC on water maze has shown significant ($p < 0.001$).

In Y- maze test, the group with induced amnesia (Negative control) denotes a reduction in the behavioural alteration by the ($p < 0.05$) when compared with the group I. The values got by treatment groups showed significance ($p < 0.05$) by improving behavioural alteration with respect to 200 mg/kg of AESC and 400 mg/kg of AESC when compared with that of the control group. The effect of Aesc on Y-Maze has shown significant ($p < 0.001$) values. See Table 1.

Table 1 *The table shows results of different parameters which are performed to evaluate neuroprotective activity.*

Group	Stats	Step Down Inhibitory Avoidance (Sec)		Water Maze	Y-Maze
		STM	LTM	Escape Latency	% Alteration
G-I	Mean	119	134.33	30.33	69.83
	SD	5.8	2.58	4.97	1.17
	SEM	2.366	1.054	2.028	0.477
G-II	Mean	85.33	96.33	60.83	50
	SD	2.16	1.63	1.47	1.41
	SEM	0.881	0.667	0.6	0.577
G-III	Mean	94.67	114.83	44.83	55.17
	SD	1.75	4.07	2.79	1.17
	SEM	0.715^{***}	1.662^{***}	1.138^{***}	0.477^{***}
G-IV	Mean	114.83	129	38.17	66.67
	SD	1.72	2.83	1.94	1.86
	SEM	2.64	1.155^{*}	0.792^{***}	0.760^{**}

Here each group consists of 6 animals (N=6), SD= Standard deviation, SEM= Standard Error Mean, STM= Short Term Memory, LTM=Long Term Memory.

3.2.2 Results of Exploratory behaviour Test. In exploratory behaviour test, the low dose of AESC has shown significant increase in the head dippings and line crossings when compared with control group. The 400 mg/kg treated group showed significant ($P<0.05$) increase in the rearings and line crossings respectively. The effect of AESC on exploratory behaviour has shown significant ($P<0.0001$) values. See Table 2.

Table 2 *The table shows results of different parameters which are performed to evaluate neuroprotective activity.*

Group	Stats	Exploratory Behaviour		
		Head Dipping	Rearing	Line Crossing
G-I	Mean	11.5	23	36.33
	SD	1.87	1.41	1.75
	SEM	0.763	0.577	0.714***
G-II	Mean	5.83	10.33	14
	SD	1.47	1.03	1.41
	SEM	0.6	0.421	0.577
G-III	Mean	7.5	14.17	28.5
	SD	1.05	1.47	1.05
	SEM	0.428***	0.600***	0.428***
G-IV	Mean	9.5	20.67	31.5
	SD	1.05	1.37	1.05
	SEM	0.509	0.557*	0.509***

Figure 1 has shown that the AESC has improved short term and long-term memories by increasing the cognitive parameters in different evaluation tests which were performed earlier.

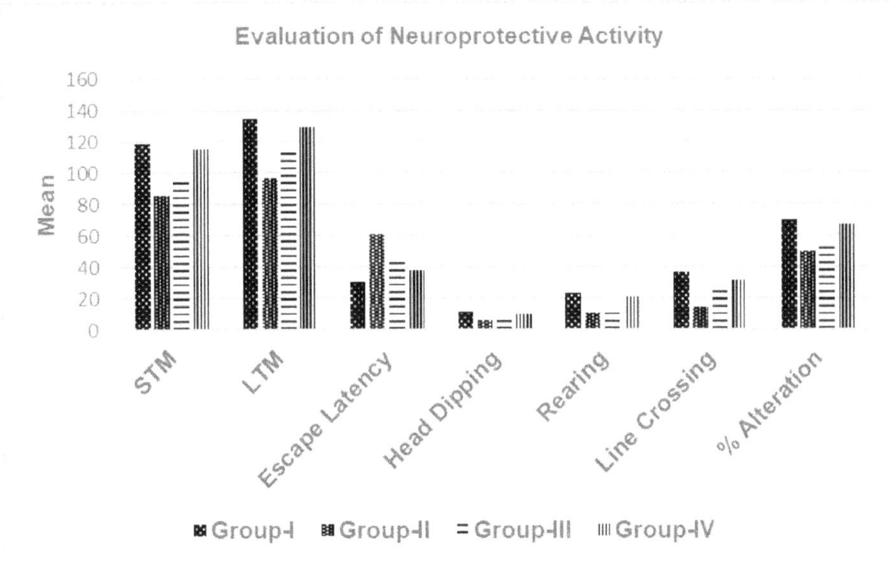

Figure 1: the above chart represents the comparison of Mean values of different Parameters measured to evaluate neuroprotective activity of the AESC, STM=Short Term Memory, LTM= Long Term Memory.

3.2.3 Results of biochemical parameters evaluated to confirm neuroprotective activity. The plant extract of *S. Cordifolia* has reduced the levels of enzymes like AchE, MAO-A, Super oxide dismutase and Glutathione peroxidase etc., and improved the levels of protective agents like Vit.c and tissue protein. See Table 3.

Table 3 *Shows results of different parameters which are performed to evaluate neuroprotective activity.*

Groups	Stats	AchE μM/ min/mg Protein	MOA-A μM/hr	T. Pro mg/dl tissue	SOD U/min/ mg protein	GPx U/min/ mg protein	GSH U/min /mg protein	Vit-C μgm/ mg Protein
1	Mean	15.58	23.51	6.56	8.65	35.45	35.42	0.9
	SD	0.36	0.14	0.39	0.33	0.42	0.69	0.07
	SEM	0.1467	0.057	0.161	0.136	0.171	0.283	0.03
2	Mean	18.45	28.61	5.14	5.75	29.61	24.35	0.46
	SD	0.33	0.18	0.04	0.44	12.3	0.63	0.05
	SEM	0.137	0.073	0.159	0.178	5.22	0.257	0.207

3	Mean	16.49	27.49	5.46	7.5	28.54	28.39	0.5
	SD	0.29	0.26	0.03	0.45	0.46	0.49	0.05
	SEM	0.116***	0.107***	0.011***	0.182**	0.187**	0.199***	0.019***
4	Mean	15.76	24.47	6.16	8.1	33	31.89	0.81
	SD	0.15	0.4	0.03	0.61	0.61	0.54	0.09
	SEM	0.062	0.164***	0.011	0.247	0.247	0.219	0.039

Here each group consists of 6 animals (N=6), SD= Standard deviation, SEM= Standard Error Mean, T. Pro=Tissue Protein.

4. DISCUSSION

Neurodegenerative disorders are often accompanied with the loss of hippocampal and cortical neurons which results in memory and cognitive impairment. Based on the Journal of Global voice of Dementia, about 65.7 million of people may suffer with neurodegenerative disorders by the year of 2030. AchE inhibitors are the most prescribed drugs in the treatment of neurodegenerative diseases, most of these are synthetic drugs and are found to be less effective that have wide range of adverse effects. So, there is a need to identify the new compounds which are more selective and are having low toxic profile.

S. Cordifolia is a plant which is having antioxidant property. The current research work showed that the neuroprotective activity of AESC on icv administration to cognitive deficient mice induced by Aβ 25-35 protein. In our study intra-cerebral ventricular injection of Aβ 25-35 induced impairment of cognition which are evaluated by considering few behavioural parameters like step down inhibitory avoidance, Morris water maze test, Y-maze and exploratory behaviour tests and biochemical parameters like SOD, Glutathione peroxidase and reductase, AchE and MAO-A etc.

In the current study we observed that there is a considerable reduction in the concentration of enzymes which can exert antioxidant activity and the elevated AchE levels, MAO-A in mice brain in negative group after administration with the Aβ 25-35. Aβ 25-35 has the capacity to create oxidative stress in the brain cholinergic hypofunction, elevation of AchE and MAO. Further continuous icv infusion of Aβ to the mice has resulted considerable reduction in expression of proteins like SOD, GPx, and glutathione- transferase in the brain of mice. We observed that the administration of AESC has reduced the cognition impairment in Aβ 25-35 treated mice, particularly it shows enhanced STM and LTM in passive avoidance test and in the behavioural parameters like head dipping, rearing, line crossing. AESC exerted significant ($p<0.001$) dose dependent increase from 5.83 ± 0.081 to 94.67± 0.715 effects in short term retention and dose dependent significant ($p<0.001$) increase in long term retention from 96.3 ±

0.667 to 114.83 ± 1.602 and 129 ± 1.155 in the present task. The effect of AESC on the open field i.e.; exploratory behaviour test is also evaluated which permitted us to study treatment influence on the locomotor activity in the animals. It was proved that rodents administered with AESC has shown a significant dose dependent elevation in the number of responses like head dipping, rearing, and line crossing in the open field test. These results of our experiments are compatible with favourable effect on locomotor activity in the water maze evaluation test. Consumption of AESC decreases the escape latency in a dose dependent tendency. 200 and 400 mg/kg doses of AESC have proved to decrease in the elevated AchE enzyme level which denotes the capacity to increase cognitive function through the reduced deterioration of acetylcholine levels. The AESC indicated reduction in the levels of MAO-A, which denotes the governing property of the extract on regulation of biogenic amines. AESC has proved to provide protection against the oxidational damage by both 200 mg/kg and 400 mg/kg doses which may be the cause for the rejuvenating property of the extract.

5. CONCLUSION

Based on the presented results the present study concluded that the whole plant extract of *S. Cordifolia* revamps cognitive deficiency caused by Aβ 25-35 and this is facilitated by the antioxidant properties of AESC. The neuroprotective potential of *S. Cordifolia* on neurodegenerative disorders may be due to the inhibiting activity against AchE, MAO, and free radical scavenging activity. However, further studies were required to prove its long-term activity on memory.

6. ACKNOWLEDGMENT

The authors are thankful to Principal and Management of Bapatla College of Pharmacy and KL College of Pharmacy for providing the facilities to carry the work.

7. REFERENCES

1. D.C. Rubinsztein, *Genome Biol.,* Aug 2006, 7(8), 1.
2. D.E. Bredesen, R.V. Rao, P. Mehlen, *Nature*, Oct 2006, 443(7113), 796.
3. D.J. Selkoe, *Nature*, Jun 1999, 399(6738), A23.
4. J. Kang, H.G. Lemaire, A. Unterbeck, J.M. Salbaum, C.L. Masters, K.H. Grzeschik, G. Multhaup, K. Beyreuther, B. Müller-Hill, *Nature*, Feb 1987, 325(6106), 733.
5. C.I. Franco, L.C. Morais, L.J. Quintans-Júnior, R.N. Almeida, A.R. Antoniolli, *J. Ethnopharmacol.*, Apr 2005, 98(3), 275.
6. S. Rejitha, P. Prathibha, M. Indira, *Redox Report*, Mar 2015, 20(2), 75.

7. M. Annie, M. Narender, B. Naga Raju, T.E.G.K. Murthy, Toxicol Int, April 2022, 29(2), 239.
8. T. Maurice, B.P. Lockhart, A. Privat, *Brain Res.*, Jan 1996, 706(2), 181.
9. K. Bromely Brits, Y. Weihong Song, J Vis Exp., Jul 2011, (53), 2920.

SCREENING OF ANTIDIABETIC ACTIVITY OF CRUDE EXTRACT OF CELASTRUS PANICULATUS IN ALLOXAN INDUCED DIABETIC RAT MODEL

FATIMA SULTANA*[1], RAMA RAO, ANUPAMA K.

Sultan ul uloom College of Pharmacy, Hyderabad, TS.

*fsultana4@gmail.com

1 INTRODUCTION

Diabetes mellitus is a commonest form of endocrine disorder that is characteristic of hyperglycaemia. This is due to a defective metabolic system of fat, proteins, and carbohydrates leading to major health challenges and lifelong microvascular and macrovascular complications. There is a destruction of beta cells of the pancreas that secrete insulin so that insulin is not produced, or defect in release of insulin, or defects in response to insulin by peripheral tissues, that cause hyperglycaemia.[1]

As of 2019, an estimated 463 million adults worldwide, or 9.3% of all adults, have diabetes. These adults range in age from 20 to 79. This number is projected to rise to 578 million, or 10.2% of the global adult population, by 2030, and to 700 million, or 10.9%, by 2045 of the world's adult population.[2]

Plant products have been largely used as drugs and cure for various symptoms. Around 800 medicinal plants have hypoglycaemic action worldwide.[3] Several of these were documented in the traditional literature and yet many of them are being used without thorough information about their phytochemicals and their action on the human body. Since these plant constituents show successful results without side effects and can be conveniently used orally, the WHO has advised assessing these conventional plant therapies.[4]

Because of their ready availability, low cost, and the fewest side effects, herbal supplements have become the mainstay of all available remedies, especially in rural areas.[5] Botanical products provide a rich source of bioactive chemicals with potent pharmacological effects and without unwanted side effects.[6, 7] For long medicinal plants have been a valuable resource for discovering novel treatments for illnesses that affect people. A variety of plants have anti-diabetic properties. These results are supported by *in vitro* data and animal models, and there is scant information about their clinical applicability.[8]

The Ayurveda treatment using *Celastrus paniculatus* seeds shows an effective change in the human body as it is helpful in curing dysentery, diarrhoea, piles, leprosy, wound healing, rheumatism, gout and Alzheimer's disease. It contains alkaloids, sterols, triterpenoids, flavonoids, saponins and tannins. The present study was carried out to show the antidiabetic property of *Celastrus paniculatus* seeds.[9, 10]

2 MATERIALS AND METHODS

2.1 Preparation of seed extract

The *Celastrus paniculatus* seeds were procured and authentified by Dr. Mohammed Ali Uddin Sheik, Secretary of Unani Research Foundation Hyderabad in Telangana State. The seeds were thoroughly dried and cleaned with muslin cloth to remove any dust, and then they are fed into a mechanical grinder in small amounts.

300 gm of seed powder was soaked in 600 to 700 ml of distilled water for at least 12 hr with stirring at constant intervals. This was heated at 60-70 degrees centigrade to reduce it to half of its volume. The extract was

thereafter filtered with the help of muslin cloth. This filtrate was concentrated to get a semi-solid extract that was refrigerated to be used as the need arises.

2.2 Phytochemical screening of *Celastrus paniculatus* Seed Extract

The *Celastrus paniculatus* seed extract (CPSE) was treated with various reagents and tests were conducted to study the phytoconstituents by standard methods.

2.3 Animals

Male Albino Wistar rats of 150-200 gm were used for the experimental study. Rats were bought via a dealer i.e. Sainath Animal Agency in Hyderabad. Institutional Animal Ethical Committee had given approval for procurement and for conducting the study using these animals, approval number IAEC/SUCP/2019/05. The animals were fed with an adequate standard pellet diet and water *ad libitum*. They were kept under a standard animal house environment in 12 hrs light and dark conditions. Fasted rats were deprived of food for 16 hrs prior to the administration of the extract.

2.4 Hypoglycaemic Study

This was a single-day study. It was conducted on normal and fasted rats. Sixteen rats were taken and divided into four groups. Group I/Normal received a dose of one ml of distilled water. Group II /Standard received a dose of the standard drug, glibenclamide 0.5 mg/kg. Group III /Test 1 received a lower dose of CPSE i.e., 800 mg/kg. Group IV/Test 2 received a higher dose of CPSE i.e., 1600 mg/kg. All the samples were administered via peroral route. Blood samples were collected to test blood sugar levels in rats at 0, 30, 60, and 120 min by tail vein method.[11]

2.5 Oral Glucose tolerance test

This was also a one-day study. Glucose tolerance study was conducted on normal rats with an adequate supply of food and water *ad libitum*. Sixteen rats were taken and divided into four groups; grouping was done as mentioned before. All the rats were fed adequately and provided free availability of water and food. The rats were administered distilled water, CPSE doses, and glibenclamide, respectively by per oral route. All the rats were given a 50% glucose solution of 2 g/kg body weight by p.o. route, half an hour after CPSE and glibenclamide administration. Collection of the blood sample was done by tail vein method. Samples were taken 30 min after glucose is given and continued at 60 min, 90 min, and 120 min.[12]

2.6 Design for Alloxan induced diabetic study

Rats fasted overnight before were administered with Alloxan a dose of 150 mg/kg. Alloxan monohydrate was dissolved in 0.9% w/v normal saline solution which was stored at a cool temperature. Single-dose of freshly prepared alloxan was given by the i.p. route. Rats were given 10-20% glucose solution to prevent hypoglycemia that may cause death, 6-8 hrs after alloxan was given, a 5% glucose solution was given for the next 24 hr to prevent hypoglycemia. Blood glucose was estimated to check diabetes-induced rats on 3rd day. Rats with blood glucose levels of more than 250 mg/dL were selected for the study. Once diabetes was induced, all doses were administered once at a specific time of the day. The study continued for three weeks after diabetes was induced. No fasting of animals was done and rats were given food and water *ad libitum*.[13]

2.6.1 Grouping. 30 numbers of animals were selected for the study and divided into five groups, each group contains six animals. I. Normal Control – This group of rats were given only distilled water and were not diabetes-induced. II. Disease/ Diabetes Control –These rats become diabetic after Alloxan (150 mg/kg b.w., i.p. route) administration. Thereafter only the vehicle was given. III. Standard /Reference group – These rats were Alloxan induced diabetic and then treated with antidiabetic standard drug, Glibenclamide 10 mg/kg b.w., p.o. inj. + 5 % w/v CMC. IV. Test 1 – In this set rats were administered with 400 mg/kg p.o. of CPSE. This was done after the rats were diabetic induced with administration of Alloxan 150 mg/kg b.w. V. Test 2 – This set has rats administered with 800 mg/kg p.o. of CPSE, after induction of diabetes with alloxan.

2.7 Estimation of body weight

Weight of all the rats were taken at the beginning of the study (i.e.0th day) and later every week during the study (i.e.7th, 14th, 21st, 28th days).

2.8 Collection of blood samples

Blood samples were collected by the tail vein method for estimation of blood glucose levels once before alloxan was given and later to confirm the induction of diabetes. Once after the study started blood samples were collected on a weekly basis. On the last day of the study, i.e. 28th day blood samples were collected by retro-orbital method for estimation of HbA1c and lipid profile.

2.9 Histopathological Study

On the 28th day, the rats were sacrificed for removal of pancreas for histopathological evaluation. Pancreas was kept in 10% Formalin solution before the histological study.

2.10 Statistical Analysis

Statistical analysis was carried out using Graph Pad Prism 8.4.0. All results were expressed as Mean ± SEM. Groups of data were compared with analysis of variance (ANOVA) followed by Dunnett's multiple comparison tests to identify significance (i.e. $p<0.001$, $p<0.01$, $p<0.05$) between groups.

3 RESULTS

3.1 Investigation of phytochemical constituents

The qualitative tests performed to detect phytochemical constituents of the aqueous extract of *Celastrus paniculatus* seeds showed the presence of alkaloids, carbohydrates, flavonoids, tannins, fixed oils, and proteins.The percentage yield of aqueous extract of *Celastrus paniculatus seeds* was observed as 13.7%

3.2 Body Weight

The body weight of the disease control group decreased (131.7 gm), whereas those of CPSE 400 mg (177.2 gm) and CPSE 800 mg (184.5 gm) showed an increase. The results of test drug are comparable to standard as Glibenclamide (192.0 gm) also showed an increase in body weight on day 21 (See Figure 1).

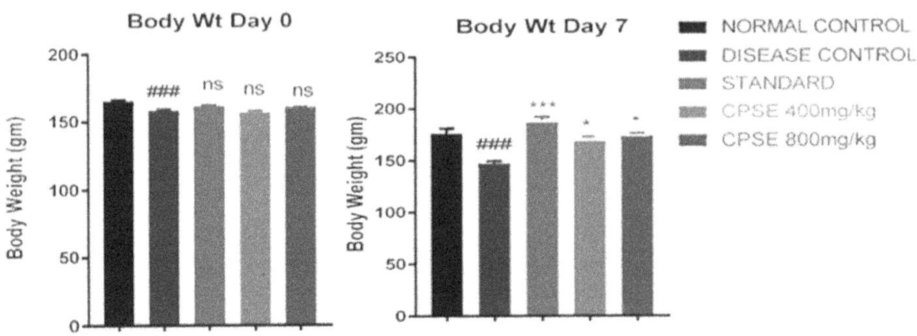

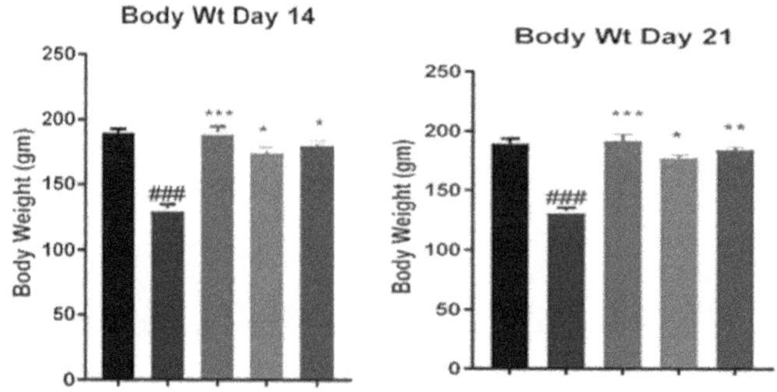

Figure 1 Effect of CPSE on Body Weight of experimental rats. Each bar in the graph indicate mean ± SEM (n=6) of the respective groups. Significant difference at ***P < 0.001, **P<0.01, *P<0.05 when compared with disease control group; ###P<0.001, when compared with normal control group; and ns-Not significant.

3.3 Hypoglycaemic Study of a single dose of CPSE in normal rats.

When a single dose is administered in the groups Glibenclamide showed an evident fall in blood glucose levels at 30 min (54.75 mg/dL). The CPSE, at 1600 mg/kg dose also showed low blood glucose levels that were comparable to the standard (60.0 mg/dL) (See Figure 2).

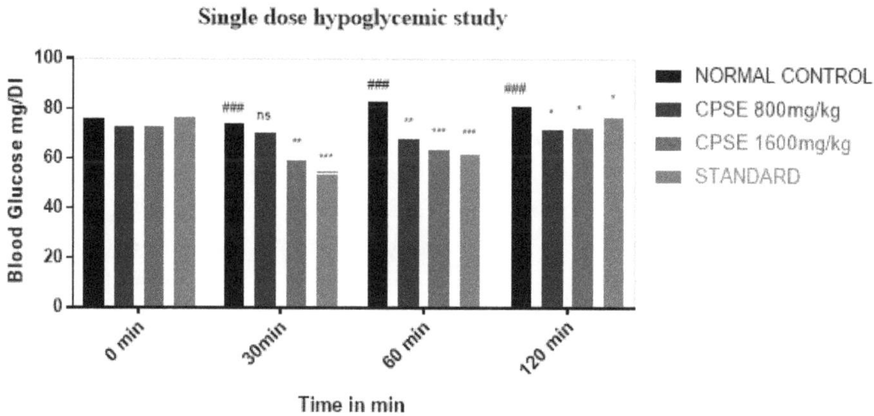

Figure 2 Effect of Single dose of CPSE on normal rats at 0 min, 30 min, 60 min and 120 min. Each bar in the graph indicate mean ± SEM (n=6). Significant difference at ***P < 0.001, **P<0.01,*P<0.05 when compared with disease control group; ###P<0.001, when compared with normal control group; and ns-Not significant.

3.4 Oral Glucose tolerance test

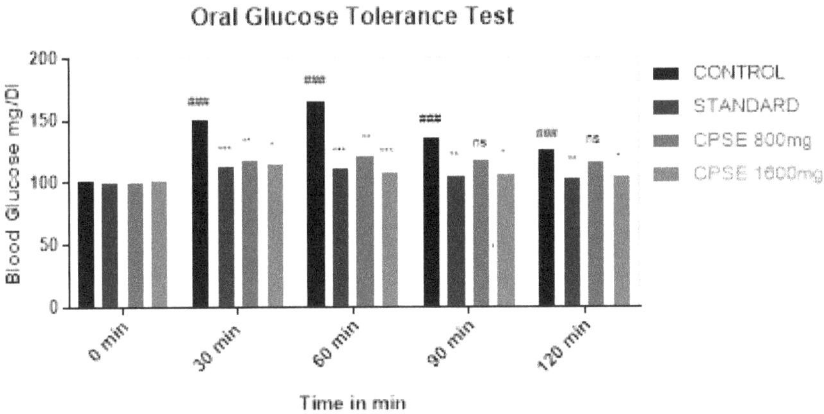

Figure 3 Effect of Single dose of CPSE on glucose loaded rats at 0 min, 30 min, 60 min, and 120 min. Each bar in the graph indicate mean ± SEM (n=6). Significant difference at ***P < 0.001, **P<0.01,*P<0.05 when compared with disease control group; ###P<0.001, when compared with normal control group; and ns-Not significant.

The control group shows a maximum glucose level at 60 min, (165.5 mg/dL). Glibenclamide administered group showed a value of 111.0 mg/dL. The CPSE 400 mg and CPSE 800 mg also showed values 121.8 mg/dL and 108.5 mg/dL, respectively suggesting that CPSE may have hypoglycaemic action (See Figure 3).

3.5 Blood Glucose

The disease control group showed the highest blood glucose level of 346.7 mg/dL and the Glibenclamide 5 mg/kg treated group showed a lowest blood glucose level of 102.8 mg/dL on day 21. There was a decrease in glucose levels of CPSE 400 mg/kg and CPSE 800 mg/kg treated groups showed 144.8 mg/dL and 123.7 mg/dL. The normal control group showed no significant change in blood glucose levels (Shown in Figure 4).

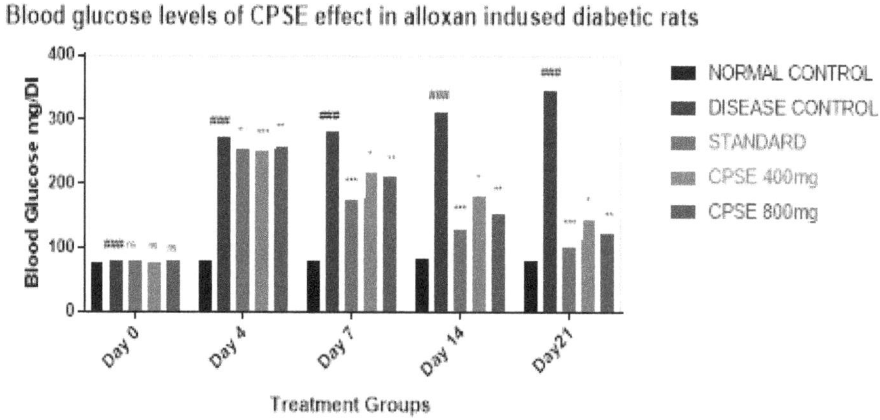

Figure 4 Effect of CPSE on Blood glucose of experimental rats, Day0, Day4, Day7, Day14 and Day21. Each bar in the graph indicate mean ± SEM (n=6). Significant difference ***P < 0.001, **P<0.01,*P<0.05 when compared with disease control group. ###P<0.001, when compared with normal control group.ns-Not significant

3.6 Histopathology Studies

The tissue sections of the pancreas of the rats from the normal control group showed normal size and shape of pancreatic islet cells. Glibenclamide treated group showed a slight degree of hypertrophy of beta cells. Rats treated with CPSE 400 mg/kg, showed pancreas with considerable hyperplasia of beta cells. Samples from CPSE 800 mg/kg showed no apoptosis or necrosis and normal acinar cells whereas samples from the disease control group showed a high degree of hypertrophy and hyperplasia of pancreatic cells (See Figure 5).

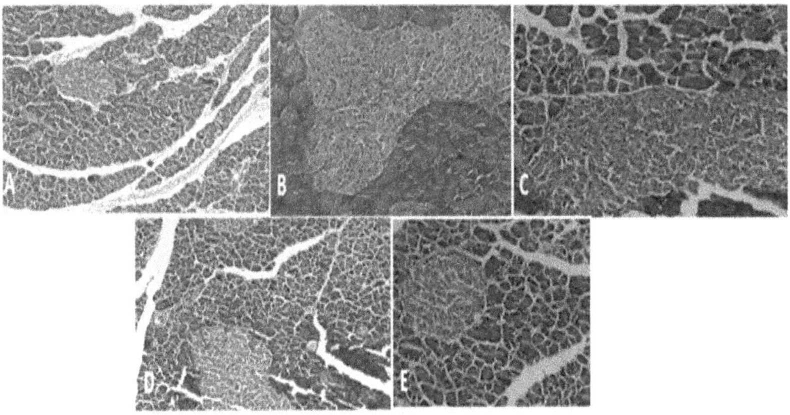

Figure 5 Histopathological study of pancreas depicting A) Normal beta cells, size and shape of islets appear normal in normal control group, B) Severe hypertrophy of islets and hyperplasia of beta cells in disease control group, C) Moderate hypertrophy of islets and hyperplasia of beta cells in standard control group. D) Moderate hypertrophy of islets and hyperplasia of beta cells in CPSE 400 mg/kg, and E) Normal acinar cells and beta cells, no apoptosis and necrosis in CPSE 800 mg/kg.

4 DISCUSSION

As diabetes is increasing with a faster pace it has become a major public health concern worldwide. Controlling not only diabetes but also its lifelong microvascular and macrovascular complications is a challenge. Use of herbal medicines is one of the approaches to treat diabetes and its complications. The conventional treatments may give dissatisfactory effect and herbal treatment may show better glycaemic control, availability and affordability.[14]

In this study, CPSE was taken to show its hypoglycaemic effect. It was used in hypoglycemic study of a single dose of CPSE in normal rats, shows oral glucose tolerance test (OGTT) and used to treat Alloxan induced diabetic rats. The preliminary phytochemical screening showed the presence of alkaloids, carbohydrates, flavonoids, tannins, fixed oils, and proteins.

A single-dose administration of CPSE 800 mg/kg and 1600 mg/kg in normal rats show a significant fall in blood glucose levels that are directly proportional to the dose of extract. A maximum reduction in blood glucose levels was observed in one hour for 800 mg/kg and 30 min for 1600 mg/kg and for Glibenclamide (0.5 mg/kg). CPSE 800 mg/kg showed 68± 3.34 mg/dL in one hour and CPSE 1600 mg/Kg showed 60±2.08 mg/dL while that of Glibenclamide is 54.75±2.28 mg/dL in 30 min.These results were in accordance with the study conducted by Shah, N.A. and Khan, M.R.[13] Oral glucose tolerance test (OGTT) consists of prior glucose loading with 50% glucose (2 gm/kg) to experimental rats. The control group showed a rise in blood glucose level of 165.5±4.6 mg/dL in one hour. Glibenclamide values are 111 ±3.49 mg/dL, CPSE 1600 mg/Kg showing 121.8±1.49 mg/dL and CPSE 800 mg/kg showing 108.5±4.49 mg/dL at one hour. The glucose clearing action of the extract was as good as that of research done by Shalini Adiga et al.[11]

Alloxan induced diabetic rats have shown a significant reduction in the weight of rats. The treatment of these animals with Glibenclamide caused restoration of weight loss. CPSE extract also showed prominent control of weight loss that is comparable with the values of the standard group. The body weight of Alloxan induced disease control group on 21st day was 131.7±2.26 gm. Upon treatment with Glibenclamide, the values were reduced to 192±2.74 gm. The body weight of the CPSE 400 mg/Kg administered group was 177.2±1.49 gm and that of the CPSE 800 mg/Kg treated group was 184.5±1.58 gm. These values were in line with the study conducted by Nora A. AlFaris et al.[15]

Glycated hemoglobin is a diagnostic test to check the blood glucose levels over a period of a few months. Alloxan induced disease group showed 6.98±0.04 % treatment with CPSE 400 mg/Kg and 800 mg/Kg showed 6.32±0.09 % and 6.1±0.13 % values, respectively. The standard group showed 5.760.11 %. These results are in accordance with G. Rajiv Gandhi et. al.[16] Diabetes induction with Alloxan 150 mg/kg in cold 0.9 % normal saline, caused hyperglycemia in the disease control group. Oral administration of Glibenclamide 5 mg/Kg and CPSE doses during the study period of 21 days lowered the values of blood glucose level. The Alloxan induced diabetic control group showed 346.7±3.89 mg/dL on 21st day of study. The CPSE treated groups of 400 mg/Kg and 800 mg/Kg also showed a marked reduction in blood glucose values of 144.8±2.89 mg/dL and 123.7±1.77 mg/dL on 21st day of study. The results are close to the glucose levels of the standard drug that has a value of 102.8±2.65 mg/dL. Such outcomes are in line with the study conducted by R Ghosh.[17]

In the 21 day study, the normal group showed no change in blood glucose level whereas the CPSE and Glibenclamide groups exhibited a notable decrease in blood glucose values; this asserts the antidiabetic action of CPSE. More distinct antidiabetic action was seen at the higher dose of CPSE i.e. 800 mg/Kg. One of the implications followed by diabetes is increased total cholesterol levels and triglycerides. This may be contributed to the fact that insulin deficiency, a reason behind metabolic disturbances, causes a build-up of lipids and augmentation of cholesterol and triglycerides. In the experimental study, the effect of CPSE on total cholesterol levels on 21 days, on Alloxan induced diabetic rats, the total cholesterol levels of the disease control group was 154±2.3 mg/dL. Glibenclamide treated group showed low values of total cholesterol i.e. 106.33±0.95 mg/dL. CPSE treated rats with 400 mg/Kg and 800 mg/Kg also showed such values on lower side i.e. 143.2±1.13 mg/dL and 125.8±1.44 mg/dL, respectively.

Similarly, triglycerides and LDL values of the disease control group increased, whereas that of the standard group decreased. The action of CPSE on triglycerides and LDL has similar actions to standard and showed satisfactory results. On the other hand, values of HDL of disease group decrease and standard drug show high levels. The result of CPSE treated rats is complementary to the values of the standard drug. This is in accordance with the study of Shaik Abdul Nabi.[18] Hence, CPSE also showed an anti-hyperlipidaemia action. Further, The histopathological studies showed that necrosis of beta cells in the disease control group and on treatment with CPSE 400 mg/Kg and 800 mg/Kg showed moderate hypertrophy and normal beta cell with no apoptosis.

5 CONCLUSION

The aqueous extract of seeds of *Celastrus paniculatus* was chosen for the study of antidiabetic action on Alloxan induced diabetic rats. The administration of CPSE to experimental rats showed hypoglycemic action when one dose of CPSE is administered to normal and fasting rats. Additionally, a good glucose clearing capability is seen in the Oral glucose Tolerance Test. CPSE further showed a gain in body weight of rats in comparison to diabetic rats that show weight loss. Blood glucose levels over the study time period of twenty one days were also reduced in CPSE administered groups. Total cholesterol, Triglycerides, and LDL levels were augmented and HDL levels are lessened. This is opposed to the values shown by diabetic rats. Histopathological studies conjointly indicate protecting results of CPSE on pancreatic cells. *Celastrus paniculatus* has shown antidiabetic medicinal action, yet its mechanism is not known. Further study may be encouraged that will add to our knowledge and facilitate diabetes treatment.

6 REFERENCES

1. J. Khawandanah, *Nutrition & diabetes*, 2019, **9**, 33.
2. M. Z. Banday, A. S. Sameer and S. Nissar, *Avicenna journal of medicine*, 2020, **10**, 174-188.
3. D. Patel, R. Kumar, D. Laloo and S. Hemalatha, *Asian Pacific Journal of Tropical Disease*, 2012, **2**, 239-250.
4. S. B Gaikwad, G. Krishna Mohan and M. Sandhya Rani, *Pharmaceutical Crops*, 2014, **5**.
5. R. Sharma and V. Arya, *Journal of Chemical and Pharmaceutical Research*, 2011, **3**, 204-212.
6. A. N. Singab, F. S. Youssef and M. L. Ashour, *Med Aromat Plants*, 2014, **3**, 2167-0412.
7. B. Salehi, A. Ata, N. V. Anil Kumar, F. Sharopov, K. Ramirez-Alarcon, A. Ruiz-Ortega, S. Abdulmajid Ayatollahi, P. Valere Tsouh Fokou, F. Kobarfard and Z. Amiruddin Zakaria, *Biomolecules*, 2019, **9**, 551.
8. A. Ghorbani, *Brazilian Journal of Pharmaceutical Sciences*, 2013, **49**, 413-422.
9. K. Deodhar and N. Shinde, *Ind J of Adv in Plant Research*, 2015, **2**, 18-21.
10. V. Dwivedi and H. Maurya, *Indian Journal of Pharmaceutical and Biological Research*, 2018, **6**, 37-42.
11. S. Adiga, K. Bairy, A. Meharban and I. Punita, *International Journal of diabetes in Developing countries*, 2010, **30**, 38.
12. H. V. Doan, S. Riyajan, R. Iyara and N. Chudapongse, *BMC Complementary and Alternative Medicine*, 2018, **18**, 1-10.
13. N. A. Shah and M. R. Khan, *BioMed research international*, 2014, **2014**.
14. H. Choudhury, M. Pandey, C. K. Hua, C. S. Mun, J. K. Jing, L. Kong, L. Y. Ern, N. A. Ashraf, S. W. Kit and T. S. Yee, *Journal of traditional and complementary medicine*, 2018, **8**, 361-376.
15. N. A. AlFaris, G. M. Alshammari, M. M. Alsayadi, M. A. AlFaris and M. A. Yahya, *Saudi Journal of Biological Sciences*, 2020, **27**, 925-934.
16. G. R. Gandhi and P. Sasikumar, *Asian Pacific journal of tropical biomedicine*, 2012, **2**, 281-286.
17. R. Ghosh, I. Dhande, V. Kakade, R. Vohra, V. Kadam and S. Mehra, *Int J Pharmacol*, 2009, **6**, 25-30.
18. S. A. Nabi, R. B. Kasetti, S. Sirasanagandla, T. K. Tilak, M. V. J. Kumar and C. A. Rao, *BMC complementary and alternative medicine*, 2013, **13**, 1-9.

IN-VITRO ANTI-INFLAMMATORY, MITOTIC, OXIDANT ACTIVITIES OF POLYHERBAL EXTRACT OF *LAURUS NOBILIS, ILLICIUM VERUM, AND MYRISTICA DACTYLOID*

Jahasultana Mohammed, Risy Namratha Jamullamudi*, Niharika P, Sathish kumar Konidala

Department of Pharmacy, Koneru Lakshmaiah Education Foundation, Vaddeswaram, AP, India.

*Corresponding email address: r.namratha747@gmail.com

1 INTRODUCTION

Plant sources are a valuable gift to humans in this era. Nowadays, in the drug design and development process, the framework was also surrounded by Natural chemistry and it is considered as the backbone for the growth of the pharmaceutical industry. Some natural products tend to have different biological impacts as they produce secondary metabolites which act as a starting material in drug discovery.[1] These secondary metabolites may also have biological activities like anti-inflammatory, antiviral, antiprotozoal, immunomodulatory, antimalarial, antitumour, antipyretic, and antimicrobial activities.[2] *Laurus nobilis, Illicium verum, Myristica dactyloids* are such plants which are known for their pharmacological activities.[3-11] *Laurus nobilis* (*Lauraceae*), commonly known as Bay, Sweet Bay, True Laurel, or Roman Laurel, is a widely herb spread in the Mediterranean area and Europe. The leaves of this plant are used as a folk medicine in the treatment of arthritis, rheumatic pains, skin inflammation, and asthma. *Illicium verum Hook*. f. (*Illiciaceae*) is an aromatic evergreen tree bearing purple-red flowers and anise-scented star-shaped fruit. It grows almost exclusively in southern China and Vietnam. Its fruit (star anise) is an important traditional Chinese medicine as well as a commonly used spice. The characteristically shaped fruit is listed in Chinese Pharmacopoeia (2010 edition) and has been applied as a traditional Chinese medicine to treat vomiting, stomach aches, insomnia, skin inflammation, and rheumatic pain. *Myristica dactyloides* belong to the *Myristicaceae* family. *Myristica dactyloides* fruits of the plant contain many volatile oil compounds. These oils are used to treat diseases like flatulency, diarrhoea, nausea, vomiting, chronic bowel complaints, spermatorrhoea, impotency, amenorrhoea, menorrhagia, dysmenorrhea, ulcers, splenic disorders, rheumatism, asthma, colic, flatulence, and dyspepsia.

The dried parts of these plants like a leaf (*Laurus nobilis*), fruit (*Illicium verum*), and flower (*Myristica dactyloids*) are well known for their utility as spices in food preparation.[12-15] The complete study of these natural products involves various isolation methods hydro-distillation, Soxhlet extraction, chromatographic procedures in a pure state, and investigation of their structure and function in live organisms. This study was intended to investigate the pharmacological effects of PHE derived from these plants parts which includes, anti-inflammatory, oxidant, mitotic and microbial activities.

2 MATERIALS AND METHODS

2.1 Materials

The dried parts of *Laurus nobilis, Illicium verum, and Myristica dactyloids* were collected. The green gram seeds were collected from the local market in Vijayawada, Andhra Pradesh, India. All the chemicals were obtained from the local vendors and are of analytical grade.

2.2 Preparation of polyherbal extract (PHE)

The dried leafs of *Laurus nobilis*, fruits of *Illicium verum*), and flowers of *Myristica dactyloids* were grounded separately to make coarse powders. These three were mixed in equal quantities and kept for maceration in methanol for seven days. The brown-coloured supernatant liquid was filtered and condensed with a minimum heat to yield a glassy brown PHE.

2.3 Preliminary phytochemical screening

The PHE was tested for identification of phytochemicals like alkaloids, carbohydrates, flavonoids, tannins, triterpenes, and steroids according to the standard procedures.[15]

2.4 *In vitro* studies

2.4.1 Anti-inflammatory activity. The crude PHE was screened for anti-inflammatory activity by *in vitro* method using the protein denaturation technique. Anti-inflammatory activity was determined by the percentage of albumin denaturation inhibition.[15]

Stock solutions of 1000 μg mL^{-1} were prepared for PHE and the reference drug, Ibuprofen. From the stock solutions, different concentrations were prepared (100 μg mL^{-1}, 10 μg mL^{-1}, 1 μg mL^{-1}, and 0.1 μg mL^{-1}) for the extract and Ibuprofen. To these dilutions, 2.8 mL of phosphate-buffered saline (pH

6.4) was mixed with 2 mL of egg albumin collected from a fresh hen's egg. The reaction mixture was incubated at room temperature (27±1°C) for 15 min. After incubation, the reaction mixture was kept at 70°C in a water bath for 10 min. Without a test sample, the blank was also prepared. After cooling, absorbances of all dilutions were measured by a spectrophotometer at 660 nm. The % inhibition was determined by using the below formula.

$$\% \text{ Inhibition} = \frac{\text{controls(Abs)} - test\ \text{(Abs)}}{\text{controls(Abs)}}\ x\ 100 \tag{1}$$

2.4.2 Antimitotic activity. The crude extract was screened for antimitotic activity by using Mung beans. The equal quantities of beans for the test, standard (Methotrexate) and blank were chosen and taken in beakers. The extract and standard were dissolved in water and are taken in separate beakers. In another beaker, only water was taken. The separated beans are weighed and noted. The weighted beans are then transferred to the appropriate beakers and soaked overnight. Later, the water was drained and then each set of beans was tied in muslin cloth. After 24 hr to 48 hr, beans were observed for germination in all three sets for determining antimitotic activity.

*2.4.3 Antioxidant activity***.** The *in vitro* DPPH scavenging activity screening method was used to test the antioxidant activity of the extract. The extract (0.1 μg mL^{-1}, 1 μg mL^{-1}, 10 μg mL^{-1}, 100 μg mL^{-1}) and standard compound (Ascorbic acid) ability to neutralize free radicals was determined by treating with a stable DPPH radical in an ethanol solution. 0.5 mL of extract, 3 mL of absolute ethanol, and 0.3 mL of 0.5 mM DPPH radical solution in ethanol were the the components of th reaction mixture. After 30 minutes of reaction with DPPH, the absorbance was measured at 517 nm using a UV–VIS spectrophotometer for all the samples. The blank solution was made by mixing ethanol (3.3 mL) and extract (0.5 mL), while the control solution was made by mixing ethanol (3.5 mL) and DPPH radical solution (0.3 mL).[16] Percentage antioxidant (AA) was calculated with the help of below mentioned formula

$$\% \text{ AA} = \frac{100 - \text{ Abs sample-Abs blank}}{\text{Control (Abs)}}\ x\ 100 \tag{2}$$

*2.4.4 Antimicrobial activity***.** Using the Agar disc plate technique, the crude extract was tested for antibacterial activity.[6] Nutrient Agar medium and Sabouraud's Dextrose Agar (SDA) medium were utilized for antibacterial and antifungal activities, respectively.

Test microorganisms

Gram-positive bacteria- *Staphylococcus aureus* (NCIM 2122), gram-negative bacteria-*Escherichia coli* (NCIM 2137), fungal strains-*Candida Albicans*

(NCIM 3102), and *Aspergillus niger* (NCIM 652) were the microorganisms used.

Antibacterial activity

The bacterial suspension was aseptically distributed on a solid plate with a sterile swab and allowed to dry. The extract residue-loaded discs were gently pressed on top of the seeded medium to ensure contact. As a pre-diffusion time, plates were held at room temperature for 30 minutes. Ampicillin (10 μg mL^{-1}) and blank discs soaked with the relevant solvents were the positive and negative controls, respectively. At 37°C, the plates were incubated for 24 hours. The inhibition zones that formed on the medium were measured in millimetres, and each test was repeated three times.

Antifungal activity

The solid plate was soaked with a sterile swab and allowed to dry after being moistened by the fungal suspension. The plant extract disc was placed on the seeded plate aseptically, and the attachment was confirmed. For compound diffusion, the plates were maintained at room temperature for 30 minutes. The negative control was a blank disc impregnated with each solvent and dried off, whereas the positive control was conventional anti-fungal disc Ketoconazole (10 μg mL^{-1}). After 72 hours of incubation at 37°C, the activity was measured. The diameter of the extract-induced inhibitory zone was measured in millimetres. The antifungal activity of each sample was estimated from three different observations.

3 RESULTS AND DISCUSSION

3.1 Preliminary phytochemical screening

The preliminary phytochemical screening was performed for PHE and the evaluated results are tabulated in Table 1.

Table 1 *Phytochemical constituents of methanolic extract.*

S. No.	Chemical Constituents	Methanolic extract
1.	Alkaloids	+
2.	Glycosides	-
3.	Flavonoids	+
4.	Tannins	-
5.	Triterpenes	-
6.	Carbohydrates	+
7.	Steroids	-

Where, + indicate presence of respective phytochemical constituent, and – indicate absence of the same.

3.2 Anti-inflammatory activity

The percentage of albumin denaturation inhibition increased with the decrease in absorption means an increase in anti-inflammatory activity. In the present study, the extract had shown dose dependant anti-inflammatory activity which was comparable to that of the standard, Ibuprofen at 0.1 µg mL^{-1}. The results are included in Table 2.

Table 2 *% Inhibition of protein denaturation for PHE and Ibuprofen.*

Conc. of dilutions (µg mL^{-1})	**% Inhibition of protein denaturation**	
	PHE	Ibuprofen
0.1	2.06	7.92
1	4.13	9.90
10	4.82	11.88
100	12.87	13.79
Control	-	-

3.3 Antioxidant activity

From the stock solution (1 mg mL^{-1}) of PHE and standard, various concentrations were prepared and screened for antioxidant activity by DPPH scavenging activity. Results showed that the percentage of scavenging activity increased with an increase in the concentration. The PHE had shown antioxidant activity similar to that of the standard at 100 µg mL^{-1}. The results were Tabulated and included in Table 3.

Table 3 *Antioxidant activity using DPPH against methanolic PHE.*

Conc. of dilutions (µg mL^{-1})	**% Scavenging activity**
0.1	-
1	-
10	16.77±0.35
100	79.25±0.57
Ascorbic acid	81.58±1.63

3.4 Anti-mitotic activity

The germination of beans experiments indicate whether cell division had taken place or not. The observations in the current study, showed antimitotic activity of the PHE. The beans treated with methanolic extract did not produce any sprouts after allowed for germination. This effect was similar to that that of the standard drug. Whereas, the beans subjected to blank had shown germination. This indicates that there is no solvent effect on beans.

3.5 In-vitro antimicrobial activity

Methanolic extract with different concentrations were made and subjected to antibacterial and antifungal activity by disc diffusion method.

Table 4 *Antimicrobial activity of Methanolic extract.*

Concentration (μg mL^{-1})	**Zone of inhibition (mm)**			
	S. aureus	***E. coli***	***C. albicans***	***A. niger***
0.1	-	-	-	-
1	-	-	-	-
10	-	-	-	-
100	19±0.58	18±0.34	15±0.26	18±0.55
Ampicillin (10 μg mL^{-1})	33±0.18	35±0.30	-	-
Ketoconazole (10 μg mL^{-1})	-	-	24±0.29	26±0.32

Ampicillin was used as a standard in antibacterial study against *S. aureus* and *E. coli* and Ketoconazole was used as a standard in antifungal study against *C. albicans* and *A. niger*. The PHE at 100 μg mL^{-1} concentration depicted moderate antimicrobial action which was comparable to the respective standards. The results are presented in Table 4.

4 CONCLUSION

The three plants are well known for their various biological activities individually. The present work was done to identify the synergistic/antagonistic effect of the methanolic PHE (*Laurus nobilis, Illicium verum,* and *Myristica dactyloids)*. The PHE had shown promising synergistic effect at 100 μg mL^{-1} for anti-inflammatory and antioxidant activities. Furthermore, exploratory studies are suggested for the analysis of detailed chemical composition of extracts from various solvents. Studies involving other *in vivo* tests are needed in order to draw firm conclusion regarding the biological role of PHE.

5 ACKNOWLEDGEMENTS

The authors thank the Management of KL College of Pharmacy, Vaddeswaram, Guntur (Dt.), Andhra Pradesh for providing the research facilities.

6 CONFLICT OF INTEREST

The authors declare no conflict of interest.

7 REFERENCES

1. M. J. Balunas and A. D. Kinghorn, *Life sciences*, 2005, **78**, 431-441.
2. C. Jain, S. Khatana and R. Vijayvergia, *Int. J. Pharm. Sci. Res*, 2019, **10**, 494-504.
3. S. Dall'Acqua, R. Cervellati, E. Speroni, S. Costa, M. C. Guerra, L. Stella, E. Greco and G. Innocenti, *Journal of medicinal food*, 2009, **12**, 869-876.
4. I. Mssillou, A. Agour, A. El Ghouizi, N. Hamamouch, B. Lyoussi and E. Derwich, *Journal of Food Quality*, 2020, **2020**, 1-8.
5. S. Karakaya, S. N. El, N. Karagözlü and S. Şahin, *Journal of medicinal food*, 2011, **14**, 645-652.
6. H. TUSEEF, M. L. RAZA and T. ASSAD, *Walailak Journal of Science and Technology (WJST)*, 2021, **18**, 9456 (9412 pages)-9456 (9412 pages).
7. H. N. Khan, S. Rasheed, M. I. Choudhary, N. Ahmed and A. Adem, *BMC Complementary Medicine and Therapies*, 2022, **22**, 1-15.
8. R. S. Habeballa, E. I. Ahmedani, N. S. Awad and M. A. Abdein, *Med. Sci*, 2020, **24**, 3469-3480.
9. W. Guo-Wei, H. Wen-Ting, H. Bao-Kang and Q. Lu-Ping, *Journal of Ethnopharmacology*, 2011, **136**, 10-20.
10. F. Sofa, D. Akhmad, Megawati, H. Sumi, K. Soleh and H. Muhammad, *Phytochemistry Letters*, 2017, **20**, 36-39.
11. D. K. a. P. R. Bhat, *int. journal of Phar. Sci. and Res.*, 2020, **11**, 6413-6424.
12. H. A. B. Shagufta Khan*, Ghulam Abbas, Muhammad Ali Versiani, Shaheen Faizi, Salman Khalid Ahmed, Saad Jawaid, Anita Naushir Akbar Ali, Rabia Jamy, Fatima Shahid and Ahsana Dar Farooq, *Letters in Drug Design & Discovery*, 2018, **15**, 678-686.
13. E. Ovidi, V. Laghezza Masci, M. Zambelli, A. Tiezzi, S. Vitalini and S. Garzoli, *Plants*, 2021, **10**, 707.
14. S. Dowek, S. Fallah, R. Basheer-Salimia, M. Jazzar and A. Qawasmeh, *International Journal of Pharmacy and Pharmaceutical Sciences*, 2020, **12**, 12-16.
15. A. N. Tamfu, O. Ceylan, S. Kucukaydin, M. Ozturk, M. E. Duru and R. M. Dinica, *Foods*, 2020, **9**, 1768.
16. L. L. Mensor, F. S. Menezes, G. G. Leitão, A. S. Reis, T. C. d. Santos, C. S. Coube and S. G. Leitão, *Phytotherapy Research*, 2001, **15**, 127-130.

ETHANOLIC PLANT EXTRACTS AS EFFECTIVE ANTIUROLITHIATIC AGENTS AGAINST OXALATE INDUCED HYPEROXALURIA IN WISTAR RATS

Ch. Malathi Suvarna[1,2*], Subhakar Raju R[1], Narender Malothu[1], Chakravarthi Guntupalli[1]

*[1]Department of Pharmacy, Koneru Lakshmaiah Education Foundation, Vaddeswaram, AP, India;

[2]Department of Pharmacology, Gland Institute of Pharmaceutical Sciences, Shivampet, Narsapur, Telangana.

*Corresponding Author: suvarnasarma31@gmail.com

1 INTRODUCTION

Plants have always been an ample source of nutritional and medicinal products. *Lantana camara* is a large strong smelling evergreen shrub with opposite leaves, small flowers of varied colours and fruits with two cotyledons.[1] It has high medicinal value with constituents like Caryophyllene, Lantedene A, B, Lantinine, Lancamarone quinine and also rich polyphenols among many others thus it's leaves were selected for its efficacy against Urolithiasis.[2] Belonging to the family Malvaceae *T. pop* is a tropical, subtropical evergreen medium sized shrubby tree with yellow flowers and five ridged pods as fruits containing elliptical brown and hairy seeds leaves are glossy and heart shaped. Kaempferol and its glycosides, glucosides lie populneol, populin, populnatin, sitosterol, lupeol, sesquiterpenoidal quinines, thespone, mansonone C, D, E, F present in leaves, bark, seeds of *T. pop* are responsible for activities such as astringent, antidiarrhoeal, hepatoprotective, diuretic activity, thus selected for this study.[3] Elevated urinary oxalate also known as Hyperoxaluria leads to the formation of kidney stones termed as Nephrolithiasis.[4,5] Patients with inflammatory bowel disease are more prone to nephrolithiasis. Treatment strategies include dietary restriction, of oxalate but it was found to be not always entirely effective.[6,7] Present study aims the evaluation of Anti-urolithiatic activity of Ethanolic extracts of *L. cam* and *T. pop* on Oxalate induced Hyperoxaluria through *in-vitro* models.

2 METHODS

2.1 Plant collection and extraction

L. cam leaves are collected from campus area of Gland Institute of Pharmaceutical Sciences, Kothapet, Medak district, Telangana. ***T. pop*** leaves are collected from loamy soil of Hyderabad, Telangana. Plants are cleaned, leaves are separated, washed, shade dried. Dried leaves are powdered to attain fine powder which is extracted through soxhlation using Ethanol as solvent. Vacuum dried extract was stored in properly labelled container.

2.2 Screening for Oxalate induced hyperoxaluria and nephrolithiatis

Four groups of rats of wistar strain are taken each group containing six animals. The rats are feed ad. libitum for a week, and acclimatized to the laboratory conditions. Initially blood is collected from rats through Retro orbital puncture and the serum is collected by centrifuging the blood samples at 600 rpm. The initial urine samples of rats are collected. The initial Calcium, oxalate & creatinine levels in the serum and urine samples are estimated. After blood collection the animals are given rest for two days. 1st group is taken as blank and the rest are treated with 0.5% of calcium (calcium carbonate) and 1.5% of oxalate (oxalic acid) for a week, for inducing calcium oxalate crystals. 2nd group is taken as control.[8] 3rd group of animals were treated with 400 mg/kg b.wt of *L. cam* Ethanolic leaf extract. 4th group of animals were treated with 400 mg/kg b.wt of *T. pop* Ethanolic leaf extract. Throughout the duration of study samples as required were collected and tested accordingly. After the completion of the treatment urine is collected, blood is collected by cardiac puncture was centrifuged for serum. The final calcium, oxalate & creatinine levels in serum and urine are estimated. These are compared with control group animals.[9, 10]

3. RESULTS

After extraction 23.5 g of *L. cam* ethanolic extract and 37 g of *T. pop* Ethanolic extract were obtained.

3.1. Urinary Calcium

It is observed that the urinary calcium levels increased after the oxalate administration in all group when compared with normal especially in control (1.2483). In group administered with *L. cam* extract the 56th day (0.95) reading shows that the extract was successful in inhibiting hyperoxaluria as shown in Table 1. In group administered with *T. pop* extract the 56th day reading (0.6833) shows that the extract was successful in inhibiting hyperoxaluria. Anti urolithiatic effect of group 4 is more than that of group 3.

Table 1 *Effect of Ethanolic extracts on Urinary Calcium (mg/dl).*

Groups/ Days	0th	28th	42nd	56th
Group 1	0.7833±0.18**	0.7433±0.51**	0.755±2.36**	0.76±0.05ns
Group 2	0.7766±0.08**	1.1466±0.35**	1.2183±1.25***	1.2483±0.36**
Group 3	0.7466±0.75**	0.9716±0.48***	0.9866±1.03**	0.95±0.6**
Group 4	0.7066±0.17**	0.95333±0.706**	0.8216±0.256***	0.6833±0.3**

The values are expressed as Mean ± SEM (n=6), Group 1 vs 3, group 1 vs 4, group 2 vs 3, group 2 vs 4, group 3 vs 4 were compared. ANOVA was followed by Dunnet' t test for statistical comparisons, p values *** p<0.001 **p<0.01 *p<0.05, ns non significant.

3.2. Urinary oxalate

It is observed that the urinary oxalate levels increased after the oxalate administration in all group when compared with normal especially in control (7.68). In group administered with *L. cam* extract the 56th day (6) reading shows that the extract was successful in inhibiting hyperoxaluria. In group administered with *T. pop* extract the 56th day reading (0.942) shows that the extract was successful in inhibiting hyperoxaluria. Anti urolithiatic effect of group 4 is more than group 3 as can be observed from Table 2.

Table 2 *Effect of Ethanolic extracts on Urinary Oxalate (mg/dl).*

Groups/ Days	0th	28th	42nd	56th
Group 1	0.762±0.28**	0.778±0.03**	0.766±0.04**	0.77±0.1**
Group 2	0.758±0.18***	6.92±0.12**	7.62±0.108***	7.68±0.43***
Group 3	0.838±0.06**	6.6±0.028**	6.8±0.328**	6±0.35**
Group 4	0.818±0.73**	4.94±0.18***	2.82±0.221***	0.942±0.1**

The values are expressed as Mean ± SEM (n=6), Group 1 vs 3, group 1 vs 4, group 2 vs 3, group 2 vs 4, group 3 vs 4 were compared. ANOVA was followed by Dunnet' t test for statistical comparisons, p values *** p<0.001 **p<0.01 *p<0.05, ns non significant.

3.3. Urinary Albumin

It is observed that the urinary albumin levels increased after the oxalate administration in all group when compared with normal especially in control (1.674). In group administered with *L. cam* extract the 56th day (1.208) reading shows that the extract was successful in inhibiting hyperoxaluria. In group administered with *T. pop* extract the 56th day reading (0.352) shows that the extract was successful in inhibiting hyperoxaluria as mentioned in Table 3. Anti urolithiatic effect of group 4 is more than that of group 3.

Table 3 *Effect of Ethanolic extracts on Urinary Albumin (mg/dl).*

Groups/ Days	**0th**	**28th**	**42nd**	**56th**
Group 1	0.284±0.30**	0.282±0.19**	0.272±0.08**	0.264±0.122**
Group 2	0.268±0.28***	1.44±0.753**	1.55±0.785***	1.674±0.28**
Group 3	0.286±0.15**	1.168±0.36**	1.094±0.26**	1.208±0.65**
Group 4	0.286±0.05**	0.598±0.13***	0.524±0.125**	0.352±0.08**

The values are expressed as Mean ± SEM (n=6), Group 1 vs 3, group 1 vs 4, group 2 vs 3, group 2 vs 4, group 3 vs 4 were compared. ANOVA was followed by Dunnet' t test for statistical comparisons, p values *** p<0.001 **p<0.01 *p<0.05, ns non significant.

3.4. Urinary pH

It is observed that the urinary pH decreased after the oxalate administration in all group when compared with normal especially in control (5.5666). In group administered with *L. cam* extract the 56th day (6.3333) reading shows that the extract was successful in inhibiting hyperoxaluria. In group administered with *T. pop* extract the 56th day reading (7.15) shows that the extract was successful in inhibiting hyperoxaluria by restoring the pH to almost normal as shown in Table 4. Anti urolithiatic effect of group 4 is more than that of group 3.

Groups/ Days	**0th**	**28th**	**42nd**	**56th**
Group1	7.1±0.61*	7.2±0.088**	7.1333±0.15**	7.0833±0.26**
Group2	7.1833±0.27**	5.6333±0.156**	5.5333±0.173**	5.5666±0.22***
Group3	7.15±0.043*	5.5833±0.013**	5.9333±0.6**	6.3333±0.2*
Group4	7.0666±0.61*	6.6166±0.06*	6.7333±0.256*	7.15±0.89NS

Table 4: Effect of Ethanolic extracts on Urinary P^H.

The values are expressed as Mean ± SEM (n=6), Group 1 vs 3, group 1 vs 4, group 2 vs 3, group 2 vs 4, group 3 vs 4 were compared. ANOVA was followed by Dunnet' t test for statistical comparisons, p values *** p<0.001 **p<0.01 *p<0.05, ns non significant.

3.5. Serum oxalate

It is observed that the serum oxalate levels increased after the oxalate administration in all group when compared with normal especially in control (24). In group administered with *L. cam* extract the 56th day (18.1666) reading shows that the extract was successful in inhibiting hyperoxaluria. In group administered with *T. pop* extract the 56th day reading (12.3) shows that the extract was successful in inhibiting hyperoxaluria. Anti urolithiatic effect of group 4 is more than that of group 3 from Table 5.

Table 5: Effect of Ethanolic extracts on Serum Oxalate (µm).

Groups/ Days	**0th**	**28th**	**42nd**	**56th**
Group 1	11.35±0.45*	11.2166±0.38*	11.3666±0.26*	11.5±0.02*
Group 2	11.7±0.86***	23.05±0.47**	23.65±0.14*	24±0.061**
Group 3	11.7l±0.25**	20.7833±0.23***	17.9166±0.69**	18.1666±0.246*
Group 4	11.75±0.36*	16.0833±0.90***	14.3333±0.72**	12.3±0.58*

The values are expressed as Mean ± SEM (n=6), Group 1 vs 3, group 1 vs 4, group 2 vs 3, group 2 vs 4, group 3 vs 4 were compared. ANOVA was followed by Dunnet' t test for statistical comparisons, p values *** p<0.001 **p<0.01 *p<0.05, ns non significant.

3.6. Serum creatinine:

It is observed that the serum creatinine levels increased after the oxalate administration in all group when compared with normal especially in control (6.42). In group administered with *L. cam* extract the 56th day (3.98) reading shows that the extract was successful in inhibiting hyperoxaluria. In group administered with *T. pop* extract the 56th day reading (0.778) shows that the extract was successful in inhibiting hyperoxaluria. Anti urolithiatic effect of group 4 is more than group 3 as shown in Table 6.

Table 6: Effect of Ethanolic extracts on Serum Creatinine (mg/dl).

Groups/ Days	**0th**	**28th**	**42nd**	**56th**
Group 1	0.63±0.55**	0.636±0.28**	0.68±0.032**	0.698±0.06*
Group 2	0.772±0.62*	2.74±0.156*	4.22±0.38**	6.42±0.29**
Group 3	0.612±0.33**	3.28±0.011**	24±0.22***	3.98±0.38**
Group 4	0.592±0.128**	2.28±0.31*	2.24±0.76**	0.778±0.045**

The values are expressed as Mean ± SEM (n=6), Group 1 vs 3, group 1 vs 4, group 2 vs 3, group 2 vs 4, group 3 vs 4 were compared. ANOVA was followed by Dunnet' t test for statistical comparisons, p values *** p<0.001 **p<0.01 *p<0.05, ns non significant.

3.7. Serum Uric Acid:

It is observed that the urinary calcium levels increased after the oxalate administration in all group when compared with normal especially in control (8.68). In group administered with *L. cam* extract the 56th day (05.54) reading shows that the extract was successful in inhibiting hyperoxaluria. In group administered with *T. pop* extract the 56th day reading (2.24) shows that the extract was successful in inhibiting hyperoxaluria. Anti urolithiatic effect of group 4 is more than that of group 3 can be observed from Table 7.

Table 7: Effect of Ethanolic extracts on Serum Uric Acid (mg/dl).

Groups	0 Day	28th day	42nd day	56th Day
Group 1	1.86±0.19**	2.06±0.38*	2.08±0.37**	2.1±0.97**
Group 2	1.88±0.203*	5.74±0.61**	7.16±0.302***	8.68±0.66**
Group 3	2.02±0.033*	5.62±0.95**	5.52±0.064**	5.54±0.82*
Group 4	2.04±0.57**	3.82±0.25**	3.08±0.58**	2.24±0.015***

The values are expressed as Mean ± SEM (n=6), Group 1 vs 3, group 1 vs 4, group 2 vs 3, group 2 vs 4, group 3 vs 4 were compared. ANOVA was followed by Dunnet' t test for statistical comparisons, p values *** p<0.001 **p<0.01 *p<0.05, ns non significant.

3.8. Blood Urea Nitrogen:

It is observed that the urinary calcium levels increased after the oxalate administration in all group when compared with normal especially in control (40.5). In group administered with *L. cam* extract the 56th day (15.2) reading shows that the extract was successful in inhibiting hyperoxaluria. In group administered with *T. pop* extract the 56th day reading (12.4) shows that the extract was successful in inhibiting hyperoxaluria as shown in Table 8. Anti urolithiatic effect of group 4 is more than that of group 3.

Table 8: Effect of Ethanolic extracts on Blood Urea Nitrogen (BUN- mg/dl).

Groups/ Days	**0th**	**28th**	**42nd**	**56th**
Group1	11.125±0.18**	11.75±0.26**	12.25±0.02*	14±0.37***
Group2	9.75±0.69**	28.5±0.125**	35±1.5**	40.5±0.48**
Group3	9.875±0.3*	22.75±0.38*	20.125±0.16*	15.2±0.82*
Group4	9.95±0.055*	20.75±0.018ns	14.875±0.96**	12.4±0.731ns

The values are expressed as Mean ± SEM (n=6), Group 1 vs 3, group 1 vs 4, group 2 vs 3, group 2 vs 4, group 3 vs 4 were compared. ANOVA was followed by Dunnet' t test for statistical comparisons, p values *** p<0.001 **p<0.01 *p<0.05, ns non significant.

3.9. Total Plasma protiens

It is observed that the quantity of total plasma protiens decreased after the oxalate administration in all group when compared with normal especially in control (2.1). In group administered with *L. cam* extract the 56th day (5.475) reading shows that the extract was successful in inhibiting hyperoxaluria. In group administered with *T. pop* extract the 56th day reading (6.2) shows that the extract was successful in inhibiting hyperoxaluria and can be referred from Table 9. Anti urolithiatic effect of group 4 is more than that of group 3.

Table 9: Effect of Ethanolic extracts on Total Plasma Proteins (mg/dl).

Groups	0 Day	28th day	42nd day	56th Day
Group 1	6.625±0.07**	6.775±0.56***	6.95±0.06**	7.35±0.29^{ns}
Group 2	6.525±0.81*	3.775±0.11**	2.775±0.84**	2.1±0.44***
Group 3	6.85±0.102**	5.25±0.02**	4.925±0.25**	5.475±0.63**
Group 4	6.825±0.35*	6.05±0.28*	6.4±0.131*	6.2±0.52**

The values are expressed as Mean ± SEM (n=6), Group 1 vs 3, group 1 vs 4, group 2 vs 3, group 2 vs 4, group 3 vs 4 were compared. ANOVA was followed by Dunnet' t test for statistical comparisons, p values *** p<0.001 **p<0.01 *p<0.05, ns non significant.

4. DISCUSSION

Our body has a mechanism of excreting all by products and end products of several metabolic reactions for homeostasis. Oxalate excretion is a part of such excretion through kidney failing or decrease in which leads to the formation of calcium oxalate crystals in kidneys. Oxalate is usually present in the diet with varying content thereby bioavailability. Small fraction (6-15%) of dietary oxalate absorbed makes dietary restriction less reliable for stone recurrence prevention there by creating an opening for alternate methods. Personalized medicine with probiotics is one such approach for patients with stone disease, specific agents targeting specific aspects of oxalate transport are administered which might contribute to hyperoxaluria and stone disease. The use of these plant extracts like *T. pop* and *L. cam* Ethanolic extracts have showed a significant effect on induced hyperoxaluria and nephrolithiasis. The study showed the groups administered with extracts showed considerable increase in urinary pH, calcium, oxalate, albumin levels. They also exhibited a high degree of decrease in serum levels of oxalate, creatinine, urea-nitrogen and plasma protein. These extracts being from natural resources are less likely to have higher adverse effects than that of the antibiotics which are cause of disturbance in the bioenvironment of gut where friendly bacteria which help by decreasing the oxalate concentration by metabolising the excess oxalate.

5. CONCLUSION

This study established that the ethanolic extracts of *L. cam* and *T. pop* 400mg/kg b.wt of had statistically significant effect in reducing hyperoxaluria created by inducing the rats with 0.5% of calcium (calcium carbonate) and 1.5% of oxalate (oxalic acid) out of both.

6. ACKNOWLEDGEMENTS

We would like to thank Dr. N. Siva Subramanian, Principal, Gland Institute of Pharmaceutical Sciences for his immense support for the completion of work.

7. REFERENCES

1. Venkatachalam, T., Kumar, V. K., Selvi, P. K., Maske, A. O., & Kumar, N. S. (2011). Int. J. Pharm. Sci., *3*(1), 52-54.
2. Mallappaumara Swamy, Uma Rani Sinniah & Mohd. Sayeed Akhtar (2015).Evid Based Complement Alternat med. doi: 10.1155/2015/506413. Epub 2015 Dec 9.
3. Sasikala, V., Radha, S. R., & Vijayakumari, B. (2013). . Pharm. Res., *6*(3), 378-382.
4. Suvarna, C., Sriya, P., Arshad, M., & Pavan, K. (2018). J. drug deliv. ther. *8*(4), 1-4. https://doi.org/10.22270/jddt.v8i4.1718
5. Adie Viljoen, Rabia Chaudhry & John Bycroft (2018). Ann. Clin. Biochem. 56(1):15-27. doi.org/10.1177/0004563218781672
6. Benoit Buysschaert, Selda Aydin, Johann Morelle, Valentine Gillion, Michel adoul, Nathalie Demoulin (2020). Kidney Int. Rep. (2020) 5, 1503–1509. doi.org/10.1016/j.ekir.2020.06.021
7. Huang Y, Zhang Y, H, Chi Z, P, Huang R, Huang H, Liu G, Zhang Y, Yang H, Lin J, Yang T, Cao S, Z. Urol Int (2020);104:167-176. doi: 10.1159/000504417
8. Holmes RP, Goodman HO, Assimos DG. Kidney Int. (2001); 5(1):270-6.
9. Bensatal A, Ouahrani MR. Urol Res.(2008); 36:283-287.
10. Atmani F, Khan SR. Br J Urol Int. (2000); 85:621-625.

A SYSTEMATIC APPROACH OF RESEARCH SCREENING MODELS OF NEURODEGENERATIVE DISEASE USING ZEBRAFISH

Kondeti. Madhu*, Dr.T.VinayKumar

Department of Pharmacology, Nirmala College of Pharmacy, Atmakuru-522503, Andhra Pradesh, India.

Email: kondeti.madhumounish@gmail.com

Phone no: 8106429298.

1. INTRODUCTION

Now a day's Alzheimer's disease is most common neurodegenerative disease. which effects the neuron in the brain that causes memory loss, anxiety, moodiness, depression and agitation. Currently NDD is swiftly in aged society, so there is fine allege all for the growth of neurodegenerative disorder. Zebrafish is considered as an emerging model due to its rapid development ,optical transparency genetic studies ,since the embryo form and grows externally following fertilization due to insight genetic control development offspring can be studied in detail.[3]

Alzheimer's disease is the most familiar motor disorder that causes the brain to reduce in size and brain cells to end due to abnormal protein, which forms plaques around the brain. The other protein is called tau which forms tangles within the brain cells. NDD has tortured millions of people world wide.[1] Among them 1 in 9 people age at 65 and older has AZ the percentage of people with AZ dementia increases with 5.0% of people ages 65 to 74.[2] Dementia is currently seventh leading cause of death among all disease the chief challenge in the learning of NDD antidote development are the multiplicity of analytic heterogeneity among human beings and in addition due to deficiency of biomarkers form it tough to identify the illness in premature phase by 2040[15] Alzheimer may be on top 2.

During starting zebrafish models these models used for identification of water contamination. Later In 1981 streisinger introduced the zebrafishes an effectual animal models mainly in genetics and others. First its small size and the maintenance is simple and breeding make ideal for high-rise through put showing for genotypes and drugs.[4] Second swift growth cycle. Third zebrafish consists of transparency embryos for optical manipulation. Zebrafish animal models is valuable to detail study on human disorder in pharmacology, developmental genetic makeup and further fields. There is a difference in organization structure of the central nervous system of zebrafish brain.[5] But there is a highsamenessin brain maturing between zebrafish and human and also shows many similarities in neurotransmitters like noradrenergic, serotonergic, dopaminergic and histaminergic systems. For example, D1 and D3 receptors in dopamine system shows complete amino acid homology in binding sites between zebrafish and human. Among these receptors D2 and D 4 receptors show 85-95% homology.[5] So, zebrafish model were used to unexpected study models.

Zebrafish is very useful tool for drug discovery in neurologic diseases. which is cost efficient high throughput screening with time saving. In this we discuss about pharmacology of AD zebrafish models.

2. BEHAVIORAL ASSESSMENT OF ZEBRAFISH IN NEURODEGENERATIVE DISEASE

Now a day's zebrafish models are used in impossible to possible neuroanatomical investigations. Due to its visible zebrafish behavior plays a vital role in disease model competence. Within 72 h post fertilization zebrafish embryos develop to larvae. Complete body develops with in 48hpf and go through metamorphosis and are prepubescent at 30days post implantation and grown up phase emerge from 90dpf.[8] Zebrafish having visual ,auditory and olfactory functions were more active during day time .Due to different advantages in larvae stage it is very beneficial to researchers for high through put screening in their behavior ,while adult zebrafishes avantgarde than larvae .experiments carried on zebrafish are very comfortable and facile in aqueous environment.[5,8] the bioavailability and harmful of the drugs can estimate in profuse steps.[9] Drug absorption cannot be eliminated when absorbed through any organ like skin and water .The study on the zebrafish models using larvae and adult gives a definite neurological improvements .

2.1 Locomotor activity

To check the swimming behavior of zebrafish, singles placed on a light with infrared light source using optical stage. Digital video recording used to identify the movements of zebrafish vertically or horizontally during the time period. the pattern of zebrafish swimming was observed automatically by using specialized software systems by monitoring the parameters, such as total moving distance, how much angle rotated and velocity are observedfrequently.[10] In the bottom vertical lines drawn with equal distance no of lines that zebrafish cross is counted to measure the zebrafish movements .[11]

2.2 Touch-evoked escape response

In larvae form zebrafish has a tendency to respond to visual, touch response suddenly increasing theirmovements. The automatic response gets by touching the head, tile, of the larvae several times by gently touching the larvae using a fine needle.[13] The movements are recorded from the time of touch up to tail gets to resting state video recorded.

2.3 Light-dark test

Altering in the light shows effect on movement and anxiety behavior in zebrafish .When light transition from light to dark rapidly more action in zebrafish ,while dark to light movements were less action .In pupa forms the movements are observed by placing larvae on 96-well plate observed using digital control which can alter the light and dark tests and observation the movements of zebrafish larvae.[15] Second method using adult zebrafish to analyze that how much time spend in tank and that tank was equally separated into two compartments using barrier into light and dark. zebrafish released to swim easily in those light and dark compartments. At first zebrafish released in light compartment there by counting that number of entries, times spent and movements in light area and dark area were observed.[18]

2.4 T/Y-maze test

Using T-mage to enhance memory and learning ability in zebrafish which is similar to rodents. T-maze consists of one long arm which is a starting zone and two short arms and two chambers to train the zebrafish. Two days zebrafish were not fed and then released them at starting point feed were placed in one arm by identifying the area of feed by zebrafish shows positive result.[19] This whole test was taken place under video recording and time spent by zebrafish in finding the feed were analyzed. This is a simple procedure developed on T-maze in this one short arm is red zone which is inconvenient zone to zebrafish and the other arm is green zone which is a convenient zone the food is provided 5 seconds after taping in center of tank. [18] Zebrafish enter the green zone within 5seconds means correct comeback.

2.5 Novel tank diving test

Zebrafish were released into novel tank to observe there swimming and locomotors activities were recorded including their swimming path. tank was divided into three equal parts which can be identified at bottom, middle andsurface horizontally and in vertically the number of arrival and time spend in the top to basenoted. Anxiety related action also observed by time and irregular movements of zebrafish in the procedure.[8] See Figure 1.

NDD Zebrafish Models

ALCL3	APP/Aβ
Okadaic acid	MAPT/Tau
Scopolamine	PSEN 1
	PSEN2

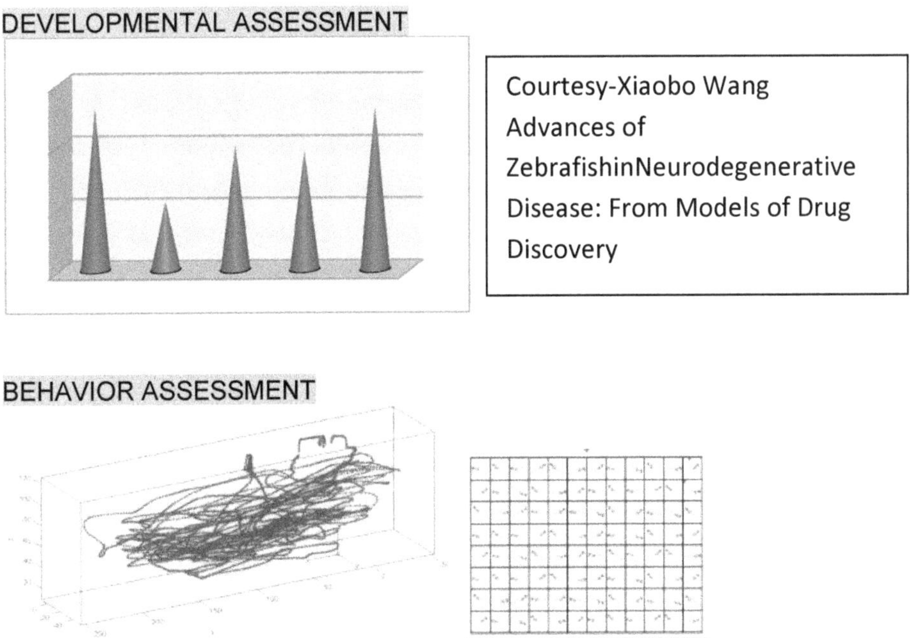

Figure 1. Behavior Assessment and Developmental Analysis in NDD studies.

3. ALZHEIMER'S DISEASE

In 1907 first AD was reported about mostly 95% of cases observed periodically in people over 65 years.[11] AD was about to spread through genes only 1% remaining cases are majorly due to natural ,genetic sensitive and some other elements .Major mentality of AD are memory dropping ,confusion

,difficulty in identification and related co-occurring conditions like depression or anxiety .Due to essential pathological modifyin AD causes passing of some neurons in brain region in control for analytical activity ,The existence of tangles and β-amyloid .On several hot theories of scientists may inspire the initiation of AD: The amyloid fountain, tau hyperphosphorylation, neuroinflammation and metal ions .There are many ways in pharmacological models to investigate AD pathological process by using zebra fish .

4. NEUROTOXIC ZEBRAFISH MODELS OF ALZHEIMER'S DISEASE

Here some neurotoxins used that triggers the AD-like progression is a good sign for investigating the AD etiology .Cholinergic dysfunction and neuronal loss were observed in the AD patients brain .In one hypothesis of AD ,acetylcholine neurotransmitter synthesis is declined ,scopolamine which is an muscarinic agonist has been examine to the zebrafish neurotoxin incite AD animal model .Scopolamine causesacademic workdebt and that effects can be changed by acetylcholinesterase inhibitors such acetylcholinesterase (AChE) and butyrylcholinesterase (BuChE) these are enzymes used in the AD progression.[12]

GABAergic system is associated with AD, some mechanism unclear. Neuronal loss of glutamergic in the hippocampal and temporal lobe in AD appearexcitotoxicity of glumergic system with the disease[28].GABAergic compounds shows allowance of the GABAergic structure to AD pathogenesis process [9].

Protein phosphate 2A(PP2A) is inhibited by okadaic acid, tau is dephosphorylating which is associated with insanity. Following zebrafish models are identifiedby inducing Aβ misfolded protein aggregate. Senileplaque, enlarge p-tau and kinase glycogen synthase 3 β(GSK-3β). [9] Deficiency in memory and learning observed in the histological and pathological modification. By using this kind of model, it is easy to represent the neuropathology relate to AD and it is not difficult for drug discovery.

Here aluminum and lead are used as neurotoxins. By inducing aluminum as AD increases the AChE activity.[10] Whereas Pb exposure outcome in de novo DNA duplicate number alteration in alethal dosecontrolledmanner. mostly gene alteration was related to amyloid precursor protein (APP) molecularpathway, Pb induced to neurological disease needs for further explanation, especially in AD.[11,12]

5. GENETIC ZEBRAFISH MODELS OF ALZHEIMER'S DISEASE

Aβ which is generated from APP WHICH IS a type 1 transmembrane glycoprotein .APPβ to form aβ between formation these are involved β and γ-secretase .APPα is produced by involving **α** and **γ** secretase .plaques are formed by the Aβ due to accumulation ofAβ.This accumulation can induced by the microinjection of Aβ and sometimes cause apoptotic cell death , activation of microglia.[12]To study the amyloidosis on animal model injecting Aβ microinjection into the embryos this method is very cost effective which provide acute AD and also used for study of neuropathology and further drug screening[36] .Tg Swedish mutant APP successfullydemonstrate the disease causing protein particularly to the eyes, heart , brain ,pulmonary vasculature when induced in zebrafish .shows symptoms like AD in this model and cerebral β amyloidosis , neuronal loss .potentially these are lethal to the mammals.[13,14]

A neurofibrillary tangle isa model in AD induce byhighly phosphorylated tau.[15] Neurofibrillary tangles ARE formed due to hyperphosphorylated tau protein and effects microtubule normal function. Thealterationis found in AD patients with frontotemporal lunacy by encoding GENE MAPT such as tau-P301L and Tau –A152T, has been accepted to copy AD key pathological feature of taupathie.[13] HuC promoter used to study the rule of a neuronal – specific in Tg induce zebrafishmodel. Tauhyperphosphorylation, aggregation, decline and behavioral deficit in zebrafish showed when inducing Ga14/UAS-based human tau P301L-Tg.[10] Same but human based Tau-A152T-Tg showed high-rise in phosphorylation, neurodegeneration and damaged proteosome act in zebrafish. Sign of tauopathies spot only tau hyperphosphorylation.[13] In development stage of Tg influences the zebrafish model traits.Six isoforms of tau due to the splicing has been identified in an adult human brain. Binding repeats when there is lack of microtubule, there by three isoforms with three repeats(3R) another three isoforms produce four repeats(4R). Additionally, 29 or 58 amino acids encoded by exon 2.[16]Here two paralogues one is MAPT ortholog,mapta which encodes the tau isoforms four, five or six. Mapta and maptbdisplays certain brain regions during different expression development, after duplication there will be sub functionalization role of MAPT orthologue. in ganglion neurons there is presence of mapta mRNA but maptb is not. teratogenicity is observed byseek to overexpress zebrafish 3R or 4R in human zebrafish. [13] by finding both wild type and short forms are immune to blend and increased in the neuronal demise.more and more effects are required to mature such AD-like models by changing different expression of tau isoforms . by inducing PSEN1 OR PSEN2 identified early onset AD patients having fibrillogenic Aβ initiation in the brain but when deletion of psen1 causing neuronal loss .by blocking psen2 translation, which effects on unborn childgrowth in the CNS with notch signaling. [10]

6. DRUG DISCOVERY IN ZEBRAFISH MODELS OF ALZHEIMER'S DISEASE

GSK-3β present in zebrafish is matching to the human GSK3β AND IT important for abnormal tau hyperphosphorylation .the lithium chloride can send back the theoretical deficits and tau hyperphosphorylation in vivo AD model by the GSK3β inhibitor caused by administer the Aβ42 into the hind brain ventricles of zebrafish fetus .to protect zebrafish locomotor lithium chloride is given for positive control for tauopathy in Tau-P301-Tg zebrafish model .Lithium and lithium carbonate form identified prevention of memory impairment and AChE activity in zebrafish by inducing scopolamine. There is a lack of evidence that GSK3β is one of the effective Inhibitor. Thecommercially presence compound is TDZD-8 which can reduce the mortality rate and increases cognitive impairment by OKA.[16]

A number of drugs have been recognized and look in this system that in act of cholinergic system can suppress cholinesterase activity. When scopolamine induced this cotinine and 6-hydroxy-L-nicitine found to be reduce tension like actionsand memory impairment in this model and also stress and AChE activity were also reduced in zebrafish. Thymus vulgaris L. essentialoil (TEO) reduces the AChE activity and antioxidant capacity in brain. thymol and p-cymene are the compounds present in the TEO known by gas chromatography. here ALCL3-induced zebrafish dyskinesia is inhibited the AChE activity by active ingredient Linarin.[17] In molecular docking Linarin is binding with AChE active sites. compound 4m has designed to reduce behavioral symptoms of AD by blocking AChE and BuChE activity which is an important lead for drug model in AD treatment.

Anti-tauopathy drugs against AD by preventing tau hyperphosphorylation .LDC8 and cyclin-dependent kinase 5(cdk5) which can shieldcounter neurodegeration ,neuroinflammation and synapse density in Aβ42 induced model by modulating the phosphorylation of microtubule- associated protein ,heparan sulphate is a highly sulfated polysaccharide which related with tau accumulation in AD .surfen or oxalyl surfen, both having antagonists properties can decrease tau hyperphosphorylation and rescues neuronal in Tg zebrafish encourage p-tau inhibition probably a alleviating AD.

Monoamine oxidase and cholinesterase inhibitors derivatives of 3-arylcoumarin which is coumarin skeleton compound showed development in zebrafish movements in ALCL3 induced zebrafish model. Anti-oxidant ferulic acid is consists of variety of functions including amyloid structure elimination. When ALCL3 induced zebrafish the erratic movements are recovered by administering ferulic acid derivative which was synthesized .Tm-10 shows a improvement in memory loss when Aβ aggregation and scopolamine induced zebrafish model. Cinnamic acid hybrid, 4e, can slow down Aβ 42 accumulation which were newly synthesized.AlCl3 induced zebrafish model is improved dyskinesia by compound 4e.sulfur amino acid metabolite relieve cognitive defects in OKA-induced AD zebrafish model. Which could be very useful in neurodegenerative improvement models.

Present all looking towards nanomedicineresistance to AD or human amyloid diseases due to economic and high through put screening .casein coated gold nano particles were designed and tested .moreover Aβ toxicity zebrafish model was rescued there mobility and cognition function by βCas AuNPs.[19] 1(BACE1)DAPT , γ-SECRETASE INHIBITOR (GSI) was educated to control Aβ PRODUCTION interfere with PSEN function , toxic impact are discovered in this GSI not only inhibiting Aβ aggregation and also the notch signaling pathway .Where JLK is coumarin inhibitors Aβ aggregation without notchinterfering.The drugs designed for AD should decrease their side effects without triggering other pathways.[20]

7. CONCLUSION

There are various of pharmacological drugs designed for NDD development haddoomed in clinical litigations over the endeda number of decades due to the clinical heterogeneity of the disease. Infection-refine treatments can only be helpful in the early stage of this kind disease because here most NDD pathologies are unrepairable. To bring about this, the zebrafish model now a days plays a key role. Zebrafish have a tremendous advantage that provide an important foundation for the elimination of NDD pathogenesis and drug identifications. In vitro chemical studies provide much higher throughput than that of vivo, while in vivo models can duplicate complex systems in the human body such as the BBB. Here Zebrafish have been used in both in vitro and in vivo drug showing, and number of new syntheses have been thoroughly checked to be resistant to NDD. Zebrafish will be beneficent in the carry of drug from the labs to the clinic. See Table 1.

Table 1. *Drugs identified that are used in NDD developmental zebrafish model.*

Compound	**Classification**	**Model**	**Effects of Phenotype**	**Behavioral method**	**Molecular mechanism**
Surfen &oxalyl surfen	Heparin sulfate	Tau-p301L-TG	Rescue Impaired Escape	Touch-Evoked Escape Response	Decrease P-Tau
AR-534	GSK-3β inhibitor	Tau-p301L-TG	Not Applicable	Touch-Evoked Escape Response	Decrease P-Tau
Linarin	flavonoid	AlCl3	Protect against dyskinesia	Light-dark test,	Inhibit AChE Activity
TDZD-8	GSK-3β inhibitor	OKA	Reduce mortallyrate	Spatial discrimination test	Decrease P-Tau
Compound 4m	3ARYLcoumarin	AlCl3	Alleviated behavior symptom	Locomotion test with trajectory recording	Inhibit AChE Activity
Li2C03	GSK-3β inhibitor	scopolamine	Reverse cognitive deficits	Novel tank test, Y-maze	Decrease P-Tau

Compound 4e	Cinnamic acid hybrid derivatives	AlCl3	Improve dyskinesia recovery rate	Locomotion test	Inhibit Aβ42 Accumulation
TEO	Thymus Vulgaris	Scopolamine	Reduce Amnesia and Anxiety	Novel tank test, amnesia, discrimination test	Reduce AChE Activity

8. REFERENCES

1. J.Vaquer-Alicea, M. Diamond. *M. I. Propagation of Protein Aggregation in Neurodegenerative Diseases. Annu. Rev. Biochem,* 2019, **88**,785.

2. Y.Hou, X.Dan, M.Babbar. *Ageing as a Risk Factor for Neurodegenerative Disease.* 2019,**15,** 565.

3. G.Streisinger, C.Walker, N.Knauber, F.Singer. *Production of Clones* of *Homozygous Diploid Zebra Fish (BrachydanioRerio),* 2007, **291,** 293.

4. G.J.Lieschke, P.D.Currie. *Animal Models of Human Disease: Zebrafish Swim into View,* 2007, **8,** 353.

5. R.M.Basnet, S.Taweedet, D.Finazzi, M.Memo. *Zebrafish Larvae as a Behavioral Model in Neuropharmacology* 2013 **7,** 23.

6. M.Cosacak, P.Bhattarrai, L.Dzewas, V.Mashkaryan. *Overexpression Results in TAU Hyperphosphorylation without Neurofibrillary Tangles in Adult Zebrafish Brain,* 2017, **7** , .

7. H.Pan, J.Zhang, Y.Wang, K.Cui, L.Wang, *Linarin Improves the Dyskinesia Recovery in Alzheimer's Disease Zebrafish by Inhibiting the Acetylcholinesterase Activity,* 2019, **222,** 112.

8. F.Wang, J.Chen, *Sleep Deprivation Caused a Memory Defects and Emotional Changes in a Rotenone-Based Zebrafish Model of Parkinson's Disease,* 2019, **372** , .

9. M.E.Nunes, T.E.Muller, M.M.Braga, B.D.Fontana, V.Quadros, *Chronic Treatment with Paraquat Induces Brain Injury, Changes in Antioxidant Defenses System, and Modulates Behavioral Functions in Zebrafish,* 2017, **54,** 3925.

10. S.E.Nada, F.E.Williams, Shah, *Development of a Novel and Robust Pharmacological Model of Okadaic Acid-Induced Alzheimer's Disease in Zebrafish,* 2016, **15,** 86.

11. M.R.Senger, K.J.Seibt, R.D.Dias, *Aluminum Exposure Alters Behavioral Parameters and Increases Acetylcholinesterase Activity in Zebrafish* (Daniorerio*) Brain, 2011,* **27**,199.

12. J.Lee, J.L.Freeman, *Exposure to the Heavy-Metal Lead Induces DNA Copy Number Alterations in Zebrafish Cells,* 2020, **33**,2047.

13. P.Bhattarai, A.K.Thomas, Y.Zhang, C.Kizil, *The Effects of Aging on Amyloid-B42-Induced Neurodegeneration and Regeneration in Adult Zebrafish Brain,* 2017, **4**, e1322666.

14. A.Abramsson, P.Kuttunen, R.K.Banote, M.Arner, *The Zebrafish Amyloid Precursor Protein-B Is Required for Motor Neuron Guidance and Synapse Formation,* 2013, **381**, 377.

15. M.S.Uddin, M.Kabir, T.Behl, P.Jeandet, G.M.Ashraf, *Revisiting the Amyloid Cascade Hypothesis: From Anti-aβ Therapeutics to Auspicious New Ways for Alzheimer's Disease,* 2020, **21,** 5858.

16. S.KSonawane, S.Chinnathambi, *S. Prion-Like Propagation of Post-Translationally Modified Tau in Alzheimer's Disease: A Hypothesis,* 2018, **65** ,480.

17. L.R.Nery, N.S.Eltz, C.Hackman, R.Fonseca, *Brain Intraventricular Injection of Amyloid-β in Zebrafish Embryo Impairs Cognition and Increases Tau Phosphorylation, Effects Reversed by Lithium,*2014, **9,** e105862.

18. Z.Sang, K.Wang *Synthesis and Evaluation of Novel Ferulic Acid Derivatives as Multi-Target-Directed Ligands for the Treatment of Alzheimer's Disease,* 2019, **10,** 1008.

19. K.Wang, J.Shi, Y.Zhou, *Synthesis and Evaluation of Cinnamic Acid Hybrids as Multi-Target-Directed Agents for the Treatment of Alzheimer's Disease,*2021, **112,** 104879.

20. A.Petit, A.Pasini, C.Alvesda Costa, *Selective Gamma-Secretase Inhibitors That Do Not Interfere With Notch Pathway In Vitro Or In Vivo,* 2003, **74**, 370.

PROTECTIVE ACTIVITY OF *BIOPHYTUM* AND LYCOPENE IN CCl_4 INDUCED HEPATOTOXICITY IN RATS

B. Nagaraju[a], N. Jyothirmai[b*], T. Daveedu[c]

[a] School of Pharmaceutical Sciences(SOPS), Sandip University, Nasik, Maharashtra, India; [b] Department of Pharmacology, Siddartha institute of Pharmaceutical Sciences, Jonnalagadda, Narsaraopeta, Guntur; [c] Department of Pharmacy, Koneru Lakshmaiah Education Foundation, Vaddeswaram, AP, India.

*Corresponding author email: bnrajupharma@gmail.com

1 INTRODUCTION

India is a diverse nation. Despite the diversity of its natural resources matches the richness of its culture and tradition. The most potent of these resources is the forest resource. Our country is blessed with an abundance of forests that are home to a large range of medicinally useful plants. There are over 45,000 different plant species in India, and thousands of them have been said to have medical properties and to be used in various ways.[1] As long as human civilization has existed, people have used medicinal plants. The usage of minerals, plant, and animal parts as a source for drugs dates back thousands of years, according to numerous ancient references on medicinal plant tradition. ancient Egyptian writings.[2] The *Biophytum* plant looks like a palm plant. Especially in wet lands of Africa, Asia and India this plant was found. It was grown at low and medium altitudes.[3] Reports regarding the efficacy of these plants for hepatoprotective were very limited, which had driven me to take up the task. Besides, we have also used the lycopene extract in combination with the plant extract since lycopene is already a proven antioxidant. In present study we focus on the Protective activity of *Biophytum and* Lycopene in CCl_4 induced liver damage in rats.

2 MATERIALS AND METHODS

2.1 Drugs and Chemicals

Silymarin is procure from sigma Aldrich, CCl4, methanol, formaldehyde, Petroleum ether and all other chemicals and reagents procure from Siemens Heal thinners Ltd. All drugs and reagents used were analytical grade only.

2.2. Plant material Collection

The whole plant of *Biophytum* were picked form rampachowdavaram area forest and Dr Prasanna Kumari authenticated the plant. A Specimen voucher was kept at Department of Pharmacology, KLEF Andhra Pradesh, India

2.3 Extract preparation

After collection of plant material, it was dried under normal temperature. After drying by using petroleum ether waste materials were removed. Now the defatted materials were subjected to solvent like methanol by using Soxhlet apparatus to get extract. Now by using reduced pressure the extract was dried, brownish colour was obtained and it was stored in desiccator.

2.3.1 Lycopene (LE) preparation. Tomato sauce was purchased from the local market, for removal of moisture content it is subjected to drying, after drying remaining solid content is considered as lycopene. To get required dose of 50 mg kg^{-1} it is dissolved in required amount of lycopene in purified water. [5]

2.3.2 Phytochemical analysis. The MEBV was evaluate for determination of different phytochemicals such as alkaloids, glycosides, phenols, tannins, saponins, carbohydrates, fatty acids and steroids. [6]

2.4 Animals

Male Wistar rats were used, animals weigh approximately 140-160gms were kept in animals house and maintained under standard quality conditions (RH 44-56%, temperature 27±2°C) and feed with standard diet and drinking water *ad libitum for* 1 week before and during study. Protocol of this study was approved by IAEC of Shri Vishnu college of Pharmacy and with the permission from CPCSEA Reg No. 439/PO/01/a/CPCSEA.

2.5 Design of Experiment

Experimental animals were divided into five groups, each group contain six animals (n=6). Except Group I remaining group animals received CCl4 at a dose of 0.1ml kg^{-1} i.p.it is already dissolved in olive oil (50% v/v). [7]

Group I: Normal Group received olive oil p.o

Group II: Toxic Group received carbon tetrachloride 0.1 ml kg^{-1} i.p

Group III: Standard Group treated with carbon tetrachloride 0.1 ml kg^{-1} i.p + 50 mg kg^{-1} silymarin p.o

Group IV: Test Group treated with carbon tetrachloride 0.1 ml kg^{-1} i.p + MEBV 200 mg kg^{-1}p.o

Group V: Test Group treated with carbon tetra chloride 0.1 ml/kg i.p 50 mg kg^{-1} LE p.o + MEBV 200 mg kg^{-1} p.o

2.6 Assessment of liver function parameters

After end of the study by using retro orbital puncture blood samples was collected. Serum was separated by centrifuged the blood sample at 4000 rpm for 10 minutes. This serum can be used for estimation of SGOT, SGPT, ALP, total cholesterol, total bilirubin and total proteins by using commercially available kits (Roche Diagnostics) and using autoanalyzer. After collection of blood samples rats were sacrificed by using euthanasia, liver was isolated; one of the lobes was fixed in 10 % formalin solution for histopathological studies and second lobe was used for estimation of antioxidants like SOD, catalase, GSH and MDA.[8]

2.7 Statistical Analysis

The statistical analysis was performed by using one-way ANNOVA followed by Tukey's post hoc test using graph pad prism 5.0. All date was expressed as mean ± SEM. Significance was evaluated at p value less the 0.0013.

3 RESULTS

Phytochemicals tests reveals the presences of Phyto substances like glycosides, tannins, phenols, flavonoids and steroids.

3.1 MEBV and Lycopene on SGOT in CCl4 induced hepatotoxicity

In this Study, CCl4 treated animal's shows high levels of Serum glutamate oxaloacetate transaminase (SGOT) 421.2±0.02 IU/L when compared to normal group animals (see Figure 1). Whereas silymarin treated animals decreases SGOT levels 146.2±0.45IU/L. MEBV with lycopene treated animals significantly reduced the SGOT levels 202±0.78 IU/L when compared to CCl4 treated group.

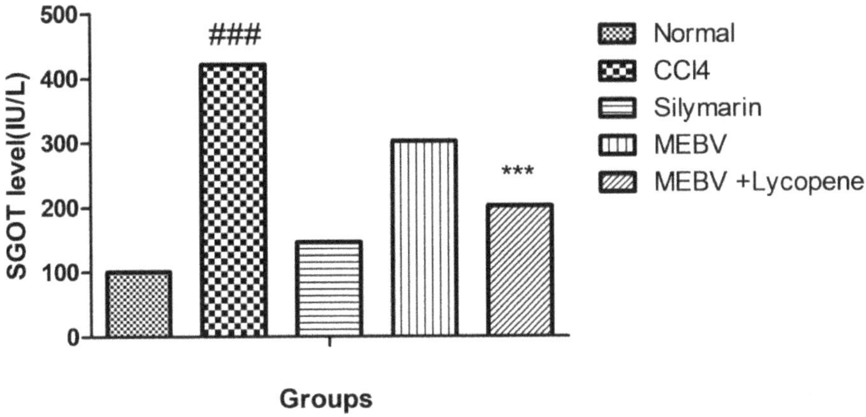

Figure 1: All values are expressed as mean ± SEM. CCl4 treated animals shows elevated levels of SGOT (### *p*<0.01) when compared to normal group, whereas silymarin and MEBV and Lycopene treated animals (*** p<0.001) significantly decreased levels of SGOT when compared to toxic group.

3.2 Effect of MEBV and Lycopene on SGPT in CCl4 induced hepatotoxicity

In this Study, CCl4 treated animal's shows elevated levels of Serum glutamate pyruvate transaminase (SGPT) 304.5±0.06 IU/L when compared to normal group animals (see Figure 2). Whereas silymarin treated animals decreases SGPT levels 102.4±0.87 IU/L. MEBV with lycopene treated animals significantly reduced the SGPT levels 135±0.88 IU/L when compared to CCl4 treated group.

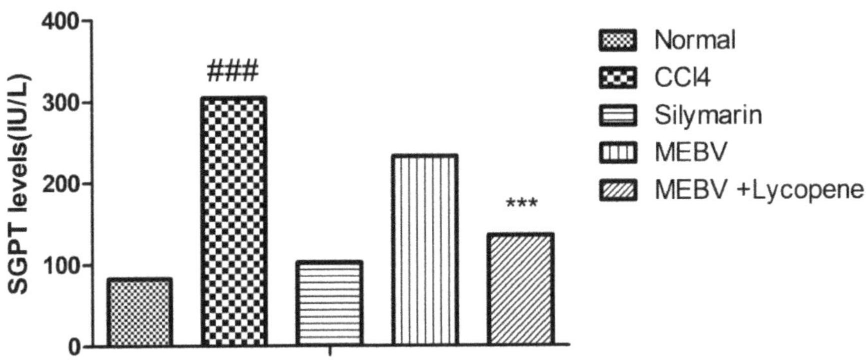

Figure 2: Values are shown as mean ± SEM and n=6. CCl4 treated animals shows elevated levels of SGPT (### *p*<0.01) when compared to normal group,

whereas silymarin and MEBV and Lycopene treated animals (*** p<0.00) significantly decreased levels of SGPT when compared to toxic group.

3.3 Effect of MEBV and Lycopene on serum alkaline phosphatase in CCl4 induced hepatotoxicity

In this Study, CCl4 treated animal's shows elevated levels of serum alkaline phosphatase 333.2±0.07 IU/L when compared to normal group animals (see Figure 3). Whereas silymarin treated animals decreases serum alkaline phosphatase levels 202.1±0.87 IU/L. MEBV with lycopene treated animals significantly reduced the serum alkaline phosphatase levels 242.4±0.45 IU/L when compared to CCl4 treated group.

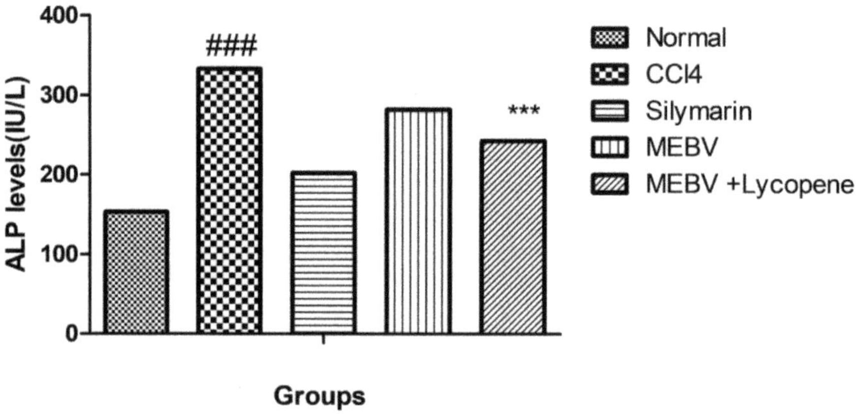

Figure 3: All values are expressed as mean ± SEM. CCl4 treated animals shows elevated levels of ALP (### *p*<0.01) when compared to normal group, whereas silymarin and MEBV and Lycopene treated animals (*** p<0.001) significantly decreased levels of ALP when compared to toxic group.

3.4 Effect of MEBV and Lycopene on serum total bilirubin in CCl4 induced hepatotoxicity

In this Study, CCl4 treated animal's shows elevated levels of serum total bilirubin 1.9 mg/dl when compared to normal group animals (see Figure 4). Whereas silymarin treated animal's decreases serum total bilirubin levels 0.3 mg/dl. MEBV with lycopene treated animals significantly reduced the serum total bilirubin levels 0.4 mg/dl when compared to CCl4 treated group.

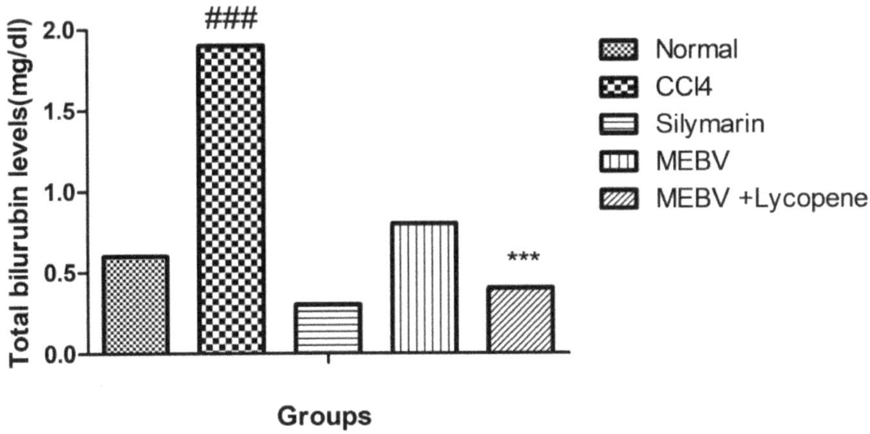

Figure 4: All values are expressed as mean ± SEM. CCl4 treated animals shows elevated levels of TB (### *p*<0.01) when compared to normal group, whereas silymarin and MEBV and Lycopene treated animals (*** p<0.00) significantly decreased levels of TB when compared to toxic group.

3.5 Effect of MEBV and Lycopene on total protein levels in CCl4 induced hepatotoxicity

In this Study, CCl4 treated animal's shows elevated levels of total protein 0.58±0.08 mg/dl when compared to normal group animals (see Figure 5). Whereas silymarin treated animals increase total protein levels 0.88±0.78 mg/dl.MEBV with lycopene treated animals significantly increased the total protein levels 0.78±0.23 mg/dl when compared to CCl4 treated group.

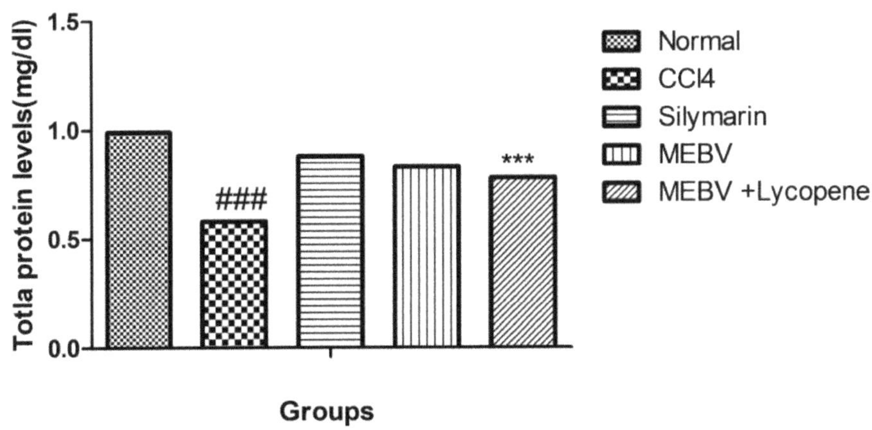

Figure 5: All values are expressed as mean ± SEM. CCl4 treated animals shows low levels of TP (### *p*<0.01) when compared to normal group, whereas silymarin and MEBV and Lycopene treated animals (*** p<0.00) significantly elevated levels of SGOT when compared to toxic group.

3.6 Effect of MEBV and Lycopene on total cholesterol levels in CCl4 induced hepatotoxicity

In this Study, CCl4 treated animal's shows elevated levels of serum total cholesterol levels 138.2±0.12 mg/dl when compared to normal group animals (see Figure 6). Whereas silymarin treated animal's decreases serum total cholesterol levels 80.8±0.56 mg/dl. MEBV with lycopene treated animals significantly reduced the serum total cholesterol levels 70.2±0.89 mg/dl when compared to CCl4 treated group.

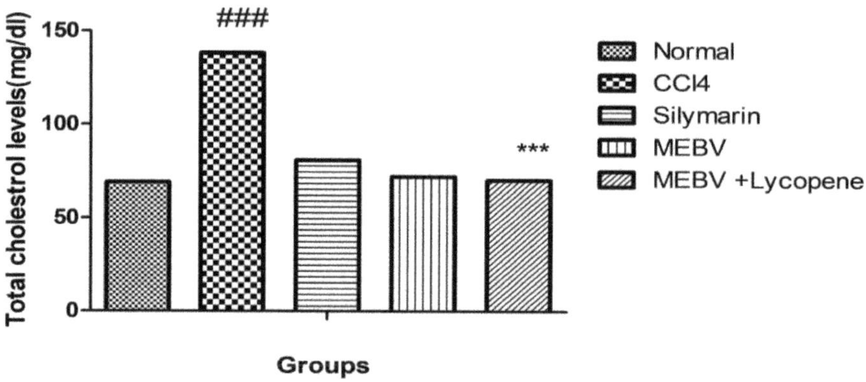

Figure 6: All values are expressed as mean ± SEM. CCl4 treated animals shows elevated levels of total cholesterol (### *p*<0.01) when compared to normal group, whereas silymarin and MEBV and Lycopene treated animals (*** p<0.00) significantly decreased levels of total cholesterol when compared to toxic group.

3.7 Effect of MEBV and Lycopene on SOD levels in CCl4 induced hepatotoxicity

In this Study, CCl4 treated animal's shows decreased levels of SOD 4±0.24 U/mg protein when compared to normal group animals (see Figure 7).Whereas silymarin treated animal's elevated SOD levels 9.4±0.24 U/mg protein. MEBV with lycopene treated animals significantly reduced the SOD levels 8.2±0.24 U/mg protein when compared to CCl4 treated group.

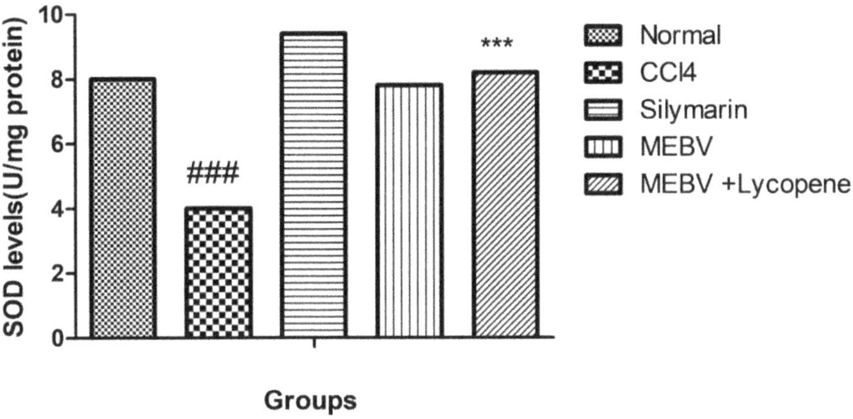

Figure 7: All values are expressed as mean ± SEM. CCl4 treated animals shows low levels of SOD (### *p*<0.01) when compared to normal group, whereas silymarin and MEBV and Lycopene treated animals (*** p<0.00) significantly elevated levels of SOD when compared to toxic group.

3.8 Effect of MEBV and Lycopene on Catalase levels in CCl4 induced hepatotoxicity

In this Study, CCl4 treated animal's shows decreased levels of catalase 5.9±0.42 uM of H_2O_2 consumed /mg protein when compared to normal group animals (see Figure 8). Whereas silymarin treated animal's elevated catalase levels 11.2±0.68 uM of H_2O_2 consumed /mg protein. MEBV with lycopene treated animals significantly reduced the catalase levels 10.2±0.86 uM of H_2O_2 consumed /mg protein when compared to CCl4 treated group.

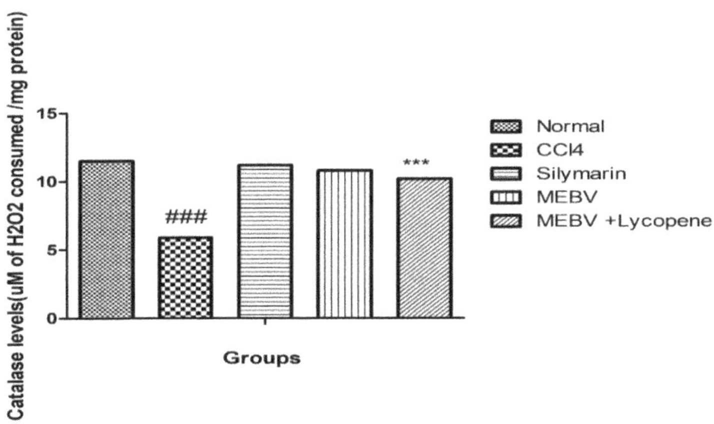

Figure 8: All values are expressed as mean ± SEM. CCl4 treated animals shows elevated levels of catalase (### *p*<0.01) when compared to normal group, whereas silymarin and MEBV and Lycopene treated animals (*** p<0.001) significantly decreased levels of catalase when compared to toxic group.

3.9 Effect of MEBV and Lycopene on GSH levels in CCl4 induced hepatotoxicity

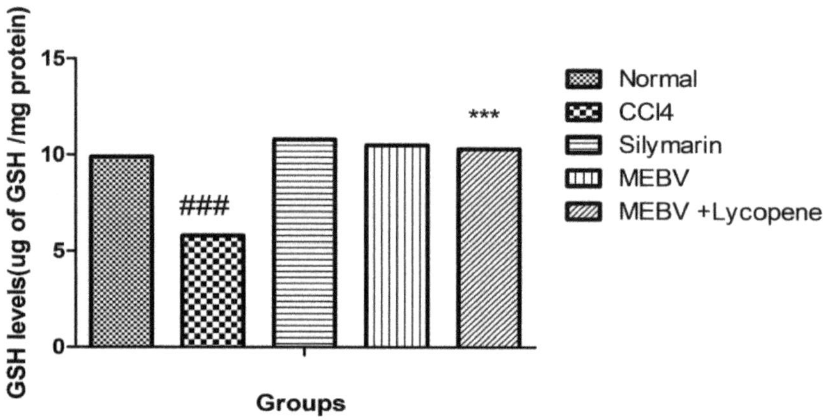

Figure 9: All values are expressed as mean ± SEM. CCl4 treated animals shows decreased levels of GSH (### *p*<0.01) when compared to normal group, whereas silymarin and MEBV and Lycopene treated animals (*** p<0.001) significantly elevated levels of GSH when compared to toxic group.

In this Study, CCl4 treated animal's shows decreased levels of GSH 5.8±0.68 ug of GSH /mg protein when compared to normal group animals (see Figure 9). Whereas silymarin treated animal's elevated GSH levels 10.2±0.46 ug of GSH /mg protein. MEBV with lycopene treated animals significantly reduced the catalase levels 10.3±0.54 ug of GSH /mg protein when compared to CCl4 treated group.

3.10 Effect of MEBV and Lycopene on MDA levels in CCl4 induced hepatotoxicity

In this Study, CCl4 treated animal's shows elevated levels of MDA 1.38±0.04 nm of MDA/ mg protein when compared to normal group animals (see Figure 10). Whereas silymarin treated animals decreases MDA levels 0.52±0.12 nm of MDA/ mg protein. MEBV with lycopene treated animals significantly reduced the MDA levels 0.58±0.48 nm of MDA/ mg protein when compared to CCl4 treated group.

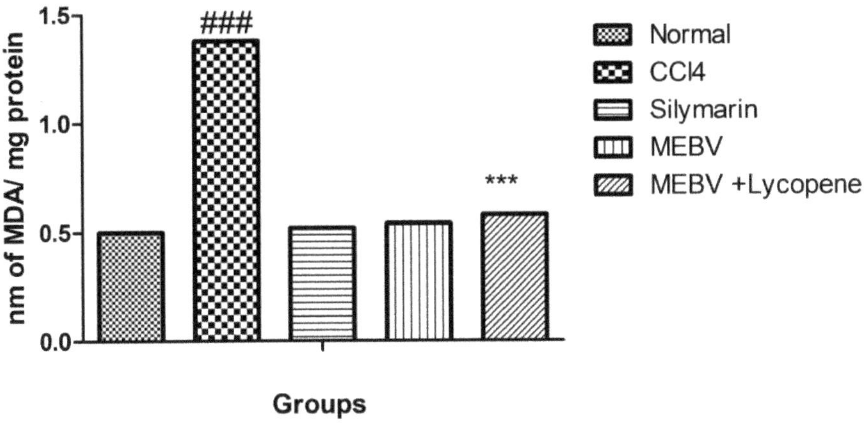

Figure 10: All values are expressed as mean ± SEM. CCl4 treated animals shows elevated levels of MDA (### *p*<0.01) when compared to normal group, whereas silymarin and MEBV and Lycopene treated animals (*** p<0.001) significantly decreased levels of MDA when compared to toxic group.

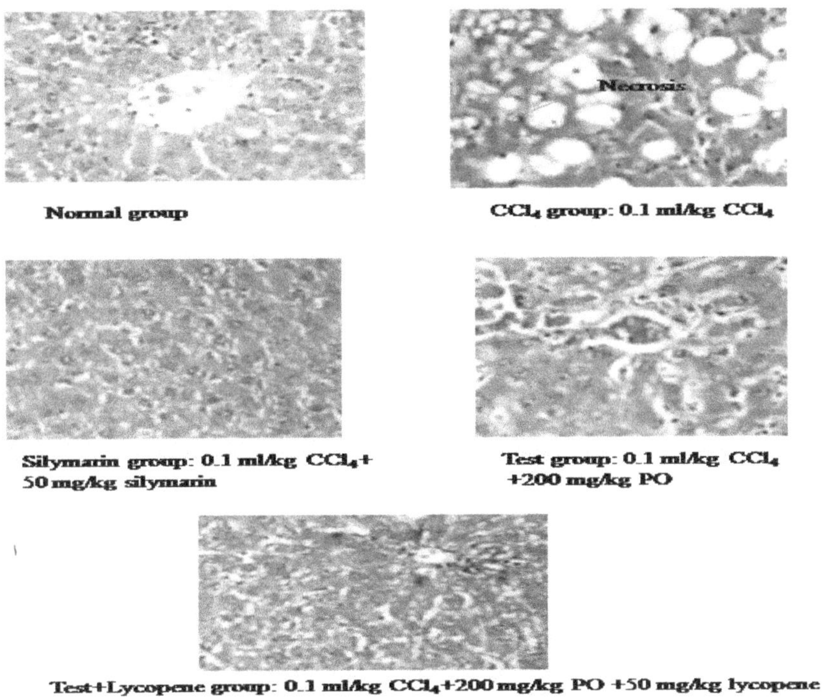

Figure 11: Histopathology of livers of various groups.

The result of liver histopathological studies discloses, CCl4 treated animals shows substantial hepatic damage by centrilobular necrosis by infiltration of liver cells with fat, associated with disturbance of metabolism, whereas Standard, and test compounds like silymarin, MEBV with lycopene treated animals shows protective effect of liver by less centrilobular necrosis and steatosis when compared to carbon tetra chloride treated animals. (see Figure 11)

4 DISCUSSION

The metabolism and excretion of xenobiotics occurs in liver, any damage or altered in liver functions leads to chronic health issues. [9,10] Traditional usage of drugs in the management of liver problems is often very rare so requisite to explore possible drugs for the management and treatment of liver problems and also displace the drugs having sub marginal efficacy and safety in market. In worldwide hepatitis is the major reason for morbidity and mortality in liver diseases. Lipidaemia, peroxidation and loss of integrity of plasma membrane are abnormalities associated with hepatitis. Better understanding of pathophysiology's of hepatitis is very important for discovery of safe and less toxic hepatoprotective drugs. [11]

Wistar rats are most commonly used experimental animals for screening of hepatoprotective drugs because of structural similarity of CYP 450 enzymes with humans. In this study we are used male rats because female rats are less prone to carbon tetrachloride induced liver damage.[12] In this research carbon tetra chloride (CCl4) used for induction of liver damage. CCl4 cause the liver damage by destruction of hepatocellular membrane due to release of different types of free radicals like trichloromethyl, hexachloroethane and COCl2. These free radicals also alter the chain reactions of lipids peroxidation, macromolecules covalent bonding, and metabolic mechanisms in mitochondria leads to liver necrosis.

Enzymes like amino transferases play an important role in the conversation of amino acids to alpha ketoacids by transfer of amino group. These enzymes are very sensitive and reliable indicates for hepatic damage and also hepatoprotective effect of various agents. Elevated levels of SGOT, SGPT and ALP are observed in CCl4 treated animals it indicates CCl4 causes liver damage. Whereas test compound with lycopene treated animals' exhibits low levels of enzymes it indicates this compound exhibits protective effect on liver. The liver is also playing a crucial role in the production of proteins like albumin, fibrinogen and globulins. These proteins are playing a very important role in human normal physiological function. Reduced levels of total protein is observed in CCl4 induced liver damage due to damage of endoplasmic reticulum which leads to loss of CYP 450 function resulting in decrease protein synthesis by liver and leads to accumulation of triglycerides results to fatty liver.

In case of test compound and test compound with lycopene significantly increases the total protein levels by increasing the regeneration process it indicates these compounds shows protective action. CCl4 also increase the cholesterol levels by inhibiting synthesis of bile acids from cholesterol leads to fatty liver. Significant reduction of cholesterol observed in test and test with lycopene treated animals. Lower levels of SOD, catalase, GSH and elevated levels of MDA is observed in CCl4 treated animals and increases levels of SOD, catalase, GSH and decreased levels of MDA observed in test and test with lycopene treated animals it indicates test compounds protect the liver by enhance endogenous antioxidants.

5 CONCLUSION

Based on the above results MEBV alone and MEBV with lycopene significantly prevents the carbon tetrachloride induced hepatotoxicity in rats by free radical scavenging action due to presence of flavonoids, phenolic compounds. Future studies need isolation of active compounds and establishment of molecular mechanism for hepatoprotection.

6 ACKNOWLEDGEMENT

The authors are thankful to K L College of Pharmacy, Koneru Lakshmaiah Education Foundation for their constant encouragement and support for conducting this study.

Conflict of interest statement

The authors declare that they have no conflict of interests.

7 REFERENCES

1. Grover JK, Yadav S, Vats V. Medicinal plants of India with anti-diabetic potential. *Journal of Ethnopharmacology* 2002;81(1):81-10.

2. Asima Chatterjee, Satyesh Chandra, Pakrashi, *Publication and information directorate the treatise on Indian medicinal plants* ,1956, 1, 243.

3. Bandaru N, Ramu A, Vidhyadhara S. Neuroprotective and antioxidant activities of fraction isolated from methanolic extract of Sapinduslaurifolia on scopolamine induced dementia. *Phcog Res* 2020;12:119-26.

4. Di Mascio P, Kaiser S, Sies H. Lycopene as the most efficient biological carotenoid singlet oxygen quencher. *Arch Biochem Biophys.* 1989; 1;274(2):532-8

5. Thilagavathi.T, Arvindganth.R, Vidhya.D, Dhivya .R , *Int. Res. J. Pharm preliminary phytochemical screening of different solvent mediated medicinal plant extracts evaluated,* 2015;6:246.

6. Mohammad Ali. *Pharmacognosy and Phytochemistry*, CBS Publishers &Distribution, New Delhi.
7. Zeashan H, Amresh G, Singh S, Rao CV. Hepatoprotective activity of Amaranthus spinosus in experimental animals. *Food Chem Toxicol.* 2008;46(11):3417-21.
8. Lahon K, Das S. Hepatoprotective activity of Ocimum sanctum alcoholic leaf extract against paracetamol-induced liver damage in Albino rats. *Pharmacognosy Res.* 2011;3(1):13-8
9. Kirtikar, K.R. and Basu, B.D,*International Book Distributors In Indian Medicinal Plants, 3rd Edn., Dehradun*,1995, 2, 1526.
10. Meena B,Anbin Ezhilan R, Rajesh R, Sheik Hussain A, Ganesan B, Anandan R*, Afr J Biochem* Res Antihepatotoxic potential of Sarassum polycystum on antioxidant defense status in D-galactosamine induced hepatitis in rats, 2008; 2:51-56.
11. Siegers CP, Reichl W, Younes M. Sex differences in the susceptibility of rats to carbon tetrachloride--alcohol-induced liver injury. Agents Actions. 1984;14(1):121-3
12. SEAKINS A, ROBINSON DS. The effect of the administration of carbon tetrachloride on the formation of plasma lipoproteins in the rat. *Biochem J*. 1963;86(3):401-7.

PHYTOCHEMICAL ANALYSIS AND CHARACTERIZATION OF ETHANOLIC EXTRACT OF *PSEUDERANTHEMUM MALABARICUM* WHOLE PLANT

K. Sujana[a], *, A. V. Surendra[c], K.V. Kishore[a], SK.Reehana[b], C. B. Himakalyani[a]

[a]University College of Pharmaceutical Sciences, Acharya Nagarjuna University ,Guntur-522510, Andhra Pradesh, India; [b]Department of Pharmaceutical Analysis, KVSR Siddhartha College of Pharmaceutical Sciences, Vijayawada, Krishna-520010, Andhra Pradesh, India; [c]Department of Pharmacy, Koneru Lakshmaiah Education Foundation, Vaddeswaram, Andhra Pradesh, India.

*Corresponding author email: drksujana36@gmail.com

1 INTRODUCTION

Pseuderanthemum belongs to the family of Acanthaceae.[1] *Eranthemum malabaricum* is the synonym for this plant. It is a woody plant, grows about 2 ft tall. Leaves are elliptic or ovate, narrowed at both ends, hairless (Figure 1.1A). These plants can grow upto 5 inch in length and 2.5 inches width, with somewhat wavy toothed margin. Nerves are 6-7 pairs, and the stalk is about 8 mm long. Flowers are borne in 4-9 inches long spikes. They are white or yellowish, 2.5-4.5 cm across, somewhat two-lipped. Sepals are about 4 mm long, nearly linear. Capsule is about 2 cm long, and cylindrical (Figure 1B).

Figure 1 *Pseueranthemum malabaricum* plant leaf (A); and flower (B).

The objective of the present study was to identify the bioactive ingredients in the ethanolic extract of the entire plant, *Pseuderanthamum malabaricum*. The information on the phytoconstituents present in the plant extract may be useful in identifying them as bioactive chemicals of natural origin and can be developed further as lead molecules for pharmacological screening.

2 MATERIALS AND METHODS

2.1 Plant

Pseuderanthemum malabaricum was collected from Tirumala hills and was authenticated by a professor (Dr Madhava Chetty K) in Botany Department of Sri Venkateswara University, Tirupati. A specimen was deposited with reference number, PM/0671. The plant was air dried in well ventilated shaded place, powdered and stored in a desiccator for research.

2.2 Physico-chemical analysis

Physicochemical data such as the ash value, extractive value and loss on drying were performed as per the standard procedures.[2]

2.2.1 Extractive values. Alcohol, water and ether-soluble extractive values were determined in relation to the air dried plant material according to the official methods prescribed in IP 1996.

2.2.2 Ash values. 3 g of air dried powder of *Pseuderanthemum malabaricum* was taken for estimating the total ash values. Acid soluble ash value and water insoluble ash values were determined in relation to the air dried plant material according to the official methods prescribed in IP 1996.

2.2.3 Loss on drying. Pseuderanthemum malabaricum powder (air dried), was precisely weighed into a thin, flat porcelain dish. It was oven dried at 105°C and cooled in a desiccator. The weight loss was reported as a decrease in moisture content.

2.3 Extraction of Plant Material

A weighed amount of air dried powder of *Pseuderanthemum malabaricum* was macerated. Also, another weighed quantity was subjected to extraction with ethanol as a solvent using Soxhlet extractor. After soxhlation, the extract was vacuum-concentrated in a rotary evaporator and % yield was calculated.[3-5]

2.4 Phytochemical screening

Ethanolic extract of *Pseuderanthamum malabaricum* powdered material was screened for its phytochemicals constituents using the standard methods.[6, 7]

2.5 Fourier transform infrared (FTIR) analysis.

FTIR technique was utilised to identify the distinctive functional groups in the ethanolic plant extract. A little portion of the extract was scanned in the wavenumber ranging from 4000 cm^{-1} to 400 cm^{-1}.

2.6 Finger print profiling of ethanolic plant extract by GC-MS/MS

GC-MS/MS analysis of *Pseuderanthamum malabaricum* ethanolic extract was done using ITQ 900 (Thermo Scientific) Model AI 1310 autosampler (for split/splitless injections) and a Model Trace 1300 gas chromatograph were both included in the GC-MS/MS equipment. The 10 l syringe was used in the big volume injection technique. Software called ThermoXcalibur was used to control the instruments and analyse the data. For analysis, a TG-5MS column (30 m length, 0.25 mm internal diameter, and 0.25 m film thickness) with a mass selective detector was employed in gas chromatography. As a carrier gas, helium was employed at a constant flow rate of 1 ml min^{-1}. With the injector temperature at 270°C and the interface temperature at 290°C, sample injection was carried out in the splitless mode. The oven's temperature was programmed to start at 75°C for 0 minutes, ramp to 170°C at 20°C min^{-1} for 0 minutes, then to 211°C at 8°C min^{-1} for 2 minutes, then to 230°C at 3.5°C min^{-1} for 2 minutes, and finally to 290°C at 18°C min^{-1} for 8 minutes. For the detection of various analytes, electron ionisation was employed at -70 eV in full-scan and selective ion monitoring modes between 50 m/z and 500 m/z.

3 RESULTS AND DISCUSSION

3.1 Physico-chemical analysis

3.1.1 Ash values. Ash values of the plant were determined and are shown in Table 1

Table 1 *Ash value of plant.*

S. No	Test type	% Ash
1	Total ash	23.66%
2	Acid insoluble ash	7%
3	Water soluble ash	1%

3.1.2 Loss on drying. Loss on drying which indicates % moisture content of the plant was determined and is shown in Table 2.

Table 2 *Loss on drying value of plant.*

S. No	Weight before drying (with china dish)	Weight after drying (with china dish)	Weight loss on drying	% Moisture content
1	65.85g	65.60g	0.25g	12.5%

3.2 Extraction by maceration

Plant was subjected to maceration using three different solvents (Ethanol, Methanol and Water) and % yield was determined and exhibited in Table 3.

Table 3 *Extractive results of plant using different solvents.*

S. No	Solvent used for extraction	Plant powder taken for maceration (g)	Amount of solvent used (ml)	Yield	
				In grams	Percentage
1.	Ethanol soluble extraction	5	100	3	60
2.	Water soluble extraction	5	100	0.18	3.6
3.	Ether soluble extraction	5	100	0.09	1.8

When the powder was macerated with ethanol, 60 % of the yield indicated that majority of the phytochemical constituents have been extracted in this alcoholic solvent. Further, using soxhlation technique the phytoconstituents in the whole plant powder were extracted with ethanol as the solvent and % yield was calculated and is shown in Table 4.

Table 4 *% yield of extraction of plant Pseuderanthemum malabaricum. Using Soxhlet extraction (Ethanolic extract)*

S. No.	weight of raw material taken for extraction (g)	weight of extract obtained (g)	% yield
1	75	10.88	14.8%
2	80	16.88	18.75%
3	90	25.2	28%

3.3 Phytochemical screening

Phytochemical screening of ethanolic extract was done to determine the constituents such as steroids, terpenoids, saponins, flavonoids, proteins, chalcones and coumarins were shown in Table 5.

Table 5 *Phytochemical screening of ethanolic extract of Pseuderanthemum malabaricum.*

S. No.	Phytoconstituents	Result
1	Steroids	Present
2	Quinones	Absent
3	Terpenoids	Present
4	Cardiac Glycosides	Absent
5	Saponins	Present
6	Phenols	Absent
7	Carbohydrates	Absent
8	Tannins	Absent
9	Flavonoids	Present
10	Proteins	Present
11	Alkaloids	Absent
12	Chalcones	Present
13	Coumarins	Present

3.4 Fourier transform infrared (FTIR) analysis.

IR interpretation of plant ethanolic extract

The presence of chemicals like phenols and flavonoids is indicated by the H-bonded -OH stretching vibration at 3347 cm^{-1} in the IR spectrum (Figure 2). Aromatic CH-stretching is visible in the peak at 2931.22 cm^{-1}. The carbonyl group's CO-stretching may be seen in the peak at 1725 cm^{-1} The fact that many peaks were found in the 900–600 cm^{-1} region of the C–H bending vibrations indicates the presence of an aromatic structure. Inferring from all of these peaks that coumarin is present, we can deduce that the peak at 1041.96 cm^{-1} is due to C-H bending vibrations. It's possible that the two bands at 1335.14 and 1397.45 cm^{-1} are CH_3 and CH_2 bending. Band at 1621.48 cm^{-1} is connected to aromatic molecule C=C stretching (flavonoids).

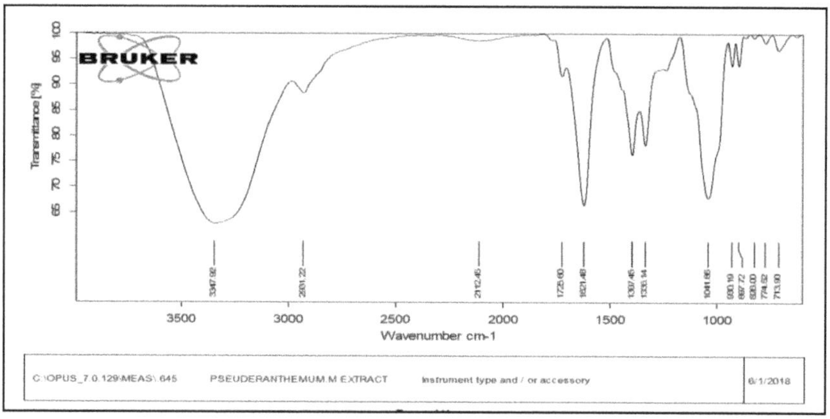

Figure 2 FTIR spectrum of ethanolic extract of Pseuderanthemum malabaricum plant.

3.5 Finger print profiling of ethanolic plant extract by GC-MS/MS

The bioactive chemicals included in ethanol crude extract were characterised using a typical GC-MS/MS chromatogram analysis (Figure 3) and the Phytocomponents were identified in the ethanolic extract are listed in the Table 6.

Table 6 *Phytocompounds identified in the ethanol extract by GC-MS/MS.*

S. No.	Name of the Substance	Molecular formula	Molecular weight (g/ mole)	Retention time (min)
1	2, 6 Dimethylhex 4 enamidine, 2 (3methylbut2enyl)	$C_{13}H_{24}N_2$	208	3.14
2	Benzene,1 chloro 3, 5 bis (1,1dimethylethyl) 2 (2 propenyloxy)	$C_{17}H_{25}ClO$	280	3.50
3	4HPyran4one,2, 3 dihydro 3, 5 dihydroxy 6 methyl	$C_6H_8O_4$	144	3.99
4	5Hydroxymethylfurfural	$C_6H_6O_3$	126	4.36
5	Pentanedioic acid, (2, 4 ditbutylphenyl) Monoester	$C_{19}H_{28}O_4$	320	6.08
6	l(+) Ascorbic acid 2,6 dihexadecanoate	$C_{38}H_{68}O_8$	652	13.86
7	Octadecanoic acid	$C_{18}H_{36}O_2$	284	15.04
8	Indazol4one, 3, 6, 6 trimethyl 1 phthalazin 1 yl 1, 5, 6, 7 tetrahydro	$C_{18}H_{18}N_4O$	306	15.47

9	(5á,13á)3,19 Epoxyandrost8ene,17 acetoxy3methoxy4, 4, 7 trimethyl	$C_{25}H_{38}O_4$	402	43.14
10	7aHCyclopenta[a]cyclopropa[f] cycloundecene2,4,7,7a,10,11h exol,1,1a,2,3,4,4a,5,6,7,10,11, 11adodecahydro1,1,3,6,9pent amethyl,2,4,7,10,11pentaaceta te	$C_{30}H_{44}O_{11}$	580	43.33

Pharmacognostical standards of the plant was done to investigate the physico-chemical parameters of *Pseuderanthamum malabaricum* whole plant and was found to be within the limits. Extraction of plant was performed by soxhlet extraction technique using ethanol as solvent and then concentrated using rotary evaporator and % yield was calculated, finally stored in a desiccator. Preliminary phytochemical screening of ethanolic extract was carried out and found that steroids, terpenoids, saponins, coumarins, flavonoids, proteins and chalcones were present. Characterization of the phytoconstituents was performed by FTIR and GC-MS/MS to determine extract characterization and for the phytoconstituent identification. Thus, the NIST database and GC-MS/MS analysis, which revealed the existence of 10 compounds, were used to identify the bioactive chemicals in the entire *Pseuderanthamum malabaricum* plant.

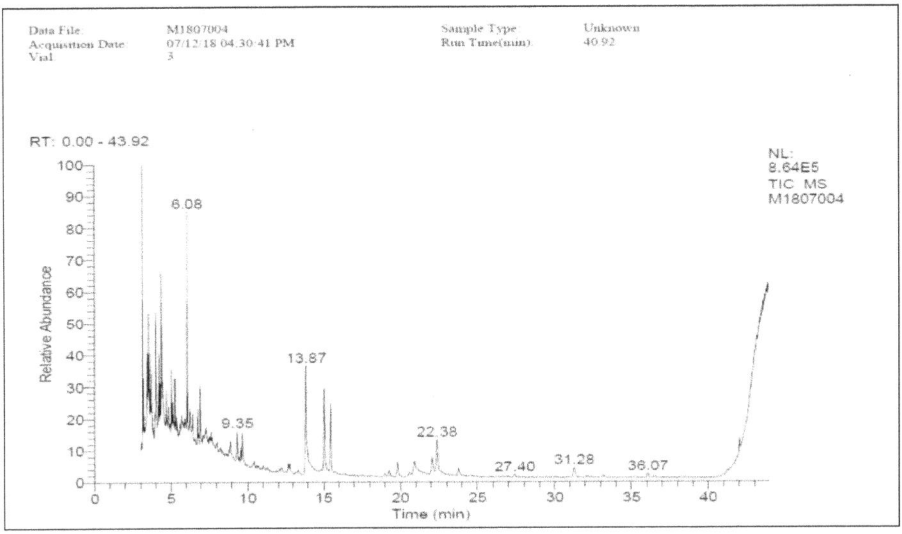

Figure 3 *GC-MS/MS chromatogram of the phytocompounds in ethanol crude extract.*

Among the identified compounds, 4H-Pyran-4-one-2,3-dihydro-3,5-dihydroxy-6-methyl-(DDMP), 2-Furancarboxaldehyde, 5-(hydroxymethyl) - (5HMF), 1-(+)-ascorbic acid-2-6- dihexadecanoate were the phytoconstituents which possess the antioxidant, and anticancer properties.

4h-Pyran-4-One-2,3-Dihydro-3,5-Dihydroxy-6-Methyl- (DDMP) which is depicted in Figure 4 is a flavonoid compound and known to have antioxidant activity.

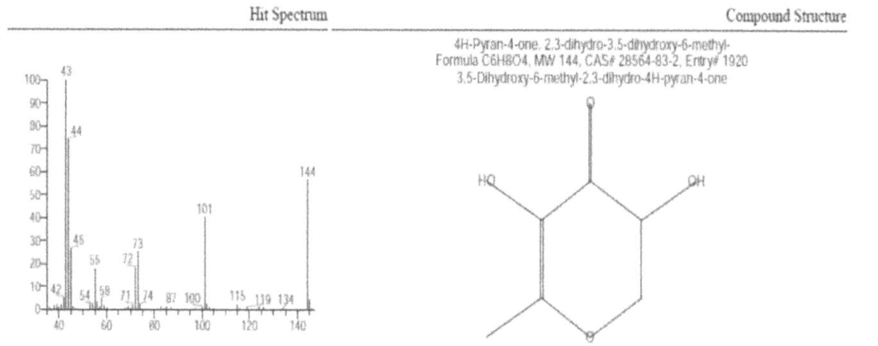

RT	Scan #	Probability	Compound Name	SI	RSI	Cas #
3.99	110.000000	22.00	4H-Pyran-4-one, 2,3-dihydro-3,5-dihydroxy-6-methyl-	638	734	28564-83-2

Figure 4 *4HPyran4one-2,3 dihydro3,5dihydroxy6methyl.*

The antioxidant capacity of edible prunes was due to the release of DDMP during drying process. This compound is also having anti-cancer activity. Further this compound was reported as a strong antioxidant in the glucose–histidine Maillard reaction products.[8]

Another compound, *2-Furancarboxaldehyde-5-(Hydroxymethyl)-(5HMF)* (Figure-3.4) exhibited the cytotoxic effect in a brine shrimp bioassay. Also it has been found to be an insecticidal compound by bioassay- guided fractionation.[9]

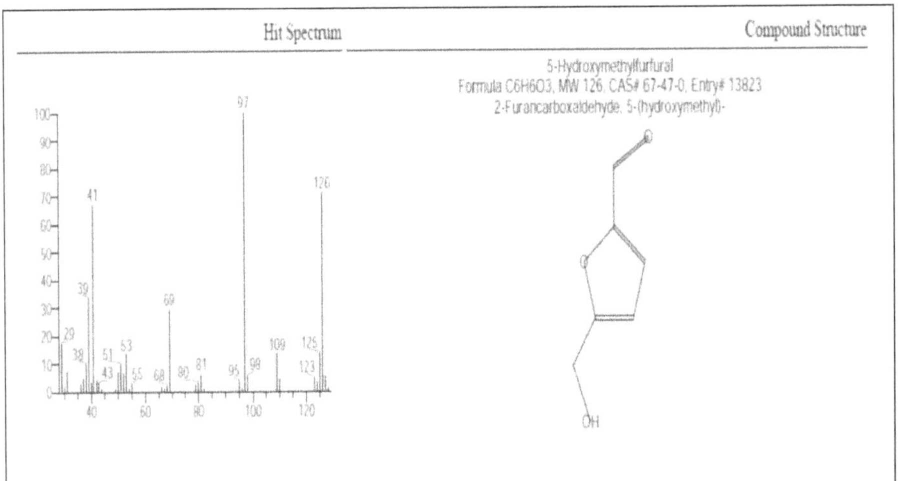

RT	Scan #	Probability	Compound Name	SI	RSI	Cas #
4.36	150.000000	74.40	5-Hydroxymethylfurfural	697	856	67-47-0

Figure 5 *5Hydroxymethylfurfural.*

1-(+)-Ascorbic Acid- 2-6- Dihexadecanoate (Figure 6) is a form of vitamin C which was known as ascorbic acid. As an antioxidant, vitamin C is among the substances known to improve sperm quality, decrease sperm agglutination, and increase sperm motility over time which may indirectly promote the male fertility.[10]

Other novel entities found in the ethanolic extract were

- 2,6 Dimethylhex4enamidine, 2 (3methylbut2enyl) (Figure 7);
- Benzene,1chloro3,5bis(1,1dimethylethyl) 2(2propenyloxy) (Figure 8);
- Pentanedioicacid (Figure 9)
- (2, 4ditbutylphenyl) Monoester, Octadecanoic acid (Figure 10);
- Indazol4one,3,6,6trimethyl1phthalazin1yl1, 5, 6, 7 tetrahydro (Figure 11);
- (5á,13á)3,19Epoxyandrost8ene,17acetoxy3methoxy4,4,7trimethyl (Figure 12);
- 7aHCyclopenta[a]cyclopropa[f]cycloundecene2,4,7,7a,10,11hexol,1,1a,2,3,4,4a,5,6,7,10,11,11adodecahydro1,1,3,6,9pentamethyl,2,4,7,10,11pentaacetate (Figure 13).

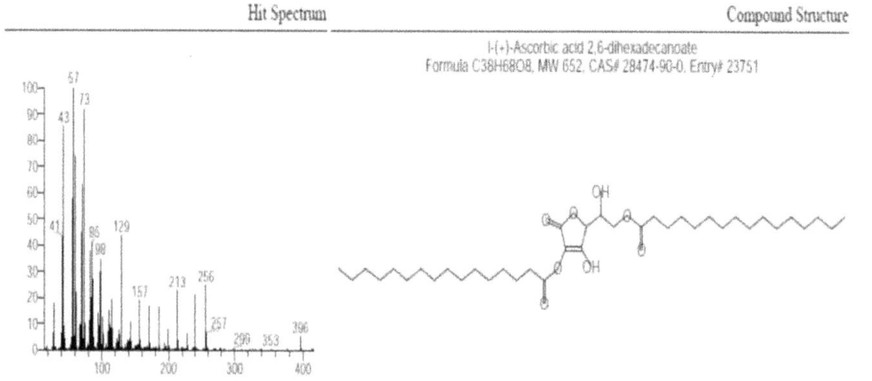

RT	Scan #	Probability	Compound Name	SI	RSI	Cas #
13.86	1157.0000 00	27.13	l-(+)-Ascorbic acid 2,6-dihexadecanoate	692	698	28474-90-0

Figure 6 *l (+) Ascorbic acid 2, 6 dihexadecanoate.*

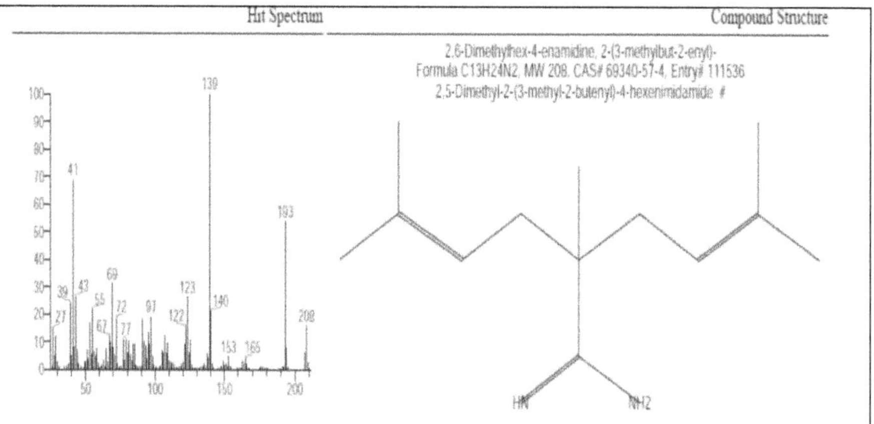

RT	Scan #	Probability	Compound Name	SI	RSI	Cas #
3.14	16.000000	5.36	2,6-Dimethylhex-4-enamidine, 2-(3-methylbut-2-enyl)-	415	532	69340-57-4

Figure 7 *2, 6 Dimethylhex4enamidine, 2 (3methylbut2enyl).*

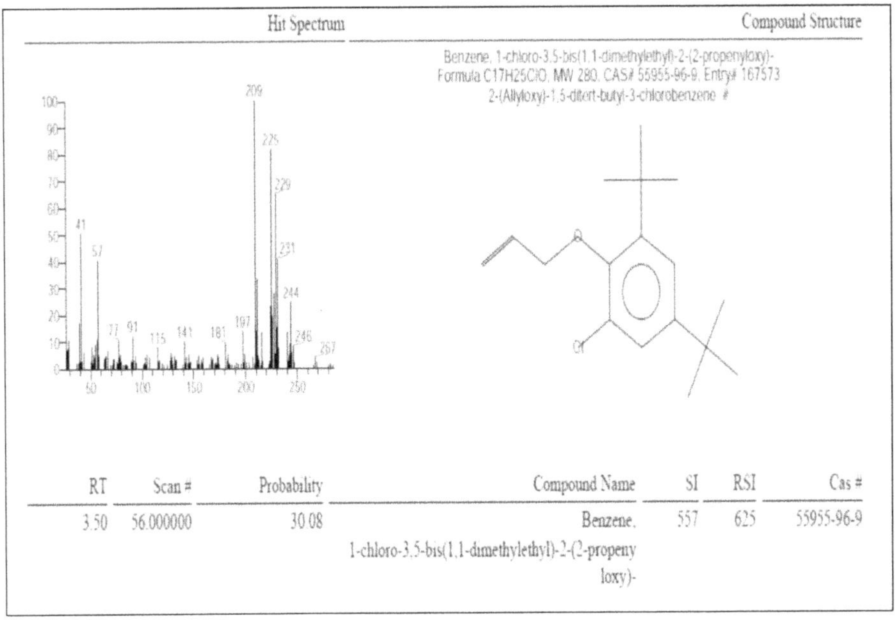

Figure 8 *Benzene, 1chloro3, 5 bis (1,1dimethylethyl) 2 (2 propenyloxy).*

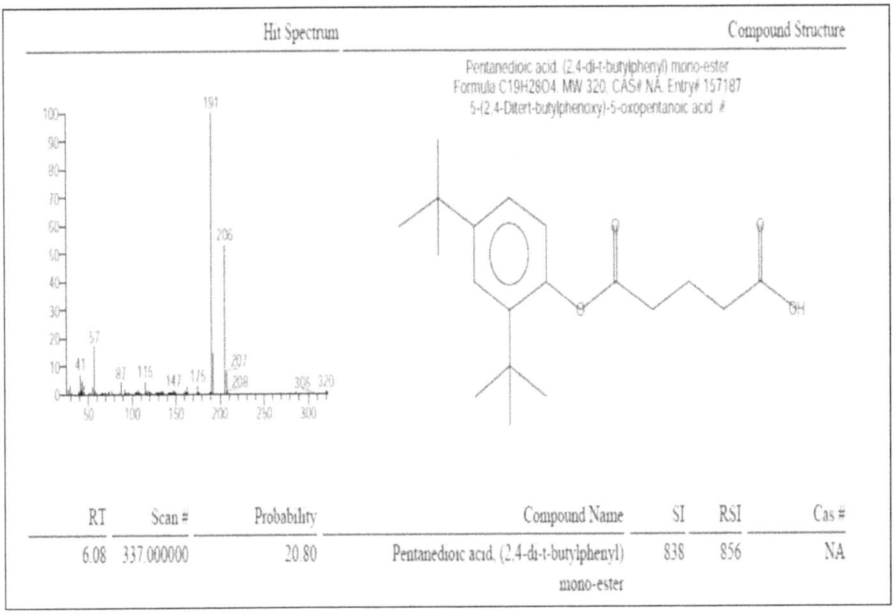

Figure 9 *Pentanedioic acid, (2,4ditbutylphenyl).*

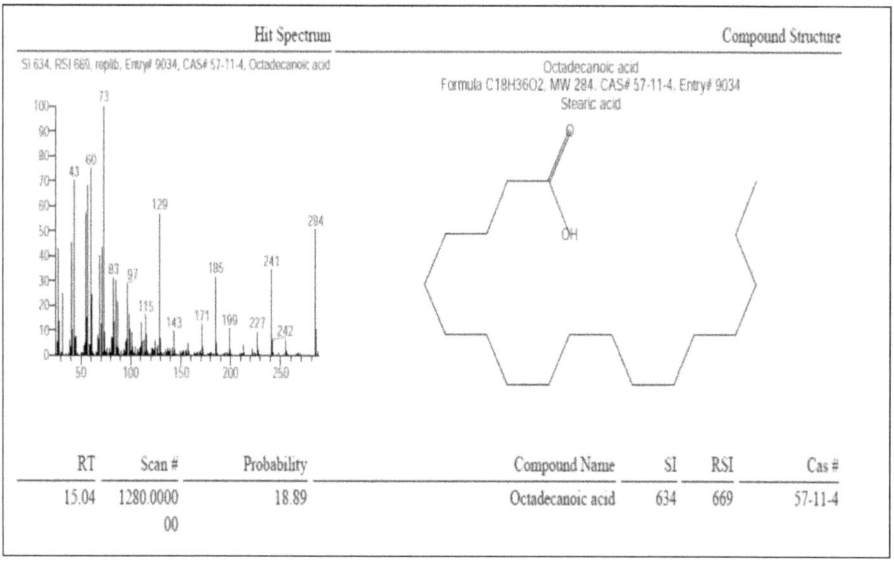

Figure 10 *Octadecanoic acid*

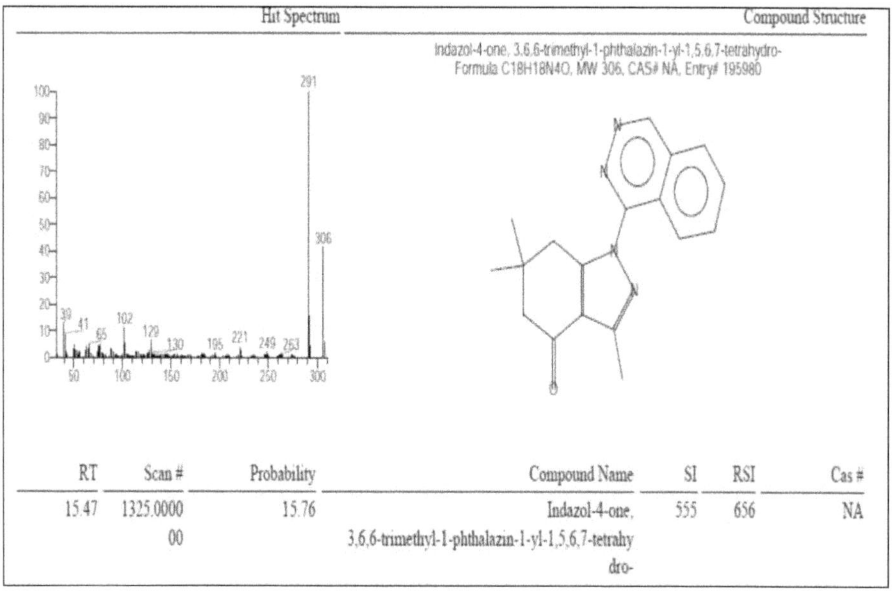

Figure 11 *Indazol4one, 3, 6, 6 trimethyl1phthalazin1yl1, 5, 6, 7 tetrahydro.*

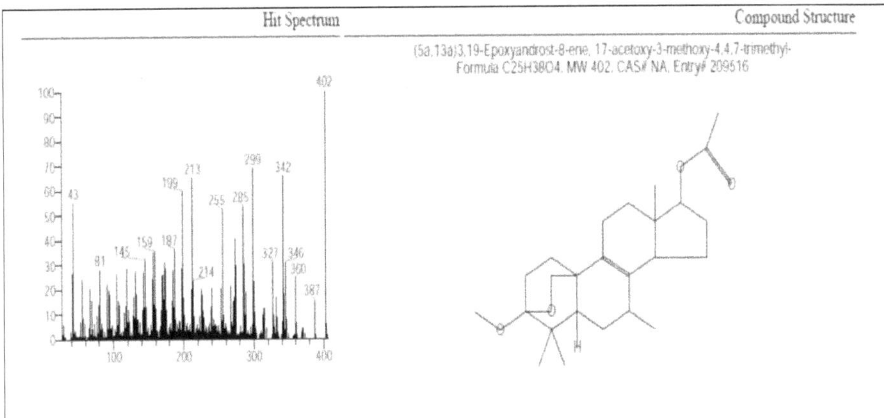

RT	Scan #	Probability	Compound Name	SI	RSI	Cas #
43.14	4097.0000 00	9.08	(5á,13á)3,19-Epoxyandrost-8-ene, 17-acetoxy-3-methoxy-4,4,7-trimethyl-	458	540	NA

Figure 12 *(5á, 13á) 3, 19 Epoxyandrost8ene,17 acetoxy3methoxy 4, 4,7 trimethyl.*

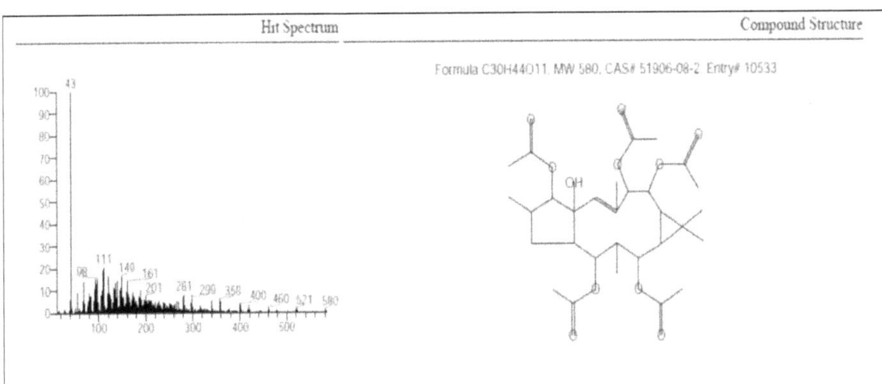

RT	Scan #	Probability	Compound Name	SI	RSI	Cas #
43.33	4117.0000 00	11.29	7aH-Cyclopenta[a]cyclopropa[f]cycloundecene-2,4,7,7a,10,11-hexol, 1,1a,2,3,4,4a,5,6,7,10,11,11a-dodecahydro-1,1,3,6,9-pentamethyl-, 2,4,7,10,11-pentaacetate	488	529	51906-08-2

Figure 13
7aHCyclopenta[a]cyclopropa[f]cycloundecene2,4,7,7a,10,11hexol,1,1a,2,3,4, 4a,5,6,7,10,11,11adodecahydro1,1,3,6,9pentamethyl,2,4,7,10,11pentaacetat.

4 CONCLUSION

Among the identified compounds, 4H-Pyran-4-one-2,3-dihydro-3,5-dihydroxy-6-methyl-(DDMP), 2-Furancarboxaldehyde-5-(hydroxymethyl) - (5HMF), 1-(+)-ascorbic acid-2-6- dihexadecanoate were the phytocompounds reported with antioxidant, and anticancer effects. Other phytoconstituents that are identified as novel entities needs to go through a series of *in vitro* and *in vivo* experimental studies for depicting their beneficial effects. From this study, it can be presumed that *Pseuderanthamum malabaricum* whole plant could be used as a source of novel molecular entities and has future scope for pharmacological studies.

5 REFERENCES

1. N. Dhatchanamoorthy, T. Arumugam and P. Raja, 2015.
2. P. K. Mukherjee, *Quality control of herbal drugs: an approach to evaluation of botanicals*, Business horizons, 2002.
3. N. Azwanida, *Med Aromat Plants*, 2015, **4**, 2167-0412.
4. P. S. Radha and S. Puri, *Plant Arch*, 2019, **19**, 397-403.
5. K. P. Ingle, A. G. Deshmukh, D. A. Padole, M. S. Dudhare, M. P. Moharil and V. C. Khelurkar, *Journal of Pharmacognosy and Phytochemistry*, 2017, **6**, 32-36.
6. A. Pandey and S. Tripathi, *Journal of Pharmacognosy and Phytochemistry*, 2014, **2**, 115-119.
7. M. Yadav, S. Chatterji, S. K. Gupta and G. Watal, *Int J Pharm Pharm Sci*, 2014, **6**, 539-542.
8. X. Yu, M. Zhao, F. Liu, S. Zeng and J. Hu, *Food Research International*, 2013, **51**, 397-403.
9. J. Mani, J. Johnson, H. Hosking, B. E. Hoyos, K. B. Walsh, P. Neilsen and M. Naiker, *Plants*, 2022, **11**, 2886.
10. X. Zhou, H. Shi, S. Zhu, H. Wang and S. Sun, *International Urology and Nephrology*, 2022, **54**, 1793-1805.

PART 5
CLINICAL CASE STUDIES

EVALUATION OF DRUG USAGE IN LACTATION IN THE COMMUNITY SETTINGS OF VIJAYAWADA

Dr. Ghanta Vijaya Kumar[a*], K. Jaahnavi Lakshmi Akhila [b],

M. Pravallika [c], R. Satya Sree [d], K. Surendra [e].

[a-e] Department of Pharmacy Practice, KVSR Siddhartha College of Pharmaceutical Sciences, Vijayawada, Andhra Pradesh – 520008.

*ghanta.vijay@gmail.com

1 INTRODUCTION

Certain vulnerable populations, such as geriatrics, children, pregnant women, and nursing mothers are not included in clinical trials of drugs due to ethical and regulatory reasons. The adverse effects of medications on lactating mothers and infants are therefore unknown. Even with all of the information available, there are still unknown risks that may apply to a small number of people, emphasising the importance of not using drugs unless absolutely necessary. When recommending drugs to a breastfeeding mother, one must consider the benefits of drug use for the mother against potential risk of drug exposure for the baby. A medicine that is acceptable to use while pregnant might not be appropriate for the mother who is nursing. A concentration gradient that permits passive diffusion of nonionized, non-protein-bound medicines is necessary for drug absorption into breast milk.[1] Because many drugs taken by breastfeeding mothers end up in breast milk, all medications must be administered with the utmost care. Thousands of women who take medications decide not to breastfeed owing to a misconception about the risk, even though the majority of drugs are excreted in amounts that are too little to have a harmful impact on the health of new-borns. [2] The most frequently reported adverse drug reactions between January 1985 and June 2011 according to the Pharmacovigilance Network were gastrointestinal (28.1%) and neurological (20.3 percent). Of the recorded adverse drug reactions, 65 were found to be severe (37.4%), 2(1.1%) obtained a causality score of "certain" (14), and 13 (7.5%) received a "probable" score (13). Drugs that are suspected include benzodiazepines, opiate analgesics, and antiepileptic drugs.[3]

Some drugs like antianxiety drugs, anti-depressants and antipsychotic drugs require a doctor's supervision during their use by adjusting the dose, limiting the amount of time a person uses a medication, timing of medicine administration in relation to breastfeeding.[4] Regular use of some drugs can cause neonatal abstinence syndrome (NAS), in which the baby goes through withdrawal upon birth. The risk of sudden infant death syndrome (SIDS) is twelve times higher in children born to women who smoked and consumed alcohol at the same time.[5] Certain drugs may occasionally be required for breastfeeding mothers. Our medical community needs to guard breastfeeding more carefully because it is so important to a baby's health and growth. Hence, when providing drugs to breastfeeding mothers, health care professionals should assess the risks and benefits by taking into mind the specific characteristics, including mother's need for the medication, possible negative effects of the medication on milk production, how much of the medication is secreted in human milk, a breastfed infant's level of oral absorption, the infant's age and the possibility of negative effects when breastfeeding. The lactation risk categories developed by Dr. Hale are used to evaluate the safety of drugs while nursing.

1.1 Dr. Hale`s Lactation Risk Categories[6]

L1- SAFEST: Drug that has been consumed by a lot of nursing women without any rise in detrimental effects on the baby being detected.
Example: Amoxicillin, Acetaminophen, Magnesium sulphate, Levothyroxine, Vit-B_{12} etc.

L2- SAFER: Drug that hasn't increased the risk of ill effects on the new-born despite only being evaluated in a small number of breastfeeding mothers. And/or there is little evidence of a danger associated with the use of this medicine in a nursing mother.

Example: Cefotaxime, Fentanyl, Prednisolone, Acetazolamide, spironolactone, Hydroxychloroquine etc.

L3- MODERATELY SAFE: There are no controlled research on breastfeeding mothers, however it is plausible that there could be unfavourable consequences on a breastfed child; alternatively, controlled studies might only reveal minor, non-threatening harmful effects. Only if the potential benefit outweighs the possible risk to the infant should medication be used.
Example: Ciprofloxacin, Ceftriaxone, Aspirin, Hydrocodone, Triamcinolone, Estradiol, Isoniazid etc.

L4- POSSIBLY HAZARDOUS: Although there is evidence to suggest a risk to a breastfed infant or to the production of breastmilk, the advantages of use for breastfeeding moms may outweigh the risk to the child (e.g., if the drug is needed in a life-threatening situation or for a serious disease for which safer drugs cannot be used or are ineffective).
Example: Colchicine, Cisplatin, Dapsone, Nadolol, Pyrimethamine, Naproxen, Lithium etc.

L5- CONTRAINDICATED: Studies on breastfeeding women have shown that there is either a medicine that has a high chance of seriously harming a new-born or that there is a considerable and established risk to the infant based on human experience.
Example: Enalapril, Captopril etc.

2 METHODS AND RESULTS

2.1 METHODOLOGY

2.1.1 Type of study design. Cross-sectional, Observational study.

2.1.2 Study site. The study was conducted in localities in and around Vijayawada which included Rani Gari Thota, Krishna Lanka, Bharathi Nagar, Loyola Gardens, Madhura Nagar, Siddhartha Nagar, Moghalrajpuram, Poranki, Ayyappa Nagar, Auto Nagar and Ajith Singh Nagar.

2.1.3 Sample size. The study comprised 60 lactating mothers who met the inclusion criteria.

2.1.4 Study duration. The study was conducted for a period 3 months out of which 2 months were for data collection and one month for data analysis.

2.2 DESIGNING OF STUDY RELATED DOCUMENTS:

2.2.1 Data collection form.

A. General Information including patient's name, age, occupation.

B. Demographic data: Age, weight, height, marital status, residential address, education level and employment.
C. Lactation status: Period of lactation.
D. Social and medical History or any other significant detail.
E. Medication history: Name of the drug, dose, route of administration, frequency, class, indication.
F. Concomitant condition (if any).
G. Current medication chart: Name of drug (generic/brand name), formulation, dose, and route of administration, frequency.

2.2.2 Informed Consent Form. English and Telugu Language. Before enrolling patients in the study, participants were given a patient information sheet in English and their vernacular language (Telugu), which explained in detail the type of study, the purpose of the study, confidentiality, and the study procedure.

2.3 ELIGIBILITY CRITERIA:

Inclusion criteria:

- Healthy Lactating mothers with infants less than 1 year and willing to participate.

Exclusion criteria:

- Mothers who were reluctant to give their consent.

2.4 DATA COLLECTION:

The study enrolled a total of 60 patients who met the inclusion criteria. A suitable data collection form was created. All relevant data for the study were collected and documented in an adequately defined data collection form, including patient demographic characteristics (age, gender), socioeconomic, social history (smoking and alcohol usage), medical and medication histories, and co-morbidities. All drugs were classified using Dr. Hale's Lactation Drug Risk Category.

2.5 DATA EVALUATION:

The data was assessed and evaluated using Dr.Hale's Lactation Drug Risk Category.

Statistical Analysis was performed using MS Excel.

3 RESULTS

To protect the safety of the newborn during nursing, it is imperative to use sensible medication therapy. The purpose of this cross-sectional observational study was to assess drug use among nursing mothers in the local area. The use of medications by a total of 60 individuals was examined.

The majority of the study's participants examined under each category were between the ages of 18-22 (31.7%), postgraduates (31.7%), and non-smokers/non-alcoholics (98.3%) (see Table 1). The L1 Category accounted for the majority of the drugs prescribed (36.53%), followed by the L2 (34.06%), L3 (26.09%), L4 (0.27%), and L5 (0%) categories. 3.02% of the drugs were not classified to any category. Out of 364 drugs prescribed, a total of 133 drugs belonged to L1 category. Folic acid was the most commonly used drug, accounting for 32.33% (43) of all drugs in the L1 category (the safest). 124 of all the prescribed drugs were in the L2 category (Safer), and Iron ascorbate accounted for the highest percentage (32.25%) of those medications (40). 95 drugs come under the L3 category, with multivitamin formulations making up 43.15 percent of the 95 drugs (41). L4 category medications represented 1 out of 364 total drugs (Possibly hazardous). Only one drug fell into this classification, and that was clopidogrel 100% (1). This study, which involved 60 patients in and around Vijayawada, found no drugs that fell under category L5 (Contraindicated). Trypsin and chymotrypsin accounted for 63.63% of the unclassified medications (11 out of 364), whereas proanthocyanidin extract accounted for 36.36% (4) (see Table 2)

The following tables show the findings of the data analysis from the study population.

Table 1 *Socio-Demographic data of Lactating Study Population.*

	Frequency (n)	**Percentage (%)**
	Age (Years)	
18 – 22	19	31.7%
23 – 27	18	30%
28 – 32	12	20%
33 – 37	9	15%
38 - 42	1	1.7%
>42	1	1.7%
	Education Level	
Illiterate	6	10%
Primary	6	10%
SSC	10	16.7%
HSC	7	11.7%
Graduate	12	20%
PG	19	31.7%

Social History		
Non-smokers / Non-alcoholics	59	98.3%
Smokers	0	0
Alcoholics	1	17

Table 2 *Drug Distribution data of Lactating study population based on Generic Name (n=364).*

S.No	Drug Name	Total No (%)
Category L1(n=133)		
1	Folic Acid	43 (32.33%)
2	Vitamin D3	38 (28.57%)
3	Pantoprazole	17 (12.78%)
4	Levothyroxine Sodium	15 (11.27%)
5	Acetaminophen	11 (8.27%)
6	Metformin Hydrochloride	5 (3.75%)
7	Ibuprofen	4 (3%)
Category L2 (n=124)		
1	Ferrous Ascorbate	40 (32.25%)
2	Metronidazole	19 (15.3%)
3	Ranitidine	12 (9.67%)
4	Serratiopeptidase	7 (5.64%)
5	Diclofenac Sodium	7 (5.64%)
6	Nifedipine	6 (7.83%)
7	Warfarin	4 (3.22%)
8	Clotrimazole	4 (3.22%)
9	Cefotaxime	3 (2.41%)
10	Sertraline	3 (2.41%)
11	Diphenhydramine	3 (2.41%)
12	Labetalol	3 (2.41%)
13	Verapamil	2 (1.61%)
14	Carbamazepine	2 (1.61%)
15	Fluoxetine	2 (1.61%)
16	Phenytoin	2 (1.61%)
17	Enoxaparin	2 (1.61%)
18	Insulin	1 (0.8%)
19	Prednisolone	1 (0.8%)
20	Metoprolol	1 (0.8%)
Category L3 (n=95)		
1	Multivitamin	41 (43.15%)
2	Calcium	39 (41.05%)
3	Progesterone	6 (6.31%)
4	L-Arginine	4 (4.21%)

5	Tramadol	2 (2.1%)
6	Aspirin	1 (1.05%)
7	Atorvastatin	1 (1.05%)
8	Lactulose	1 (1.05%)
	Category L4 (n=1)	
1	Clopidogrel	1 (100%)
	Category L5 (n=0)	
	Category Not Assigned (n=11)	
1	Trypsin Chymotrypsin	7 (63.63%)
2	Proanthocyanidin Extract	4 (36.36%)

4 DISCUSSION

4.1 Socio-Demographic data of the study population:

In the bulk of the 60 participants, 31.7% (19) belonged to the 18 to 22-year-old age group, followed by 30.7% (18) of the 23 to 27-year-old group, 20.7% (12) of the 28- to 32-year-old group, 15.7% (9) of the 33 to 37-year-old group, and 1.7% (1) of both the 38 to 42 and above 42-year-old age groups. Regarding the participants' educational background, postgraduates make up the majority of the population (approximately 31.7%), followed by graduates (about 20%), SSC graduates (16.7%), HSC graduates (11.7%), primary school graduates (11.7%), and illiterates (about 10%). 59 of the 60 study participants, or 98.3 percent, were not smokers or drinkers. We found only 1 alcoholic in our study population.

4.2 Drug Distribution data of the study population:

A total of 364 different medications were seen in our study with 60 participants. Folic acid was the most commonly used drug, accounting for 32.33% (43) of all drugs in the L1 category (the safest), followed by vitamin D3 28.57% (38), Pantoprazole 12.78% (17), levothyroxine sodium 11.27% (15), acetaminophen 8.27% (11), Metformin hydrochloride 3.75% (5), Ibuprofen 3% (4).

A total of 133 drugs belonged to L1 category. Through breast milk, folic acid can continue to nourish a child. The body's ability to produce new, healthy cells is encouraged by folic acid. It is crucial that the child continues to get folic acid, Vit-D3 outside of the womb because it has been related to optimal brain development throughout childhood. A prior prospective study found that low levels of maternal pantoprazole in milk had no negative impact on nursing infants. Treatment with levothyroxine has been demonstrated to enhance breastfeeding, and it is advised for lactating hypothyroid women who wish to breastfeed their child. In nursing mothers, acetaminophen is a useful option for pain and fever control. In diabetic patients, metformin increases insulin sensitivity and decreases hepatic glucose production. Although the incidence of neonatal hypoglycemia is relatively minimal, as shown in a previous study, metformin does enter into breast milk in minute amounts. Ibuprofen is frequently used by nursing women as an analgesic or anti-inflammatory drug.

Out of the total 364 medications prescribed, 124 were in the L2 group (Safer). Iron ascorbate made up roughly 32.25% (40), followed by Metronidazole 15.3% (19), Ranitidine 9.67% (12), Serratiopeptidase 5.64% (7), Diclofenac sodium 5.64% (7), Nifedipine 7.83% (6), Warfarin and Clotrimazole 3.22% (4), Cefotaxime, Sertraline, Diphenhydramine and Labetalol 2.41% (3), Verapamil, Carbamazepine, Phenytoin, Fluoxetine and Enoxaparin 1.61% (2). Insulin, Prednisolone and Metoprolol 0.8% (1). Most babies have enough iron in their bodies to last them for the first six months or so of life. However, new-borns need an additional source of iron beyond breast milk by the age of 6 months as little iron is present in breast milk. Breastfed new-borns receive metronidazole in lower doses than those required to treat baby infections when administered intravenously and orally by the mother. Animal data in a study suggested that metronidazole may be carcinogenic if it is given in a large single dose and the baby should feed on expressed breast milk that has been discharged in advance after breast-feeding has been stopped for 12 hours. The infant's stomach would neutralize any small amount of Ranitidine that might enter the breast milk after administration by mothers. Most often, Serratiopeptidase is used to relieve pain and the ensuing inflammation. Diclofenac has a short half-life and produces very little glucuronide metabolite. It is often thought to be safe to use when nursing when prescribed at the proper dosage. Nifedipine is used for treating painful nipple vasospasm like Raynaud phenomenon in nursing mothers but Nifedipine is excreted into human milk and there is insufficient data on long-term adverse reactions, cautionary use of nifedipine is needed as it could cause serious adverse reactions in nursing infants.

Although nursing and warfarin are often compatible, a very high dose may be more hesitant to advocate it. Since clotrimazole has a poor oral bioavailability and is therefore unlikely to have a negative impact on a breastfed new-born, it is a widely accessible over-the-counter therapy for fungal infections. Low quantities of cefotaxime in milk are not anticipated to have any negative impact on nursing infants. Cephalosporins have occasionally been associated with alteration of the infant's gut flora, which has led to diarrhoea or thrush, although these effects have not been sufficiently studied. It's acceptable to use cefotaxime while nursing. When treating postpartum depression, sertraline is among the most secure antidepressants to take while nursing. Diphenhydramine in small, infrequent doses would not be anticipated to have any negative effects on breastfed new-borns. Larger doses or longer use may have negative effects on the infant or lessen milk production. Chest discomfort and excessive blood pressure are both treated with labetalol. Because of the low quantities of labetalol in breastmilk, infants who consume it shouldn't experience many negative consequences because the amounts they consume are modest. But a study stated that while breastfeeding, the infant should be periodically examined for potential adverse effects, such as a slowed heartbeat and low blood pressure. Verapamil serum levels in new-borns may be measurable, but they are modest. Verapamil shouldn't be anticipated to have any negative effects on nursing infants. According to one study, the majority of women taking anticonvulsants while nursing was taking a combination of them, such as phenytoin and carbamazepine, which were shown to be effective. Compared to the majority of other SSRIs, fluoxetine has a greater average drug concentration in breastmilk. In some breastfed babies, negative consequences like colic, fussiness, and sleepiness have been documented. The blood thinner enoxaparin is used to treat diseases that cause blood clots. Enoxaparin may be used to treat chest discomfort, pulmonary embolism, and deep vein thrombosis. Insulin is considered to be safe for usage during lactation for those women with diabetes. Prednisolone is highly coupled to plasma proteins and only a little amount of it enters breastmilk. Metoprolol is a cardio selective beta blocker that is used to treat Hypertension. Metoprolol levels in baby plasma have been shown to be negligible or very low.

95 medications out of a total of 364 fall under the L3 group, including Multivitamin 43.15% (41), Calcium 41.05% (39), Progesterone 6.31% (6), L-Arginine 4.21% (4), Tramadol 2.1% (2), Aspirin, Atorvastatin and lactulose1.05% (1). Multivitamins and calcium with vitamin C, vitamin D, vitamin B complex, probiotics, and zinc can be used to help strengthen immunity. The risk of using progesterone while lactating is that it occasionally causes the breast milk to stop or produce less than usual. Excessive sleepiness in new-borns, trouble nursing, respiratory issues and limpness are associated with usage of tramadol in infants. One breastfed baby may develop metabolic acidosis as a result of a mother's prolonged, high-dose aspirin consumption. A baby using aspirin has a higher risk of developing Reye's syndrome. Statins prevent the production of cholesterol. Due to concerns over the modification of baby lipid metabolism, statin using women should not breastfeed. Lactulose has been used as a laxative commonly.

L4 category medications made up a total of 1 out of 364 drugs (Possibly hazardous). Clopidogrel 100% (1) was the only medication found in this category. Clopidogrel is an anti-platelet drug. Infant's bruising and bleeding should be monitored while clopidogrel is being used by the nursing mother.

There have been 0 reported medications under category L5 (Contraindicated) in this study which included 60 patients in and around Vijayawada.

Eleven medicines out of 364 fell into the unclassified category, with trypsin and chymotrypsin accounting for 63.63% (7) and Proanthocyanidin for 36.36% (4).

According to the aforementioned information, the sample population primarily utilized medicines in categories L1 followed by L2, L3, L4 and L5 with L4 being the least frequently used and L5 drugs being not found. (See Figure 1)

Figure 1: Distribution of drugs according to Dr. Hale`s Lactation Drug Risk Category

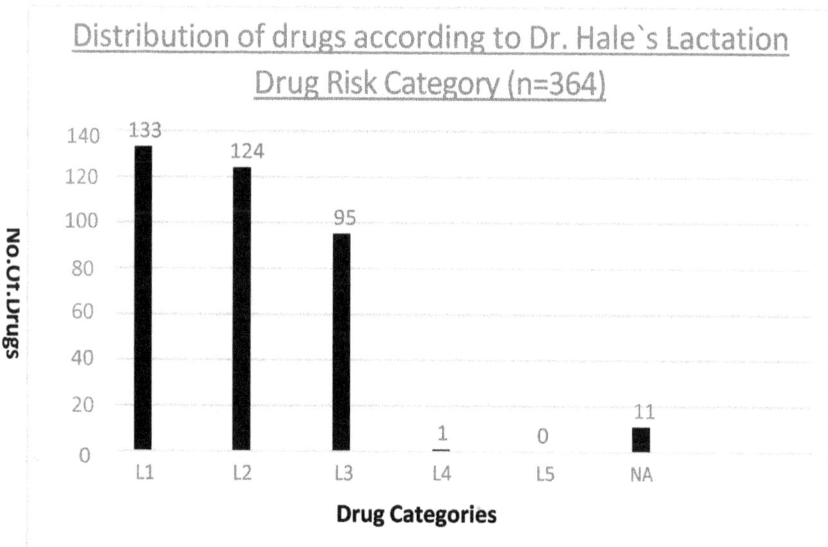

5 CONCLUSION

Lactation is a unique physiological condition in which the use of a few medications is extremely concerning due to the potential teratogenicity of certain medications. The thalidomide tragedy compelled American and International regulatory bodies to establish systematic toxicity testing protocols, a significant step forward in toxicity testing. For the vast majority of prescription drugs on the market, there is a scarcity of reliable research on the safety and effectiveness of medications used during breastfeeding. Hence, while prescribing drugs during lactation, the clinician must review the current literature concerning the drug's safety and choose the drugs after evaluating the safety parameters.

This cross-sectional observational study was conducted in community settings in and around Vijayawada to collect comprehensive information on the prescribing patterns of lactating women. This study helped us understand how common drug use was in lactation.

Dr. Hale's lactation risk category was used to evaluate the medication use of 60 nursing women. According to the data, the sample population primarily used drugs from categories L1, L2, L3, L4, and L5, with L4 medications being the least commonly used and L5 medications being absent. Prescription patterns in breastfeeding were found to be rational, with a few notable exceptions. Doctors were concerned about using safer drugs during breastfeeding, as well as having more faith in evidence-based medicine. Post-natal counselling was provided to participants.

The findings of our study imply that the Lack of knowledge about use of numerous medicines during lactation, especially newer ones, is a barrier to nursing. This emphasizes the need for further research and regulatory attention in lactating women with large sample size in various strata.

STUDY LIMITATIONS:

- Only a small sample was used for the investigation.
- The study and results are applicable only to some specific communities in and around Vijayawada.
- Our study was based only on the information from prescriptions and brief interviews with patients.

Acknowledgment

The authors are thankful to the management of KVSR Siddhartha College of Pharmaceutical Sciences for providing the facilities to carry the work. We are also thankful to the patients who agreed to participate in this study.

6 REFERENCES

1. Spencer, J. P., Gonzalez, L. S., 3rd, & Barnhart, D. J. (2001). Medications in the breast-feeding mother. American Family Physician, 64(1), 119–126.

2. Burkey BW, Holmes AP. Evaluating medication use in pregnancy and lactation: what every pharmacist should know. J Pediatr Pharmacol Ther. 2013 Jul;18(3):247-58. doi: 10.5863/1551-6776-18.3.247. PMID: 24052789; PMCID: PMC3775560.

3. CDC. About breastfeeding. Centers for Disease Control and Prevention. 2021 [cited 2022 Sep 28].

4. Gunatilake R, Patil AS. Drug Use During Breastfeeding [Internet]. MSD Manual Consumer Version. [cited 2022 Sep 28]. Available from: https://www.msdmanuals.com/home/women-s-health-issues/drug-use-during-pregnancy/drug-use-during-breastfeeding.

5. National Institute on Drug Abuse. Substance use while pregnant and breastfeeding [Internet]. National Institute on Drug Abuse. [cited 2022 Sep 28]. Available from: https://nida.nih.gov/publications/research-reports/substance-use-in-women/substance-use-while-pregnant-breastfeeding

6. Hale TW. *Medication and Mother's Milk.* 15th ed. Amarillo, TX: Hale Publishing; 2012.

ASSESSMENT OF MEDICATION ADHERENCE FOR CONTROLLING DIABETES IN ANDHRA PRADESH

CH K V L S N Anjana Male[a], P. Dharani Prasad[b], A. Santhi Sri[c*], Snehitha Kasu[d], Mary Asha Prathi[d] and Anusha Nakka[d]

[a]Dept of Pharmaceutical Chemistry and Phytochemistry, Nirmala college of pharmacy, Atmakuru, Mangalagiri, Guntur district, A.P.

[b]Dept of Pharmacology, Sree Vidyaniketan College of Pharmacy, Tirupati, A.P, India.

[*c]Acharya Nagarjuna University College of Pharmaceutical Sciences, Nagarjuna Nagar, Guntur- 522002, Andhra Pradesh, India.

[d]Nirmala college of pharmacy, Atmakuru, Mangalagiri, Guntur district,A.P.

Corresponding author: santhisriakkiraju20@gmail.com

1 INTRODUCTION

Millions of individuals worldwide are impacted by the considerable and rapidly spreading public health issue known as diabetes mellitus (DM). It quite often coexists with other chronic conditions, and its prevalence is rising year after year to epidemic proportions.[1] China (20.8 million) and the United States (17.7 million) came in second and third, respectively, in terms of the number of individuals with diabetes mellitus in the globe in 2000, at the rear India (31.7 million) and China. 90% of cases of type 2 diabetes, which affects 387 million people in 2009, are anticipated to rise to 483 million and 592 million by the years 2030[2] and 2035[3], respectively. More than 77% of morbidities and 88% of fatalities in impoverished nations are caused by DM.[4, 5]

In India, diabetes is quickly becoming recognized as a possible epidemic.[6, 7] According to the ICMR's National Diabetes Study Report, there were 62.4 million type 2 diabetics and 77 million prediabetics in India in 2011. By 2030, these projected numbers are anticipated to rise to 101 million.[8, 9]

Although effective screening tools and diabetes treatments are available to the urban Indian population, people in rural areas do not frequently have access to these medical benefits. Between urban and rural communities, health resources are unevenly shared, and there are various different types of rural poverty. Food hardship, illiteracy, scarce sanitation, and the occurrence of communicable diseases may all be contributing factors, which raise the possibility that local governments and policymakers are underestimating and under prioritizing the diabetes threat that is rapidly approaching. The ultimate goals of diabetes therapy are glycemic control and preventing early complications, which depend on patient lifestyle modifications and regimen compliance.[10]

Reduced exercise, an unhealthy diet (characterised by a penchant for fatty foods, carbohydrates, and an excessive use of dates, etc.), increasing attractiveness of soft drinks and artificially sweetened fruit juices, and smoking are all contributing to Saudi Arabia's culture. Each of those elements has a direct influence on the risk of obesity, coronary artery disease (CAD), hypertension, and type 2 diabetes mellitus.[11] Unfortunately, chronic disease patients typically fail to change their eating habits and improve their physical activity, despite their doctors' recommendations after the diagnosis has been made. Complications are more likely as a result, especially for people with type 2 diabetes and hypertension.

Diabetes counselling, whether it takes the form of written or verbal patient education that is specifically tailored to the patient or involvement in health education seminars on general topics, has a significant impact on the patient's acquaintance and understanding of the risk coupled with disregarding their health. The key to extending survival and enhancing quality of life is adhering to the doctor's instructions about the pharmaceutical management procedures as well as making basic and long-term modifications in lifestyle. The best method for spreading knowledge and influencing individuals to adopt healthy lives is patient counselling. Adopting a good lifestyle reduces not only the risk of developing illnesses, but also the possibility of complications caused by these illnesses.[12]

Despite the acknowledged advantages of contemporary treatment regimens, medication non-adherence is becoming a bigger worry for healthcare professionals, with one study reporting that it affects more than 50% of diabetic patients. The expenses of medical emergency treatments as a result of non-compliance exceed the total cost of a drug regimen that is followed, so the consequences of non-adherence extend beyond those directly related to health (such as treatment failure, rehospitalisation, and death).[13]

There was a marked improvement in patient compliance in diabetics who engaged in an educational session centered on the importance of anti-diabetes medication and adherence to it.[14]

2 METHOD OF STUDY

2.1 Methodology

For six months, from August to January (2018-19), a prospective observational study was carried out in a rural area of Mangalagiri, Andhra Pradesh. According to the inclusion and exclusion criteria, subjects were hired.

2.2 Inclusion criteria

Individuals older than 18 years who have diabetes mellitus or diabetes and co-morbid ailments.

2.3 Exclusion criteria

- Patients under age of 18.
- Pregnant women and nursing mothers.
- The individuals who are hesitant to take part in the study..

Subjects who meet the requirements of the study are admitted, and each participant signs an informed consent form. The data on demographics, diagnoses, lifestyle factors (eating habits, physical activity, social history), and currently prescribed medications were obtained through a data collecting form, and the details of the subjects were acquired. Every subject gets patient counselling through each review. Each participant was assigned the MMAS-8 (Morisky Medication Adherence Scale) both before and after counselling.

3 RESULTS

A total of 1872 patients were screened over the course of six months, from August to January, and 189 of them were enrolled in the study. We discovered that patients aged 50-60 were more prone to Diabetes than other age groups (Table 1); Diabetics along with Hypertension was more in number than other co morbidities (see Table 2). Surprisingly there was also improvement in the social behaviour and life style parameters which are depicted in Tables 3&4. Medication adherence of patients was evaluated using Morisky Medication adherence scale (MMAS) which was filled by every subject. The obtained result of MMAS was given below in the Table 5.

Table 1 *Age wise distribution of patients.*

Age groups of patients	No. of patients
20-30	12
30-40	24
40-50	42
50-60	64
60-70	36
70-80	11

Table 2 *Patients with Diabetes and co-morbidities.*

List of Co morbidities with Diabetes	No. of Co morbidities
DM + HTN	64
DM +Hypothyroid	21
DM+HTN+Hypothyroidism	14
DM+Others	18

Table 3 *Lifestyle parameters and social history among the patients.*

Lifestyle parameters			
Before counselling		**After counselling**	
Yes	No	Yes	No
56	133	72	117
Social History			
Before counselling		**After Counselling**	
Yes	No	Yes	No
73	116	49	140

Table 4 *Exercise habits of patients.*

Exercise				
	Before Counselling		**After Counselling**	
	Yes	**No**	**Yes**	**No**
Brisky walking	9	180	36	153
Normal walking	26	163	57	132
Others	23	166	23	166

Table 5 *Medication adherence score.*

S.no	QUERY	Before		After	
		Yes	**No**	**Yes**	**No**
1.	Do you sometimes forget to take your medications?	63	126	29	160
2.	Occasionally people skip taking their medications for reasons other than simple ignorance. Have you failed to take your medication on any days within the last two weeks, in your memory?	50	139	38	151
3.	Have you ever altered your medication intake or stopped taking it entirely without telling your doctor because you felt worse from taking it?	24	165	24	165
4.	Do you frequently fail to bring your medicine when you travel or leave home?	73	116	36	153
5.	Did you take all your medication yesterday?	189	0	180	9
6.	Have you ever stop your treatment when you feel your manifestations are under control?	0	189	0	189
7.	For some people, the necessity of taking medication every day is a big hindrance. Do you ever struggle to follow your medication regimen?	0	189	0	189
8.	How regularly do you forget to take all of your medication? A. Hardly ever B. Rarely C. Usually D. Every time	0	189	0	189

Graph pad prism software was used to analyze the obtained results. When compared to before counselling, patient counselling significantly improved their lifestyle modifications and social habits; the p value was 0.139, which is significant. We are also pleased to see that their exercise routine has improved.

Our investigation made it abundantly evident that people consume more carbohydrates (mostly rice), thus we advised the patients to make dietary changes. The p value was discovered to be 0.013, which is extremely significant.

3.1 Medication adherence score

With a mean difference of 4.6 0.47 and a p-value of 0.0001 when compared to before counselling, paired t-test statistics show a significant improvement in medication adherence after counselling.

4 DISCUSSION

In 2016, the estimated prevalence of diagnosed type 1 and type 2 diabetes was 0.55% (1.3 million Indian adults) and 8.6% (21 million Indian adults), respectively.

We gathered information during the course of our six-month study on the prevalence and trends of diabetes mellitus in a rural area (Atmakuru village), and we discovered that it was 10.9%. According a population-based study with 1892 individuals aged (34% of people aged 41 to 50 and 54% of women) showed that older adults were more prevalent in our study. Our results show that patient counselling was successful in helping diabetes patients in this remote location make a few simple but important lifestyle modifications. The dietary standards improved, but some unexpected outcomes emerged: there was a decline in alcohol consumption and smoking, which was significant at a p value of 0.139; there was less exercise, which was reported to be significant at a p value of 0.139. Less exercise was performed, which was proven to be irrelevant at a p value of 0.808, and dietary patterns changed, which has been found to be very significant at a p value of 0.013.

Based on the MMAS-8 score, medication adherence was evaluated in this study. It is an extremely efficient and validated scale that is designed to make it easier to recognize the activities that provide a barrier to medication adherence.

Based on the findings of the study, medication adherence was low (78.8%) prior to counselling and lowered to 19% after counselling.Medication adherence enhanced as percentage values risen from moderate 3.7%, 16.9% to high 57.6%, 25.3%.

5 CONCLUSION

This study led to the conclusion that enhancing the status of health education in rural areas has a substantial impact on expanding public awareness and education, primarily among diabetics. As a result, uniquely designed patient education programmes presented in rural areas will have a significant impact.

Diabetics who involved in an educational session focusing on the importance of anti-diabetes medication and adherence to it showed a significant rise in adherence to treatment in this study. There was also a dramatic change in the patient's social conduct and lifestyle modifications, which can reduce further risk factors in patients. Organizing educational programmes and enhancing patient awareness can certainly assist diabetic patients prevent disease peaks and prevent Diabetes Mellitus from evolving into an epidemic in India.

6 ACKNOWLEDGEMENT

We acknowledge all the participants in the study.

7 CONFLICTS OF INTREST

The authors declare that there was no conflict of study.

8 REFERENCES

1. Aymen Elsous, Mahmoud Radwan, Hasnaa Al-Sharif, *Front. Endocrinol*, 2017, **8**,100.
2. Seema Abhijeet Kaveeshwar, Jon Cornwall, *Australasian medical Journal,* 2014, **7**, 45.
3. Anjana RN, Ali NK, Pradeepa R, *Indian J Med Res*, 2011, **4**, 369.
4. Harries AD, Satyanarayana S, Kumar AMV, *Public Health Action,* 2013, **3**, 3.
5. FarsaeiS, Sabzghabaee AM, Zargarzadeh AH, Amini M, Iran. *Iran J Pharm Res*, 2011, **10**,933.
6. Joshi SR, Parikh RM. *J Assoc Physicians India,* 2007, **55**,323.
7. Kumar A, Goel MK, Jain RB, Khanna P, Australas Med J, 2013,**6**,524.
8. Whiting DR, GuariguataL, Weil C, Shaw *J. Diabetes Res Clin Pract*, 2011, **94**,311.
9. Ponnusankar S, Surulivelrajan M, Ananda moorthy N, Suresh B, *J Patient Education and Counselling,* 2003,**137**,1.
10. Hazzaa. HM, *East Mediter Health J*, 2004, **10**, 663.
11. Aladin Alwan, *East Mediter Health J*, 1997, **3**, 616.
12. HoPM, BrysonCL, Romfield JS. *Journal of the American Heart Association*, 2009, **119**, 3028.
13. Al-Hazzaa HM, *East MediterHealth J*, 2004, **10**, 4.4.
14. OsterbergL, Blaschke T. *N Engl J Med,* 2005, **353**,487.

A STUDY ON PATTERN OF UTLISATION OF BLOOD AND ITS COMPONENTS IN A TEACHING HOSPITAL

Zareen Unnisa[a], Mohammed Fareedullah[a*], Sadia Siddiqua[a], Sheema Jabeen[a], Roshan S[b], Risy Namratha Jamullamudi[c]

[a] Department of Pharmacy Practice, Deccan School of Pharmacy, Hyderabad, TS
[b] Department of Pharmacognosy, Deccan School of Pharmacy, Hyderabad, TS
[c] Department of Pharmaceutical Chemistry, KL College of Pharmacy, Koneru Lakshmaiah Education Foundation, Vaddeshwaram, Guntur, India

*Corresponding Author Email: fareedullah29@gmail.com

1 INTRODUCTION

Blood is a rare, expensive, and valuable resource. The risk of transfusion must therefore always be evaluated before a patient is prescribed blood and blood products.[1] A crucial component of the current healthcare system is blood transfusion. When used properly, blood transfusions can lower the death rate and improve health status.[2] Similar to other medications, blood transfusions can have short-or long-term side effects like transfusion reactions and can spread infections like HIV, hepatitis, syphilis, and others.[3]

Blood parts include packed red blood cells, plasma that has just been frozen, cryoprecipitate, platelet-rich plasma, and whole blood.[1]

It is important to circumvent the improper usage of this source as it is costly and the factors affecting the cost of blood transfusion are as follows:

- Collection cost.
- Cost of laboratory.
- Cost of handling.
- Blood administration cost.[1,3]

Risk of transfusion can only be decreased by close association between the blood bank services and doctors in the management of transfusion process in which they are liable for:

- Enough blood and blood products are available.
- Proper use of blood and its derivatives.[2]

Numerous procedures have been evolved to curtail the improper use of blood components. It includes guidelines, conferences, education, and self-audit to monitor transfusion practice.[2]

A periodic review of blood components and use is required to ascertain the pattern of blood usage in any hospital or health centre.[4] Hence this study was conducted in Owaisi Hospital & Research Centre (OHRC) after obtaining the ethical clearance.

2 AIMS AND OBJECTIVES

The main goal was to find out how blood and its components were used in a teaching hospital.

The study's objectives include:

1. To assess the transfusion procedures used in our hospital over the course of a year
2. To analyse the utilisation pattern of each blood and its products based on age group, gender, disease condition requiring transfusion.
3. To figure out how blood products are used in different hospital departments.

3 MATERIALS AND METHODS

A retrospective study was carried out at the blood bank of a teaching hospital in Telangana. During this study, the number of units of blood and its components (Fresh frozen plasma, packed cells, platelet-rich plasma, and whole blood) issued for period of one year was documented (January 2019 to December 2019) from the record books of blood bank. Each recipient's details were noted down. This included their age, gender, ABO and Rhesus blood groups, the wards or departments where blood units were given, and instructions for using different blood components.

4 STUDY SELECTION

Inclusion criteria

- Transfusion of whole blood, packed cells, platelet-rich plasma, and fresh frozen plasma.
- This study included all the requests for blood and its products sent to our blood bank for the period mentioned above from all clinical departments.

Exclusion criteria:

- Blood or blood components supplied outside our hospital
- This study excluded all incomplete request forms submitted to our blood bank.

Results:

Our blood bank distributed 7638 units of blood and blood components from January 2019 to December 2019. 3594 units of whole blood, 1978 units of packed cells, 985 units of fresh frozen plasma, 998 units of platelet-rich plasma, and 83 units of single donor platelets (SDP) were transfused out of the total transfusions (Table 1).

Month	Whole blood	Packed cells	Fresh frozen plasma	Platelet Rich Plasma	Single Donor Platelet (SDP)
January	332	144	76	79	2
February	286	146	77	73	2
March	300	168	78	84	2
April	299	163	83	80	0
May	317	164	83	83	0
June	282	165	86	80	0
July	306	177	84	88	4
August	299	175	78	88	8
September	286	162	86	90	27
October	297	172	82	92	25
November	290	172	94	85	11
December	300	170	78	76	2
Total	**3594**	**1978**	**985**	**998**	**83**
Percentage	**47.05%**	**25.89%**	**12.89%**	**13.06%**	**1.08%**

Table 1 *Blood and its components issued in one year.*

Substantial number of units of blood was supplied to the department of General Medicine accounting for 37.8% followed by Gynaecology and Obstetrics and General Surgery. (18.6% and 15.2%) respectively and small number of units were supplied to operation theatre accounting for only 0.21%. (Figure 1)

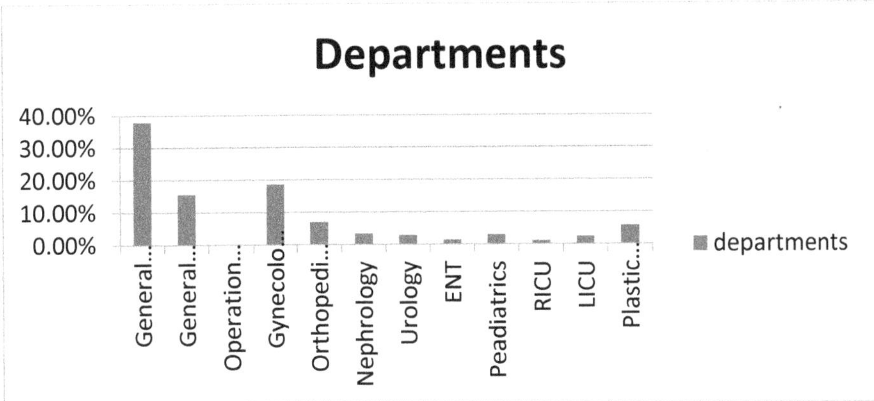

Figure 1: Bar Graph Showing Supply of Blood to Different Departments

A greater percentage of patients (n=3128/7638, or 40.9%) were between the ages of 41 and 64. A smaller percentage (n=116/7638, or 1.5%) were in the age group of <1. Male patients outnumbered female patients by a ratio of 1.02:1. (Table 2)

Age	Male	Female	Total No. of Patients
< 1	79	37	116
1-16	281	218	499
17-40	1050	1930	2980
41-64	1890	1238	3128
≥ 65	569	346	915
Total	**3869**	**3769**	**7638**

Table 2 *Distribution of Patients According to Age and Gender.*

The various blood types of 7638 patients are shown in Table 3. Blood group O positive patients made up the largest percentage of patients (32.4%), followed by blood group B positive patients (31.3%). Blood group AB negative patients made up the smallest percentage (0.2%). (Table 3)

Blood group	Rh positive	Rh negative	No. of patients	Percentage
A	2134	48	2182	28.5%
B	2400	80	2480	32.4%
O	2476	91	2567	33.6%
AB	389	20	409	5.35%
Total	**7399**	**239**	**7638**	**100%**

Table 3 *Different Blood Group of Patients.*

Anaemia was the most frequent sign for all blood components combined (48.6%), followed by elective surgeries (31.6%). (Table 4)

Table 4 *Indications for which blood products were issued.*

Indications	Whole blood	Packed cells	Fresh Frozen plasma	Platelet Rich Plasma	SDP	Total	Percentage
Surgery	1274	1140	-	-		2414	31.6%
Anaemia	1580	1381	753	-		3714	48.6%
Internal bleeding	-	-	254	-		254	3.3%
Thrombo-cytopenia	-	-	-	623	38	661	8.6%
Dengue	-	-	-	294	45	339	4.4%
Malaria	3	-	-	59		62	0.81%
Dialysis	64	130	-	-		194	2.53%

The total number of units discarded in a period of one year were 656 and reasons for discarding was date of expiry, HbsAg, HCV and HIV. (Table 5)

Table 5 *Number of units discarded.*

Reason	**No. of units discarded**
HbsAg	6
HIV	1
HCV	1
VDRL	0
Malarial parasite	0
Expiry of shelf life	648
Total	**656**

5 DISCUSSION

The provision of blood transfusion services is an essential component of modern healthcare. The blood transfusion system's main objective is the efficient and less wasteful use of blood and its products.[5, 6]

In blood banks, vital part of the quality control & assurance are internal audits. The goal is to give patients' blood components that are efficient and secure. The Food and Drug Administration (FDA) considers blood and its products to be drugs because their use is intended to help patients therapeutically. The FDA uses pharmaceutical industry standards to collect, test, store, and supply safe blood. This serves to reiterate the quality of blood and its constituent parts.[4, 7]

The blood bank must be able to supply the demand for this life-saving product while simultaneously examining and evaluating the current trends of blood ordering systems in order to prevent misuse of blood and its components, which could result in a shortage of blood for someone in a life-threatening situation. Numerous authors have emphasised the significance of internal audits and education campaigns that emphasise the proper selection of blood and its products. They claimed that following such audits, inappropriate patient requests for blood and its components and their misuse significantly decreased.[4, 8, 9, 10, 11, 12, 13]

After the proper cross-matching and screening, 7638 units of whole blood and its derivatives were distributed for this study. Substantial amounts of whole blood were used out of the total units issued, and this result was consistent with studies by Joshi et al. [1] who discovered that whole blood was used more frequently than other components. Packed red cells were in substantial number when compared to other components like FFP, PRP and SDP which is correlating with the findings of Anshoo et al. Venkatachalapathy and Subhashish also documented increased utilisation of packed red cells among blood components.[14, 15]

In our study males and females received almost same units of blood transfusion which contrasts with the studies done by Mathew et al., and Bansod et al. They noted males received a greater number of units of transfusion when compared to females.[16, 17] In this study, majority of blood units were issued to General Medicine department followed by Gynaecology and Obstetrics and General Surgery which was in contrast with the finding of Venkatachalapathy & Subhashish & Giriyan SS.[6, 15] Anemia was the most frequent cause of blood transfusions, followed by elective surgery, which was in line with research by Mallikarjuna Swamy CM and Gaur DS.[3, 4]

We discovered that the blood groups O positive and B positive were more prevalent in the study subjects. According to studies by Mallikarjuna Swamy

CM, AB negative blood made up 0.2% of the patients, making it the least prevalent blood type.[3]

We further investigated the reasons for the blood component discards and found that 656 units of blood components were discarded annually due to issues such shelf-life expiration (648 units), HbsAg positivity in 6 samples, HIV and HCV in 1 unit each, and 648 units due to shelf-life expiration. It is important to study the method of using blood and its blood components for monitoring transfusion procedures for quality and to study different part need to enhance better component separation to avoid wastage and shortage.[18]

6 CONCLUSION

This study, carried out in a tertiary care centre, provides new information on the functions of blood and its constituent elements. Therefore, it is crucial to regularly review how blood is used and what goes into it to ensure that it isn't being used in a way that might leave someone in a situation where they don't have enough blood to survive.

7 ACKNOWLEDGEMENT

Authors are thankful for the services provided by Mr. Ibrahim, In charge Blood Bank, OHRC, for his consistent support and cooperation during the data collection process. Authors also thank Mr. Mohammed Misbah Ul Haq & Mr. Mohd Munaf ur Razzak, PharmD Interns, Deccan School of Pharmacy for their valuable inputs in editing the manuscript.

8 REFERENCES

1. R. J. Anil, J. A. Rakesh and S. K. Anjali, *International Journal of Health Sciences & Research*, 2014, **4(8)**: 38.

2. P. A. Vijay, M. Akhtar and S. D. Mahore, *International Journal of Biomedical and Advance Research,* 2013, **4(9)**: 657.

3. S. C. M. Mallikarjuna, P. Shashikala and G. U. Kavita, *International Journal of Current Research and Review*, 2012, **4(12)**: 47.

4. S. G. Dushyant, N. Gita and C. Neena, *Indian Journal of Hematology and Blood Transfusion,* 2009, **25(3)**: 91.

5. E. A. Pozo, P. M. Rosales, Cd. Almeidaa-Neto, *Pan American Journal of Public Health,* 2015, **37(6)**: 435.

6. S. S. Giriyan, H. D. Chethana and N. Sindhushree, *Journal of Blood & Lymph,* 2017, **7(2)**: 1.

7. C. Hillyer, L. Silberstein, P. Ness, *Blood Banking and Transfusion Medicine,* Churchill Livingstone, New York, 2nd Edition, 2006.

8. C. J. Yeh, C. F. Wu and W. T. Hsu, *Transfusion Audit of Fresh Frozen Plasma in Southern Taiwan. Vox Sanguinis,* 2006, **91(3)**: 270.

9. T. E. Hawkins, J. M. Carter and P. M. Hunter, Transfusion Medicine, 1994, **4(1)**: 45.

10. E. Greeno, J. McCullough and D. Weisdorf, *Transfusion*, 2007 **47(2)**: 201.

11. B. Diedrich, M. Remberger and A. Shanwell, *Transfusion,* 2005, **45(7)**:1064.

12. R. Zimmerman, M. Buscher and C. Linhardt, *Transfusion,* 1997, 37(10): 1075 – 1083.

13. C. Vishwanathan, R. Jain and M Kamath, *Indian Journal of Hematology and Blood Transfusion,* 1999, **17**: 26.

14. A. Anshoo, B. Saidunnisa and C. Meghna, *International Journal of Science and Research,* 2013, **2(8)**: 56.

15. T. S. Venkatachalapathy and D. Subhashish, *Journal of Blood and Lymph,* 2012, **2(3)**: 1.

16. P. N. Bansod, N. Jethani and G. Pachori, *International Journal of Medical Sciences and Public Health*, 2015, **4(6)**: 787.

17. A. S. Mathew, S. S. Kurian, N. P Sundaresan, *Academic Medical Journal of India,* 2014, **2(1)**: 28.

18. A. B. Hulwan, S. V. Jagtap, S. R. Kanetkar, *Journal of Dental and Medical Sciences,* 2017, **16(6)**: 90.

A STUDY OF QUALITY OF LIFE IN PATIENTS WITH BREAST CANCER ON CHEMOTHERAPY IN TERTIARY CARE HOSPITAL

Kunta Naresh*, CH Roshini, CH Chandasree, D Kiranmai, G Satya Sharanya

Department of Pharmacy Practice, Malla Reddy Pharmacy College, Dhulapally, Secunderabad – 500100 Telangana, India.

*Corresponding Author E-Mail ID: kuntanaresh785@gmail.com

1 INTRODUCTION

Breast cancer is the most common type of cancer in women, it accounts for 27% of all cancers in women in India. The survival of cancer becomes more typical in the higher stages. In India, more than 50% of the women suffer from stage 3 & 4 of breast cancer.[1] Post cancer survival for women with breast cancer was 60%. In oncological researches quality of life survey is important in health care. Cancer affects different aspects of Quality of life. Therefore, it is important for the healthcare professionals to know about the impact of breast cancer treatment on patients QoL. The purpose of this study is to assess the quality of life in breast cancer patients.[2]

Now-a-days Quality of Life is important aspect to study. It is defined as a multidimensional concept of both positive and negative aspects of dimensions such as Physical, Social, Functional and Emotional domains as well as the negative aspects of physical discomfort and other symptoms caused by a diseases or its treatment.[3] To survey QoL, we interviewed the patients at the beginning of the chemotherapy and on completion of the chemotherapy cycle by using FACT-B Questionnaire having 0-4 (0-fully ambulatory without symptoms; 1-fully ambulatory with symptoms; 2-requires bed rest < half of the walking day; 3-requires bed rest > half of the walking day; 4- bedridden) scoring depending on severity.

2 METHODOLOGY

A Hospital based observational, prospective study was conducted on patients (n=111) with breast cancer on chemotherapy for 6 months in Oncology Department. The data was collected regarding Demographic details, Complaints, Menstrual history, Family history, Marital status and educational status; interviewed the patient by using FACT-B Questionnaire and scores were allotted from 0 to 4 depending on severity.[4]

2.1 Statistical methods

The statistical tests like "t-test: paired two samples for mean" and Kruskal Walli's test were conducted to estimate the Quality of Life of the patients.

2.2 Inclusion and exclusion criteria

We included female patients of age $\geq$ 18 years, histologically confirmed breast cancer and who are under chemotherapy and are willing to give informed consent for the study.

We excluded males, pregnant women, and women < 18 years, patients under Radiotherapy, Hormonal therapy and Psychiatric disorders and patients with other previous or concurrent malignancies and who are not willing to participate in the study.

3 RESULTS

In our study 111 patients were enrolled as per our study criteria. The roles of several factors were studied and assessed. This study helps clinicians in decision making.

3.1 Age

Among 111 patients the prevalence of breast cancer is higher in the age group of 31- 40 years with 31 patients (28%) followed by 27 patients (24%)

were between 51-60 years, followed by 25 patients (23%) were between 41-50 years (See Table 1).

Table 1 *Distribution of patients according to age groups.*

Range	No. of patients	Percentage (%)
21-30	8	7
31-40	31	28
41-50	25	23
51-60	27	24
61-70	14	13
71-80	6	5

3.2 Complaints

Among 111 patients, lump in the right breast is the most common complaint in 32 patients (28.8%) (See Table 2 and Figure 1).

Table 2 *Distribution of patients according to Complaints.*

Complaints	No. of patients	Percentage (%)
Change in size of breast	5	4.5
Lump in the left breast	25	22.5
Lump in the right breast	32	28.5
Pain & discharge from the nipples	6	5.4
Pain & lump over breast	5	4.5
Pain & swelling over breast	30	27.02
Swelling over breast	8	7.2

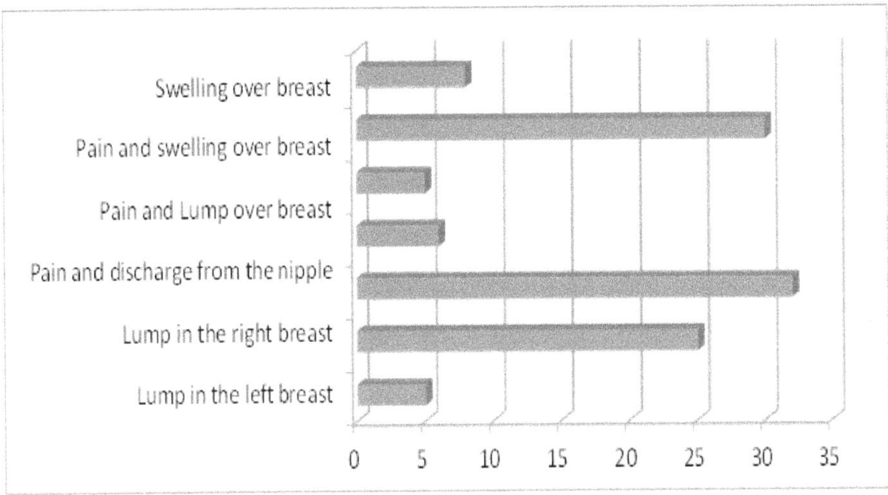

Figure 1 Graphical representation of patient's complaints.

3.3 Menstrual History

Menstrual history was assessed in 111 patients (See Table 3 and Figure 2).

Table 3 *Distribution of patients according to Menstrual history.*

Menstrual History	No. of patients	Percentage (%)
Regular menstruation	26	9
Irregular during chemo	19	17
Menopause	54	48
Hysterectomy	26	26

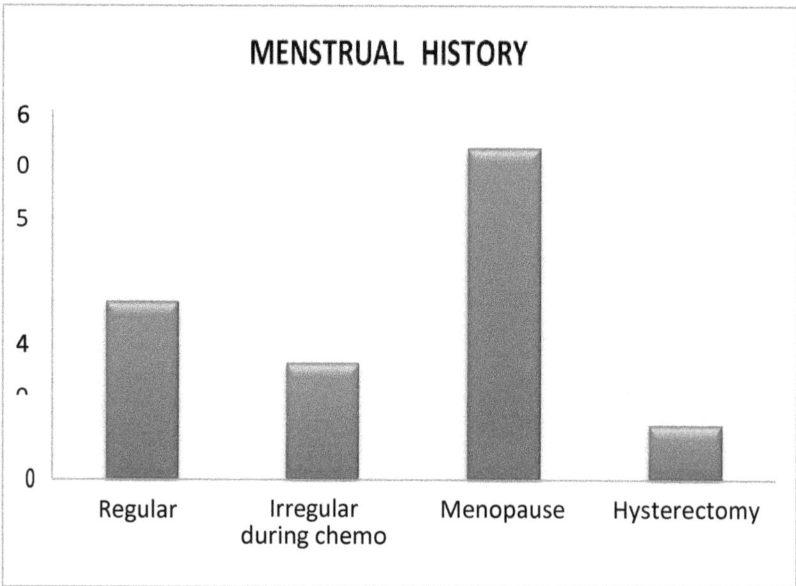

Figure 2 *Graphical representation of patients according to Menstrual History.*

Breast cancer risk was found more in patients with menopause (48%) and found least in patients with hysterectomy.

3.4 Family History

Among 111 patients, the familial history of breast cancer was found in 12 patients(11%) and 99 patients (89%) do not have family history (See Table 4 and Figure 3).

Table 4 *Distribution of patients according to Family History.*

Family History	No. of Patients	Percentage (%)
Yes	12	11
No	99	89

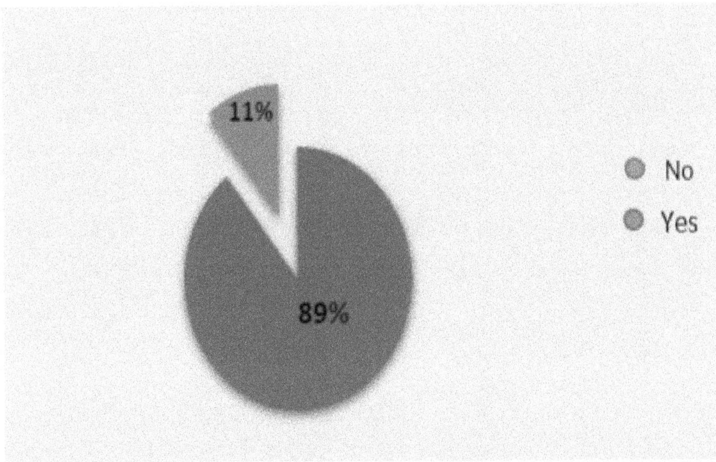

Figure 3 *Graphical representation of patients based on Family History.*

3.5 Quality of Life

3.5.1 Physical well-being (PWB). The comparison of Quality of life before and after the chemotherapy in 111 patients was assessed. The questionnaire was asked at the beginning of the chemotherapy and on completion of chemotherapy cycle (See table 5).

Comparison of factors effecting Physical well-being

When we assessed factors effecting Physical well-being before and after chemotherapy, we found that all the factors got worsen after chemotherapy, among them the most influential factor that is affecting the quality of life of breast cancer patients was nausea followed by bothering about side effects (See Figure 4).

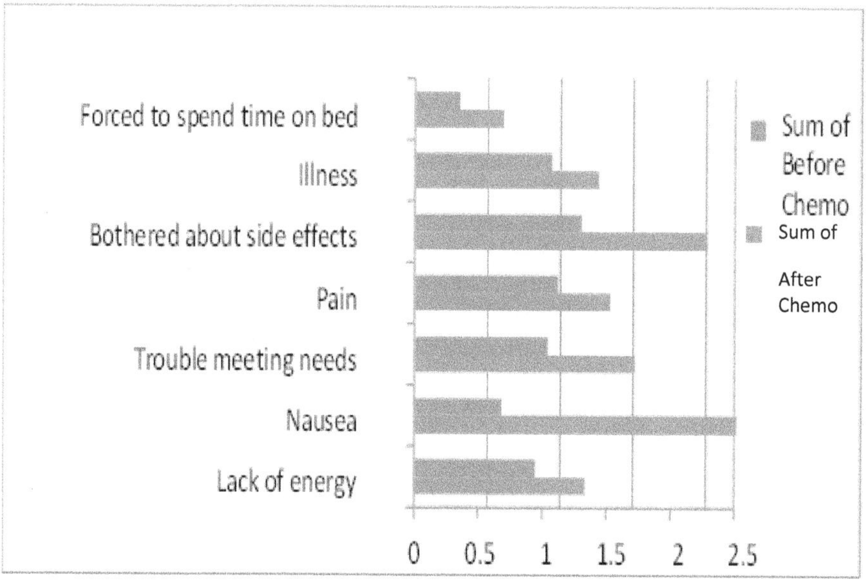

Figure 4 *Graphical representation of factors affecting Physical well-being.*

3.5.2 Social well-being (SWB). The questionnaire was asked about support to the patient from their family members, and/or friends and related questions evaluated for the score maximum of 28 (See table 5).

Comparison of factors affecting Social well-being

When we assessed factors effecting Social well-being before and after chemotherapy. We found that all the factors except the sex life got better after chemotherapy (See Figure 5).

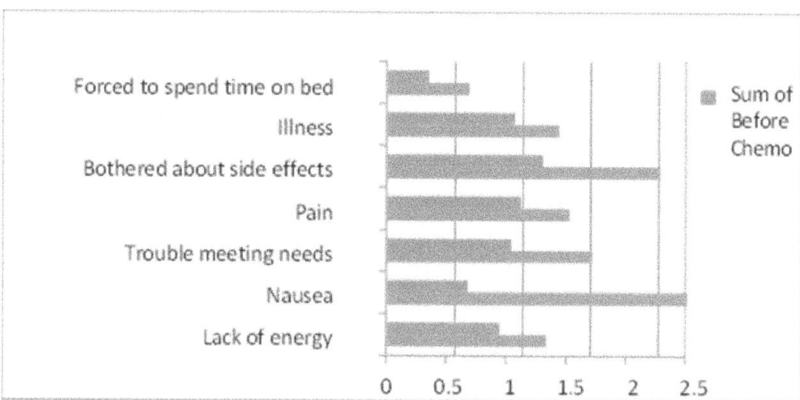

Figure 5 *Graphical representation of factors effecting Social well-being.*

3.5.3 Emotional well-being (EWB). The questionnaire was asked about their emotional status, disease condition, their feelings towards future.

Other related questions were additionally evaluated for the score maximum of 24 (See table 5).

Comparison of factors affecting Emotional well-being

When we assessed factors effecting Emotional well-being before and after chemotherapy, we found that after chemotherapy the factors like worrying about dying, nervousness and hopelessness were reduced (See Figure 6).

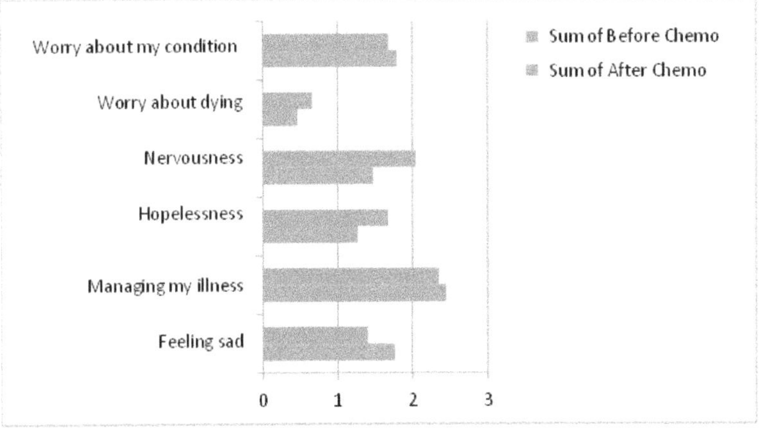

Figure 6 *Graphical representation of Emotional well-being.*

3.5.4 Functional well-being (FWB). The questions asked about functional well-being like their working environment, acceptance of illness and other associated questions evaluated for the score maximum of 28 (See table 5)

Comparison of factors affecting Functional well-being

When we assessed factors effecting Functional well-being before and after chemotherapy, we found that after chemotherapy all the factors got worsen. The most influential factors were the ability to enjoy work and things they do (See Figure 7).

Figure 7 *Graphical representation of Functional well-being.*

Breast cancer subscale (BCS)

Additional questions like shortness of breath, hair loss, swollen limbs and body pains, change in weight were asked.

Comparison of factors affecting Breast cancer subscale

When we assessed factors effecting Breast cancer subscale before and after chemotherapy, we found that among all factors pain in certain parts of the body has reduced after chemotherapy. The most influential factor is bothering about hair loss (See Figure 8).

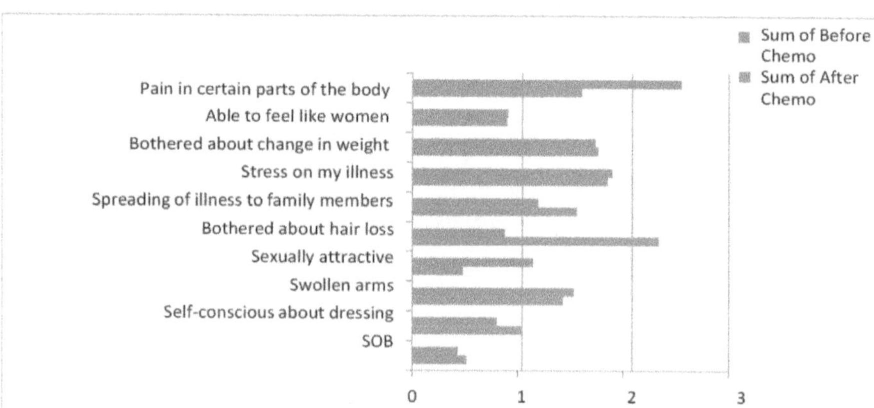

Figure 8 *Graphical representation of breast cancer subscale(additional)*

Table 5 *Comparison of Quality-of-life domains before and after chemotherapy*

	PWB		**SWB**		**EWB**		**FWB**		**BCS**	
	Before	After	Before	After	Before	After	Before	After	Before	After
Mean	5.7	9.01	17.18	17.8	9.82	9.23	19.56	16.9	1.45	12.9
Variance	8.7	25.9	15.27	10.8	11.97	19.4	15.12	42.7	16.23	19.9
Observations	111	111	111	111	111	111	111	111	111	111
Pearson Correlation	--0.117		0.27		-0.072		0.27		0.003	
Hypothesized MD	0		0		0		0		0	
df	110		110		110		110		110	
t-Stat	-5.64		-1.52		1.07		4.22		-2.58	
P(T<=t) one-tail	6.41		0.06		0.14		2.15 E-05		0.005	
t Critical one tail	1.65		1.65		1.65		1.65		1.65	
P(T<=t) two-tail	1.28 E-07		0.12		0.28		5.03 E-05		0.011	
t Critical two-tail	1.98		1.98		1.98		1.98		1.98	

p-value compared at α=0.05 by student t-test for paired samples.

The scores were calculated by using "Student t-test for paired samples" and the meanvalues of each Quality-of-Life domain before and after chemotherapy were found. In PWB, FWB and BCS domains the values

show significant difference at α=0.05, it indicates that Quality of life has decreased after chemotherapy. In SWB and EWB domains the values show no significant difference at α=0.05, it indicates that Quality of Life remained unchanged before and after chemotherapy (See table 5).

Comparison of Quality-of-life domains

Table 6 *Comparison has been done among the QoL parameters before and after chemotherapy*

Domains	Mean Rank	
	Before Chemo	**After Chemo**
PWB	105.1	179.0
SWB	403.73	410.75
EWB	262.52	230.92
FWB	451.81	380.98
BCS	166.42	182.44

Before chemotherapy, it has been identified that Quality of life is poor in Functional well-being (FWB), followed by Emotional well-being (EWB). Comparatively the least effected parameter is Physical well-being (PWB). The patients are getting good support from family and/or friends *i.e.,* social well-being (SWB) that also complies with the Kruskal-Walli's result.

After chemotherapy, it has been identified that Quality of life is poor in Functional well-being (FWB) followed by Emotional well-being (EWB). Comparatively the least effected parameter is Physical well-being (PWB). The patients are getting good support from family and/or friends *i.e.,* Social well-being (SWB) that also complies with the Kruskal-Walli's result (See table 6).

4. DISCUSSION

In this study the prevalence of breast cancer was seen more in the age group between 31-40 years (28%) followed by age group between 51-60 years (24%). A Safaee *et al (2008).,*[5] reported that 48.27% of the patients were in the age group of 31-40 years.

Among 111 patients enrolled 32 patients (28.8%) had lump in the right breast. It reveals that strong relationship exists between breast cancer and lump in the breast.

Menopause is one of the risk factors for developing breast cancer. In our study among 111 patients 54 patients (48%) are menopausal women. Maria Lavdaniti *et al (2019).*,[6] study showed there is a significant effect of menopausal status on physical role.

Family history is one of the risk factors for incidence of breast cancer. In our study we found that 12 patients (11%) have familial history of breast cancer. MR Ataollahi *et al.*,[7] reported that family history is fixed and unchangeable risk factor for developing breast cancer.

Literacy is important contributing factor for quality of life in breast cancer patients. Educational status was recorded in our study; we found that 55% of the women were illiterate. B Moghimi *et al(2008).*,[5] the level of education in their study was relatively low (17.6%).

Quality of life was assessed, in physical well-being the most affected factor was nausea, in social well-being family acceptance of disease was quite good, in emotional well-being after chemotherapy the factors like worrying about dying, nervousness and hopelessness were reduced, in functional well-being most influential factors were the ability to enjoy work and things they do and in breast cancer subscale. We found that the most effected factor is bothering about hair loss.

When we assessed QoL before and after chemotherapy by using "Student t-test method" we found that there is a statistical difference between mean values before and after chemotherapy in the following domains PWB, FWB and BCS and no statistical difference between mean values before and after chemotherapy in SWB and EWB.

All Quality-of-life domains were compared by with Kruskal-Walli's test by using IBM SPSS software and found that Functional well-being domain is most affected. Social well-being was quite good.

5. CONCLUSION

Quality of life is important issue in healthcare. After chemotherapy quality of life of breast cancer patients will get affected. Our study was based on methodological approach of collecting data, interviewing patient using FACT-B questionnaire, calculating means using Student t-test paired samples and for comparing QoL domains we used Kruskal-Walli's test.
Higher prevalence of breast cancer is seen in age group between 31-40 years. The most common complaint is lump in the breast. Menopausal women are at higher risk of developing breast cancer. By using student t-test we found that there is a statistical difference between mean values

before and after chemotherapy in PWB, FWB and BCS. We utilized Kruskal-Walli's test to compare all the domains and found that Functional well-being is poor.

From the findings of our study we conclude, it is important that healthcare providers should inform patients about both long-term and short-term side effects of chemotherapy and medical issues.

REFERENCES

1. Akshay kudpaje, Aparna Subraya Gangoli, *Cytecare Statistics of Breast Cancer in India,* 2018

2. Safaee A, Moghimi-Dehkordi B, Zeighami B, Tabatabaee, *Predictors of quality of life in breast cancer patients under chemotherapy.*, 2008, **45,** 107-111.

3. Skevington SM, Lotfy M, O'Connel KA, WHOQOL Group, *The World Health Organization's WHOQOL-BREF quality of life assessment: psychometric properties and results of the international field trial*, 2004, **13(2)**, 299-310.

4. Brady M.J, Cella D.F, Mo F, Bonomi A.E, *Reliability and validity of the Functional Assessment of Cancer Therapy – Breast quality-of-life instrument. Journal of Clinical Oncology* 1997, **15(3)**, 974-986.

5. Maria Lavdaniti, Dimitra Anna Owens, Polixeni Liamopoulou, Kalliopi Marmara, *Factors influencing quality of life in breast cancer patients six months after the completion of chemotherapy,* 2019, **7**, 26.

6. MR Ataollahi, J Sharifi, MR Paknahad, and A Paknahad, *Breast cancer and associated factors: a review* ,2015, **8,** 6–11.

PERCEPTION KNOWLEDGE AND PRACTICES REGARDING MENSTRUATION AMONG ADOLESCENT GIRLS AND REPRODUCTIVE AGE GROUP WOMEN

Sukeerthi Katari*, Naresh Dumala*, Piyush Bandela, Shaik Mohammad Noor, Maha Lakshmi Kammili, Fathima Asra

Department of Pharmacy, Koneru Lakshmaiah Education Foundation, Vaddeswaram, AP, India

* Corresponding email address: sukeerthikatari@gmail.com;

nareshpharma2020@gmail.com

1 INTRODUCTION

Menstruation is defined as a cycle of a series of natural changes in hormone production and the structures of the uterus and ovaries of the female that make pregnancy possible. Menstruation is a phase in every female's life, especially in adolescent girls to middle-aged women. The menstrual cycle is an every-month recurring cycle where a female goes through multiple mood alterations, spasmodic pains, and serious bodily hormonal changes.[1] Menstruation in females (menarche) starts at a mean age of 13 years. The menstrual cycle has separate phases in it, such as menstruation, follicular phase, ovulation, and the luteal phase, usually, the cycle longs up to 28 days, where the menstrual flow might occur every 21 days to 35 days, and may last two to seven days.[2] Menstruation is still considered unholy or women during which period may be isolated in rural villages and isolated parts of India, and the response to menstruation is dependent on awareness and understanding of the issue.[3, 4] Teenage pregnancies and unsafe abortions have already emerged as important issues that can exacerbate current issues like, reproductive tract infections /sexually transmitted infections/Human immunodeficiency virus /Acquired immune deficiency syndrome among adolescent women.[5] It has been observed that 40-45 percent of adolescent females experience menstruation issues.[6] They are caused mostly by psychosocial stress and emotional changes. Moreover, poor personal cleanliness and dangerous sanitary conditions have predominantly contributed to gynaecological issues in teenage females.[7] There has also been high evidence of reported cases of infections due to lack of hygiene during menstruation.[3, 8] The main objective of the study was to identify the basic knowledge, perceptions of women about menstruation and also identifying the practices of them during menstruation and with the help of collected data from the participants. The study intended to provide the right knowledge, hygiene practices and also helps to overcome the myths regarding menstruation.

2 METHODOLOGY

2.1 Study location

This is a cross-sectional and observation based study, conducted in the localities in and around the Vijayawada city and Mangalagiri town of Andra Pradesh state in India. The study comprised a sample size of 306 female participants who met the requirements of inclusion criteria. The study was performed for a period of six months, out of which five months were taken for data collection and one month for data analysis. The participants were given with a questionnaire, the need of study and contents of it were clearly explained in English and local languages. Data of participants who accepted the Informed consent was subjected for analysis and rest was excluded. The entire data was kept confidential.

2.2 Design

2.2.1 Data collection form. The data collection form used for collecting the data for the present work include below mentioned points as,

A. Demographic Information including human subject name, age, occupation, socio-economic status
B. Age at the time of menarche
C. Persons who informed the girls about menstruation
D. Fear experienced by the person after knowing about menstruation.
E. Girls for whom teacher was the source of information
F. Frequency of emotional changes experienced by them at menarche
G. Bathing frequency during menstruation
H. Are girls being isolated at home during menstruation
I. Usage of napkin/clothes during menstruation
J. Are they preserving their clothes during menstruation
K. Number of sanitary pads used during menstrual period
L. Frequency of cleaning private parts during menstruation
M. Behavior of family members during menstruation
N. Do they discuss their menstrual problems with others
O. Person with who they discuss the problems

P. Provision of adequate water in the school toilets
Q. Do the onset of menarche impacts the capacity of attainment of women to conceive
R. Food habits during menstruation

2.2.2 Informed consent. Before involving the participants in the study, they were clearly explained about the contents and need of the study. Data of participants who accepted to use their data was only used to extract the results. All the data was kept confidential.

2.3 Eligibility Criteria

Inclusion criteria: Female participants who attained menarche and willing to participate.
Exclusion criteria: Female participants who have not attained the menarche and those in post-menopausal phase were excluded from the study.

2.4 Data Collection

The study enrolled a total of 306 participants who met the inclusion criteria. Data collection form with the appropriate questionnaire as mentioned in the previous section was given to the female participants. The contents of the data collection form were clearly explained to the participants.

2.5 Data Evaluation

The data of only 300 participants who accepted the informed consent form was only used in the evaluation procedure. Data collected was handled and analyzed for comparative interpretational studies using MS Office tools for windows.

3 RESULTS

3.1 Age distribution and socio-economic status

The Cross-sectional observational study was performed on female participants of both adolescent and reproductive age groups. The basic knowledge, perception, and practices during menstruation were collected in the study. Ddistribution of age and Socio- economic status data among all the participants of the study is presented in Table 1.

Table 1. *Age distribution and socio-economic status of participants (n=300).*

Variables	Frequency (n=300)	Percentage (%)
Age distribution		
16-20	169	56.3
21-25	98	32.7
26-30	29	9.7
31-35	4	1.3
Socio- economic status		
Lower- Middle Class	20	6.7
Middle Class	240	80.0
Rich	1	0.33
Upper –Middle Class	39	13.0

Among the participants, 56.3% (192) fall under the age group of 16-20, 32.7% (98) are in the age group 21-25, 9.7% (29) belong to the 26-30 age group, and 1.3% (4) belong to the 31-35 age group. The socio-economic status of the participants revealed that majority of the participants, 80% (240) fall under middle class category, whereas 13% (39) belong to upper middle class, 6.7% (20) belong to lower middle class and 0.33% (1) belong to rich category.

3.2 Source of information about menstruation

Date of information about menstruation and the person who provided the same is presented in Table 2. Out of which most participants 77.3% (232) responded as their mother was their source of information, 12.6% (38) received the information about menstruation from their friends and 6.7% (20) got the education on menstruation by the teachers whereas 3.3% (10) received the same information from other sources. Also, data about the teacher being the source of information to clarify the doubts 49.7% (149) participants clarified their doubts, and 50.3% (151) were responded negatively. The same is also included in the data table 2.

Table 2. *Person who informed them first about menstruation and teacher being source of information (n=300).*

Variables	Frequency (n=300)	Percentage (%)
Person Informed them first about menstruation		
Friends	38	12.6
Mother	232	77.3
Others	10	3.3
Teacher	20	6.7
The teacher is the first source of information		
Yes	149	49.7
No	151	50.3

3.3 Subjects feared after knowing about the menstruation

Information about the fear experienced by them after knowing about menstruation is presented in Table 3. Out of all, 26.0% (78) participants experienced more fear, 25.7% (77) less fear, and, 48.3% (145) did not fear as such. Also in the table 3, data on the frequency of emotional changes experienced by them during the phase of menarche is available. Majority of the participants, 47.3% (142) experienced more emotional changes and 41.7% (125) experienced less emotional changes. However, 11.0% (33) did not experience any kind of emotional changes during menarche.

Table 3. *Fear experienced after knowing about menstruation and frequency of emotional changes at menarche (n=300).*

Variables	Frequency (n=300)	Percentage (%)
Fear experienced after knowing about menstruation		
Less	77	25.7
More	78	26.0
Nothing as Such	145	48.3
Frequency of emotional changes at menarche		
Less	125	41.7
More	142	47.3
No	33	11.0

3.4 Menstrual hygiene practices

The data on different menstrual hygiene practices, usage of either napkins or cloth during menstruation are presented in Table 4. Among the participants, 91.7% (275) use napkins whereas 8.3% (25) use cloth during menstruation. When enquired on the information about preserving cloth for re-use during menstruation, majority 96.7% (290) of the participants responded like they do not preserve cloth for reuse, whereas, 3% (10) of the participants responded as they do preserve the cloth for reuse purpose. Table 4 also depicts several pads used by participants, during the menstrual period. 43.3% (130) of the participants use three in a day whereas 33.7% (101) use two in a day. 19.3% (58) use four pads in a day, and 3.7% (11) use only one in a day. The bathing frequency during menstruation showed that 8.3% (25) of the subjects take bath three times a day and the 64.0% (192) of the subjects two times in a day. However, 27.0% (81) of all the subjects take bath once in day and 0.7% (2) do not take bath in a day. Table 4 also shows the information on hygiene practices like cleaning their private/intimate parts during menstruation. Out of all, 93.3% (280) participants follow the hygiene practices and remaining were not. 81.3% (244) of the participants have the provision of adequate water in school/college toilets s and 18.7% (56) do not have adequate provisions.

Table 4. *Usage of napkin, clothes the frequency of changing napkin and preserving, bathing habits and cleaning private parts and provision of water at college toilets (n=300).*

Variables	Frequency (n=300)	Percentage (%)
Usage of napkin/cloth during menstruation		
Napkin	275	91.7
Cloth	25	8.3
Preserving the cloth during menstruation for re-use		
No	290	96.7
Yes	10	3.3
Number of pads used during the menstrual period		
One in a Day		
Two in a Day	11	3.7
Three in a Day	101	33.7
Four in a Day	130	43.3
Bathing frequency during menstruation	58	19.3
Three times a day		
Two times a day	25	8.3
Once a day	192	64.0
None in a day	81	27.0
Cleaning private parts more often during menstruation	2	0.7
No		
Yes	20	6.7
Provision of adequate water in your college toilets	280	93.3
No		
Yes	56	18.7
	244	81.3

3.5 Behavioural responses during menstruation

When the participants were enquired about the behavioural responses like whether they were being isolated at home during menstruation or not interesting data was recorded and the same is shown in the Table 5.

Table 5. *Isolation at home, behaviour of family members, discussion of problems with others and food habits during menstruation (n=300).*

Variables	Frequency (n=300)	Percentage (%)
Isolated at home during menstruation		
No	200	66.7
Yes	100	33.3
Does family behave differently with you during menstruation		
No	249	83.0
Yes	51	17.0
Do you discuss your menstrual problems with others		
No	54	18.0
Yes	246	82.0
Food habits during menstruation		
Sweet cravings	181	60.3
Spicy food	59	19.6
Normal /Regular food	32	10.6
Specific Food	28	9.3

66.7% (200) of the women did not isolate themselves and there were no changes during menstruation when compared with other days. However, 33.3% (100) women isolated themselves and they have also noticed the change in the behaviour of other family members. 82.0% (246) of the women discuss their menstrual problems with others confidently, whereas 18.0% (54) of the women in study population faced difficulty in discussing the same with others. The information about persons with whom these women discuss their menstrual problems revealed that 68.7% (169) of the participants discuss with their mother, and 28.0% (69) of the participants with their friends, 2.8 % (7) with others, and 0.4% (1) with the teachers in their school. Information about food habits during menstruation revealed 60.3% (181) of the study population have sweet cravings, 19.6% (59) have spicy food as their major intake during menstruation, while 10.6 % (32) of them will take normal and regular food. 9.3% (28) of the study population have restricted their food habits during menstruation.

Onset of menarche and attainment of capacity of women to conceive

Information about their basic knowledge on the onset of menarche compiles with the attainment of the capacity of women to conceive, showed that 65.3% (196) of study population have no idea about the relation between menarche and ability to conceive, while 25.3% (76) of the study population have the awareness. When compared with the percent of women who had the awareness about menarche and conceiving ability, the percentage of women with no idea was highly significant among the studied population. Furthermore, 9.3% (28) of the study population have no awareness on the ability to conceive from the onset of menarche.

4 DISCUSSION

The perception, knowledge, and practices during menstruation of all the 300 participants were examined and analyzed for understanding the real world situations pertaining to the menstruation in the southern part of the India. This study revealed that there was a significant impact of education of mothers on their female child[9]. It is well known fact that mother is the first teacher and for every child she is the person to whom they can approach[10], this study data exhibited the same. The analysed data clearly showed that the socio-economic status of the study participants has a direct impact on certain parameters like the usage of napkins/ cloth, preserving them and number of napkins they use per day. A recent study highlighted the socio economic inequalities and impact on hygiene management during menstruation in the in eight low-and middle-income countries[11]. This study also revealed that most of the participants are hesitant to discuss menstruation with others including teachers and most discussions are with mother and friends. The main problem for limiting the girls or women in discussing about the menstrual concerns is, they do not have idea on sources for accessing information. Furthermore, the limited available sources is another barrier to initiate discussions regarding the problems of menstruation.[12]

Fear of menarche and menstruation was noticed because of a lack of proper education regarding menstruation.[13] Hygiene practices were followed by most of the study participants. The data on bathing frequency and cleaning private/intimate regions indicated that majority of the study population are practicing hygienic procedures. Food habits indicated that, sweet cravings were noticed more, healthy food habits were comparatively less. A study conducted on young women individuals indicated that there is an alteration in appetite and food craving during menstrual cycle. Increase in appetite was observed in 85.8 % before menstruation 70.4 % of individuals during menstruation out of 311 young students. Further, it was highlighted that there was a significant association of increase in appetite with depression and sleepiness during menstruation. The same study concluded that the association between rise in appetite and intake of sweets and junk food followed by premenstrual symptoms indicate the need for promoting proper menstruation-related eating habits.[14]

Most of the participants are having lack of knowledge between menarche and the attainment of women's capacity to conceive. An extensive awareness and education about the facts may improve their understanding on conceiving capability once after the age of menarche is reached. A report on awareness level of adolescents regarding menarche indicated that in rural societies where parents do not discuss these matters with their children, and can be achieved by enabling family life education/sex education in the curricula of school.[15]

CONCLUSION

In the current study, the collected Data was analysed and correlated with various aspects like the educational status of a mother, hygiene practices they follow, and basic knowledge of them regarding menstruation. The analysed data helped us in educating the participants to break out the myths of menstruation, open themselves to talk about menstruation, and the right information regarding healthy food intake, hygiene practices to be followed during menstruation. The importance and need of frequent changing of napkin during menstruation was explained to the participants. The study revealed that there is a need for family members to break the myths and behave normally with the women during menstruation. It is further suggested to educate women about menstruation for betterment of the situation.

ACKNOWLEDGEMENT

All the authors are thankful to Principal, K L College of Pharmacy, and Dr. Bapuji Maringanti, Advisor K L University for the support during this work.

REFERENCES

1. T. S. Thakre BS, Reddy M, Rathi N, Pathak K, Ughade S., *DC Dutta's Textbook gynaecology*, 2013, **8**, 82.
2. S. S. T. Subhash B. Thakre, Monica Reddy, Nidhi Rathi, Ketaki Pathak, Suresh Ughade *Journal of Clinical and Diagnostic Research*, 2011, **5**, 1027-1033.
3. I. P. Damayanti, *Journal of Midwifery and Nursing*, 2021, **3**, 30-33.

4. D. Shanbhag, R. Shilpa, N. D'Souza, P. Josephine, J. Singh and B. Goud, *International Journal of Collaborative Research on Internal Medicine & Public Health*, 2012, **4**, 1353.
5. K. Jain, S. K. Garg, J. B. Singh, M. Bhatnagar, H. Chopra and S. K. Bajpai, *Health and population; perspectives and issues*, 2009, **32**, 204-209.
6. J. C. Bhatia and J. Cleland, *Studies in family planning*, 1995, **26**, 203-216.
7. A. Mallick, S. Shaikh and P. C. Dhara, *International Journal of Physiology, Nutrition and Physical Education*, 2022, **7**, 304-309.
8. J. Pane, R. Pakpahan and S. Nadeak, *Indonesian Journal of Global Health Research*, 2022, **4**, 921-928.
9. A. J. Hawkey, J. M. Ussher and J. Perz, *The Palgrave Handbook of Critical Menstruation Studies*, 2020, 99-113.
10. R. Power, K. Wiley, M. Muhit, E. Heanoy, T. Karim, N. Badawi and G. Khandaker, *BMC Women's Health*, 2020, **20**, 1-9.
11. L. Rossouw and H. Ross, *International Journal of Environmental Research and Public Health*, 2021, **18**, 2571.
12. N. Pandey, *Mass Communicator: International Journal of Communication Studies*, 2020, **14**, 26-30.
13. A. L. DeMaria, C. Delay, B. Sundstrom, A. L. Wakefield, Z. Naoum, J. Ramos-Ortiz, S. Meier and K. Brig, *Women & Health*, 2020, **60**, 87-98.
14. M. Yukie, I. Aoi, K. Mizuki and Y. Toshiyuki, *International Journal of Nutrition and Metabolism*, 2020, **12**, 25-30.
15. P. Mahajan and N. Sharma, *Journal of Human Ecology*, 2004, **16**, 215-218.

A CASE STUDY ON RELAPSED MANTLE CELL LYMPHOMA

K. Chandrakaladar[a*], L. Bhavana[a], A.M.V. Priyanka[a*], G. Kishore[a], B. Devi[a], K. Sreya[a]

[a]Department of pharmacy practice, Nirmala college of pharmacy, Atmakur, Mangalagiri, Guntur, Andhra Pradesh.

*corresponding author details:

K. Chandrakaladar, A.M.V. Priyanka

EMAIL ID - kchandrakaladar9@gmail.com Ph. No: 7671871935

EMAIL ID – akurathipriyanka16@gmail.com

1 INTRODUCTION

Mantle Cell Lymphoma (MCL) is a type of non-Hodgkin lymphoma (NHL) distinguished by clonal expansion of malignant B-type lymphocytes originating in the mantle zone of lymphoid follicles.[1] It is a well-known hematopoietic malignancy,[2] with the gastrointestinal tract being the most common site of involvement other than lymph nodes.[3] Extranodal involvement in the bone marrow, spleen, liver, and gastrointestinal system (GIS) has been observed in MCL.[4] See Table 1 for the Ann Arbor classification for Lymphona.

Table 1 *Ann Arbor Classification for Lymphoma.*

I (I_E)	One lymph node region or extra lymphatic site (I_E)
II (II_E)	Two or more lymph node regions or at least one lymph node region plus a single localized extra lymphatic site(II_E) on the same side of the diaphragm
III (III_E, III_S)	Lymph node regions or lymphoid structures (e.g. thymus, Waldeyer's ring) on both sides of the diaphragm with the optional localized extranodal site (III_E) or spleen (III_S)
IV	Diffuse or disseminated extra lymphatic organ involvement

Although mantle cell lymphoma as a primary gastrointestinal tumor is uncommon. It either produces multiple intestinal polyps, for which the term multiple lymphomatous polyposis has been coined, or it can present as a dominant tumor, most commonly in the ileocecal region. [5,6,7]

Gastrointestinal lymphomas are uncommon, accounting for only 1-4% of all gastrointestinal cancers. [8] In general, the stomach is the most commonly affected site (56.4%), followed by the small intestines and ileocecal region (16.4%).[9] MCL patients were mostly middle-aged and elderly men, with a median age of 61 years. [3]

This lymphoma is linked to the chromosomal translocation t(11,14)(q13;q32), which causes cyclin D1 overexpression.[10] The translocation of the bcl-1 gene from chromosome 11 to chromosome 14, which contains the immunoglobulin heavy-chain gene, is common in this lymphoma and results in increased bcl-1 protein expression.[11] Recent studies have revealed that intestinal mantle cell lymphomas typically present at an advanced stage and have a poor prognosis, with a median survival of 3-4 years.[12,13] Although traditional treatment regimens include R-CHOP and R-DHAP,[14] recent research has suggested that bendamustine and rituximab may play a role, especially if the patient's functional status precludes a more aggressive chemotherapy regimen.[15]

2 CASE SUMMARY

A 45-year male patient was admitted with complaints of sleep deprivation, enlargement of the abdomen, abdominal pain, and generalized weakness, as well as a history of chemotherapy 2nd cycle 3 weeks prior. Now scheduled for the third cycle of chemotherapy on July 25, 2022 (See Table 2). He has a past history of type-2 diabetes mellitus for 10 years, and hypertension for 7 years on medication.

According to colonoscopy and histopathology reports, he was diagnosed with MCL one year ago. After 6 cycles of R- CHOP chemotherapy and after bone marrow aspiration and biopsy, the results show normal values.

After one year, on June 12th, 2022, the patient was admitted with complaints of abdominal pain, painful urination, and generalized weakness.

Table 2 *Lab investigations.*

Hematology	**Observed values**			**Normal values**
	25/7/22	29/7/22	5/8/22	
Haemoglobin	10.9	9.8	10	14-18 g/dl
WBC	5200	9200	5600	4400-11000 cells/cu.mm
Platelets	3,56,000	4,07,000	1,40,000	1.5-4.0 lakh/ml

As the patient is a known case of hypertension, ECG and 2DECHO were done and the report showed that the ECG study was normal and 2DECHO showed normal cardiac chambers, no RWMA of LV, normal LV systolic function, trivial MR/TR, no clot/veg/effusion.

The patient suggested for colonoscopy for any abnormalities as he complaints enlargement of the abdomen and abdominal pain - Colonoscopy report showed that the clinical diagnosis – was K/C/O Mantle cell Lymphoma, post-chemotherapy status, currently, there appears to be an active disease in the ileocecal region, biopsies were obtained for evaluation.

Terminal ileum - Large nodules were seen in the last 5 cm

A biopsy was done and sent for histopathology for evaluation, **Histopathology report showed that the** Method - Light microscopy with interpretation**,** Clinical diagnosis - Mantle cell lymphoma**. Impression –** Features are suggestive of a mantle cell lymphoma involving the small intestine in a diagnosed case of chemotherapy.

Based on the subjective and objective evidence the patient was diagnosed as **RELAPSED – MANTLE CELL LYMPHOMA**

3 TREATMENT

This patient is treated with the Chemotherapy treatment regimen of **R-DHAP.**

Chemotherapy drugs are prescribed for 3 days.

On the first day (26/7/2022) the patient was treated with - T. DOLO - 650mg – PO, INJ. AVIL - 2cc – IVB, INJ. RANTAC – 50mg – IVB, **INJ. RITUXIMAB – 600mg** in 500ml NS IV over 3-4hrs.

On the second day (28/7/2022) the patient was treated with - INJ. FOSOPREPITANT – 150mg in 150ml iv, INJ. EMESET – 8mg – IVB – BD, INJ. RANTAC – 50mg – IVB – BD, **INJ. DEXA – 40mg in 100ml** NS IV over 30mins, INJ. LASIX – 10mg - IVB, **INJ. CISPLATIN – 165mg** – IV in 500ml NS over 24hrs (6 am).

On the third day (29/7/2022) the patient was treated with – **INJ. CYTARABINE – 3000mg** in 500ml NS IV and **T. DEXA 40mg –** OD for 3 days.

Along with the chemotherapy regimen T. ZINCOVIT 1tab – PO – OD, T. PAN 40mg – PO – OD, T. GLIMAS – 1tab – PO – BD, T. GRAFEEL – 300mg – SC – OD were given for 13 days.

4 DISCUSSION

Non-Hodgkin lymphoma is a diverse group of malignant lymphatic tumors with distinct biological, morphological, and clinical characteristics. The majority of non-Hodgkin lymphoma is caused by tumorous analogs of B-lymphocytes, with T-cell and NK-cell tumors accounting for the remainder. MCL is a B-cell tumor composed of monomorphic small or medium lymphoid cells with an uneven nucleus contour and CCND1 gene translocation. K. Lennert, a German pathologist, described the most common "classical" variant of mantle cell lymphoma in 1973.

MCL accounts for 6% of all non-Hodgkin lymphoma in adults. MCL affects 2-3 people per 100,000 in the United States and Europe each year. Patients are typically between the ages of 60 and 65. Men are 2-6 times more likely than women to be diagnosed with the disease. The average survival time is 3-5 years. Lymphoma leukemization denotes a widespread impairment of the hematopoietic organs. MCL bone marrow damage and leukemization occur in 20-80% of cases.

The most important pathogenetic stage of MCL development is the translocation of genes encoding cyclin D1 (CCND1) and immunoglobulin heavy chains - t(11; 14). (q13;q32). As a result, the main diagnostic criteria for MCL are the presence of cyclin D1 in immunohistochemical studies or the presence of t(11; 14)(q13;q32) by D-FISH.

Some lab tests, such as histopathology and colonoscopy, were performed on this patient to confirm the clinical diagnosis. According to this report, the cancer progression is in the last 5 cm of the ileum region, and extranodal involvement of the ileocecal region was observed and confirmed as MCL.

The tumor is aggressive in the vast majority of cases, and at the time of diagnosis, disseminated lesions of lymphatic and extranodal organs are determined, which corresponds to stage IV according to the Ann Arbor classification, as was observed in our patient. More than 70% of patients with MCL are diagnosed with terminal stage IV of the disease. The presence of stage IV in our patient indicates the presence of lymphatic and extranodal involvement.

Despite the fact that MCL is sensitive to first-line chemotherapy, relapses and multidrug resistance to cytostatic effects determine a poor prognosis. If complete remission is not achieved after the start of therapy, the patient's survival time is limited to 1.5 years. Consolidation of the first remission with allogeneic bone marrow or allogeneic hematopoietic stem cell transplantation may be the only potential cure for MCL, but it is limited due to procedure toxicity and the patient's age. The median overall survival without hematopoietic stem cell transplantation is three years.

Until the 1970s, the diagnosis was solely based on morphological (histological and cytological) findings. When the reproducibility of the WHO classification was investigated, it was discovered that when only histological criteria were used, the conclusion of hematology experts on the diagnosis coincided in 41-93% of cases. Pathologists were able to reproduce their own diagnosis in approximately 85% of cases after "blind" re-evaluation. As a result, at this point, morphological research and tumor immunophenotyping are required diagnostic procedures in all lymphoproliferative diseases.

The accumulated information on the use of immunophenotypic and cytogenetic techniques allowed for a more precise definition of a number of lymphoproliferative diseases, including those whose nosological independence was not thought to be proven. Morphological examination remains a critical component of diagnostics because it reveals the cellular substrate of a tumor and determines its tissue architecture. The list of CD antigens introduced into the nomenclature of differentiating antigens of human leukocytes and determined by immunophenotyping is constantly updated, and it now contains approximately 400 CD antigens and subtypes. As a result, a number of new antibodies are introduced into immunohistochemical diagnostic panels, allowing tumor immunophenotypes to be determined and the specificity of the morphological study to be increased.

The National Comprehensive Cancer Network (NCCN) guidelines for MCL treatment discussed stage-based chemotherapeutic treatment strategies. Induction therapy with several chemotherapeutic options, most commonly R-DHAP - rituximab, dexamethasone, cytarabine - and platinum-based agents such as carboplatin, cisplatin, or oxaliplatin, is used in stage 1 and stage 2 disease treatment. Rituximab, cyclophosphamide, doxorubicin, and prednisone are examples of alternative therapies (R–CHOP). R-DHAP, clinical trials, or high dose consolidation with autologous stem cell rescue, as well as maintenance rituximab, are used to treat bulky stage 2, stage 3, and stage 4 diseases.

5 CONCLUSION

This case of ileocecal region intussusception as a single mass lesion adds to the literature by describing another variable clinical presentation of primary gastrointestinal MCL. Primary ileocecal MCL, though uncommon, should be considered in the differential diagnosis of a single mass lesion in the ileum. As demonstrated in this case, adequate tissue sampling of colonic masses is critical for determining the diagnosis and directing treatment. This case serves as a reminder to keep a broad differential in mind, which includes uncommon diseases and atypical pathology. Maintaining and applying a thorough understanding of medical principles during work-up, diagnosis, and treatment is critical for optimal patient care and, as shown in this case, avoiding an unnecessary surgical procedure that could delay the appropriate chemotherapeutic treatment.

6 DECLARATION OF COMPETING INTEREST

The authors declared that there is no conflict of interest in their article.

7 SOURCES OF FUNDING

This study was conducted by releasing self-funding, no external funds are availed for the study.

8 ETHICAL APPROVAL

This study has got approval from IRB Nirmala college of pharmacy with Ethical approval number – **10/IRB/NRML/2022-2023**

9 INFORMED CONSENT FORM

Written informed consent was obtained from the patient for the publication of this case report and accompanying images.

10 ACKNOWLEDGMENT

We acknowledge our thanks to Nirmala College of Pharmacy and Manipal Hospitals for providing the necessary facilities.

11 REFERENCES

1. Agha R., Borrelli M., Farwana R., Koshy K., Fowler A., Orgill D., *Int. J. Surg.* 2018;**60**:132–136.
2. Banks PM, Chan J, Cleary ML, Delsol G, Wolf-Peeters C, Gatter K, *et al. Am J Surg Pathol* 1992; **16**: 637–640.
3. Hou WH, Wei P, Xie JL, Zheng YY, Zhang YL, Zhou XG. *Chinese journal of pathology.* 2018;47:417–422.
4. N. L. Harris, E. S. Jaffe, J. Diebold et al., *Journal of Clinical Oncology*, vol. 17, no. 12, pp. 3835–3849, 1999.
5. Isaacson PG, MacLennan KA, Subbuswamy SG. 1984; **8**: 641–656.
6. O'Briain DS, Kennedy MJ, Daly PA, O'Brien AA, Tanner WA, Rogers P, *et al. Am J Surg Pathol* 1989; **13**: 691–699.
7. Ruskone-Fourmestraux A, Delmer A, Lavergne A, Molina T, Brousse N, Audouin J, Rambaud JC . *Groupe D'etude Lymphomes Digestifs Gastroenterol* 1997; **112**: 7–16.
8. Howell JM, Auer-Grzesiak I, Zhang J, et al. Can J Gastroenterol 2012;26:452-6.
9. Chen Y, Chen Y, Chen S, et al. Medicine (Baltimore) 2015;94:e2119.
10. P. M. Banks, J. Chan, M. L. Cleary et al., *American Journal of Surgical Pathology*, vol. 16, no. 7, pp. 637–640, 1992.
11. Banks PM, Chan J, Cleary ML, Delsol G, Wolf-Peeters C, Gatter K, *et al. Am J Surg Pathol* 1992; **16**: 637–640.
12. Ruskone-Fourmestraux A, Delmer A, Lavergne A, Molina T, Brousse N, Audouin J, Rambaud JC . *Groupe D'etude Lymphomes Digestifs Gastroenterol* 1997; **112**: 7–16.
13. Lavergne A, Brouland JP, Launay E, Nemeth J, Ruskone-Fourmestraux A, Galian A . *Cancer* 1994; **74**: 3042–3050.
14. Lenz G, Dreyling M, Hoster E, Wörmann B, Dührsen U, Metzner B, et al. J Clin Oncol. 2005 Mar;23(9):1984–92.
15. Flinn IW, van der Jagt R, Kahl B, Wood P, Hawkins T, MacDonald D, et al. J Clin Oncol. 2019 Apr;37(12):984–91.

www.ingramcontent.com/pod-product-compliance
Lightning Source LLC
LaVergne TN
LVHW070827220926
848486LV00027B/38
* 9 7 8 1 8 3 7 6 7 0 8 3 3 *